Future Bioenergy and Sustainable Land Use

Members of the German Advisory Council on Global Change (WBGU)

(as of 31 October 2008)

Prof Dr Renate Schubert (chair), Economist
Director of the Institute for Environmental Decisions, ETH Zurich (Switzerland)

Prof Dr Hans Joachim Schellnhuber CBE (vice chair), Physicist
Director of the Potsdam Institute for Climate Impact Research and visiting professor
at Oxford University, UK

Prof Dr Nina Buchmann, Ecologist
Professor of Grassland Science, Institute of Plant Sciences, ETH Zurich (Switzerland)

Prof Dr Astrid Epiney, Lawyer
Professor of International Law, European Law and Swiss Public Law, Université de Fribourg (Switzerland)

Dr Rainer Grießhammer, Chemist
Director of the Institute for Applied Ecology, Freiburg/Breisgau

Prof Dr Margareta E. Kulessa, Economist
Professor of International Economics, University of Applied Science, Mainz

Prof Dr Dirk Messner, Political Scientist
Director of the German Development Institute, Bonn

Prof Dr Stefan Rahmstorf, Physicist
Professor for Physics of the Oceans at Potsdam University and head of the Climate System Department
at the Potsdam Institute for Climate Impact Research

Prof Dr Jürgen Schmid, Aerospace Engineer
Professor at Kassel University, Chairman of the Executive Board of the Institute for Solar Energy
Technology

WBGU is an independent, scientific advisory body to the German Federal Government set up in 1992 in the run-up to the Rio Earth Summit. The Council has nine members, appointed for a term of four years by the federal cabinet. The Council is supported by an interministerial committee of the federal government comprising representatives of all ministries and of the federal chancellery. The Council's principal task is to provide scientifically-based policy advice on global change issues to the German Federal Government.

The Council:
• analyses global environment and development problems and reports on these,
• reviews and evaluates national and international research in the field of global change,
• provides early warning of new issue areas,
• identifies gaps in research and initiates new research,
• monitors and assesses national and international policies for sustainable development,
• elaborates recommendations for action,
• raises public awareness and heightens the media profile of global change issues.

WBGU publishes flagship reports every two years, making its own choice of focal themes. In addition, the German government can commission the Council to prepare special reports and policy papers.
For more information please visit www.wbgu.de.

GERMAN ADVISORY COUNCIL ON GLOBAL CHANGE

Future Bioenergy and Sustainable Land Use

Routledge

Taylor & Francis Group

LONDON AND NEW YORK

German Advisory Council on Global Change (WBGU)
Secretariat
Reichpietschufer 60-62, 8th Floor
D-10785 Berlin, Germany

http://www.wbgu.de

German edition published in 2009, entitled
Welt im Wandel: Zukunftsfähige Bioenergie und nachhaltige LandnutzungWBGU, Berlin 2009

First published by Earthscan in the UK and USA in 2010

ISBN 978-1-84407-841-7 hardback

Translation by Christopher Hay, Seeheim-Jugenheim, ecotranslator@t-online.de

Pictures for cover design with kind permission of CLAAS Germany (combine harvester Lexion 600) and Schmack Biogas AG, photographer Herbert Stolz (biomethane plant). All other pictures Prof Dr Meinhard Schulz-Baldes, WBGU.

For a full list of publications please contact:
Earthscan
2 Park Square, Milton Park, Abingdon, Oxon OX14 4RN
711 Third Avenue, New York, NY 10017

Earthscan is an imprint of Taylor & Francis Group, an informa business

A catalogue record for this book is available from the British Library

Library of Congress Cataloging-in-Publication Data has been applied for.

For Product Safety Concerns and Information please contact our EU representative:
GPSR@taylorandfrancis.com.
Taylor & Francis Verlag GmbH, Kaufingerstraße 24, 80331 München, Germany.

ISBN 13: 978-0-415-84786-5 (pbk)
ISBN 13: 978-1-84407-841-7 (hbk)

Council Staff and Acknowledgments

This Special Report builds upon the expert and committed work performed by the WBGU Secretariat staff and by the WBGU members and their assistants.

Scientific Staff at the Secretariat

Prof Dr Meinhard Schulz-Baldes
(Secretary-General)

Dr Carsten Loose
(Deputy Secretary-General)

Dr Karin Boschert

Dr Oliver Deke

Dipl Umweltwiss Tim Hasler

Dr Nina V. Michaelis

Dr Benno Pilardeaux
(Media and Public Relations)

Dr Astrid Schulz

Administration, Editorial work and Secretariat

Vesna Karic-Fazlic (Accountant)

Martina Schneider-Kremer, MA (Editorial work)

Margot Weiß (Secretariat)

Scientific Staff to the Council Members

Dipl Phys Jochen Bard (Institute for Solar Energy Technology, ISET Kassel, until 30.06.2007)

Steffen Bauer, MA (German Development Institute, DIE Bonn)

Dipl Volksw Julia E Blasch (Institute for Environmental Decisions, ETH Zurich, Switzerland)

Dr Georg Feulner (Potsdam Institute for Climate Impact Research, PIK)

Dr Sabina Keller (ETH Zurich, Switzerland)

Dipl Geogr Andreas Manhart (Institute for Applied Ecology, Freiburg, until 30.04.2008)

Dr Martin Scheyli (University Fribourg, Switzerland)

MSc Dipl Ing Michael Sterner (Institut für Solare Energieversorgungstechnik, ISET Kassel, from 01.07.2007)

Dr Ingeborg Schinninger (ETH Zurich, Switzerland, until 31.05.2007)

Dr Jennifer Teufel (Institute for Applied Ecology, Freiburg, from 01.05.2008)

WBGU owes a debt of gratitude to the important contributions and support provided by other members of the research community. This report builds on the following expert studies:

- Dipl.-Umweltwiss. Tim Beringer, Prof. Wolfgang Lucht (Potsdam Institute for Climate Impact Research, PIK): 'Simulation nachhaltiger Bioenergiepotentiale'.
- Dr Göran Berndes (Department of Energy and Environment, Physical Resource Theory, Chalmers University of Technology, Gothenburg, Sweden): 'Water demand for global bioenergy production: trends, risks and opportunities'.
- Dr André Faaij (Utrecht University, Copernicus Institute): 'Bioenergy and global food security'.
- Dr Uwe R. Fritsche, Kirsten Wiegmann (Öko-Institut, Darmstadt Office): 'Treibhausgasbilanzen und kumulierter Primärenergieverbrauch von Bioenergie-Konversionspfaden unter Berücksichtigung möglicher Landnutzungsänderungen'.
- Dr Les Levidow, PhD (The Open University, Development Policy and Practice (DPP) Group, Milton Keynes, UK), Helena Paul (EcoNexus, Oxford, UK): 'Land-use, Bioenergy and Agro-biotechnology'.
- Dipl.-Ing. Franziska Müller-Langer, Anastasios Perimenis, Sebastian Brauer, Daniela Thrän, Prof. Dr-Ing. Martin Kaltschmitt (German Biomass Research Centre – DBFZ, Leipzig): 'Technische und ökonomische Bewertung von Bioenergie-Konversionspfaden'.
- Mark W. Rosegrant, Anthony J. Cavalieri (International Food Policy Research Institute – IFPRI, Washington, DC): 'Bioenergy and Agro-biotechnology'.
- Mark W. Rosegrant, Mandy Ewing, Siwa Msangi, and Tingju Zhu (International Food Policy Research Institute – IFPRI, Washington, DC): 'Bioenergy and Global Food Situation until 2020/2050'.
- Dr Ingeborg Schinninger (ETH Zürich, Institut für Pflanzenwissenschaften): 'Globale Landnutzung'.
- Dr oec. troph. Karl von Koerber, Dipl. oec. troph. Jürgen Kretschmer, Dipl. oec. troph. Stefanie Prinz (Beratungsbüro für Ernährungsökologie, Munich): 'Globale Ernährungsgewohnheiten und -trends'.

For help in creating the graphics we are indebted to Danny Rothe, Design Werbung Druck, Berlin.

During its intensive conference held in May 2008 in Schmöckwitz, Berlin, WBGU drew valuable input from the papers on 'THG-Emission Bio-Prozesse mit LUC' by Dr Uwe R. Fritsche (Öko-Institut, Darmstadt Office) and on 'Technischen und ökonomischen Bewertung von Bioenergiekonversionspfaden' by Dipl.-Ing. Franziska Müller-Langer (German Biomass Research Centre – DBFZ, Leipzig). We should also like to thank Tim Beringer (Potsdam Institute for Climate Impact Research, PIK) for presenting the results of his 'Modellierung zu nachhaltigem globalen Bioenergiepotenzial'.

WBGU also wishes to thank all those who promoted the progress of this report through discussion, comments, advice and research or by reviewing parts of the report:

Prof. Dr Markus Antonietti (Max-Planck-Institut für Kolloid- und Grenzflächenforschung, Potsdam); Ing. Michael Beil (Institut für Solare Energieversorgungstechnik – ISET Hanau); Verena Brinkmann (Sector Project HERA – Household Energy Programme, GTZ Eschborn); Qays Hamad, Advisor to the Executive Director for Germany (The World Bank, Washington, DC); Peter Herkenrath and Dr Lera Miles (UNEP-WCMC, Cambridge); DirProf. Dr Christian Hey and Dr Susan Krohn (German Advisory Council on the Environment – SRU, Berlin); Holger Hoff (Potsdam Institute for Climate Impact Research and Stockholm Environment Institute); Philipp Mensch (ETH Zürich); Gregor Meerganz von Medeazza, PhD (Sustainable Energy and Climate Change Initiative – SECCI, Washington, DC); Ritah Mubbala (Institut für Solare Energieversorgungstechnik – ISET, Kassel); Dipl.-Volksw. Markus Ohndorf (ETH Zürich); Dr Alexander Popp (Potsdam Institute for Climate Impact Research, PIK); Dr Timothy Searchinger (Princeton University, Princeton, NJ); Dr Karl-Heinz Stecher (KfW Bankengruppe, Berlin); Dr-Ing. Alexander Vogel (German Biomass Research Centre – DBFZ, Leipzig) and Dr Tilman Altenburg, Dr Michael Brüntrup, Dr Matthias Krause, Christian von Drachenfels, Dipl.-Ing. agr. Heike Höffler, Julia Holzbach and Kathrin Seelige (German Development Institute – DIE, Bonn).

WBGU is much indebted to the persons who received the WBGU delegation visiting India from 5 to 17 February 2008, and to the organizers of the visit. The German Embassy in New Delhi provided extensive support in making the necessary arrangements. WBGU proffers warmest thanks to Ambassador Mützelburg and all the embassy staff for their invaluable assistance. WBGU is particularly indebted to Dr von Münchow-Pohl and Ms Subhedar, who planned the different parts of the itinerary and arranged meetings and discussions. Thanks are also due to Ms Holzhauser, Mr Wirth and Ms Tiemann, who accompanied WBGU to meetings in Delhi. We should also like to thank the GTZ team: Ms Kashyap, Mr Glück, Dr Bischoff, Dr Porst and Mr Babu.

Many local experts from politics, administration and science offered guided tours, prepared presentations and were available for in-depth discussions and conversations. WBGU proffers them all its warmest thanks.

Contents

Boxes

Tables

Figures

Acronyms and Abbreviations

ACP	African, Caribbean and Pacific Group of States
ADB	Asian Development Bank
AfDB	African Development Bank
BEFS	Bioenergy and Food Security Project (FAO)
BMELV	Bundesministerium für Ernährung, Landwirtschaft und Verbraucherschutz [Federal Ministry of Food, Agriculture and Consumer Protection, Germany]
BMU	Bundesministerium für Umwelt, Naturschutz und Reaktorsicherheit [Federal Ministry for the Environment, Nature Conservation and Nuclear Safety, Germany]
BMZ	Bundesministerium für wirtschaftliche Zusammenarbeit und Entwicklung [Federal Ministry for Economic Cooperation and Development, Germany]
BtL	Biomass-to-Liquid
CAP	Common Agricultural Policy (EU)
CBD	Convention on Biological Diversity
CCS	Carbon Capture and Storage
CDM	Clean Development Mechanism (Kyoto Protocol)
CGIAR	Consultative Group on International Agricultural Research
CHP	Combined Heat and Power
CITES	Convention on International Trade in Endangered Species of Wild Fauna and Flora (UN)
COP	Conference of the Parties
CO_2	Carbon Dioxide
CRIC	Committee for the Review of the Implementation of the Convention (UNCCD)
CPD	Centers of Plant Diversity (IUCN)
CSD	Commission on Sustainable Development (UN)
CST	Committee on Science and Technology (UNCCD)
DALY	Disability Adjusted Life Years
dLUC	Direct Land-Use Change
DM	Dry Matter
EEG	Renewable Energy Sources Act (Germany)
EGS	Environmental Goods and Services (WTO)
EMPA	Swiss Federal Laboratories for Materials Testing and Research
ETI	Ethical Trading Initiative
ETS	Greenhouse Gas Emission Trading Scheme (EU)
EU	European Union
EUGENE	European Green Electricity Network
EUIE	EU-Initiative Energy for Poverty Reduction and Sustainable Development
FATF	Financial Action Task Force on Money Laundering
FAO	Food and Agriculture Organization of the United Nations
FLO	Fairtrade Labelling Organizations International
FSC	Forest Stewardship Council
GATT	General Agreement on Tariffs and Trade
GBEP	Global Bioenergy Partnership (FAO)
GDP	Gross Domestic Product

GEF	Global Environment Facility (UNDP, UNEP, World Bank)
GHG	Greenhouse Gas
GIS	Geographical Information System
GLASOD	The Global Assessment of Human Induced Soil Degradation (ISRIC)
GSP	Generalized System of Preferences (EU)
GSPC	Global Strategy for Plant Conservation (CBD)
GTZ	Deutsche Gesellschaft für Technische Zusammenarbeit [German Society on Development Cooperation]
GuD	Gas-steam Power Plant
GMO	Genetically Modified Organisms
HANPP	Human Appropriation of Net Primary Production
HCVA	High Conservation Value Areas
IBEP	International Bioenergy Platform (FAO)
IAASTD	International Assessment of Agricultural Knowledge, Science and Technology for Development
IADB	Inter-American Development Bank
ICRISAT	International Crops Research Institute for the Semi-Arid Tropics (CGIAR)
ICSB	International Conference on Sustainable Bioenergy (recommended)
ICSU	International Council for Science
IDA	International Development Association (World Bank)
IEA	International Energy Agency (OECD)
IFAD	International Fund for Agricultural Development
IFC	International Finance Corporation (World Bank)
IFOAM	International Federation of Organic Agriculture Movements
IFPRI	International Food Policy Research Institute (FAO)
IGBP	International Geosphere Biosphere Program (ICSU)
IHDP	International Human Dimensions Programme on Global Environmental Change (ISSC, ICSU)
ILO	International Labour Organization (UN)
iLUC	Indirect Land-Use Change
IPCC	Intergovernmental Panel on Climate Change (WMO, UNEP)
IRENA	International Renewable Energy Agency
ISCC	International Sustainability and Carbon Certification (BMELV)
ISRIC	International Soil Reference and Information Centre
ISSC	International Social Science Council (UNESCO)
ITTO	International Tropical Timber Organization
IUCN	World Conservation Union
IMF	International Monetary Fund
KfW	German Development Bank
LDC	Least Developed Countries
LIFDC	Low Income Food Deficit Countries (FAO, WFP)
LULUCF	Land Use, Land-Use Change and Forestry
MA	Millennium Ecosystem Assessment (UN)
MDG	Millennium Development Goals (UN)
MERCOSUR	Mercado Común del Sur (Argentina, Brazil, Paraguay, Uruguay)
MESA	Multilaterales Energiesubventionsabkommen (recommended)
MODIS	Moderate Resolution Imaging Spectroradiometer
NaWaRo	Nachwachsende Rohstoffe
NEDC	New European Driving Cycle
NGO	Non-governmental Organization
OECD	Organisation for Economic Co-operation and Development
PEFC	Programme for the Endorsement of Forest Certification Schemes
PIK	Potsdam Institute for Climate Impact Research
PSA	Programm Pagos por Servicios Ambientales (Costa Rica)
REC	Renewable Energy Certificates
REDD	Reducing Emissions from Deforestation and Degradation (UNFCCC)

REEEP	Renewable Energy and Energy Efficiency Partnership (UK)
REN21	Renewable Energy Policy Network for the 21st Century
RIL	Reduced-impact Logging
RSB	Roundtable on Sustainable Biofuels
RSPO	Roundtable on Sustainable Palmoil
RTRS	Roundtable on Responsible Soy Association (Switzerland)
SAFE	Silvorable Forestry for Europe Project
SAI	Social Accountability International
SAN	Sustainable Agriculture Network (Rainforest Alliance)
SRF	Short-rotation Forestry; or: Short-rotation Coppice
SRU	Sachverständigenrat für Umweltfragen [Council of Environmental Experts, Germany]
UBA	Umweltbundesamt [Federal Environment Agency]
UNCCD	United Nations Convention to Combat Desertification in Countries Experiencing Serious Drought and/or Desertification, Particularly in Africa
UNCTAD	United Nations Conference on Trade and Development
UNDP	United Nations Development Programme
UNEP	United Nations Environment Programme
UNESCO	United Nations Educational, Scientific and Cultural Organization
UNFCCC	United Nations Framework Convention on Climate Change
UNIDO	United Nations Industrial Development Organisation
WBGU	Wissenschaftlicher Beirat der Bundesregierung Globale Umweltveränderungen [German Advisory Council on Global Change]
WCD	World Commission on Dams (World Bank, IUCN)
WCMC	World Conservation Monitoring Centre (UNEP)
WDPA	World Database on Protected Areas (UNEP, IUCN)
WFP	World Food Programme (UN)
WHO	World Health Organization (UN)
WSSD	World Summit on Sustainable Development
WTO	World Trade Organization
WWF	World Wide Fund for Nature

Summary for policy-makers

Global bioenergy policy for sustainable development: WBGU's guiding vision

The incipient global bioenergy boom is giving rise to vigorous and strongly polarized debate. Different underlying aims, such as reducing dependence on imported oil and gas or using biofuels to reduce the CO_2 emissions of road traffic, predominate in different quarters and shape the political agenda. Supporters of bioenergy argue that, at a time of sharply increasing demand for energy, bioenergy can help to secure energy supply and to mitigate climate change as well as create development opportunities, particularly in the rural areas of industrialized and developing countries. Critics, on the other hand, maintain that growing energy crops will heighten land-use conflicts as food cultivation, nature conservation and bioenergy production compete for land, and that bioenergy is likely to impact negatively on the climate. Because of the dynamics and huge complexity of the issue, as well as the considerable scientific uncertainty and the multiplicity of interests involved, it has not as yet been possible to carry out an integrated assessment of the contribution bioenergy can make to sustainable development. WBGU aims to show that the sustainable use of bioenergy is possible and to outline how to exploit opportunities while at the same time minimizing risks.

To that end, WBGU presents an integrated vision that will provide policy-makers clear guidance for the deployment of bioenergy. The principle behind the change of direction that is required must in WBGU's view be the strategic role of bioenergy as a component of the global transformation of energy systems towards sustainability. The guiding vision is inspired by two objectives:

- *Firstly* the use of bioenergy should contribute to mitigating climate change by replacing fossil fuels and thus helping to reduce greenhouse gas emissions in the world energy system. The fact that bioenergy carriers can be stored and used to provide control energy in power grids can make a strategically important contribution to stabilizing electricity supplies when there is a high propor- tion of wind and solar energy in the energy systems of industrialized, newly industrializing and developing countries. In the long term, bioenergy in combination with carbon dioxide capture and secure storage can even help to remove some of the emitted CO_2 from the atmosphere.

- *Secondly* the use of bioenergy can help to overcome energy poverty. In the first place this involves substituting the traditional forms of bioenergy use in developing countries that are harmful to people's health. The modernization of traditional bioenergy use can reduce poverty, prevent damage to health and diminish pressures placed on natural ecosystems by human uses. Some 2.5 billion people currently have no access to affordable and safe forms of energy (such as electricity and gas) to meet their basic needs. Modern yet simple and cost-effective forms of bioenergy can play an important part in significantly reducing energy poverty in developing and newly industrializing countries.

WBGU's central message is that use should be made of the global sustainable potential of bioenergy, provided that risks to sustainability can be excluded. In particular, the use of bioenergy must not endanger food security or the goals of nature conservation and climate protection.

If this ambitious guiding vision is to be realized, politicians must play their part in shaping the processes involved. It is essential to avoid undesirable developments that could prevent proper use being made of the available opportunities. Some of the political measures that are currently in place – such as inappropriate incentives under the Framework Convention on Climate Change or the European Union's quota specifications for biofuels – actually promote bioenergy pathways that exacerbate climate change. It is also important that bioenergy does not trigger competition for land use in a way that puts food security at risk or leads to the destruction of rainforests or of other natural and semi-natural ecosystems. When assessing the use of energy crops it is important to take account of both direct and indirect land-use changes, since these changes have a crucial impact on the greenhouse gas balance and on the risks to biological diversity. By contrast, the use of biogenic wastes and residues entails far fewer risks for land use.

On account of the many possible bioenergy pathways, their different characteristics, and the global linkages among their effects, it is not possible to arrive at a single sweeping assessment of bioenergy. The analysis must be more specific, and in its report WBGU therefore considers bioenergy from an interdisciplinary, systemic and global perspective. WBGU has created an analysis matrix; this involves defining ecological and socio-economic sustainability criteria for the use of bioenergy, conducting an innovative global analysis of the potential of bioenergy on the basis of these criteria, and finally evaluating specific bioenergy pathways in terms of their greenhouse gas balance and environmental impacts over the entire life cycle, taking account of objectives and costs in the process.

Building on that analysis, WBGU develops strategies showing how bioenergy can be deployed as part of sustainable energy systems in industrialized, newly industrializing and developing countries. In the process it becomes evident that the modern forms of bioenergy that are currently in use are insufficiently geared towards the goals of sustainability and climate change mitigation. This applies in particular to the use of annual energy crops grown on agricultural land in order to produce liquid fuels for transport purposes. It would be better to give priority to bioenergy pathways that generate electricity and heat from residues or from perennial crops. WBGU therefore calls for a rapid end to the promotion of biofuels in the transport sector by means of a progressive reduction in the blending quotas for fossil fuels and for the scheme to be replaced by an expansion of electromobility.

With an appropriate regulatory framework, the sustainable use of fuels derived from energy crops can be an important component in the transformation towards sustainable energy systems, with the potential to function as a bridging technology until around the middle of the century. By then the growth in wind and solar energy production is likely to be so far advanced that sufficient energy will be available from these sources. At the same time the pressures on global land use will have increased significantly, principally as a result of three factors: the growth in a world population whose food consumption patterns are increasingly land-intensive, the increasing demand for land to cultivate biomass as an industrial feedstock, and, not least, the impacts of climate change. As a result, the cultivation of energy crops will probably have to be reduced in the second half of the century, while the use of biogenic wastes and

residues will be able to continue. In view of these escalating trends, the problem of competing land use is a potential source of future conflict with implications ranging far beyond the field of bioenergy. Global land-use management is therefore a key task of future international policy-making and an essential requirement for a sustainable bioenergy policy.

For steering the use of bioenergy, WBGU proposes a global regulatory framework for a sustainable bioenergy policy. The key elements of such a framework are a revised UN climate regime with corrected incentives, the setting of sustainability standards, and accompanying measures to safeguard sustainability by strengthening and developing international environmental and development regimes (such as the biodiversity and desertification conventions). Within this framework WBGU formulates promotion strategies with the aim of furthering efficient, innovative technologies and increasing investment in necessary infrastructure – thus contributing to attainment of the guiding vision's two objectives.

By supporting country-specific sustainable bioenergy strategies, development cooperation can help to mobilize sustainable bioenergy potential in developing and newly industrializing countries, to significantly reduce poverty and to build climate-friendly energy systems. An important condition for developing countries, if they are to start using modern forms of bioenergy, is the strengthening of their capacities to take action (such as governance capacities in relation to developing and implementing a sustainable bioenergy policy; monitoring capacities in relation to land-use conflicts; application-oriented research into bioenergy). In addition, for such countries it is essential that bioenergy strategies are linked with food security strategies. This applies in particular to the low-income developing countries who are net importers of food.

In view of the major opportunities and risks associated with it, and the complexity of the subject, bioenergy policy has in a short time become a challenging political task for regulators and planners – a task which can only be accomplished through worldwide cooperation and the creation of an international regulatory framework. In this flagship report WBGU provides decision-makers with guidance to help them in this process of crafting a differentiated and coherent global bioenergy policy.

1
Present use and future potential of bioenergy

To acquire a comprehensive perspective on bioenergy it is necessary to look beyond the narrow focus on the cultivation of energy crops for the production of liquid fuels for transport purposes and to consider the full potential. For this purpose it is in WBGU's view useful to divide bioenergy use analytically into the following areas: (1) traditional bioenergy use, (2) use of biogenic wastes and residues, (3) cultivation of energy crops.

MOST PRESENT BIOENERGY USE IS TRADITIONAL BIOMASS USE

Modern bioenergy plays only a small part in present global bioenergy use, representing about 10 per cent of the total. Biofuels for transport purposes, while much discussed, account for a mere 2.2 per cent of all bioenergy. The lion's share of global bioenergy use – almost 90 per cent of the total, or around 47 EJ per year – is accounted for by traditional bioenergy: this represents around one-tenth of current global primary energy use. This traditional usage involves burning wood, charcoal, biogenic residues or dung, mainly on inefficient three-stone hearths. Around 38 per cent of the world's population, mostly in developing countries, depend on this form of energy, which is harmful to health. More than 1.5 million people a year die from the pollution caused by these open fires. Simple technical improvements to stoves can to a large extent prevent the health risks posed by biomass use while at the same time doubling or even quadrupling its efficiency. The process of modernizing traditional bioenergy use or replacing it with other – preferably renewable – forms of energy can therefore provide important leverage for poverty reduction worldwide, a fact that has been often neglected in the debate on bioenergy and development policy.

THE SUSTAINABLE POTENTIAL OF BIOGENIC WASTES AND RESIDUES

WBGU estimates the technical potential of biogenic wastes and residues worldwide to be around 80 EJ per year. However – for soil protection and other reasons – the sustainably usable potential can be set at only about 50 EJ per year, of which around a half may be economically viable. The scientific basis for estimates of the sustainable global potential of wastes and residues is very slim; WBGU recommends that further studies be carried out so that more precise estimates can be made.

A NEW MODELLING OF THE GLOBAL SUSTAINABLE POTENTIAL OF ENERGY CROPS

Since the available estimates of potential are based on different methods and deliver widely varying results, WBGU has undertaken a new analysis of the global sustainable potential of energy crops. This estimate is based on a dynamic global vegetation model. Scenarios of the potentially available areas of land incorporated those sustainability requirements that

must in WBGU's view be met if a globally integrated perspective is adopted. Future land requirements for food security and nature conservation were estimated and excluded from energy crop cultivation. Areas of land were also excluded if the greenhouse gas emissions arising from the conversion to agricultural land would take more than ten years to be compensated for by the carbon removed from the atmosphere by the cultivation of energy crops; these areas were primarily forests and wetlands. Different scenarios relating to climate, emissions and irrigation were also examined, although set against food security and nature conservation the influence of these three factors is relatively small. These different scenarios result in figures for the global sustainable technical potential from energy crops of between 30 EJ and 120 EJ per year.

Figure 1 shows a scenario that represents an average estimate of potential. It describes the technical potential that can be produced in a sustainable manner. However, considerations of economic viability and political conditions in the different parts of the world impose further restrictions on this technical potential. WBGU therefore conducted a further analysis of the regions in which the modelling identifies significant sustainable bioenergy potentials. The preconditions for rapid realization of these potentials include a minimum level of security and political stability in the countries and regions concerned: significant investment activity cannot be expected in fragile states or those embroiled in civil war. Infrastructure-related and logistical capacities are also required, together with a basic level of regulatory competence, if sustainability requirements are to be formulated and implemented.

In the light of these factors five regions were considered in more detail; in the other regions it was either the case that the estimated bioenergy potentials are relatively low (e.g. the Middle East and North Africa), or that economic and government capacity can be regarded as given in the foreseeable future (e.g. North America, Europe). As the results of the modelling show, there are considerable potentials for the sustainable cultivation of energy crops in tropical and subtropical latitudes. Central and South America alone account for 8–25 EJ per year. The political and economic conditions there are also particularly favourable for realizing the sustainable bioenergy potential compared to the other regions. In addition, good prospects for harnessing the sustainable potential to the extent of 4–15 EJ per year exist in China and its neighbouring countries; there, too, it would be possible to secure the necessary investment and develop the required capacities. There is also considerable potential on the Indian subcontinent (2–4 EJ per year) and in South-East Asia (1–11 EJ per year).

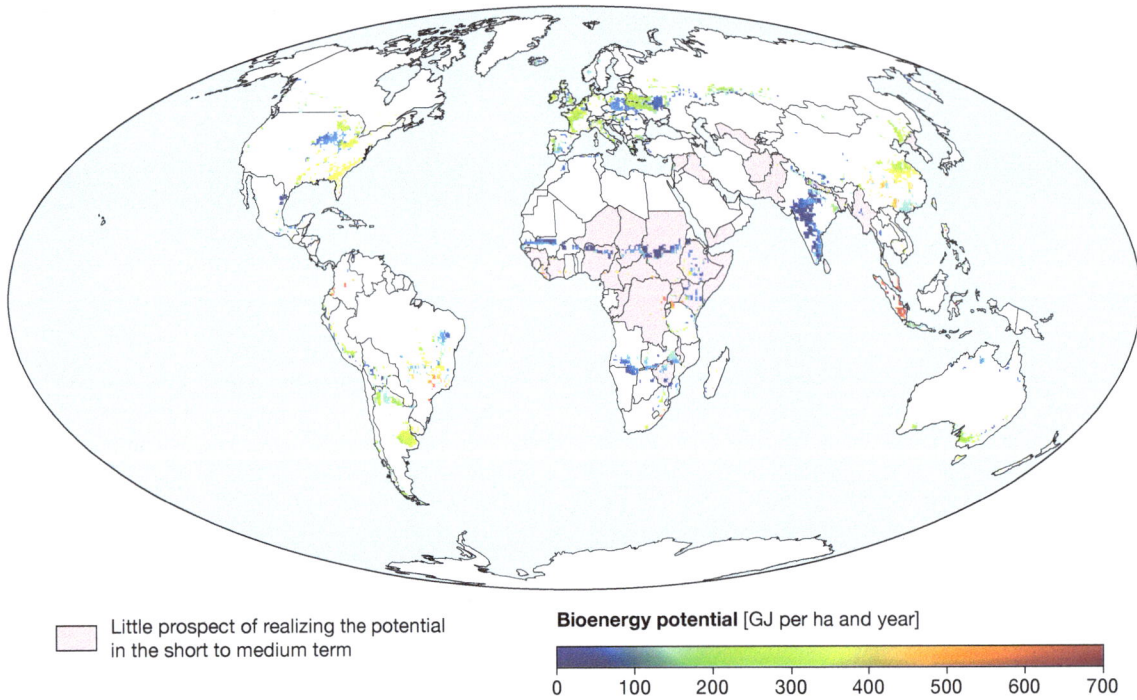

Little prospect of realizing the potential in the short to medium term

Bioenergy potential [GJ per ha and year]

0   100   200   300   400   500   600   700

Figure 1
Regions with potential for sustainable bioenergy from crops and countries that are affected by state fragility or collapse of the state. The map shows the distribution of possible areas for the cultivation of energy crops and the potential production in the year 2050 for a WBGU scenario involving a low level of need for agricultural land, high level of biodiversity conservation and non-irrigated cultivation. One pixel corresponds to 0.5° x 0.5°. In order to assess whether the identified sustainable bioenergy potentials are likely to be realizable, the quality of governance in individual countries was rated using the Failed States Index (FSI). The countries coloured light red have an FSI > 90, indicating that in the short to medium term the prospect for realizing bioenergy potentials can be regarded as poor.
Source: WBGU, drawing on data from Beringer and Lucht, 2008 and from Foreign Policy, 2008

These regions, however, face particular challenges in the form of high land-use density, risks to food security, deforestation and the need to conserve biological diversity. On account of state fragility or the collapse of government, in many African countries it is unrealistic to expect that the full potential of around 5–14 EJ per year in sub-Saharan Africa will be realized. In African countries where the economic and political situation is more favourable the options for tapping the potential should be explored in more detail.

THE SUSTAINABLE POTENTIAL OF BIOENERGY IS SIGNIFICANT!
Including the potential from wastes and residues (ca. 50 EJ per year), WBGU estimates the total sustainable technical potential of bioenergy in the year 2050 to be 80–170 EJ per year. This represents around a quarter of current global energy use and less than one-tenth of the expected level of global energy use in 2050. However, this range represents the upper limit; some of this technical sustainable potential will not be viable, for example for economic reasons or because the area in question is one of political con-

flict. The economically mobilizable potential may amount to around a half of the sustainable technical potential. In view of these figures the importance of bioenergy should not be overestimated, but the expected scale is nonetheless significant. Considering the strategic merits of bioenergy, it should not be neglected in the future development of energy systems. The challenge for policy-makers is to make full use of the sustainable bioenergy potential that is economically mobilizable while at the same time ensuring through suitable regulation that undesirable developments are avoided and sustainability limits observed.

2

Risks and undesirable developments arising from unregulated bioenergy expansion

Against the potentials and opportunities must be set the risks of unregulated bioenergy development. The increased cultivation of energy crops couples the rapidly growing worldwide demand for energy to global land use. This increases the demand for agricultural land, which is already becoming scarcer, and increases the likelihood of land-use conflicts in the future. Some ecosystem services and products are inextricably linked to land use and the production of biomass and cannot be substituted by other means. These include, for example, the conservation of biodiversity, biogeochemical cycles, biomass as food and feed and to some extent the use of biomass as feedstock in industrial production processes. In contrast, renewable energy can also be produced in ways that are unlikely to trigger land-use conflicts, such as through the generation of wind power or solar energy. Risks arise when the cultivation of energy crops triggers direct or indirect competition for land, with the result that non-substitutable uses of biomass are displaced and hence jeopardized. These risks were taken into account in WBGU's analysis of potential, but in the practical mobilization of this potential it is a major challenge for a sustainable bioenergy policy to avoid them.

RISKS TO FOOD SECURITY
If the food requirements of the world's growing population are to be met, global food production will need to increase by around 50 per cent by 2030. The amount of land needed for future food production is also influenced by the land-intensive food consumption patterns of the industrialized countries, which are spreading to the growth regions of emerging economies such as China. This demand can only partly be met by increasing productivity per unit of land; in consequence the FAO estimates that the amount of land used for agriculture will need to be increased by 13 per cent by 2030. It is therefore likely that there will be a significant increase in competition for the use of agricultural land and, consequently, a trend towards rising food prices. Furthermore, a significant increase in the cultivation of energy crops implies a close coupling of the markets for energy and food. As a result, food prices will in future be linked to the dynamics of the energy markets. Political crises that impact on the energy markets would thus affect food prices. For around one billion people in the world who live in absolute poverty, this situation poses additional risks to food security, and these risks must be taken into account by policy-makers.

RISKS TO BIOLOGICAL DIVERSITY
The increased demand for agricultural products that arises from the expansion of bioenergy use can be met by intensifying existing production systems, at the expense of the biological diversity of the land thus farmed. The other option is to claim new agricultural land at the expense of natural ecosystems; this process is at present regarded as the most important driver of the current global crisis of biological diver-

sity. This impact on biodiversity may occur directly, such as when tropical forests are cleared and the land used for energy crops. Indirectly triggered land-use changes are more difficult to assess: when agricultural land is given over to the cultivation of energy plants, the production that previously took place on this land must now take place elsewhere. Through the world market for agricultural goods these indirect displacement effects often acquire an international dimension. Thus, an uncontrolled expansion of energy crop cultivation would further exacerbate the loss of biological diversity.

RISKS TO CLIMATE CHANGE MITIGATION
The conversion of natural ecosystems into new agricultural land releases greenhouse gases. Whether and to what extent greenhouse gas emissions can be reduced by using bioenergy from energy crops depends to a large extent on the land-use changes involved. Emissions created by the conversion of ecosystems that contain a high proportion of carbon (such as forests and wetlands, as well as some natural grasslands) generally negate the climate change mitigation effects that bioenergy use might have. In such cases the use of energy crops may even exacerbate climate change. Both direct and indirect land-use changes must therefore be taken into account in evaluating the greenhouse gas balance of bioenergy use.

RISKS TO SOIL AND WATER
Forms of bioenergy that focus on the use of annual energy crops on agricultural land are insufficiently compatible with the goals of soil protection. Perennial cultivation systems, on the other hand, may actually help to restore degraded land. Whether the cultivation of energy crops is acceptable in terms of soil protection also depends on agro-ecological conditions in the region. In addition, the removal of residues from agriculture- or forestry-based ecosystems must be restricted, as the soil may otherwise be depleted of organic substances and mineral nutrients. Uncontrolled expansion of energy crop cultivation and inappropriate cultivation systems may also greatly increase the pressure of use on the available water resources. Energy crops are a new driving force in the land-use sector; the major effects that they may have on future water use have as yet barely been explored.

3

Sustainable bioenergy pathways: WBGU's findings

On the basis of the two objectives of its guiding vision, WBGU explores a number of important bioenergy pathways. The use of bioenergy only has a climate change mitigation effect if the greenhouse gas emissions arising from the land-use changes and from the cultivation and use of the biomass are lower than the emissions that would arise if fossil fuels were used. Bioenergy can best contribute to overcoming energy poverty when its advantages are exploited by locally adapted technology: biomass can decentrally store and provide energy without the need for major financial or technical investment.

Production of biomass for use as energy: What are the key issues?

In producing biomass for use as energy a fundamental distinction needs to be made between wastes and residues on the one hand and energy crops on the other.

PRIORITY FOR THE USE OF WASTES AND RESIDUES
The use of biogenic wastes and residues has the advantage of causing very little competition with existing land uses. It involves no greenhouse gas emissions from land-use changes and cultivation, so that the contribution to climate change mitigation is determined primarily by the conversion into bioenergy carriers and their application in energy systems. When using residues, care must be taken to meet soil protection standards – and hence ensure climate change mitigation – and that pollutant emissions are avoided. Overall, WBGU attaches higher priority to the recycling of biogenic waste for energy (including cascade use) and to the use of residues than to the use of energy crops.

LAND FOR ENERGY CROP CULTIVATION
Where specially cultivated energy crops are used, it is essential to take account of land-use changes. While emissions from direct land-use changes can be quantified using standard values, much greater uncertainty attaches to indirect land-use changes. WBGU uses a provisional method for calculating these indirect effects, enabling an initial rough estimate to be made.

WBGU is strictly opposed to the direct or indirect conversion of woodland, forests and wetlands into agricultural land for energy crops; such conversion is usually accompanied by non-compensatable greenhouse gas emissions and its impacts on biological diversity and soil carbon storage are invariably negative. The cultivation of energy crops should preferably be restricted to land for which the change of use to bioenergy production does not involve indirect land-use changes. The total greenhouse gas emissions initially caused in the context of cultivation should not exceed the quantity of CO_2 that can be re-sequestered by the cultivation of energy crops on the land in question within ten years.

The cultivation of biomass on marginal land (that is, land with a limited productive or regulatory function) has the significant advantage that land-use competition, for example with food security, is unlikely; in consequence, indirect land-use changes will probably not be induced. WBGU therefore concludes that marginal land should be preferred for the cultivation of energy crops and this type of land use should be encouraged, provided that the interests of local population groups are taken into account and the implications for nature conservation are assessed before cultivation commences.

CULTIVATION SYSTEMS FOR ENERGY CROPS
The principal criteria used by WBGU to assess the sustainability of cultivation systems are the effects on biological diversity and soil carbon storage. Bioenergy can only be classed as sustainable energy if the land on which it is grown continues in the long term to produce as much biomass as is used for energy – in other words, if long-term soil fertility is ensured. Only in this situation can it justifiably be assumed that the carbon that is removed from the atmosphere and stored by the energy crops and that is re-released in the form of CO_2 when the crops are used for energy does not lead to an increase in the concentration of CO_2 in the atmosphere and therefore does not need to be regarded as an emission. In addition, differing yields per unit of land must be taken into account. From this point of view perennial crops such as *Jatropha*, oil palms, short-rotation plantations (fast-growing timber) and energy grasses score better than annual crops such as rape, cereals or maize; the former group should therefore always be preferred. If suitable cultivation systems are chosen, additional organic carbon can be incorporated into the soil; this improves both the greenhouse gas balance and soil fertility.

Conversion, end-use application and system integration: What are the best ways of using bioenergy?

Once the biomass has been made available, the climate change mitigation effect is mainly determined by two factors: the way in which biomass is converted into usable products such as gas, plant oils, biofuels or wood pellets, and the way in which it is used and integrated into the energy system – for example, into transport or into the generation of heat or electricity. On the whole, however, these influences carry less weight than the effect of direct or indirect land-use changes in connection with the cultivation of energy crops. Much depends on what energy carrier the biomass replaces and on the magnitude of the energy losses in the conversion pathway. In industrialized countries, in the rapidly developing urban and industrialized regions of newly industrializing countries and in some cases also in developing countries the way in which bioenergy is used should be geared towards its climate change mitigation effect. In relation to overcoming energy poverty the primary tasks are modernization of traditional bioenergy use and provision of access to modern forms of energy such as electricity and gas. Both are challenges that are of particular importance in the rural regions of developing countries. Here, too, the use of bioenergy can have a positive effect on climate change mitigation.

MITIGATING CLIMATE CHANGE

From the point of view of climate change mitigation the most attractive application areas for bioenergy are, firstly, those in which bioenergy can replace fossil fuels with high CO_2 emissions, predominantly coal.

Roughly similar reductions in greenhouse gases can be achieved by various conversion pathways producing electricity, such as co-combustion in coal-fired or cogeneration plants, the use of biogas from fermentation or crude gas from gasification in cogeneration (combined heat and power, CHP) plants, or the use of biomethane in small-scale CHP plants or combined-cycle power plants. Where biomethane is used, however, a greater climate change mitigation effect can be achieved if the CO_2 which must in any case be captured during the production process can be securely stored. The conversion of biomass into electricity has the additional advantage that, unlike liquid fuels for transport, it eases the shift towards electric mobility. The current greenhouse gas (GHG) abatement costs of these pathways vary widely: while the simple co-combustion of solid biomass and the use of biogas or biomethane from fermentation already represent cost-efficient climate change mitigation options, this is not yet the case for gasification technologies (although significant reductions in costs can be expected). The use of biomethane is also particularly attractive for technological and system-related reasons, since it can be collected and distributed over natural-gas grids and converted very efficiently into electricity in small-scale CHP units or combined-cycle power plants near where it is needed. The biomethane route can already be recommended for industrialized countries; for industrialized regions in newly industrializing and developing countries it is a promising option for the future.

On account of its high energy efficiency combined heat and power production is to be preferred to the generation of electricity alone, provided that demand for the heat exists. In regions where this is appropriate CHP can also be used to generate cooling, a factor which is of interest for many developing and newly industrializing countries. Where bioenergy is used exclusively for the production of heat (e.g. pellet stoves) GHG abatement costs are relatively high and the potential for reducing greenhouse gas emissions is only about half that which can be achieved in the electricity sector; such use is therefore only worthwhile as a transition measure where alternative renewable energy sources are not available. With direct generation from renewables (wind, solar) constituting an ever-larger proportion of production, there will in future be a significant increase in the overall energy efficiency of electric heat pumps, so that in the medium term they will represent a viable alternative for heat generation. Overall CHP pathways are to be preferred both to pure electricity and pure heat use pathways.

From the point of view of climate change mitigation the first-generation biofuels (such as biodiesel from rape or bioethanol from maize), which involve the cultivation of temperate, annual crops on agricultural land, score very badly. When emissions from indirect land-use changes are taken into account, they frequently result in higher emissions than would arise from the use of fossil fuels. Where residues are used (e.g. timber waste, liquid manure, straw) the impact on the greenhouse gas balance is indeed positive, but the reduction in greenhouse gases is only about half that of applications in the electricity sector. Second-generation biofuels are not on the whole any better.

A different picture emerges for the use of perennial tropical plants such as *Jatropha*, sugar cane or oil palms that are grown on degraded land and result in carbon being stored in the soil there. In this situation a major climate change mitigation effect can be achieved at low cost. However, if these crops are grown on freshly cleared land or on agricultural land and thus are associated with direct or indirect land-use changes, the greenhouse gas balance becomes negative; in some cases emissions will be substantially larger than would be the case using fossil fuels. Ensuring sustainability in the cultivation of energy crops is therefore the deciding factor in evaluating the climate change mitigation effect of these pathways.

Since there are as yet no established sustainability standards for biofuels, their import and use pose problems. Once relevant minimum standards have been introduced, it may be appropriate to import plant oils and bioethanol – perhaps produced in tropical regions – for power and heat applications. During the transition period, however, care should be taken to avoid any promotion of biofuels that fail to meet the envisaged minimum standards.

For the future of mobility on the roads, WBGU considers the most appropriate solution to be the generation of electricity from renewables in combination with the use of electric vehicles. This means of utilizing bioenergy achieves a significantly higher climate change mitigation effect than blended biofuels. If electric vehicles were to be introduced on a large scale, it is likely that the costs could be drastically reduced within 15–20 years, enabling the GHG abatement costs – which at present remain very high – to also be reduced. Through the use of smart grids, electromobility can also contribute as control energy to the stabilization of power grids. WBGU recommends a swift phase-out of the promotion of biofuels for transport purposes. The quotas for blending biofuels with fossil fuels should be frozen and should then be completely removed within the next three to four years.

Overall, the substitution of bioenergy for fossil fuels, making use of the sustainable bioenergy potential estimated by WBGU, can achieve a global reduction in greenhouse gas emissions of 2–5 Gt CO_2eq per year. However, this would require all the biomass to be used in such a way that the greenhouse gas reduction amounts to 60 t CO_2eq per TJ of raw biomass used. This corresponds to roughly a doubling of the mitigation efforts currently under discussion in the EU as a standard for biofuels in the transport sector. WBGU proposes this level as a necessary precondition for promotion of bioenergy use. From a very optimistic viewpoint it might be possible to achieve a reduction in greenhouse gases of up to 4–9 Gt CO_2eq per year. By way of comparison: global anthropogenic greenhouse gas emissions currently amount to around 50 Gt CO_2eq per year, and a hypothetical stop to global deforestation would reduce these emissions by up to 8 Gt CO_2eq.

Leaving aside bioenergy pathways that involve the use of marginal land in the tropics or are based on established technologies such as co-combustion in coal-fired power plants or the production of biogas through fermentation, the GHG abatement costs of many bioenergy pathways in 2005 were significantly more than € 60 per t CO_2eq; in WBGU's view they cannot therefore be currently considered to be cost-efficient climate change mitigation options.

The cultivation of energy crops must therefore be carefully weighed against other climate change mitigation options, such as afforestation or the avoidance of deforestation. It is particularly important that energy crop cultivation does not undermine the politically very complex endeavours to reduce emissions from deforestation.

If exploitation of the sustainable bioenergy potential is combined with the capture and secure storage of CO_2, it is possible for "negative" CO_2 emissions to be produced. By this means around 0.2 ppm CO_2 could be removed from the atmosphere per year. This corresponds to around one-tenth of the current annual increase in the concentration of CO_2 – hence even over quite lengthy periods of time this technology can counteract only a relatively small proportion of the human-induced increase in the concentration of CO_2.

Until a global system of mandatory limits to greenhouse gas emissions is put in place that encompasses all relevant sources, WBGU recommends emissions standards for bioenergy.

Overcoming energy poverty

In the rural regions of developing countries, and to some extent also in their urban areas, overcoming energy poverty is an important precondition for tackling poverty in general. As a first step WBGU recommends as an international objective the complete phase-out of traditional forms of bioenergy use that are harmful to health by the year 2030.

To achieve this, some technologies can even now be implemented rapidly and at low cost. The use of improved cooking stoves can cut fuel consumption by between one-half and three-quarters while at the same time drastically reducing the risks to health. Greater emphasis should also be placed on the promotion of small, decentralized biogas plants for residues and wastes, and on the use of plant oil – produced from oil plants grown locally on marginal land – for lighting, electricity generation and mechanical energy use. These technologies also help to reduce the pressure of use on natural ecosystems and to tackle poverty, because the time and money required to procure the fuel is significantly reduced. They provide an important lever for significantly improving the quality of life of many hundred millions of people in a short time and at low cost. It is important, though, to ensure at all stages of development cooperation in this field that the technologies are accepted and that they can be maintained by the individuals who use them.

Further down the track to reducing energy poverty, access to modern forms of energy, particularly electricity and gas, is a priority. In developing countries medium-scale use of modern bioenergy to generate electricity in CHP or gasification plants can be an important means to this end, particularly if biomass such as that from residues or from timber plantations on marginal land is used. The use of liquid fuels for stationary applications (e.g. electricity generation, water pumps, cooking) may be appropriate in rural regions of developing countries, if these regions are at a disadvantage in terms of infrastructure on account of their remoteness.

The larger-scale production and use of modern bioenergy, which can likewise contribute to the tackling of energy poverty in developing countries, should always also be considered from the point of view of its climate change mitigation effect. For those bioenergy pathways that are associated with low GHG abatement costs, new sources of funding can be accessed through international climate protection instruments.

Energy crops as bridging technology

The sustainable use of bioenergy from energy crops can be an important bridging technology during the transformation from existing fossil energy systems to future systems based predominantly on wind and solar energy. It can fulfil this function only until approximately the middle of the century, for two reasons:

Firstly, demands on global land use will increase markedly in the coming decades as a result of dynamic trends such as a growing world population with increasingly land-intensive patterns of food consumption, increased soil degradation and water scarcity. In addition, for reasons of climate change mitigation, among others, there will be a growing tendency for petrochemical products to be produced from biomass. The non-substitutable land use for the manufacture of textiles, chemical products, plastics, etc. is likely to require around 10 per cent of world agricultural land. After use some of the biomass-based products will be able to be recycled as biogenic waste for purposes of energy recovery ("cascade use"). These increasing pressures on land use take place against a backdrop of increasingly manifest anthropogenic climate change. Because of all this, the cultivation of energy crops will probably have to be cut back in the second half of the century.

Secondly, in forthcoming decades there will be a growing trend for renewable energy in the form of electricity to be produced directly by wind and water power, as well as by solar energy on a large scale from the middle of the century; by this time, therefore, energy crops will largely have fulfilled their function of bridging the way to sustainable energy provision. This will not affect the part of bioenergy use that centres on the use of wastes and residues which, together with the remaining use of fossil fuels, will increasingly take on the task – as control energy in power grids – of balancing fluctuations in the output of directly generated electricity from renewables. In combination with smart electricity grids, electromobility can also make an important contribution to control energy.

4
Research recommendations for sustainable bioenergy use

While WBGU highlights in this report viable corridors for sustainable bioenergy use in some areas, gaps in knowledge remain that need to be filled through further research. WBGU identifies a particular need for research in six fields:

1. *Broadening the knowledge base on global land use:* In order to create the scientific basis for setting up a global land register supported by a Geographical Information System (GIS), the state of global land use and land cover as well as the dynamics of global land-use changes must be studied and evaluated in more detail than has so far been the case. This needs to include the collection of high-resolution data on vegetation cover, hydrology and soil condition, agricultural usage and surface sealing in the different regions of the world.

2. *Determining more precise greenhouse gas balances for different bioenergy pathways:* The greenhouse gas balance is the crucial indicator of the climate benefit (or in some cases harm) of a particular use of bioenergy. It has to date only been possible to calculate it imprecisely, for example with regard to indirect effects such as the displacement of previous land use onto other land.

3. *Determining the potential, the greenhouse gas balances and the economic deployment pathways of residue use:* Biogenic residues, such as those from agriculture and forestry, represent a still virtually untapped potential for energy generation. The opportunities for making use of them in future should be researched.

4. *Analysing the role of bioenergy in a future energy system at national, regional, and global levels:* The strategic importance of bioenergy and its integration in particular energy systems (e.g. as control energy in power grids) should be explored in more detail. These factors play an important part in the selection of preferred bioenergy pathways.

5. *Clarifying the links between food security and bioenergy:* The complex local, national and global cause-effect chains that link bioenergy use and food security urgently need to be examined from a socio-economic perspective. This research needs to take geopolitical factors into account: could the "primacy of securing energy supply" of the western world and other powerful political players in a world energy system of which bioenergy is an important component result in increased food-security problems in poor and politically less influential countries? How could such scenarios be avoided through international cooperation?

6. *Analysing international land-use competition and developing the components of a global land-use management system:* As a result of various driving forces, land will in the forthcoming decades become a scarce resource worldwide. Land use will in consequence become a matter of global governance. Research should explore interest structures relating to global land use and help to develop an effective global regime for managing land resources and preventing land-use conflicts.

5
Recommendations for action: Components of a sustainable bioenergy policy

The competition between farming biomass as a resource for energy production and growing food on increasingly scarce agricultural land links two fundamentals of human societies: energy and food. Adoption of a systemic perspective further reveals that the emerging bioenergy policy involves complex issues that are not restricted to matters of energy, agriculture and climate policy; transport policy and foreign trade policy as well as environmental, development and security policy all play an important role. Because non-sustainable bioenergy strategies can harm the climate, exacerbate food-security problems and drive land-use conflicts, policy-making must establish a framework that addresses all the matters mentioned above. Furthermore, bioenergy policy cannot be formulated solely within the national context; it requires collective, transboundary action and effective multi-level governance. To render bioenergy use sustainable, complex regulatory measures need to be taken; this represents a major challenge for a policy-making system that is structured mainly along "departmental" lines. Competing goals need to be reconciled at both national and international level.

In the light of these considerations and in view of the urgency to redirect global policy, WBGU has developed a differentiated mix of policy instruments for a sustainable global bioenergy policy. The considerable risks attached to energy crop cultivation – risks for climate change mitigation and from land-use competition – must be countered by institutional regulation. The first task is to ensure that the expansion of bioenergy use contributes to climate change mitigation. The accounting rules under the UN climate protection regime must be adjusted to remove any incentives to engage in a bioenergy energy policy that is counterproductive for climate change mitigation. Since this will not be accomplished in the short term and cannot guarantee that other sustainability criteria (food security, conservation of biological diversity, etc.) will be met, work on drawing

up and applying bioenergy standards must be undertaken simultaneously. WBGU proposes a demanding minimum standard in combination with additional criteria to be met as a pre-condition for any kind of bioenergy promotion (promotion criteria). Accompanying measures to secure global food production and biological diversity and to protect soil and water resources are also necessary. Existing UN institutions such as the Food and Agricultural Organization (FAO), the Biodiversity Convention (CBD) and the Convention to Combat Desertification (UNCCD) can contribute to these processes. In conclusion WBGU assesses which forms of bioenergy use should be explicitly promoted through national policies and international development cooperation.

5.1
Making bioenergy a consistent part of international climate policy

REFORM ACCOUNTING PROCEDURES FOR CO_2 EMISSIONS FROM BIOENERGY
The existing provisions in the United Nations Framework Convention on Climate Change (UNFCCC) and the Kyoto Protocol create false incentives in relation to bioenergy production and use; they distort the picture of the contribution made by bioenergy to climate change mitigation and may even promote bioenergy use that is harmful to the climate. In WBGU's view the modalities for determining contributions to commitments under the Kyoto Protocol and its successor regime must therefore be corrected. The correction needs to involve the following elements: firstly, the use of bioenergy must no longer be counted en bloc as free of CO_2 emissions ("zero emissions") in the energy sector. However, WBGU is not advocating replacement of the presumed zero emissions by cumulated emissions from a life-cycle analysis of the bioenergy, since this would not be compatible with the other allocation modalities within the UNFCCC and would lead to double counting. Instead, within the energy sector the actual CO_2 emissions arising from the combustion of the biomass should be counted and included. In return, the uptake of CO_2 from the atmosphere by energy crops in the land-use sector should also be counted. This correction would align the way in which bioenergy is treated with the principle used elsewhere of always allocating emissions to the place and time of their creation. Secondly, the existing rules, under which only selected CO_2 emissions and absorptions from land use and land-use change are or can be set against the commitments made by states, should be replaced by full accounting of all emissions from these sectors. Ideally this accounting would form part of a wider agree-

ment on the conservation of the carbon stocks of terrestrial ecosystems within the UNFCCC. Thirdly, there need to be supplementary regulations regarding trade between countries that have and countries that have no binding commitments to limit emissions. In addition, for those emissions from the life cycle of bioenergy use for which there is already an appropriate allocation to the inventories (e.g. non-CO_2 emissions from agriculture), the countries that have committed to limit emissions should systematically introduce incentives for limiting emissions at stakeholder level (e.g. for farmers and foresters).

CONSIDER BIOENERGY IN THE CDM IN MORE
SPECIFIC DETAIL
The Clean Development Mechanism (CDM) involves only a small number of bioenergy projects and these have as yet had only a limited influence on overall bioenergy use in newly industrializing and developing countries. Any expansion of CDM projects that include the cultivation of energy crops should be viewed with scepticism unless it can be ensured that the use of land for this purpose will not give rise to the well-known displacement effects and result in terrestrially stored carbon being released elsewhere. The scope for CDM projects to improve or replace inefficient traditional biomass use should be utilized without damaging the integrity of the CDM. As a matter of principle, CDM projects in the area of bioenergy should be certain of meeting the minimum standard called for by WBGU.

LIMIT EMISSIONS CAUSED BY LAND-USE CHANGES
IN DEVELOPING COUNTRIES
Since the present expansion of the cultivation of energy crops can contribute to an increase in tropical deforestation, an effective regime for reducing the emissions from deforestation and forest degradation in developing countries (REDD) under the UNFCCC is extremely important. An appropriate REDD regime should provide effective incentives for rapidly generating real emissions reductions by reducing deforestation, and it should mobilize international funding transfers at a sufficient level. The regime should consist of a combination of national targets to limit emissions and project-based procedures in order (i) to prevent leakage effects and (ii) to permanently protect the natural carbon reservoirs such as tropical primary forests from deforestation and degradation as well as limit emissions from grassland conversion. The REDD regime would ideally form part of a comprehensive agreement on the conservation of the carbon stocks of terrestrial ecosystems within the UNFCCC.

MOVE TOWARDS A COMPREHENSIVE AGREEMENT
ON THE CONSERVATION OF TERRESTRIAL CARBON
RESERVOIRS
CO_2 emissions arising from land use, land-use change and forestry (LULUCF) should be fully and systematically included in the post-2012 regime in order to ensure that the incentive given to bioenergy use by the UNFCCC is based on the actual contribution to climate change mitigation made by this use. However, the absorption and release of CO_2 by the biosphere differs from the emissions of fossil energy sources in a number of fundamental respects, including measurability, reversibility, long-term controllability and interannual fluctuations. Since the different sectors also have very different characteristics in terms of time-related dynamics and amenability to planning, it would seem more appropriate – from the point of view of remaining within the 2°C guard rail – to define separate reduction targets rather than one overarching target. WBGU therefore recommends that a comprehensive separate agreement on the conservation of the carbon stocks of terrestrial ecosystems be negotiated. This agreement should (i) take up the debate on REDD, (ii) replace the existing rules on offsetting reduction commitments in the sectors listed in Annex A to the Kyoto Protocol against sinks (including through CDM activities) and (iii) fully include all CO_2 emissions from LULUCF. Despite separate target agreements, WBGU considers it appropriate from the point of view of economic efficiency to aim for a certain level of fungibility; however, on account of measurement problems and other uncertainties attaching to LULUCF emissions, this fungibility should be clearly demarcated and associated with deductions.

5.2
Introducing standards and certification for bioenergy and sustainable land use

In order to ensure sustainable production of bioenergy carriers within WBGU's guard rails for sustainable land use, it is necessary to introduce sustainability standards for bioenergy. A minimum standard for bioenergy carriers should be met before bioenergy products are allowed onto the market.

GRADUALLY INTRODUCE A MINIMUM STANDARD
FOR BIOENERGY AND SUSTAINABLE LAND USE
As a first step, a statutory minimum standard for all types of bioenergy should be introduced promptly at EU level. The sustainability criteria for liquid biofuels for transport contained in the planned EU directive on the promotion of renewable energies should be further developed and applied as a minimum stand-

ard for all types of bioenergy in the EU. In addition to provisions relating to soil, water and biodiversity conservation, the standard should include impacts of indirect land-use changes and criteria for restricting the use of genetically modified organisms (GMOs). Certain core labour standards of the International Labour Organization (ILO) should also be made mandatory. With regard to greenhouse gas emissions, WBGU recommends to request a specific absolute emissions reduction in relation to the quantity of raw biomass used, rather than a relative emissions reduction based on the final energy or useful energy. The use of bioenergy carriers should reduce life-cycle greenhouse gas emissions by at least 30 t CO_2eq per TJ of raw biomass used in comparison with fossil fuels.

The cultivation of energy crops and the supply of biomass resources should only be promoted if this gives rise to a demonstrable reduction of energy poverty or to demonstrable advantages for climate change mitigation, as well as soil, water and biodiversity conservation, and if such cultivation also rates positively with regard to social criteria. Another precondition for promotion should be that the use of bioenergy carriers can achieve a reduction in life-cycle greenhouse gas emissions of at least 60 t CO_2eq per TJ of raw biomass used in comparison with fossil fuels. Bioenergy pathways considered particularly worthy of promotion are the use of biogenic wastes and residues and the cultivation of energy crops on marginal land, if the above-mentioned promotion criteria are met.

In order to attain the goal of globally sustainable land use there is a need in the medium term for a global land-use standard to regulate the production of all types of biomass for a wide range of uses (food and feed, use for energy and use as an industrial feedstock, etc.) across national borders and cross-sectorally. The EU member states should therefore prepare suitable provisions for extending the bioenergy standards to all types of biomass.

Until a globally agreed land-use standard is created, the anchoring of bioenergy standards in bilateral agreements remains an effective instrument for increasing sustainability. WBGU recommends that the European states include binding sustainability criteria in future agreements with countries that are important producers and consumers of bioenergy. Existing bilateral agreements should be amended to this end. In return, trading partners who adhere to the minimum standard should be accorded free market access for bioenergy carriers.

With a view to WTO rules and in order to limit recourse to alternative markets for bioenergy products that fail to meet the minimum standard, the German government should also endeavour to ensure

that international consensus on a minimum standard for sustainable bioenergy and on a comprehensive international bioenergy strategy is achieved as quickly as possible. During the transition period efforts must be made to rapidly dismantle all promotion of non-sustainable bioenergy use.

ESTABLISH CERTIFICATION SCHEMES FOR SUSTAINABLE BIOENERGY CARRIERS

To enable adherence to the minimum standard to be demonstrated, corresponding certification systems must be created promptly. WBGU recommends the development of an internationally applicable certification scheme for all types of biomass. This makes it easier for the bioenergy standards to be extended at a later stage to other uses of biomass. The International Sustainability and Carbon Certification system drawn up on behalf of the German Ministry of Food, Agriculture and Consumer Protection (BMELV) or a comparable certification system should be put in place at an early stage.

The duty to furnish proof that the standards have been adhered to could lie in the first instance with the entity marketing the end product. This would remove the need for a duty to certify the origin of bioenergy feedstocks that could also be used for non-energy purposes. While the certification should be carried out by private companies, institutions capable of imposing sanctions must be created by the state to monitor actual implementation of the standards. Developing countries, and in particular the least developed countries, should be offered technical and financial assistance in setting up certification systems and monitoring bodies, and in implementing the certification.

ENSURE WTO CONFORMITY OF ENVIRONMENTAL AND SOCIAL STANDARDS

The World Trade Organization (WTO) conformity of a unilateral European standard can be justified in law, particularly with regard to criteria for the reduction of greenhouse gas emissions and the protection of global biodiversity, because the necessity of protecting climate and biodiversity is laid down in multilateral environmental agreements in international law. In general the acceptance of environmental and social standards in the WTO regime needs to be further improved. In addition, the intended liberalization of trade in relation to what are known as "environmental goods and services" (EGS) must not run counter to the goal of sustainable production and use of such goods and services. In the context of the relevant negotiations the German government should therefore work to ensure that goods are not classified as EGS unless they are guaranteed to meet the minimum standard called for by WBGU and/or result from sustainable bioenergy pathways.

5.3
Sustainably regulating competition among uses

ENSURE PRIORITY FOR FOOD SECURITY

Unless action is taken, the degree of scope for food production will in future come under increasing pressure, partly as a result of the emerging bioenergy boom. In order to prevent a crisis situation developing, there is a need for action in the following areas:

- *Develop an integrated bioenergy and food security strategy:* Over and above the measures specified by the departmental working party on world food affairs in its report to the German Federal Cabinet, WBGU recommends including the cultivation of energy crops in an integrated bioenergy and food security strategy in which food security has priority. This is particularly important for those low-income developing countries that are net importers of food (Low-Income Food-Deficit Countries, LIFDCs). Any controlled expansion of bioenergy must be accompanied by global efforts to strengthen farming. For this to happen, the food situation in affected regions must first be improved, for example by distributing free seed for the next growing season. At the same time the conditions for food security and food production must be improved over the long term and consistently incorporated into other policy areas such as climate protection and nature conservation. Cultivation of energy crops should be promoted primarily on marginal, in particular degraded land.

- *Take greater account of increasing pressure on land use as a result of changing food consumption patterns:* The sharply increasing pressure on land use as a result of land-intensive food consumption patterns in industrialized countries, and the replication of these patterns in large and dynamically growing newly industrializing countries, is exacerbating global competition for land use. This is a major challenge and one that remains largely underestimated: it is assessed that by the year 2030 around 30 per cent of necessary food-related production increases will be attributable to this. This relationship between individual eating habits, global land use and food security is insufficiently well known; it should be brought to the attention of consumers through educational campaigns. Priority should be given to creating awareness of the issue, particularly in the industrialized countries, and encouraging people to change their behaviour. Initiatives at international level, for example in connection with the UN organizations, could also play a part. These initiatives should be supported by international cooperation on the land required for the per capita consumption of food. Measures of sustainability such as the ecological footprint can illustrate the fact that on a global scale natural resources are currently being used at a rate that exceeds their capacity for regeneration.

- *Promptly identify risks posed by land use to food security:* An effective early warning system is needed if societies are to be better prepared for future crises. Existing monitoring capacities, such as those of the FAO and the World Food Programme, should be strengthened and more efficiently networked. In addition, as pressure on global land use increases WBGU recognizes an increasing need for risks to food security arising from competing use to be identified at an early stage. In this connection global monitoring and early warning systems are extremely important.

- *Take account of the coupling of land use, food markets and energy markets:* The challenges of global food security must today be dealt with against the backdrop of increasing pressure on global land use; they can no longer be addressed through national endeavours alone. In a globalized world, policy-making must take account of the ever-closer links between land use and agricultural commodity price trends on the one hand and the energy market on the other. Policy-makers must therefore create regulatory mechanisms to deal with situations such as trends in the energy markets that have undesirable consequences for food security. In the long term it is important for food security that the world agricultural markets should generate an impetus for production increases, particularly in the poorer developing countries. To this end import barriers for agricultural goods should be further dismantled and export subsidies and other production-promoting measures worldwide, but particularly in the industrialized countries, should be reduced. Any liberalization of trade must, however, take account of the fact that developing countries vary in their circumstances and needs. For example, LIFDCs are directly and adversely affected by price rises on the world market. Exceptions to a general liberalization should therefore be made for a group of the predominantly poorer developing countries.

BIODIVERSITY CONSERVATION: UTILIZE THE OPPORTUNITIES PRESENTED BY THE CBD

The expansion of bioenergy must not result in the directly or indirectly induced conversion of natural ecosystems. To prevent this, an effective system of protected areas is essential. WBGU recommends that a global, ecologically representative and effectively managed system of protected areas with adequate financing should be set up on 10–20 per cent of the world's terrestrial surface by 2010. The Convention on Biological Diversity (CBD) is the key international agreement for implementing this guard rail for biosphere conservation.

- *Close the funding gap that affects protected areas:* To this end WBGU recommends mobilizing a sum of € 20–30 per capita per year in the high-income countries. In the first instance use should be made of the LifeWeb Initiative, which was set up and provided with considerable funds at Germany's instigation, so that tangible bilateral projects move swiftly forward. At the same time other donor countries should be persuaded to give financial support to LifeWeb. If this is successful, the initiative can in the medium term become a nucleus for a protected area protocol to the CBD that links implementation of measures relating to protected areas with funding instruments. The practical and political feasibility of the protocol and possible links with the emerging REDD regime under the UNFCCC should be researched and evaluated as options. In addition, WBGU supports an expansion of international compensation payments for foregone agricultural and forestry income, in order to make the transition to sustainable land use financially viable for developing countries. Pilot projects are to be used to assess whether national-level habitat banking systems in industrialized countries can be opened to providers of ecosystem services in developing countries. Countries with economies in transition, newly industrializing countries and countries rich in raw materials should also be more closely involved in the financing of international nature conservation. Plans should be being made now for a market-like mechanism in which the assurance that previously certified areas are protected is traded for money.
- *Use the CBD to develop biodiversity guidelines for sustainability standards:* In the light of the outcomes of COP-9 it cannot be assumed that rapid progress will be made, but nonetheless this process should be promoted by the German presidency of the CBD and as far as possible moved rapidly forwards. In order to build the necessary monitoring capacities, the development of the world database on protected areas should be promoted at the same time. The impetus for sustainability standards in the bioenergy sector should be used in the medium term to arrive at general guidelines for all forms of biomass production.

IMPROVE WATER AND SOIL PROTECTION THROUGH THE CULTIVATION OF ENERGY CROPS
Present trends in global water and soil use are tending in the wrong direction. Without policy change this will result in a worsening water crisis and increasing soil degradation in many areas.

- *Make analysis of regional water and soil availability a requirement:* Since water and soil are highly endangered resources in many regions, any large-scale promotion of bioenergy cultivation systems should be preceded by an integrated analysis of regional water and soil availability. Non-adapted bioenergy cultivation systems and the globally mounting demand for energy can significantly increase the pressure of use on soil and water resources. The cultivation of energy crops should therefore be aligned with regional strategies for sustainable soil and water management.
- *Use the cultivation of energy crops to restore marginal land:* If the proper cultivation system is chosen, the cultivation of energy crops on marginal land (such as degraded land) can actually result in an improvement in soil fertility. The cultivation of energy crops on degraded land is therefore a strategic option – it can be used to restore land at least some of which could later be available for food production. This could play a part in reducing the increasing pressure on land use.

5.4
Making targeted use of bioenergy promotion policies

It is important that, in principle, only those bioenergy pathways are promoted that contribute to climate change mitigation in a particularly sustainable way. In WBGU's view this means that not only is the minimum standard adhered to but that, taking account of total life-cycle emissions, the use of bioenergy is able to avoid emissions of at least 60 t CO_2eq per TJ of raw biomass used. Since for practical reasons promotion needs to be provided at the various stages of the production process (cultivation, conversion and end-use application systems), it is usually necessary to work with default values regarding the emissions of the other stages.

Particularly in connection with the promotion of energy crop cultivation, WBGU regards it as important that, in addition, ecological and social promotion criteria are met. Likewise, where biogenic residues are mobilized, ecological limits should also be observed so that soil fertility is maintained. Finally, promotion of conversion and end-use application systems should be undertaken in such a way as to ensure that they fit with the vision of the transformation towards sustainable energy systems. Undesired lock-in effects should be avoided and promising technologies such as electromobility should be promoted.

Alongside the focus on climate change mitigation, sustainability of energy systems involves addressing energy poverty. Modernizing off-grid or traditional uses of bioenergy can play a valuable part in this, particularly in the rural regions of developing countries. In such situations WBGU regards promotion of bioenergy-based projects as justified even if climate change mitigation and promotion criteria are not fully met.

REMODEL PROMOTION IN THE AGRICULTURAL SECTOR

Sustainable biomass production for energy purposes should ideally only be promoted if the land use contributes to nature or soil conservation. At the very least, instances of the promotion of biomass production that do not meet the WBGU minimum standard should be brought to an end within the next few years, and transferred to sustainable methods of production wherever possible. In general, production subsidies in the agricultural sector should as far as possible be removed; this would bring an end to inefficient competition for subsidies between countries and remove market distortions in world agricultural trade. Subsidies that yield substantial benefits in development-related or environmental terms form an exception; they should be explicitly permitted.

PHASE OUT PROMOTION OF LIQUID BIOFUELS AND PROMOTE ELECTROMOBILITY

Technology policy on the use of bioenergy in the transport sector must be re-directed. From the point of view of sustainability, promotion of liquid biofuels for road transport – particularly in industrialized countries – cannot be justified. The reasons for this include the high GHG abatement costs, low or negative GHG reduction potentials per unit of land or per unit of biomass used, and the lock-in effects on an inefficient infrastructure based on the combustion engine. Blending quotas should not be increased any further, and the current blending of biofuels should cease completely within the next three to four years. The road-traffic-related emissions reductions that have been agreed at EU level will then have to be achieved by other means. In the transport sector the highest energy efficiency of biomass is achieved through the generation of electricity and its use in electric vehicles. An appropriate framework for the expansion of electromobility should be developed. Promotion policies can assist businesses in their technological development by helping to expand opportunities for connection to the electricity grid. Demand for electric or hybrid vehicles can be stimulated through taxation policies.

PROMOTE BIOENERGY PATHWAYS FOR ELECTRICITY AND HEAT PRODUCTION

Greater incentives for utilizing the potential of organic wastes and residues are created primarily through the promotion of renewables in electricity and heat production. The aim must be to promote the use of biogenic wastes and residues in such a way as to ensure that it is distinctly more attractive than the generation of electricity from energy crops. In tandem with this there is a need for appropriate regulation on the extraction of residues from agriculture and forestry, the dumping of waste and cascade uses. In some countries there is already promotion of the direct combustion of biomass (primarily wood chips and pellets from residues) in coal-fired and cogeneration plants and of the use of biogas, crude gas and biomethane; this should be continued and introduced as a priority in all regions in which coal plays a major part in electricity generation. However, it is essential to ensure that the biomass used meets the minimum standard with regard to sustainability. The production of electricity from biomass that meets the promotion criteria should be particularly encouraged. In addition, particular emphasis should be placed on promoting the use of biomethane if the CO_2 which is captured during the production process can be removed to secure storage. If at the same time the international scaling-up of cogeneration and combined-cycle power plants accelerates as a result of appropriate climate and energy policy measures and suitable promotional approaches, it will be possible to utilize highly efficient bioenergy pathways and hence achieve globally significant reductions in emissions. In WBGU's view it is entirely appropriate to promote the combustion of wood chips or pellets for electricity generation, but state subsidies for pure heat use should be provided in industrialized countries at most for a transition period, until a transformed energy system is in place in which this need is met from CHP plants or from heat pumps running on renewable electricity.

INITIATE AN INTERNATIONAL AGREEMENT ON (BIO)ENERGY SUBSIDIES

In order to cut back energy subsidies that harm the environment and give a higher priority to sustainability criteria, states need to coordinate their policies at international level. They should enter into agreements whereby non-sustainable energy subsidies are removed in all countries and guidelines for permissible subsidies, based on the principle of sustainability, are established. This could occur in the context of a Multilateral Energy Subsidies Agreement (MESA), which at the outset might involve only the most important energy producers and consumers. In the long term the agreement could form part of the WTO regime.

STRATEGICALLY MANAGE THE USE OF BIOMASS AS
AN INDUSTRIAL FEEDSTOCK

In order to pave the way for strategies for the use
of biomass from agriculture and forestry as a feedstock in industrial production processes, material
flow analyses and land-use inventories should be
drawn up both globally and nationally. The scenarios
should describe likely developments (competition
for land use, substitution processes, etc.) and options
for action. For key categories of materials and products (cellulose, paper products, etc.) sustainability
standards for the cultivation and extraction of feedstocks should be set and product standards with high
recycling quotas should be specified. Through suitable measures it should be possible for high levels
of resource and product consumption to be greatly
reduced.

5.5

Harnessing the sustainable bioenergy potential in developing and newly industrializing countries

MAKE TACKLING ENERGY POVERTY A PRIORITY OF
DEVELOPMENT POLICY

As a target WBGU recommends endeavouring to
ensure that traditional forms of bioenergy use that
are harmful to health are replaced by 2030. Facilitating access to modern forms of energy does not
have to be included as a stand-alone goal in the Millennium Development Goals (MDGs), but it should
be explicitly included in the MDGs as a means of
tackling poverty and, moreover, should be more
strongly anchored in the energy policy portfolios of
stakeholders involved in international development
cooperation. As a first step, tackling energy poverty
should be systematically included in Poverty Reduction Strategy Papers (PRSPs). The international
community should particularly promote bioenergy
projects that advance rural off-grid energy supply in
developing countries.

BASE STRATEGIES FOR REDUCING ENERGY POVERTY
ON RELIABLE DATA

So that alternative ways of providing energy services can be examined and obstacles to implementation can be better understood, actors involved in
international development cooperation must work
with national actors to draw up strategies for tackling
energy poverty. These approaches should be based
on reliable empirical findings and must be embedded in suitable policy strategies. WBGU therefore
recommends carrying out multi-country cross-sec-
tional evaluations and nationally, regionally and
locally specific studies in order to obtain information
on best practices.

SUPPORT DEVELOPING COUNTRIES IN DRAWING UP
NATIONAL BIOENERGY STRATEGIES

So that the opportunities and development potentials
of bioenergy can be realistically assessed and risks
can be minimized, WBGU recommends that strategic issues be discussed in the country context and
with as broad a range of stakeholder groups and
affected sections of the population as possible, and
that decisions then be taken on the priority goals of
any promotion of bioenergy. Development cooperation actors should support partner countries in
developing these strategies, examining all the forms
in which bioenergy and its alternatives can be used
and applied, as well as evaluating the suitability of
these forms in the context of the local situation. They
should also seek to ensure that the minimum standard and promotion criteria are met and that the necessary governance capacities, such as land-use planning and certification, are strengthened. In addition,
it is essential that bioenergy strategies be linked to
food security strategies.

PROMOTE PILOT PROJECTS THAT INVOLVE
PARTICULARLY SUSTAINABLE CULTIVATION SYSTEMS
AND THE USE OF WASTES AND RESIDUES

Cultivation methods that are particularly sustainable and that help to combat soil erosion, conserve
biodiversity, reduce energy poverty and advance rural
development should be promoted in pilot projects.
Such methods include, for example, the socially
acceptable cultivation of perennial energy crops on
degraded land, or agroforestry. WBGU also recommends that the country-specific potentials of wastes
and residues be assessed and then utilized in electricity generation, particularly in agro-industrial biogas
plants and cogeneration plants where the waste heat
is used. Pilot projects can improve the mobilization
of the potential of residues and wastes.

CREATE BIOENERGY PARTNERSHIPS

Multilateral cooperation for purposes of sustainable bioenergy use can be supplemented by intergovernmental partnerships. Technology agreements
are appropriate in this context, for example for scaling up technologies for processing and using biomethane. These technologies can be linked to aspects of
sustainable land-use policy or to trade partnerships.

PROMOTE THE RESTRUCTURING OF THE WORLD
ENERGY SYSTEM

In order to increase the purchasing power of people affected by energy poverty, development cooperation should continue its financial support of
microfinancing systems. Cooperation between the
private and public sectors should be encouraged in
order to mobilize private capital. Greater use can be

made of CDM projects for the large-scale substitution of fossil fuels. The technologies recommended by WBGU in connection with the sustainable use of bioenergy in the energy systems of developing countries serve not only to tackle energy poverty; the majority of them also address the issue of climate protection. For instance, making projects that aim to improve the efficiency of traditional uses of bioenergy eligible as small-scale CDM activities is justifiable and can contribute to financing. In addition, the international community should coordinate and support the restructuring of the world energy system. WBGU recommends that the German government should position itself at the forefront of such a process at European level and in the supervisory bodies of the international organizations involved, so that it can continue to maintain its pioneering role in climate change mitigation.

5.6
Building structures for a sustainable global bioenergy policy

SET UP A GLOBAL LAND-USE REGISTER
To be able to monitor direct and indirect land-use changes when introducing standards and the requisite certification systems, it is important that a global, GIS-supported land-use register is set up. As a key element of this, rapid further development of the world database of protected areas managed by the World Conservation Monitoring Centre of the United Nations Environment Programme (UNEP-WCMC) is recommended. However, the global land-use register must go beyond this database; for each imported bioenergy carrier it must be able to provide information on the land on which it was produced (geographical coordinates, manner of cultivation, commitment to adherence to sustainability criteria, etc.).

CREATION OF AN INSTITUTIONAL FRAMEWORK FOR THE GLOBALIZATION OF STANDARDS
The Global Bioenergy Partnership (GBEP) should be used as a forum for developing a uniform international bioenergy standard and accelerating multilateral policy formulation. This partnership brings together key stakeholders and includes newly industrializing countries. Efforts should, however, be made to ensure that relevant civil society stakeholders have greater involvement in the dialogue. GBEP or the Task Force on Sustainability should be helped to channel, in their capacity as an intergovernmental forum, the formal and informal processes involved in drawing up global sustainability standards and to work towards the creation of global standards and

guidelines. The proposals of WBGU, which has taken up important ideas put forward by the Roundtable on Sustainable Biofuels, could provide a basis for this.

PROMOTE BIOENERGY THROUGH IRENA
The International Renewable Energy Agency (IRENA) is being set up with the aim of promoting worldwide use of renewable energies through policy advice, technology transfer and knowledge dissemination; this is an appropriate step towards the streamlining and institutional strengthening of international energy policy. Nevertheless, in addition to promotion of renewable energies IRENA should include all aspects of the transformation towards sustainable energy systems in its remit. It should be enabled to address aspects of energy demand and issues relating to energy, the environment and development in a comprehensive and integrated manner.

CONVENE AN INTERNATIONAL CONFERENCE ON SUSTAINABLE BIOENERGY
In order to arrive at a shared global understanding of the opportunities and risks of bioenergy and a consensus on appropriate standards in relation to the production and use of different forms of bioenergy, WBGU recommends that an International Conference on Sustainable Bioenergy be convened at an early stage. This conference could be modelled along the lines of renewables 2004. It could be used to formulate objectives and general promotion principles, exchange ideas for best-practice approaches and draw up agreements on international bioenergy partnerships and on the importance of bioenergy for a sustainable global energy system. It is important that it should bring together actors from the policy areas of agriculture, energy, the environment and development.

5.7
Conceiving of global land-use management as a challenge of the future

Inherent in the problem of competing land use, in WBGU's view, is a potential for future conflict that reaches far beyond the sphere of bioenergy. Critical trends in world food security are even now becoming apparent, and they will become more acute as the world population increases to around 9 billion and land-intensive food consumption patterns become ever more widespread. Global land-use management is therefore set to become a key future task if land-related conflicts are to be avoided.

SET UP A GLOBAL COMMISSION FOR SUSTAINABLE
LAND USE

The increasing pressure on land use is a global challenge of an extent and complexity which is as yet little understood. This calls for the development of a
complex new field of global governance in which
issues of food, energy, development, environmental
and climate policy mesh. On account of the diverse
global interactions and linkages involved, it will no
longer be possible to see land use solely as an issue
for action at individual country level. This is power-
fully illustrated by the example of the worldwide
effects of indirect land-use changes associated with
the expansion of bioenergy, and by the issue of equitable per-capita land use in connection with global
food security. A new global commission for sustainable land use should be set up to start these processes at international level and to organize how to
approach the issue. The commission's task should
be to identify the key challenges arising from global
land use and to assemble the scientific state-of-the-
art. Drawing on this groundwork, the commission
should then elaborate the principles, mechanisms
and guidelines required for global land-use management. The commission could be located within UNEP
and work closely with other UN organizations such
as the FAO. The findings should be regularly placed
on the agenda of the UNEP Global Ministerial Environment Forum or the strategically important G8+5
gatherings of heads of state and government.

Introduction

It is better to be approximately right than
precisely wrong.
JOHANN WOLFGANG VON GOETHE
1749–1832

Bioenergy in the form of open fire was the first source of energy used by humankind. Even today a quarter of the world's population still depends on this traditional form of bioenergy use. During the last 150 years fossil fuels – initially coal, but then oil and natural gas – have replaced wood as the dominant source of primary energy. The emergence of markets for modern bioenergy is by contrast a relatively recent phenomenon, with dynamics that are driven by varying motives in industrialized and developing countries. Liquid biofuels for the transport sector occupy a central position in the current public debate.

Bioenergy presents opportunities for climate and environmental policy, security of energy supply and rural or economic development and has in consequence been the subject of policy measures and programmes in many countries, sometimes on a vast scale. As food prices have risen in recent years, however, it has become clear that bioenergy poses risks due to competing interests in using land to grow crops for food or biomass for energy. In addition, the conversion of natural or semi-natural land for energy crop cultivation leads to the release of greenhouse gases as vegetation and soil carbon decompose; biological diversity is also lost.

The rising price of oil has intensified the search for fuels that can be used in combustion engines to replace petrol and diesel and has thus accelerated the expansion of bioenergy. Many stakeholders in bioenergy policy focus first and foremost on the production and utilization of energy crops, despite the lack of adequate scientific evidence for many of the assumptions on which policy-makers base their decisions. For example, it is not yet sufficiently clear which energy crop deployment pathways under which conditions of cultivation and use can make a significant contribution to climate change mitigation, how direct and indirect land-use change can be inventoried and how competition for land use can be avoided. On the other hand, the generation of energy from wastes and residues is a potentially sustainable method of using biomass for energy and entails relatively few problems; tapping that potential has not yet received the attention it deserves.

Since agro-ecological and socioeconomic conditions vary widely and national energy supply structures also differ from country to country it is impossible to make universal recommendations on the use of bioenergy. In addition to establishing global guard rails and standards to safeguard the sustainability of bioenergy use, the local situation must always be appraised from case to case. However, the rapid pace of current bioenergy expansion with the concomitant risks to sustainability highlight the need to establish national and international frameworks for the use of biomass as an energy source.

In this setting marked by uncertain knowledge and conflicting political interests, the present report by WBGU seeks above all to map routes for sustainable bioenergy use, reveal the associated opportunities, determine the prevailing uncertainties, identify risks and highlight the scope and need for regulation in both the short and long term.

This WBGU report examines the issues surrounding bioenergy from a global perspective and depicts the differing motives of the industrialized, newly industrializing and developing countries in connection with the use of biomass for energy. Bioenergy is a major issue of much wider scope than the debate on liquid biofuels suggests. The report therefore distinguishes between traditional biomass use, biogenic wastes and residues and energy crops. It includes an assessment of the globally sustainable potential for energy crop cultivation, which is limited by the WBGU guard rails for food security, climate change mitigation and nature conservation. Within this context competition for land use is also explored and evaluated. In addition WBGU assesses more than 60 bioenergy pathways, from resource extraction to energy service, in terms of their (positive, neutral or negative) contribution to the global shift towards sustainable energy systems.

WBGU sees bioenergy as having a sustainable global potential amounting to around one-quarter of present primary energy use. The challenge is to utilize these opportunities while minimizing the risks, taking account at the same time of increasingly globalized markets, highly divergent political interests and accelerating climate change. Policy-makers have to develop a framework for the sustainable use of biomass for energy quickly, before technological developments lock in that would do the climate more harm than good. The present report will, it is hoped, provide some guidance.

The increased production and use of biomass for energy purposes and the creation of a market for modern bioenergy is actively pursued for disparate reasons and with diverse policies in different parts of the world (GBEP, 2008). Promotion policies and programmes – some of which are on a large scale – are based on arguments such as climate change mitigation, conservation of the environment, energy and supply security and rural or national development. In its analysis WBGU focuses on the role of bioenergy in a sustainable global energy system and thus has a specific perspective on the global bioenergy discussion. In order to highlight the potentials and limits of bioenergy and the parameters of policy-making, it is important that the dimensions and dynamics of the overall debate are first understood.

The most important current discourses on bioenergy are briefly summarized below. Consideration of the different discourses, which tend to be conducted in parallel, reveals the commonalities and contradictions of present bioenergy policies. It will also become apparent that a wide range of political and economic interests come into play at both national and transnational level in industrialized and developing countries; it is essential to be aware of these if one is to understand the current debate and the predicted prospects of a sustainable policy of the future. This chapter thus describes the wider context within which WBGU frames the formulation of its own objectives and priorities for a sustainable global bioenergy policy.

2.1
Current discourses on bioenergy

In the recent past at least three different bioenergy discourses have emerged, underpinned by diverse motives and stakeholder constellations. It is due to the dynamics of these different discourses that no predominant view of the benefits and disadvantages of bioenergy has as yet become established.

Firstly there is a discourse centred on environmental policy, which focuses on the contribution of bioenergy to climate change mitigation and resource conservation. Bioenergy is regarded as a 'green', climate-friendly form of energy. It is therefore seen as playing an important role, particularly in the industrialized countries, in enabling the Kyoto commitments to be met. In the long term it is thus envisaged that bioenergy will contribute to the transformation of energy systems towards a low-carbon economy. This discourse is currently supported by the IPCC guidelines, which classify the use of bioenergy as in principle carbon neutral (IPCC, 2006).

Typical of policies that are based on this discourse is the EU's Biofuels Directive (2003/30/EC), which aims to reduce traffic-induced CO_2 emissions through the blending of biofuels. That the traffic sector's contribution to climate change mitigation should involve liquid biofuels can be explained in part by the vested interest of a major stakeholder in European economic policy: the automobile industry. The use of biofuels in conventional combustion engines requires only minor technical modifications; by using biofuels, extensive and costly technologi-

---

Box 2.1-1

Terminology: Bioenergy, biofuels, agrofuels

Many of the bioenergy-related terms that are bandied about in the public debate are not used in a standardized manner. Bioenergy is the final energy or useful energy that is converted and made available from biomass. Biofuels are liquid or gaseous fuels of biogenic origin; they can be used

as transport fuels or deployed in the stationary applications of power generation or cogeneration.

The prefix 'bio' has a positive connotation, but biofuels may also be derived from the non-sustainable cultivation of energy crops. Because of this, the term 'agrofuels' is now often used, or – less frequently – 'agri-ethanol', 'agro-energy' or 'agrogas'. WBGU continues to use the original terminology, because 'bioenergy', 'biofuels' and 'biogas' are the more familiar terms.

cal change can therefore be avoided, and the industry can at the same time claim to be making a serious contribution to climate change mitigation. This provides companies with a low-cost means of demonstrating their commitment to tackling climate change, and relieves consumers of the need to change their behaviour directly, for example by reducing their car usage. Since the actual effectiveness of these biofuels in mitigating climate change was initially not seriously questioned, this approach appealed to decision-makers in politics and industry and appeared to engender little opposition. The assumed effectiveness in mitigating climate change became a pivotal argument in favour of subsidizing biofuels from energy crops, in the industrialized countries and elsewhere. Now that greenhouse gas balances are better understood and interdependencies with food production and nature conservation have become apparent, supporters of biofuels face growing criticism. As a result first steps towards a correction of policy are already being taken, while some are going so far as to call for a moratorium on the cultivation of energy crops (Umwelt Aktuell, 2008).

A second discourse on resource scarcity, rising energy prices and energy security regards bioenergy as an alternative to the fossil energy carriers – coal, oil and natural gas. It builds on the assumption that the use of biomass can contribute to greater energy and supply security and to reduced dependence on fossil and nuclear fuel imports.

Sharp price rises and the predicted scarcity of fossil fuels, particularly oil ('peak oil'), and the growing demand from newly industrializing countries, have in recent years kindled a new debate on security of supply (Worldwatch Institute, 2007; Economist, 2008a). Since the production of mineral oil is concentrated in a small number of regions, many of which are politically unstable, security-related and geostrategic motives for the substitution of oil imports are also coming to the fore (Mildner and Zilla, 2007; Adelphi Consult and Wuppertal Institut, 2007). This combination of reasons plays a particularly important role in the USA (White House, 2006). In the European Union, too, dependence on Russian gas and oil is seen as a serious risk to the security of supply (EU Commission, 2005c). In both cases these arguments have been used to support ambitious plans for expanding the use of biomass for energy, and in particular the use of liquid biofuels.

Reducing dependence on imports is, however, also an explicit goal of the bioenergy programmes of many newly industrializing and developing countries. The main aim of such a reduction is to circumvent rising procurement costs for fossil resources. The high oil prices of recent years have significantly worsened the balance of trade of many countries, and

import substitution through bioenergy is extolled as a possible way out of this situation (UN-Energy, 2007b). For example, high crude oil prices and the goal of self-sufficiency of supply were major determinants of the biofuel policy used by the Brazilian government in 2006 to achieve its goal of self-sufficiency in crude oil (IEA, 2006a). Other newly industrializing and developing countries, such as India and Indonesia, also point to import substitution as an important motive for their biofuel strategies (e.g. Planning Commission, 2003).

This discourse, too – which is encountered in equal measure in developing, newly industrializing and industrialized countries – is frequently confined to liquid biofuels and the transport sector. Alongside the mineral oil suppliers and small and medium-sized businesses that perceive major market openings for biofuels in industrialized countries (Economist, 2008a), there are also powerful arguments in developing and newly industrializing countries for a development pathway that involves crude oil substitution. These include the growing affluent consumer groups and the rapidly rising demand for motor cars. Within such specific national supply discourses the argument of improved access to energy in rural areas plays only a subordinate role. In consequence, priorities and support policies are insufficiently geared to the needs of countries and regions affected by energy poverty.

In a third discourse centring on rural development and economic potential the fresh opportunities for growth and employment in agriculture are emphasized. In industrialized countries the increased use of biomass for energy is seen as an opportunity to revitalize the sectors of the economy based on agriculture and forestry and secure jobs in these areas (DBV, 2004). This combination of reasons plays an important role in both the USA and the European Union, not least because it can be used to legitimize new or continuing agricultural subsidies (Koplow, 2007; Kutas et al., 2007).

Many newly industrializing and developing countries likewise support the expansion and promotion of specialized energy crop farming. Many of these countries are predominantly agrarian and they particularly stress the opportunities for national development that arise from employment effects in agriculture and the possible growth potentials of the export-oriented production of energy crops and biofuels (Lula da Silva, 2007). It is argued that natural geographical conditions, regional climate, the availability of agricultural and forestry land and low wage costs result in comparative cost advantages on the world market; these would open up global sales opportunities, perhaps extending to specific trade partnerships with industrialized countries where demand is located (Mildner and Zilla, 2007; Mathews, 2007). In particular, the major agricultural producers among the newly industrializing and developing countries – such as Brazil, Indonesia, Malaysia, South Africa and Argentina – have high hopes of an emerging world market in biofuels. Even though this development discourse has recently been put on the defensive as the possible social and ecological consequences have been pointed out (slogans have included: 'food, not fuel', 'tortilla crisis' and 'destruction of the rainforest'), the argument centred on new development opportunities arising from bioenergy continues to play an important role.

Beyond these primarily macro-economic considerations, multinational companies also see significant commercial potential in the areas of agrochemistry and plant biotechnology (Bayer CropScience, 2006; Economist, 2008a). As a result, the interests of agricultural policy and energy policy stakeholders sometimes coincide, a situation that amplifies the impact and forcefulness of this discourse.

When viewed as a whole there is a great deal of overlap between the individual discourses and the interests of the stakeholders pushing them. It is often suggested that bioenergy as such – without differentiating further – could have positive effects in a number of issue areas ('win-win-win'). Interactions, conflicting objectives and risks are overlooked – partly out of ignorance, partly as a calculated strategy. Different interest groups compete to dominate the discourse on bioenergy and thus assert their influence on relevant policy-making processes.

It is noticeable that in the public debate on alternative energies there is still little attempt to distinguish between different production and deployment forms of bioenergy. In particular, liquid biofuels are often equated with bioenergy in general. It is even rarer to encounter any differentiation between the use of bioenergy in fundamentally different energy sectors such as power, heat and transport. The same attitude is revealed in the narrow focus of bioenergy policy to date on the transport sector and biofuels. Otherwise unheard-of alliances of different interest groups – such as the automobile industry and environmental conservationists, or groups representing agricultural interests and energy companies – have been able to state their case with particular forcefulness. In consequence, a policy of promoting bioenergy appeared to everyone involved to be a worthwhile strategy. But it is an open question whether the bioenergy policy that is currently being pursued is meaningful and effective in the sense of involving coherent promotion of climate change mitigation and energy security while also heeding the principles of sustainable development.

2.2
Sustainable global energy systems and land-use systems

When seen in terms of its multiple interlinkages, bioenergy is the most complex of all the known forms of renewable energy. The potential benefits and the risks of extensive undesirable effects are both high. This makes it all the more urgent to question the globally sustainable deployment of bioenergy: what should the use of biomass for energy achieve, what can it achieve and what are the associated risks and limits?

Bioenergy is in the first place a form of energy. As WBGU has already shown in previous reports, it is essential to turn energy systems towards sustainability worldwide – both in order to protect the natural life-support systems on which humanity depends, and to overcome energy poverty in developing countries (WBGU, 2004a). The increased use of bioenergy must therefore be evaluated in terms of whether and to what extent it contributes to this global shift towards sustainable energy systems.

A sustainable energy system must be anchored in a general process of sustainable development in order to ensure that the use of bioenergy is not at the expense of other sustainability dimensions. Furthermore, conversion into energy is not the only use of biomass. The issue of the sustainable use of bioenergy is therefore just one aspect of a wider question – in view of the fact that biomass, while it is renewable, is not available in unlimited quantities, in what way and for what purposes should it be used in order to facilitate globally sustainable development?

The following sections elucidate the areas of bioenergy use in which WBGU considers a significant contribution to sustainable development to be possible and which therefore form the core of the report's analysis.

2.2.1
Bioenergy, energy system transformation and climate change mitigation

Effective climate change mitigation is essential if there is to be any prospect of globally sustainable development (WBGU, 2007). In order to avoid dangerous climate change, within the next ten years the emissions trend must be reversed and by 2050 global greenhouse gas (GHG) emissions must be cut to half their 1990 level. Currently (2004), 56.6 per cent of global greenhouse gas emissions are CO_2 emissions from the combustion of fossil energy carriers. Overall, central energy generation contributes 25.9 per cent of global GHG emissions, transport contributes 13.1

per cent and industry 19.4 per cent (IPCC, 2007c). A transformation of energy systems is therefore indispensable for attainment of the climate change mitigation targets (WBGU, 2003, 2004a).

Two other sectors that are highly relevant to climate change mitigation are forestry and agriculture, which contribute respectively 17.4 per cent and 13.5 per cent to global greenhouse gas emissions. Emissions from the forestry sector are predominantly CO_2 emissions from ongoing deforestation; those from agriculture are attributable in approximately equal proportions to emissions of methane and nitrous oxide (IPCC, 2007c). Whether the climate change mitigation targets can be achieved therefore depends not only on the transformation of energy systems but also to a significant extent on the future development of global land use.

Bioenergy, provided that it is not limited to the use of wastes and residues, is directly linked to land use and therefore has the potential to lead to a change in emissions in the agriculture and forestry sectors. It thus forms an interface between the two most significant drivers of climate change – global energy systems and global land use.

2.2.2
Bioenergy, energy system transformation and energy poverty

A further goal of the global reconfiguration of energy systems is to overcome energy poverty in developing countries. Energy poverty involves a lack of adequate options for access to affordable, reliable, high-quality, safe and environmentally sound energy services to meet basic needs (WBGU, 2004a). Access to modern energy is an important element in tackling poverty and a condition for attainment of the Millennium Development Goals (WBGU, 2004a). Some 2500 million people presently depend on biomass as the primary source of energy for cooking. In many countries, particularly in sub-Saharan Africa, biomass accounts for more than 90 per cent of household energy consumption (IEA, 2006b). The majority of this bioenergy is used in traditional form; it therefore frequently involves inefficient technologies and major risks to health (Section 8.2). In WBGU's view, further development of existing bioenergy use or its replacement by low-emission forms of energy is key to overcoming energy poverty.

2.2.3
Specific properties of biomass

Since the amount of biomass that is annually renewed in the biosphere is limited and conversion into energy is only one of a number of ways in which biomass can be used, any expansion of energy crop cultivation needs to be evaluated in the context of competing demands. In particular, the area of land available for energy crop cultivation is limited by the absolute necessity of ensuring an adequate level of food production. Similarly, the energy yield achievable per unit of land cannot be increased indefinitely, since there is a theoretical upper limit to the efficiency of photosynthesis in converting incident solar energy into biomass. This makes it all the more important not simply to regard bioenergy as a mere quantitative contribution to overall energy, but to conduct a general evaluation of the qualitative properties of biomass in order to identify how they might contribute to the objectives of a sustainable energy system.

PROPERTIES OF BIOMASS AS AN ENERGY CARRIER
Plants are able to absorb and store solar energy without technological intervention. Humans can utilize this property by burning biomass in various forms. Conversion and storage of bioenergy requires only the simplest of technology; bioenergy has therefore been utilized since the dawn of human history. Today bioenergy is used predominantly by the poor, for whom it represents an affordable and easily manageable form of energy. From the point of view of utilization, biomass and fossil fuels – which are ultimately stored biomass from prehistoric times – share similar properties. In particular, biomass can be used upon demand. This means that even in complex energy systems it can play an important part in securing the energy supply: as the proportion of renewable energies increases, biomass can balance and supplement the intermittent feed-in of wind and solar energy in electricity supply systems.

PROPERTIES OF BIOMASS AS A CARBON SINK AND CARBON RESERVOIR
At the same time as storing energy, plants also store carbon, which is removed from the air in the form of CO_2. If the biomass is used as energy, the stored CO_2 is again released. As with the use of fossil fuels, it is technologically possible – although not straightforward – to separate and store the CO_2 in the course of energy generation. In generating biomethane, some of the CO_2 must in any case be separated in order to make the gas usable. In the case of biomass, however, it is also possible and technologically fairly simple to store CO_2 temporarily if some or all of the utilization for energy is foregone or delayed. Depend-

ing on the use of the biomass and the way in which it is kept, CO_2 can be stored in this way for several centuries (wood) or even millennia (charcoal). There is at present no relevant technical process that can be used to remove CO_2 directly from the atmosphere in a way similar to that used by plants.

PROPERTIES OF BIOMASS AS AN INDUSTRIAL FEEDSTOCK

Biomass is also used as a raw material and feedstock by the manufacturing and chemical industries, and in the building trade. In developing countries, in particular, wood is an easily available building material and resource. The material use of biomass is also relevant to climate policy because, as well as storing carbon, it also enables the use of emission-intensive materials such as cement to be avoided.

SUBSTITUTABILITY OF BIOMASS

It is clear from the various qualitative properties of biomass that appropriate use of biomass can in principle make a positive contribution to climate change mitigation and the overcoming of energy poverty. However, it is also possible to pursue these objectives by other means, without using biomass – for example, by improving efficiency in energy supply, using other renewable energies or employing technology to separate and store CO_2 from fossil sources. There are, though, properties and uses of biomass that cannot be substituted. This applies, for example, to biomass as a key component of ecosystems, or biomass as a food- and feedstuff. From the point of view of sustainability, therefore, biomass should only be used for climate change mitigation and energy supply in ways that do not jeopardize its non-substitutable uses and properties. The following chapter sets out the guard rails and guidelines that should in WBGU's view be observed if bioenergy use is to be globally sustainable.

Sustainability constraints upon bioenergy

WBGU derives a sustainable corridor for bioenergy use primarily from its own guard rail concept (WBGU, 1995b). The Council uses the term 'guard rail' to refer to quantitatively defined damage limits, exceedance of which is intolerable or would have catastrophic consequences. An example of such a limit is an increase in global mean temperature by more than 2°C from pre-industrial levels. Sustainable development pathways follow trajectories that lie within the range delimited by the guard rails. This approach is based on the realization that it is virtually impossible to define a desirable, sustainable future in positive terms – that is, in terms of a goal or state to be achieved. It is, however, possible to agree on the boundaries of a range that is acknowledged to be unacceptable and that society seeks to avoid. If the system is on course for collision with a guard rail, steps should be taken to change direction.

Adherence to the guard rails described in this chapter is, however, only a necessary and not a sufficient criterion for sustainability (WBGU, 2001a). The constraints presented by both the socioeconomic and the ecological dimensions of sustainability cannot always be precisely formulated as guard rails. In the socioeconomic arena, for example, many of the requirements of a sustainable bioenergy policy are not quantifiable. Furthermore, the majority of the socioeconomic requirements that are in principle quantifiable cannot be converted into a global guard rail, because they are country- or situation-dependent. Ecological damage limits, too, cannot always be formulated as guard rails – perhaps because regional differences are too great or because no satisfactory global indicator can be specified. For these reasons WGBU specifies, in addition to the guard rails, other sustainability requirements; these provide additional criteria for the sustainable use of bioenergy that cannot be defined in terms of guard rails. They involve, for example, various aspects of land use or adherence to social standards.

3.1
Ecological sustainability

3.1.1
Guard rail for climate protection

In WBGU's view, climate change impacts are intolerable if they are associated with a mean global rise in near-ground air temperatures of more than 2°C from pre-industrial levels, or a rate of temperature change of more than 0.2°C per decade. This guard rail has been explained in detail in earlier WBGU reports (WBGU, 1995b, 2006). Adherence to this guard rail requires the concentration of greenhouse gases in the atmosphere to be stabilized below 450 ppm CO_2eq. To achieve this, global greenhouse gas emissions need to be at least halved by the middle of the century.

A considerable proportion of the CO_2 released by human activities dissolves in seawater and causes acidification there. In order to avoid undesired or dangerous changes in marine ecosystems, the pH level of the uppermost ocean layer should not fall in any major ocean region by more than 0.2 units against the baseline of pre-industrial levels. Adherence to the 2°C guard rail would automatically result in adherence to the acidification guard rail, provided that there is a sufficient reduction not only in the overall 'basket' of greenhouse has emissions but also in CO_2 emissions as such (WBGU, 2006).

A use of bioenergy for which climate change mitigation effects are claimed should be judged by the contribution it makes to adherence to the climate protection and acidification guard rails. For adherence to the 2°C guard rail it is of no consequence whether a particular sector (such as transport) achieves a particular reduction in emissions. The only deciding factor is the development over time of global emissions and of greenhouse gas removals by sinks across all sectors. A realistic assessment of the contribution of bioenergy use to climate change mitigation must take account of the development of emissions in all sectors. To gauge adherence to the acidification guard rail, the effect of bioenergy use on the global carbon cycle must also be considered.

3.1.2
Guard rail for biosphere conservation

The Council has proposed the following guard rail for biosphere conservation: 10–20 per cent of the global area of terrestrial ecosystems (and 20–30 per cent of the area of marine ecosystems) should be designated as parts of a global, ecologically representative and effectively managed system of protected areas (WBGU, 2001a, 2006). In addition, approximately 10–20 per cent of river ecosystems including their catchment areas should be reserved for nature conservation (WBGU, 2004a).

This guard rail is based in part on the realization that ecosystems and their biological diversity are crucial to the survival of humanity, because they provide a variety of functions, services and products (MA, 2005a). Protected areas, in particular, are an indispensable instrument of sustainable development (CBD, 2004b; Section 5.4). It should be noted that the conservation and sustainable use of biodiversity are by no means mutually exclusive: they can be combined in various ways, depending on ecological circumstances (WBGU, 2001a). The World Conservation Union (IUCN, 1994) has accordingly drawn up a graduated category system for protected areas that allows specific relationships between conservation and sustainable use.

A particularly pressing issue is conservation in the hotspots of biological diversity. These are areas in which a large number of wild species is found within a small area or which contain a large number of endemic species or unique ecosystems; they are therefore particularly valuable for the conservation of biological diversity (Mittermeier et al., 1999; Myers et al., 2000). Conservation should in addition include species that are particularly worthy of protection and areas that still contain undisturbed ecosystems on a large scale (wilderness areas, e.g. tropical and boreal forests). For global food security it is also important to maintain the 'gene centres' in which a wide genetic range of crops or related wild plants is found (Vavilov, 1926; Stolton et al., 2006).

The international community has agreed to establish a protected area system of this sort by 2010 (Section 10.5; CBD, 2004b). A positive sign is that the number of protected areas and the proportion of the world's surface covered by them has risen sharply in recent years, so that protected areas now cover around 12 per cent of the global land surface (Box 5.4-1). However, closer examination reveals many of these protected areas to be 'paper parks' (Dud-

ley and Stolton, 1999a) – that is, they are indeed protected by ordinance, but local management is so inadequate that it often cannot even halt the destructive exploitation of biological resources (e.g. illegal logging, predatory fishing). Furthermore, strictly speaking only the areas in IUCN categories I–IV should be included in the tally, since in categories V and VI the emphasis is more on sustainable use than on the conservation of biological diversity. The call for an effectively managed system of protected areas is therefore met by only a fraction of the 12 per cent (Box 5.4-1).

In the same way as the objectives of the Global Strategy for Plant Conservation (GSPC) agreed in the context of the Convention on Biological Diversity, this global guard rail needs to be differentiated and operationalized on a regional basis (CBD, 2002a; Section 10.5). The GSPC's 16 global targets for 2010 include the following:

– at least 10 per cent of each of the world's ecological regions effectively conserved.
– protection of 50 per cent of the most important areas for plant diversity assured. Criteria for the selection of these areas would include species richness, endemism, and uniqueness of habitats and ecosystems.
– 60 per cent of the world's endangered species conserved in situ (e.g. through protected areas).
– 70 per cent of the genetic diversity of socio-economically valuable plant species conserved (gene banks and on-farm conservation).

However, even a perfectly functioning system of protected areas cannot halt the loss of biological diversity. It must be complemented by two processes: integration of the protected areas or protected area systems into the surrounding landscape (CBD, 2004b) and mainstreaming of conservation through differentiated application of the principle of sustainable land use to all land used for agriculture or forestry. The objective is the 'integrated, sustainable management of land, water and living resources' (Ecosystem Approach: CBD, 2000, 2004a). This means that ensuring the sustainability of land use calls for additional ecological sustainability requirements that take account of the nature conservation dimension (Section 3.1.4).

3.1.3
Guard rail for soil protection

In view of the importance of soil protection measures for future food security, it is worth elaborating guard rails for global soil conservation in the form of quantitative values which if exceeded would be irreversible and endanger human livelihoods (WBGU, 2004a; UBA, 2008a). Schwertmann et al. (1987) set the tolerance limit for human-induced soil degradation at a level at which there is no significant deterioration in the natural yield potential of the soil over a period of 300–500 years. In concretizing this guard rail a distinction needs to be made between the two greatest risks to which soil is exposed: degradation through erosion and through salinization.

WBGU has proposed tolerance limits for these two factors (WBGU, 2004a). For soil erosion this means that, strictly speaking, the quantity of soil removed or otherwise degraded must not exceed the quantity newly created, as this would reduce the yield potential in the long term. However, since soil formation takes place on geological timescales, this can be only a distant objective. In the temperate zone, for example, WBGU sets a tolerance limit of 1–10 tonnes per hectare per year, depending on soil depth. As a tolerance limit for soil salinization in irrigation farming WBGU (2004a) proposes that over a period of 300–500 years the saline concentration and composition should not exceed the level that can be tolerated by crops in common use.

3.1.4
Additional ecological sustainability requirements

Not all ecological sustainability dimensions can be formulated as globally valid guard rails. This may be because regional differences are too large or because no satisfactory global indicator is available. In this section WBGU therefore describes additional requirements for sustainable bioenergy use.

For example, in considering the sustainable use of water resources in connection with bioenergy the main issue is the management of water used for irrigation when there is a threat of competition with the use of water for food production. In WBGU's view the water stress indicators found in the literature are not suitable for quantifying a globally valid guard rail. Even in regions with high levels of water stress, many of the adverse effects of irrigation can be avoided and sustainability attained if systematic measures are put in place. A further consideration is that the indicators fail to take account of 'green' water – the water from precipitation that is available to plants in the form of soil moisture.

Even where guard rails are formulated in global terms – for example, in connection with biodiversity conservation or soil protection – their application must be considered in the specific context of local and agro-ecological conditions. In accordance with the ecosystem approach of the CBD (2000) the objective must be the 'integrated, sustainable management of land, water and living resources', which includes humans as an integral component of many

ecosystems. The Addis Ababa principles and guidelines for the sustainable use of biological diversity can be referred to in this regard (CBD, 2004d; Section 10.5), as can the FAO's definition of sustainable land use: 'Sustainable land management combines technologies, policies and activities aimed at integrating socio-economic principles with environmental concerns so as to simultaneously: (1) maintain or enhance production/services (Productivity); (2) reduce the level of production risk (Security); (3) protect the potential of natural resources and prevent degradation of soil and water quality (Protection); (4) be economically viable (Viability); (5) and socially acceptable (Acceptability)' (Smyth and Dumanski, 1993).

Rules and regulations at European and German level are formulated in significantly more specific and concrete ways. In the EU, direct payments to farmers are linked to adherence to mandatory standards of environmental conservation, food and feed security, animal health and animal protection – a system known as cross-compliance (BMELV, 2006; UBA, 2008a). This attachment of conditions to subsidy payments constitutes an environmental policy instrument (SRU, 2008). For German agriculture various laws and ordinances define 'good farming practice' in terms of ecological and safety standards that farmers must adhere to. However, many provisions relating to good farming practice are still formulated in highly indeterminate terms in statutes and ordinances (SRU, 2008).

The existing regulations and available landmark studies should be used as a basis for drawing up specific, internationally recognized management rules or standards for sustainable land use (Section 10.3). Such standards should also take account of the greenhouse gas balance of the various farming systems, because – for example – intensification of land use results in nitrous oxide emissions as a consequence of nitrogen fertilizer use and CO_2 emissions as a consequence of processes such as the conversion of grassland.

3.2
Socioeconomic sustainability

3.2.1
Guard rail for securing access to sufficient food

ACCESS TO FOOD FOR ALL
The expansion of bioenergy use can have an adverse effect on food production and – particularly in low-income developing countries that are net importers of food (Low-Income Food-Deficit Countries, LIFDCs) – on food security, because land, water resources and

agricultural resources (such as machinery, fertilizers, seed, feed, fuel) are withdrawn from food production and used instead to grow energy crops. In WBGU's view securing the word food supply must take precedence over all other uses of those areas of the world's land surface that are suitable for farming. While bioenergy can be substituted by other sources of fuel, there is no substitute for food. According to the FAO definition, food security exists when all people, at all times, have physical, social and economic access to sufficient amounts of safe and nutritious food that meets their dietary needs and food preferences for an active and healthy life (FAO, 2008b). WBGU therefore proposes here as a guard rail that access to sufficient food should be secured for all people.

A necessary but not a sufficient requirement for this is that enough food is produced to meet the calorie needs of all people. For operationalization of the guard rail it can thus be deduced that the amount of agricultural land available globally must at least be sufficient to enable all people to receive food with an average calorie content of 2700 kcal per person per day (equivalent to approximately 11.3 MJ per person per day) (Box 3.2-1). According to FAO figures (FAO, 2003b) global food production currently amounts to approximately 2800 kcal per person per day (Beese, 2004). On a global scale, therefore, enough food energy is currently produced, so that hunger and malnutrition are primarily problems of access and/or distribution.

LAND NEED DEPENDS ON NUTRITION STYLE AND LAND PRODUCTIVITY
Factors that are important for the extent of the potential for providing the world's population with sufficient and nutritious food are people's nutrition habits and the productivity of the land. The nutrition

potential of existing agricultural land depends to a large extent on the way in which the crop is used. For example, the majority of the maize harvest in North America and Europe is fed to animals. This means that the maize provides food for people only via the production of meat and milk. In the course of this 'refinement' a large proportion of the food calories originally present in the maize is lost. Around one-third of the world's grain yield is currently used as animal feed. Overall, global food production must be increased by 50 per cent by 2030 and by around 80 per cent by 2050. This will need to be achieved mainly through increases in productivity per unit area of land (Section 5.2).

3.2.2
Guard rail for securing access to modern energy services

In WBGU's view (WBGU, 2004a), securing elementary energy services must involve access to modern forms of energy. WBGU therefore proposes the following guard rail: access to modern energy for all people should be ensured. In particular, this must entail ensuring access to electricity and replacing the use of biomass that is harmful to health with modern fuels. In the medium term WBGU considers the minimum quantity of final energy for basic individual needs to be 700–1000 kWh per capita per year.

There are considerable difficulties – from both normative and methodological/technical points of view – in calculating minimum per capita energy needs. Climatological and geographical considerations must be taken into account, as must cultural, demographic and socioeconomic factors. In addition, converting energy services into required energy quantities

Box 3.2-1

A person's calorie requirements

In the run-up to the World Food Summit of 1996 there was extensive discussion of minimum levels of calorie availability. The original plan was to specify the availability of 2700 kcal per person per day (equivalent to roughly 11.3 MJ per person per day) as a target. However, this was abandoned on the grounds that an average per capita calorie level conceals inequalities of provision within a country and provides no information on food quality. Nonetheless, it is scarcely feasible to operationalize a 'nutrition guard rail' without recourse to a figure of this sort.

A person's energy requirements are made up of components relating to basic energy consumption (basic metabolism, depending on age, gender and weight), physical activity and individual life circumstances (pregnancy, lactation) (FAO, 2004). Physical activity accounts for a significant pro-

portion of a person's energy consumption and is measured by the Physical Activity Level (PAL). Normal PAL scores range from 1.2 for people whose life is entirely sedentary to 2.4 for those who carry out the heaviest types of work (DGE, 2007). Guideline values for energy intake required by people aged 19–25 are 3000 kcal per person per day for men and 2400 kcal per person per day for women. For men engaged in heavy physical work this figure can rise to just under 4000 per person per day. For men and women aged 25–51 the guideline values for average energy intake are 2900 and 2300 kcal per person per day respectively, while for people aged 51–65 they are 2500 kcal per person per day for men and 2000 kcal per person per day for women.

In industrialized countries the actual average calorie intake is around 3400 kcal per person per day, while in many developing countries the figure is under 2000 kcal per person per day (Ethiopia, at 1600 kcal per person per day, is at the bottom of the scale) (Meade and Rosen, 1997; FAO, 2006a).

involves making assumptions about the technologies used. For these reasons the literature contains little in the way of differentiated specification of such a minimum requirement. Despite these difficulties, the calculation can be justified (WBGU, 2004a), because this minimum requirement is defined not as a target but as an absolute minimum, and failure to meet it must be classed as non-sustainable.

Where efficient technologies in accordance with the state of the art are deployed, WBGU estimates the absolute individual minimum energy requirement to be approximately 450 kWh per person per year (in a 5-person household) or 500 kWh per person per year (in a 2-person household; WBGU, 2004a). Those figures lie in the range of 300–700 kWh per person per year that is usually quoted in the literature. 450 or 500 kWh per person per year can represent only an absolute minimum, since this figure takes no account of heating, transport and support for domestic and subsistence activities. It is for this reason that WBGU's guard rail of 700–1000 kWh per person per year is above this level.

3.2.3
Guard rail for avoiding health risks through energy use

The International Covenant on Economic, Social and Cultural Rights (UN Social Covenant) defines health as a fundamental human right (Art. 12). The right to a reasonable standard of living (Art. 11) is also defined as such a right; this includes access to energy for purposes such as cooking and heating. In many countries and regions these two rights do not match, because energy that is 'clean' or adapted to the form of use is not available. The forms of energy used in these areas can cause significant harm to human health. In particular, the burning of fossil fuels and biomass produces gases and particles that cause air pollution, and this harbours major risks to health (WBGU, 2003a).

To assist formulation of guard rails in the form of non-tolerable limits to health impacts associated with the production and use of energy, the concept of Disability Adjusted Life Years (DALYs) can be used. DALYs are a measure of the impact on health expressed in reduced life expectancy. They are made up of life years that are lived with health impairments or disease and life years that are lost through premature death (Murray and López, 1996). In large parts of the world urban air pollution and indoor air pollution already account for less than 0.5 per cent of regional DALYs. As a guard rail WBGU therefore proposes that the proportion of regional DALYs attributable to these two risk factors should be reduced to below

0.5 per cent for all WHO regions and sub-regions (WBGU, 2004a).

3.2.4
Additional socioeconomic sustainability requirements

In producing and using bioenergy, a number of socioeconomic factors need to be taken into account if the requirements for sustainable development are to be met.

WBGU therefore explores measures by which these factors may be addressed (standards: Section 10.3). Socioeconomic sustainability criteria are relevant in the context of bioenergy in both industrialized and developing countries. However, there are three reasons for paying particular attention to developing countries. Firstly, the problems associated with traditional biomass use in developing countries are widespread and a major obstacle to development (Box 8.2-1; Section 10.8); traditional biomass use in industrialized countries, on the other hand, does not represent a significant problem. Secondly, the same applies to access to energy and to sufficient food. Thirdly, the agricultural sector in developing countries, in contrast to that in most industrialized countries, plays a key role in economic and social development. In low-income countries around 20 per cent of GDP is generated in the agricultural sector; in high-income countries the figure is only 2 per cent (World Bank, 2008c). In industrialized countries only a small percentage of the workforce is employed in the agricultural sector; in some developing countries proportions over 40 per cent or even over 60 per cent are found (World Bank, 2008c). To this must be added the major significance of the agricultural sector in developing countries in overcoming extreme income poverty: some 700 million people, or three-quarters of all those who live on less than US$ 1 per day, are rural dwellers in developing countries (World Bank, 2004).

Local working conditions, viewed from the perspective of social sustainability, are an important aspect of the production of biomass for use as energy. For example, conditions are not sustainable if pesticides are used in large quantities and the health of plantation workers and local residents suffers as a result. In addition, at least the most basic core standards of the International Labour Organization (ILO) should be observed (safety, prohibition of exploitative child labour, prohibition of slave labour, elementary employee rights, etc.). Another clear criterion of non-sustainability of bioenergy cultivation is if smallholders or indigenous groups are deprived of their livelihood by being displaced to make way for plantations.

Economic aspects of sustainability are also particularly important for poorer countries. Many developing countries hope that bioenergy will bring development opportunities – perhaps by tackling rural poverty directly, by reducing dependence on imports of fossil fuels or by increasing energy supply security. They also perceive opportunities in relation to the export of modern energy, which can further a country's economic development. The extent to which such hopes are fulfilled does not depend solely on whether cultivation is ecologically sustainable: national and local political and socioeconomic conditions are also key factors (Section 10.8). Another crucial issue is whether an expansion of the bioenergy sector is economically sustainable in the sense of being able to continue operations in the long term even without subsidies; if ongoing subsidy of the sector is required, funds will no longer be available for projects of greater social and economic promise.

3.3
Conclusion

WBGU bases its assessment of a sustainable bioenergy policy on a range of guard rails in the form of non-tolerable damage limits, supplemented by additional sustainability requirements that cannot strictly be described as guard rails. In WBGU's view this provides a basis for operationalizing sustainability requirements in the field of bioenergy, such as through standards or agreements in international law. Building on this concept, WBGU has in addition modelled the global energy crop potential (Chapter 6).

Bioenergy, land use and energy systems: Situation and trends

This chapter analyses the trends of present-day land use and the global bioenergy sector. It forms the starting point for addressing the question of the extent to which increasing use of energy crops is compatible with sustainable land use worldwide. Section 4.1 explores the present position of bioenergy in the various sectors of the global energy system. It includes discussions of the current contribution of bioenergy to meeting the world's primary energy needs, and of the trade in bioenergy carriers (Section 4.1.1), as well as outlining current global policies promoting bioenergy (Section 4.1.2).

Subsequent sections describe global land cover (Section 4.2.1) and land use (Section 4.2.2). Interventions in the landscape also have a major impact on ecosystem factors such as biodiversity and the carbon cycle – the latter also being relevant to energy crop use (Section 4.2.3). The growing global demand for energy crops combined with limited availability of land exacerbates competition between different forms of land use. It is therefore likely that the increased use of energy crops will affect the extent of natural and semi-natural ecosystems; these consequences will be both direct and – via displacement effects – indirect. Finally, the effects of direct land-use changes are evaluated in Section 4.2.4.

4.1
Bioenergy in the global energy system

Bioenergy plays an important role in today's global energy supply. Alongside traditional biomass use, more and more energy is being generated from wastes and residues and from energy crops grown especially for the purpose. This section describes how the bioenergy sector is currently developing, which technologies are available or likely to be available in the future, how the trade in biogenic resources and processed bioenergy carriers develops and how national policy measures influence global demand for bioenergy.

4.1.1
Current bioenergy use

4.1.1.1
Bioenergy in the global energy system

Around 10 per cent of the global demand for primary energy is currently met by energy from biomass and waste. In 2005 this amounted to approximately 47.2 EJ out of a total of 479 EJ (GBEP, 2008). The lion's share of this is attributable to traditional biomass use (e.g. firewood); in comparison with this the contribution of modern bioenergy and waste use is small. According to the International Energy Agency the proportion of primary energy consumption met by bioenergy in 2005 was 4 per cent in the OECD countries, 13 per cent in China, 29 per cent in India, 18 per cent in Latin America and the developing countries of Asia and 47 per cent in Africa (IEA, 2007a). In many countries of sub-Saharan Africa bioenergy accounts on average for more than 90 per cent of the primary energy supply (MEMD, 2004; IEA, 2007a).

The figures quoted here were calculated by the physical energy content method. In comparison with the methodologically more appropriate substitution method, this undervalues the contribution of renewables in the electricity sector and thus distorts the overall result (Box 4.1-1; Figure 4.1-1).

Today renewables meet around 16.7 per cent of the world's primary energy needs. Bioenergy accounts for 60 per cent of renewable energy and is thus the most important of the world's renewable energy sources (BP, 2008; OECD, 2008; REN21, 2008). 86 per cent of bioenergy is used in the heat sector, mostly for cooking and heating; this is the 'traditional' use of biomass. For around 2500 million people (38 per cent of the world population) in more than 80 newly industrializing and developing countries biomass in the form of firewood, charcoal and animal dung is still the most important source of energy (IEA, 2006b). In comparison with traditional biomass use, the use of modern biomass in the form of power, heat and fuel plays only a small part, representing 14.5 per cent of the

Box 4.1-1

Applying the substitution method

METHODOLOGICAL ISSUES WHEN CALCULATING PRIMARY ENERGY REQUIREMENT
The primary energy requirement of a nation's economy is an important indicator for economic and climate change policy. It is therefore extremely important that it is calculated correctly. However, identifying precisely the primary energy contribution of electricity generated from renewables in particular poses a methodological challenge. At present two methods are used to calculate electricity balances: the physical energy content method and the substitution method.

The physical energy content method is used predominantly internationally and was also introduced in Germany in 1995 for political reasons according to the VDI (Association of German Engineers) guideline 4661 (VDI, 2003). Until then the substitution method was used. The latter is based on the assumption that electricity which is not produced from fossil fuels replaces corresponding electricity production from fossil-fuelled power stations. With both methods conversion factors are used which indicate the amount of primary energy required to produce one energy unit of electricity.

Producing 1 kWh of electricity (1 kWh = 3.6 MJ) in a conventional fossil-fuelled power plant with 38 per cent energy efficiency (global average: BP, 2008) requires around 2.63 kWh of primary energy. This involves conversion losses of about 1.63 kWh (62 per cent). In contrast, with hydropower, solar and wind energy electricity is generated without thermal conversion losses (100 per cent efficiency). Thus 1 kWh of electricity obtained from wind corresponds to 'only' 1 kWh of primary energy, but replaces 2.63 kWh of primary energy from fossil fuels. Whether and how this difference is taken into account in the balance sheet is extremely important. The physical energy content method uses 1 kWh as the conversion factor for electricity directly produced from renewables, whereas the substitution method uses the substituted value 2.63 kWh.

The physical energy content method equates 1 kWh of electricity (final energy) from renewables to 1 MJ of fossil chemical primary energy or 1 kWh of thermal energy; this contradicts the laws of thermodynamics and is wrong in purely physical terms. Only 0.38 kWh of electricity can be produced from 1 kWh of fossil chemical energy with the reference generation mix. These figures underscore that the two forms of energy are not equivalent and cannot be rated as such. This is corroborated by the Association of German Engineers (VDI) who state in their guideline VDI 4661 that the conversion factors for renewables, nuclear power and electricity from waste used in the physical energy content method were determined "by means of political decision-making and partly without taking the physical and technical boundary conditions into account" (VDI, 2003). This statement could well have alluded to the use of nuclear power which is clearly seen in a better light in terms of primary energy according to this method. With the physical energy content method the proportion of renewables in the electricity sector is systematically falsely represented: according to the present method of calculation, even given total electricity provision from renewables there would remain the need for fossil-based primary energy to make up the 62 per cent shortfall in electricity production. That this fact has not yet become glaringly obvious is merely because the proportion of renewables in the electricity supply is still

comparatively small. Should there be a steady increase in this proportion, then the substitution method is the only method capable of giving a correct undistorted calculation of primary energy.

For this reason WBGU favours adopting this method and assumes a reference value of 2778 $kWh_{\text{primary energy}}$/kWh$_{\text{electricity}}$ for the substitution of fossil fuels which corresponds to an efficiency rating of 36 per cent for the global thermal generation mix. This value is derived from the average efficiency rating of 38 per cent for the OECD thermal generation mix which is somewhat higher than the global value and is also used by the Renewable Energy Policy Network (REN21, 2008). With the aid of the OECD reference value global primary energy demand is calculated annually in the BP Statistical Review of World Energy (BP, 2008) according to the substitution method.

Given expansion in renewables in line with the BMU Lead Scenario 2006 (Nitsch, 2007) renewables will cover 45 per cent of the primary energy requirement for electricity generation in 2030 according to the substitution method, but only 24 per cent according to the misleading physical energy content method currently employed (Sterner et al, 2008). This systematic difference becomes particularly clear in the global calculation of nuclear energy and hydropower. In 2005 both energy sources produced roughly the same volumes of electricity: nuclear energy 2770 TWh; hydro-

a) The physical energy content method

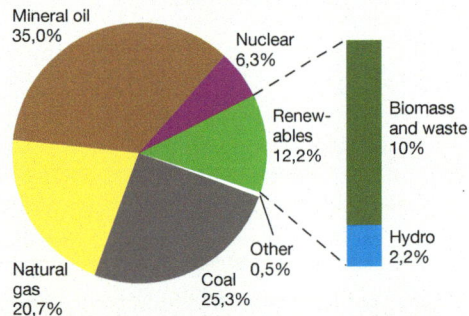

b) The substitution method

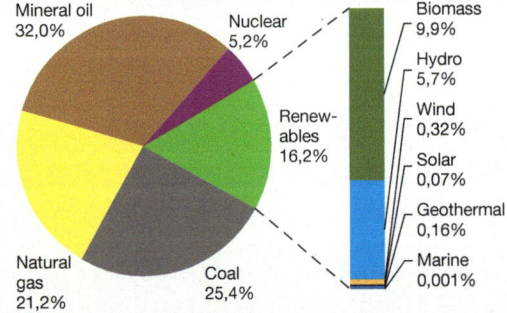

Figure 4.1-1
Shares of different energy carriers in global primary energy requirement. a) by the physical energy content method in 2005, *'Other' comprises other renewables; Total primary energy requirement: 479 EJ. b) by the substitution method in 2006; Total primary energy requirement: 509 EJ.
Sources: BP, 2008; REN21, 2008; GWEC, 2008

power 2934 TWh (IEA, 2006b; BP, 2008). According to the physical energy content method used by the IEA nuclear energy contributes 30.2 EJ of primary energy (efficiency rating: 0.33) and thus almost three times as much as hydropower which contributes 10.5 EJ of primary energy (IEA, 2006b). However, according to the substitution method

their contributions would be roughly the same: 26.7 EJ or 5.2 per cent for nuclear energy and 28.8 EJ or 5.7 per cent for hydropower (BP, 2007). The differences in the calculation of primary energy become clear in the diagrams showing the shares of primary energy carriers in the global energy supply (Fig. 4.1-1).

total. The much-discussed biofuels for the transport sector constitute only 2.2 per cent of global bioenergy use; however, their use has increased very markedly over the last decade (GBEP, 2008; OECD, 2008). Around 34.5 per cent of bioenergy worldwide is converted into electricity (Figure 4.1-2).

In terms of quantity the largest consumer of bioenergy is China at approximately 9 EJ per year, followed by India (6 EJ per year), the USA (2.3 EJ per year) and Brazil (2 EJ per year). In the major European countries the quantity is lower; in France and Germany it amounts to approximately 0.45 EJ per year. In the major newly industrializing countries the quantity is declining, because biomass for heat generation is increasingly being replaced by natural gas and liquid gas (GBEP, 2008). In the industrialized countries, on the other hand, the quantity is rising, largely on account of the increased use of biofuels in the transport sector but also because of increased use in power generation (co-combustion of woody biomass in coal-fired power plants, biogas systems). The generation of energy from waste is also classed as bioenergy. This includes energy from landfill gas and sewage gas, black liquor from the paper industry, forest timber waste, organic waste and other municipal waste.

4.1.1.2
Use of bioheat and bio-electricity in the energy system

THE CONTRIBUTION OF BIOHEAT
At present bioenergy makes its greatest contribution in the heat sector. Forty-four per cent of felled timber is used as firewood (FAO, 2006b). According to FAO figures, global use of firewood peaked in the 1990s and is now falling. Global use of charcoal doubled between 1975 and 2000; increasing urbanization is a driving factor in this (MA, 2005c). Eighty-nine per cent of global bioenergy use takes the form of traditional biomass use; 71 per cent thereof is attributable to households in developing countries (GBEP, 2008).

The Renewables Global Status Report (REN21, 2008) states that in 2006 installed biomass heating capacity amounted globally to around 235 GW_{th}. According to IEA estimates, the quantity of modern bioenergy used for heating in buildings and industry is around 3 EJ per year. This includes heat from cogeneration and heat used for drying agricultural and forestry products (IEA, 2007c). Modern biomass heating is found mainly in countries in which extensive biomass resources are available, and in particular where central district heating systems exist.

THE CONTRIBUTION OF BIO-ELECTRICITY
Biomass is at present used less in the electricity sector than in the generation of heat. Globally, grid-connected biopower capacity in 2006 amounted to an estimated 45 GW, accounting for 0.4 per cent of global power consumption (REN21, 2008). This corresponds to around 21 per cent of the generating capacity of renewables (excluding large hydropower). Biomass power plants are used both in developing countries and in Europe and the USA. One scenario calculates that their global power generation capacity could reach 306 GW by 2030 and 505 GW by 2050 (Greenpeace, 2007).

Almost all types of biomass can be used to generate power through combustion, gasification or fermentation. Combustion is used mainly for the production of steam in electricity and cogeneration plants; the steam is then used in conventional steam turbines to generate electricity. Normally compressed

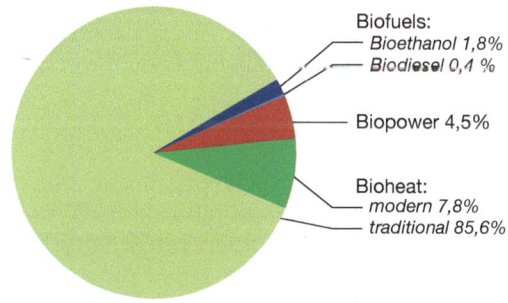

Figure 4.1-2
Breakdown of global bioenergy use (primary energy, total 50.3 EJ) into provision of electricity, heat and fuel. Sources: BP, 2008; OECD, 2008; REN21, 2008

biomass is used in the form of wood pellets or bri-quets, which have a calorific value similar to that of lignite. Alternatively, biomass can be burned with another fossil fuel (e.g. coal). This co-firing in large coal-fired power plants has the advantage of higher overall power efficiency (up to 45 per cent) than is obtained in small biomass power systems (30–35 per cent) (IEA, 2007b).

Bio-electricity is also generated through the com-bustion of biogas in gas and combustion engines. Biogas is produced decentrally through fermen-tation of liquid and solid biomass; the use of waste such as animal dung offers major ecological advan-tages in this context. In Europe gaseous and solid types of biomass contribute in roughly equal propor-tions to electricity generation: for example, in Ger-many in 2006, biogas systems met 0.9 per cent of electricity needs while solid biomass met 1.2 per cent (BMU, 2007a). Alongside biogas systems, gasification and gas power generation in combined-cycle power plants provides a particularly efficient means of con-verting waste-based biomass into electricity.

THE CONTRIBUTION OF BIOENERGY FROM COGENERATION (COMBINED HEAT AND POWER, CHP)

Thermal power generation processes will ideally also use the waste heat that arises. In southern countries waste heat from cogeneration (combined heat and power generation, CHP) is used for industrial pro-cesses such as drying. In northern countries it is used mainly for space heating and hot water provision, either directly or indirectly via local and district heat-ing grids. Global data on cogeneration is difficult to gather because its applications are very diverse (pro-cess heat, space heating), the need for it is seasonal (heating), and in warmer countries cogeneration is only occasionally used for cooling purposes. In 2005 in Germany 58PJ of biomass was used for power and heat production in cogeneration at an efficiency level of 86 per cent; this amounts to 0.4 per cent of the pri-mary energy used (Nitsch, 2007).

TRADE IN BIOENERGY CARRIERS IN THE POWER AND HEAT SECTOR

Bioenergy carriers are often produced and used in separated locations. In particular, the final use of modern bioenergy often occurs at a considera-ble distance from the place of production. There is, therefore, a trans-regional trade in pre-products of bioenergy production such as biogenic solid fuels (raw wood, roughly sawn wood, pellets), raw mate-rials used in conversion (energy crops, timber waste, etc.) and bioenergy as an end product (biofuels, elec-tricity from bioenergy). The character and extent of this trade are determined by the availability of raw

materials and conversion technologies and by inter-national price and cost structures (Schlamadinger et al., 2005).

As far as end use in the power and heat sector is concerned, national and international trade in bioen-ergy is linked to the logistic availability of electric-ity and district heat grids that provide sufficient capacity. Physical and technological constraints at present restrict the economic attractiveness of such trade. Over medium distances, such as within Europe, trade in bio-electricity can be cost-effective (Schla-madinger et al., 2005; Schütz and Bringezu, 2006). At the level of pre-products that are used by com-bustion technologies in the electricity and heat sec-tor (energy wood), however, trade has so far taken place only on a limited scale worldwide. Thus of the 1770 million cubic metres of wood that was used as firewood in 2005 (of a global removed volume of around 3000 million cubic metres), only 3–4 million cubic metres or 0.2 per cent was traded internation-ally (FAO, 2007a). High transport costs in relation to the value of the goods often render exports uneco-nomic (Thrän et al., 2005).

Despite this, there is evidence of expanding inter-national markets for some biogenic solid fuels that are processed industrially (wood chips, pellets). Driven by various national climate and energy-policy measures, demand for pellets is growing in Europe, North America and Asia. Brazil, Argentina, Chile and New Zealand are also planning to develop the infrastructure for pellet production. The develop-ment and spread of modern pelletizing technology reinforces this trend (Thrän et al., 2005; Peksa-Blan-chard et al., 2007). Usage has so far been dominated by pellets from wood production. The primary source is wood from short-rotation plantations, but also used are woody biomass arising as a residue of forestry (timber waste), of agriculture (in particular straw), of wood processing (including industrial wood and in particular wood shavings), and end-of-life wood (bulky waste, demolition) (IZT, 2007). Pelletizing of other residues (e.g. press cakes from oil plants) is still under development. Globally, paper and cellulose production accounts for most of the material use of forests; it also gives rise to the energy-rich byproduct black liquor, almost all of which is used directly in the generation of power and process heat.

Alongside to solid materials, biogas is also of inter-est as a pre-product for heat generation. In order to be traded trans-regionally in the future, biogas can be processed into biomethane and fed into existing gas supply grids (Bringezu et al., 2007; Thrän et al., 2007). In 2007, in its Integrated Climate and Energy Pro-gramme (IKEP), the German government pledged to increase the amount of biomethane in the natural gas grid from the current figure of 0 per cent to 6 per

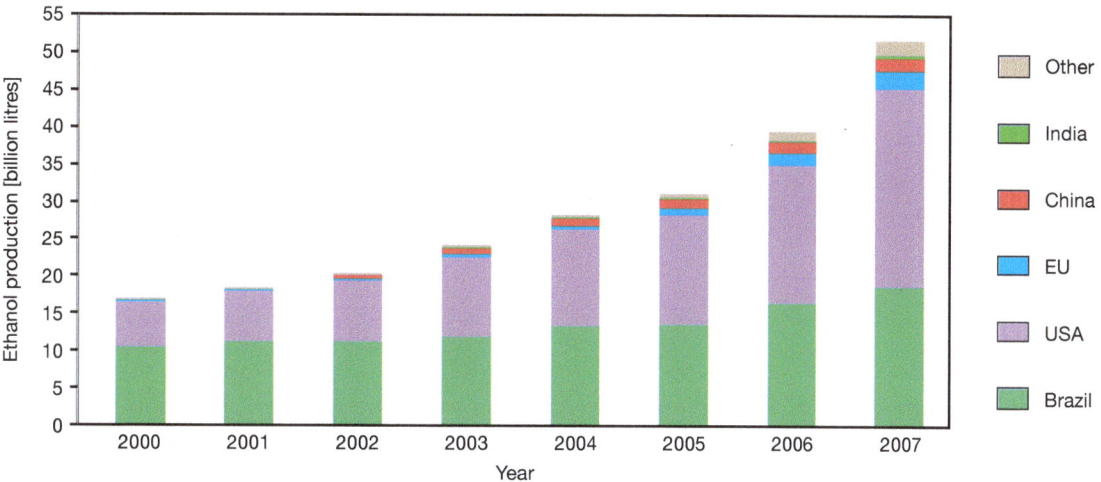

Figure 4.1-3
Global production of ethanol for use as fuel (2000–2007).
Source: Licht cited in OECD, 2008

cent by 2020 and to 10 per cent by 2030 (BR, 2007). These 6,000–10,000 million cubic metres of biogas per year will bring corresponding trade flows in their wake.

4.1.1.3
Use of biofuels

CONTRIBUTION OF BIOFUELS
Viewed in absolute terms, the use of biomass as biofuel in the transport sector is still at a low level. In recent years, however, it has expanded rapidly as a consequence of political decisions and targetted state promotion policies (Section 4.1.2).

Use of bioethanol
Global bioethanol production in 2007 totalled 52 million litres, equivalent to 1.2 EJ (OECD, 2008); output has thus trebled since 2000 (Figure 4.1-3). The largest bioethanol producers are Brazil and the USA, which between them have a market share of almost 90 per cent (Table 4.1-1). The raw materials used in production differ from region to region: in the USA bioethanol is produced mainly from maize, Brazil uses sugar cane and Europe uses, among other crops, sugar beet and wheat. The sugar contained in the plants is fermented with the aid of yeast and enzymes to form bioethanol and CO_2. It is then dehydrated in a multi-stage distillation process and brought to an ethanol content of 99.5 per cent (FNR, 2007a).

Bioethanol is used for transport purposes by being blended at a low percentage with petrol. The proportion of bioethanol is normally either 5 per cent (E5) or 10 per cent (E10). Also available are 'flexible fuel' vehicles, which can run on E85 (85 per cent bioetha-

nol, 15 per cent petrol). The energy content per litre of ethanol is, however, only 65 per cent of that of fossil petrol, which means that the quantities produced cannot be directly compared: the quantity of bioethanol used by a vehicle will be around one and a half times the quantity of petrol needed to travel the same distance.

Use of biodiesel
Global biodiesel production in 2007 totalled 10.2 million litres (equivalent to 0.32 EJ). The annual figure has increased more than tenfold since 2000 (OECD, 2008; Figure 4.1-4).

Biodiesel (fatty acid methyl esters, FAME) is produced by esterification from plant oils, principally rape oil, soya oil and palm oil. In Europe rape is the chief crop that is grown and processed into biodiesel. By contrast, almost 90 per cent of global palm

Table 4.1-1
Production of fuel ethanol in the main production countries and worldwide (figures for 2007).
Source: Licht cited in OECD. 2008

Country / region	Production	
	Amount [1,000 million litres]	Proportion [%]
United States	26.5	51.0
Brazil	19.0	36.5
European Union	2.3	4.4
China	1.8	3.5
India	0.4	0.8
World	52.0	100.0

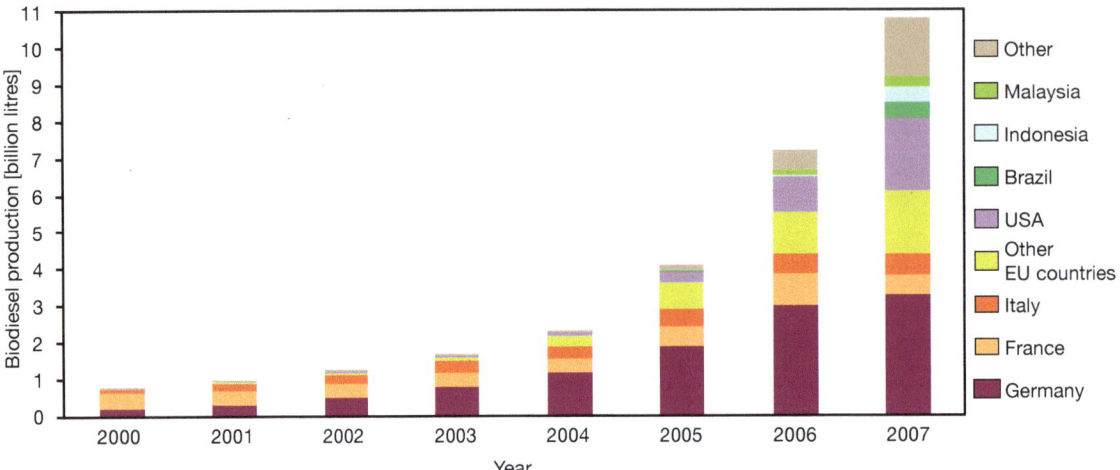

Figure 4.1-4
Global production of biodiesel (2000–2007).
Source: OECD, 2008

oil is produced in Malaysia and Indonesia; most of it is exported as food, but an increasing proportion is being refined locally into biodiesel. In 2007/08 the largest soya-producing countries were the USA with 71.3 million tonnes, Brazil with 61 million tonnes and Argentina with 47 million tonnes (Toepfer International, 2007). While cultivation in the USA has declined in recent years, it is increasing in South America. While the majority of soya production continues as before to be processed into food and feed, soya is also being used increasingly for biodiesel production. Argentina is in addition developing its production capacity for export. In the past year production of plant oils and fats totalled 9.5 million tonnes, of which 2.1 million tonnes of soya oil were used for biodiesel production (Ronneburger, 2008). An analysis by Greenpeace showed that in Germany 20 per cent of blended plant diesel is produced from soya oil (Greenpeace, 2008).

In comparison with bioethanol, the energy yield of biodiesel is still relatively low, amounting in 2007 to 0.32 EJ. The main producer of biodiesel is the European Union, which accounts for 60 per cent of the world market (OECD, 2008), and in particular Germany and France (WI, 2007). While global production has increased in recent years, it is partly declining as a result of current high raw material prices or changes in national tax concessions. In addition the production capacity of some plants has been reduced, and some plants have closed completely.

Biodiesel, like bioethanol, is blended with fossil fuels. A 5 per cent blend with traditional diesel (B5) is already the norm in Europe. New high-performance diesel engines can also use 100 per cent biodiesel. Use of B100 has been widespread for some years, particularly in Germany, where it is available at more

than 1900 filling stations. At 96 per cent biodiesel has roughly the same energy content as traditional diesel, but it has better physical properties (viscosity, cetane rating); volume-specific quantities can therefore be compared (IEA, 2006b; FNR, 2007a).

Use of plant oil
Plant oil from crops such as rape, soya, sunflower or oil palm can also be used directly as fuel in a combustion engine. Since, however, such use normally requires modification of the engine, plant oil differs from bioethanol and biodiesel in that its direct use in the transport sector is not yet of any significance on a global scale.

Use of the second and third generation of biofuels
Techniques for producing synthetic biofuels (second generation: biomass-to-liquid, BTL) are in development. They hold the promise of better fuel characteristics as well as higher hectare yields and greenhouse gas reduction potentials, because – contrary to first generation fuels – the entire plant can be used. However, it is not certain that these expectations can be fulfilled (Sections 7.2 and 7.3). The expected advantages are offset by the need for significantly more complex production plants with higher investment costs. The techniques are based on thermochemical gasification of woody biomass and residues. This method enables the production of fuels such as Fischer-Tropsch diesel, biogenic hydrogen, biomethane, dimethyl ether, methanol, biokerosine and ethanol (Sterner, 2007). The third generation of biofuels is still at the stage of basic research. Research focuses in essence on the production of hydrogen with the aid of microalgae. It will be some years before BTL fuels, in particular synthetic diesel, are ready for the market. The first

Table 4.1-2
Global biodiesel production in selected production countries and worldwide (figures for 2007).
Source: OECD, 2008

Country / region	Production	
	Amount [1.000 million l]	Proportion [%]
European Union	6.1	59.9
United States	1.7	16.5
Brazil	0.2	2.2
China	0.1	1.1
India	0.05	0.4
Malaysia	0.3	3.2
Indonesia	0.4	4.0
World	10.2	100.0

Table 4.1-3
Global cultivation area, production and net trade for grain and sugar. The trade quantities for sugar relate to processed raw sugar. n. a. = not available. Data for 2006.
Source: FAOSTAT, 2007; trade data (for 2004/05) from Thrän et al., 2005

	Cultivation area [million ha]	Production [million t]	Net trade [million t]
Grain			
Wheat	216.1	607	89.9
Maize	144	784.8	76.6
Barley	55.5	136.2	13.8
Rye	5.9	15.7	22.9
Triticale	3.6	12.6	n.a.
Oats	11.3	26	n.a.
Sugar			
Sugar cane	20.4	1.557.7	33.2
Sugar beet	5.4	247.9	n.a.

commercial plant, producing 340 barrels of diesel per day, is planned for 2008, with a second plant producing 4500 barrels per day planned for 2012 (Choren, 2007). The latter figure is equivalent to 0.12 per cent of present diesel consumption in the EU.

TRADE IN BIOFUELS FOR THE TRANSPORT SECTOR
An analysis of trade flows for biofuels can only be an approximation, because processed bioenergy carriers rarely feature in official trade statistics. For example, the Harmonized System Commodity Description and Coding System (HS) of the World Customs Organization covers trade in bioethanol (HS 2207 10) and biodiesel (HS 3824 90) but does not distinguish between their use as biofuel and use in other industrial applications (Zarrilli, 2006). In addition, raw materials present an identification problem, because crops such as maize, sugar cane and certain oil plants can be used for different purposes (energy, food, as an industrial feedstock). Classifying raw materials at the first level of processing as being intended for bioenergy production is therefore difficult (Zarrilli, 2006) and presupposes a precise data collection system. Nevertheless, such classification is necessary in order to estimate land-use displacements (Section 4.2) that may be attributable to the use of crops for bioenergy but that may in some circumstances be undesirable.

Trade in bioethanol
International trade in ethanol takes place at present only on a small scale. Only 10 per cent of global ethanol production – including ethanol not used for energy purposes – is traded internationally. Brazil accounts for more than half of the export market (5000 million litres in 2006, without intra-EU trade; OECD,

2008). Pakistan, the USA, South Africa, Ukraine and Central American states are other exporting countries, although they play a significantly smaller part. The target countries of Brazilian exports are India, the USA, South Korea, Japan and various European states (WI, 2007). The 720 million litres that were imported into the USA in 2005 met 5 per cent of that country's domestic demand (Zarrilli, 2006). Since bioethanol is often not produced in the country of cultivation, trade in the raw materials of ethanol production – at present primarily grain and sugar – is also of interest (Table 4.1-3).

Trade in biodiesel
In 2007 international exports of biodiesel amounted to 1300 million litres, representing 12 per cent of global production. The principal exporters were Indonesia and Malaysia, each with around 400 million litres; the main importer, with more than 1100 million litres, was the EU. The USA also imported significant quantities, but on account of its re-exports to the EU it was a net exporter (OECD, 2008). The raw materials of biodiesel production – oils and fats, and oil plants – were, however, traded internationally in considerable quantities. Globally, though, the energy sector is only a sub-segment of the trade in plant oils. On account of the classification problems mentioned above, precise data is hard to obtain, but it can be assumed that around 80 per cent of traded oils and fats are used in the food sector (Thrän et al., 2005). Table 4.1-4 provides an overview of selected oil seeds used in biodiesel production.

Table 4.1-4
Global cultivation area, production and net trade for selected oil seeds and plant oils. n.a. = not available. Trade data for 2006.
Source: FAOSTAT, 2007, 2008b; Production data for palm oil. Trade data (for 2004/05) from Thrän et al., 2005
* Trade data palm oil = global exports 2005; Source: Pastowski et al., 2007

	Cultivation area [million ha]	Production [million t]	Net trade [million t]		
			Seed	Meal	Oil*
Soya	93.0	216.1	57.2	45.0	9.0
Rape	27.8	49.5	5.7	2.2	1.3
Sunflower	23.7	27	1.3	2.3	1.6
Oil palm	13.3	192.5	n.a.	n.a.	26.3*
Jatropha	n.a.	n.a.			

Palm oil currently accounts for the greater part of the trade in plant oils. It is followed by soya oil, with a net trade volume of 9 million tonnes, and sunflower oil with 1.6 million tonnes. The net trade volume of rape oil is at present only 1.3 million tonnes, of which around 70 per cent comes from Canada; rape oil is exported primarily to the USA and China (Table 4.1-4; Thrän et al., 2005). Trade in *Jatropha* oil has so far been negligible. In the industrialized countries the use of plant oils as a raw material for biodiesel production is increasing demand rapidly. For example, demand for plant oils in the European Union has increased noticeably since the introduction of directive 2003/30/EC on the promotion of biogenic fuels (Thrän et al., 2005). Many developing countries have ambitious expansion plans for plant oil production (Section 4.1.2). Taking the average of 2004–2006 as a baseline, the OECD-FAO forecasts that global production of oil seeds will increase by 25 per cent by 2016. It is assumed that South-East Asian countries such as Malaysia, Thailand, Indonesia and the Philippines will further develop their export potential (including for biodiesel) in the near future; a number of African and South American countries in which climate conditions are favourable for the cultivation of energy crops are likely to do the same (Table 4.1-5; Zarrilli, 2006; WI, 2007).

BIOENERGY FOR THE TRANSPORT SECTOR – PRICE DEVELOPMENT
Prices for biofuels on national and international markets are determined in the short term by regional conditions of supply and demand, which are in turn significantly influenced by the different promotion policies (Section 4.1.2). In the past prices of biofuels have in the long term developed in the same direction as prices of crude oil and fossil fuels (OECD, 2008). In general, though, rising energy prices do not necessarily make biofuels more competitive. In most countries it is still the case that biofuels are only able to retain their market position if the difference between

biofuel production costs and the price of petrol or diesel is supported by subsidies or other forms of promotion. Rising production costs for biofuels have been driven in the main by the increased price of energy crops, which are simultaneously in demand for food production (Section 5.2.5.2; IMF, 2007). These trends, however, evolve differently in different regions and for different energy crops (Figure 4.1-5).

4.1.2
Current bioenergy promotion policy

The current increase in the use of biomass for energy in many countries is the result of specific state promotion measures that serve a range of climate-related, energy-related and economic goals (Chapter 2). State intervention in the markets for biomass and energy changes market prices and these price changes give incentives for increased use or production of bioenergy. Because there are many possible ways in which biomass can be used as food, feed, or fuel, the production and use of bioenergy is influenced by policy measures in the agriculture and forestry sector as well as by those in the energy sector. Instruments of national and international environmental policy are also relevant for the bioenergy sector; an example is the opportunity for developing countries to finance bioenergy projects by selling emissions credits through the Clean Development Mechanism (CDM) (Section 10.2).

Promotion measures for the generation of electricity from renewables are currently planned in at least 60 countries, of which 23 are developing countries. In addition, some states have promotion policies specifically for bioenergy; measures aimed at expanding the production and use of biofuels are particularly widespread. At least 17 countries already have mandatory blending quotas for biofuels in place (Table 4.1-5; REN21, 2008). Analysis of the particularly relevant bioenergy promotion measures in the energy

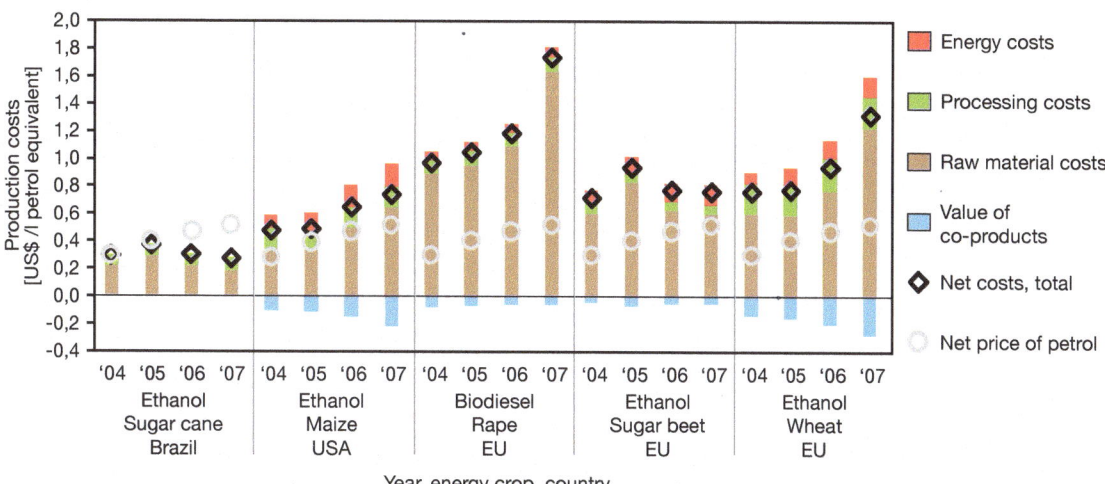

Figure 4.1-5
Production costs for selected biofuels 2004–2007 in chief production countries. The columns depict the proportion of costs by input factors. The distance between net total costs (rhombus) and the net price of petrol (circle) represents the level of competitiveness. It becomes apparent that, despite the high price of crude oil and fossil fuels, the competitiveness of these biofuels has not improved consistently in recent years.
Source: OECD, 2008

sector shows that it is possible to distinguish various types of subsidy at various stages of the product life cycle (Table 4.1-5; GBEP, 2008; SRU, 2007; REN21, 2008).

STAGE 1: PROCUREMENT AND USE OF RAW MATERIALS AND OTHER FACTORS OF PRODUCTION

Promotion measures at the first stage of processing aim to favour domestic production of raw materials as energy carriers. At present the focus in many countries is on promoting the cultivation of energy crops. Typical instruments in this area are agricultural subsidies or import tariffs on agricultural goods from abroad. Subsidies targeted at agricultural producers include, for example, guaranteed minimum prices, output-related payments, area based payments and direct income transfers to producers. Promotion of this type can also be aimed at wood or residues. Under the Common Agricultural Policy (CAP), a subsidy of € 45 per hectare has since 2004 been payable in the EU for the cultivation of energy crops, provided that producers can demonstrate that they have a contract with the processing industry. For cultivation of perennial bioenergy crops on set-aside land a subsidy for start-up costs is also payable; the size of this payment is regulated at national level (EU, 2003).

There are in addition indirect promotion measures that relate to the use of auxiliary resources in crop cultivation, such as water and energy. Specific promotion measures, perhaps in the form of tax concessions or state financing assistance, encourage the use of further value-adding factors of production – such as equipment, machinery, land, or labour – in

connection with bioenergy production (Doornbosch and Steenblik, 2007; Steenblik, 2007).

STAGE 2: EXPANSION OF INFRASTRUCTURE, RESEARCH AND DEVELOPMENT

The expansion of infrastructure for storing, transporting and marketing bioenergy carriers, particularly biofuels, is also often promoted with state funds. In some cases the state itself provides the necessary infrastructure; in other cases the state helps private stakeholders develop and operate such infrastructure by offering tax exemption, low-cost loans or subsidies. Some developing and newly industrializing countries (e.g. India, the Philippines) have in addition set up state-run bioenergy pilot projects that contribute to the development of market infrastructure or to research into and development of production techniques. Proceeds from the sale of certificates under the CDM can help to finance such demonstration projects in developing countries (Section 10.2). Many countries are also continuing to invest in state research into renewable energy technologies; this is the case in China, the USA and Peru, but also in Germany (Jull et al., 2007; GBEP, 2008).

STAGE 3: PRODUCTION THROUGH TO FINAL PRODUCT

Methods of supporting the process of bioenergy production include promoting production facilities through state investment grants, or granting tax breaks to producing businesses. In addition, parallel promotion structures can be established for goods – such as protein feed, glycerine or rapeseed meal

Table 4.1-5
Examples of bioenergy promotion policy in selected countries. As at August 2008. RE = renewable energies
Sources: Biopact, 2006; Lindlein, 2007; REN21, 2006; Reuters, 2007; UNCTAD, 2006a; Zarrilli, 2006; GTZ, 2007a; Steenblik, 2007; WI, 2007; Doornbosch and Steenblik, 2007; Jull et al., 2007; Economist, 2008b; GBEP, 2008; IEA and JREC, 2008; MME, 2008

Country/ Country group	Motivation	Promotion policy (divided into three categories) (1) General expansion targets for renewables (2) Goals/measures relating to power and heat (3) Goals/measures relating to mobility (biofuels)
Industrialized countries	.	
EU-27	Climate change mitigation, security of supply, agricultural diversification, rural development	(1) RE expansion targets (12% RE from 2010, planned: 20% RE final energy from 2020); tax relief for RE (national); area payments for growing energy crops on set-aside land (€45 per ha); promotion of research: EU's 7th Framework Programme (2) Expansion target for electricity from RE (21% from 2010); fixed feed-in tariffs for bioenergy (national, e.g. in Germany through EEC); trade in RE certificates (national) (3) Mandatory blending quotas for biofuels (5.75% from 2010, planned: 10% from 2020); (in some cases) tax exemptions or relief for biofuels (national); import tariffs for biofuels (€0.102 for denatured ethanol, €0.192 for undenatured ethanol; 1.9% ad valorem for palm oil; 6.5% ad valorem for biodiesel)..
USA	Security of supply, energy autonomy, rural development, environmental protection	(1) Investment subsidies for RE technologies; tax breaks for RE; biorefinery demonstration projects; Renewable Portfolio Standards in individual states (2) Fixed feed-in tariffs for electricity from RE; programme for using forest timber waste (woody biomass grants) (3) Expansion targets for biofuels (56,000 million litres from 2012 / 136,000 million litres alternative fuels from 2022, equivalent to 20% of national fuel requirements in 2022); Import tariffs on ethanol (2.5% ad valorem plus US$0.1427 per litre); tax relief for biofuels (US$0.135 per litre); tax breaks for cars using fuel, hybrid, or flex-fuel technology; favourable financing conditions for farmers and biofuel producers in connection with the development of infrastructure and production facilities; use of 20% blended biodiesel in public transport and state-owned vehicles; tax breaks for ethanol filling stations; state promotion of biofuel research (Bioenergy Research Centers).
Canada	Climate change mitigation, environmental protection, energy security, technological progress	(1) State grants and tax breaks for RE; Renewable Portfolio Standards in four provinces; national feed-in premium for electricity from RE amounting to CAN$0.01 per kWh; state procurement policy: meeting 20% of the government's electricity requirement through electricity from RE; state subsidies for systems for generating heat from RE (2) Grants for R&D in bioenergy (3) Government grants for the development of biofuel production and necessary infrastructure; mandatory blending of 50% of 5% ethanol from 2010 and 2% biodiesel after 2012; after 3 years progressively reducing production subsidies for ethanol (CAN$0.10 per litre) and biodiesel (CAN$0.20 per litre); import tariffs on ethanol amounting to CAN$0.0492 per litre; promotion of research into 2nd-generation biofuels.
Australia	Rural development, diversification of transport energy sources	(1) RE expansion target (planned: 20% RE or 45,000 GWh from 2020); government grants and favourable financing conditions for investment in RE technologies (2) Renewable Portfolio Standards (9.5 TWh electricity from RE p.a. from 2010); tradeable RE certificates; fixed feed-in tariffs for electricity from RE in South Australia (3) Expansion target for biofuels (350 million litres by 2010); import tariffs on undenatured ethanol (5% plus AUS$0.381 per litre); tax rebates for domestically produced ethanol; production subsidies for biofuels; promotion of the biofuel infrastructure through subsidies.
Japan	Reduced dependency on fossil fuels, climate change mitigation, environmental protection, agricultural diversification	(1) National biomass strategy (Biomass Nippon Strategy) (2) Expansion target for electricity from RE (1.63% from 2014) and specific expansion targets from 2010 for electricity from biomass and landfill gas (5860 million litres crude oil equivalent) and for heat from biomass (3080 million litres crude oil equivalent); mandatory grid feed-in for electricity from biomass; Renewable Portfolio Standards (3) Expansion target for biofuels (50 million litres domestic production by 2011; 500 million litres crude oil equivalent; 6000 million litres per year from 2030); substitution of 20% of fossil fuels as from 2030 by alternative fuels; import tariffs on palm oil (3.5% ad valorem); tax relief for biofuels; preference given to filling stations that offer biofuels.

Newly industrializing countries

Brazil	Independence of oil imports, economic development through the export of bioenergy, rural electrification, climate change mitigation, environmental protection	(1) PROINFA promotion programme for RE; bioenergy research under the National Agroenergy Plan (2) Expansion targets for electricity from RE (3.300 MW from wind, biomass, micro-hydropower from 2006); fixed feed-in tariffs for electricity from RE and reduced transmission and distribution tariffs; electricity purchase on preferential terms from RE system operators; promotion of RE for rural electrification (3) Blending quotas for biofuels (20–25% for ethanol, from July 2008 3% and from 2013 5% for biodiesel); tax relief for biofuels; tax relief and production subsidies for flex-fuel vehicles; national biodiesel programme (incl. Social Fuel Seal); import tariffs on palm oil (11.5% ad valorem); use of ethanol for state vehicle fleet.
China	Security of supply, climate change mitigation, environmental protection, rural development	(1) Expansion targets for RE (15% of primary energy from RE from 2020, of which 30 GW bioenergy); creation of a fund for promoting research into RE; low-interest loans for infrastructure development; tax relief for producers and consumers of bioenergy; US-China Memorandum of Understanding on Biomass Development (research & technology cooperation) (2) Fixed feed-in tariff for electricity from biomass; tax relief for biogas; promotion of small biogas systems in rural areas (3) Expansion targets for biofuels (15% of transport energy from 2020, i.e. 13,000 million litres bioethanol p.a. and 2300 million litres biodiesel p.a. from 2020, 50,000 million litres fuels from solid biomass from 2020); ad valorem import tariff on ethanol (30%); refund of VAT on ethanol; 10% blending quota for ethanol in nine test regions; state *Jatropha-* and ethanol-based model projects and demonstration systems.
India	Energy autonomy, security of supply, rural electrification	(1) Tax relief and low-interest loans for RE system operators; discounts on transport and distribution of RE; bioenergy projects under the CDM (2) National expansion targets for electricity from bioenergy (10% by 2012; planned: 15% from 2032); fixed feed-in tariffs for electricity from RE; subsidizing of biomass power plants and biogas systems; promotion of small biogas systems for rural electrification (Remote Village Electrification Programme) (3) National blending targets for biofuels (10% ethanol from 2008; 20% ethanol and 20% biodiesel from 2017); ambitious promotion policies are pursued in some of the individual states; fixed purchase price for ethanol; state *Jatropha* model projects.
Mexico	Energy autonomy, rural energy supply, climate change mitigation, environmental protection	(1) Tax relief for investment in RE; accelerated writing down for RE projects (2) Expansion targets for electricity from RE (1 GW from 2006; 8% from 2012; 4 GW from 2014); discounts on transport and distribution of electricity from RE (3) Expansion targets for biofuels (production of 454 million litres bioethanol p.a. from 2012; 20% biodiesel blending from 2011/12); mandatory ethanol blending quota of 10% in urban areas; import tariffs on ethanol (ad valorem tariff of 10% plus US$ 0.36 per litre); biodiesel demonstration projects.
South Africa	Rural development, energy autonomy, climate change mitigation	(1) Expansion targets for RE (4% from 2013); subsidizing of technology development; (2) Expansion target for electricity from RE: 10TWh from 2013 (3) Tax relief for biofuels; voluntary blending of biofuels (9%); mandatory blending quotas for biofuels (planned: 8% for ethanol and 2% for biodiesel from 2008), combined with 50% tax exemption for biodiesel and 100% tax exemption for ethanol; state biofuel pilot projects (incl. *Jatropha*).

Developing countries

South-East Asia (Philippines, Thailand, Indonesia, Malaysia)	Security of supply, rural development, rural electrification, meeting energy need	(1) RE use targets (Indonesia: 15% of primary energy from RE from 2025; Thailand: 8% of primary energy from 2011); tax relief for RE projects (Philippines); bioenergy projects under the CDM (Indonesia); (2) Expansion targets for electricity from RE (Thailand: 8% from 2011; Malaysia: 5% from 2005; Philippines: 4.7 GW from 2013); fixed feed-in tariffs for electricity from RE (Indonesia, Thailand); favourable financing conditions for bioenergy producers (Philippines); promotion of RE for rural electrification (Indonesia, Philippines, Thailand) (3) blending quotas for biofuels (Malaysia: 5% biodiesel from 2008; Thailand: 10% ethanol from 2007, 3% biodiesel from 2011/10% biodiesel and bioethanol from 2012; Philippines: 1% biodiesel and 5% ethanol from 2008 / 2% biodiesel and 10% ethanol from 2010; Indonesia: at present 3% ethanol and 2.5% biodiesel); expansion targets for biodiesel production/ use (Thailand: 3100 million litres biodiesel p.a. from 2012 and 1100 million litres ethanol p.a. from 2011(production); Indonesia: 1300 million litres biofuels p.a. year from 2010 (production); 10%/20% of national fuel consumption from 2010/25 (usage, planned)); tax relief for biofuel projects or components.

		(Philippines, Malaysia, Thailand); mandatory blending for government vehicle fleet/public transport (Thailand: 10% ethanol in government vehicle fleet; Philippines: 1% biodiesel in government vehicle fleet; Malaysia: 5% biodiesel in public transport); Jatropha pilot projects and biodiesel pilot systems (Thailand).
West Africa (incl. Senegal, Mali, Ghana, Nigeria)	Energy autonomy, rural development, agricultural diversification	(1) Use targets for RE (Mali: 15% of primary energy from RE from 2020, Senegal: 15% of primary energy from 2025) (2) Expansion targets for electricity from RE (Nigeria: 7% from 2025); (3) Expansion targets for biofuels (Nigeria: production of up to 140 million litres p.a., up to 10% ethanol blending planned; Senegal: biodiesel from *Jatropha* and ethanol from sugar cane, with target of biodiesel self-supply from 2012; Mali: decentral use of *Jatropha*; (state) *Jatropha* pilot projects and research projects (Mali, Senegal, Ghana, Nigeria, Burkina Faso).
(South-) East Africa (incl. Kenya, Tanzania, Malawi, Mozambique, Zimbabwe)	Rural development, rural electrification	(1) and (2) not known (3) Mandatory biofuel blending (Malawi: mandatory ethanol blending of 10–20% since 1982; Mozambique: planned blending of biodiesel and bioethanol); expansion targets (Zimbabwe: production of up to 50 million litres ethanol p.a. planned) *Jatropha* and ethanol demonstration projects (Kenya, Malawi, Mozambique, Tanzania, Zimbabwe); tax exemption on biofuels (Mozambique: planned).
South America (incl. Argentina, Bolivia, Columbia, Guatemala, Peru)	Rural electrification; rural development; energy autonomy	(1) State promotion of research into RE (Peru) (2) Tax exemption for purchase of electricity from RE (Columbia); state demonstration projects (Bolivia); expansion targets for electricity from RE (Argentina: 8% electricity from RE from 2016; Chile: 5% electricity from RE from 2010); premium payments on RE, tax reliefs for investment in bioenergy (Argentina, Guatemala); fixed feed-in tariffs for electricity from RE (Argentina); promotion of RE for rural electrification (Argentina, Bolivia, Guatemala) (3) Mandatory biofuel blending (Columbia: 10% ethanol in cities >500,000 inhabitants, from 2008 5% biodiesel; Bolivia: 2.5% biodiesel and 10% ethanol from 2007, 20% biodiesel from 2015; Peru: 7.8% ethanol and 5% biodiesel from 2010, Argentina: 5% ethanol and 5% biodiesel from 2010; Guatemala: up to 20% ethanol); tax exemptions for raw material and biofuel production and components (Columbia, Argentina, Bolivia); operation of state vehicle fleet and public transport with biofuels (Argentina).

– that arise as co-products in connection with the production of biofuels, thus providing an additional incentive to production (GBEP, 2008; SRU, 2007).

At the centre of production promotion are measures that promote broad use of bioenergy in the market and that stimulate or guarantee market demand, thus giving producers, as providers, a degree of investment security. National production targets or use targets, either for renewables in general or for bioenergy in particular, have this function. In the electricity sector in some countries, facilities for generating bioenergy or the operators of such facilities are supported by means of fixed feed-in tariffs and feed-in to the national electricity grid is guaranteed for a certain period of time. In addition, in the area of power and heat generation production quotas for various types of renewable energy (renewable portfolio standards) are sometimes used in connection with a certificate system (renewable energy certificates) (GBEP, 2008).

In the mobility sector many countries have introduced mandatory blending quotas or expansion targets for biofuels. For example, the United States has set ambitious targets for expansion of the biofuel sector; it plans usage of 56,000 million litres per year by 2012 and to increase this figure by 2022 to 136,000 million litres, which would be equivalent to 20 per cent of the country's projected annual fuel consumption (Doornbosch and Steenblik, 2007; EERE, 2008; GBEP, 2008; REN21, 2008; Box 4.1-2). Government procurement policies mandating the purchase of vehicles with alternative fuel options (e.g. flex-fuel technology) or the use of biofuels in state and municipal vehicles constitute another form of promotion. This approach has been adopted in countries such as the USA, Thailand and the Philippines (Doornbosch and Steenblik, 2007; GBEP, 2008).

STAGES 4 AND 5: TRANSPORT AND DISTRIBUTION / USE AND CONSUMPTION

In some countries the transport and marketing of electricity from bioenergy is promoted through reductions in transmission and distribution tariffs; this is the case in, for example, Brazil, South Africa and India. In industrialized countries the use of electricity from bioenergy is often encouraged through expenditure tax rebates or expenditure tax exemptions. In the heating sector households and businesses may be offered financial assistance for the conversion of heating systems to renewable energy sources. In addition, biofuels are wholly or partly exempt from excise duty in many countries. Table 4.1-5 gives examples of

Box 4.1-2

Current bioenergy use and promotion policy in the USA

Bioenergy is playing a more and more important role in the policies of the USA. However, the country's bioenergy policy has so far focused mainly on fuels for transport; promotion of bioenergy for heat and electricity has been minor. In expanding its ethanol production the USA's main aim is to reduce its dependence on imports of mineral oil. Environmental and climate change mitigation considerations are also taken into account, but are of secondary importance. At present bioenergy accounts for around 3 per cent of primary energy consumption in the USA, with biofuels constituting 25 per cent of this. The remaining 75 per cent is used for heat and electricity generation and is produced from wood and wood residue. In terms of bioenergy consumption the USA ranks in third place, after China and India (Zarrilli, 2006; GBEP, 2008).

As a result of the continuing promotion policy of recent years, the USA is now the world's leading producer of ethanol, closely followed by Brazil (Section 4.1.1). In the USA ethanol is produced chiefly from maize. In 2005 14.6 per cent of American maize production was processed into ethanol; in 2007 this figure had risen to more than 17 per cent (Zarrilli, 2006; GBEP, 2008). In 2007 the USA imported additional ethanol totalling approximately 1600 million litres from Brazil, Costa Rica, El Salvador and individual countries of the Caribbean Basin Initiative (CBI); of this, 714 million litres came from Brazil (RFA, 2008). Biodiesel production in the USA takes place on a much smaller scale. In 2006 1700 million litres of biodiesel were produced, mostly from soya beans; representing 16.5 per cent of global production. This means that the USA is the world's second-largest producer of biodiesel, after the EU (Licht, cited in OECD, 2008).

The government has put forward ambitious plans for future biodiesel production. According to the Renewable Fuel Standard, which was increased by the Energy Independence and Security Act (EISA) in 2007, 56,000 million litres of non-fossil fuels will have to be blended with fossil fuels in 2012, and 10 years later the figure will rise to 136,000 million litres. This would correspond in 2022 to around 20 per cent of total US fuel consumption (REN21, 2008; EERE, 2008). The overall targets are subdivided into annually increasing targets for first-generation biofuels and increasing targets for second-generation biofuels ('advanced biofuels') and biofuels from cellulose ('cellulosic biofuels'). There are also requirements regarding the greenhouse gas reduction potential of the biofuels over their entire life cycle. Under these requirements facilities for producing ethanol from grain that started production after the EISA came into force must achieve at least a 20 per cent reduction in life-cycle greenhouse gas emissions as compared with GHG gas emissions of the fossil reference in the base year of 2005. Biofuels classed as advanced biofuel or cellulosic biofuel must be able to demonstrate reductions of at least 50 per cent and 60 per cent respectively (EIA, 2008; GBEP, 2008; EERE, 2008). According to estimates of the US Ministry of Agriculture, biodiesel production will increase as a result of the various government promotion measures (Table 4.1-5) to 7500 million litres annually by 2010 and to 12,600 million litres by 2015. The government will in future focus more on second-generation biofuels, particularly fuels from wastes and residues (NGA, 2008; GBEP, 2008).

The government has set no targets for the generation of power and heat from biomass (GBEP, 2008). Despite this, regulation provides some incentives to the use of power and heat from biomass. Some states have set Renewable Portfolio Standards for the feed-in of electricity (REN21, 2008). Furthermore, from 1 January 2008 facilities that produce electricity from biomass will benefit from a Renewable Electricity Production Tax Credit; they will thus receive US\$ 0.0019 per kWh for electricity from biomass from a closed circuit and US\$ 0.001 per kWh for electricity from an open circuit. Through the Renewable Energy Production Incentive an additional US\$ 0.0015 per kWh will be paid for electricity from renewables for the first 10 years. The government and electricity companies also issue Clean Renewable Energy Bonds, which support projects in the area of regenerative electricity generation. Biofuels nevertheless remain the focus of bioenergy promotion in the USA. Producing bioethanol from cellulose on a competitive basis is seen as a particular challenge (GBEP, 2008).

---

bioenergy promotion measures in selected countries. Within these countries individual states or provinces may have adopted additional measures that cannot be discussed in this context.

CONCLUSIONS AND OUTLOOK

There is little difference between national promotion policies for bioenergy in terms of the instruments used: in the electricity sector fixed feed-in tariffs are the preferred instrument. In the fuel sector blending quotas and national expansion targets predominate. Subsidies or tax concessions for producing or using bioenergy are also frequently used strategies. Particularly large tax concessions for biofuel blending are granted by the US government, which is providing a tax credit of 0.14 US\$/litre for ethanol blending until the end of 2010, and a tax credit of 0.12 US\$/litre for biodiesel until the end of 2008 (REN21,

2008). High import tariffs on bioenergy carriers are more commonly found in industrialized countries (e.g. EU, USA). These levies are intended to rectify the competitive disadvantage of domestic producers, who incur higher production costs for biofuels than providers from developing and newly industrializing countries.

It is evident that if support for biofuels were removed demand for biodiesel and bioethanol would be significantly lower than it is with support in place. For example, it is estimated that without support demand for biodiesel would fall by 87 per cent in the European Union and by 55 per cent in the USA. Globally, demand for biodiesel would be reduced by about 50 per cent. Demand for ethanol, on the other hand, is less dependent on promotion policies. If all support were removed it is likely that demand for ethanol would fall by only 14 per cent, since ethanol production in Brazil – one of the most important producing countries – remains largely competitive even without support (OECD, 2008).

In some countries production of biofuel in the private sector is uneconomic; correspondingly large financial incentives for market participants are required in order to boost production and demand. Hence estimated expenditure on state promotion measures for biofuels in the USA, the EU and Canada in 2006 amounts to around US$ 11,000 million. Continuation of promotion policies could result in costs rising to US$ 27,000 million per year in the next 5–10 years (Steenblik, 2007; OECD, 2008). Since state promotion measures for biofuels, in these countries and elsewhere, places increasing financial pressure on national budgets, a change in policy can currently be observed: there is movement away from active fiscal promotion policy towards promotion through usage targets such as blending quotas for biofuels (GBEP, 2008). Such targets create guaranteed sales markets for businesses that produce biofuels or the associated raw materials. At the same time the costs of promotion – that is, the additional costs of production – are shifted to the consumer. Production and usage targets are applied in the OECD countries, in the major newly industrializing countries of Brazil and India, and to an increasing extent also in developing countries such as the Philippines and Indonesia (Steenblik, 2007; GBEP, 2008).

In some developing and newly industrializing countries, however, a growing number of voices are questioning the use of national blending quotas on the grounds that they initiate market effects that can ultimately jeopardize national food security (Section 5.2). Some countries have in consequence enacted regulations excluding particular raw materials from biofuel production or promotion. For example, in South Africa maize is not eligible for ethanol promotion, and in China grain may not be used to manufacture fuel (Box 5.2-2; Reuters, 2007; Weyerhaeuser et al., 2007).

Pressure is also being placed on the EU's planned 10 per cent blending target for biofuels on the grounds that it incurs high macroeconomic costs and may not in fact contribute to the main aim of bioenergy promotion policy, which is to mitigate climate change. Thus the London-based consultancy Europe Economics estimates that meeting the EU's 10 per cent target for biofuels will require total annual transfers to the wider biofuels industry of € 11,000–23,000 million annually by 2020 (Europe Economics, 2008). Various studies, including those carried out by the European Commission's Joint Research Centre (JRC) and by the OECD, agree in concluding that attainment of the 10 per cent blending target will most likely not contribute to a significant reduction in greenhouse gas emissions in the transport sector. In relation to climate change mitigation and the costs per tonne of greenhouse gas abatement, biofuels have so far tended to yield below-average results: with abatement costs at present significantly over € 100 per tonne, biofuel use is considerably more expensive than alternative abatement options (Doornbosch and Steenblik, 2007; de Santi, 2008; OECD, 2008; Section 7.3; Box 4.1-3). With regard to the policy goals of increasing energy security and promoting rural development, the JRC also sees no significant benefit in the EU's biofuel policy. Instead it calculates that there is an 80 per cent probability that biofuel promotion for the EU-25 between 2007 and 2020 will be associated with overall net costs to the economy amounting to around € 33,000–65,000 million at present-day values (de Santi, 2008).

Some developing countries, particularly in Africa and South America, have considerable potential for producing biomass for energy purposes but may not yet have formulated corresponding policy targets. In some countries targets have been formulated but concrete promotion policies have not been implemented; for example, this is the case in Chile, El Salvador and Panama (Jull et al., 2007). The necessary investment in infrastructure and technology presents an obstacle for these countries in developing a bioenergy market. Export is also difficult for developing countries on account of the subsidy and trade policies of many industrialized countries. Developing countries face additional trade barriers arising from the fact that technical standards are not harmonized internationally (Mathews, 2007; Jull et al., 2007; von Braun, 2007; Lindlein, 2007). For these reasons the preferred option for many developing countries in Africa, Latin America and Asia is to focus on small-scale production of bioenergy for domestic use (Section 10.8).

4.2
Global land cover and land use

Among the most significant impacts of human society on the environment are changes in land cover and in land use. The former relate to changes in the biophysical features of the Earth's surface, while the latter are determined by the purposes for which people use the land (Turner et al., 1990; Lambin et al., 2001; Schinninger, 2008). More than three-quarters of the world's ice-free land surface has already undergone changes as a result of human use (Ellis and Ramankutty, 2008). General understanding of the causes of changes in land cover and land use is unfortunately often dominated by over-simplifications. Neither population growth nor poverty can be regarded

Box 4.1-3

Current bioenergy policy and use in the EU

In early 2008 the European Commission presented its draft of an EU Directive on the promotion of energy from renewable sources, providing a framework for implementation of the targets and strategies for expansion of the use of renewables that were drawn up by the Commission and the Council at the beginning of 2007. Under the Directive the proportion of total energy consumption met by renewables is set to reach 20 per cent by 2020, and the proportion of total EU fuel consumption met by biofuels in that year should be 10 per cent. The EU's main aim in this is to make a contribution to climate change mitigation (EU Commission, 2008a).

Bioenergy at present constitutes around 4 per cent of primary energy consumption in the EU (EEA, 2007b). Biofuels account for around 1 per cent of total fuel consumption (EU Commission, 2006a; REN21, 2008). Biofuel production within the EU is promoted by tax relief that applies in many member states. In 2007 total biodiesel production in the EU amounted to 6100 million litres, corresponding to 59.9 per cent of global production. Within the EU, Germany, France, Italy, the Czech Republic and Spain were the most important producing countries (Licht cited in OECD, 2008). Ethanol is produced in the EU in significantly smaller quantities, mainly from grain and sugar beet. The principal producing countries in the EU are Germany and Spain. In 2007 the EU produced a total of 2300 million litres of ethanol, representing 4.4 per cent of global production (Zarrilli, 2006; Licht cited in OECD, 2008; REN21, 2008).

For any further expansion of the contribution of biofuel to total fuel consumption to meet the 10 per cent quota laid down in the draft guidelines, the EU is dependent – particularly as far as ethanol is concerned – on imports (REN21, 2008). The most important countries from which ethanol is imported are Brazil and Pakistan. Palm oil for biodiesel production is important primarily from Malaysia (Zarrilli, 2006). However, there are growing doubts about the ecological and economical sustainability of a 10 per cent blending quota for biofuels. It is uncertain whether blending will actually achieve the intended goals, which include climate change mitigation, security of supply and job creation; at the same time, the estimated costs of government promotion measures are highly likely to exceed the hoped-for benefits (de Santi, 2008). The Industrial Committee of the European Parliament has therefore invited discussion of a modification of the 10 per cent quota. Under the amended plan, at least 40 per cent of the quota should be met through the use of second-generation biofuels, hydrogen or electro-mobility (EU Parliament, 2008). A final decision on the quota is expected by the end of 2008.

Electricity production from biomass is another important pillar of the EU's 20 per cent target. Promotional measures used in individual members states include feed-in tariffs, renewable energy certificates (RECs), tax incentives and subsidies for production capital; of these, feed-in tariffs and RECs have been found to be the most effective. Twenty-one member states participate in an inter-European transfer systems for RECs, the European Energy Certificate System (EECS; EU Commission, 2005b; REN21, 2008). In contrast to electricity and fuels, there are as yet no concrete targets for heat production from biomass. Plans include improving the energy efficiency of buildings and promoting district heating systems (EU Commission, 2005a). Increased utilization of the potential of high-efficiency combined heat and power, CHP in the member states is also planned (GBEP, 2008).

---

as the sole cause of global changes in land cover, which involve in particular the conversion of forest into cropland (Lambin et al., 2001). These changes arise primarily in response to economic opportunities, which in turn are closely linked to social, political and infrastructure-related conditions. The effects of changes in land cover and land use (Sections 4.2.1 and 4.2.2) impact in turn on the carbon storage capacity, greenhouse gas emissions and fertility of the soil (Section 4.2.3), as well as on the local climate and hence also on local land cover.

4.2.1
Global land cover

Land cover includes not only topographical features of the Earth's land surface but also structures such as buildings and roads and aspects of the natural environment such as soil type, vegetation type, biodiversity, surface water and ground water (Meyer, 1995). Discussion of land use issues usually focuses on agricultural and forestry use (intensity, type), even though settlement use and water use also fall under this heading.

Direct and immediate effects of human-induced changes in land cover and land use on the environment (conversion, loss, fragmentation, eutrophication) lead to changes in the nutrient cycle and in the hydrological and thermal regime; often they also result in increased erosion of the converted surface. These effects must be taken into account in the discussion of bioenergy use. Indirect effects at ecosystem level, on the other hand, are often manifested as loss of biological diversity (Jarnagin, 2004).

DATA SOURCES

For describing land cover and land cover changes, whether on a regional or global scale, the most important source of information is remote sensing data (DeFries and Townsend, 1999; Figure 4.2-1). However, classification of land cover varies widely between data sources, which also use different definitions for the various categories. Thus according to Lepers et al. (2005) around 90 different definitions of 'forest' are used in different parts of the world. The FAO (1997) defines forest as a vegetation unit with

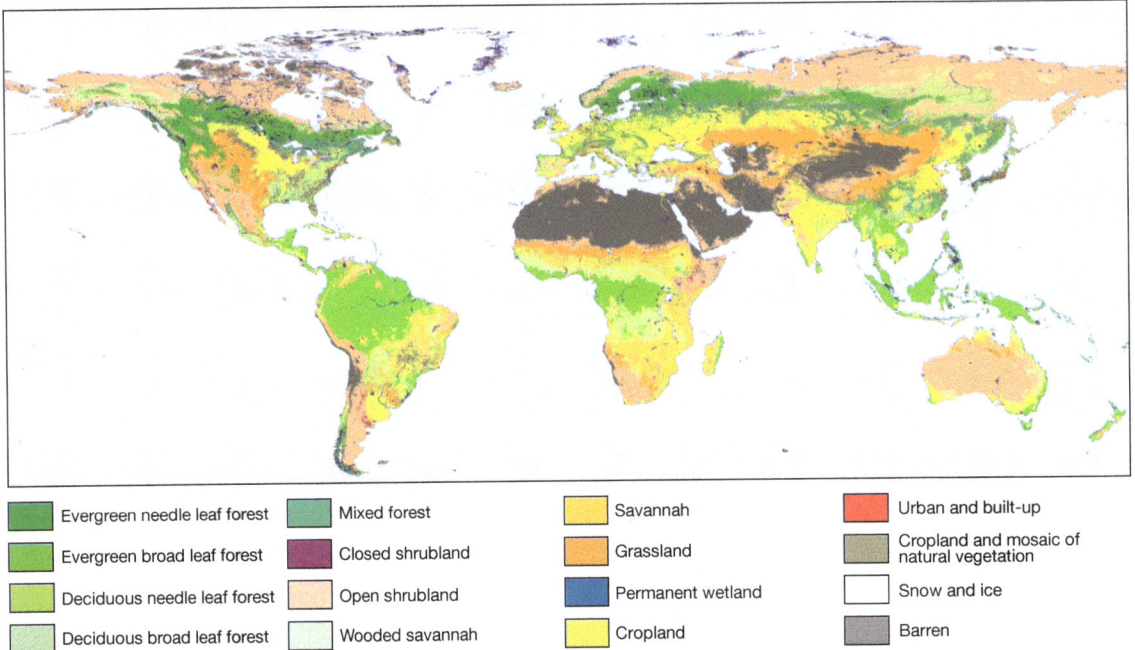

Figure 4.2-1
Global distribution of land cover types, based on MODIS satellite data (Land cover science data set of the IGBP;
0.05° resolution, year 2001).
Source: U.S. Geological Survey – Earth Resources Observation and Science Center, 2008

tree crown cover of ≥10 per cent, an area of ≥0.5 ha and potential growth of >5 m. The International Geosphere Biosphere Programme (IGBP), by contrast, defines forest as predominantly woody vegetation (>60 per cent) with a growth height of >2 m.

Similar problems arise for pasture land. In most cases national data are inconsistent with data in the FAO's statistical database (FAOSTAT). For example, Ramankutty et al. (2008) specify the world's total area of pasture land as 28.0 million square kilometres; this is 18 per cent less than FAOSTAT's estimate of 34.4 million square kilometres. The largest discrepancies occur in Saudi-Arabia, Australia, China and Mongolia and result from different definitions of 'pasture land' – a problem that is also highlighted by FAOSTAT. In addition to differing definitions of arable and pasture land, multi-functional land use also presents a problem: for example, in some countries, particularly in Africa and Asia, arable land is used for grazing animals after the harvest (Figure 4.2-4). Other problems arise from the data mixing of different remote sensing data sets; from the data provided it is not possible to obtain any information about land use below the topmost vegetation cover identified by the remote sensor. There are also differences in the time scales used in some inventory data; in most industrialized countries surveys are carried out every 5–10 years. The spatial resolution of surveys also varies. As a result, the African continent and the former Soviet Union tend to be under-represented in the data.

Alongside land cover and population density, Ellis and Ramankutty (2008) also include other land use. This new classification (Figure 4.2-2) could prove very useful for integrative modelling approaches, because ecosystems are affected by a wide range of human-induced factors, including climate change, nitrogen input, pollution and above all land-use changes. Measuring the drivers and the outcomes of their synergetic effects and integrating them into standard climate and ecosystem models presents a challenge (Fischlin et al., 2007). In addition this classification is still so new that its suitability for the task cannot yet be evaluated.

SITUATION AND TRENDS
The greatest changes in land cover were brought about in the past through the conversion of forests and grassland into arable and pasture land (Schinninger, 2008). In the last 300 years arable land has increased by 460 per cent and pasture land by 560 per cent (Klein Goldewijk, 2001). Today only 19 per cent of land with cultivation potential still forms part of forest ecosystems (Fischer et al., 2002); of this land, the largest areas suitable for cultivation are in South and North America, followed by Central America and Africa.

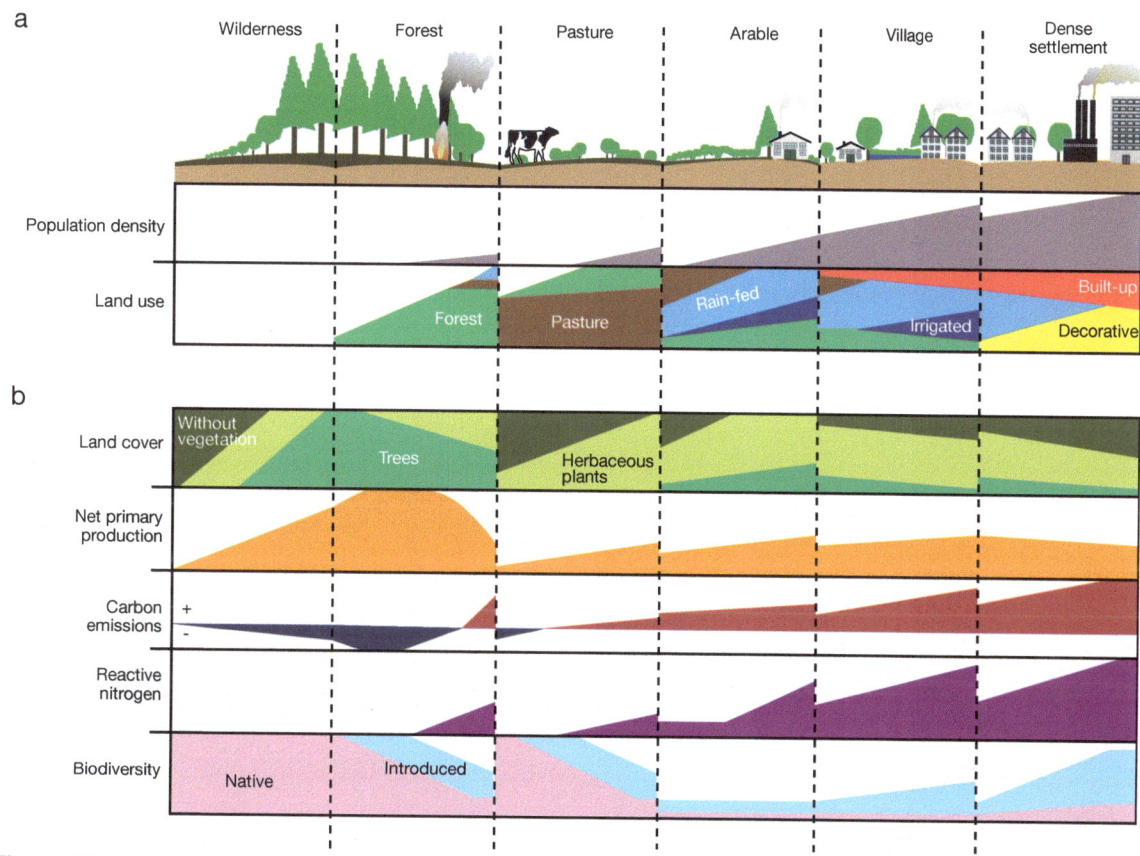

Figure 4.2-2
Conceptual model of habitats subject to different levels of anthropogenic influence. From left to right: increase in anthropogenic influence. (a) Habitats structured according to population density (logarithmic scale) and land use (percentage scale). Within these habitats these factors form patterns of (b) ecosystem structures (land cover), processes (net primary production, carbon emissions, availability of reactive nitrogen) and biodiversity (native vs. introduced and domesticated biodiversity; quoted relative to the original biodiversity; white areas represent the net reduction in biodiversity). Quelle: Ellis und Ramankutty, 2008

The quantity of land used for agriculture has increased by nearly 500 million hectares over the last four decades. This trend is likely to continue in future (Fedoroff and Cohen, 1999; Huang et al., 2002; Trewavas, 2002; Green et al., 2005). Rosegrant et al. (2001) forecast the conversion of a further 500 million hectares into agricultural land by 2020; much of this land will be in Latin America and sub-Saharan Africa. The FAO expects that the global area of land used for food production will need to be increased by 120 million hectares by 2030 in order to secure the food supply of the growing world population (FAO, 2003a).

Lepers et al. (2005) summarize changes in land cover during the period 1981–2000, basing their description on remote sensing data, expert opinions, land area surveys and statistics on land cover and land use at regional, national and international level. Difficulties in synthesizing existing sets of data on land-use changes arose from the absence of standardized definitions, differences in the spatial resolu-

tion of different remote sensing data, and differences in the time spans and spatial coverage of the data collected. See Box 4.2-1 on the definition of 'marginal land'. The results of the study show that the largest areas of rapid land-cover change were located in Asia. In South-East Asia there was a rapid increase in the area of agricultural land between 1981–2000, often linked with large-scale land clearance. The Amazon basin continues to be a focal point of rainforest clearance. In Siberia there was a rapid increase in forest degradation as a result of non-sustainable use and the increasing frequency of forest fires. By contrast a fall in the area of land used for agriculture was recorded in the south-east of the USA and eastern China. In addition many of the most densely populated and fastest growing cities were located in the tropics.

Dynamic global vegetation models (DGVMs) are used to simulate changes in land use – particularly in the distribution of vegetation – over time, and thus to assess changes in ecosystem functions and services

Box 4.2-1

Defining the concept of 'marginal land'

Discussion of the potential of bioenergy often involves considering the cultivation of bioenergy crops on agricultural and forestry land whose yields are relatively low. In this connection the terms 'marginal land', 'degraded land', 'unproductive land', 'set-aside land', 'wasteland' and 'fallow land' are often used in parallel or even synonymously, usually without further differentiation. In the present report WBGU uses the term 'marginal land' as an umbrella term for (1) areas with little capacity for fulfilling a production or regulation function, and also for (2) areas that have lost their production and regulation function, sometimes to a significant extent. (1) includes areas whose productivity for agriculture or forestry is considered low. Also in this category are arid and semi-arid grasslands, desert fringes and areas of steep ground and structurally weak or erosion-prone soils, particularly in mountainous regions. (2) covers

formerly productive areas; they may have lost their yield potential as a result of human-induced soil degradation (e.g. overused, degraded and therefore unproductive land, including both forests and pasture and arable land), or the land may have been deliberately taken out of production (e.g. set-aside land in central Europe that has been taken out of production for economic or political reasons). Marginal areas are generally highly susceptible to soil degradation.

WBGU avoids using the term 'wasteland' on account of its associations with neglected, unused land. The term 'fallow land' refers in the strict sense to an unworked field in a rotational farming system (arable, meadow/pasture, fallow); the fallow stage is needed to enable the soil to recover when the land is farmed without artificial fertilizers. The word 'fallow' is also used to describe land taken out of production that still exhibits signs of human use. This term, too, is deliberately avoided by WBGU in this report, because it leaves the reasons for non-use unclear and is therefore imprecise.

(Fischlin et al., 2007). This option is also used in the modelling carried out by Beringer and Lucht (2008) and described in Chapter 6. Much progress has been made in recent years in merging DGVMs and climate models, enabling researchers to study feedback effects between the biosphere and processes in the atmosphere (Fischlin et al., 2007).

Changes in land cover and land use in the temperate zone in recent decades have, however, affected not only vegetation cover but also the albedo; the increase in albedo has probably had a cooling effect (Govindasamy et al., 2001; Bounoua et al., 2002). The feedback effects of land-use changes on climate are influenced by a complex interplay of various local factors (evaporation rate, ground water storage capacity, albedo). These effects are, however, also dependent on large-scale air circulation movements and may therefore operate differently in different regions. Historically, human-induced land cover changes may have reduced temperatures in agricultural areas at medium latitudes by 1–2°C (Feddema et al., 2005). On the other hand, it is projected from simulations that future human-induced influences on land cover, taking account of the further deforestation of the tropics, will result in further warming by 1–2°C (Feddema et al., 2005). Depending on the type of future land-use changes that are incorporated into two IPCC-SRES scenarios (A2 = Regionalized economic development and B1 = Global sustainability; Nakicenovic and Swart, 2000), there are significant differences in the results of the climate simulation for 2100 as a result of the feedback between the land surface and the energy balance of the atmosphere (Feddema et al., 2005). For example, the conversion of the rainforest in the Amazon region into arable land in Scenario A2 leads to a temperature increase of over

2°C in 2100, which in turn would affect the Hadley and monsoon circulation.

Other model-based studies highlight future land-cover changes, particularly in the tropics and subtropics (DeFries et al, 2002; Voldoire, 2006). However, it remains extremely difficult to forecast future land cover and describe the future distribution of vegetation, despite the fact that these models usually ignore socioeconomic factors (DeFries et al., 2002). Land-use dynamics have only recently been integrated into dynamic vegetation models (Voldoire et al., 2007); this represents a major step forward. Land-use changes undoubtedly exert a major influence on future regional and global climate (Feddema et al., 2005; Pitman et al., 1999; Pielke et al., 2002; Voldoire et al., 2007); depending on geographical location they may exacerbate or diminish the resulting climate change (DeFries et al., 2002; Feddema, 2005; Voldoire, 2006).

4.2.2
Global land use

Land use – which is closely connected to land cover – describes the type, manner and purpose of land use by humans, and/or the use of existing resources; for example, use for agriculture, mining or forestry (Meyer, 1995). The term 'land-use changes' (Figure 4.2-3) refers both to the human-induced replacement of one type of land use by another – for example, the conversion of forest into agricultural land – and to changes in management practices within a land-use type – for example, intensification of agriculture. Land-use changes and the associated loss and fragmentation of habitats are important drivers of past and present ecosystem changes and of the loss

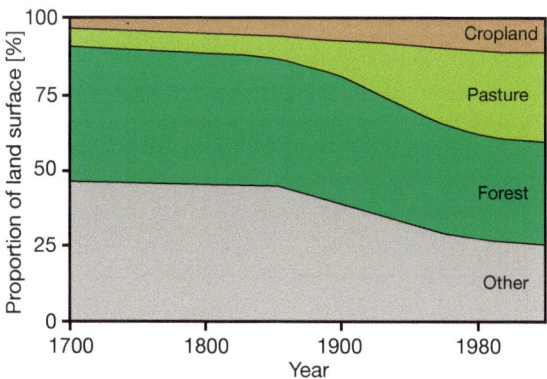

Figure 4.2-3
Estimated land-use changes between 1700 and 1995.
Source: Klein Goldewijk and Battjes, 1997, cited in Lambin et al., 2001

of biological diversity. This means that studies of the effects of climate change that fail to consider land-use changes may yield incorrect estimates of ecosystem responses (Fischlin et al., 2007).

Agricultural activities are a major cause of land-use changes. According to the FAO, the global land area used for agricultural purposes in 2005 amounted to 49.7 million square kilometres (FAOSTAT, 2006); 69 per cent of this total, equivalent to 34.1 million square kilometres, was in use as pasture land and 31 per cent, or 15.6 million square kilometres, was arable land or land under permanent cultivation (Figure 4.2-4). A new study by Ramankutty et al. (2008)

with a resolution of 10 km combines national and sub-national statistics with data on agricultural land use and remote sensing data on land cover; for 2000 it identifies 15 million square kilometres of arable land and 28 million square kilometres of pasture land. According to these statistics, humans are using around 34 per cent of the global ice-free land area for agricultural purposes.

Ramankutty et al. (2008) also investigated the extent to which potentially natural vegetation has been affected by agricultural use. If up-to-date maps of the distribution of agricultural land are merged with the maps of potentially natural vegetation drawn up by Ramankutty and Foley (1999), the results show that around 30 per cent of temperate, deciduous forests have been converted to arable land and 50 per cent of grasslands have been converted to pastures. Although Ramankutty and Foley (1999) analysed the global distribution of arable land for as far back as 1992, it is not possible to draw any conclusions about changes during this period from comparison of the two studies, since the methods and sources used were changed.

The greatest effects of changed land use are observed in the net primary production (NPP) of plants – that is, in the production of biomass by primary producers, taking cellular respiration into account. Monfreda et al. (2008) modelled the global NPP of arable land for the year 2000. The regions with the largest NPP of over 1 kg C per m^2 per year were Western Europe, East Asia, the central USA,

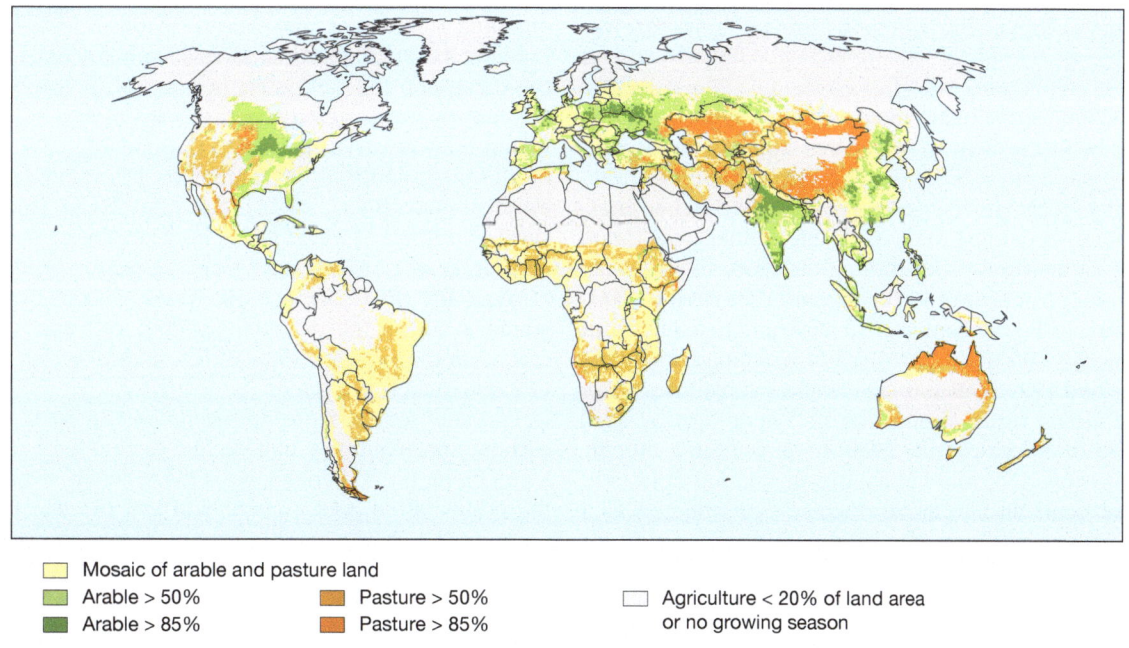

Figure 4.2-4
Current global extent of arable and pasture land. The colours show the ratio of arable land to pasture in each area.
Source: UNEP, 2007a

Brazil and Argentina. Approximately 13 per cent of global arable land is planted with perennial vegetation, which stores more carbon in its roots than annual vegetation, and around 24 per cent of cultivated agricultural land is planted with crops that have the more efficient C4 photosynthesis mechanism (e.g. maize, sorghum, millet and sugar cane; Monfreda et al., 2008). At a global NPP of 56.8 Gt C per year, humans, who make up just 0.5 per cent of the biomass of heterotrophic organisms, appropriate 15.6 Gt C or almost 24 per cent of global NPP (Haberl et al., 2007). Of this biomass used by humans (see also HANPP: Human Appropriation of Terrestrial Net Primary Production; Imhoff et al., 2004), 58 per cent is used as feed and only 12 per cent directly as food. A further 20 per cent is used as raw materials and 10 per cent as firewood (Krausmann et al., 2007).

Water consumption in agriculture is also altered significantly as a result of land-use changes; use of water in agriculture is already higher than in all other sectors of the economy (MA, 2005b). In low-wage countries 87 per cent of the water removed is used for agricultural purposes; in countries with medium wage levels the corresponding figure is 74 per cent and in countries with high income levels it is only 30 per cent (World Bank, 2003). There are at present 276 million hectares of irrigated agricultural land (FAOSTAT, 2006); this represents a five-fold increase since the beginning of the 19th century. With the increasing need for irrigation, water management becomes an important issue. In addition other problems, particularly in connection with food production, are foreseeable as a result of climate change. Globally around 3600 million hectares (approximately 27 per cent of the land surface) are too dry for rain-fed agriculture. Seen in the light of water availability, only around 1.8 per cent of these dry zones is suitable for growing cereals under irrigated conditions (Fischer et al., 2002). According to the FAO, the annual growth rate of agricultural production is therefore likely to fall from 2.2 per cent to 1.6 per cent in the period 2000–2015, to 1.3 per cent in 2015–2030 and to 0.8 per cent in 2030–2050 (FAO, 2006b). When compared with the period 1999–2001 this nevertheless constitutes a rise in global cereal production of 55 per cent by 2030 and of 80 per cent by 2050. To achieve this, though, a further 185 million hectares of rain-fed land (+19 per cent) and 60 million hectares of irrigated land (+30 per cent) must be brought into production for cereal growing. However, on account of the projected decline in water availability in some regions as a result of climate change, these regions (e.g. the Mediterranean basin, Central America and the sub-tropical regions of Africa and Australia) could become too dry for rain-fed agriculture (Easterling et al.,

2007). Alongside climate-induced, regional problems of water availability, Scanlon et al. (2007) point out that, because of delayed ecosystem responses (such as ground water replenishment, water quality), the effects of past land-use changes on water levels have not yet become apparent; these effects could lead to future competition for water use.

4.2.3
The influence of land-use changes on ecosystem services

Human intervention in the environment brings about changes in biological diversity and hence also changes in ecosystem services, which range from greenhouse gas emissions and carbon storage in the soil and vegetation to erosion control and aesthetic aspects. In particular biological diversity (or biodiversity) is strongly influenced by the land-use changes that are relevant in connection with bioenergy use (Section 5.4). The following section examines the effects of land-use changes of various ecosystem types on carbon storage, greenhouse gas emissions and biological diversity.

4.2.3.1
Conversion of forest

For the use of bioenergy – as for agricultural use in general – the necessary land was and is frequently obtained from areas that were previously forested (Section 4.2.1). In addition, forests are a key issue in the debate on bioenergy use because of the potential loss of the largest carbon reservoirs and sinks within the terrestrial biosphere. Around 20 per cent of all human-induced CO_2 emissions arise from forest clearance (IPCC, 2007b). The rate of CO_2 emissions as a result of forest losses in the 1990s is put at an average of 1.6 Gt (0.5–2.7 Gt) C per year (Cramer et al., 2004; IPCC, 2007a). The FAO (2006c) even assumes that in the period 1990–2005 the stock of carbon in the living biomass of forests is diminished at a rate of 4 Gt per year.

Globally forests store an estimated 638 Gt C of carbon; approximately half of this carbon is in the living biomass and in dead wood (MA, 2005b), while the other half is bound in the soil and the leaf-litter layer (FAO, 2006c). This is equivalent to roughly 40 per cent of the carbon present in the terrestrial biosphere (Matthews et al., 2000). When a change of use takes place and forest is cleared to make way for the cultivation of bioenergy crops, the consequent carbon losses must first be compensated before the effect of the bioenergy on the greenhouse gas balance can be described as positive (Section 6.4.3.3). In this connection it must be borne in mind that carbon is stored not only in the biomass but also in the soil. In South America, particularly in the tropical rainforest, around one-third of the total carbon is stored in the soil, but in European forests the proportion of carbon stored in the soil is around two-thirds (FAO, 2006c).

In the last 20 years there have been major losses of carbon as a consequence of land-use changes involving the clearance of tropical forests in order to create arable and pasture land (IPCC, 2007b). In Brazil, for example, large areas of tropical forest have been sacrificed to soya production (Tollefson, 2008). Carbon losses through clearance are particularly critical because the carbon reservoirs that are destroyed are not only large but also very old. The carbon in wood and dead wood may have been stored for several decades or centuries; in humus the reservoir may have existed for centuries or millennia (Vieira et al., 2005). Since carbon reservoirs interact continuously with the environment, the effects of clearing forests for bioenergy use can vary widely from region to region. For example, the change in the carbon reservoir in the soil of cleared tropical forests that are converted to pasture land depends largely on the soil type (Figure 4.2-5; Bormann and Likens, 1979; López-Ulloa et al., 2005). Thus after 25 years the net quantity of organic carbon lost per unit of surface area was almost five times greater for pasture land on former rainforest soil with a clayey subsoil than it was for similar pasture land with a sandy soil, where the carbon stock remained almost constant over the same period (van Dam et al., 1997). In addition, the type and intensity of the land-use change in connection with bioenergy use play a major part in determining how much carbon is lost by the ecosystem. The direct effects of the conversion on carbon losses depend crucially on whether only the annual new growth is used for traditional bioenergy use, whether whole trees or large areas of land are cleared for a bioenergy plantation and whether the soil is damaged or even eroded by harvesting machinery or fire.

In addition, the remaining carbon stock, whether in the soil or in the residual vegetation, has a noticeable influence on the further course of carbon loss. If a great deal of dead biomass is left on the land after clearance, or if the soil is rich in organic substances, these materials will be broken down by micro-organisms, provided that new vegetation does not overshadow the soil (Bormann and Likens, 1979). This means that, depending on tree type and region, young tree plantations can continue to be sources of carbon for several years or decades, as a result of increased soil respiration, despite the growth of the

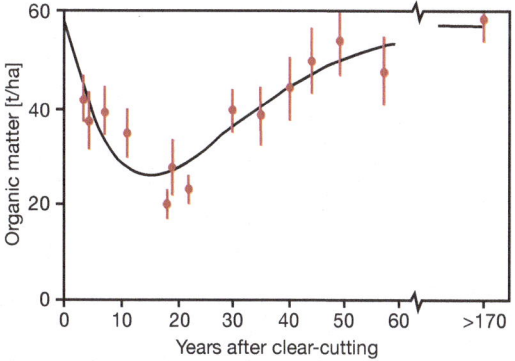

Figure 4.2-5
Change in the quantity of organic matter on the forest floor after clear-cutting Nordic broad-leaved woodland. The points are average values of 30 samples per area of forest with 95% confidence intervals.
Source: Bormann and Likens, 1979

trees (Figure 4.2-6; Harcombe et al., 1990; Buchmann and Schulze, 1999; Baldocchi, 2008).

The nitrogen balance is also affected by land-use changes. Clearance of tropical forests causes nitrous oxide and nitrogen oxide emissions to rise by 30–350 per cent, depending on the nitrogen input, temperature and moisture levels (IPCC, 2007b). If the land is cleared by burning rather than felling, the carbon from the biomass and soil is immediately released into the atmosphere as CO_2 during the combustion process; if combustion is incomplete, CO and methane (CH_4) are also released. Forest and bush fires (excluding clearance by burning) release 1.7–4.1 Gt C annually, which is equivalent to around 3–8 per cent of total terrestrial net primary production (IPCC, 2007b). Approximately 14 per cent of human-induced CH_4 emissions are attributable to the burning of biomass (Wuebbles and Hayhoe, 2002). Here, too, carbon

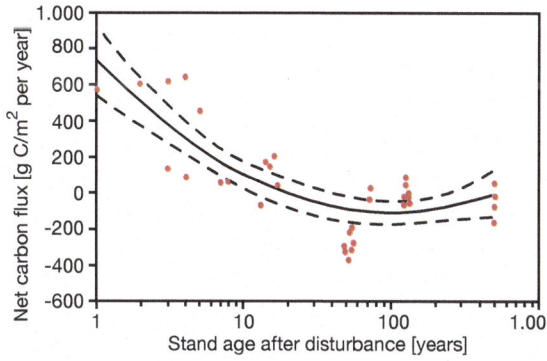

Figure 4.2-6
Relationship between net carbon flux and stand age after disturbance. The data originate from a number of chronosequence studies of coniferous stands in central and western Canada and the Pacific north-west coast of the USA.
Source: Baldocchi, 2008

losses vary widely from region to region, depending on land cover; they are also dependent on the intensity of the fire. Smouldering fires with incomplete ashing give rise to particularly large methane emissions. In addition, the effects persist long-term. The low organic content of the soil reduces soil fertility and impairs the soil structure; this frequently leads to increased surface run-off of water and to erosion. On the other hand, the formation of charcoal significantly increases the length of time for which the carbon is retained in the soil (Section 5.5).

In tropical rainforests the almost completely closed nutrient cycles with a very large number of participating organisms result in very high productivity. Disruption of these cycles can have devastating effects on biological diversity (Section 5.4; WBGU, 2000). The loss of forest areas and fragmentation of the landscape destroys important habitats. Tropical rainforests are the most species-rich of all terrestrial ecosystems: 15 per cent of all plant species live in tropical rainforests with extraordinarily high species density on an area that occupies only 0.2 per cent of the global land area (Mooney et al., 1995). Although some species are well able to adapt to new habitats, many endemic species have an extremely local range and are therefore particularly likely to be exterminated by the conversion of forest into agricultural land. For example, many plant species of the tropical cloud forest in Latin America occur in an area less than 10 km² in size (Mooney et al., 1995).

In conclusion it can be stated that the conversion of forest into pasture or arable land is always associated with considerable carbon losses and therefore does not represent an option for efficient climate change mitigation.

4.2.3.2
Conversion of wetlands

The productivity of wetlands and the extent to which carbon is stored in organic soils also vary very widely from region to region. Wetlands are among the most productive of all locations. Moors cover only 3–4 per cent of the terrestrial land surface, but they store around 25–30 per cent of the global carbon that is bound in plants and soils; this is equivalent to around 540 Gt C (MA, 2005b). Siberian raised bogs store up to 2 kg C per m² per year (Peregon et al., 2008).

Large quantities of carbon are released as wooded peatlands in South-East Asia are drained and cleared. The quantity of carbon stored by such peatlands in Indonesia, Malaysia, Brunei and Papua New Guinea is estimated at 42 Gt (Hooijer et al., 2006). Forty-five per cent of these forests have already been cleared and the subsoil drained, often for the cultivation of oil palms. The drained organic soils are susceptible to fire, which increases the loss of carbon from the ecosystem. Hooijer et al. (2006) estimate the CO_2 emissions arising from the loss of peatlands at 632Mt (megatons) per year, with a possible increase during the period 2015–2035 to an emissions maximum of around 823Mt per year. In contrast to carbon, which is released as CO_2, CH_4 emissions in tropical wetlands are very low (Jauhiainen et al., 2005).

Change of use also has a serious impact on biological diversity (Section 5.4). Drainage of wetlands causes abrupt changes in the ecystem as result of the intrusion of oxygen; this leads to the extinction of many animal species that are adapted to this specific habitat. In addition, the radically different hydrological conditions endanger many higher plants (MA, 2005c) and affect the water regime and the local water cycle. For these reasons further steps to convert wetlands for energy crop cultivation should be rejected.

4.2.3.3
Conversion of grassland

Grassland – or pasture land (usually degraded), which is the dominant usage form of grassland worldwide – is frequently mentioned in the bioenergy debate in connection with potential land reserves. Grassland ecosystems cover between 20 per cent and 40 per cent of the continental land area, depending on the definition and the method of data collection used (White et al., 2000; Scanlon et al., 2007). In the last 40 years the area of pasture land has increased globally by 10 per cent to around 3500 million hectares, representing 69 per cent of land in agricultural use (FAO-STAT, 2006; IPCC, 2007c). The carbon stock in grassland is smaller than in forest, because it exists almost exclusively in the soil; nevertheless, the loss of biological diversity when grassland is converted for the cultivation of energy crops is an important issue.

On account of their extent, grassland soils are important global carbon reservoirs. They contain around 34 per cent of the carbon present in terrestrial ecosystems (White et al., 2000). While grassland ecosystems store on average less carbon per unit of area than forests do, the amount stored is considerably more than in arable land (Kirby and Potvin, 2007). In tropical savannahs significant quantities of carbon are released by wildfires (White et al., 2000; IPCC, 2007c). Tylianakis et al. (2008) found that the production of underground biomass correlates positively with plant richness in temperate grasslands; the more heterogeneous the site, the more significant the correlation. Human-induced changes in the carbon cycle of grassland ecosystems are the result of arable farming, urbanization, soil degradation, grazing, fragmentation and the introduction of non-native organisms (White et al., 2000). In comparison with extensive grassland use, intensive use in temperate latitudes leads to greater carbon input to the soil on account of the increased root production and does not result in substantially higher environmental stress per hectare yield (Kägi et al., 2007).

The afforestation of pasture land increases the amount of organic carbon in the soil, depending on the age of the pasture. De Koning et al. (2003) found, however, that the amount of carbon stored in secondary forest was lower than in young pasture land that was less than ten years old; for 20–30-year-old pasture land the afforested areas accumulated up to 20 per cent more carbon per year. In addition, afforestation with trees is to be clearly preferred to arable use involving perennial plants because the lower frequency and intensity of use usually results in less carbon being lost through soil respiration. In semi-arid regions the conversion of grazed bush steppe into forest can lead to a considerable increase in carbon storage in the soil within 35 years (Grünzweig et al., 2007).

On the other hand, when grassland is converted into arable land, CO_2 is released as a result of the increase in soil respiration. The soil loses its year-round and perennial vegetation cover and thus becomes more susceptible to erosion. Similar mechanisms operate when grassland is overused through being overstocked with animals. This type of grassland conversion occurs most often in arid and semi-arid regions (Sahel, Central Asia), where biomass production is in any case low on account of climatic conditions. Overgrazing and the associated loss of vegetation lead to soil erosion and the release of carbon, thus accelerating desertification (Steinfeld et al., 2006). In addition, grassland ecosystems harbour significant species diversity. Forty of the world's 234

Centres of Plant Diversity (CPD) identified by the IUCN are located in grasslands. A further 70 of these CPDs contain some grassland habitats. Grassland ecosystems thus feature in almost half of the identified CPDs (White et al., 2000). Many of these grassland hotspots of plant diversity are also home to a large number of endemic bird species (White et al., 2000). The more intensive the human interaction with these ecosystems, the greater the loss of biological diversity (Mooney et al., 1995). When grassland is converted into arable land, biodiversity diminishes rapidly.

Grassland ecosystems, particularly those with significant plant diversity, thus represent an important opportunity for carbon sequestration, while the biomass harvested from them can be used for energy in climate-friendly ways (Rösch et al., 2007; Tilman et al., 2006). Conversion to arable rotation should therefore be rejected, while conversion through afforestation can be regarded as a positive step in terms of carbon sequestration and in some cases in terms of biological diversity. The situation with regard to the conversion of existing, intact grassland or of grassland on degraded land for the cultivation of perennial crops needs to be considered with a closer eye to specific detail; the effects (negative or positive) on the carbon reservoir in the soil and the (negative) effects on biological diversity must be carefully weighed against each other.

4.2.3.4
Conversion of arable land

The carbon reservoir in the soil of arable and pasture land is very variable and depends on farming methods, climate and the crop that is grown (Figure 4.2-7).

Intensive agriculture usually restricts itself to a narrow range of crops with low genetic diversity – crops

Figure 4.2-7
Organic carbon at two soil depths in relation to vegetation cover. Switchgrass 5 years: the field was burned in a 5-year cycle. Switchgrass 1 year: the field was burned annually. *Measurement depth 0–15 cm and 30–45 cm. Sources: Lemus and Lal, 2005; Al-Kaisi and Grote, 2007; Kirby and Potvin, 2007

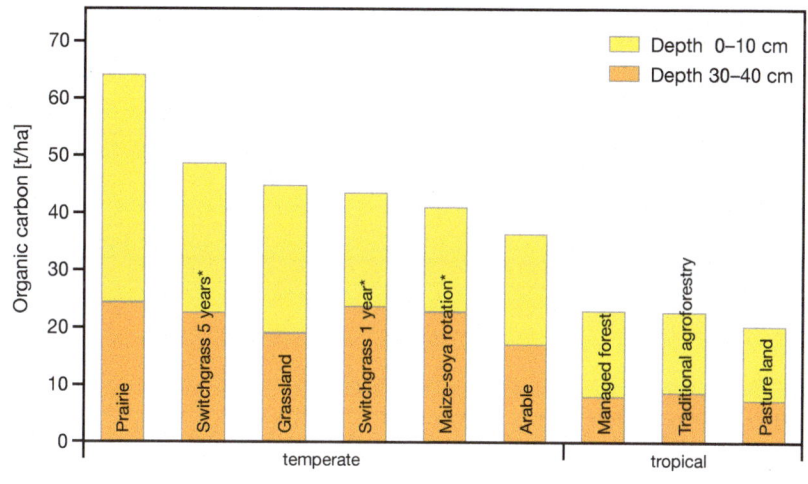

that are bred for high yields and that require the use of artificial fertilizers and pesticides; in some cases irrigation is also needed. Less intensively farmed arable land usually displays greater biodiversity (Mooney et al., 1995), although species diversity remains low in comparison with other ecosystems.

When arable land is converted to grassland the size of the carbon reservoir in the soil increases. The process is a slow one and takes decades. The higher rate of carbon storage in grassland is usually attributable to the fact that the ground is usually covered with vegetation throughout the year, with greater underground productivity and less disturbance of the soil than in arable land; this reduces erosion and loss of CO_2 through soil respiration (Yimer et al., 2007).

Afforestation with short-rotation woody crops provides a means of storing carbon on arable land (e.g. Hansen, 1993; Mann and Tolbert, 2000; Grogan and Matthews, 2002; Section 7.1.2). In addition, perennial crops are effective in protecting the soil against erosion and hence against the loss of organically bound carbon in the soil (Lewandowski and Schmidt, 2006). Overall, therefore, the conversion of arable land for perennial crops or even for re-afforestation can be evaluated positively.

4.2.4
Summing up

The conversion of forestry areas and wetlands into agricultural land invariably has negative consequences for biological diversity and carbon storage in the soil. The conversion of such land is associated with large-scale emissions of greenhouse gases. Biodiversity and carbon storage in the soil are greater in forests, grassland or pasture land than in arable and degraded land; perennial crops have a more positive effect on both factors than annual crops. When crops are grown in rotation, greenhouse gases are usually released: on an annual average through cultivation the soil loses more CO_2 through cultivation than is input through litterfall, and intensive N fertilization can lead to the release the greenhouse gas N_2O (Table 4.2-1).

Table 4.2-1
Qualitative rating of the effects of direct land-use changes on biological diversity, the quantity of carbon in the soil and vegetation (time scale: >10 years) and greenhouse gas losses during conversion. In assessing conversion only the effects of direct land-use changes were considered.
Crops, 1–3 years = cultivation in 1–3-year rotation; crops, perennial = min. 5-year cultivation, e.g. short-rotation plantation, *Jatropha*, oil palms.
Source: WBGU

Conversion of	to	Rating
Forest	Crops, 1–3 years	negative
	Crops, perennial	negative
	Grassland, pasture	negative
Wetland	Crops, 1–3 years	negative
	Crops, perennial	negative
	Grassland, pasture	negative
Grassland, pasture	Crops, 1–3 years	negative
	Crops, perennial	unclear
	Forest	positive
Degraded land	Crops, 1–3 years	negative
	Crops, perennial	unclear
	Grassland, pasture	positive
Arable land	Crops, perennial	positive
	Grassland, pasture	positive
	Forests	positive

Effects

▮ (red)	negative
▮ (yellow)	unclear
▮ (green)	positive

Competing uses

5.1
Introduction

Throughout the world fertile land is scarce and subject to a wide range of claims on its use. In view of the growth in world population it is impossible to meet all usage claims to the extent that would be desirable. Instead, careful decisions must be made to give precedence to certain claims and reduce the dominance of others. Human society currently uses around 34 per cent of the world's land surface for agricultural purposes, primarily for food and feed production (Sections 4.2.2 and 5.2). Added to this is the growing importance of cultivating plant biomass for feedstock uses in products (Section 5.3). The potential availability of agricultural land is restricted by the need to conserve the natural environment, particularly semi-natural and natural areas (Section 5.4) and the need to mitigate climate change (Section 5.5). Other limitations arise from overuse – particularly advancing soil degradation – and problems of increasing scarcity and pollution of freshwater resources (Section 5.6). Human-induced changes in natural surface run-off, such as occur when large reservoirs are built, add to the shortage of land suitable for growing crops. The global spread of urbanization and expansion of the associated infrastructure have a similar effect. Cities, urban agglomerations and their infrastructure tend to be concentrated in the most fertile regions of the world (such as river deltas, alluvial fans, riverbank areas and places where rivers divide or converge). This tendency is primarily at the expense of agricultural land. Further expansion of urban structures therefore competes directly with the use of land in the vicinity of settlements for agriculture. Towns, cities and urban agglomerations currently cover – depending on the method of calculation used – between 1.5 per cent and 2 per cent of Earth's terrestrial surface (calculated from data from Salvatore et al., 2005; Girardet, 1996). Urban structures occupy 4.8 per cent of the land area in Germany (UBA, 2003a) and around 5 per cent in the EU-24 (EEA, 2006). On a global scale, therefore, the direct effects of urbanization in terms of land take are rela-

tively small and so will be ignored here. The following section describes the background to globally competing uses and their future dynamics in the light of the increasing importance of energy crop cultivation.

5.2
Competition with food and feed production

5.2.1
Introduction

More than 923 million people worldwide – most of them in developing countries – are affected by food insecurity (FAO, 2006a; FAOSTAT, 2006). Food insecurity exists when people lack access to sufficient amounts of safe and nutritious food that meets their dietary needs and food preferences and permits an active and healthy life (FAO, 2001a). The majority of people affected by food insecurity do not have the income that would enable them to buy the food they need reliably and throughout the year (FAO, 2006b). Between 2006 and 2008 food insecurity increased significantly as a result of the sharp rise in food prices worldwide (UN, 2008). The World Food Programme and the World Bank estimate that if food prices remain high at least another 100 million people will fall further into poverty and be threatened by hunger (UN, 2008). The majority of these people will be in Low-Income Food-Deficit Countries (LIFDCs). On the other hand, rising food prices could alleviate poverty as the income of many agricultural producers rises and as higher prices create incentives for the expansion of production, thereby generating income (Box 8.2-3). Many factors have contributed to the present sharp rise in food prices (2007/2008); the significance of the individual factors and the long-term trend of prices are disputed and need to be more fully researched (Section 5.2.5.2). Around 5000 million hectares of farmland, including pasture, are available globally; of this, 1500 million hectares are arable land (Section 4.2.2). At present some 20 million hectares globally – a relatively small proportion of the total – are used for the cultivation of energy crops (Faaij, 2008). However, if the worldwide bioenergy boom claims an ever-increasing quantity of farmland it could become a critical factor for global food and feed production.

5.2.2
Growing food supply and rising demand

In the past, worldwide population growth has been the most significant factor driving demand for food and feed. The world population is at present estimated at 6600 million, of whom some 80 per cent live in developing countries (FAOSTAT, 2006). It is estimated to grow to around 8300 million by 2030 and to around 9200 million by 2050 (UNPD, 2006). Global food production will need to increase by some 50 per cent by 2030 if an increase in food insecurity is to be avoided (OECD, 2008).

Another important factor is the change in food consumption habits as a consequence of urbanization, rising incomes and associated lifestyle changes (von Koerber et al., 2008). In industrialized countries around three-quarters of the population now live in urban areas; in developing countries the proportion is just under one half. By 2030 the urban population will grow further; 60 per cent of the total world population will by then be living in towns and cities, and the proportion is expected to rise still further (UNPD, 2006). The diet of the urban population tends to contain more white flour, fat, sugar and processed foods than that of rural dwellers (Mendez and Popkin, 2004). The structures through which food is typically made available in cities (e.g. supermarkets, fast-food restaurants) support these trends (Popkin, 2006).

The level of available income is also very important in relation to demand for food. In the next 30 years real incomes in developing countries are likely to rise by on average 2 per cent per year; in the least developed countries this growth rate may reach 4 per cent (Schmidhuber and Shetty, 2005). Higher incomes are usually associated with a more varied diet and increased consumption both of high-quality foods and of highly processed products and convenience foods (FAO, 2007b). In particular, demand for meat and other animal-based foods tends to rise (Keyzer et al., 2005). Once a consistently high income level is reached, growth in the consumption of animal products stagnates (if population growth remains constant) and the market becomes saturated (Delgado et al., 1999; Keyzer et al., 2005).

On a global average, calorie availability has increased in recent decades, primarily as a result of increased land productivity (Section 5.2.4.1). However, this has not solved the major problem of inadequate distribution. Between 1970 and 2000 the average quantity of food energy available rose from around 2400 to 2800kcal per person per day (Table 5.2-1). In the 1960s 57 per cent of the world population lived on less than 2200kcal per person per day; today only around 10 per cent of the population does so (FAO, 2003b). The greatest progress in this respect has been made in the developing countries and is heavily influenced by success in some densely populated regions such as East Asia. For example, in China calorie availability has increased dramatically within a short time and is now approaching the level of the industrialized world (FAO, 2006b; FAOSTAT, 2008a).

In other regions such as sub-Saharan Africa food supply has not improved significantly since the 1970s. Only a few countries (e.g. Nigeria, Ghana and Benin)

Table 5.2-1
Average available food energy in different world regions (kcal per person per day), [1]Mean for the 3-year span.
Source: FAO, 2006b

	Food energy [kcal/person/day]						
	1969/71[1]	1979/81[1]	1989/91[1]	1999/01[1]	2015	2030	2050
Developing countries	*2,111*	*2,308*	*2,520*	*2,654*	*2,860*	*2,960*	*3,070*
Sub-Saharan Africa	2,100	2,078	2,106	2,194	2,420	2,600	2,830
North Africa/Middle East	2,382	2,834	3,011	2,974	3,080	3,130	3,190
Latin America	2,465	2,698	2,689	2,836	2,990	3,120	3,200
South Asia	2,066	2,084	2,329	2,392	2,660	2,790	2,980
East & South-East Asia	2,012	2,317	2,625	2,872	3,110	3,190	3,230
Transition countries	*3,323*	*3,389*	*3,280*	*2,900*	*3,030*	*3,150*	*3,270*
Industrialized countries	*3,046*	*3,133*	*3,292*	*3,446*	*3,480*	*3,520*	*3,540*
World	**2,411**	**2,549**	**2,704**	**2,789**	**2,950**	**3,040**	**3,130**

have increased food provision to more than 2400kcal per person per day (FAO, 2006b).

The FAO estimates that in 2050 around 90 per cent of the world's population will live in countries with an average calorie availability of more than 2700kcal per person per day. Today around 51 per cent of the population is in this position; 30 years ago the figure was only around 4 (four!) per cent (FAO, 2006b). However, the FAO's calculated calorie availability is a purely statistical figure compiled from national data on food production, food trade and population figures. This conceals the fact that access to food can vary widely within a country; the problem of malnutrition among some sections of the population therefore persists even in developing countries where average calorie availability is apparently adequate (FAO, 2006b).

5.2.3
Challenges arising from changed dietary habits

As the availability of food calories increases, the composition of people's diets also changes. In the course of economic progress a carbohydrate-rich diet based on plant foods (such as cereals, roots, tubers, legumes) is in many developing countries gradually replaced by a diet containing more fat and protein. As already mentioned, the proportion of animal-based foods, sugar and plant oils in total food calories will increase further in the coming decades (FAO, 2006b; Popkin, 2006).

5.2.3.1
A summary of individual foods: Global trends

CEREALS

Cereals currently account for 50 per cent of total food consumption, making them the most important food group worldwide. In developing countries up to 80 per cent of people's diet is based on cereals (FAO, 2006b). Per-capita consumption of cereals peaked in the 1990s and has been falling continuously since the turn of the century. Only in sub-Saharan Africa and Latin America did cereal consumption continue to rise in the 1990s (FAO, 2006b). The future development of cereal consumption will be influenced by two opposing trends. On the one hand, the range of food on offer is tending to include more animal products; this is particularly the case in countries that have attained a medium to high level of food consumption. On the other hand, cereal consumption is increasing in countries in which the food supply continues to be relatively limited or in which diet is shifting from roots and tubers to cereals. It is likely that the quantity of cereals used for direct consumption will gradually decrease worldwide. However, if all potential uses are taken into account – including food, feed and other uses such as seed and the production of ethanol and starch – consumption is likely to increase from 309 kg per person in 2000 to an anticipated 339 kg by 2050 (FAO, 2006b). In the light of the rising demand for meat, feed grain is a particularly important factor in the future development of the cereals sector. In 2020 the developing countries will use an estimated 65 kg of feed grain per person per year, while the industrialized nations will – at 374 kg per person per year – use roughly six times that amount (Delgado et al., 1999). Keyzer et al. (2005) comment in this con-

Table 5.2-2
Consumption of meat, milk and milk products in various world regions.
[1]Without butter; [2]Mean for the 3-year span
Source: FAO, 2006b

	Meat [kg/person/year]				Milk and milk products[1] [kg/person/year]			
	1969/1971[2]	1999/2001[2]	2030	2050	1969/1971[2]	1999/2001[2]	2030	2050
Developing countries	*10.7*	*26.7*	*38*	*44*	*28.6*	*45.2*	*67*	*78*
Sub-Saharan Africa	10.2	9.5	14	18	29.6	28.3	34	38
North Africa/Middle East	12.6	21.7	35	43	68.1	73.2	90	101
Latin America	33.5	58.5	79	90	84.0	108.8	136	150
South Asia	3.9	5.5	12	18	37.0	67.6	106	129
East & South-East Asia	9.2	39.8	62	73	3.7	11.3	21	24
Transition countries	*49.5*	*44.4*	*59*	*68*	*185.7*	*160.2*	*179*	*193*
Industrialized countries	*69.7*	*90.2*	*99*	*103*	*189.1*	*214.0*	*223*	*227*
World	**26.1**	**37.4**	**47**	**52**	**75.3**	**78.3**	**92**	**100**

text that commonly used forecasts of demand for feed grain are often significant underestimates. For developing countries it is often assumed that large quantities of material unsuitable for human consumption (such as household waste and harvest residues) will be used as animal feed. In future, however, a shift from traditional to cereal-intensive feeding methods is to be expected.

MEAT, MILK AND MILK PRODUCTS

The changing nutrition patterns in developing countries can be seen most clearly in the increasing consumption of animal products. Further rises in the consumption of meat, milk, milk products and eggs in these countries are forecast. There are large regional and national differences not only in the quantity but also in the type of products (Table 5.2-2).

The slow growth of meat consumption is strongly influenced by India, which is home to around 70 per cent of the population of South-East Asia and where traditionally very little meat is eaten. India currently has the lowest rate of meat consumption in the world. In South Asia a slow but steady rise in consumption of animal-based foods is evident, involving in particular milk and milk products, but also poultry meat (FAO, 2003a). Demand for animal products is likely to increase significantly in the region as incomes rise and urbanization progresses (Rosegrant et al., 2001; Keyzer et al., 2005). Sub-Saharan Africa is also a region in which relatively few animal-based foods are eaten. Slow but steady growth in demand for animal-based foods is expected there. Meat consumption in Latin America is traditionally relatively high; it is expected to rise further, as will milk consumption (FAO, 2003a). In East Asia consumption of animal products – particularly pork and to a lesser extent poultry – and milk is rising rapidly (FAO, 2003a). By 2050 East Asia is expected to have the second-highest per-capita consumption of meat in the developing world, second only to Latin America. Meat consumption is projected to grow more slowly in future than it did between 1960 and 2000 since the countries that have in the past exhibited rapid growth (principally China and Brazil) will experience saturation of their demand (FAO, 2006b).

SUGAR

Sugar consumption in industrialized and transition countries has fallen from around 40 kg per person per year in the 1970s to 33–37 kg per person per year in 2000. In a reversal of the trend of recent decades, the transition countries are forecast to increase sugar consumption to 41 kg per person per year by 2050. In industrialized countries sugar consumption will remain almost constant (FAO, 2006b). The developing countries, by contrast, are exhibiting a steady upward trend; they increased their consumption from 15 kg per person per year in 1970 to 21 kg per person per year in 2000.

PLANT OILS

The cultivation of oil plants has in recent years been one of the most vigorously growing sectors of agriculture, with an annual growth rate that has overtaken even that of livestock management. The greatest influence on cultivation has been the increasing consumption of plant oils and meat. However, consumption of plant oils in developing countries will grow more slowly in the years to 2050 than it has done in recent decades (FAO, 2006b). The use of oil plants for the production of cleaning agents, lubricants and biodiesel will increase significantly faster in the same period (FAO, 2006b).

5.2.3.2
Land requirements of dietary habits and foods

LAND AVAILABILITY AND LAND USE

Around 34 per cent of the world's existing land surface is available for agricultural use (Section 4.2.2). The majority of this land, amounting to 3408 million hectares (69 per cent) is extensive pasture. If land used for feed production is also included, around 80 per cent of the world's agricultural land is found to be used for cattle rearing (Steinfeld et al., 2006). This contrasts with the fact that animal-based foods play only a small part in the world food supply (17 per cent in 2003; FAOSTAT, 2006).

The total area of farmland has increased by nearly 460 million hectares in the last 40 years (1963–2003). However, the increase in farmland has slowed since the mid-1990s and occurs almost exclusively in the developing countries (Table 5.2-3; Steger, 2005).

Per-capita availability of farmland is decreasing worldwide (von Koerber et al., 2008). This is mainly a result of strong population growth, the rate of which has overtaken the moderate expansion of farmland. In the industrialized countries modest population growth is accompanied by a small loss of agricultural land, which thus leads to a relatively small reduction in per-capita land availability. For economic and ecological reasons, opportunities for future expansion of arable land and permanent crops will be limited. The FAO (2003b) assumes that the area of land may increase by 13 per cent between 1997–1999 and 2030. However, the world population is likely to rise by 22 per cent in the same period (UNPD, 2006). This means that the productivity of the existing or newly acquired land must be increased if deterioration of the food situation is to be avoided. The situation is exacerbated by the fact that the industrialized coun-

Table 5.2-3
Farmland per person in various world regions (ha/person).
Source: von Koerber et al., 2008 based on FAOSTAT, 2008a

	Farmland [ha/person]			Farmland and permanent crops [ha/person]		
	1962	1982	2002	1962	1982	2002
Developing countries						
Africa	3.60	2.15	1.32	0.56	0.37	0.26
Asia	0.64	0.46	0.39	0.26	0.18	0.15
Latin America/ Caribbean	2.5	1.8	1.4	0.50	0.38	0.30
North America	2.4	1.9	1.5	1.1	0.92	0.87
Oceania	29.0	21.2	14.6	2.2	2.2	1.7
EU-15	0.51	0.43	0.37	0.31	0.25	0.22
Industrialized countries	*1.90*	*1.58*	*1.38*	*0.68*	*0.57*	*0.48*
Developing countries	*1.22*	*0.83*	*0.64*	*0.33*	*0.23*	*0.18*
World	**1.43**	**1.0**	**0.80**	**0.44**	**0.32**	**0.25**

tries use more agricultural land than they themselves possess. These virtual areas of land are, however, not included in Table 5.2-3. For example, by means of agricultural imports the EU-15 countries increase the amount of land available per person by about 20 per cent. This is mainly attributable to imports of animal feed – especially soya beans and press cake made from them – for the intensive livestock management that is carried out in Europe (Steger, 2005).

FOOD-RELATED LAND REQUIREMENTS
In addition to the area of agricultural land available it is useful to consider the specific amount of land needed for selected food crops (Table 5.2-4).

The amount of farmland needed for the production of different food crops varies greatly in different parts of the world, depending on local conditions and intensity of cultivation, as influenced by factors such as soil quality, climate, and use of fertilizers and crop treatments (von Koerber et al., 2008).

In a case study of New York state in the USA, the land requirement of different foods was expressed in terms of their energy content (land needed per 1000kcal). This method has the advantage that it takes account of the varying energy densities of different foods (Table 5.2-5).

It reveals that animal-based foods require a significantly greater area of land than do plant foods: 31 m^2 of land are required to produce 1000kcal of food calories from beef (using predominantly extensive grazing methods), while the same calorific value can be produced from cereals on just 1 m^2 of land (exclusively arable land). Of the plant foods studied, oil plants require the largest land area (Peters et al., 2007).

Table 5.2-4
Land requirement in m^2/kg of food in various countries (2006, m^2/kg yield).
Source: von Koerber et al., 2008 from FAOSTAT, 2008a

	Land requirement [m^2/kg]						
	Germany	Brazil	Ethiopia	China	India	Ukraine	World
Oil plants	2.8	4.1	15.3	4.1	8.7	7.6	**3.9**
Wheat	1.4	6.3	5.5	2.2	3.8	4.6	**3.6**
Rice	–	2.6	5.3	1.6	3.2	2.9	**2.4**
Maize	1.3	3.0	4.5	1.9	5.2	2.5	**2.1**
Fruit	0.66	0.64	0.88	1.10	0.91	2.20	**0.98**
Potatoes	0.27	0.45	1.4	0.70	0.59	0.75	**0.60**
Vegetables	0.34	0.49	2.8	0.52	0.86	0.67	**0.59**

Table 5.2-5
Land requirement of foods in relation to the energy content of the consumable product (based on yields in the USA, case study of New York state).
Source: Peters et al., 2007

	Land requirement [m²/1.000 kcal]
Animal-based foods	
Beef	31.2
Poultry	9.0
Pork	7.3
Eggs	6.0
Full-cream milk	5.0
Plant-based foods	
Oil fruits	3.2
Fruit	2.3
Pulses	2.2
Vegetables	1.7
Cereals	1.1

Germany requires 17.2 million hectares of land to secure its current food consumption. This is approximately equal to the existing area of farmland, which thus means that the food supply could in theory be maintained without imports. In Germany 39 per cent of food calories come from animal-based foods and 61 per cent from plant-based ones. For a population of approximately 80 million this represents a land requirement of 0.22 ha per person per year (Seemüller, 2001). To meet this level of food consumption through organic farming alone would – on account of lower yields per unit area – require 22.5 million hectares, or roughly 24 per cent more. This would correspond to a land requirement of 0.28 ha per person per year (Seemüller, 2001; Badgley et al., 2007).

5.2.3.3
Additional land requirements as a result of changing dietary habits

The optimistic assessments of future food security put forward by international organizations are questioned by Keyzer et al. (2005) for a number of reasons. They are particularly critical of the estimates of the land needed for grain for meat production. It is usually assumed that consumption of animal-based foods increases linearly with rising income. This ignores the fact that many people who acquire more purchasing power will eat disproportionately more meat as they engage in catch-up consumption. Once a particular level of affluence is reached, demand for

greater quantities of meat is replaced by demand for meat of higher quality. Despite this, the growth in demand for meat is often underestimated, because in many developing countries large sections of the population are on the threshold of greater consumption opportunities (von Koerber et al., 2008).

The 'affluent diet' involves not only greater meat consumption but also increased consumption of items such as food oil, beverages, fruit, cheese, biscuits and ice cream, which further increases the need for land (Gerbens-Leenes et al., 2002). While agriculture is in principle able to meet the food needs of a growing world population, it is unable to support the global expansion of an affluent diet containing a large amount of meat. If developing countries were to adopt the nutrition habits of the western world, the global land requirement would double or treble (Gerbens-Leenes et al., 2002). A similar conclusion is reached by Balmford et al. (2005). Such an adaptation of nutrition habits is already apparent in China and Brazil and is forecast to occur in other regions in the coming decades (FAO, 2003a, 2006b). This will exacerbate the pressure on land.

5.2.4
Limits to potential food production

The FAO estimates that by 2030 some 50 per cent more food will be needed in order to feed the world's population, which will by then number more than 8000 million. Since opportunities for expanding the total area of agricultural land are very limited, not least because of water availability, soil degradation and the requirements of nature conservation (Sections 5.4 and 5.6), 80 per cent of this increase will have to be achieved through agriculture that is more intensive while being at the same time sustainable and environmentally friendly (FAO, 2003a). For the last 40 years increases in global food production have been attributable primarily (to around 80 per cent) to increases in land productivity. The modernization of farming and the change from draught animals to machines has also freed up land that was previously used for traditional 'fuel production' – that is, to feed the animals. In the developing countries these advances in productivity have varied widely from country to country and from region to region. Sub-Saharan Africa, in particular, has been bypassed by these productivity increases (Brüntrup, 2008). Increasing area yields will continue to be a key factor in production increases; growth rates will, however, decline as a consequence of high crude oil and energy prices, declining increases in yields as the limits of present technology are reached, soil erosion and the overuse of freshwater resources (Brüntrup,

2008). For example, the FAO anticipates that annual growth in world grain production, currently at 1 per cent, will increase to 1.4 per cent by 2015 and then fall to 1.2 per cent (for comparison: 1970s: 2.5 per cent per year; 1980s: 1.9 per cent per year; 1990s: 1 per cent per year; FAO, 2003a; OECD and FAO, 2005, 2006). However, these estimates do not allow for the influence of climate change, which will have a detrimental effect on agricultural production in the medium term (Section 5.2.4.2).

5.2.4.1
Potentially available land and soil degradation

The FAO projects that it will be possible to increase the worldwide area of arable land by around 13 per cent between 1997–1999 and 2030 – although much of this increase will be the result of deforestation (FAO, 2003a). This projection does not take account of the adverse effects of land degradation as a result of soil erosion, deforestation and climate change (WBGU, 2008). The productivity of newly converted farmland is likely to be lower than that of existing agricultural land, particularly if the land that is taken into production is marginal land that may in addition be remote from markets (Rosegrant et al., 2001; Balmford et al., 2005). Potentially cultivable land is very unevenly distributed. More than half of it belongs to only seven countries – Angola, Congo and Sudan in Africa and Argentina, Bolivia, Brazil and Columbia in South America. This can be contrasted with the fact that in the Middle East 87 per cent and in South Asia 94 per cent of suitable land is already cultivated. This means that in these countries, a well as in East Asia and North Africa, any increases in agricultural output must result almost entirely from increases in land productivity (Beese, 2004).

An indication of the pressure under which ecosystems are already placed as a result of human use is provided by the HANPP index (human appropriation of net primary production; Section 4.2). The index describes the proportion of potential net primary production (NPP) that is appropriated through human activities involving use and change. Haberl et al. (2007) estimate that globally this index stands at around 25 per cent. Agricultural and forestry yields account for around 53 per cent of this HANPP, land-use-related production changes account for 40 per cent, and around 7 per cent is lost through fire. The consequences of a further rise in HANPP are ecosystem overuse, soil degradation, additional species under threat and the accumulation of carbon dioxide in the atmosphere. Table 5.2-6 shows how the HANPP index is distributed according to region.

5.2.4.2
Climate change impacts on production potential

In the event of a rise in temperature of 1–3°C (against a 1990 baseline), it is likely that global agricultural production will initially increase, because decreases in many developing countries could be more than offset by higher yields in regions at higher latitudes (WBGU, 2007). Particularly marked decreases will occur in Africa, because the area of agriculturally useful land in arid and semi-arid regions will be reduced, the growing period will be shortened and potential yields will fall. In some sub-Saharan countries the yield of rain-fed farming could decline by up to 50 per cent by 2020 (Lal et al., 2005; IPCC, 2007b). If the rise in global mean temperatures reaches 2–4°C, agricultural productivity is likely to decline worldwide. An increase in temperature of more than 4°C is likely to cause significant damage to global agriculture (IPCC, 2007b). This means that if climate change is allowed to advance without any countermeasures, pressure on usable agricultural land will be significantly increased. Almost all projections assume that world market prices for cereals will rise if temperatures increase by 2°C, if not before (e.g. Adams et al., 1995; Fischer et al., 2002; IPCC, 2007b).

5.2.5
Impacts of the bioenergy boom on food security

The cultivation of biomass for energy purposes competes with food and feed production for land and other factors of agricultural production. Where bioenergy cultivation is an alternative land use, prices of agricultural inputs rise (if other conditions remain equal), so that staple foods become more expensive (FAO, 2008c). Rising prices increase the burden on consumers but provide farmers who produce for the market with wider income-generating opportunities. The cultivation of energy crops and the use of biomass for energy can also contribute to rural development, for example through the improved supply of decentralized energy or income-generating employment effects. The overall effect will vary from country to country and from case to case (country studies: Boxes 4.1-2, 4.1-3, 5.2-2, 5.4-2, 6.7-2, 8.2-2, 8.2-4 and 10.8-1) and will depend on regional factors such as natural, agricultural and social conditions, the type of bioenergy to be used and the development of global food markets.

5.2.5.1
The four dimensions of food security

The FAO definition of food security distinguishes the four dimensions of availability, access, stability and utilization (Faaij, 2008). Availability of food refers to the capacity of an agro-ecological system to produce sufficient food. Access to food refers to the ability of households to economically access food, defined in terms of enough purchasing power or access to sufficient resources. Stability refers to the time dimension of food security and describes whether the food supply is consistently secure or whether it is either temporarily or permanently jeopardized by factors such as price fluctuation or falls in production. Utilization of food refers to people's ability to take up nutrients and is closely linked to health factors (in particular access to clean water and the general level of rural development) and to the way in which food is prepared. All four dimensions are significantly influenced by the growing of energy crops.

Food availability can be threatened to the extent that land, water, fertilizers, etc. are diverted from food and feed production and used instead for the cultivation of energy crops. The degree of competition will hinge on a variety of factors, including the development of agricultural yields in food production, the development of meat consumption and the pace at which second-generation energy crops are introduced. These next-generation crops would significantly reduce the competition with food for land resources. However, when the whole plant is used it is important to ensure that sufficient biomass remains to ensure the fertility of the soil. Particularly where second-generation crops are grown, the market for energy crops could provide an additional opportunity for improving the incomes of farming households (Faaij, 2008). But even first-generation biofuels have similar potential in some situations, since improving the efficiency of traditional biomass use contributes to the general improvement of productivity in agri-

Table 5.2-6
Human appropriation of the net primary production of natural ecosystems (HANPP): regional distribution. Source: Haberl et al., 2007

Region	HANPP [%]
North Africa & West Asia	42
Sub-Saharan Africa	18
Central Asia & Russia	12
East Asia	35
South Asia	63
South-East Asia	30
North America	22
Latin America & Caribbean	16
Western Europe	40
South-East Europe	52
Oceania & Australia	11

culture and other sectors. Each case must be examined on its own merits to identify whether conflicts arise with the use of land for grazing.

Inadequate access to sufficient food is at present the main reason for food insecurity. This is likely to increase, at least in the short term, if the global expansion of bioenergy production causes food prices to rise faster in real terms than incomes. Parts of the population that are already affected by food insecurity or vulnerable to it as a result of inadequate purchasing power or lack of land-use rights will be put at further risk (Faaij, 2008). The expansion of energy crops is an important factor contributing to the rise in food and feed prices. Price rises in 2007 and 2008 have driven up the prices of energy crops such as maize, sugar, palm oil, rape and soya as well as feed prices and the prices of staple foods such as cereals (see below). Depending on local conditions, higher prices for agricultural products could in the long term boost the economic power of rural areas and the incomes of their inhabitants. Three-quarters of

Box 5.2-1

Has 'peak phosphorus' already been reached?

Phosphorus (P), with nitrogen (N) and potassium (K), is one of the three main constituents of artificial fertilizers, which are therefore always referred to by their N-P-K percentages. While nitrogen can be obtained from the air in almost unlimited quantities using the Haber-Bosch process, phosphorus is a finite resource and one which, unlike oil, cannot be replaced by other fuels or substances. As a nutrient, phosphorus is an essential contributor to the increase in land productivity that a growing world population requires.

Déry and Anderson (2007) take the view that global extraction of phosphorus, usually as phosphate, passed its peak ('peak phosphorus') in 1989. As the debate on peak oil has made clear, problems start to arise not when a resource begins to run out but when the extraction peak is reached. From this point on extraction becomes more difficult and more costly.

In contrast to oil, however, phosphates can be 'reclaimed'. One response to reaching peak phosphorus must therefore be to close nutrient cycles in agricultural production, in particular by fertilizing the land with organic fertilizers. Other options are to reclaim nutrients from sewage sludge and to make more efficient use of fertilizers in farming.

the word's poor live in rural areas; they could benefit from rising agricultural prices, because 80 per cent of the income of the rural population is derived from farming (Brüntrup, 2008). The expansion of bioenergy could therefore help to improve development opportunities (Müller, 2008). Food security would then increase. However, the possible long-term positive income effects for agricultural producers must be weighed against the short-term negative effects for those who must purchase some or all of their food. Unrest in many countries at the beginning of 2008 highlighted the dramatic effects of escalating food prices (cf. the conflict constellation relating to food production set out in: WBGU, 2007).

The stability of the food supply can be temporarily or permanently affected by the consequences of extreme weather events, market turbulences, civil conflict or environmental degradation. The increased coupling of the agricultural and energy markets can contribute to the destabilization of food prices, since it causes price volatility from the petroleum sector to be transmitted more directly and more strongly to the agricultural sector (Faaij, 2008). This increases the risk of temporary food insecurity.

The utilization of food can also be affected by the cultivation of energy crops. For example, if the cultivation of energy crops reduces the availability and quality of water, this has an adverse effect on health and reduces food security, since sick people utilize food less well. On the other hand, improved efficiency in traditional biomass use can bring about a significant reduction in potential health risks (indoor air pollution, time-consuming collection of firewood; Box 8.2-1) and make cooking cheaper and cleaner, thus helping to improve the nutrition situation and the utilization of food.

The income- and price-related effects of the bioenergy boom and their effects on food security will be analysed in the following section.

5.2.5.2
The influence of the bioenergy boom on prices and incomes

Since food security is largely an issue of distribution and hence of purchasing power, consideration of the ways in which the bioenergy boom affects the food situation must take account not only of price effects but also of income effects. Private landowners' decisions on land use hinge primarily on the profits to be made from the different forms of land use. Unless societal or state regulation dictates otherwise, the landowner will usually practice the form of land use or produce the agricultural product that is expected to yield the highest profit. The level of profit depends mainly on the costs of the factors of production, the costs of inputs (fertilizers, pesticides, machinery use) and marketing and the prices of the end products. If the price of energy crops rises in comparison with the price of other agricultural products as a result of rising demand for energy crops, more energy crops will be grown and the supply of food and feed will fall, so that food prices will also rise. One of the factors influencing demand for energy crops and the price obtainable for them is the price of the fossil fuels that can be replaced by bioenergy. This means, for example, that a rise in oil prices leads to rises in food prices (Figure 5.2-1).

However, dynamic effects must also be considered. Rising oil prices raise the price of fossil inputs in agriculture and hence also the price of biofuels, thus reducing the price advantage that bioenergy has over oil. The high price of crude oil, which for a time

Figure 5.2-1
Development of food and oil prices since 1980. Food price index 2005 = 100; includes: cereals, plant oil, meat, fish, sugar, bananas and oranges, alongside the price index for crude oil (2005 = 100).
Source: Wiggins and Levy, 2008

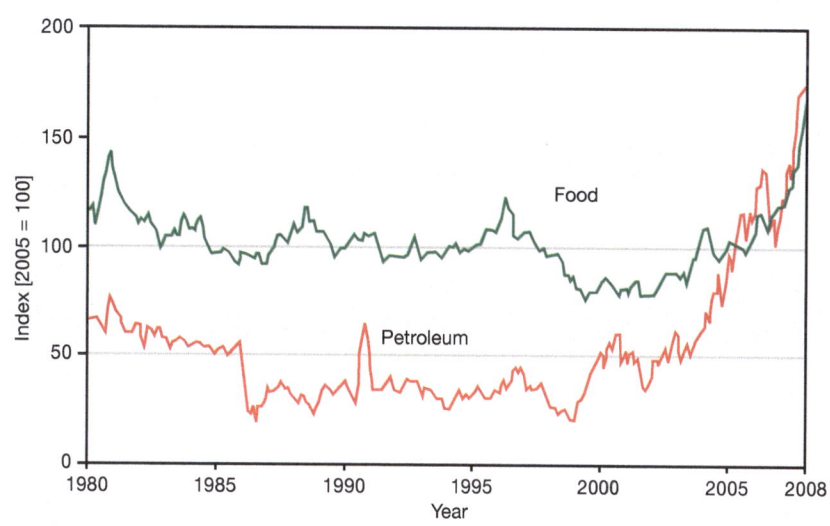

in mid-2008 exceeded US$ 140 per barrel (Figure 5.2-1), and the present blending quotas for biofuels in the USA and the EU are currently causing an increase in the use of grain, sugar and palm oil to produce bioethanol or biodiesel (Section 4.1). This creates particular difficulties for developing countries that import food and whose balance of trade deteriorates because of the higher prices. The problem is particularly severe if the income obtained from the cultivation or export of biofuels is insufficient to cover the purchase of food, or if prices fluctuate markedly. Between 2005 and 2008 food prices rose on average by 83 per cent (World Bank, 2008d). Further rises, in part attributable to bioenergy, may occur. They further restrict the food-purchasing options of consumers with low incomes (Faaij, 2008).

THE CONTRIBUTION OF THE BIOFUEL BOOM TO FOOD PRICE RISES

The price rises occurring in the food sector are only partially attributable to the biofuel boom (Ressortarbeitsgruppe 'Welternährungslage', 2008; von Braun, 2008). Other causes are the growing global demand for food, changing nutrition habits in the aspiring newly industrializing countries, and the growth in world population (Section 5.2.2). On the supply side, production costs have risen as a result of higher input prices for energy, transport and fertilizer, extreme weather events such as droughts and floods, the low US dollar price, insufficient investment in rural infrastructure and in agriculture, particularly in developing countries, and the decline in food storage. In addition, recent price rises and fluctuations have been fuelled by speculation on the international commodity markets and the walling off of markets in producer countries through the imposition of export levies and bans, as has recently occurred in Argentina, Vietnam, China, and Cambodia. Figure 5.2-2 summa-

rizes the price development of wheat, rice and maize since 1990 (Wiggins and Levy, 2008).

The extent of the influence that the biofuels boom has on food prices is rated very differently in different studies. For example, the United States Department of Agriculture estimates that biofuels have caused a 2–3 per cent rise in food prices (USDA, 2008), while a report of the World Bank comes to the conclusion that biofuels are responsible for 75 per cent of the recent price rises (Mitchell, 2008); both figures are disputed. IFPRI (2008) puts the influence of the increased demand for biofuels on average grain prices during the period 2000–2007 at 30 per cent; the OECD estimates the influence to be 5 per cent for wheat, 7 per cent for maize and 19 per cent for plant oils (OECD, 2008). The large degree of uncertainty about the relative importance of bioenergy production for the food price hampers any assessment of future developments. It can, however, be assumed that expansion of bioenergy production will cause food prices to rise further. The future price of oil will also be a major influence on price changes, since it will increase demand for biofuels and thus put the food supply under further strain (IFPRI, 2008). There is a need for further research into the effect of bioenergy use on food prices (Section 11.4.4).

The sharp rise in prices is a short-term reaction. In the medium term the markets will react with an expansion of the food supply, and the currently high prices are likely to fall again. However, prices are unlikely in the medium term to return to the low level of the start of this century (Ressortarbeitsgruppe 'Welternährungslage', 2008). The OECD and the FAO estimate that prices of agricultural goods will fall again from their present record high but will for the next 10 years remain above the average level of the past decade; they will also remain very volatile (OECD and FAO, 2008). In WBGU's view the glo-

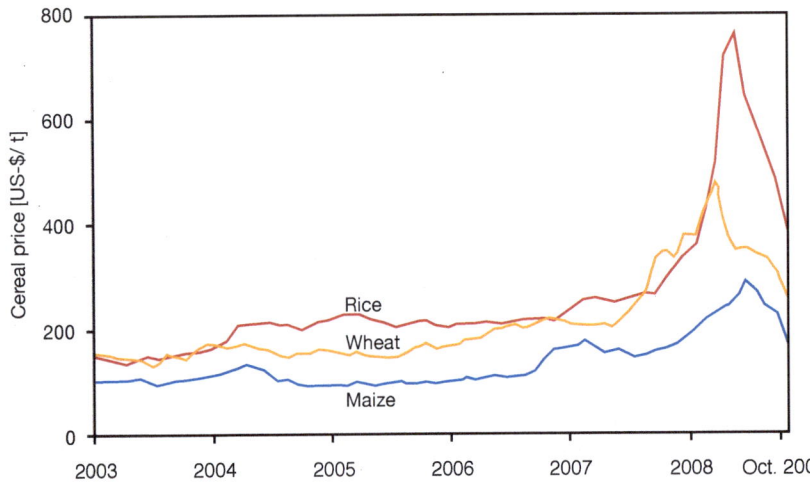

Figure 5.2-2
Development of cereal prices (2003–2008).
Source: von Braun, 2008b drawing on data from FAO, 2008f

bal trends described in this chapter (Sections 5.2.2–5.2.4) make it likely that this dynamic will persist long-term.

EFFECTS OF THE PRICE RISE

Higher agricultural prices are regarded as necessary in the long term for poverty reduction and development of the world's poorest countries (Constantin, 2008). The World Bank estimates that in 2008 900 million people in the rural areas of developing countries were living on less than US$ 1 a day; the majority of these people are involved in farming and could benefit from rising prices (World Bank, 2008c). In considering the effects of the rise in food prices a distinction must be made between the macroeconomic and the microeconomic level, and between short-term and long-term consequences.

At the macroeconomic level price rises will, at least in the short term, tend to benefit those countries that have a well-developed agricultural infrastructure. Capital-intensive farms in Latin America particularly stand to benefit (Constantin, 2008). Net importers of food and energy, on the other hand, will be particularly hard hit by price rises. The FAO names 22 developing countries that are particularly vulnerable on account of high levels of malnutrition combined with heavy dependence on oil

imports (Table 5.2-7). The FAO estimates that the cost to developing countries of food imports rose by 33 per cent in 2007 (FAO, 2008a). A similar situation affected the group of least developed countries (LDCs) and the low-income food-deficit countries (LIFDCs). The continuing rise in cost of imported food has severe implications for both groups of countries: the imported food now costs more than twice what it cost in 2000. The increase in the amount paid by the LIFDCs for imported cereals between the financial years 2006/2007 and 2007/2008 is particularly high at an estimated 56 per cent (FAO, 2008a). This has a negative impact on the balance of trade in these countries (Figure 5.2-3).

This threat is, however, tempered by the fact that in the countries of sub-Saharan Africa, the majority of which are counted among the LIFDCs, high import taxes on staple foods and high transport costs mean that there is little connection between local and international markets. The influence of high international wheat, maize and rice prices remains slight. By 2005 cereal imports to sub-Saharan Africa amounted to US$ 3400 million per year, which is equivalent to half of one per cent of the region's gross domestic product (Ng and Aksoy, 2008).

In the long term higher food prices would lead to an increase in supply even in countries with poorly

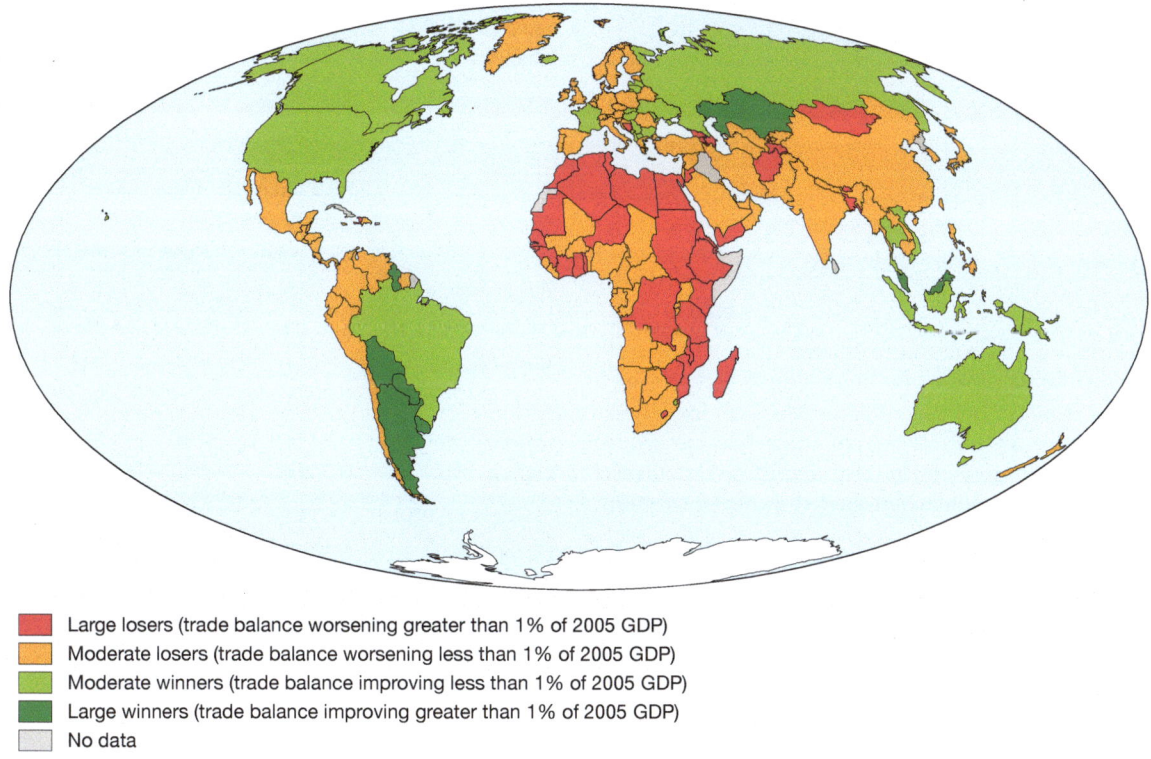

🟥	Large losers (trade balance worsening greater than 1% of 2005 GDP)
🟧	Moderate losers (trade balance worsening less than 1% of 2005 GDP)
🟩	Moderate winners (trade balance improving less than 1% of 2005 GDP)
🟢	Large winners (trade balance improving greater than 1% of 2005 GDP)
⬜	No data

Figure 5.2-3
Impact of projected food price increases (2007–2008) on trade balances.
Source: Maxwell, 2008

Table 5.2-7
Countries with high food insecurity which as net importers of oil and cereals are particularly vulnerable to price rises.
Source: FAO, 2008a

Country	Proportion of petroleum imported [%]	Proportion of major grains imported [%]	Proportion of population undernourished [%]
Eritrea	100	88	75
Burundi	100	12	66
Comoros	100	80	60
Tajikistan	99	43	56
Sierra Leone	100	53	51
Liberia	100	62	50
Zimbabwe	100	2	47
Ethiopia	100	22	46
Haiti	100	72	46
Zambia	100	4	46
Central African Republic	100	25	44
Mozambique	100	20	44
Tanzania	100	14	44
Guinea-Bissau	100	55	39
Madagascar	100	14	38
Malawi	100	1	35
Cambodia	100	5	33
Korea	98	45	33
Rwanda	100	29	33
Botswana	100	76	32
Niger	100	82	32
Kenya	100	20	31

developed agricultural infrastructure; this would involve expansion of the agricultural sector, which could increase economic power and incomes. Higher prices also alter the terms of trade between countries in favour of agricultural and semi-agricultural (rural) sectors, which can lead to positive development effects.

However, the macroeconomic perspective alone is not sufficient for assessing a country's food security. The microeconomic level must also be considered. The higher the disposable income of a household, the greater the quantity and quality of the food that can be purchased. Food prices are an important aspect of this, but the connections between food security and food prices are very complex. Firstly, it is important to distinguish between net producers and net consumers of food. The latter group includes in partic-

ular the urban poor, the landless and many subsistence farmers. Higher food prices invariably impact severely on net consumers, as is even now apparent ('bread riots', 'tortilla crisis'). Secondly, farmers who are net producers of food can profit from the higher prices and thus increase their income if other conditions remain unchanged. However, this can only happen if the price rises reach the local markets and are not erased by national price policies and transport costs (Wiggins and Levy, 2008). Furthermore, the profits accruing to small farmers as a result of higher food prices also depend on how these profits are distributed along the national value chain and on the extent to which input prices have risen in relation to food prices (Constantin, 2008).

It is thus impossible to make general statements about the net effect of higher food prices on food security. The effect will depend on socio-economic and agro-ecological conditions within a country and on the specific product whose price has risen. For example, poor farmers in a developing countries may be net sellers of a product that has risen in price and at the same time net purchasers of a product that has also risen in price (Faaij, 2008). Table 5.2-8 shows the percentage of net sellers of staple foods in selected countries in three important development regions.

Table 5.2-8
Proportion of households in selected countries which produce food above the subsistence level and are therefore net sellers of staple foods.
Source: FAO, 2008a

	Proportion of households [%]		
	Urban	Rural	All
Bangladesh, 2000	3.3	18.9	15.7
Pakistan, 2001	2.8	27.5	20.3
Vietnam, 1998	7.1	50.6	40.1
Guatemala, 2000	3.5	15.2	10.1
Ghana, 1998	13.8	43.5	32.6
Malawi, 2004	7.8	12.4	11.8
Madagascar, 1993	14.4	59.2	50.8
Ethiopia, 2000	6.3	27.3	23.1
Zambia, 1998	2.8	29.6	19.1
Cambodia, 1998	15.1	43.8	39.6
Bolivia, 2002	1.2	24.6	10.0
Peru, 2003	2.9	15.5	6.7
Maximum	*15.1*	*59.2*	*50.8*
Minimum	*1.2*	*12.4*	*6.7*
Unweighted average	*6.8*	*30.7*	*23.3*

Box 5.2-2

Country study: China – competition of 'food versus fuel'

With 1300 million inhabitants China is the most populous country in the world and, after the USA, the second-largest consumer of energy. Within the next decade this energy-hungry newly industrializing country will overtake the USA in energy consumption. In 2006 China accounted for around 15.6 per cent of worldwide primary energy consumption – and most of this energy came from fossil resources (BP, 2007). In 2005 around 70 per cent of China's primary energy was obtained from coal and somewhat more then 20 per cent from oil. Since China has the third-largest coal reserves in the world – after the USA and Russia – the majority of its energy will for the foreseeable future continue to be generated from coal. Gas and nuclear power play only a small part in primary energy consumption, accounting respectively for just under three per cent and one per cent. Renewable energies, too, make little contribution, except for hydropower (ca. five per cent of primary energy consumption) and biomass. Biomass use, however, is largely attributable to continuing widespread use of traditional biomass (BP, 2007; GBEP, 2008).

With rising economic growth China's energy demand will more than double by 2030, despite measures to increase energy efficiency. Growth rates in energy demand are projected to reach up to 15 per cent per year (GBEP, 2008). The reason lies not least in the growing volume of traffic (IEA, 2007d; Weyerhaeuser et al., 2007). China will continue to seek to meet its increasing need as far as possible from domestic production. Apart from for oil this is possible, since China has large stocks of coal and other potentials such as wind and hydropower are far from being fully exploited.

Electricity supply – 80 per cent of which comes from coal and 16 per cent from hydropower – varies widely for structural reasons, and grid losses are high. Rural regions, in particular, have an insufficient electricity supply. The rural energy supply is based to a large extent on small hydropower systems and traditional biomass use. To improve the rural supply situation, around 17 million biogas systems that can operate on biological waste have been installed in rural regions since 1975 (GTZ, 2006, 2007a). But the use of modern bioenergy is also making progress in China. There are already some facilities for generating electricity from biomass; in 2006 they produced in total 2 GW of electricity (REN21, 2008). This electricity was generated mainly from bagasse and in many cases was used within the sugar industry to meet its own needs. There are also facilities for producing bioethanol from grain. Overall China is the largest user of bioenergy in the world (9 EJ in 2005), ahead of India, the USA and Brazil (GBEP, 2008). At the same time the country makes use of only a fraction of its bioenergy potential. Ways in which this potential could be more fully exploited include in particular the use of organic materials in combined heat and power plants (CHP) with steam turbines and the generation of electricity from biogas in gas turbines (GTZ, 2007a).

The government's official target is to generate 15 per cent of primary energy from regenerative sources (excluding traditional biomass) by 2020. Electricity from biomass is planned to contribute 20 GW to this target (GBEP, 2008). In addition, the country aims to produce 13,000 million litres of bioethanol and 2300 million litres of biodiesel per year by 2020 in order to reduce dependence on imported petroleum. China produced around 1000–3000 million litres of bioethanol in 2006, making it the third-largest ethanol producer in the world, behind the USA and Brazil. In 2006 far less biodiesel was produced (approx. 70–100 million litres), most of it from waste oil (GBEP, 2008; REN21 2008). Most of the ethanol (more than 80 per cent) is produced from maize, but manioc, rice, sugar and cellulose waste are also used as source materials. Biodiesel can be produced from rapeseed oil, sunflower oil, soya oil and groundnut oil, as well as from waste oil (GTZ, 2006). There are also plans to grow Jatropha for biodiesel production on up to 15 million hectares of land in the south-west of the country, in the provinces of Guizhou, Sichuan and Yunnan (Weyerhaeuser et al., 2007). An additional 3 million tonnes of biodiesel could also be produced in future from waste oil or low-quality by-products of food oil production, although in the short term logistics for the use of these source materials are insufficiently developed (GTZ, 2006). China also plans to produce biodiesel in future from woody biomass (GBEP, 2008).

In terms of area China is the fourth-largest country in the world. However, only 10 per cent of the land is usable for farming. Twenty-seven per cent of the country is desert, and marginal mountain regions constitute a further substantial area. Forests cover 16.5 per cent of the country. Desertification is continuously increasing as a result of overuse. Irrigation efficiency in China is still around 45 per cent, compared with 70 per cent in industrialized countries (GTZ, 2006). In addition, bioenergy promotion can conflict with food provision and water availability. At present China is still able to supply its own need for grain, but it could soon have to resort to imports, as it already has to do –because of the increasing demand – with meat. The primary goal of China's bioenergy policy is therefore food security, since although productivity has risen demand is increasing and supply bottlenecks are likely to occur (GTZ, 2006).

Following sharp rises in grain prices in China, the government banned the production of ethanol from grain in 2007 in order to prevent further price rises (Weyerhaeuser et al., 2007). The aim is to develop alternative sources for ethanol production, such as millet, manioc and cellulose. However, the rise in grain prices has helped to improve incomes in the agricultural sector. In addition it is projected that the deployment of agricultural and forestry products for energy could create 9.2 million jobs in China, which would improve incomes among the rural population and hence also improve access to food (GTZ, 2006). According to estimates of GTZ (2006), land is available for biofuel production that does not compete with food production. An optimistic scenario estimates that 7.6 million hectares are suitable for bioethanol production and 67.5 million hectares are suitable for biodiesel production. Bioenergy can therefore go at least some of the way towards meeting China's increasing energy need. However, competition between bioenergy and food production for land use must be carefully monitored.

Calculation of the unweighted average across all countries shows that only 23 per cent of all households and 31 per cent of rural households are net sellers of food. This means that the majority of households in the countries concerned are net purchasers of food and thus likely to be adversely affected by high food prices. Poor households will be disproportionally affected in both urban and rural regions (FAO, 2008a). In both rural and urban areas positive effects at household level result in the long term from the relatively high labour intensity of the agricultural sector in developing countries and from links with upstream and downstream sectors (Brüntrup, 2008).

5.2.6
Summary: Ways to defuse competition for land use

Food production in the coming decades faces major challenges on account of the dynamically growing demand for food and feed worldwide. The main drivers of growing demand are world population growth and increasing affluence in the developing and newly industrializing countries that aspire to economic growth. The change in nutrition habits that accompanies increasing affluence adds a new dynamic to the process. At the same time, opportunities for increasing food production are limited by land scarcity, climate change and soil degradation. The growing importance of bioenergy further increases the pressure on agricultural land. Against this backdrop it should be borne in mind that rising prices both for food and feed and for energy can represent not only a threat to food security but also an opportunity to reduce it. Key factors are on the one hand conducive policy settings and on the other the socio-economic and agro-ecological situation at local level. The development of world market prices for food is influenced by a wide range of factors and can only to a limited extent be attributed to the global increase in energy crop cultivation.

From a global perspective the key to defusing the competition between food and energy crops, as discussed above, consists in (1) giving preference to the cultivation of energy crops on degraded and marginal land (if the land contributes to the subsistence of local population groups, their interests must be considered), (2) the development of integrated bioenergy and food security strategies at country level, (3) increasing land productivity, coupled with reform of international agricultural and trade policy and (4) promoting nutrition styles low in meat. In some poor developing countries there is also a place for measures to stem population growth and to prevent postharvest losses (Section 10.4). The recommendations made in Chapter 12 spell out these options.

5.3
Using biomass as an industrial feedstock

Biomass from plants and animals is not used only for food and feed (Section 5.2) and for the production of bioenergy; it is also used as a material in products. Material uses of biomass in products take many different forms (Figure 5.3-1) and vary widely from region to region. Resources of animal origin – such as skins and leather, wool, fat – are also used as materials.

The literature on feedstock uses of biomass shows that it is not usual, or even possible, to distinguish between 'stalk' biomass and 'woody' biomass. In German-speaking countries, feedstock crops and raw timber consigned to use as feedstock are grouped together as *'nachwachsende Rohstoffe'* (NaWaRo); the term covers all materials derived from living matter that are specifically used for purposes other than human food or animal feed (FNR, 2006c).

When considered across their entire life cycle, products manufactured from biomass (such as bioplastics or soap) are not necessarily associated with fewer CO_2 emissions than products derived from petroleum. As a result of direct or indirect land-use changes in the course of cultivation or of energy-intensive processing, it is in some cases possible for a greater quantity of greenhouse gases to be released than occurs with petroleum-based products.

The amount of biomass deployed globally as an industrial feedstock or in material form and the area of land used to grow it has not been assessed. Contrary to data on production or consumption, the data for individual material uses is often unclear or contradictory. This is because often no clear distinction is made between deployment for energy and for material purposes, or between the use of fossil and biogenic sources, and many products (such as detergents) contain materials from mineral, biogenic and fossil sources. In addition, there is often no clear distinction between primary and secondary (recycled) resources, and processed wood products are not always included. Furthermore, there is major trade in raw materials, semi-finished products and finished products, which further increases the difficulty in delineating the data. For example, Germany is a net exporter of some processed semi-finished cotton products.

5.3.1
Feedstock use of plant raw materials (excluding wood) in Germany

In order to identify how much land might be available for the deployment of biomass as energy, it is

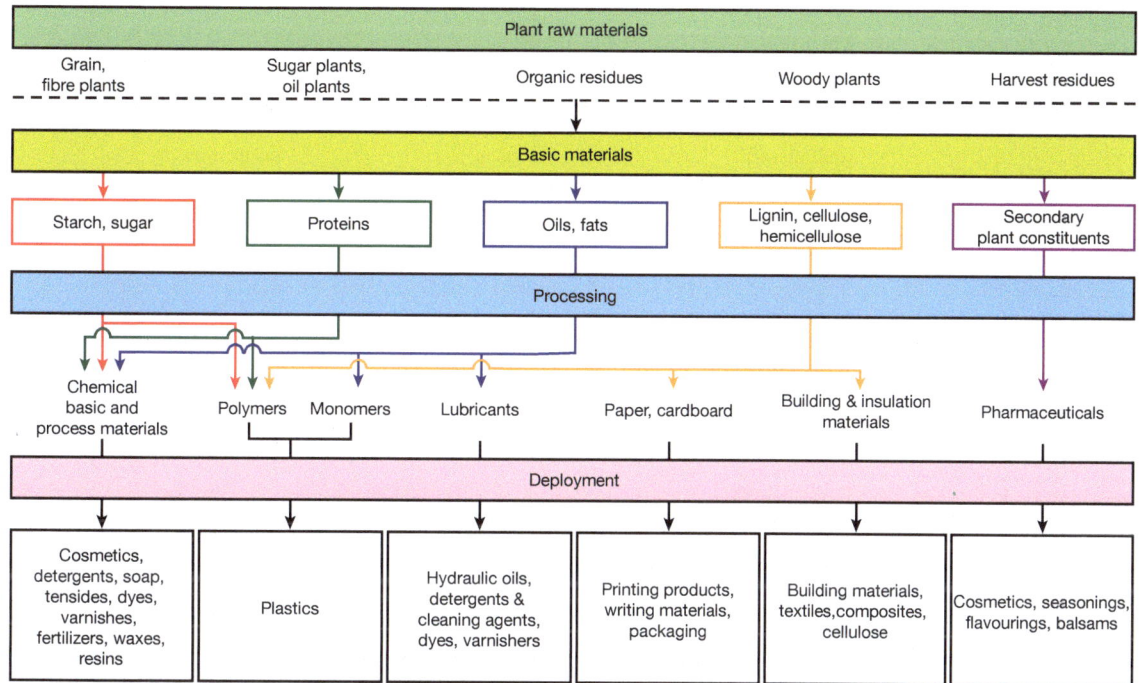

Figure 5.3-1
Deployment chains for use of biomass as an industrial feedstock.
Source: SRU, 2007, expanded

important to at least understand the scale of the use of biomass as an industrial feedstock. Such an assessment is therefore attempted below, with a distinction made between wood and cellulose products from forested land and plant and animal products from arable land and grassland. In the absence of any studies of the global land area dedicated to the use of biomass as an industrial feedstock, the extent of such land use is first calculated using an industrialized country (Germany) as an example and then extrapolated under simplifying assumptions to global requirements. Raw materials of animal origin (e.g. animal fats, leather, wool) are not included in the calculation.

The land requirement is calculated by considering the raw materials that in terms of quantity are the most important, namely wood/cellulose, natural fibres, oils and fats, starch and sugar (TAB, 2007). Other raw materials are grouped under an overall figure. Even for the most important raw materials it is not easy to calculate the figures for per-capita consumption, since the quantities of biogenic raw materials are seldom quoted when semi-finished goods are imported and exported.

According to figures provided by the Office of Technology Assessment at the German Parliament (Büro für Technikfolgenabschätzung, TAB), around 120,000 tonnes of natural fibres are processed industrially in Germany each year, which corresponds to around 1.45 kg per capita (TAB, 2007). For cot-

ton alone it is likely that annual per-capita consumption is around 10 kg, which would mean that total consumption in Germany is around 825,000 tonnes annually. No statistical data is available on the per-capita consumption of (processed) oils, fats, starch and sugar in products, so consumption must be estimated on the basis of the simplified assumption that it is equivalent to the annual quantity processed in Germany: 805,000 tonnes of plant oils, 345,000 tonnes of animal fats, 492,000 tonnes of chemical starch and 240,000 tonnes of sugar (TAB, 2007). Other plant raw materials such as medicaments are used in negligible quantities (although their land requirement could be relatively greater).

In Germany in 2005 1.4 million hectares of land were used for growing renewable raw materials for use as energy and as an industrial feedstock. According to figures provided by the TAB (2007), renewable raw materials for use as an industrial feedstock were grown on only 0.28 million hectares (0.23 per cent of all arable land). The total area of farmland used in Germany amounts to 17 million hectares, made up of 12 million hectares of arable land and 5 million hectares of pasture. The area of land actually used for growing plants that form an industrial feedstock for products consumed in Germany is likely to be several times higher than the figure quoted above if land used abroad (e.g. for cotton and tropical oils) is taken into account; it would be higher still if all the

Table 5.3-1
Production of and world trade in forest products. Trade figures are the mean of import and export from official statistics.
Source: FAO, 2007a

	Global production (2005) [million m³]	Trade (2005) [million m³]	Trade as a proportion of production [%]
Fuelwood	1,766.9	3.6	0.2
Roundwood, industrial timber	1,644.3	120.8	7.3
Sawnwood	421.8	132.2	31.3
Wood materials	220.1	79.1	36.0
	[million tonnes]	[million tonnes]	[%]
Cellulose	189.7	41.0	21.9
Paper and cardboard	353.4	111.8	31.6

plastics, bitumen and lubricants that are used annually were to be produced from biomass. For example, the annual German per-capita consumption of cotton of around 10 kg requires almost 1.2 million hectares of land for cultivation (10 per cent of the arable land in Germany). Worldwide, cotton-growing occupies around 2.5 per cent of arable land.

Plastics have until now been produced almost exclusively from crude oil and gas; bioplastics account for less than 1 per cent of total production. If in the distant future all plastics, bitumen and lubricants (which together account for around 8 per cent of crude oil consumption) were to be produced from biogenic raw materials, a very large quantity of land would be required. For per-capita consumption at half the level currently prevailing in Germany and a world population of 9000 million, WBGU calculates that around 10 per cent of world agricultural land would be needed for products that have traditionally been of agricultural origin (textiles, chemical products from oils, sugar and starch) and for the biogenic production of plastics, bitumen and lubricants.

5.3.2
Feedstock use of forestry products

In 2005 the world's forests covered less than 4000 million hectares. Their overall area is declining, despite regional variations in trends. Net gains in forest area are being reported for parts of Europe and Asia as a result of afforestation, but in other regions, particularly in Africa and Latin America, forested areas are shrinking in size (FAO, 2005).

Forest ecosystems provide a combination of services and are therefore often subject to multiple uses; that is, a forest can be used for different purposes simultaneously. In 34 per cent of forests the production of timber products and non-woody forest prod-

ucts is the primary form of use. Overall more than 50 per cent of forests are used productively in combination with other ecosystem services, such as soil and water conservation, nature conservation or leisure activities (FAO, 2005). In structural terms timber plantations are gaining in importance.

Worldwide removal of wood in 2005 was estimated at around 4000 million m³ (Table 5.3-1). The quantity of wood produced worldwide is thus significantly larger than the total quantity of steel, aluminium and concrete. According to the FAO's figures, 44 per cent of felled timber is used as fuelwood (FAO, 2005). Fifty-six per cent is used as roundwood and industrial timber (including for cellulose products), trimmed timber and finished products (e.g. furniture), in other words primarily for material purposes. However, wood residues that arise in the course of processing are often used for energy, so that the quantity actually used in material form may be somewhat less than 56 per cent.

Fuelwood has in the past seldom been traded internationally as most of it is used traditionally for heating and cooking. High transport costs relative to the value of the goods tend to make export uneconomical (Thrän et al., 2005). Wood pellets, however, are already traded internationally and this trade is likely to increase markedly in future. Competition between the use of wood for energy and for material purposes arises if the wood is used immediately for energy. When wood is used as a material, the wood products can – with a few exceptions, such as toilet paper – be used for energy at the end of their product life (cascade use, Section 5.3.3).

The trade in industrial timber and roundwood is difficult to delimit, since some of the wood is processed and then exported as cellulose or paper. The main exporters of processed timber are the Russian Federation, Canada and the USA (Thrän et al., 2005). Europe and Asia are the main purchasers of such tim-

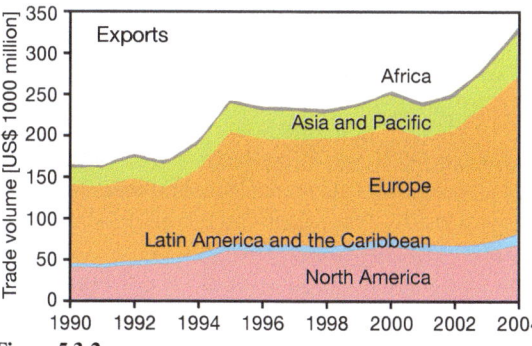

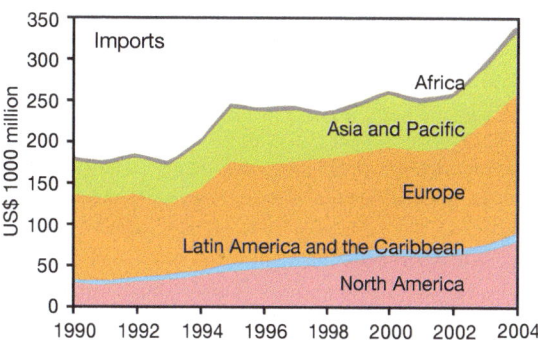

Figure 5.3-2
Trade in forest products – regional trends since 1990.
Source: FAO, 2007a

ber from Russia (FAO, 2007a; Figure 5.3-2). Internal trade in Europe and North America accounts for the majority of the volume traded worldwide (Thrän et al., 2005).

There is considerable potential for savings in the fuelwood used for heating and cooking in developing countries, because much cooking in those countries is still done over open fires (Section 8.2). On the other hand, population increase and rising prices for alternative fuels could increase demand for fuelwood. A significant increase in demand for wood used as a material (cellulose products such as paper; construction timber; wood for furniture) is also to be expected. For example, in its European Forest Sector Outlook the UN anticipates that in Europe (where consumption is already very high) wood consumption will increase annually from 2000 to 2020: by 1.8 per cent for sawnwood, 2.6 per cent for panels and 2.9 per cent for paper products (UNECE, 2005).

Around one-fifth of wood production is used for paper manufacture (Worldwatch Institute, 1999). Per-capita consumption of wood products used materially is significantly higher – in the case of paper products some twelve times higher – in the industrialized countries than in newly industrializing and developing countries. Fifteen per cent of the world population has an annual per-capita consumption of paper of 240 kg, while for 85 per cent the per-capita figure is 19 kg (Edelbrock, 2005). The world average is 57 kg per person per year. If this average doubles to 114 kg per person per year, assuming 50 per cent of paper is recycled and the world population grows to 9000 million, the additional quantity of raw wood needed to cover increased paper consumption alone would amount to around 1000 million m³ (approximately 25 per cent of the quantity currently used for energy and as a material). If per-capita consumption rose to that of Germany's population the additional quantity needed would be around 2700 million m³ (approximately 70 per cent of the quantity currently used for energy and as a material).

Even extensively forested industrialized countries such as Germany import large quantities of wood products – in particular to meet their need for cellulose – and utilize a great deal of forested land worldwide. It is estimated that the use of wood products in Germany results in 23,300 million hectares of land worldwide being needed for such products. Germany's own forests, covering 10,100 million hectares, meet only 43 per cent of this requirement (Wuppertal-Institut and RWI, 2008).

5.3.3
Cascade use

The production of biomass for use as an industrial feedstock (fibres, plastics, technical oils, etc.) is likely to require around 10 per cent of world agricultural land (Section 5.3.1). This is a significant quantity of land, although it would be reduced to the extent that the biomass-based products could be deployed for energy purposes at their end of their useful life (cascade use). In this context, however, it should be remembered that energy losses are often incurred in the course of production, in use in open or semi-open applications (e.g. detergents) and in collection and recycling.

5.3.4
The outlook for material production without oil, gas and coal

Even today raw materials such as fibres, technical oils, etc. are produced wholly or partly from cultivated plants. Nevertheless, those products that are most important in terms of quantity (such as plastics) are still produced predominantly from petroleum. If use of all fossil fuels were to be abandoned, these products would also have to be manufactured from biomass. In the long term, therefore, there will be radical changes in the ways in which material production takes place. In particular:

- Crops selected for cultivation should as far as possible be those that have natural high-performance qualities, for example natural products such as medicaments, fragrances and soap constituents.
- The majority of such raw materials should be produced in 'biorefineries' (such as lignocellulose biorefineries). Such biorefineries would be analogous to today's petrochemical refineries in that they could synthesize a range of typical base materials, which would yield 'family trees' of feedstocks and chemicals of biogenic origin (TAB, 2007; IFEU, 2007). Important base materials would be derived from carbohydrates or lignocellulose. Production could – but would not have to be – coupled with the manufacture of biofuels.
- In the feedstock sector, as in the energy sector, efficiency would have to be an important consideration. Product design and waste management need major reconfiguration to ensure high material and energy recovery ratios without complex and costly collection and processing.

5.4
Competition with biological diversity

The Earth's biological diversity is in crisis: plant and animal species are today becoming extinct at a rate 100–1000 times higher than the average over the world's history, and this rate is set to increase further (MA, 2005a). The loss of biological diversity threatens important ecosystem services (e.g. coastal protection, water supply, pollination, etc.; MA, 2005e). Moreover, as plant and animal species become extinct their genetic and physiological blueprints are irretrievably lost – blueprints that could be of major value in areas such as crop development or medical research (WBGU, 2001a; Chivian and Bernstein, 2008). For these reasons biodiversity is recognized as a key element of sustainable development (WEHAB-Framework des WSSD; WEHAB Working Group, 2002). In the Convention on Biological Diversity (CBD, 2002b) and at the World Summit on Sustainable Development (WSSD, 2002) the world community agreed to bring about a significant reduction in the rate of biodiversity loss by 2010 (Chapter 3 and Section 10.5). A number of signs indicate, however, that this target will not be met (MA, 2005a): of the 15 indicators used by the CBD to assess attainment of the 2010 target, 12 show an unbroken downward trend. Only one indicator – the area of designated protected areas – shows an upward trend (CBD, 2006a; Box 5.4-1).

The most important reason for the current global crisis of biological diversity is habitat loss as a result of the conversion of natural and semi-natural ecosystems for agricultural and forestry purposes (Section 4.2.3) and of the intensive use of the resulting production systems. The cultivation of energy crops adds an additional type of land use to the mix, with the potential to exacerbate the competition for usable land. The increasing competition for land, and hence also the risk of significantly increasing the loss of biological diversity (UNEP, 2007a) has two main dimensions: first, the conversion of natural ecosystems and, second, the intensification of farming and forestry on existing land. Conversion of natural ecosystems can also occur indirectly if the cultivation of energy crops displaces the previous form of land use. That use must then transfer to other land, which may result in natural ecosystems being converted for this purpose (Searchinger et al., 2008). These indirect displacement effects often have an international dimension: expansion of the cultivation of energy crops on farmland in Germany can result in changes beyond its borders, perhaps in the form of increased clearance of tropical rainforests or ploughing up of savannah.

Analysis of the issues underlying this competition for land use is hampered by the fact that scientific study of the connections between greatly expanded bioenergy use and ecosystem effects have only recently acquired momentum (Fritsche et al., 2006). For example, the comprehensive report of the Millennium Ecosystem Assessment makes only passing reference to the problem of bioenergy (MA, 2005a, b, c, d). In a survey of British nature conservation experts just three years later the increasing need for bioenergy was classified as a major risk for the further loss of semi-natural habitats and biological diversity (Sutherland et al., 2008).

5.4.1
Competition between energy crop cultivation and existing protected areas

The cultivation of energy crops can compete directly with the conservation of biological diversity if it involves the conversion of land that forms part of existing protected areas. Since protected areas are one of the most important instruments for conserving biological diversity and securing ecosystem services (Box 5.4-1; CBD, 2004b; MA, 2005c), protected area status should largely rule out other usually harmful uses.

This is, however, not always the case. Even the traditional forms of bioenergy use (fuelwood, charcoal) can compete with protected areas if the interests of

Box 5.4-1

Protected areas: Situation and trends

The World Conservation Union (IUCN) defines protected areas as follows: 'A clearly defined geographical space, recognised, dedicated and managed, through legal or other effective means, to achieve the long-term conservation of nature with associated ecosystem services and cultural values.' (Dudley, 2008). Protected areas have increased significantly in recent decades in both number and area (Figure 5.4-1). There are now some 115,000 protected areas worldwide (WDPA, 2008), with an area of more than 20 million km²; this represents more than 12 per cent of the terrestrial land area (Chape et al., 2005). This figure includes areas whose primary purpose is sustainable use rather than the conservation of biological diversity (IUCN categories V and VI; UNEP-WCMC, 2008).

However, the figures for the number and extent of protected areas are not reliable indicators of effective nature conservation (Pressey, 1997) since many of the protected areas exist only on paper (IUCN, 2003) and the protected status of many areas is not effectively managed and applied (UNEP, 2007a; Dudley et al., 2004). Furthermore, the selection of areas has frequently not been based on scientific consideration of how the greatest contribution to the global protected area system can be achieved; in many cases preference has been given to protecting land that is not of major economic interest.

EFFECTIVENESS AND THREAT

Protected areas – particularly those in the tropics – are under threat, primarily through overuse (hunting, gathering) and human settlement, through land conversion and fragmentation within protected areas and through their increasing isolation as a consequence of the conversion of surrounding land (Carey et al., 2000; IUCN, 2003). They are also at risk from invasive species (Box 5.4-3) and climate change (Section 5.4.4). A study of ten tropical countries found that only one per cent of their forest protected areas can be regarded as secure and that many are suffering from degradation and area loss (Dudley and Stolton, 1999b). By contrast, in the majority of the tropical protected areas considered by von Bruner et al. (2001) further land clearance was being prevented and other detrimental activities were at least being held in check. Protected areas that are better

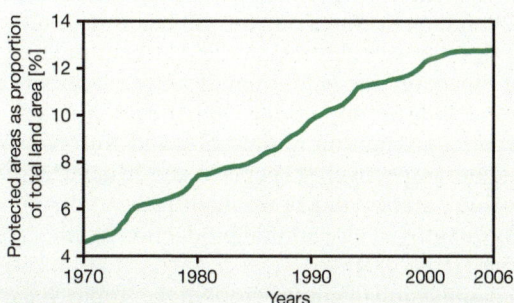

Figure 5.4-1
Increase in the extent of protected areas worldwide (1970–2000). The proportion of the terrestrial area under protection has increased markedly in recent decades and is now more than 12%.
Source: UNEP, 2007a

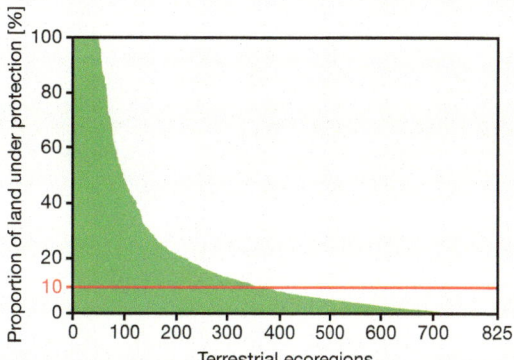

Figure 5.4-2
The representativity of ecoregions in the existing protected area system. Frequency distribution of the 825 terrestrial ecoregions according to the percentage under protection. In fewer than half of the ecoregions is more than 10% of the land under protection, and in 140 ecoregions less than 1% is protected.
Source: UNEP, 2007a

equipped and run, in financial and management terms, are more likely to be effective in conserving their biodiversity (IUCN, 2003). Protected areas in countries with weak governance structures and limited capacity for the management of existing protected areas are particularly at risk (Brandon et al., 1998). In view of the increasing pressure on land use, it is likely that many existing protected areas will require additional investment if they are to stem the erosion of biological diversity.

REPRESENTATIVITY

Accurate information on the extent to which biological diversity is being conserved in the existing protected area system is not available. For birds, however, it has been found that 20 per cent of species do not occur in any protected area (Rodrigues et al., 2004). Comparison of the 825 ecoregions listed by the WWF (Olson et al., 2001) with the coverage of existing protected areas (Figure 5.4-2) shows that fewer than half of the ecoregions have more than 10 per cent of their area protected, and in 140 ecoregions less than one per cent of the area is protected (CBD, 2006a). Crop genetic diversity is particularly important for human food security, yet crop diversity centres are clearly underrepresented in existing protected areas (Stolton et al., 2006). The protection system is thus insufficiently representative either of ecosystem types or of species and genetic diversity.

CLIMATE CHANGE

Future planning and management of protected areas must take account of climate change, because around 50 per cent of protected areas are likely to be affected by climate zone shifts (Halpin, 1997). If it is to adapt to climate change, the protected area system must therefore be more flexible, larger, better networked and better integrated into the surrounding farming areas (UNEP, 2007a; Hannah et al., 2007).

Overall the existing protected area system is thus a necessary but not a sufficient instrument for preventing further loss of biodiversity (MA, 2005c; McNeely, 2008). Internationally there is broad agreement that the existing global network of protected areas needs to be expanded and better financed if it is to fulfil its purpose (Section 10.5).

Existing protected area systems are neither big enough nor sufficiently well planned and managed (CBD, 2004b). The necessary expansion includes corridors for networking protected areas, additional protected areas in ecosystem types or ecoregions in which conservation is still under-represented, areas that are important for the conservation of endangered species or genetic diversity, and buffer zones between protected areas and intensively farmed land. The Convention on Biological Diversity stipulates that this extension of the terrestrial protected area network should be completed by 2010. In view of the expected increase in pressure on land use this is a highly ambitious goal which needs institutional improvements and stronger political support (CBD, 2004b; Section 10.5).

the local population with regard to access to the biological resources of these areas collide with their natural conservation goals. Conflicts with traditional use can, however, usually be resolved through appropriate planning and participation (MA, 2005c).

Large-scale, modern bioenergy projects harbour greater potential for conflict. In Ethiopia, for example, large areas of land adjacent to the Babile Reserve, a protected area for elephants, have been cleared for the production of ricinus. It is possible to find examples of direct competition through encroachment and expansion of areas of land used for energy crops into protected areas. For example, in Uganda plans to allocate more than 7000 ha of a protected area of tropical rainforest (Mabira Forest) to the development of sugar cane plantations were halted only after vehement local protest (ABN, 2007). Another example concerns the Indonesian province of Riau, which has lost 65 per cent of its natural forest cover in the last 25 years and still has very high rates of clearance and degradation (2005–2006: 11 per cent loss). Once the forest has been cleared, oil palms or acacia plantations are often planted. Even within the protected areas deforestation has not been prevented, although it has taken place more slowly than outside these areas. The locally managed protected areas in Riau (covering around 22 per cent of the province) are in this respect significantly less effective than the national areas (6 per cent of the province): in the former, primary forest cover has fallen from 81 per cent to 47 per cent, while in the latter it has fallen from 90 per cent to 70 per cent (Uryu et al., 2008).

In addition to direct competition with bioenergy plantations, indirect effects must also be considered, since the land use displaced by energy crop cultivation can increase the pressure on existing protected areas. This is particularly relevant to protected areas in the tropics, which are already under considerable threat (Carey et al., 2000; IUCN, 2003; Box 5.4-1).

However, within protected area systems there are potentials for sustainable use for the purpose of energy crop cultivation. The International Union for Conservation of Nature (IUCN) divides protected areas into categories according to their conservation goal and intensity of use (IUCN, 1994). Some of these categories permit the conservation purpose to be combined with sustainable use. For example, in farmland that has been in use over a long period the conservation of biological diversity can often only be ensured if the historical type of land use is continued or its effects are simulated. For example, this applies to a great deal of marginal land in Central Europe that has been used for extensive grazing or coppicing and that now has great biological diversity. In these areas certain bioenergy uses (wood chips, grass and shrub-cutting) can be combined with landscape preservation measures as a type of 'conservation through use' (WBGU, 2001a; Wiegmann et al., 2007). If suitable cultivation systems are chosen (Section 7.1) it may also be possible to combine use for bioenergy with the conservation purposes in transition and buffer zones between protected areas and surrounding intensively used land.

5.4.2
Competition between energy crops and natural ecosystems outside protected areas

The majority of terrestrial biological diversity is found outside protected areas, particularly in natural or semi-natural ecosystems that are not at present used intensively by humans. The conversion of such ecosystems for agricultural use is currently the most important direct driver of the loss of biological diversity (Baillie et al., 2004). This pressure will increase further: according to the FAO (2003b), an additional 120 million hectares of land will be brought into use in developing countries by 2030. A major expansion of bioenergy use would significantly exacerbate this trend. Substitution of only 10 per cent of petrol and diesel consumption in the USA and Europe would require respectively 43 per cent and 38 per cent of the region's arable land, or would displace a corresponding quantity of agricultural production to other countries (IEA, 2004).

Cultivation of energy crops in the tropics is particularly attractive, because land is available cheaply in these regions and if conditions are favourable hectare yields are high (Doornbosch and Steenblik, 2007). However, among the areas that are converted for cultivation is a high concentration of localities with high biological diversity (case studies in BirdLife International, 2008). For example, the increased demand

Box 5.4-2

Country study: Indonesia – competition with nature conservation

Beside Malaysia, Indonesia is one of the most important producers of palm oil, accounting for 42.6 per cent of world output. Oil palms are currently grown on 41,200 km² of land, or 2.3 per cent of the country's surface – and the figure is rising sharply (Figure 5.4-3). The palm-growing areas are located mainly in the moist tropical regions of Kalimantan, Sumatra and Sulawesi. Three-quarters of the palm oil produced is exported; in the past most of it has been used in the food and cosmetics industries (FAOSTAT, 2008b).

At present Indonesia meets 28.5 per cent of its primary energy requirement from the combustion of biomass and waste (IEA, 2008c). The Indonesian government's biomass strategy has the specific aim of expanding the use of biomass for energy in order to help meet the national energy requirement, expand exports in the bioenergy sector and create jobs in rural areas. On the domestic front a key target indicator is an increase in biofuel use to at least 10 per cent of oil consumption by 2010 (Setyogroho, 2007). The blending quota for petrol is currently 3–5 per cent, with the blended bioethanol being produced mainly from sugar cane and cassava. Diesel, most of which is biodiesel from palm oil, is blended at a rate of 2.5 per cent. It is due to be supplemented by increasing use of *Jatropha* (Setyogroho, 2007; Butler, 2008). There are also other quantitative targets at national level. For example, 5.25 million hectares of 'unused' land are due to be planted with oil palms, *Jatropha*, sugar cane and cassava by 2010 (Butler, 2008).

Rising prices for energy and palm oil make Indonesia's exports of palm oil profitable. In no other country can palm oil be produced in such large quantities at such favourable prices (FAOSTAT, 2008b). While production costs are lower in Thailand and China than in Indonesia, neither Thailand nor China has sufficiently large areas of cropland suitable for oil palms. With its sparsely populated, moist tropical regions of Kalimantan, Sumatra and Sulawesi and a large pool of cheap labour, Indonesia thus has significant advantages. Outside the main centres Indonesia's economy is largely restricted to farming, fishing, local trade and in some regions mining; it is therefore hoped that development of bioenergy will significantly strengthen the rural regions. For example, Djaja (2006) calculates that the cultivation of energy crops alone could involve 7.4 million jobs, the majority of them new ones, by 2010.

Indonesia's forest cover declined from 64 per cent to 49 per cent in the period 1990–2005. An average of 18,715 km² were cleared per year – an area about the size of Saxony (FAO, 2006c). The causes of deforestation are legal and illegal logging and periodic forest fires, to which non-intact forest ecosystems are particularly vulnerable. There is a cor-

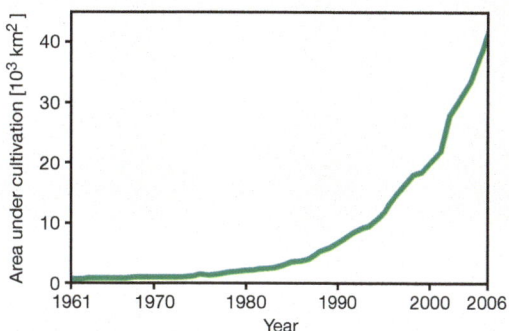

Figure 5.4-3
Development of land under oil palm cultivation in Indonesia (1961–2006).
Source: FAOSTAT, 2008b

relation between the expansion of palm oil plantations and the clearance of primary and secondary forest in Indonesia (Glastra et al., 2002; Reinhardt et al., 2007; UNEP, 2008). Koh and Wilcove (2008) conclude from their evaluation of FAO data that more than half of the expansion of palm oil cultivation in Indonesia and Malaysia during the period 1990–2005 was at the expense of primary and secondary forest. Since these forests are among the global hotspots of biological diversity, their destruction brings with it irreversible loss of biological diversity and ecosystem services (Section 4.2.3).

It should be noted that this deforestation for the cultivation of oil palms is relevant not only from the point of view of biotope and species conservation, but also from that of climate change mitigation. Annual greenhouse gas emissions from land-use changes and forestry amount to 2,565 million tonnes of CO_2eq, which amounts to around 84 per cent of all Indonesian emissions (WRI, 2008). Indonesia is thus the fourth-largest emitter of greenhouse gases, after the USA, China and the EU. A particular contributor to the problem is the cultivation of oil palms on peat soils, in the course of which enormous quantities of carbon stored in the soil escape into the atmosphere as CO_2 (Hooijer et al., 2006).

The Indonesian government exerts at present only very limited influence on deforestation; in consequence it is estimated that 70–80 per cent of logging activity is carried out illegally (World Bank, 2006c). Enforcement of forestry laws is hampered by unclear responsibilities and personnel shortages. The problems are exacerbated by the fact that elite members of the army and administration, together with politicians and industrialists, are often involved in illegal logging activities, with the result that regulatory measures are often blocked at high level (World Bank, 2006c).

for biofuels drives the expansion of palm oil plantations (Box 5.4-2). In South-East Asia the connection between palm oil plantations and the clearance of natural forests is evident (Malaysia, Indonesia; Reinhardt et al., 2007; UNEP, 2008; Koh and Wilcove, 2008). Conversion for palm oil plantations is very profitable and governance in some provinces in this region is weak, so that existing statutory regulations on conservation of the forests are not always

enforced (Glastra et al., 2002). The consequence is not only the emission of large quantities of greenhouse gases (Hooijer et al., 2006), but also a significant threat to biological diversity, including endangered megafauna such as the Sumatra tiger, the Asiatic elephant, the Sumatra rhinoceros and the orang utan. Box 5.4-2 explores the palm oil boom in Indonesia in more depth.

Furthermore, a change of farmland use from food production to bioenergy production can trigger displacement effects, which can lead indirectly to the conversion of natural ecosystems (Wissenschaftlicher Beirat Agrarpolitik beim BMELV, 2007; Searchinger et al., 2008). These effects can thus significantly accelerate the loss of biological diversity. A striking example of this is the expansion of land for the cultivation of sugar cane for bioethanol production in Brazil; this displaces other uses (soya, pastoral farming) into natural biodiversity-rich ecosystems including those in the Amazon region (tropical rainforest) or in the Cerrado (Klink and Machado, 2005; Sawyer, 2008). The Cerrado is the Brazilian savannah, which is a major hotspot of biological diversity: it is home to 50 per cent of all endemic Brazilian species and 25 per cent of endangered Brazilian species. Around 45 per cent of the Cerrado is still covered by natural vegetation, but less than 2 per cent currently has protected status. The conversion rate in the Cerrado is at least twice as high as in the Amazon region (Sawyer, 2008). More and more of the land is being given over to large-scale monocultures (sugar cane, soya), with a resulting loss of biological diversity and ecosystem services (Kaltner et al., 2005; Section 5.4.3).

On the American continent displacement effects are observable on an international scale (Searchinger et al., 2008). State promotion of ethanol in the USA results in more land being given over to maize growing; in addition, less maize is exported, since more of the crop is retained within the country in order to be processed into biofuel. As a result of maize growing, less land in the USA is used for growing soya, reducing the quantity of soya available on the world market. The result is a rise in soya prices, which in turn leads to faster clearance of Brazilian rainforests so that more soya can be grown there (Morton et al., 2006).

This conversion threatens to affect high conservation value areas (HCVAs) that would have been candidates for an expanded system of protected areas or even essential constituents of such a system (Section 5.4.1). Particularly alarming is the continuing loss of ecologically important areas in regions that are hotspots of biological diversity (Mittermeier et al., 1999; Myers et al., 2000) or that are ecologically underrepresented in the existing protected areas (MA, 2005a). Only a fraction of HCVAs currently have protected area status. In the context of spatial planning in relation to sustainable agriculture and forestry the HCVA concept can be an important instrument for the conservation of biological diversity. It was first introduced by the Forest Stewardship Council (FSC): identification and maintenance of conservation value in these areas is a requirement for FSC certification (FSC, 1996). The follow-

ing areas deserve particular protection in this regard (WWF, 2007):

- Hotspots of biodiversity (e.g. endemism, endangered species, genetic diversity);
- Large natural ecosystems in which populations of most wild species are still to be found in their natural distribution patterns;
- Areas of rare or endangered ecosystems;
- Areas that provide important ecosystem services (e.g. protection against landslides, floods or erosion);
- Areas that provide important ecosystem products for the local population (e.g. for subsistence or health) or that are of importance for their traditional cultural identity (e.g. areas with religious or spiritual significance).

5.4.3
Competition between energy crops and the conservation of biological diversity in agricultural areas

INTENSIFICATION OF FARMING SYSTEMS

More than three-quarters of the ice-free terrestrial surface of the world consists of biomes heavily influenced by humans; these contain a mosaic of anthropogenic ecosystems – involving, for example, agriculture or forestry – and natural or semi-natural ecosystems which account for a large part of the biomes' biological diversity (Ellis and Ramankutty, 2008; Section 4.2). Increased demand for agricultural products is met in part by expanding the areas of land used and in part by intensifying the use of existing land (Tilman et al., 2002).

The increased use of energy crops will exacerbate this trend towards intensification. Many of the bioenergy cultivation systems in use today are intensive monocultures which aim to achieve high yields through the extensive use of agrotechnology, agrochemicals (fertilizers, pesticides), energy and – increasingly – irrigation (Section 7.1). At the same time the additional demand for farmland for bioenergy crops can create displacement effects which increase the intensity of land use in other places; this is similar to the way in which energy crop cultivation can lead indirectly to the expansion of land use (Section 5.4.2). Intensification often means that small-scale, diverse, in the main extensively farmed land with comparatively high biological diversity becomes an area of large-scale, biologically impoverished monoculture. Within this process loss of biological diversity, loss of genetic varietal diversity and loss of cultural traditions are often closely linked (FAO, 1996). Even the most well-functioning and well-developed system of protected areas cannot halt this loss of biological diversity if land use in the surrounding areas of farmland is not sustainable (MA, 2005a).

The various risks to biological diversity posed by intensification can arise directly or indirectly through increased cultivation of energy crops:

- *Destruction and fragmentation of natural ecosystems:* The transition to large-scale agricultural monocultures often involves the destruction of areas (which may in themselves be small) of high conservation value (e.g. field borders and structural elements of the agricultural landscape, protected area buffer zones and natural ecosystems); this poses an additional threat to biological diversity (MA, 2005a).
- *Risks arising from the loss of agrobiodiversity:* Agrobiodiversity provides important ecosystem services for sustainable agriculture (pollination, nutrient recycling, erosion protection, etc.), but agrobiodiversity may be lost in the conversion of small-scale, biodiverse farming systems into large-scale monocultures. This form of intensification is linked with the genetic erosion of varietal diversity (Phillips and Stolton, 2008).
- *Risks arising from over-fertilization and eutrophication:* Increased tillage, erosion and sediment removal can pose a risk to natural ecosystems even at a considerable distance. For example, water run-off from US-American farmland is heavily polluted with nutrients that are carried by the Mississippi to the Gulf of Mexico, where they result in large anoxic 'dead' zones on the ocean floor (Donner and Kucharik, 2008; Diaz and Rosenberg, 2008).
- *Risks arising from pesticide pollution:* The input and accumulation of pollutants can pose a significant risk to biological diversity unless limits are set to pesticide use and integrated plant protection is pursued within a framework of sustainable agricultural practice (SRU, 2007).
- *Risks arising from the overuse of water resources:* Annual energy crops grown under irrigation have a large water requirement (Section 5.6). Overuse of local water resources for agriculture often goes hand in hand with the loss of wetlands. Wetlands harbour above-average diversity but at the same time they are at particular risk from conversion and degradation (IWMI, 2007).
- *Risks arising from invasive alien species:* The risks posed by the spread of invasive alien species are explored in Box 5.4-3.
- *Risks arising from the spread of genetically modified material:* The use of genetically modified organisms entails the risk that genetically modified material will spread in wild populations (Box 7.1-3).

These effects of intensification apply both to energy crop cultivation systems and to other intensive farming systems. However, there is a difference between the bioenergy farming systems in use today, which have ecological impacts very similar to those arising from the intensive production of food (e.g. cereals), feed (e.g. soya) or feedstocks (e.g. cotton) (SCBD, 2008), and the energy crop cultivation systems that are expected to proliferate in the future which will enable the whole plant to be used (Doyle et al., 2007). In terms of some of these ecological impacts the latter type score more positively if perennial, biodiverse cultivation systems are used in which only above-ground biomass is harvested and little tillage takes place (for more on the sustainability of bioenergy cultivation systems see Section 7.1). The yields obtained from these cultivation systems will also be improved if the crops are well supplied with nutrients and water through fertilization and irrigation; much will depend on whether these additional inputs are economically feasible and applied in a sustainable manner. For the moment, however, these considerations remain theoretical. Since use of these new farming systems is not yet widespread, there is as yet little concrete evidence of their positive or negative impacts on biological diversity (SCBD, 2008).

MARGINAL LAND

Depending on the definition and method of calculation used, between 1000 and 3000 million hectares or 7–20 per cent of Earth's terrestrial surface can be described as marginal land – that is, land whose soil fertility is so low that it is either unsuitable for agricultural production or capable only of supporting very limited production (Box 4.2-1; Worldwatch Institute, 2007). There are many reasons why land may be classified as marginal: some areas have shallow soil are or situated on steep slopes that are threatened by erosion; others are acidified or salinated. Inadequate drainage and waterlogging, or alternatively lack of water, are other factors that can limit the productivity of marginal land. In many cases the situation is the result of degradation caused by human activity – that is, non-sustainable land use, perhaps involving inappropriate irrigation or overuse, has turned productive land into marginal land (degraded land; Smeets et al., 2004).

It is often suggested that cultivation of bioenergy crops should be permitted only as an intensification of land use on marginal 'unused' land, in order to avoid displacement of existing land use and hence competition with food production (Box 6.7-2; Section 5.2). However, marginal land is in fact often used by the local population for food production; for example, it may be used for extensive pasturing of animals. There may also be competition with bio-

Box 5.4-3

Invasive alien species

Invasive alien species are an important reason for biodiversity loss (MA, 2005a). While the likelihood of a newly introduced plant species becoming invasive and thus causing harm is small, the harm done can be extensive, especially as invasion is usually irreversible (Mack et al., 2000). New risks also arise in connection with the cultivation of energy crops. Bioenergy cultivation systems of the future in which the entire above-ground biomass is used (such as grasses, wood) usually require crops with properties that differ from those that are desirable in crops for food or use as an industrial feedstock; this means that the choice may fall on other species or varieties that have been little cultivated in the past and about whose invasive potential little is known.

In this connection it should be noted that there is a great deal of overlap between the list of desirable ecological properties of energy crops (Heaton et al., 2004) and the properties that are often found in invasive plant species (Table 5.4-2). The likelihood of an introduced species becoming invasive increases with the frequency with which it is planted (Mack et al., 2000). Species that have caused no trouble for decades may reveal their invasive potential when used on a large scale.

Raghu et al. (2008) warn that grass species such as *Miscanthus* and switchgrass (*Panicum*), whose use as energy crops is currently under discussion (Section 7.1), also have properties indicative of enhanced invasive risk. The problem must be taken very seriously because invasion, which is usually irreversible, brings with it ongoing costs for the agriculture and forestry sector and is seriously detrimental to biological diversity (MA, 2005a). There are only a few examples of cases in which an invasive plant has been successfully brought under control again or ultimately exterminated. Using biological measures which involve introducing a natural enemy of the invasive species is a risky process, particularly in relation to grasses. Many essential crop plants are themselves grass species (rice, wheat and other cereals; feed grasses for animal production, etc.), which means that

Table 5.4-1
Desirable ecological properties of energy crops and their relevance to the risk of invasiveness.
Source: Raghu et al., 2008 and literature quoted there

Desirable characteristics of energy crops	(1) Characteristics present in invasive species (2) Contribute to success
C4 photosynthesis	(1), (2)
Long ground-cover period	(1), (2)
Perennial	(1)
No known pests or diseases	(1), (2)
Rapid growth in spring	(1), (2)
Sterile seeds	(1)
Redistribution of nutrients in below-ground plant parts in autumn	(1), (2)
High efficiency of water use	(1), (2)

the introduced pest may transfer to these crops and cause new damage there (Goeden and Andres, 1999).

Low and Booth (2007) list 18 species that are being used or proposed as energy crops but that also have invasive potential or have already become invasive. For example, *Jatropha curcas* was banned as an energy crop in Western Australia and Northern Territory after a study showed that the plant is considered invasive in 14 countries (Randall, 2004). Importing *Jatropha* into Australia is prohibited for this reason. *Ricinus communis*, which is used in Ethiopia as a bioenergy crop, is also regarded in Australia as invasive.

These ecological risks must be carefully examined before the species concerned are introduced for bioenergy use (e.g. Mack et al., 2000; CBD, 2002c).

---

logical diversity (Fritsche et al., 2008), because restoration of this land usually involves conversion with subsequent intensification. More intensively used, more productive land usually exhibits less biological diversity than marginal land, particularly if the marginal land has been left to succession for a lengthy period (Worldwatch Institute, 2007; the example of China: Hepeng, 2008).

Growing appropriate energy crops can, however, provide an opportunity for restoring degraded land that is species-poor and of limited usefulness (e.g. *Imperata* grassland in South-East Asia). Sustainable cultivation methods can have positive impacts on biological diversity for land that is already degraded. This is especially the case if perennial, ground-covering energy crops are used and if they form part of farming systems which involve a range of species (grasses, trees and shrubs: Section 7.1; The Royal Society, 2008). Given appropriate countryside plan-

ning, such planting systems can simultaneously deliver other ecosystem services, for example by enriching soil carbon and hence increasing soil fertility, providing protection against wind erosion, or providing a buffer zone for wetlands or nature reserves (Worldwatch Institute, 2007; Berndes, 2008). These potentials are to a large extent dependent on the particular landscape type and on the type of bioenergy farming system that is planned.

5.4.4
The cross-cutting issue of climate change

Climate change is already causing visible displacement of populations (Parmesan and Yohe, 2003; Moritz et al., 2008) and will in future present a significant additional threat to biological diversity (Thomas et al., 2004). Some 20–30 per cent of the plant and animal species studied to date will be exposed to an increased risk of extinction if the global mean temperature exceeds its pre-industrial level by more than 2–3°C (IPCC, 2007b). The greatest threat is faced by the tropical regions which harbour the highest biological diversity but which can also expect the highest rates of loss (Colwell et al., 2008; Deutsch et al., 2008).

For this reason abatement strategies are needed to counteract the increase in greenhouse gases in the atmosphere. Since increased use of bioenergy can help to mitigate climate change (Sections 2, 5.5 and 7), it can also indirectly stem the loss of biological diversity. The benefits and disadvantages must therefore be weighed against each other: the more climate-friendly the cultivation and deployment of energy crops becomes, the more acceptance of the associated loss of biological diversity is justified. However, the effect on the greenhouse gas balance of many of today's bioenergy cultivation systems is poor or even harmful, particularly if such cultivation is accompanied by direct or indirect land-use changes (Chapter 9). The conversion of rainforest, moorland, savannah or grassland into bioenergy plantations usually results in significant greenhouse gas emissions for which even the long-term deployment of bioenergy can scarcely compensate (Sections 5.5, 7 and 9; Fargione et al., 2008).

Synergies therefore arise: the conservation of ecosystems with large natural carbon reservoirs in the vegetation and soil makes sense from the point of view of both climate change mitigation and nature conservation. Consideration of greenhouse gas balances leads to promotion of bioenergy cultivation systems that also have advantages in terms of biodiversity (use of perennial and biodiverse cultivation systems, long rotation times, accumulation of carbon in the soil, etc.: Section 7.1). It should be borne in mind that as climate change advances, energy crop cultivation may become less profitable as a result of changes in productivity.

5.4.5
Conclusions

The cultivation of energy crops can have both direct and indirect negative impacts on biological diversity. However, synergies are also possible, for example through the use of residues arising from landscape maintenance. WBGU's conclusions are as follows:

- *Expansion and management of the protected area system.* The existing global system of protected areas should be secured and extended (Section 10.5). Conversion of land within protected areas for the cultivation of energy crops should

be rejected. The gaps in representativity (ecosystem types, species) should be scientifically identified, as should candidates for new protected areas. The process should include consideration of the interlinking of protected areas and their integration into the surrounding landscape (buffer zones, areas with differing balances between conservation and sustainable use). Ensuring improved management of existing and new protected areas presents a particular challenge. In certain areas the sustainable use of biomass arising from landscape preservation may be compatible with the conservation purpose. Implementation of the internationally agreed objectives of protected areas and financing issues are discussed in Section 10.5.

- *Restrict the conversion of semi-natural and natural ecosystems for bioenergy as far as possible.* Both for the conservation of biological diversity and for mitigation of climate change it is extremely important to as far as possible limit the conversion of natural ecosystems – both within and outside a comprehensive system of protected areas – as a direct or indirect consequence of energy crop cultivation. In particular, primary forests, wetlands and savannahs, diversity-rich virgin grasslands and other natural ecosystems of high conservation value should under no circumstances be converted. Large-scale use of marginal land should not take place until the nature conservation value of the land has been assessed.

- *When growing energy crops, consider the conservation of biological diversity.* Internationally recognized environmental standards for the cultivation of energy crops should be developed and should include the conservation of biological diversity among the dimensions addressed (Section 10.5). To reduce the negative effects of intensification, international guidelines or standards for sustainable land use, going beyond the cultivation of energy crops, should also be drawn up.

5.5
Land-use options for climate change mitigation

The cultivation of energy crops is only one of a number of options for using land in ways that help to mitigate climate change. As has been explained above, the cultivation and deployment of bioenergy is controversial and only under certain conditions does it constitute an appropriate strategy for reducing greenhouse gas emissions. In addition, the cultivation of energy crops sometimes competes with other forms of land use that themselves help to mitigate climate change and may in some circumstances be better suited to this purpose. These other forms of land use involve meas-

Box 5.5-1

Land requirement of solar energy and photosynthesis compared

For around 3000 million years some life forms on Earth have had the ability to use the sun's energy to synthesize organic compounds. This process of photosynthesis is the basis for the conversion of solar energy into plant biomass and hence ultimately also the basis of bioenergy production. Nevertheless, in terms of its land requirement photosynthesis is not the most efficient method of using solar energy. Considered over the whole vegetation period and the entire plant stock, even the most productive plant communities are able to utilize on average no more than two per cent of the solar energy to which they are exposed (Kaltschmitt and Hartmann, 2003). The reasons for this are twofold. Firstly, a significant amount of solar radiation lies outside the wavelength range that can be used for photosynthesis and, secondly, unavoidable energy losses arise from reflection, absorption, plant respiration and the conversion of light into chemical energy.

A comparison of the land requirement for bioenergy use with that of the direct technical use of solar energy illustrates the differing efficiencies. The comparison is based on the world's current primary energy requirement of around 500 EJ per year (Chapter 6).

If one assumes typical yields for the cultivation of energy crops of around 10 tonnes of dry mass per hectare per year (Doornbosch and Steenblik, 2007) and a biomass energy content of around 19 kJ per gram of dry mass (Section 6.3), to meet the current global need for primary energy entirely from bioenergy would require 2500 million hectares of land. For comparison: the global area of land identified in Chapter 6 as being available for the sustainable cultivation of energy crops amounts to around 400 million hectares.

For the technical use of solar energy the equation works out as follows: the mean solar irradiance at the edge of Earth's atmosphere is 1367W per m² (Bishop and Rossow, 1991). However, on account of the diurnal variation of the sun, the seasons and processes of reflection and absorption in the atmosphere, only some of this reaches the ground. Mean annual solar irradiance is also dependent on geographical latitude and on local climate. In desert regions, though, a mean of around 250W per m² can be assumed (Bishop and Rossow, 1991). If one further assumes an efficiency of 15 per cent for the conversion of the irradiated solar energy into usable final energy (which roughly corresponds to the efficiency of a modern photovoltaic system), the world's current primary energy need could be generated through the technical use of solar energy on an area of only 40 million hectares. This is equivalent to a square with sides just over 600 km in length and is one-sixtieth of the land needed to produce the same quantity of bioenergy.

In the long term, therefore, the direct technical use of solar energy is clearly superior to bioenergy on account of its lesser land requirement. In addition, solar energy can make use of land that is not in competition with food production or nature conservation (e.g. deserts or built-up areas).

---

ures in the forestry and agricultural sector including the feedstock use of forestry and agricultural products. The following sections describe the potential of these land-use options for climate change mitigation, identify land-use competitions and synergies and discuss their effectiveness and efficiency in comparison with the deployment of bioenergy. In addition, Box 5.5-1 compares the land needed for the deployment of bioenergy from energy crops with the land needed for the deployment of solar energy.

5.5.1
Forests and climate change mitigation

5.5.1.1
Avoiding deforestation and forest degradation

A currently intensely debated form of climate change mitigation is the reduction of tropical deforestation. In 2004 around 17 per cent of global emissions of long-lived greenhouse gases arose from deforestation and the decomposition of biomass (Rogner et al., 2007). These emissions originate predominantly in the tropics. Bioenergy deployment, that presupposes the cultivation of energy crops, can through various mecha-

nisms reduce the likely success of strategies for preventing tropical deforestation.

Clearly the cultivation of energy crops can lead directly to deforestation if forests are converted for this purpose. In addition, deforestation can be accelerated by indirect effects.

Figure 5.5-1 shows emissions as a result of deforestation for the year 2000 as calculated by Houghton (2003). According to this source, Brazil and Indonesia alone were responsible for more than 50 per cent of emissions from deforestation. The available

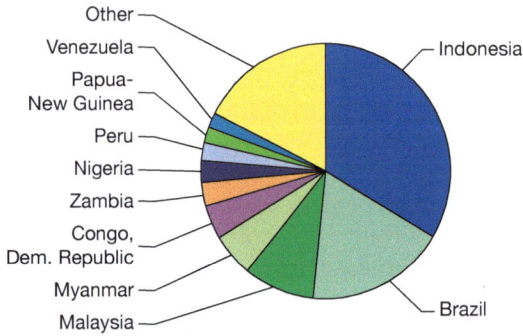

Figure 5.5-1
Global emissions from deforestation, by country. The emissions depicted here amount in total to 2 Gt C.
Source: based on Houghton, 2003 cited in
Schulze et al., 2007

sets of data on emissions from deforestation differ in their description both of regional distribution and of total quantity and are not consistent. Nevertheless it remains clear that Indonesia, Brazil, Malaysia, Congo, Myanmar and Venezuela have particularly high levels of emissions from deforestation (Schulze et al., 2007). At the same time an expansion of bioenergy production is to be expected in some of these countries, e.g. Brazil, Indonesia and Malaysia (UNCTAD, 2006b). It is therefore likely that considerable effort will be required to reduce deforestation rates in these countries, since the increasing cultivation of energy crops will put increased pressure on land use.

Geist and Lambin (2002) investigated the reasons for deforestation on the basis of 152 case studies; they distinguish between direct causes and underlying influences. The findings show that deforestation is usually attributable to a number of direct causes, but the expansion of agricultural land played a part in almost all the cases they studied. The underlying influences, too, seldom operated in isolation. In one-third of the cases studied rising product prices for cash crops were an influential factor. It can be assumed that increasing demand for bioenergy will result in price rises for agricultural products (Section 5.2). Using the Mato Grosso region of the Amazon as an example, recent studies by Morton et al. (2006) show that deforestation in favour of arable land has increased by comparison with deforestation in favour of pasture, with the extent of deforestation correlating with the price of soya. Searchinger et al. (2008) show in their calculations that price rises triggered by bioenergy use promote the conversion of forest and grassland into arable land and do so irrespective of whether other unused arable land is available. Typical quantities of CO_2 that are released in the process amount to around 604–1146 tonnes of CO_2 per hectare of forest that is converted to farmland, and 75 305 tonnes of CO_2 per hectare when grassland or savannah is converted (Searchinger et al., 2008). It is very possible that global expansion of bioenergy production will significantly impede attempts to stem tropical deforestation, even if the direct conversion of tropical forests into land for the cultivation of energy crops can be avoided.

The costs of avoiding deforestation vary from region to region and depend on the alternative uses that are foregone. The growing demand for bioenergy pushes agricultural prices upwards. Taking account only of compensation for opportunity costs – that is, compensation for foregone proceeds from the land use that would lead to deforestation – Grieg-Gran (2006) estimates that it would cost US$ 5000 million to prevent emissions from deforestation in several important newly industrializing and developing countries which between them account for half of all

these global emissions. This corresponds to US$ 483–1050 per hectare to which administrative costs of between US$ 4 and US$ 15 per hectare must be added. Each avoided tonne of CO_2 gives rise to average costs of US$ 1–2. The costs are higher if the calculation is adjusted to include the opportunity costs of macroeconomic (growth) effects that would arise in the event of deforestation and intensive land use, and direct costs that arise in practice as a result of deviations from model conservation policy (Nabuurs et al., 2007). Various other studies have arrived at considerably higher figures. Some authors put the cost of halving deforestation at between US$ 20,000 million and US$ 33,000 million (Stern, 2008; Strassburg et al., 2008; UNFCCC, 2007b). Halting the process of deforestation worldwide would cost US$ 185,000 million (UNFCCC, 2007b).

It is doubtful whether the conservation of tropical primary forests can be combined with relevant use of these forests for bioenergy or for material feedstocks, since the ecosystem is highly sensitive to disturbance; in addition, even small-scale incursions, such as for the construction of a road, result within a few years in deforestation (Section 7.1.5.1).

Policies for reducing emissions from deforestation and policies for expanding the use of bioenergy are therefore in competition with each other. However, it is clear for climate protection reasons alone that deforestation for the purpose of expanding the amount of land cultivated for energy crops is not beneficial. In addition, the conservation of tropical primary forests has numerous other positive effects, for example on the conservation of biodiversity (Section 5.4). In WBGU's view the conservation of tropical forests should therefore always take precedence over the expansion of bioenergy use. By the same token it is important to ensure that the cultivation of energy crops does not contribute directly or indirectly to tropical deforestation (Chapter 9).

5.5.1.2
Afforestation

In the past it has been rare for climate change mitigation to be the most important driver of afforestation. However, this may change as global efforts to combat climate change gain momentum, and the rate of afforestation may rise sharply (Nabuurs et al., 2007). The effect of afforestation on reducing CO_2 (carbon sink) can vary widely, depending on the tree species used, the location and other factors. The amount of carbon stored through the accumulation of biomass after afforestation varies between 1 and 35 t CO_2 per hectare per year (Richards and Stokes, 2004, cited in Nabuurs et al., 2007). However, afforestation does not always result in a significant carbon sink: the carbon content of the soil determines whether carbon is initially absorbed or emitted (Section 4.2.3). Afforestation of farmland with a low soil carbon content usually results in an increase in soil carbon. By contrast, afforestation of land with a high soil carbon content (such as grassland ecosystems, particularly on organic soils) leads initially to a decrease in soil carbon (Nabuurs et al., 2007). Moreover, the carbon absorption capacity of the terrestrial biosphere is limited. According to estimates of House et al. (2002), even the extreme scenario (global re-afforestation, i.e. reversal of all land-use changes that have taken place historically before the year 2000) would only achieve a reduction of 40–70 ppm CO_2 in the atmosphere. This can be compared with the fact that the atmospheric concentration is already approximately 100 ppm above the pre-industrial level.

A comparison of the climate change mitigation effect of afforestation and the use of land for energy crops must therefore take account of different locations and tree species as well as of different bioenergy deployment paths. Righelato and Spracklen (2007) compare the greenhouse gas savings arising from the cultivation and use of typical first-generation biofuels in the transport sector with the storage effect of afforestation on the same land. As examples they consider the use of sugar cane, wheat, sugar beet and maize for ethanol and rape and woody biomass for diesel. The values for avoided emissions used by Righelato in this study are between 0.8 and 7.2 tonnes of CO_2 for ethanol and up to 8.1 tonnes of CO_2 for diesel from woody biomass, in each case per hectare per year. The types of afforestation considered (without annual harvesting) are the natural conversion of neglected tropical farmland (CO_2 storage: 15–29 t CO_2 per hectare per year) and the afforestation of temperate farmland with pines (11.7 t CO_2 per hectare per year). In these examples afforestation enabled storage over a 30-year period of between two and nine times as much CO_2 as was avoided through biofuel use. However, the calculations do not include possible emissions from changes in land use for the cultivation of energy crops, which would tend to make bioenergy an even more unattractive option in relation to afforestation (Righelato and Spracklen, 2007). It is open to debate whether consideration of longer time periods would shift the balance more towards biofuel use. However, new studies (Luyssaert et al., 2008) show that – contrary to present opinion – even very old forests (200–800 years) can still store considerable quantities of CO_2 each year (2.4 ± 0.8 t C per hectare per year, i.e., 8.8 ± 2.9 t CO_2 per hectare per year). This means that even over relatively long periods (more than 30 years) no shift in the balance is to be expected.

Afforestation is not to be equated with the planting of short-rotation plantations (SRPs). SRPs involve the use of fast-growing tree species with high storage potential, but regular harvesting can result in the release of considerable quantities of soil carbon, and in addition emissions arise from fertilizer use (Section 4.2). The higher the existing carbon stock in the soil, the greater the risk that this stock will be released by frequent harvesting and the associated tillage or disturbance of the soil; this risk is particularly high in the first years or decades of SRP use. For the conservation of soil carbon, afforestation for permanent forestry use is therefore clearly preferable to SRP use. However, whether the overall climate change mitigation effect of SRPs is lower or higher than that of afforestation also depends on the deployment of the SRPs in the energy system and the fossil fuels that they replace (Section 7.3). In weighing up the climate change mitigation effects of different options (e.g. annual energy crops, SRPs, afforestation) the different dynamics of the various measures over time must always be considered (Section 5.5.4).

The likely costs of afforestation or reforestation vary considerably depending on the region and the existing land use. Various studies have identified abatement costs of US$ 22 per tonne of CO_2 and less (Nabuurs et al., 2007; Benitez et al., 2005, cited in Stern, 2006). As Kooten et al. (2004) show, the costs may be significantly higher if forestry management involves further protective measures. On the basis of various CO_2 prices, estimates of the global economic potential for afforestation by 2100 range from 0.57 to 4.03 Gt CO_2 per year, which would require up to 231 million hectares of land (Canadell and Raupach, 2008).

Large-scale afforestation can have negative ecological and socio-economic effects; it may reduce food security, reduce water run-off, or result in loss of biodiversity or of income (Canadell and Raupach, 2008). Forest plantings usually need more water than grassland or arable land and can thus have a seriously detrimental effect on local water levels. Depending on the location and type of planting, however, positive effects can also be achieved (Jackson et al., 2005). As with energy crop cultivation, the sustainability of afforestation measures therefore needs to be considered in a differentiated manner and from a wider perspective than that of the greenhouse gas balance alone.

5.5.1.3
Forest management, sustainable forestry

Another way of mitigating climate change through forestry-related measures is by increasing the carbon pool in existing forests through changed management techniques, such as by lengthening harvest cycles or reducing disturbance, e.g. through better protection against forest fires (particularly in sub-tropical latitudes), or by reducing soil compaction by heavy machinery (particularly in temperate and boreal latitudes). It is possible for this to include rather than compete with the use of forest products for energy or other purposes. Hence the IPCC, in considering the various climate change mitigation opportunities in the forestry sector, concludes that: 'In the long term, sustainable forest management strategy aimed at maintaining or increasing forest carbon stocks, while producing an annual yield of timber, fibre, or energy from the forest, will generate the largest sustained mitigation benefit' (Nabuurs et al, 2007; IPCC, 2007c).

Particularly in the tropics, however, implementation of such an approach is beset by significant problems. For example, at present only seven per cent of traded tropical timber comes from sustainable forestry (Canadell and Raupach, 2008). According to the International Tropical Timber Organization (ITTO), one of the major obstacles is the fact that other forms of land use often generate higher profits than sustainable forestry management (ITTO, 2006). Sustainable extraction of biomass from tropical primary forests is invariably difficult (Section 7.1.5.1).

In temperate forests, by contrast, sustainable management – which includes the extraction of biomass – is possible (Section 7.1.5.2). However, the regeneration of forests can lead to a net release of carbon which takes many centuries to compensate (Harmon et al., 1990; Luyssaert et al., 2008). Sustainable use of boreal forests is theoretically possible but is not at present common practice.

5.5.2
Agriculture and climate change mitigation

Agriculture directly contributes 10–12 per cent of global greenhouse gas emissions (IPCC, 2007c). Only a very small proportion of these are CO_2 emissions: while very large CO_2 fluxes pass from farmland into the atmosphere, they are counterbalanced by the binding of CO_2 through photosynthesis. The net emissions into the atmosphere are estimated at less than one per cent of global anthropogenic CO_2 emissions (IPCC, 2007c). The non-CO_2 emissions from agriculture arise in the following areas:

- 38 per cent N_2O from the soil (especially after fertilizer application),
- 32 per cent CH_4 from the digestive processes of ruminants (e.g. cattle, sheep, etc.),
- 12 per cent from the combustion of biomass,
- 11 per cent from paddy cultivation,
- 7 per cent from dung.

In the majority of regions N_2O from the soil is the main source of these greenhouse gas emissions. This is predominantly attributable to the use of nitrogen fertilizers in the production of food and feed crops. Only in the countries of the former Soviet Union, the Pacific OECD countries, Latin America and the Caribbean is methane from ruminant digestion the main source; this is due in particular to the relatively large number of animals in these regions (Smith et al., 2007b).

Direct agricultural emissions rose between 1990 and 2005 by 14 per cent; N_2O emissions from the soil increased disproportionately by 21 per cent during this period (Smith et al., 2007b). It is expected that N_2O will rise further due to increased fertilizer use and increased animal manure production. Agricultural N_2O emissions are projected to increase by 35–60 per cent by 2020 and 2030 respectively. Livestock-related CH_4 emissions are expected to rise by between 20 and 60 per cent in the same period; by contrast, CH_4 emissions from paddy cultivation are expected to rise by only a few per cent or may even show a reduction (IPCC, 2007c).

Agriculture can contribute to climate change mitigation through reductions in CO_2, N_2O and CH_4 emissions, through an increase in carbon storage in the soil or biomass, and through the use of agricultural products and residues to generate bioenergy (IPCC, 2007c).

Many factors will continue to make it necessary to increase the productivity of existing agricultural land (Section 5.2). This will require increased use of fertilizer and an expansion of irrigation, and will thus entail greater energy inputs. These measures could result in increased greenhouse gas emissions (Smith et al., 2007b). If energy crop cultivation is significantly expanded, it is likely that the increased competition for land will further strengthen the need to increase the productivity of existing farmland (Section 5.2). The extent to which this would lead to a further increase in GHG emissions from agricultural land cannot be calculated by a simple formula. It is, however, clear that the level of these emissions increases – caused in part by energy crop cultivation – will be determined by a large number of individual decisions on management options (such as efficiency improvements in fertilizer use). The use of land for the cultivation of energy crops thus tends to counteract efforts to reduce emissions on existing agricultural land.

According to the IPCC, one of the most effective methods of reducing emissions in agriculture is the conversion of cropland into land with semi-natural vegetation (Smith et al., 2007a). However, this should be carefully evaluated from the point of view of food security in the same way as the use of arable land for the cultivation of energy crops (Section 5.2). The mechanism for mitigating climate change operates here not through the avoidance of emissions but through the increase in carbon storage in the soil or biomass. The reversion of cropland to semi-natural vegetation inevitably competes with the option of using it for the production of energy crops. It is, however, possible for the production of biomass for energy to be combined with increased carbon storage in the soil if suitable cultivation systems are employed (Sections 7.1 and 4.2). Similar arguments apply to the restoration of drained and degraded land. If the extension of biomass cultivation increases the pressure on available land, this can result in marginal land being brought back into use. This can increase the risk of erosion and further degradation. The various consequences in the form of possible CO_2 emissions are, however, uncertain (Smith et al., 2007b). Alternatively, though, the use of marginal land for bioen-

ergy land can contribute to the restoration of that land and hence to climate change mitigation, provided that appropriate cultivation and management systems are used (Sections 4.2 and 5.6).

5.5.3
Climate change mitigation through the use of long-lived biomass products

Biomass is a carbon reservoir: the carbon is absorbed from the CO_2 in the atmosphere through photosynthesis as the crop grows. For example, the carbon bound in a cubic metre of wood corresponds to around 0.92 t CO_2 (Nabuurs et al., 2007). Depending on the region, the terrestrial biosphere fixes 0–55 t CO_2 (0–15 t C) per hectare per year through photosynthesis; this is termed the net primary production (Figure 5.5-2). On a worldwide scale this means that around 217 Gt CO_2 (corresponding to 59 Gt C) per year is fixed in the biomass through photosynthesis; around 14 per cent (30 Gt CO_2 or 8 Gt C) is then removed through harvesting by humans (Haberl et al., 2007). This carbon fixation through photosynthesis is, however, counterbalanced by CO_2 emissions at almost the same rate from the decomposition of biomass; in consequence, net CO_2 storage by the terrestrial biosphere amounts

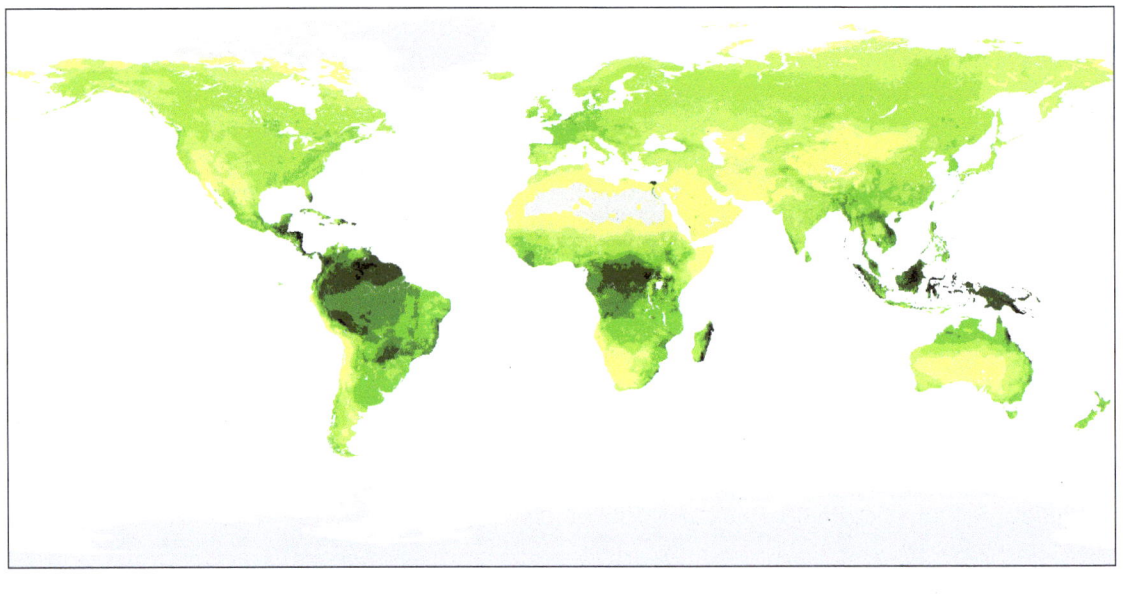

Net primary production [gC/m² and year]

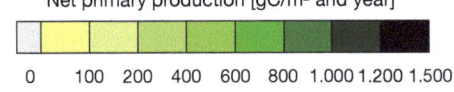

0   100   200   400   600   800 1.000 1.200 1.500

Figure 5.5-2
Present net primary production in different parts of the world. 100 gC per m² per year corresponds to 1 t C per hectare per year or 3.7 t CO_2 per hectare per year.
Source: based on Haberl et al., 2007

to only around 3.7 Gt CO_2 (corresponding to 1 Gt C), although there are major fluctuations from year to year (WBGU, 2003). Emissions from the use of fossil fuels and the cement industry amounted in 2006 to 8.4 Gt C (Canadell et al., 2007). Comparison of these figures shows that a change in the average life of harvested products is undoubtedly able to influence the carbon cycle and hence the CO_2 concentration in the atmosphere. When biomass is used for energy generation, the stored carbon is usually released relatively quickly and directly into the atmosphere; in this case, therefore, the desired climate change mitigation effect arises from the substitution of higher-emission fuels (Chapter 7). By contrast, when biomass is used as an industrial feedstock, the carbon initially remains fixated. This is in particular the case for timber products and long-lived bioplastics. A special case of long-term carbon fixation in biomass is black carbon sequestration (Box 5.5-2).

Pingoud (Pingoud, 2003, cited in UNFCCC, 2003) calculates that between 1960 and 2000 the global quantity of carbon stored in wood products rose by an average of 0.04 Gt C (corresponding to 0.15 Gt CO_2) per year, rising in this period from 1.5 Gt C (5.5 Gt CO_2) to more than 3 Gt C (11 Gt CO_2). However, this corresponds to less than one per cent of cumulative anthropogenic CO_2 emissions, which totalled approximately 264 Gt C in the same period (WRI, 2008).

As well as temporarily storing carbon directly, biomass products can also replace emissions-intensive materials such as concrete, steel, aluminium and plastics. According to Nabuurs et al. (2007), wood can achieve a particularly significant climate change mitigation effect if it is first used to replace concrete as a building material and later, after disposal, is used as biofuel. A study by Reinhardt et al. (2007) explores the effects of biomass use in the chemical industry and comes to the conclusion that replacing fossil raw materials with biomass can, in terms of the area of cropland used, avoid a quantity of greenhouse gas emissions similar to that avoided when biomass is used for energy in the transport sector.

The discussion shows that the feedstock use of biomass in manufacturing or construction has worthwhile potential for helping to mitigate climate change, and that such use does not always compete with the use of biomass for energy since the two can sometimes be combined in cascade use.

5.5.4
Conclusions

The use of biomass for energy as a climate change mitigation option is often in direct or indirect competition with other climate change mitigation options. Direct competition arises, firstly, in relation to land use. The land can either be used for the production of energy crops that replace fossil fuels, or the aim can be to maintain or increase the carbon stored on the land (e.g. by forgoing deforestation or grassland conversion for the purpose of energy crop cultivation). Secondly, there is direct competition in terms of the way the biomass is used. Its property as a carbon reservoir can be utilized by deploying the biomass as a raw material or protecting it in other ways from oxidation and decomposition (Box 5.5-2). Alternatively it can be used as energy, which releases the carbon stored in the biomass but can replace other emissions-intensive forms of energy.

In addition to direct competition, there are also indirect effects that operate via agricultural prices.

Box 5.5-2

Black carbon sequestration as a climate change mitigation option

For some time the possible contribution of black carbon sequestration to climate change mitigation has been under discussion (e.g. Marris, 2006; Lehmann, 2007). The starting point is the process of photosynthesis, in which CO_2 is absorbed from the atmosphere and stored as carbon in biomass. For biosequestration biomass is first heated in an oxygen-starved process (low-temperature pyrolysis), which produces both charcoal and other volatile substances. The volatile substances can be used as biogas or, after processing, as liquid fuel to produce energy, while the charcoal can be incorporated into agricultural soils as a means of carbon sequestration. In the ground it has a relatively long life: the exact length is the subject of research, but estimates range from centuries to millennia (Lehmann, 2007). In addition to providing long-term storage in the soil of the carbon

absorbed by the plants from the atmosphere, the charcoal also improves the structure and fertility of the soil. This effect is known from the very fertile *terra preta* soils of the Amazon basin (Denevan and Woods, 2004; Fowles, 2007). If the charcoal were to be deposited deep underground even longer storage times could be achieved, although without the positive influence on soil fertility.

If the resulting charcoal were to be used itself for energy, it could directly replace fossil fuels. However, Lehmann (2007) calculates that incorporating the charcoal into the soil leads to a reduction in emissions that is 12–84 per cent greater than when the charcoal is used for energy. The author further estimates that black carbon sequestration combined with the use for energy of the waste gases arising from pyrolysis could be worthwhile at a CO_2 price of US$ 37 or more. Evaluation of this climate change mitigation option must, however, take account of possible competing goals arising from the increasing need for energy and organic raw materials.

As expansion of energy crop cultivation increases the pressure on farmland, other climate change mitigation measures in connection with land use become more and more difficult to implement. For example, it becomes increasingly difficult to reduce deforestation or to reduce N_2O emissions arising from fertilizer use.

If bioenergy use is to play a part in climate change mitigation, the merits of these different mitigation options must therefore be weighed against each other (Figure 5.5-3).

Comprehensive greenhouse gas balances of different measures can provide guidance here (Chapter 7). However, in comparing different climate change mitigation options and evaluating how they can best contribute to adherence to the climate protection guard rail it is not enough merely to consider direct emissions and emissions reductions; the dynamics of each option over time must also be taken into account. Table 5.5-1 summarizes the time dynamics of different climate change mitigation options relating to land use. In WBGU's view, avoidance of a dangerous degree of climate change means that global temperatures must not rise by more than 2°C from their pre-industrial level. Two things are necessary if this is to be achieved. Firstly, the trend of global emissions must be reversed as quickly as possible and, secondly, the basis must be laid for further long-term, continuous and substantial emissions reductions extending to the middle of the century and beyond.

In addition to climate change mitigation options relating to land use, reducing emissions from fossil fuels is also crucial. Bioenergy is equally relevant in both areas. Unlike emissions from fossil fuels, CO_2 emissions from land use and land-use changes are

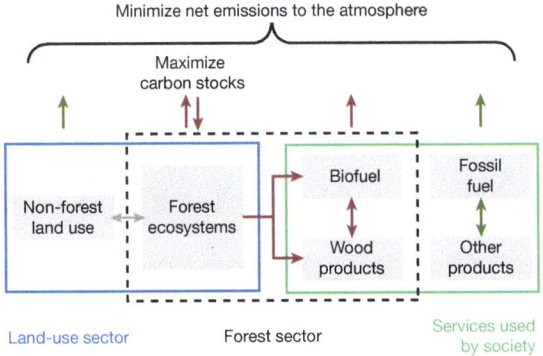

Figure 5.5-3
Climate change mitigation through appropriate land use: weighing up the options, taking the forestry sector as an example.
Source: Nabuurs et al., 2007

generally reversible in the sense that a corresponding quantity of CO_2 can in theory (if managed appropriately) be fully reabsorbed by the biosphere within a reasonable period of time (a few years or decades). At the same time carbon sequestration by ecosystems in soils or biomass is usually limited by management practices, and the annual increase in sequestration reduces over time (Smith et al., 2007a), although even very old ecosystems can still store considerable quantities of CO_2 (Luyssaert et al., 2008). Moreover, changes in management can often result in large quantities of carbon stored in biomass and in the soil being released, for example when the ecosystem becomes degraded or different farming techniques are employed.

By contrast, 20 per cent – a significant proportion – of CO_2 emissions from the combustion of fossil

Table 5.5-1
Time dynamics of climate change mitigation options in land use.
Source: adapted from Nabuurs et al., 2007

	Mitigation mechanism	Impact	Timing of impact	Timing of cost
A	Enhance site-level carbon pool: afforestation, management, etc.	↑	⌐	⌐
B	Maintain site-level carbon pool: avoid deforestation, degradation, etc.	↓	⌐	⌐
C	Enhance carbon pool in biomass products	↑	―	⌐
D	Bioenergy and substitution (if B is met)	↓	―	⌐

Impact of cost	Timing of impact	Timing
↑ Enhancement of sink	⌐ delayed	⌐ delayed
↓ Reduction of sources	⌐ immediate	⌐ up-front
	― sustained or repeated	⌐ on-going

fuels remain in the atmosphere for centuries (IPCC, 2007a). Montenegro et al. (2007) consider that as much as 25 per cent of these emissions may remain in the atmosphere for more than 5000 years. Figure 5.5-4 shows schematically the carbon reservoirs and fluxes that determine the CO_2 concentration in the atmosphere, and highlights the differences between them.

House et al. (2002) conclude from their calculations that even the use of extreme and unlikely land-use changes during the 21st century in an attempt to reduce CO_2 increases in the atmosphere would have only a small effect in comparison to the effects of various emissions scenarios for the use of fossil fuels.

It is therefore evident that efforts to reduce emissions from the use of fossil fuels must form the core of any serious climate change mitigation policy. From a climate change mitigation perspective, land-use measures cannot adequately replace the reduction of emissions from fossil fuels. Nevertheless, land-use measures must also be accorded high priority, particularly where there are synergies with other sustainability objectives such as the conservation of

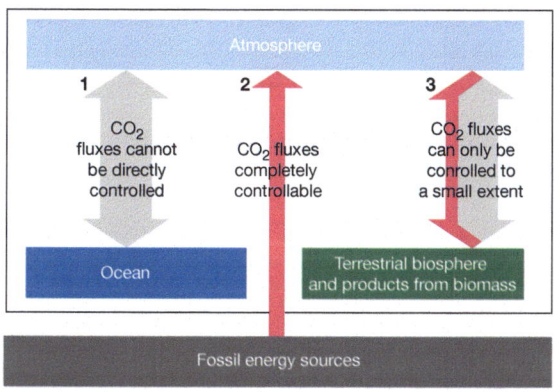

Figure 5.5-4
The global carbon cycle. The atmospheric concentration of CO_2 is determined primarily by
1: CO_2 fluxes between the ocean and the atmosphere: these are large natural fluxes that cannot be directly controlled by humans using presently available technology. However, humans influence them indirectly through the atmospheric concentration of CO_2. They act as a buffer.
2: CO_2 emissions from the use of fossil fuels: these fluxes are entirely human-induced and thus completely controllable. With present technology they are largely irreversible (technical CO_2 sequestration not yet feasible on a large scale). The volume of these fluxes can be very accurately measured.
3: CO_2 fluxes between the terrestrial biosphere (incl. biomass products) and the atmosphere: these are large, predominantly natural fluxes that humans can influence to only a small extent (mainly through land-use changes). By altering land use the distribution of carbon can within limits be transferred between the atmosphere and the terrestrial biosphere. These measures are largely reversible and the precision of flux measurement is relatively low.
Source: WBGU

biodiversity. These considerations show that even a life-cycle analysis of the greenhouse balances arising from the use of bioenergy is only of limited usefulness, particularly with regard to the contribution made by bioenergy to adherence to WBGU's 2°C guard rail. The life-cycle analysis provides a 'snapshot' that takes no account of the differing characteristics of the emissions or emissions reductions associated with land use or other factors, particularly with regard to the long-term dynamics of these emissions/ reductions over time or to their reversibility. The life-cycle analysis can therefore be only one element of the evaluation and comparison of different climate change mitigation measures as part of a comprehensive climate change mitigation strategy.

5.6
Competing use of soil and water

5.6.1
Soil degradation and desertification

While the increased cultivation of energy crops can increase the risk of soil degradation, it can also contribute to the restoration of degraded land. Whether the cultivation of energy crops is acceptable and eligible for financial support or whether it should be resisted depends ultimately on the cultivation system (Section 7.1) and on regional agro-ecological conditions. Where unadapted cultivation systems are used (e.g. inappropriate tillage or irrigation), particularly on marginal land, the risks of further soil degradation can be considerable. Research is needed to clarify which cultivation systems can reduce the risks and how much potential for soil restoration exists (Section 11.4).

Land all over the world is affected by soil degradation. Particularly at risk are the world's arid regions, which constitute 40 per cent of the terrestrial surface. More than 250 million people in these regions are directly affected; a further 1000 million live in at-risk areas. The majority of developing countries are situated in Earth's arid zones; the 50 poorest countries are therefore also those most affected by desertification, i.e. soil degradation in arid regions (UNCCD, 2008). In Africa 65 per cent of cropland, 31 per cent of pasture and 19 per cent of forests is degraded. But the situation also affects Latin America (45 per cent of cropland, 14 per cent of pasture, 13 per cent of forests) and Asia (38 per cent of cropland, 20 per cent of pasture, 27 per cent of forests), especially China (FAO, 1990 cited in WBGU, 1995a; MA 2005e). Four levels of soil degradation are distinguished (Oldeman et al., 1991): 'Light degradation' means that the terrain has somewhat reduced agricultural suitability, but restoration of full productivity is possible. In the case of 'moderate' degradation the terrain has greatly reduced agricultural productivity and major improvements are required before the soils can be used fully and productively. 'Strong' degradation means that the terrain has completely lost its productive capacity and is no longer usable for farming. Much investment and hard work over a long period is required for the land to be restored. 'Extreme' degradation means that the terrain cannot be either cultivated or restored.

WBGU's estimate of land with potential for the cultivation of energy crops (Chapter 6) disregards land in the categories of 'strong' and 'extreme' degradation. However, there remains 84 per cent of the land categorized as exhibiting 'light' to 'moderate' degradation; the productivity capacity of this land for food is greatly reduced and it is therefore not in direct competition with food production. With correct selection and suitable management the cultivation of energy crops on such terrain can even open up new opportunities if the planting serves to prevent further degradation. This can in the long term increase the amount of organic carbon in the soil and thus improve the soil quality of this marginal land. It must, however, be borne in mind that marginal or degraded soils (Box 4.2-1) are usually more vulnerable to (further) soil degradation than highly productive soils. In WBGU's view, therefore, energy crops should not be grown on marginal land until a soil protection strategy has been drawn up which specifies how cultivation and management systems are to be adapted to local conditions. Priority should be given to perennial plants, preferably grown in a mixed culture, that require little tillage and whose root biomass remains in the soil (Section 7.1).

In the long term the sustainable cultivation of suitable energy crops on marginal land opens up the following strategic option. WBGU assumes that it is only during the forthcoming transition period that the cultivation of energy crops will play an important role in the global energy mix (Section 9.2.3); the possibility therefore arises that through the cultivation of energy crops some land will be restored to the point where it can later be used if required for food production or feedstocks. This will help to ease the increasing pressure on land use.

5.6.2
Overuse of freshwater resources

The cultivation of energy crops can, however, increase competition not only for the use of land but also for available freshwater resources. Particularly at risk are regions that already suffer from water scarcity, such as Central Asia, parts of South Asia and North Africa and sub-Saharan Africa, where population growth and neglect of the water sector lead to shortages. The cultivation of energy crops can heighten competition for freshwater but it can also help improve efficiency of water use (Berndes, 2008; Lundqvist et al., 2008). Humans already use or regulate more than 40 per cent of renewable, accessible freshwater resources (MA, 2005d). The pressure on global freshwater resources is increasing at the rate of about 10 per cent per decade (for all freshwater use), particularly as a result of increasing affluence (rising per-capita water consumption) and population growth (effects include increasing need for water for irrigated agriculture). There are now 1200 million people living in regions affected by water scarcity. Water withdrawal will need to increase by about 20 per cent by 2050 simply to cover the increased need for food (de Fraiture et al., 2007). The cultivation of energy crops further increases the pressure of use on regional freshwater supplies (McCornick et al., 2008) and can contribute to overuse in a region – that is, to a situation in which water withdrawal exceeds the natural renewal rate (Figure 5.6-1). According to an estimate of the International Water Management Institute (IWMI, 2007), particularly extensive expansion of bioenergy could cause water use by energy crops (measured as agricultural evapotranspiration) to almost double by 2050. Lundqvist, too, assumes that water use will nearly double, but does not attribute this to energy crops alone. According to his study, energy crops will be responsible for only around 20–40 per cent of the increased need for water (Lundqvist et al., 2008).

According to Berndes (2008), the factors influencing the effect of energy crops on the hydrological cycle include:

- the location (and water catchment area) in which the crops are grown, and the agro-ecological situation of that location, particularly with regard to the available freshwater resource;
- the particular energy crop grown; energy crops vary widely in their efficiency of water use and hence in their need for water;
- the type of vegetation that the energy crops replace; the net change in water availability may be positive or negative. Areas with sparse vegetation may experience improved water availability as a result of energy crop cultivation (e.g. through reduced surface run-off and better infiltration), while clearance of dense forest for the purpose of cultivating crops such as soya or maize reduces water availability (e.g. through shorter ground cover periods and hence increased surface run-off).

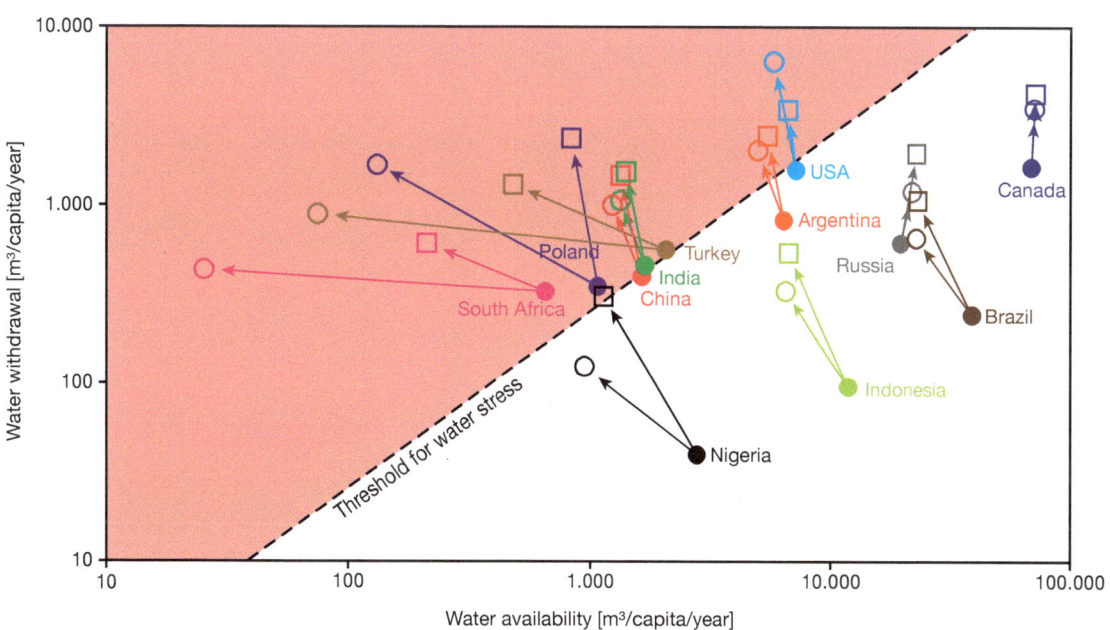

Figure 5.6-1
Development of per-capita water withdrawal and availability in a model of the impact of energy crop cultivation in selected countries to 2075. The solid circles represent the initial situation in 1995. The arrows indicate the shift to two scenarios which – besides impacts of increased food crop production and climate change – includes the impacts of substantial energy crops cultivation, either in exclusively rain-fed systems (open circles) or with partial irrigation (squares). Water scarcity is defined on the basis of the water stress indicator of Raskin et al. (1995), which describes a ratio between consumption and the available resource of less than 25% as the threshold of water stress (red area). In countries in which water resources are scarce both energy crop cultivation scenarios exacerbate the water situation.
Source: Berndes, 2008

Cultivated crops already evapotranspire around 7000 km³ of freshwater globally per year (including evaporation); in irrigated cultures this water comes from rivers, lakes or aquifers ('blue water'). Energy crops currently use an additional 100 km³ (or approximately 1 per cent; de Fraiture et al., 2007). Production of one litre of biofuel from energy crops requires on average around 820 litres of irrigation water and involves 2500 litres of water being evapotranspired. However, these global averages are difficult to interpret because there are considerable regional differences (Table 5.6-1). In Europe, where the chief crops are rape and maize grown under rain-fed conditions, very little water is used for irrigation. In the USA, where maize is usually grown under rain-fed conditions, three per cent of irrigation water is used for energy crop cultivation; this amounts to 400 litres of irrigation water per litre of bioethanol. In Brazil sugar cane is the most important energy crop; it is grown predominantly under rain-fed conditions, so that only very little irrigation water is used for energy crop cultivation. In China, by contrast, an average of 2400 litres of irrigation water are used to produce one litre of ethanol from maize. Overall, around two per cent of irrigation water in China is used for energy crop cultivation. In India sugar cane is grown predominantly under irrigated conditions, so that nearly

3500 litres of water are need to produce one litre of ethanol (de Fraiture et al., 2007). The use of irrigation water for energy crops is thus an important factor in the regional assessment of water competition. The water requirements of energy crops must therefore be an important consideration when deciding whether or not to grow such crops.

Berndes (2002) has modelled the influence of energy crop cultivation in selected countries to 2075, using two scenarios that distinguish between rain-fed and irrigated systems (Figure 5.6-1). For the model assumptions made, the results show that no critical developments in the water sector are expected as a result of the expansion of energy crop cultivation in Canada, Brazil, Russia, Indonesia and some countries in sub-Saharan Africa. However, in a number of countries that are already affected by water scarcity a deterioration of the position is likely, even if energy crops are grown only under rain-fed conditions. These countries include South Africa, Poland, Turkey, China and India. Finally there is a group of countries in which the cultivation of energy crops causes the critical threshold of water withdrawal (the point at which water withdrawal amounts to more than 25 per cent of the available resource) to be exceeded (USA, Argentina).

Table 5.6-1
Water use for energy crops for ethanol production in selected countries.
Source: adapted from de Fraiture et al., 2008

	Bio-ethanol [milllion t]	Main energy crop used	Raw material used [million t]	Land on which bioenergy is grown [million ha]	Proportion of farmland used for bioenergy [%]	Total eva-potranspir-ation [km³]	Proportion of evapo-transpira-tion from bioenergy [%]	Water withdrawal for bioenergy [km³]	Proportion of water withdrawal for irriga-tion for bioenergy [%]
Brazil	15,098	sugar cane	167.8	2.4	5.0	46.02	10.7	1.31	3.5
USA	12,907	maize	33.1	3.8	3.5	22.39	4.0	5.44	2.7
Canada	231	wheat	0.6	0.3	1.1	1.07	1.1	0.08	1.4
Germany	269	wheat	0.7	0.1	1.1	0.36	1.2	–	0.0
France	829	beet	11.1	0.2	1.2	0.90	1.8	–	0.0
Italy	151	wheat	0.4	0.1	1.7	0.60	1.7	–	0.0
Spain	299	wheat	0.8	0.3	2.2	1.31	2.3	–	0.0
Sweden	98	wheat	0.3	0.0	1.3	0.34	1.6	–	0.0
UK	401	sugar beet	5.3	0.1	2.4	0.44	2.5	–	0.0
China	3,649	maize	9.4	1.9	1.1	14.35	1.5	9.43	2.2
India	1,749	sugar cane	19.4	0.3	0.2	5.33	0.5	6.48	1.2
Thailand	280	sugar cane	3.1	0.0	0.3	1.39	0.8	1.55	1.9
Indonesia	167	sugar cane	1.9	0.0	0.1	0.64	0.3	0.91	1.2
South Africa	416	sugar cane	4.6	0.1	1.1	0.94	2.8	1.08	9.8
World	**36,800**			**10.0**	**0.8**	**98.0**	**1.4**	**30.6**	**2.0**

When integrating energy crops into land-use strategies the key challenge is to ensure that local effects and conflicts of objectives are considered. For example, a plantation of fast-growing trees (SRP) may not only increase water scarcity in the region through the high rate of evapotranspiration but also have a deleterious effect on the water supply of neighbouring users and adjacent ecosystems (Calder, 1999; Perrot-Maître and Davis, 2001; Berndes, 2008). On the other hand, the increased use of marginal land (such as pasture) for the sustainable cultivation of energy crops can help to avoid competition for the use of water and provides an opportunity for more efficient use of the water that is available to the plants as soil moisture ('green water'). If perennial energy crops are grown on marginal land, the increasing demand for bioenergy could encourage the spread of land-use systems that use water more efficiently (Berndes, 2008). A number of energy crops are drought-resistant and relatively water-efficient; they provide an opportunity for reducing competition for water between food and energy crops. Crops that cover the ground all year round not only use each precipitation event but also protect the soil against erosion and provide shade. For example, agroforestry sys-

tems can increase water productivity by reducing the amount of unproductive precipitation that is otherwise lost through run-off or evaporation (Ong et al., 2006). In order to assess the full impact of land-use practices on water availability, an integrated analysis of the water catchment area should therefore be carried out before energy crops are planted on a large scale; at present, however, this is rarely done (Rockström et al., 2007).

5.6.3
Conclusion: Integrate energy crop cultivation into sustainable soil and water management

Large-scale expansion of energy crop cultivation and inappropriate cultivation systems (Section 7.1) can significantly increase the pressure of use on available resources. This can result in competition between the cultivation of food and energy crops, both for the available land and also for the available water. This is not yet a major problem, but in at-risk regions continuing promotion of inappropriate cultivation systems can cause significant problems within a very short time. The cultivation of energy crops should not result in a region incurring water stress or in soil degradation that exceeds the soil protection guard rail (Chapter 3). If it were to do so, the expected social benefit from energy crops would be less than the harm done by increased soil degradation and inadequate water supplies.

The Comprehensive Assessment of Water Management in Agriculture (IWMI, 2007), the SIWI study (Lundqvist et al., 2008) and the GLASOD study (Oldeman et al., 1991; Oldeman, 1992) highlight the fact that current trends in global water and soil use are in the wrong direction. Without a change of policy, many regions will face an increasingly severe water crisis and greater soil degradation. In regions that are already experiencing a high level of water stress or soil degradation the cultivation of energy crops must not be permitted to exacerbate these negative environmental effects. However, if the correct techniques are used the cultivation of appropriate energy crops can actually improve the situation. In the long term the cultivation of energy crops on marginal and degraded land is a strategic option, because restored land becomes available for future food production. The choice of cultivation system is crucial, because systems can vary considerably in their water and soil quality requirements (Section 7.1). Both energy crops and afforestation for CO_2 storage are new drivers in the land-use sector; they may have major impacts on water use that are as yet poorly understood (Berndes, 2002; Jackson et al., 2005). Discussion of the expansion of bioenergy has as yet paid little heed to the water problem. Developments are now required in two directions. Firstly, the cultivation of energy crops must be integrated into a regional strategy for sustainable soil and water management. Since there is no overall template, these strategies should always be drawn up locally and in the light of local conditions. Secondly, there are still significant gaps in our knowledge of the links between energy crop cultivation and local or regional water resources that need to be filled with dedicated research.

Over the last 10,000 years large areas of Earth's terrestrial surface have undergone radical changes as the growing world population has used land for its various needs. The most important human land-use activities include the clearance or commercial use of forests, agriculture and the expansion of human settlements (Foley et al., 2005). Farmland alone, comprising both cropland and pastures, now covers around 40 per cent of the land surface (Foley et al., 2005). Almost a quarter of Earth's potential net primary production is already subject to human influence through harvesting, productivity changes resulting from land use, and fires (Haberl et al., 2007).

The human use of land thus competes directly with natural land cover, which plays an important part in the conservation of biological diversity and also functions as a carbon reservoir in the climate system. The increasing deployment of biomass for energy production increases the pressure on previously unused land; on existing farmland, it competes with the need to produce food for the growing world population (Chapter 5).

It is against this backdrop that WBGU sets out to identify the size of the sustainable global potential for energy crops until the middle of the century, using the modelling system that it has commissioned and which is described in this chapter (Beringer and Lucht, 2008). The plant primary production available for bioenergy will be determined on a region-by-region basis, taking account of the guard rails for food and environmental, climate and soil protection (Chapter 3). Using simple scenarios, the guard rails described in Chapter 3 will be used to identify exclusion areas within which the cultivation of energy crops would not be defined by WBGU as sustainable.

Whether this global sustainable potential for the deployment of bioenergy from the cultivation of energy crops can be realized depends mainly on economic and social conditions in those regions in which farmland that meets the WBGU criteria is available. WBGU's assessment of global potential at the end of the chapter is therefore preceded by a detailed socio-economic analysis of the countries in question.

Before the model commissioned by WBGU and its results are described in detail, similar appraisals of global bioenergy potential in the recent literature will first be summarized.

6.1
Previous appraisals of bioenergy potential

6.1.1
Bioenergy potentials in the recent literature

In calculating the global potential of bioenergy, the literature distinguishes – as with other energy carriers – between theoretical, technical, economic and sustainable potential (Box 6.1-1).

In its energy report WBGU estimates the global sustainable potential of bioenergy to be around 104 EJ per year (WBGU, 2004a) – i.e. around 20 per cent of the present global primary energy demand of around 480 EJ per year (GBEP, 2008). For the year 2050, calculated by the physical energy content, WBGU's exemplary energy path (WBGU, 2004a) method involves a 10 per cent contribution of bioenergy to the global primary energy requirement; calculated by the substitution method this contribution falls to 7 per cent on account of the higher proportion of wind and solar energy (Box 4.1-1). Present production of bioenergy, as measured by figures for 2005, amounts to around 47 EJ (GBEP, 2008), most of it in the form of traditional bioenergy use (Smeets et al., 2007). WBGU's appraisal takes account of the land available for biomass use on the different continents, excluding land used for food production and protected areas designated for the conservation of biodiversity and ecosystem functions.

Previous studies of the global potential of bioenergy have yielded a wide range of conclusions. A comparison of studies of the contribution of bioenergy in future energy systems shows that estimates for the year 2050 range from 47 EJ per year to 450 EJ per year (Berndes et al., 2003). The comparatively low figure put forward by WBGU arises from the

Box 6.1-1

Types of potential

Discussion of the potentials of various energy carriers usually distinguishes between theoretical potential, technical potential, economic potential and sustainable potential (WBGU, 2004a). In the context of this report, these terms are defined as follows:

THEORETICAL POTENTIAL
Theoretical potential describes the physical upper limit of the energy available from a particular source. In the case of solar energy this is the total solar radiation incident on the area in question. This potential therefore takes no account of land-use restrictions or of the efficiency of the conversion technologies used.

TECHNICAL POTENTIAL
Technical potential is defined specifically for each technology; it is derived from the theoretical potential and the annual efficiency of the respective conversion technology. Restrictions relating to the land realistically available for energy production are also taken into account. The criteria used in selecting land are not applied uniformly in the literature. Technical, structural and ecological restrictions and statutory specifications are sometimes included. The level of the technical potential of different energy sources is thus not a clearly defined value but dependent on a wide range of conditions and assumptions.

ECONOMIC POTENTIAL
Economic potential describes how much of the technical potential is economically usable under the given economic conditions (at a particular point in time). For biomass, for example, the economic potential is the quantity of biomass that it is economical to extract in the face of competition with other products and land uses. It may be possible to exert significant influence on economic conditions through policy measures.

SUSTAINABLE POTENTIAL
The sustainable potential of an energy source takes account of all the dimensions of sustainability. A range of ecological and socio-economic factors must usually be considered. Sustainable potential is not clearly delineated, since some authors already include ecological factors in their consideration of technical or economic potential.

It should be borne in mind that these terms are used in very different ways by different authors; in consequence, the sequence described above does not necessarily represent an increasingly tight progression. Hence the modelling commissioned by WBGU and described in this chapter refers to a 'technical sustainable potential', because the absence of integrated models has made it impossible to also assess economic viability.

consideration given to the competing claims of other forms of land use and from the fact that some other estimates have assumed unrealistically high yields (WBGU, 2004a). Some more recent studies of global bioenergy potential are discussed below. All figures for potentials represent the gross energy contribution – i.e. any losses that occur during conversion to final energy are not included.

Hoogwijk et al. (2003) evaluate existing studies and investigate the influence of various factors on the proportion of bioenergy from different sources in global energy production in 2050. The studies they review vary in their estimate of the world's future food requirement (influenced by population development and dietary habits), in the food and feed cultivation systems that they consider, and in the assumptions they make about productivity, land availability and requirements for biomass feedstock cultivation. Only existing conservation areas are excluded from bioenergy production. The resulting estimates for the year 2050 span a wide range of possible values from 33 EJ to 1135 EJ per year. The way in which these bioenergy potentials are distributed between sources is interesting. Estimates for bioenergy production on existing agricultural land (after the food requirements of the growing world population have been met) range from 0 to 988 EJ per year (the figure of 0 arises from the assumption that all existing agricultural land is needed for food production).

Bioenergy production on degraded land lies in the range of 8–110 EJ per year, while production from biogenic wastes and residues (agricultural and forestry residues, dung, organic waste) amounts to 62–108 EJ per year. Figures for the feedstock use of biomass range from 83 EJ to 116 EJ per year (Hoogwijk et al., 2003). These figures highlight the importance of assumptions about the area of land that will be needed in future to secure the world food supply. Very high potentials (in the range of around 1000 EJ per year) for the contribution of bioenergy to the world energy supply are only possible if it is assumed that land previously used for food production can be released, as a result either of efficiency improvements or less land-intensive dietary habits.

This is also illustrated by a study of Wolf et al. (2003) which considers the land available for food, feed and biomass and investigates the influence of agricultural production systems and dietary habits. However, the study does not distinguish between different energy crops and their yields on different soils. Restricting consideration to land currently in agricultural use and assuming medium population growth and a moderate nutrition style, global technical bioenergy potential is estimated at between 59 EJ (extensive cultivation for food, feed and biomass) and 417 EJ per year (intensive cultivation for food, feed and biomass). If not only existing farmland but also all potentially available agricultural land is used, these figures rise to 257 EJ for extensive cultivation and 790 EJ per year for intensive cultivation (Wolf et al., 2003). The influence of dietary habits is interesting. For extensive farming for food, feed and biomass on existing land the potential decreases from 59 EJ per year to 0 EJ per year if nutrition styles involve large quantities of meat and milk products and are therefore very land-intensive; if nutrition styles are less land-intensive the figure rises to 194 EJ per year.

Field et al. (2008) argue that sustainable cultivation of energy crops is only possible on abandoned agricultural lands that have previously been used as cropland or pasture, provided that they have not been converted to urban or forest areas. The authors thus implicitly exclude land used for growing food and feed, existing protected areas and wilderness areas, which they consider essential for securing the world food supply and for nature conservation. On the basis of this land appraisal and taking account of the spatially differentiated and climatologically determined net primary production on this land, they arrive at a global potential for the additional cultivation of energy crops that is sustainable according to their criteria of 27 EJ per year (Field et al., 2008).

Another study estimates that the sustainable production of bioenergy from the extensive use of high-diversity grassland on unused and degraded land could contribute around 45 EJ per year to global energy production. This type of use would in addition entail low inputs of chemical fertilizers and pesticides, good carbon storage in the soil and relatively high biodiversity (Tilman et al., 2006). Without placing restrictions on the type of energy crop grown, a new study arrives at a similar potential of 32–41 EJ per year on abandoned and degraded land (Campbell et al., 2008).

Smeets et al. (2007) explore global bioenergy potentials to 2050 for three types of biomass (bioenergy crops, agricultural and forestry residues and waste, and additional forestry yields) without taking climate change into account. In view of the need to conserve biodiversity, this study excludes existing protected areas, forests, barren land, scrubland and savannahs. Drawing on various assumptions for increasing yields in food production, the authors identify technical potentials of 215–1272 EJ per year for the cultivation of energy crops on surplus agricultural land, although the assumptions underlying even the lowest figure appear noticeably optimistic (Faaij, 2008). The higher figures assume major technological progress in food production as well as the use of irrigated agriculture. The global potential of bioenergy production from agricultural and forestry residues and wastes in 2050 was projected to be 76–96 EJ per year, while that from additional forest growth was estimated at 74 EJ per year (Smeets et al., 2007).

Hoogwijk et al. (2005) analyse the energy potential of short-rotation plantations of woody biomass for the years 2050–2100 for the four IPCC scenarios A1, A2, B1 and B2. The area of land set aside for nature conservation that cannot be used for biomass cultivation is assumed under the A scenarios to be 10 per cent and under the B scenarios to be 20 per cent of the global land area. Assumptions about world population, dietary habits and technological development are based on the storylines of the IPCC scenarios. The technical potential for bioenergy production arising from the use of abandoned agricultural land is put at 130–410 EJ per year in 2050 and at 240–850 EJ per year in 2100. The potential of land not previously used for agriculture, after deduction of grasslands, forests, urban areas and existing protected areas, is estimated at 35–245 EJ per year for 2050 and 35–265 EJ per year for 2100 (Hoogwijk et al., 2005; Smeets et al., 2007).

In its World Energy Outlook 2007 the International Energy Agency (IEA) puts the annual global primary energy use from biomass and residues in 2030 for its four scenarios at 68 EJ (Reference Scenario), 73 EJ (Alternative Policy Scenario), 69 EJ (High Growth Scenario) and 82 EJ (450 ppm Stabilisation Case; IEA, 2007a). This economic potential was calculated using an economic energy system model, taking account of various policy scenarios.

A study by the Institute for Energy and Environmental Research, Heidelberg (Institut für Energie und Umweltforschung, IFEU) commissioned by the German chemical industry association (Verband der Chemischen Industrie, VCI) estimates global bioenergy potential in 2050 to be 240–620 EJ per year. Of this, 215–420 EJ per year arises from the cultivation of energy crops on surplus agricultural land. The study takes account of the future need for feedstock use of biomass, and extreme scenarios for yield increases in agriculture have been excluded. In addition, timber growth contributes 0–45 EJ per year to the global potential, while all types of biogenic wastes and residues contribute 25–155 EJ (IFEU, 2007).

The OECD Round Table on Sustainable Development estimates that the sustainable global potential of bioenergy in 2050 totals 245 EJ per year (Doornbosch and Steenblik, 2007). Of this potential, 109 EJ per year arise from the cultivation of energy crops and 136 EJ per year from the use of agricultural and forestry residues, dung and organic waste for energy. The authors estimate the area of land available for the cultivation of energy crops at 440 million hectares. They exclude land currently used for food production, an additional 200 million hectares for securing the world food supply, and forests, but do not reserve any land for nature conservation.

Under an 'alternative scenario' for climate-friendly future energy production, a study commissioned by Greenpeace and the European Renewable Energy Council (EREC) puts the sustainable contribution of bioenergy to global energy production in 2050 at around 105 EJ per year (Greenpeace and EREC, 2007).

6.1.2
Summary and evaluation

Estimates of the potential contribution of bioenergy to global energy consumption are summarized in Table 6.1-1. Although the potentials quoted vary widely, ranging as they do from 30 EJ to 1,200 EJ per year, it is nevertheless possible to identify from this literature review some trends from which, despite some major uncertainties, a reasonably consistent picture emerges.

The greatest uncertainty results from the fact that the amount of land needed to meet the future food requirement of the world population is unknown. This land requirement depends not only on population growth, but also on the development of dietary habits, technological progress and the level of intensification of agricultural production (Section 5.2). Very high bioenergy potentials of the order of 1000 EJ per year are only technically realizable if land at present used for food production becomes available for the cultivation of energy crops as a result either of efficiency improvements or of less land-intensive dietary habits. It is of course possible that securing the food supply of a growing world population might in fact require the use of even more land, as the FAO, among others, have forecast (FAO, 2003a).

If land that has been used until now for food production is excluded, the only land that remains available for the cultivation of energy crops is marginal land (Box 4.2-1) with a very uncertain energy potential of around 30–200 EJ per year for a non-irrigated and not highly intensified farming system.

According to the studies described here, this potential from the cultivation of energy crops is supplemented by additional forestry yields at around 80 EJ per year and by biogenic wastes and residues (including agricultural and forestry plant residues, dung and organic waste) at around 80 EJ per year.

It must, however, be borne in mind that the majority of these estimates relate to technical potential; economic potential and in particular sustainable potential are likely to be less. Furthermore, competing use has not always been considered. For example, WBGU estimates that the bioenergy potential from

additional forest growth is small, because of the rising demand for wood products (Section 5.3.2). The cascade use of these material products goes only some of the way towards alleviating the problem because of the inevitable losses that are entailed (Section 5.3.3). WBGU therefore puts a figure of 0 EJ per year on the sustainable potential of additional forest growth, but notes that further research in this area is needed.

There are very few studies of the sustainable potential arising from the use of biogenic wastes and residues. In its energy report (WBGU, 2004a), WBGU estimates that this potential amounts in total to 67 EJ per year. On the basis of more recent studies, WBGU regards a realistic figure for the global technical potential from biological wastes and residues from agriculture and forestry and from dung to be 80 EJ per year. However, not all of this is usable, since these estimates do not necessarily take account of economic considerations and sustainability criteria. For example, for soil protection reasons residues from agricultural and forestry ecosystems cannot be removed completely, as this would result in too much organic material being removed from the soil (Münch, 2008). At a rough estimate it seems realistic to assume that the technical sustainable potential is around 50 EJ per year, of which approximately half is economically realizable. WBGU points out that this figure must be regarded as very uncertain, since there are issues relating to the sustainable and economic use of biogenic wastes and residues that still need to be clarified through research.

6.2
Global land-use models: The state of scientific knowledge

6.2.1
Effects and impacts of human land use

Human-induced changes in Earth's land cover influence the climate by changing the reflectivity (albedo) of Earth's surface and affecting the carbon cycle (Lambin et al., 2003). It is estimated that around 35 per cent of anthropogenic carbon dioxide emissions since 1850 are the result of human land use (Foley et al., 2005). In addition, land use and land-use changes affect the water cycle, the nutrient cycle, biological diversity and soil quality (Lambin et al., 2003).

Conversely, biogeophysical variables such as climate, water availability and soil quality, and changes in them, affect not only the natural vegetation; together with political, economic and social factors

Table 6.1-1
Technical (TP), economic (EP) and sustainable potential (SP) of bioenergy in EJ per year from various studies.
Compilation: WBGU

Sources	Potential, year	Forest increment	Energy crop cultivation			Residues			Total
			Farmland	Unused land	Degraded land	Agriculture	Forestry	Other	
Studies that consider all contributions to the bioenergy potential									
WBGU (2004a)	NP	0	37			17	42	8	104
Hoogwijk et al. (2003)	TP, 2050	0	0–988		8–110	10–32	42–48[2]	10–28	33–1135
Smeets et al. (2007)	TP, 2050	74	215–1272			76–96			365–1442
IEA (2007a)	WP, 2030								68–82[4]
IFEU (2007)	WP, 2050	0–45	200–390	15–30		15–70	5–30	5–55	240–620
Doornbosch und Steenblik (2007)	NP, 2050		109			35	91	10	245
Faiij (2008)	NP, 2050	60–100	120[6]	70[6]		40–170			430–600[6]
Studies of the potential from the cultivation of energy crops									
Wolf et al. (2003)	TP, 2050		0–790[1]						
Hoogwijk et al. (2005)	TP, 2050			130–410	35–245				
	TP, 2100			240–850	35–265				
Tilman et al. (2006)	NP			45[3]					
Campbell et al. (2008)	NP			32–41					
Field et al. (2008)	NP			27					
WBGU (2008)	NP, 2050			34–120[7]					

[1] depending on dietary habits and degree of intensification of agricultural production
[2] including 32 EJ per year from cascade use of biomaterials
[3] extensively used grassland of high biodiversity
[4] for the four IEA scenarios (Reference Scenario, Alternative Policy Scenario, High Growth Scenario, 450 ppm Stabilisation Case)
[5] Alternative Scenario
[6] an additional 140 EJ per year in energy crop cultivation as a result of technological progress in agriculture is assumed
[7] climate model HadCM3, emissions scenario A1B, depending on guard rail scenario and irrigation

they are the major drivers of land-use changes (Heistermann et al., 2006).

Models of land use and land-use change attempt to study these complex interactions by applying numerical methods. Various types of land-use model and their typical strengths and weaknesses will now be briefly described.

6.2.2
Typology of global models of land use and land-use change

A fundamental distinction must be made between the description of current land use and that of land-use changes. The modelling of current or future land use attempts, for example, to quantify the effects on the carbon and water cycle of the displacement and expansion of agricultural land, or to estimate the effects of climate change on plant productivity. In this case information on the underlying land-use changes comes from exogenous data.

Models of land-use change, on the other hand, set out to consider the processes – usually socio-economic ones – that are likely to determine the future use of the biosphere by humans.

Models of global land use and land-use change can be classified in various ways (Verburg et al., 2004); a useful approach is to categorize them on the basis of their underlying methodology (Heistermann et al., 2006):

- Geographical models attempt to depict the spatial distribution of land-use types and the interaction between them, taking account of biogeophysical variables such as soil type and quality, climate, water availability, and the material fluxes (particularly of carbon) that are important for the vegetation. This means that they are particularly good at reflecting fundamental biogeophysical constraints on the supply of agricultural products but may not adequately model land-use changes arising from socio-economic influences (e.g. change in demand for particular products).
- Economic models, by contrast, focus on the socio-economic drivers of land use and land-use change and hence on the demand side of the world economy. Demographic and cultural factors, changes in dietary habits, policies of promoting particular agricultural products and the structure of the world market are among the issues that play an important role in such models. However, economic models do not always adequately depict major biogeophysical constraints on agricultural production (e.g. relating to the soil or to climate change).

- Integrated models seek to combine the strengths of both approaches; they attempt to provide a more realistic description of changes in the human use of land, since these changes are influenced by both biogeophysical and socio-economic factors.

6.3
Description of the model

The modelling carried out for this report uses the LPJmL (LPJ managed Land) model (Bondeau et al., 2007), which is based on the LPJ dynamic global vegetation model (Lund-Potsdam-Jena; Sitch et al., 2003). It consists of a geographical model of terrestrial land use which is combined with scenarios of the land area potentially available for biomass cultivation. The model has a spatial resolution of 0.5°, which is determined by the climate models used. The economic drivers of future land use are only implicitly considered in the scenarios.

Using process-oriented descriptions of key biogeochemical, biophysical and biogeographical mechanisms, LPJmL is able to simulate the large-scale distribution of the various vegetation types. This yields a series of parameters relating to factors such as plant productivity and the distribution and dynamics of carbon and water storage in vegetation and soils. A dynamic model of this type depicts how the geographical distribution of plants reacts to changes in the prevailing weather conditions, providing a picture of the large-scale vegetation shifts that might occur as a result of progressive climate change.

6.3.1
Methods used in the model

6.3.1.1
Modelling plant productivity

LPJmL was developed for the analysis of interactions between climate and the biosphere on a global scale. The model design therefore requires certain simplifications and generalizations to be made. For example, the diversity of plant life and growth forms is reduced to nine plant functional types (PFTs), which are characterized by their photosynthetic metabolism (C3 or C4), phenology (deciduous or evergreen), growth form (woody or non-woody) and life span (annual or perennial). Local climate conditions and competition for light and water determine the dynamics of the vegetation over time and space. The calculations of gross primary production (GPP) and plant respiration are based on a modified Farquhar-Collatz approach (Farquhar et al., 1980; Collatz et al., 1992) and are linked via the stomatal conductivity direct to the water balance of the plants (Gerten et al., 2004). This enables the effect of drought on photosynthesis and transpiration to be realistically depicted. Various allometric and functional rules determine the allocation of the assimilated carbon to the four plant storage organs of leaves, heartwood, sapwood and fine roots (Shinozaki et al., 1964). Fire events have a major impact on the carbon cycle of an ecosystem and are in some biomes characteristic elements of vegetation development. In LPJmL their occurrence is estimated from the available combustible material and prevailing soil moisture (Thonicke et al., 2001). Dead biomass enters the soil, where the decay rate of organic substances is calculated using a modified Arrhenius equation (Foley, 1995), taking account of soil temperature (Lloyd and Taylor, 1994) and moisture. For the CO_2 fertilizing effect a value of 20–30 per cent was assumed. This conforms well with measured productivity increases, such as those obtained in FACE (Free-Air Carbon Dioxide Enrichment) experiments for fast-growing tree species in short-rotation plantations (Calfapietra et al., 2003; Liberloo et al., 2006; Hickler et al., 2008).

6.3.1.2
Agriculture in LPJmL

As well as simulating the distribution and dynamics of potentially natural vegetation, LPJmL is also able to depict land used for agriculture (Bondeau et al., 2007). In addition to modelling biophysical and biogeochemical processes, the model calculates the productivity and yields of the most important crops, again utilizing the approach of generic plant types. It distinguishes 13 crop functional types (CFTs), 11 arable crops and two grass types. For all CFTs the model enables cultivation to take place on either irrigated or non-irrigated land.

6.3.1.3
Modelling the cultivation of energy crops

For the depiction of biomass plantings in the model, three additional plant types were defined and parameterized: two fast-growing trees and a high-productivity grass. The trees can be divided according to their potential cultivation area into a tropical and an extra-tropical type. Parametrization of the extra-tropical tree species was based on the growth dynamics and yield of poplar and willow species whose growth properties make them suitable for use as energy crops. It was assumed that they would be planted at

a density of around 15,000 individuals per hectare. The tropical tree type is represented by a commercially grown eucalyptus species grown in the model at a density of 2000 individuals per hectare. Both tree species are grown on a short-rotation system and harvested every eight years. Harvesting involves removal of 90 per cent of the above-ground biomass; the below-ground root mass is retained in its entirety, enabling realistic growth of coppice shoots the following year.

In addition to woody biomass, the model includes the cultivation of highly productive C4 grasses such as *Miscanthus* grasses and switchgrass *(Panicum)* on a large scale. It is particularly interesting to note that certain species of these plants can maintain a high level of photosynthesis even at low temperatures. Harvesting takes place annually at the end of the growth period, again involving removal of 90 per cent of the above-ground biomass.

6.3.1.4
Comparison with measured data

The most important quality criterion for global vegetation models is the degree of similarity between their results and empirical measurements. LPJmL has been extensively tested against various independent observation data. This has shown that the model can correctly depict the large-scale distribution and dynamics of terrestrial vegetation (Lucht et al., 2002; Sitch et al., 2003; Hickler et al., 2004; Erbrecht and Lucht, 2006). Its simulation of soil moisture, run-off, transpiration and the seasonal variability of these factors also largely corresponds with measured values (Wagner et al., 2003; Gerten et al., 2004). A comparison of agricultural yields and FAO statistics shows that LPJmL can correctly represent geographical differences in yields (Bondeau et al., 2007).

Validation of the simulated biomass plantings is difficult, because the cultivation of cellulose plants is at present largely confined to special experimental areas, usually under optimal growth conditions. Comparisons show that the modelled biomass yield of switchgrass (*Panicum*) and of short-rotation plantations of fast-growing trees lies within a range that can already be achieved with progressive farming methods.

6.3.1.5
Calculation of global bioenergy potential

For each grid cell in the model, the amount of land available for the cultivation of energy crops can be determined from the combination of various exclusion criteria arising from the guard rails for food production and nature, soil and climate protection (Chapter 3). The model assumes that half of this land will be cultivated with high-productivity grasses and half with fast-growing tree species. If only one of these plant types grows on the land in question (e.g. in many regions trees cannot be grown on non-irrigated land), the entire area is planted with this plant type. The area and yield potential of both plant types are used to calculate the quantity of primary energy (i.e. the chemical energy contained in the biomass) that can be produced annually. The calculation assumes an energy content of the dry mass of 19.0 kJ/gram (Wirsenius, 2000).

6.3.2
Data sets used in the model

6.3.2.1
Climate change and climate data

Monthly data for temperature, precipitation and cloud cover and annual data for atmospheric CO_2 concentration are used to drive LPJmL. Various climate models were used to depict future climate development. It should, however, be noted that the absolute values of the data from the climate models partly differ significantly from the measured values, which can seriously affect the quality of the vegetation modelling. For the scenario calculations with LPJmL the anomalies of temperature and precipitation were therefore superimposed on the long-term means (1961–1990) of the observed data of the Climatic Research Unit (CRU; New et al., 2000) (Schaphoff et al., 2006).

6.3.2.2
Land-use data

In addition to the climatological data in the individual grid cells the model also requires information on soil characteristics, current land use and the distribution of non-irrigated and irrigated farmland (Klein Goldewijk et al., 2007; Portmann et al., 2008; Ramankutty et al., 2008). Because of the relatively low spatial resolution of the model, steep slopes were not explicitly excluded, but the climate data used in the model prevent unrealistic biomass production on such land. The exclusion of marginal soils was based on the data of the Global Assessment of Human Induced Soil Degradation (GLASOD; Oldeman et al., 1991), as shown in Figure 6.4-1. Extremely degraded land (Category 4) cannot be converted in the model into

land for growing energy crops. On strongly degraded land (Category 3) only 30 per cent of potential yield can actually be achieved. Settlement areas are not explicitly excluded from the modelling; however, they constitute only around 2 per cent of global land use (Lambin et al., 2001) and can therefore be ignored.

6.4
Model assumptions and scenarios

6.4.1
Climate models and emissions scenarios

For the scenarios used in this report, LPJmL was driven with data from various current climate models, all of which were calculated for the Fourth Assessment Report of the IPCC (IPCC, 2007d). The selection criterion for the climate models was the best possible fit of simulated and observed values for temperature and precipitation in the period 1961–1990. The models selected were ECHAM5 (Roeckner et al., 2003), HadCM3 (Pope et al., 2000), CM2.1 (Delworth et al., 2006), ECHO-G (Legutke and Voss, 1999) and CCSM3.0 (Collins et al., 2006). All the climate models were driven using three IPCC emissions scenarios (A1B, A2 and B1, IPCC, 2000).

6.4.2
Irrigation scenarios

For the purposes of modelling global bioenergy potential a distinction is made between non-irrigated and irrigated cultivation; the model assumes that 10 per cent of cultivation land is irrigated. By way of explanation: the extent of currently irrigated land as a proportion of total agricultural land (cropland and pasture) varies widely from region to region. The figure ranges from 0.5 per cent in sub-Saharan Africa, 2.6 per cent in the former Soviet Union, 4.7 per cent in North America and 6.1 per cent in Europe to 25.8 per cent in South-East Asia and India, with a global mean of 5.4 per cent (Portmann et al., 2008). The proportion of cropland that is irrigated is higher; in 1998 it was around 16.9 per cent, and the FAO expects it to rise to around 18.0 per cent by 2030 (Faurès et al., 2000). However, the model envisages that energy crops will not be grown on existing cropland, and a significant proportion of the land potentially available for energy crops is situated in developing countries. For these reasons, and also in view of the insufficient availability of water in many regions, WBGU considers 10 per cent to be a realistic maximum for

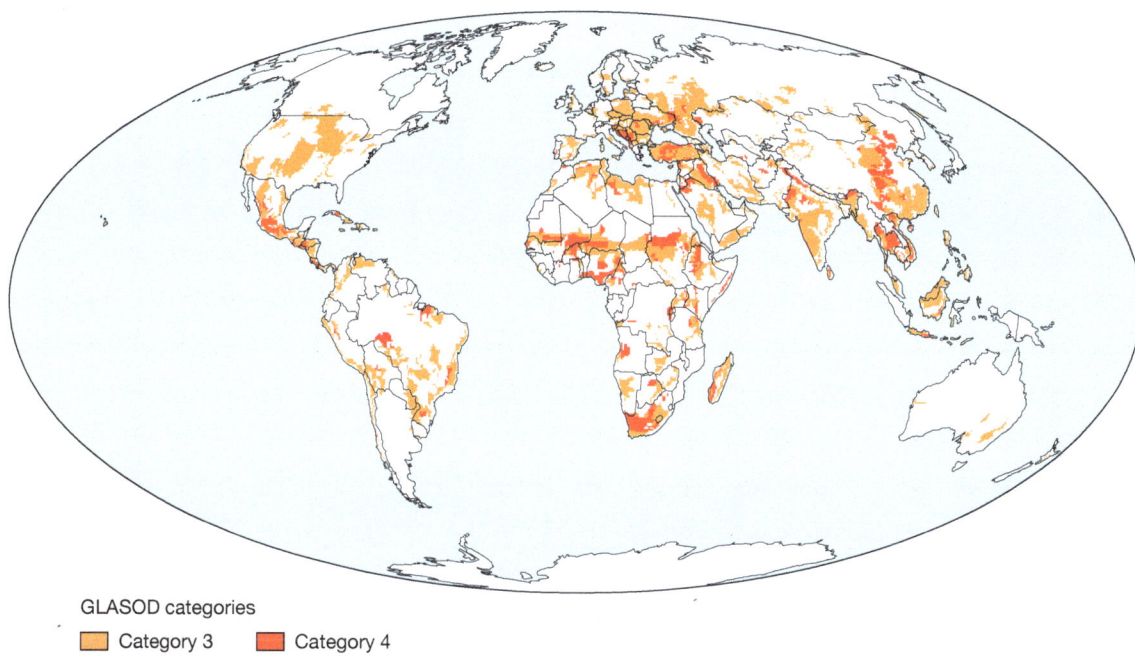

GLASOD categories

☐ Category 3   ☐ Category 4

Figure 6.4-1
Extremely degraded (Category 4, total area 680 million hectares) and strongly degraded (Category 3, total area 2400 million hectares) land excluded from bioenergy cultivation. It is assumed that yields on Category 4 land are 0% of those achieved on non-degraded land, while on Category 3 land they are 30%.
Source: Beringer and Lucht, 2008, based on Oldeman et al., 1991

the proportion of land on which irrigated cultivation takes place.

6.4.3
Scenarios for the calculation of biomass potentials

The global potential for bioenergy is calculated from the modelled potential yields and the amount of land available for the cultivation of biomass. In accordance with WBGU's guard rail approach (Chapter 3), a scenario-based approach was selected for the analysis of the potential for sustainable bioenergy production. Three main factors were viewed as key to the extent and distribution of land for energy crop cultivation in the coming decades: the land needed for food production, the land needed for nature conservation and the greenhouse gas balance of the land-use changes that would be required.

6.4.3.1
Scenarios for securing food production

It is difficult to assess the amount of additional land needed for agricultural food production, since the figure depends on population development, dietary habits and technological progress in agricultural pro-

duction; insufficient information is available on the future development of these parameters. However, it is considered unrealistic to expect that land at present used for food production could become available for the cultivation of energy crops (Section 5.2).

In the present model calculations, two scenarios for the land requirement of food production are therefore distinguished:
- *Scenario A (high demand for agricultural land):* This scenario adopts a forecast of the United Nations Food and Agriculture Organization (FAO) stating that an additional 120 million hectares of land worldwide will be required for food production by 2030 (FAO, 2003a). In this scenario, therefore, land currently used for food production and a further 120 million hectares of the most productive land is unavailable for the cultivation of energy crops.
- *Scenario B (low demand for agricultural land):* The less restrictive Scenario B assumes that the land currently used for food production will in future continue to be sufficient to meet the world's food needs and will not be used for the cultivation of energy crops.

The land excluded under the two scenarios is shown in Figure 6.4-2.

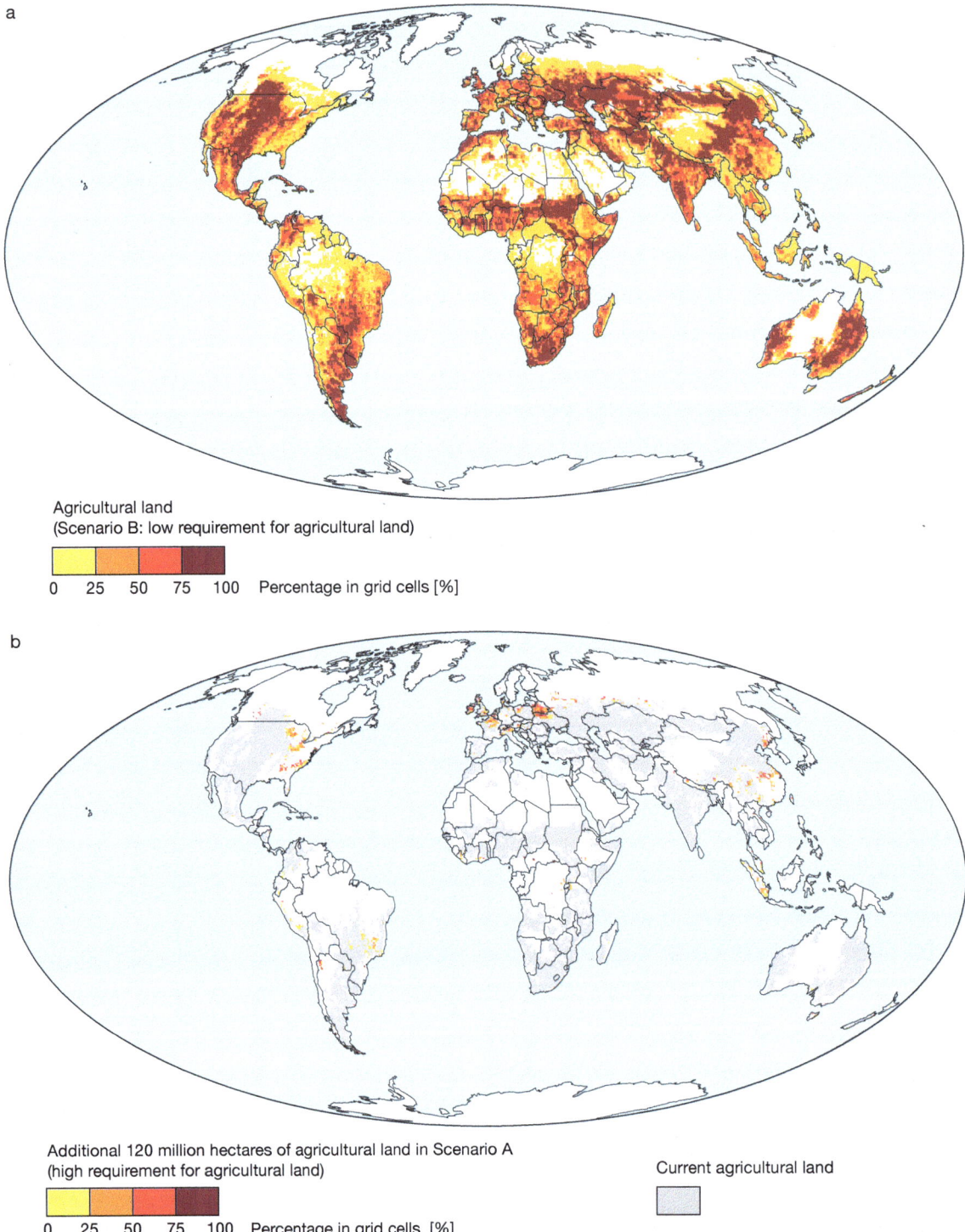

Figure 6.4-2
Land excluded in order to secure the food supply. (a) Percentage of current agricultural land in the grid cells of the model.
This land is excluded from bioenergy production in Scenario B. (b) Additional land excluded in Scenario A in order to permit
expansion of agricultural land for food production. These are the most productive 120 million hectares of the land available for
bioenergy cultivation in Scenario B.
Source: Beringer and Lucht, 2008

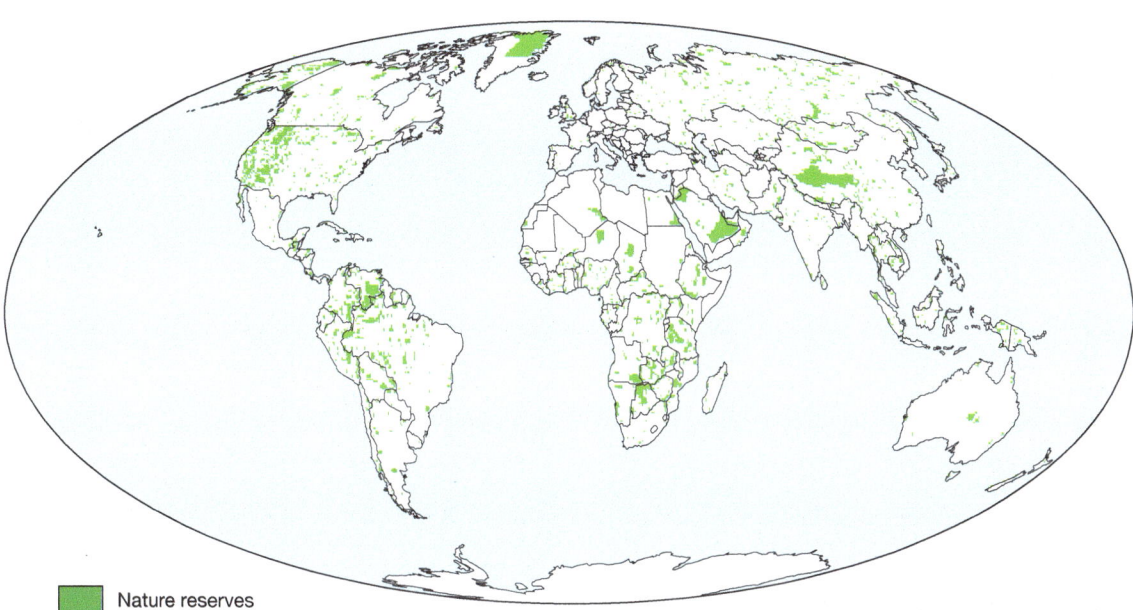

Nature reserves

Figure 6.4-3
Geographical distribution of current nature reserves, with a total area of 1330 million hectares. These areas are excluded from energy crop cultivation in the model.
Source: Beringer and Lucht, 2008, based on WDPA, 2008

6.4.3.2
Scenarios for nature conservation

The exclusion of areas with high nature conservation value is based on various scenarios for considering areas of high biological diversity and wilderness areas. Completely excluded from any use are, in the first place, existing protected areas that appear in the World Database on Protected Areas (WDPA, 2008), as shown in Figure 6.4-3.

In order to also exclude areas of high biological diversity that do not at present have protected status, the following four indicators are used:
- *Biodiversity Hotspots* (Mittermeier et al., 2004) are areas in which a particularly high concentration of endemic species is experiencing higher than average loss of habitat;
- *Endemic Bird Areas* (Stattersfield et al., 1998) are characterized by a high concentration of bird species with a small geographical distribution;
- *Centres of Plant Diversity* (WWF and IUCN, 1994) contain either a high diversity of plant species or a large number of endemic species (or both);
- *Global 2000* (Olson et al., 2001) is a list of more than 200 land, freshwater or marine ecosystems with particularly high biodiversity and that are representative of their particular ecosystem type.

Wilderness areas form a further category of types of land that should be protected. Wilderness areas are defined as large contiguous areas (e.g. tropical rainforests, boreal forests, grasslands, semi-deserts, etc.)

that, because of their remoteness from civilization or for other reasons, are still in a natural state. They do not always exhibit high concentrations of biodiversity but frequently provide very valuable ecosystem services. For areas of untouched wilderness the following three data sets were used: High-Biodiversity Wilderness Areas (Mittermeier et al., 2003), Frontier Forests (Bryant et al., 1997) and Last of the Wild (Sanderson et al., 2002).

In order to merge the data sets for biodiversity and wilderness areas of high conservation value and thus arrive at actual exclusion areas, the areas have been categorized according to the number of indicator data sets in which they occur: the greater the agreement between the various data sets, the higher the proportion of land placed under protection. Here again two scenarios are distinguished:
- *Scenario A (high nature conservation)*: In this scenario wilderness areas are always placed under 100 per cent protection, even if they are only listed in one of the wilderness area data sets. For biodiversity hotspots a graded system is used: as a starting point 10 per cent of all land is placed under protection; where an area is listed in one data set 20 per cent of the area is placed under protection, rising to 30, 50 and 80 per cent for two, three or, respectively, four data sets.
- *Scenario B (low nature conservation)*: In this less restrictive scenario wilderness areas are placed under 100 per cent protection only if the area is listed in at least two of the data sets. Biodiversity

Table 6.4-1
Proportions of protected areas for the conservation of wilderness areas and biodiversity hotspots under the two scenarios.
Source: Beringer and Lucht, 2008

Scenario	Number of concurrences in indicator data sets							
	Wilderness areas			Biodiversity hotspots				
	1	2	3	0	1	2	3	4
A: High nature conservation	100 %	100 %	100 %	10 %	20 %	30 %	50 %	80 %
B: Low nature conservation	0 %	100 %	100 %	0 %	0 %	0 %	50 %	100 %

hotspots are only placed under 50 or 80 per cent protection if they are listed in three or, respectively, four data sets.

The indicators for the scenarios are summarized in Table 6.4-1 and the resulting exclusion areas are shown in Figure 6.4-4.

6.4.3.3
Scenarios for greenhouse gas emissions from land-use changes

The release of CO_2 from vegetation and the soil through the clearing of forests or drying out of wetlands results in greenhouse gas emissions significantly greater than those that can be saved through the substitution of fossil fuels by the subsequent biomass use (Sections 4.2.3.1 and 4.2.3.2). Many such sites already form part of areas of high biodiversity and low human impact (natural forests, wetlands) and are therefore in any case not considered for energy crop cultivation.

Also excluded from human use are wetlands from the Global Lakes and Wetlands Database (Lehner and Doll, 2004) that represent a major carbon sink in the model but that have not been included in the list of reserved protected areas (Figure 6.4-5).

Finally, in view of the climate protection guard rail (Section 3.1.1), cultivation of energy crops must be excluded on land where the greenhouse gas emissions arising from conversion would only after a very long period be compensated by the carbon removed from the atmosphere. Ideally this calculation should include the emissions resulting from agricultural cultivation and the processing of energy crops (agricultural machinery, fertilizers) and the fossil fuel emissions saved through the deployment of biomass for energy. Since such data is not available for the purpose of this model, the calculations performed take account only of the extent to which the emissions arising from land-use change are compensated by the carbon subsequently removed from the atmosphere by the soil and biomass growth. This can be

used as an indicator of the minimum compensation time that can be achieved in the overall greenhouse gas balance.

With regard to the greenhouse gas balance it should be borne in mind that the carbon absorbed by the soil ideally remains there, while the carbon stored in biomass growth is released again after harvesting, although it does substitute fossil CO_2 emissions. It remains to be clarified whether the carbon stored in the biomass is a good indicator for the quantity of fossil CO_2 emissions that are mitigated when the deployment of this biomass for energy substitutes the use of fossil fuels.

If the carbon contained in different energy carriers is considered per unit of stored energy (Kaltschmitt and Hartmann, 2003), the potential CO_2 emissions associated with the use of biomass for energy correspond roughly to those of hard coal and are about 20 per cent below those of lignite. If lignite is substituted directly, it is therefore theoretically possible for the biomass use to save almost 20 per cent more fossil CO_2 than is stored in the biomass; when lignite is replaced, about as much fossil CO_2 is saved as is stored in the biomass. Other fossil fuels (oil, gas) have a lower carbon content and so offer correspondingly less scope for savings. This ignores the losses that take place during technical conversion.

The carbon stored in the biomass is hence a good indicator of the maximum CO_2 emissions that can be saved, since in addition to the emissions from the change of land use emissions from cultivation and possible conversion losses arise.

It is thus clear that substituting fossil fuels by biomass usually saves fewer greenhouse gas emissions overall than those to which the carbon stored in the biomass corresponds. The only biomass pathways that are regarded as acceptable are those for which the compensation time – i.e. the period following land-use conversion after which the real reduction in emissions begins – is a relatively short one of not more than ten years. For this to be achieved, the released carbon must therefore be re-absorbed by the soil and the biomass growth within at the most ten

a

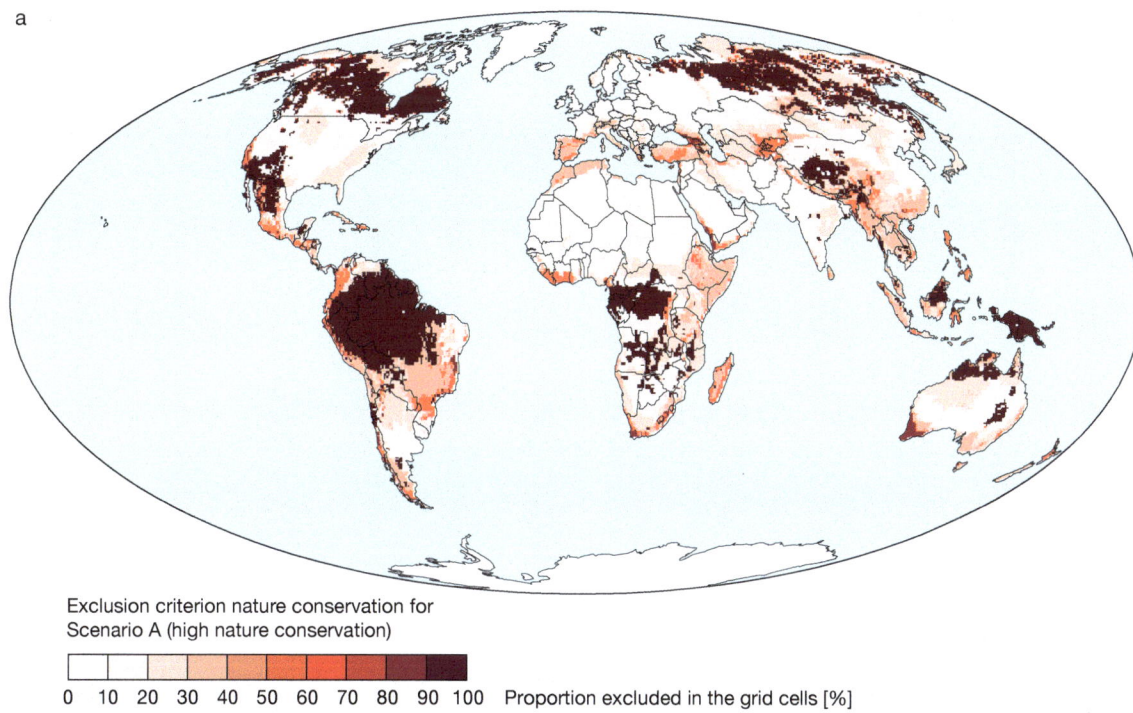

Exclusion criterion nature conservation for
Scenario A (high nature conservation)

0   10   20   30   40   50   60   70   80   90   100   Proportion excluded in the grid cells [%]

b

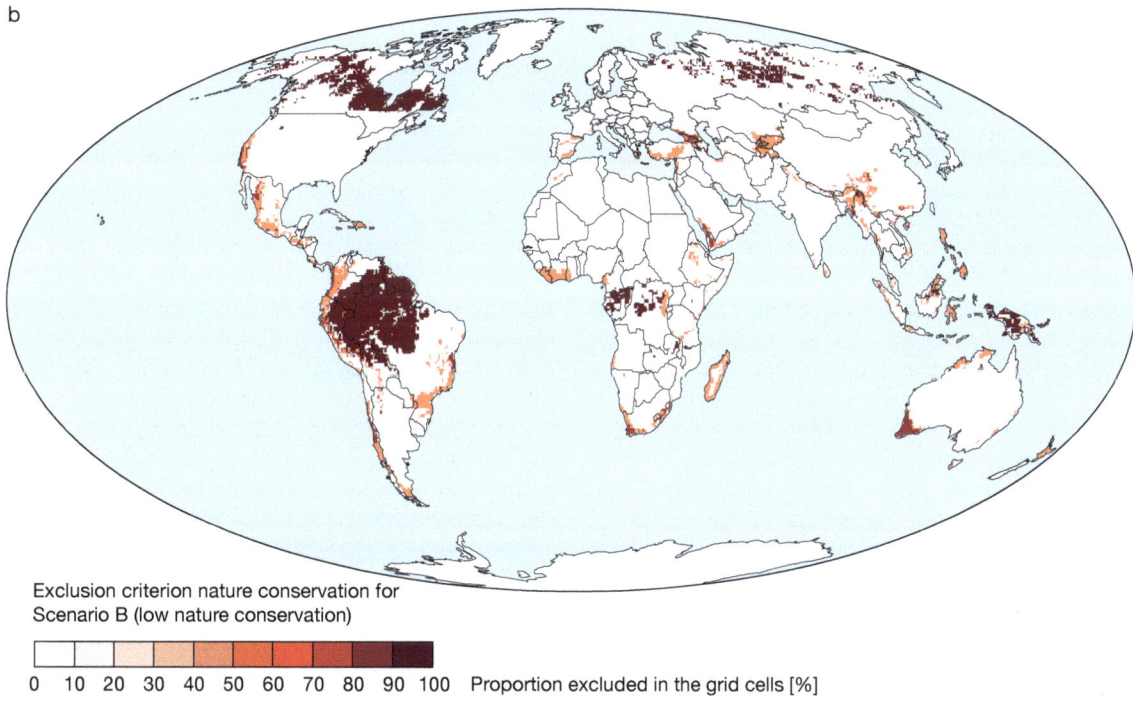

Exclusion criterion nature conservation for
Scenario B (low nature conservation)

0   10   20   30   40   50   60   70   80   90   100   Proportion excluded in the grid cells [%]

Figure 6.4-4
Areas for the conservation of wilderness and biological diversity and hence excluded from energy crop cultivation for the two
scenarios described in the text. (a) Scenario A: high nature conservation; (b) Scenario B: low nature conservation (Table 6.4-1).
Source: Beringer and Lucht, 2008

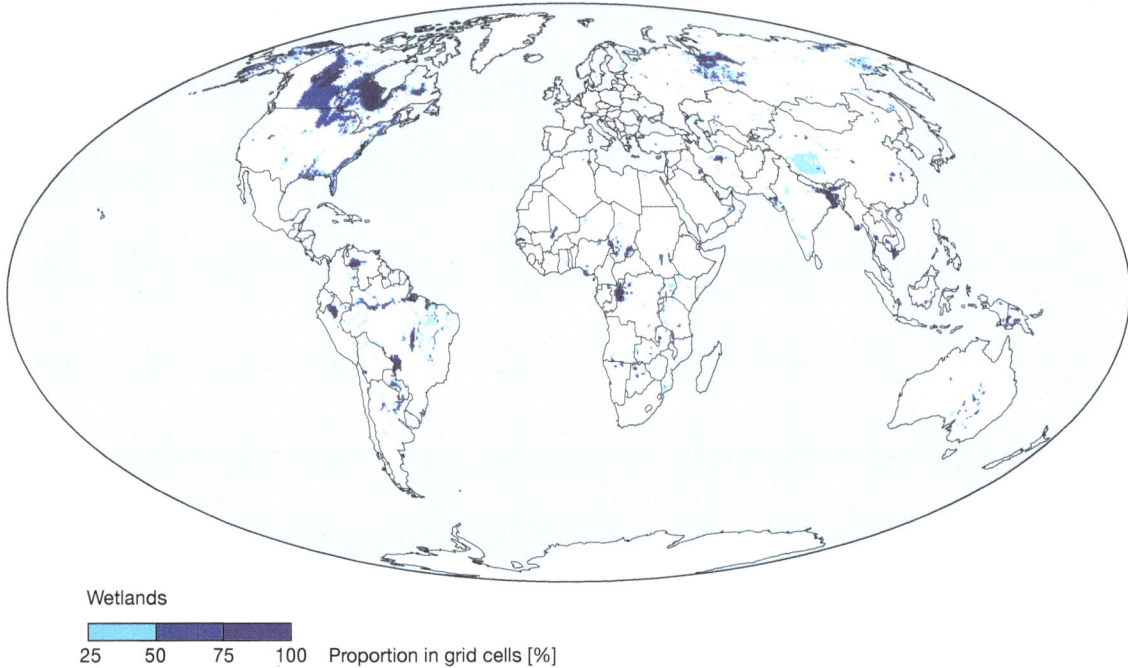

Wetlands

25   50   75   100   Proportion in grid cells [%]

Figure 6.4-5
Wetlands excluded from biomass use, with a total area of 1150 million hectares.
Source: Beringer and Lucht, 2008, based on Lehner and Doll, 2004

years. Two scenarios are therefore used, which differ in the compensation period that they envisage:

- *Five years:* In this scenario the maximum compensation period for the emissions arising from the land-use change is set at five years.
- *Ten years:* In this less restrictive scenario the maximum compensation period is ten years.

The geographical distribution of these exclusion areas is shown in Figure 6.4-6. Comparison with the geographical distribution of forested areas (FAO, 2006c) reveals that the majority of the excluded areas are forests (Figure 6.4-7). These areas are excluded from conversion into land for the cultivation of energy crops, although they might in some cases still contribute to the bioenergy potential arising from forest residues (Section 5.5).

6.5
Results of the modelling of the global potential of energy crops

6.5.1
Influence of the climate models and emissions scenarios

By comparison with the importance of the exclusion criteria for the available land, the influence of the various climate models and the two emissions sce-

narios on the modelled bioenergy potentials is only very slight. For example, the potential for non-irrigated agriculture under the A1B scenario and a particular land-use scenario is 34.5 EJ per year in the HadCM3 model and 34.1 EJ per year for ECHAM5. The corresponding figures for scenarios A2 and B1 for HadCM3 are 34 and 33 EJ per year respectively. The differences are thus noticeably below 10 per cent.

The reason for this is probably that differences in the forecast changes in climatological site conditions in the regions with potential land available for the cultivation of energy crops are only minor. For example, for the Amazon region the projections of different climate models vary widely, but land in this region has for nature conservation and climate change mitigation reasons been excluded from biomass use for the purpose of the present modelling.

The results of the modelling are thus largely independent of the climate model and emissions scenarios used. The following findings therefore relate only to calculations made with the HadCM3-Modell using the A1B scenario.

6.5.2
Influence of the compensation period

It was found that the different compensation periods of five or ten years for removal of the carbon

a

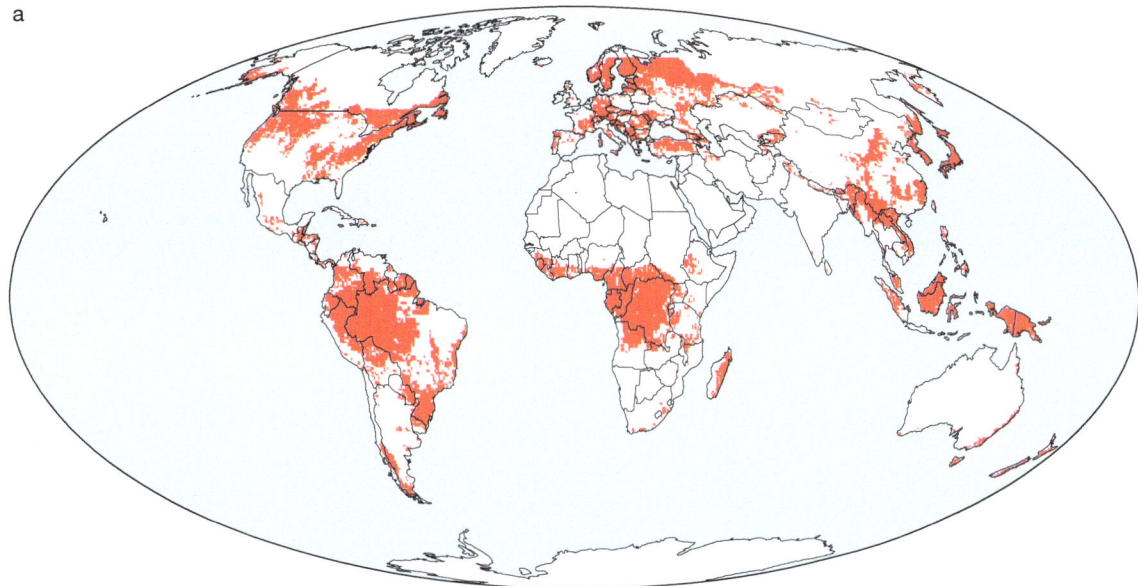

Compensation of the CO_2 released as a result of land-use change

Not possible within 5 years

b

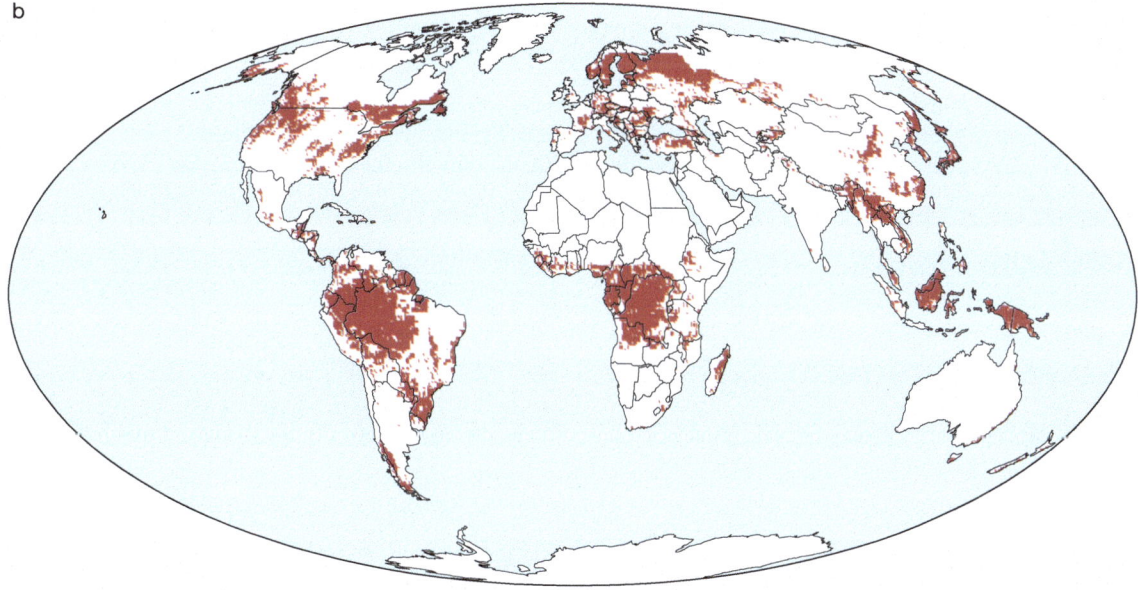

Compensation of the CO_2 released as a result of land-use change

Not possible within 10 years

Figure 6.4-6
The maps show regions in which biomass cultivation cannot compensate within (a) five years (Scenario A: total area 3713 million hectares) and (b) 10 years (Scenario B: total area 2891 million hectares) for the loss of carbon as a result of the land-use change.
Source: Beringer and Lucht, 2008

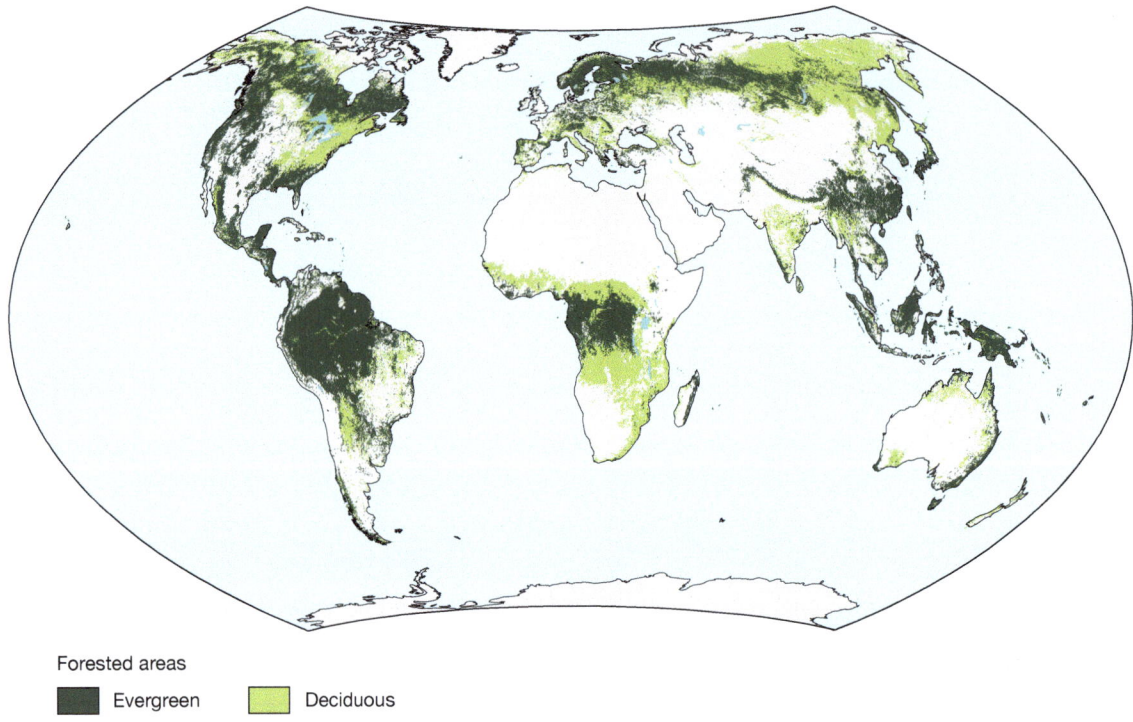

Figure 6.4-7
Global distribution of forested areas.
Source: FAO, 2006c

released during conversion of the land for the purpose of growing biomass has only a very slight influence on the simulated bioenergy potentials. The maximum variation is 10 cent. In consequence only the results for a compensation period of ten years are reported here.

6.5.3
Bioenergy potentials for four scenarios

Since the different climate models and emissions scenarios and the two compensation periods considered

have only a comparatively small influence on the calculation of the globally sustainable bioenergy potential in 2050, the following description covers only the dependence of this potential on the two remaining factors, the scenarios for food production and nature conservation.

In this section we therefore consider four scenarios that arise from combinations of the two scenarios for the land needed to secure the food supply (Section 6.4.3.1) and the two scenarios for conserving biodiversity and wilderness areas (Section 6.4.3.2). The nomenclature used in shown in Table 6.5-1.

Table 6.5-1
Definition of the four land-use scenarios used.
Source: Beringer and Lucht, 2008

Scenario	Description	Food stuff production	Nature conservation
1	High farmland requirement / high nature conservation	A	A
2	High farmland requirement / low nature conservation	A	B
3	Low farmland requirement / high nature conservation	B	A
4	Low farmland requirement / low nature conservation	B	B

Table 6.5-2
Potential cultivation areas and bioenergy potentials in 2000 and 2050 for the four land-use scenarios.
Source: Beringer and Lucht, 2008

Scenario		Cultivation area [Mha]	Bioenergy potential in 2000 [EJ per year]		Bioenergy potential in 2050 [EJ per year]	
			Non irrigated	Irrigated	Non irrigated	Irrigated
1	High farmland requirement / high nature conservation	240	35	42	34	42
2	High farmland requirement / low nature conservation	380	63	74	61	71
3	Low farmland requirement / high nature conservation	360	75	83	74	83
4	Low farmland requirement / low nature conservation	500	110	120	100	120

Table 6.5-3
Bioenergy potentials for the years 2000 and 2050 in different world regions (Figure 6.5-5) for four land-use scenarios.
Source: Beringer and Lucht, 2008

Bioenergy potentials non-irrigated for 2000 [EJ per year]									
Scenario	AFR	CPA	EUR	GUS	LAM	MEA	NAM	PAO	PAS	SAS
1	6,0	3,6	3,4	1,3	10	0,8	5,2	1,6	0,6	2,5
2	8,3	8,4	5,3	2,2	19	1,3	9,7	4,9	1,0	3,0
3	7,9	10	11	7,0	15	0,9	11	3,0	6,6	2,6
4	11	13	13	7,4	27	1,4	13,4	6,5	10	3,1

Bioenergy potentials irrigated for 2000 [EJ per year]									
Scenario	AFR	CPA	EUR	GUS	LAM	MEA	NAM	PAO	PAS	SAS
1	8,4	3,8	3,7	1,4	12	1,0	5,5	2,3	0,7	3,6
2	11	8,7	5,9	2,3	21	1,6	10	7,4	1,0	4,2
3	10	11	12	7,2	17	1,0	11	3,8	6,7	3,7
4	14	14	14	7,7	30	1,7	14	9,1	10	4,3

Bioenergy potentials non-irrigated for 2050 [EJ per year]									
Scenario	AFR	CPA	EUR	GUS	LAM	MEA	NAM	PAO	PAS	SAS
1	5,1	4,1	4,9	2,4	8,0	0,5	5,0	1,8	0,7	2,0
2	6,9	9,9	7,4	3,6	14	0,8	10	4,4	1,0	2,6
3	8,2	10	12	8,1	13	0,5	11	2,4	6,8	2,2
4	11	14	14	8,9	23	0,8	14	5,1	10	2,8

Bioenergy potentials irrigated for 2050 [EJ per year]									
Scenario	AFR	CPA	EUR	GUS	LAM	MEA	NAM	PAO	PAS	SAS
1	7,6	4,3	5,4	2,6	9,5	0,7	5,3	2,3	0,8	3,2
2	10	10,3	8,1	3,8	16	1,0	11	6,3	1,0	3,8
3	11	11	13	8,5	14	0,7	12	3,0	7,0	3,4
4	14	15	15	9,4	25	1,0	15	7,1	11	4,0

a

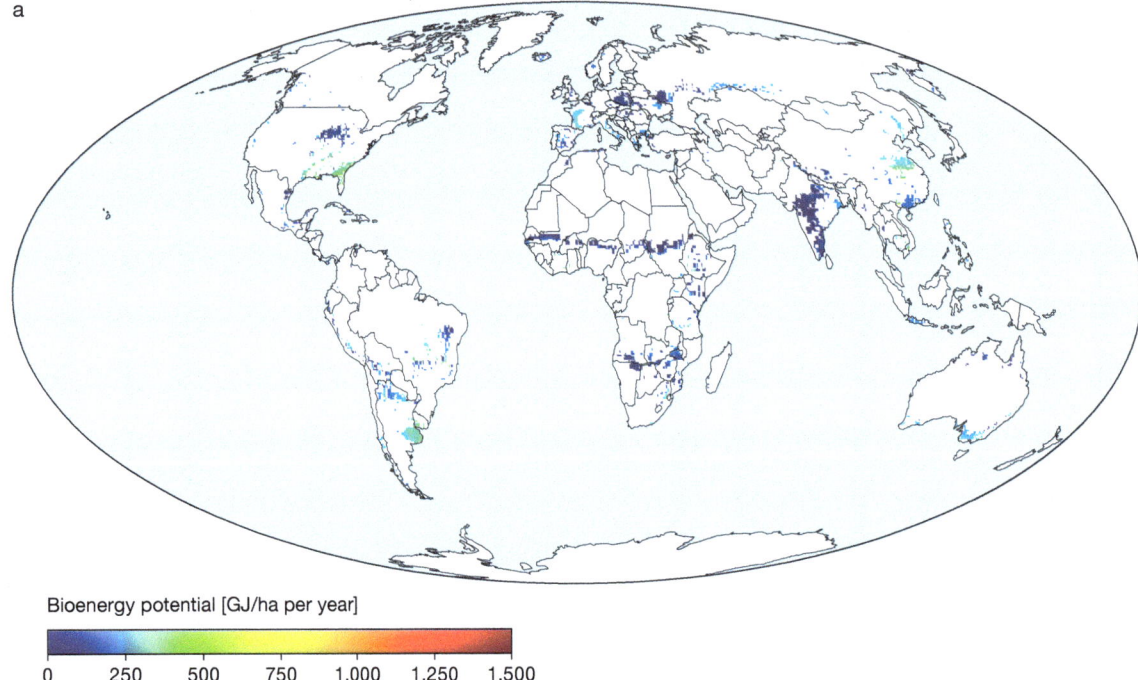

b

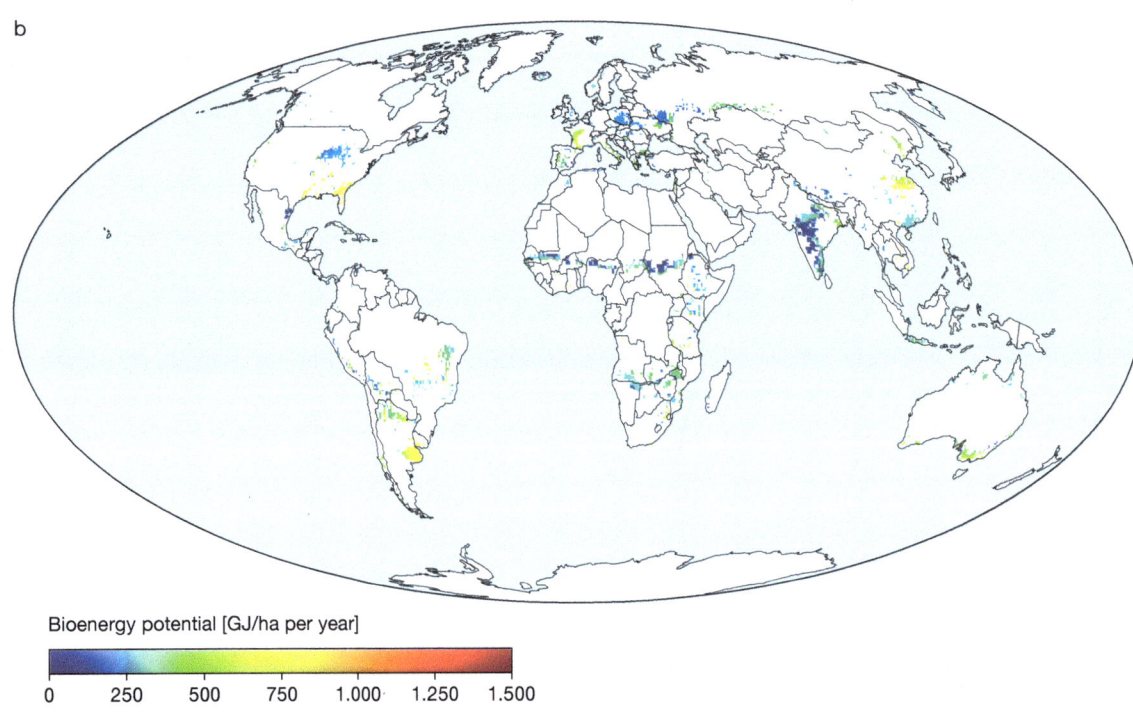

Figure 6.5-1
Geographical distribution of possible energy crop cultivation areas for Scenario 1 (high farmland requirement, high biodiversity conservation). Bioenergy potentials are shown for the year 2050 for (a) non-irrigated cultivation (totalling 34 EJ per year) and (b) irrigated cultivation (totalling 42 EJ per year).
Source: Beringer and Lucht, 2008

a

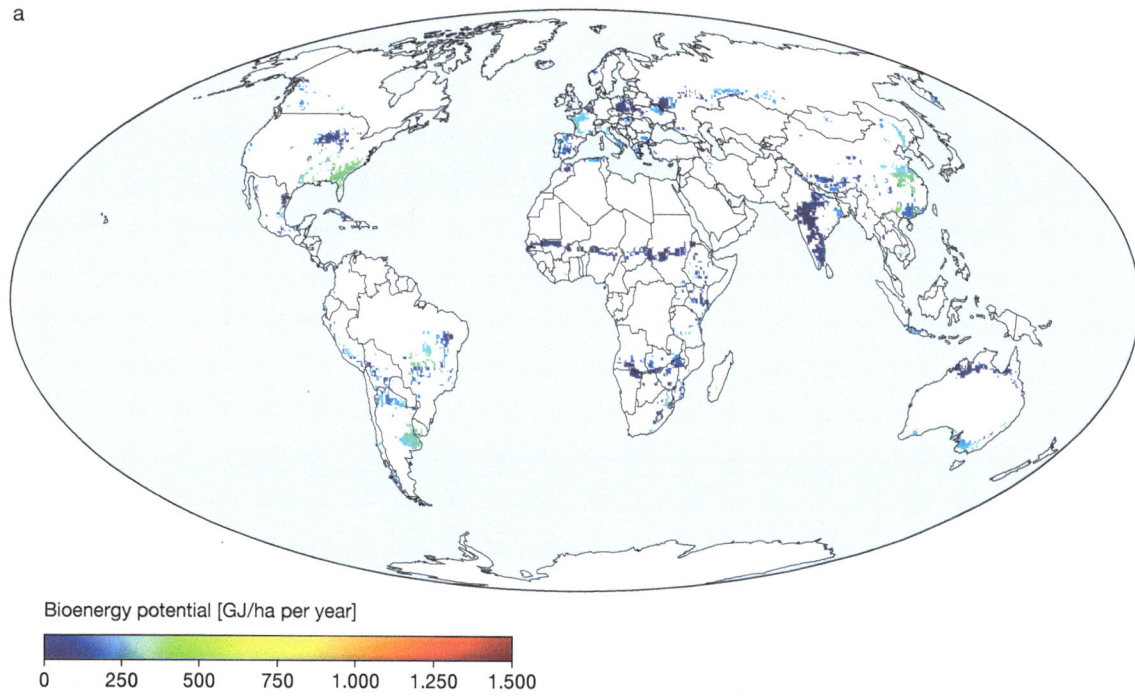

Bioenergy potential [GJ/ha per year]

0   250   500   750   1.000   1.250   1.500

b

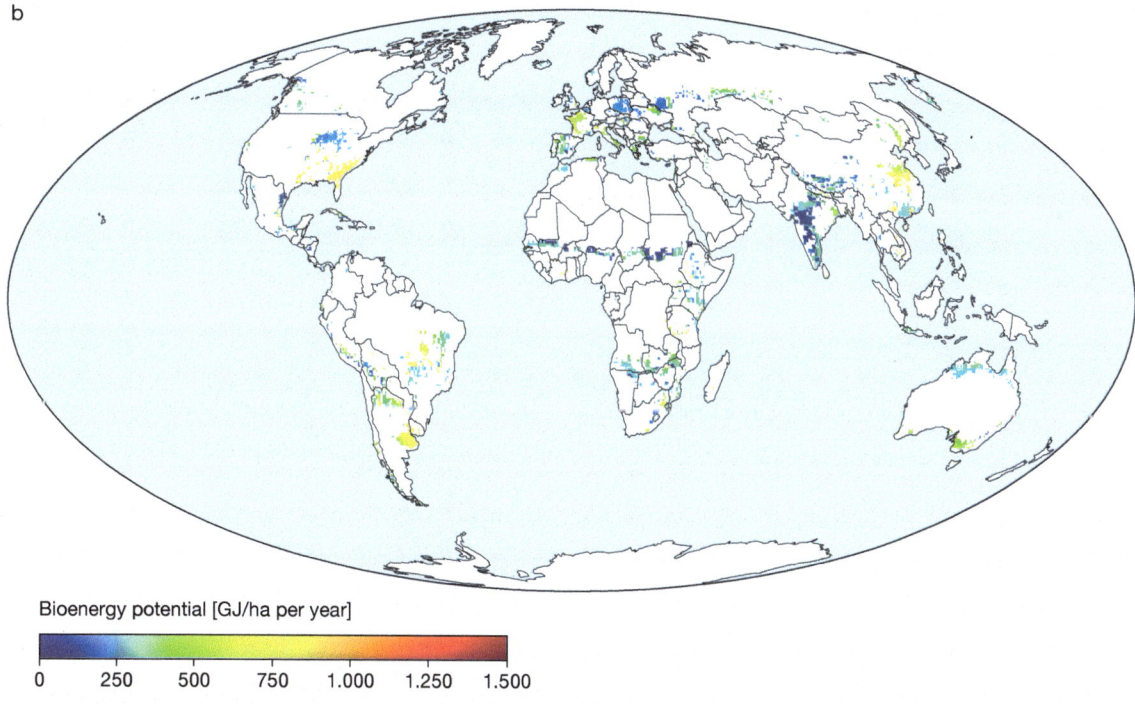

Bioenergy potential [GJ/ha per year]

0   250   500   750   1.000   1.250   1.500

Figure 6.5-2
Geographical distribution of possible energy crop cultivation areas for Scenario 2 (high farmland requirement, low biodiversity conservation). Bioenergy potentials are shown for the year 2050 for (a) non-irrigated cultivation (totalling 61 EJ per year) and (b) irrigated cultivation (totalling 71 EJ per year).
Source: Beringer and Lucht, 2008

a

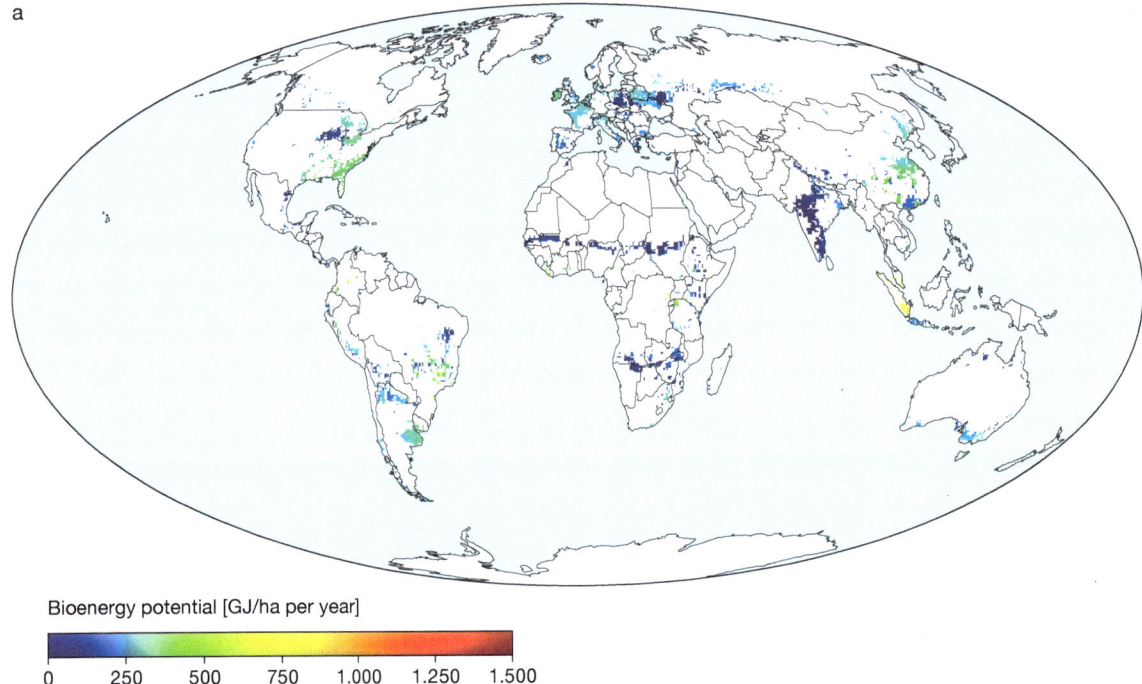

b

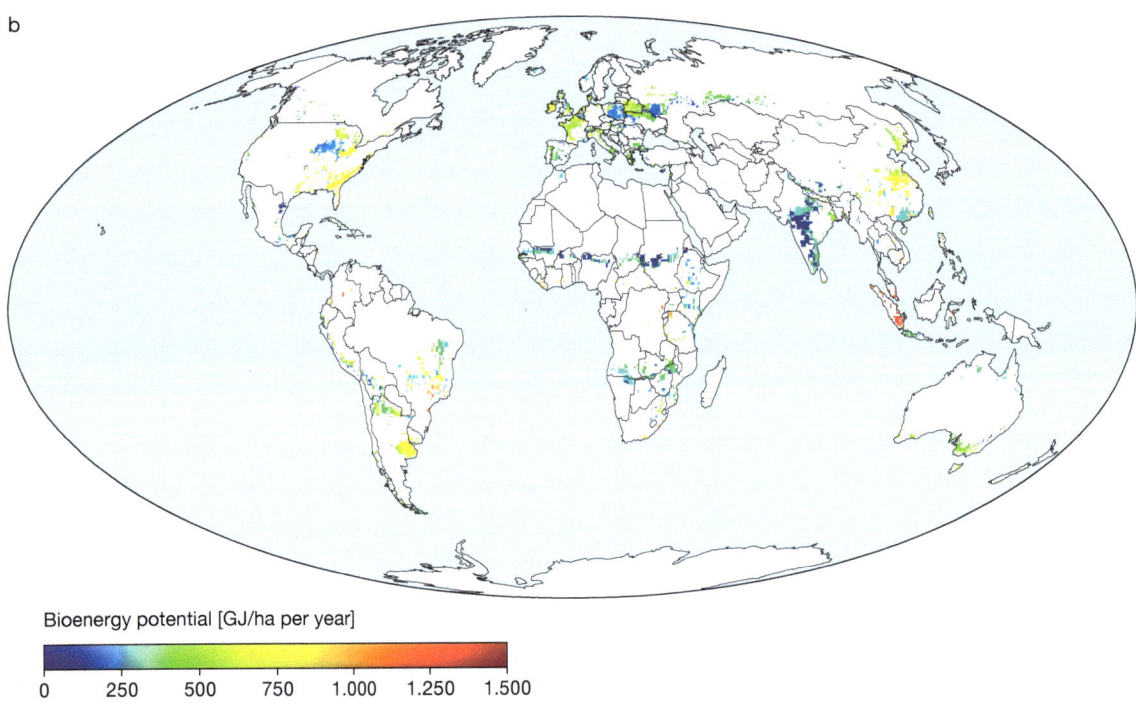

Figure 6.5-3
Geographical distribution of possible energy crop cultivation areas for Scenario 3 (low farmland requirement, high biodiversity conservation). Bioenergy potentials are shown for the year 2050 for (a) non-irrigated cultivation (totalling 74 EJ per year) and (b) irrigated cultivation (totalling 83 EJ per year).
Source: Beringer and Lucht, 2008

a

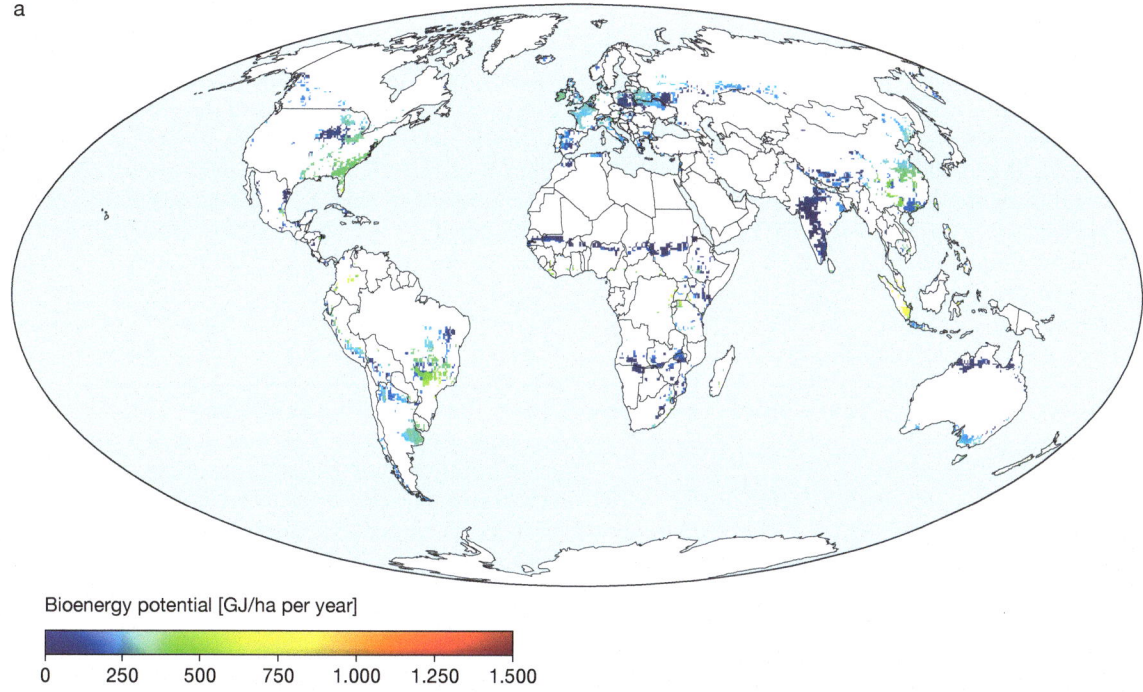

Bioenergy potential [GJ/ha per year]

0   250   500   750   1.000   1.250   1.500

b

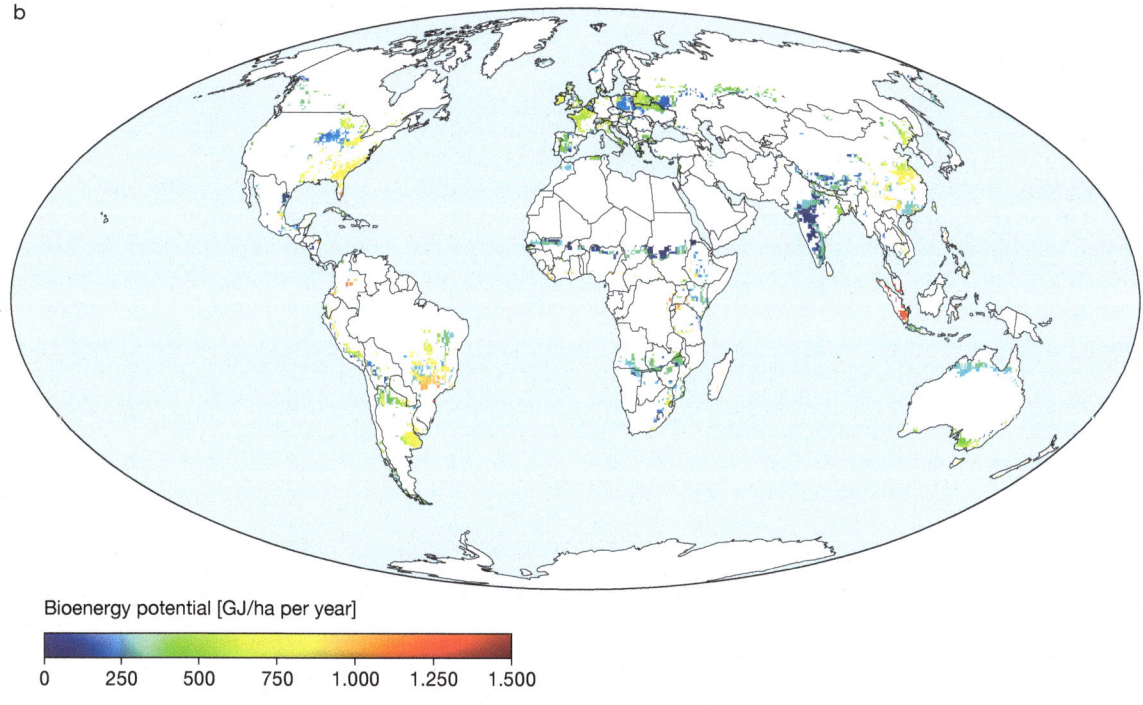

Bioenergy potential [GJ/ha per year]

0   250   500   750   1.000   1.250   1.500

Figure 6.5-4
Geographical distribution of possible energy crop cultivation areas for Scenario 4 (low farmland requirement, low biodiversity conservation). Bioenergy potentials are shown for the year 2050 for (a) non-irrigated cultivation (totalling 100 EJ per year) and (b) irrigated cultivation (totalling 120 EJ per year).
Source: Beringer and Lucht, 2008

These four scenarios yield the following results for global bioenergy potential from the cultivation of energy crops in 2050 (Table 6.5-2). Taking all guard rails into account, the land available for the cultivation of energy crops varies, depending on scenario, between 240 and 500Mha. The yields achievable on this land from non-irrigated cultivation in 2050 have an energy potential of 34–100 EJ per year, while the energy potential from irrigated cultivation is 42–120 EJ per year. Around 75 per cent of the modelled bioenergy potential comes from grasses and 25 per cent from trees. In global terms the comparative figures for potentials in 2000 differ very little from those for 2050, but the regional distribution differs (Table 6.5-3).

6.5.4
Geographical distribution of possible land for energy crop cultivation

The geographical distribution of potential land for energy crop cultivation in 2050 is shown in Figures 6.5-1 to 6.5-4. The major impact of the expanding agricultural area needed for food production in Scenarios 1 and 2 is readily apparent. As a consequence of this land requirement, the productive regions in the medium latitudes of Eastern Europe and North America are unavailable for the cultivation of biomass. Production of biomass as a raw material over relatively large, contiguous areas is therefore restricted to the transition region of the Sahel zone and African savannahs, areas in southern Africa, the Indian subcontinent and parts of northern Australia. The area of land available in the growing areas listed corresponds to between 20 and 30 per cent of the land on which field crops are currently grown.

Bioenergy potentials for ten world regions are shown in Table 6.5-3, with the following regions being used (Figure 6.5-5): sub-Saharan Africa (AFR), China and neighbouring countries (CPA), Europe (EUR), the Community of Independent States (States of the former Soviet Union, CIS), Latin America and the Caribbean (LAM), Middle East and North Africa (MEA), North America (NAM), the Pacific OECD states (with Japan, Australia and New Zealand, PAO), Pacific Asia (South-East Asia, PAS) and South Asia (with India, Pakistan and Bangladesh, SAS).

6.5.5
Biomass yields for trees and grasses

Potential biomass yields for the two crops simulated in the model – high-productivity grasses and fast-growing trees in short-rotation plantations – each grown under both non-irrigated and irrigated condition, and for the two scenarios with the largest and smallest exclusion areas, are shown in Figures 6.5-6 to 6.5-9. The modelled yields of the grasses and trees form the basis for the biomass potentials shown in Figures 6.5-1 to 6.5-4. In the model the land available for growing energy crops, after removal of the exclusion areas, is planted 50 per cent with high-productivity grasses and 50 per cent with fast-growing trees. If only one plant type grows well in a grid cell (e.g. only grasses grown under non-irrigated conditions), the whole area is allocated to this plant type. The resulting dry mass yields are then converted into energy units, assuming a conversion factor of 19.0 kJ per gram (Section 6.3.1.5). The relative contribution of the cultivation of high-productivity grasses and fast-growing trees species to the bioenergy potential of biomass can be estimated from the maps shown here.

6.6
Key uncertainties in the modelling

6.6.1
Quality of the climate data

The various climate models differ from each other, particularly with regard to the simulated changes in precipitation where some of the processes involved are not fully understood or are difficult to simulate. At the same time, the quantity of water available to plants is the most important determinant of the yield potential of biomass plantings in the simulation. However, the low impact of the varying climate data on biomass yields in LPJmL indicates that the effect of altered temperature and precipitation conditions in the areas suitable for biomass planting is relatively small.

6.6.2
Response of plants and ecosystems to climate change

The effects of altered temperature and precipitation conditions and of the increasing carbon dioxide concentration in the atmosphere on the individual plant or on whole ecosystems are still not fully understood. An example is what is known as CO_2 fertilization, which leads to increase water-use efficiency in C3 plants. CO_2 fertilization is most marked in dry areas; in the model it results in a productivity increase of around 10–20 per cent.

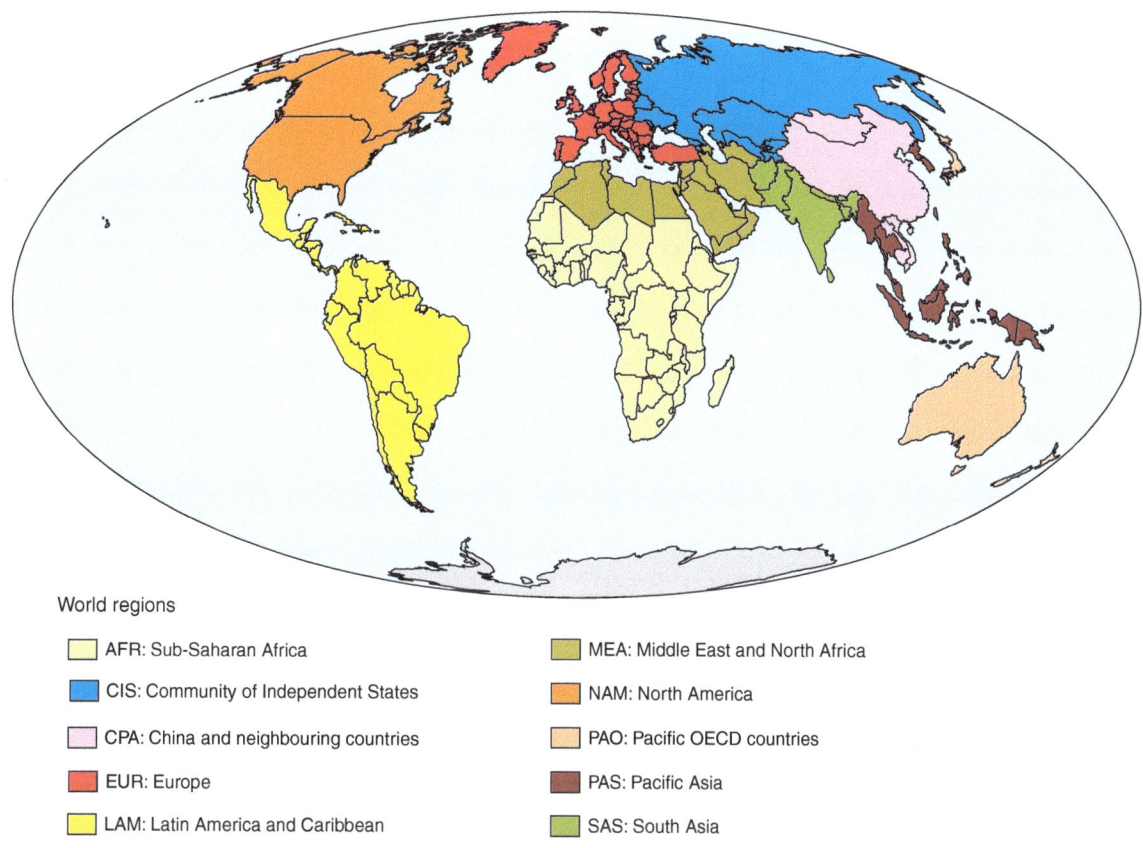

World regions

AFR: Sub-Saharan Africa

CIS: Community of Independent States

CPA: China and neighbouring countries

EUR: Europe

LAM: Latin America and Caribbean

MEA: Middle East and North Africa

NAM: North America

PAO: Pacific OECD countries

PAS: Pacific Asia

SAS: South Asia

Figure 6.5-5
The ten world regions used in this chapter.
Source: Beringer and Lucht, 2008

This effect is the main cause of the increase in plant productivity in the course of the 21st century that occurs in the model. The simulated effects of the increased CO_2 concentration accord with observations (e.g. young forests). However, it is unclear whether the rise in net primary production is permanent.

6.6.3
Availability of water and nutrients

Regional hydrology – with the exception of precipitation – is not included in the model. This means that any existing competition for scarce freshwater resources (Section 5.6.2) is not considered in the calculation of bioenergy potentials arising from the cultivation of energy crops. This is a particular problem in the case of irrigated cultivation, since it is not clear whether the quantity of water required for irrigation is actually available.

A similar problem arises in connection with nutrients that are essential for plant growth. In particular, the negative impact on the climate of greenhouse gas

emissions arising from nitrogen fertilizer use needs to be taken into account.

The application of organic and mineral nitrogen fertilizers on farmland leads to considerable nitrogen losses, because the plants absorb on average less than half of the nitrogen (N) applied (MA, 2005b). The remainder escapes into the air in the form of volatile nitrogen compounds (nitrous oxide N_2O, nitrogen oxides NO_X, ammonia NH_3), or is leached into groundwater as nitrate (NO_3). Nitrous oxide is one of the four major greenhouse gases that affect the climate (Denman et al., 2007). Almost 60 per cent of anthropogenic N_2O emissions are caused by farming (Smith et al., 2007a).

Nitrogen usage in agriculture at present amounts to 127 million tonnes per year worldwide and is forecast to increase by 1.4 per cent per year until 2011/2012 (FAO, 2008b). In the modelling of agricultural production potentials, nitrogen losses caused by farming are not negligible.

In order to avoid nitrogen losses, accurate information about the nitrogen content of the soil and the nutrient requirements of the crop must be available. Where artificial irrigation is used, a monitored irriga-

a

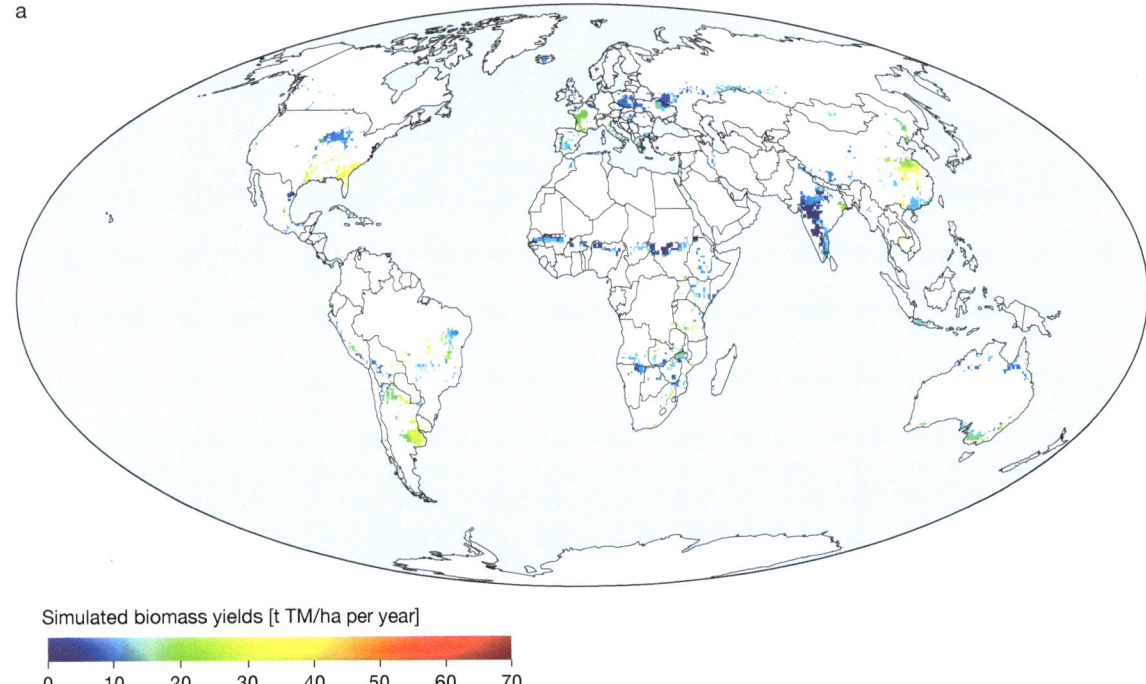

Simulated biomass yields [t TM/ha per year]

0   10   20   30   40   50   60   70

b

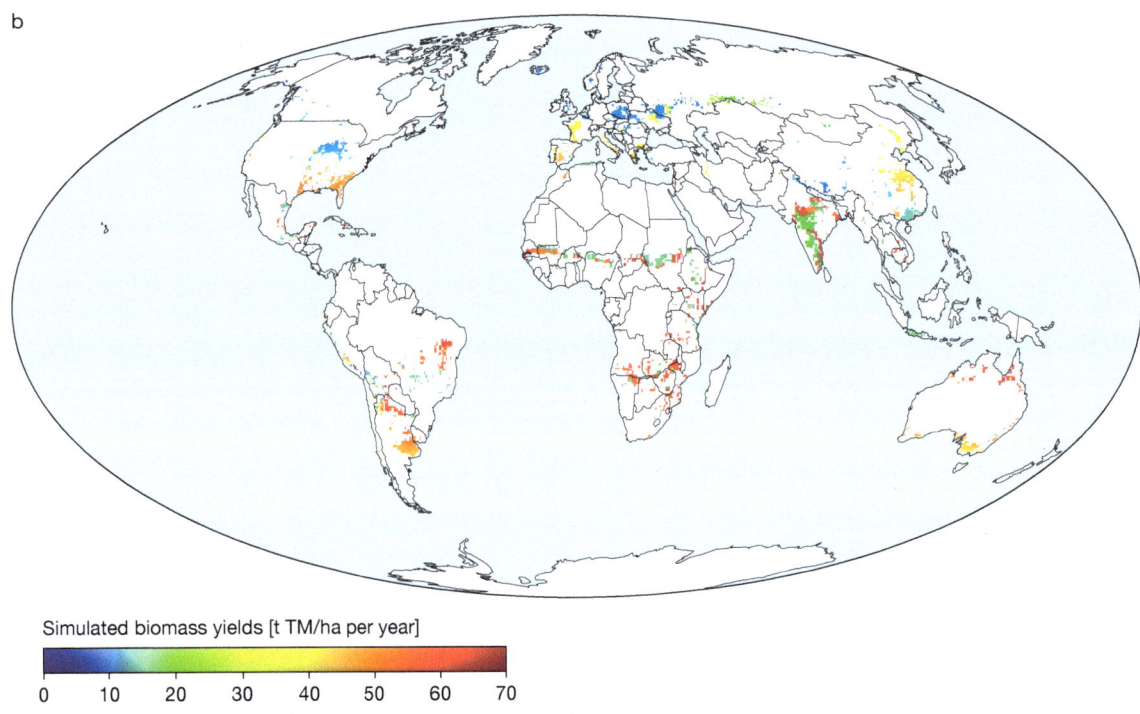

Simulated biomass yields [t TM/ha per year]

0   10   20   30   40   50   60   70

Figure 6.5-6
Simulated biomass yields in the year 2050 for grasses in (a) non-irrigated and (b) irrigated cultivation. Excluded land based on Scenario 1. Calculation of the bioenergy potential assumes that half the remaining land is planted with high-productivity grasses and half with fast-growing tree species. This can be compared with Figure 6.5-7 to provide an estimate of the relative contribution to bioenergy potential of biomass from the cultivation of high-productivity grasses and that from fast-growing tree species.
Source: Beringer and Lucht, 2008

a

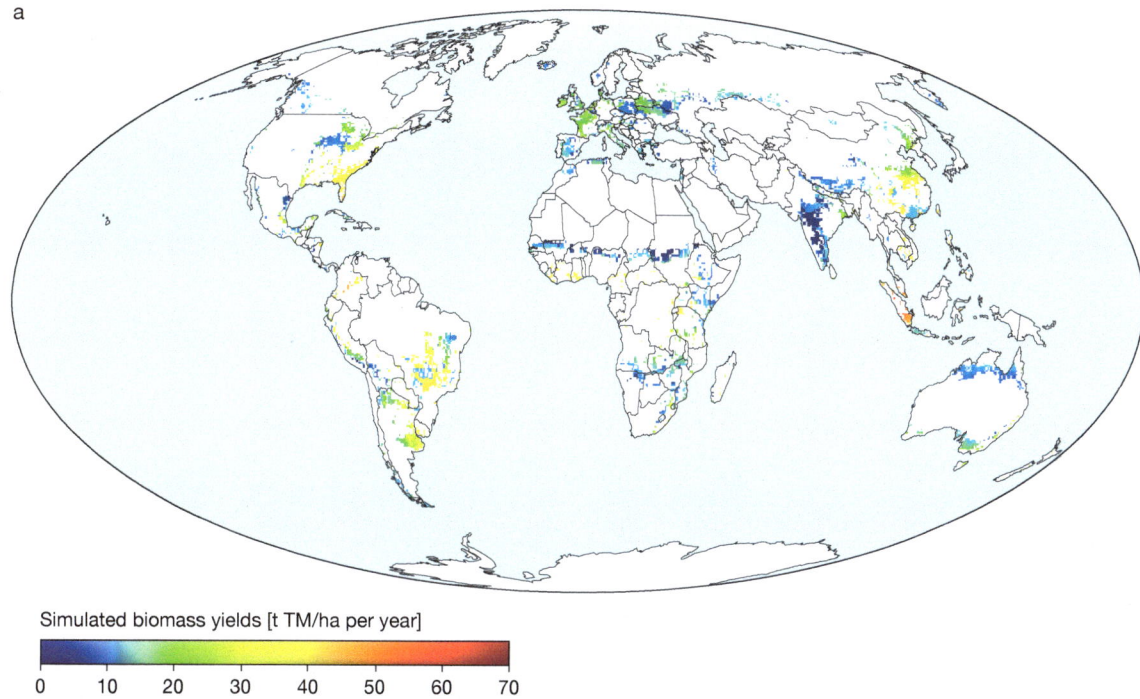

b

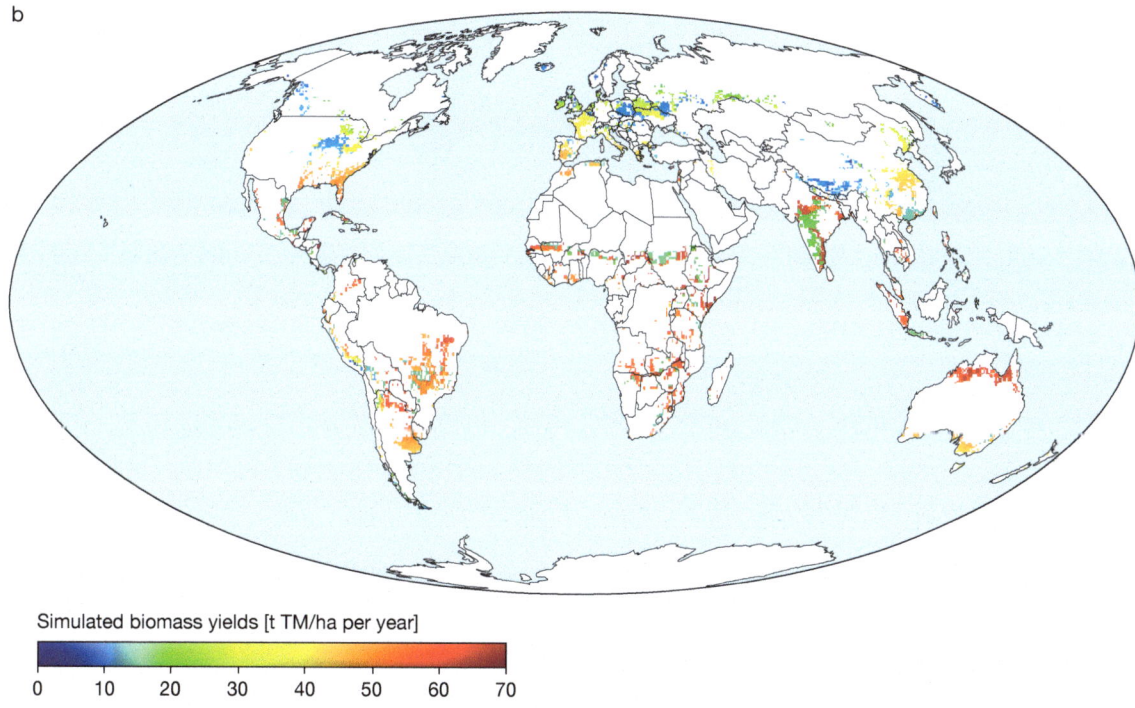

Figure 6.5-7
Simulated biomass yields in the year 2050 for trees in (a) non-irrigated and (b) irrigated cultivation. Excluded land based on Scenario 1. Calculation of the bioenergy potential assumes that half the remaining land is planted with high-productivity grasses and half with fast-growing tree species. This can be compared with Figure 6.5-6 to provide an estimate of the relative contribution to bioenergy potential of biomass from the cultivation of high-productivity grasses and that from fast-growing tree species.
Source: Beringer and Lucht, 2008

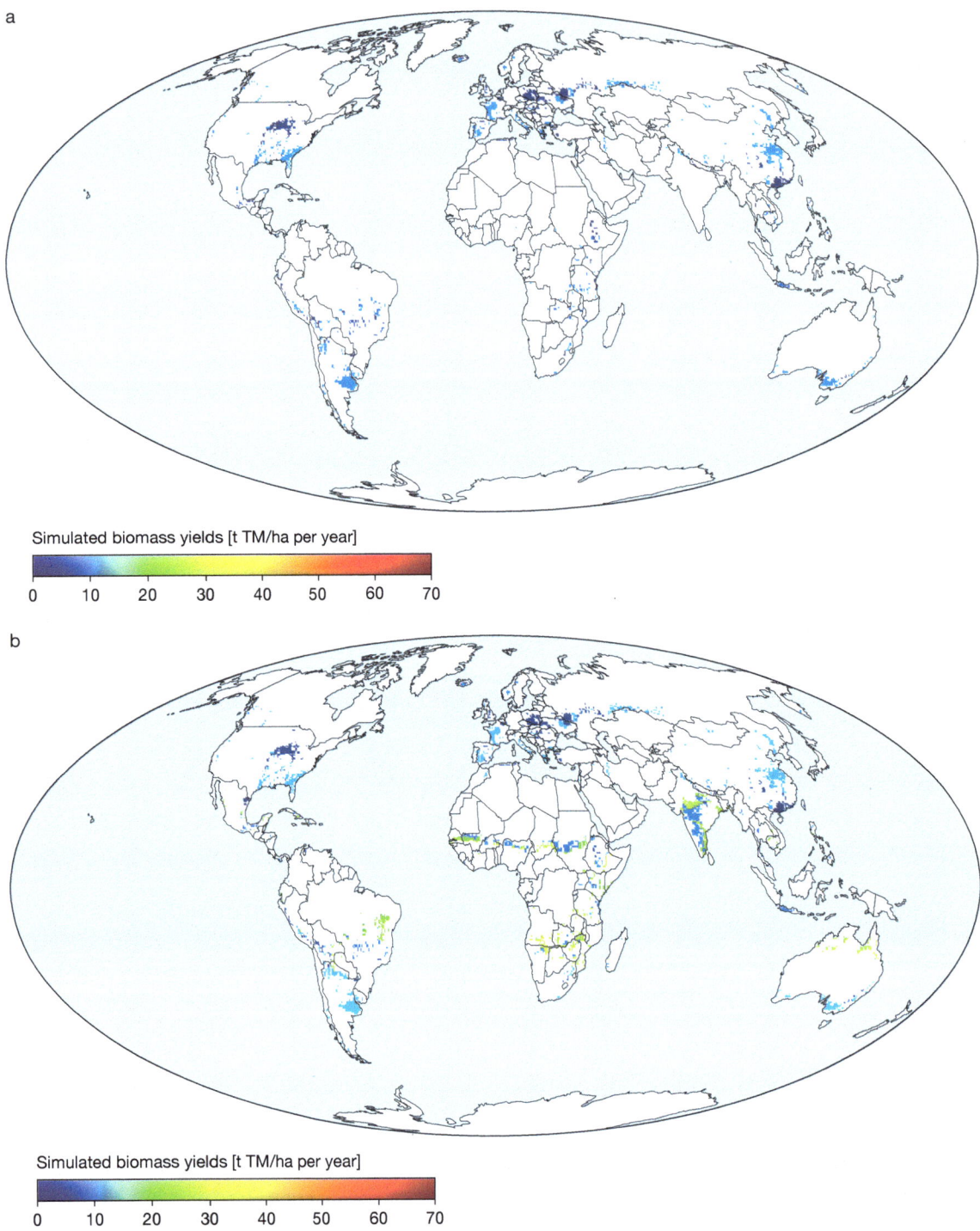

Figure 6.5-8
Simulated biomass yields in the year 2050 for grasses in (a) non-irrigated and (b) irrigated cultivation. Excluded land based on Scenario 4. Calculation of the bioenergy potential assumes that half the remaining land is planted with high-productivity grasses and half with fast-growing tree species. This can be compared with Figure 6.5-9 to provide an estimate of the relative contribution to bioenergy potential of biomass from the cultivation of high-productivity grasses and that from fast-growing tree species.
Source: Beringer and Lucht, 2008

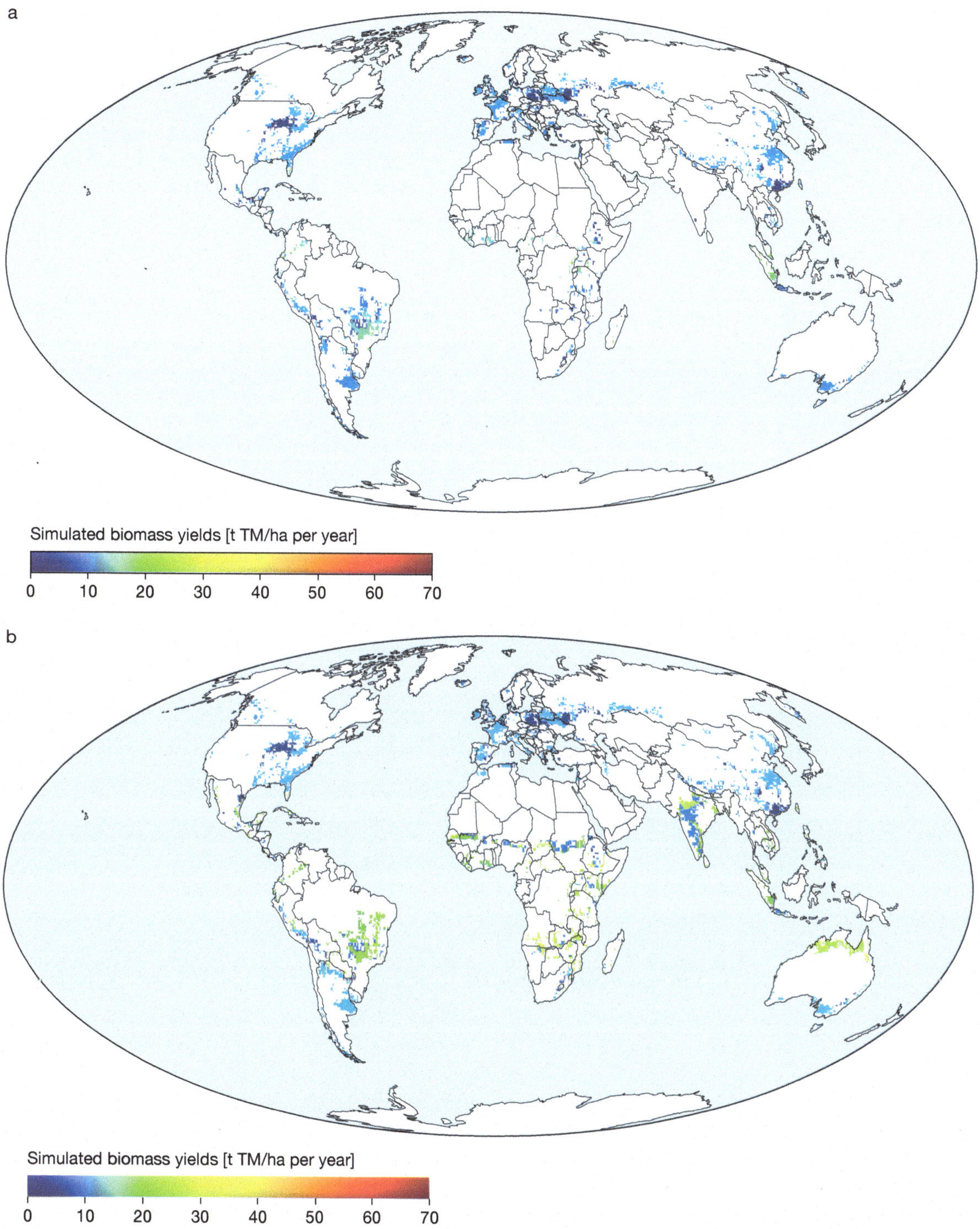

a

Simulated biomass yields [t TM/ha per year]

0   10   20   30   40   50   60   70

b

Simulated biomass yields [t TM/ha per year]

0   10   20   30   40   50   60   70

Figure 6.5-9
Simulated biomass yields in the year 2050 for trees in (a) non-irrigated and (b) irrigated cultivation. Excluded land based on Scenario 4. Calculation of the bioenergy potential assumes that half the remaining land is planted with high-productivity grasses and half with fast-growing tree species. This can be compared with Figure 6.5-8 to provide an estimate of the relative contribution to bioenergy potential of biomass from the cultivation of high-productivity grasses and that from fast-growing tree species.
Source: Beringer and Lucht, 2008

tion regime should be put in place in order to avoid nitrate leaching (Fang et al., 2006).

6.6.4
Development of energy crop yields

The simulation of biomass cultivation areas takes no account of any increase in yield levels over time as a result of breeding improvements or genetic modification (Box 7.1-11). It is entirely conceivable that yield potentials of energy crops will rise considerably as a result of further research work. The simulated yields are, however, already correspond to the levels achieved at existing experimental locations farmed under optimal conditions. In some cases they reach the potentials forecast for 2025, e.g. for the cultivation of switchgrass in North America (*Panicum*; Box 7.1-8).

6.6.5
Land-use data

The availability of land, in terms of current land use, is a major influence on the modelled energy potentials.

On this point, however, there are significant uncertainties in the underlying data sets (Section 4.2.2). Data on unused land in developing and newly industrializing countries is usually based on satellite scans. Local investigations often reveal that local people graze livestock or collect firewood on these areas of apparently unused land. These areas are therefore needed to secure the livelihood of the population and cannot be used in their entirety for the cultivation of energy crops. An indicator for the use of land for grazing is provided by the global distribution of productive livestock density, which is shown in Figure 6.6-1.

An example is the high level of potential for biomass production in India. It is very probable that the data on the distribution of agriculturally used land there is erroneous and that the area of unused, available land is therefore overestimated. The actual potential is therefore likely to be lower than shown in simulations (Box 6.7-2).

Similar problems are raised by the data on marginal land (Box 4.2-1). In this case the additional difficulty arises of estimating potential yields of energy crops on severely degraded soils.

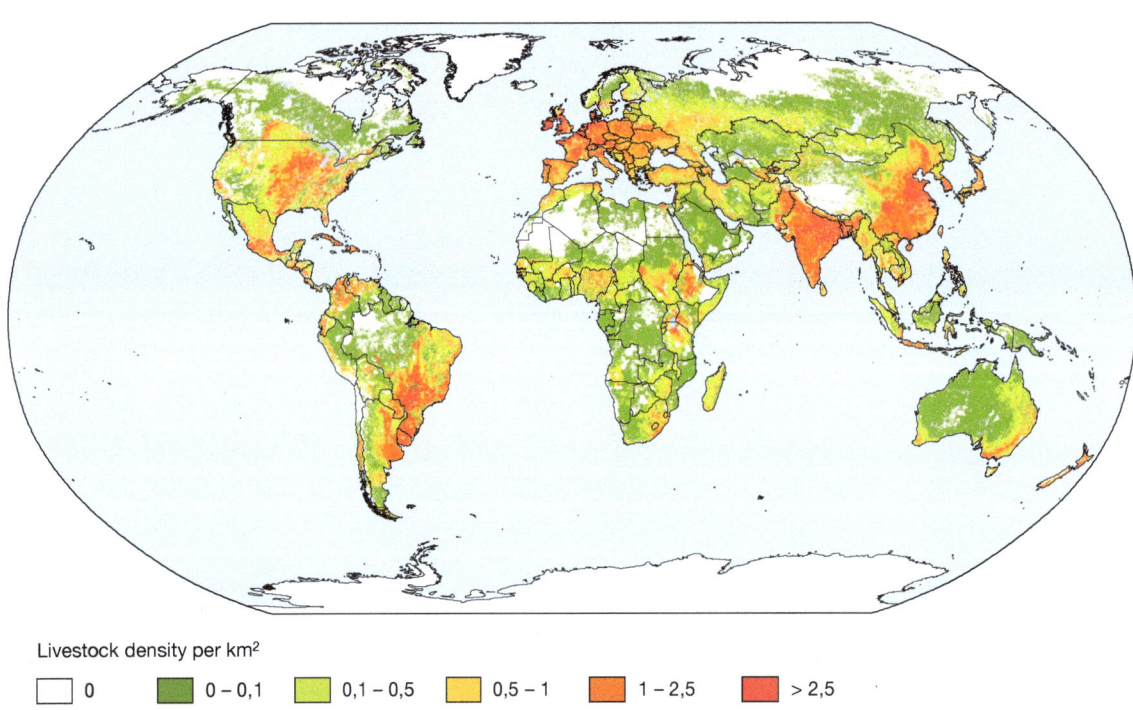

Livestock density per km²

☐ 0   ■ 0 – 0,1   ■ 0,1 – 0,5   ■ 0,5 – 1   ■ 1 – 2,5   ■ > 2,5

Figure 6.6-1
Geographical distribution of livestock density worldwide. The data covers pigs, poultry, cattle and small ruminants. The livestock density is given in livestock units (LU) per km². Smaller animals receive a lower weighting corresponding to their feed needs; for example, a sheep or goat is equivalent to 0.10–0.15 LU (may vary according to region), a pig is equivalent to 0.20–0.25 LU. Sources: FAO, 2003b; Steinfeld et al., 2006

Box 6.7-1

Socio-economic and political indicators

FAILED STATE INDEX
The Failed State Index (FSI) has been produced since 2005 by the Fund for Peace, an independent research institute based in Washington DC, and the journal *Foreign Policy*. The aim is to measure the empirical phenomena of state failure and collapse and thus obtain 'a profile of the world disorder of the 21st century' (Debiel and Werthes, 2006; WBGU, 2007). The Failed State Index comprises twelve social, economic and political indicators, each of which is given a rating of between 0 and 10 points. The twelve indicators are: mounting demographic pressures; massive movement of refugees or internally displaced persons creating complex humanitarian emergencies; legacy of vengeance-seeking group grievance or group paranoia; chronic and sustained human flight; uneven economic development along group lines; sharp and/or severe economic decline; criminalization and/or delegitimization of the state; progressive deterioration of public services; suspension or arbitrary application of the rule of law and widespread violation of human rights; security apparatus operates as a 'state within a state'; rise of factionalized elites; intervention of other states or external political actors. Theoretically, total failure on all indicators would result in a state receiving the maximum FSI rating of 120. The Failed State Index 2008 rates 177 states. *Foreign Policy* categorizes the forty states with the highest FSI ratings as 'critical' or 'endangered'. In the most recent ranking Somalia, with 114.2 points, has the highest score. Countries that are classed as 'endangered' or 'borderline' include Egypt and Laos (both 88.7) and Equatorial Guinea and Rwanda (both 88.0; Foreign Policy, 2008).

WBGU bases its country assessments on these qualitative categories; it considers the likelihood of realizing the theoretically existing bioenergy potentials in countries with an FSI index of 90 or more to be very low (Figure 6.7-1).

GLOBAL COMPETITIVENESS INDEX
The annual Global Competitiveness Reports of the World Economic Forum have since 2006 contained a ranking of the economic competitiveness of nations, compiled on the basis of the Global Competitiveness Index (GCI; López-Claros et al., 2006; Porter et al., 2007). This index rates the investment climate and competitiveness of individual countries, using productivity increases as the key indicator of sustainable economic growth. The Index, which awards ratings ranging from 1 to 7, is based on aggregated data in nine relevant areas: institutions, infrastructure, macroeconomy, health and primary education, higher education and training, market efficiency, technological readiness, business sophistication and innovation. The 131 countries assessed are ranked according to their GCI score; the Index also indicates whether the trend is upwards, downwards or stationary when compared with the previous year. On account of its complex system of indicators and assessment factors covering both qualitative and quantitative data, the GCI is regarded as more informative than many other business indices (von Drachenfels, 2007).

In rating the general business climate in the context of bioenergy production, WBGU categorizes countries as follows: GCI above 5.50 = excellent business climate; 4.50–5.49 = good business climate; 3.50–4.49 = difficult business climate; below 3.50 = unsuitable business climate. The highest score among the countries assessed is 5.67 (USA), the lowest score is 2.78 (Chad).

6.6.6
Future irrigation possibilities

The yield of irrigated biomass cropland is significantly higher than that of non-irrigated cultivation systems. The energy potentials quoted for irrigated cultivation are based on the assumption that 10 per cent of all biomass cropland is irrigated, although at present only around 5 per cent of such land, on a global average, is irrigated (Portmann et al., 2008). At the same time, regulation of soil water content in the model is highly efficient, corresponding to that achieved through trickle irrigation. The results thus presuppose that progressive agricultural technologies are available worldwide and deployed comprehensively. It must therefore be assumed that actual and sustainable biomass potential will more closely resemble that of the non-irrigated scenarios and that only slight increases through partial irrigation will be achieved. In consequence the high harvest yields under irrigated conditions in the transition zone between semi-arid and humido-arid areas of Africa are unlikely to be realized within the next few decades.

6.7
Regional survey

The geographical distribution of possible energy crop cultivation areas described in Section 6.5.4 identifies some regions that seem, on account of their prevailing biogeophysical conditions, to be suited in principle to the sustainable cultivation of energy crops. However, any serious consideration of actually realizable potentials must take account of socio-economic and political conditions, as well as biogeophysical ones, in the regions concerned. The regions that are favourably located for the cultivation of energy crops lie predominantly in tropical and sub-tropical latitudes. Some of these countries are poorly developed, characterized by weak and fragile state structures or affected by unresolved armed conflicts. In the light of these issues, over-optimistic expectations regarding the mobilization of bioenergy potentials in some regions must be corrected. Nevertheless, attempts to realize bioenergy potentials in less developed regions can in the medium to long term result in an agriculturally driven development dynamic and thus help to improve socio-economic conditions.

Box 6.7-2

Country study: India – using marginal land for biofuel production

India's biofuel strategy for biodiesel is based on oleiferous, non-edible plants and fruits. In contrast to countries such as Malaysia, Indonesia or Brazil, where biofuels are grown on fertile soils or areas of cleared forest, India relies on using marginal land ('wasteland'; Box 4.2-1). Although a standardized definition is difficult to arrive at, marginal land is taken to be land that from an agricultural point of view is poorly used, sub-standard or degraded. The Indian Wasteland Atlas classifies 17 per cent (55.2 million hectares) of the country's land as marginal and subdivides this marginal land into 13 categories according to soil type. Thirty-two million hectares are considered to be cultivable in principle (Ministry of Rural Development, 2003). For comparison: the WBGU scenarios identify between 28 and 32 million hectares of potentially usable land in India. The Indian government considers 17.4 million hectares to be suitable for *Jatropha* cultivation (Planning Commission, 2003). According to calculations of the Indian research institute TERI and the German international development corporation GTZ, as much as 38 million hectares would need to be brought under cultivation if the potential national goal of 20 per cent biodiesel blending by 2030 is to be achieved (anticipated biodiesel demand in 2030 ca. 203 million tonnes, compared with 66 million tonnes in 2010). In addition, yields would have to rise five-fold from the current level of 1–2 t per hectare to 5 t per hectare (TERI and GTZ, 2005).

Sixty-two per cent of India's land is currently used for farming. Forested areas occupy 22 per cent (data for the year 2000; Ministry of Agriculture, 2008); a target of one-third by 2012 has been set (MoEF, 2006). At the same time the availability of farmland is dwindling as a result of settlement expansion, industrial use and degradation. Degradation is further exacerbated by fuelwood use, large numbers of livestock and climatic changes. To meet the demand for food from the country's own resources, the productivity of Indian food production would have to rise by more than 50 per cent in the next two decades (TERI and GTZ, 2005). In September 2008, after lengthy internal discussions, the Indian government adopted a new biofuel policy. The blending quota of 20 per cent for biodiesel was affirmed and is due to be achieved by 2017 (Economist, 2008b). In view of the already high intensity of land use, it is questionable how such ambitious biofuel targets can be achieved in one of the most densely populated countries in the world without putting usable land under further pressure.

Furthermore, how much of India's marginal land can actually be used for growing *Jatropha* and similar oleiferous plants depends not only on biophysical parameters but also on land and usage rights. India has the largest number of rural poor and landless people in the world. The inadequate access of large parts of the rural population to usable land and the disregard of existing land and usage rights already represent an obstacle to development and a critical political factor in Indian democracy (Hanstad et al., 2008). The marginal land earmarked for growing *Jatropha* is often state-owned or common land that is used by landless people to collect fuelwood and graze livestock. Any large-scale leasing of these estates to agricultural businesses without allowing participation of the rural poor and landless is likely to lead to displacement. Moreover, land-use rights are often unclarified, because there is a wide discrepancy between the official title to land and actual use as sanctioned by customary law. There is also a lack of accurate statistics describing how many people live in these areas.

In practice there are already signs in some Indian states of conflicts arising from land issues and the generous state support provided to private companies in connection with *Jatropha* cultivation (Grain, 2008; Peoples Coalition, 2008; Shiva, 2008). Local participative procedures function poorly, increasing the risk of eviction. At local level corruption encourages the de facto privatization of common land; in some cases land that the government has specifically made available for *Jatropha* cultivation has in practice been used for other purposes (Altenburg et al., 2008). Particular attention must be paid to these social and political aspects of land use if the potentials of marginal land for energy crop cultivation are to be used and social repercussions are to be avoided (TERI and GTZ, 2005; Cotula et al., 2008). In view of the farmland actually available for growing energy crops and the existing social risks, the results of the modelling appear over-optimistic.

WBGU considers three factors to be of particular importance in this context. Firstly, realization of the theoretically possible bioenergy potentials requires at least a minimum of investment activity; this in turn cannot take place without a minimum level of security and stability. Where security and stability do not exist and cannot be ensured in the foreseeable future, there is no suitable foundation for the creation of a dynamic bioenergy farming system. Secondly, rapid realization of bioenergy potentials requires certain infrastructure and logistics capacities, which in many developing countries do not exist at the requisite level. It is because of location-specific disadvantages that many developing countries remain largely outside the global economic dynamic, despite positive initiatives by internal and external players (Collier, 2007). Thirdly, the extent to which bioenergy potentials, as defined by WBGU, can be actually realized depends in part on the capacity to observe the sustainability guard rails (Chapter 3). This in turn presupposes more than security and stability and sound infrastructure; it requires a minimum level of state regulatory competence in order to define an appropriate administrative and legal framework and to monitor and enforce adherence to it. In forested regions in which deforestation is already proceeding at a rapid rate, the absence of a strong regulatory framework is a particularly critical factor. Without effective controls, further economic or political incentives for the cultivation of energy crops in these areas could have devastating consequences for the conservation of biological diversity and the mitigation of climate change, and run counter to the sustainable realization of bioenergy potentials (Section 5.5.1.1).

In order to identify the extent to which political, institutional and socio-economic conditions within a country limit realization of the theoretically achievable bioenergy potentials, WBGU draws on the Failed State Index of the Fund for Peace and the journal Foreign Policy, the Global Competitiveness Index of the World Economic Forum (Box 6.7-1) and on expectations arising from the sustainability guard rails for individual countries and regions formulated in Chapter 3.

The regions that are of particularly interest on account of their theoretically available sustainable bioenergy potentials are evaluated below. The indicators used can only depict the present situation and at best suggest trends for the immediate future. Nevertheless, they permit regional bioenergy production conditions to be assessed from a wider perspective than that of purely biogeophysical considerations. Since major structural changes are unlikely in the short term, it must be assumed that countries that perform poorly on these indices will continue in the medium term to provide only restricted or even severely restricted opportunities for realizing bioenergy potentials (Figure 6.7-1). This must be borne in mind in the following interpretation of the modelling and assumptions about the actual potentials of bioenergy in a sustainable global energy system.

The qualitative regional assessment of the sustainable bioenergy potentials on the basis of the two above-mentioned indices is restricted to six of the ten regions identified for the preceding modelling. These regions are Latin America and the Caribbean (LAM), sub-Saharan Africa (AFR), China and neighbouring countries (CPA), the Community of Independent States (CIS), Pacific Asia (PAS) and South Asia (SAS). The remaining regions are not considered here, either because theoretical bioenergy potentials are low (e.g. the Middle East and North Africa, MEA), or because economic and governance capacities can for the foreseeable future be regarded as adequate, as for example in North America (NAM), which has theoretically possible potentials of 5–15 EJ per year, and Europe (EUR), which also has potentials of 5–15 EJ per year.

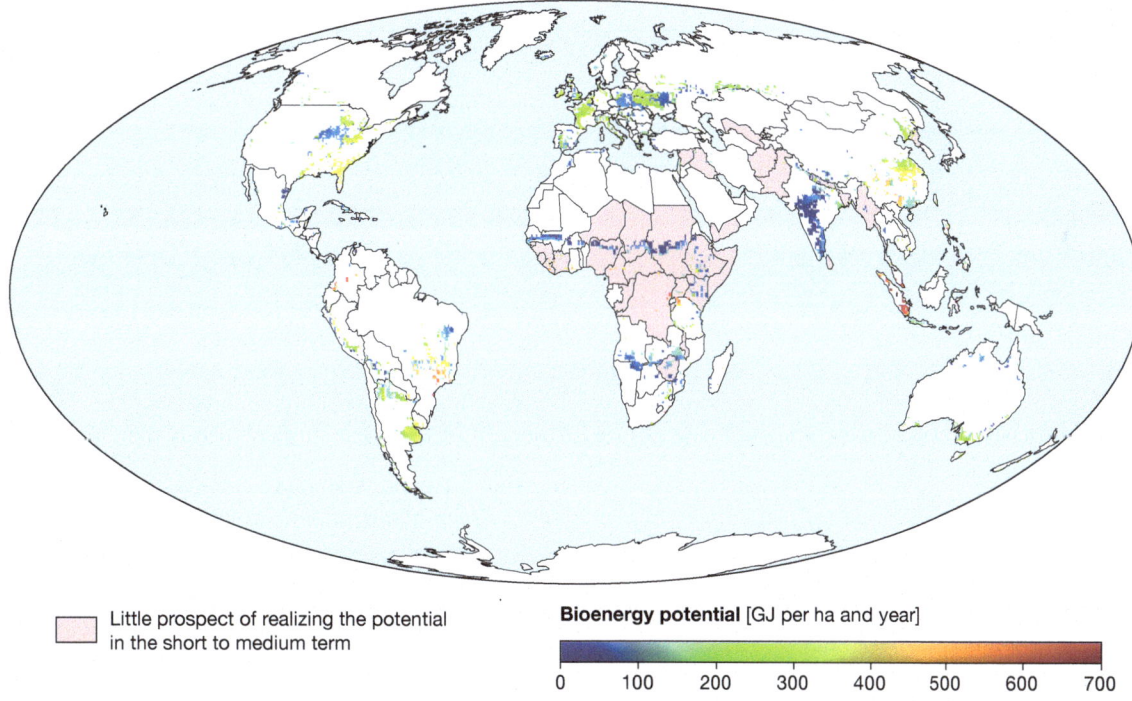

Little prospect of realizing the potential in the short to medium term

Bioenergy potential [GJ per ha and year]

0   100   200   300   400   500   600   700

Figure 6.7-1
Regions with potential for sustainable bioenergy from crops and countries that are affected by state fragility or collapse of the state. The map shows the distribution of possible areas for the cultivation of energy crops and the potential production in the year 2050 for a WBGU scenario involving a low farmland requirement, high level of biodiversity conservation and non-irrigated cultivation (Scenario 3). One pixel corresponds to 0.5° x 0.5°. In order to assess whether the identified sustainable bioenergy potentials are likely to be realizable, the quality of governance in individual countries was rated using the Failed State Index (FSI). The countries coloured light red have an FSI > 90, indicating that in the short to medium term the prospect for realizing bioenergy potentials can be regarded as poor.
Source: WBGU, drawing on data from Beringer and Lucht, 2008 and from Foreign Policy, 2008

6.7.1
Latin America and the Caribbean

The greatest hopes for large-scale bioenergy production are pinned on the greater region of Latin America; in many parts of the region biogeophysical conditions are ideal. The scale of theoretical sustainable bioenergy potential in Central and South America ranges from 8 EJ per year in the least favourable case (Scenario 1 non-irrigated) to 25 EJ per year (Scenario 4 irrigated). This is equivalent to 22–24 per cent of the modelled global potential. Except in politically unstable countries such as Columbia, Bolivia and Haiti, political, institutional and socio-economic conditions in the area are comparatively good (Faust and Croissant, 2007). Brazil and Argentina, which together represent more than half of both the area and the population of South America, have scores of 67.6 and 41.4 respectively in the Failed State Index. In terms of competitiveness the majority of countries in the region, which at the same time represent between them by far the largest part of the territory and of the population, have a GCI index of around 4; the average index of the Central and South American countries covered by the GCI is 3.87. In the World Economic Forum ratings Chile, with a GCI index of 4.77, lies above industrialized countries such as Spain (4.66), Italy (4.36) and Greece (4.08), and the regional powerhouses of Mexico (4.26) and Brazil (3.99) are significantly above the average. Of the larger Latin American countries, only Venezuela (3.63) and Bolivia (3.55) are significantly below average. Future political development in Bolivia is difficult to assess. The Failed State Index (84.2) identified the country as bordering on instability even before the violent unrest of autumn 2008.

Compared with regions in other developing countries, socio-economic and political conditions for realization of the theoretical bioenergy potentials in South and Central America are relatively favourable. Brazil's rapid rise to world market leader in bioethanol production gives further evidence for this presumption (Box 8.2-4). Nevertheless, development of bioenergy potential along sustainable lines faces two major challenges in Central and South America. Firstly, the need to conserve the rainforests gives rise to significant conflicts over land use. Brazil alone is the source of around one-fifth of global greenhouse gas emissions from deforestation and is thus the second-largest generator of such emissions in the world. Peru (in 8th place) and Venezuela (in 10th place) are also among the countries with the highest deforestation rates in the world (Section 5.5.1.1; Figure 5.5-1). Strategies must therefore be developed to reduce deforestation rates in these countries and prevent the increasing cultivation of energy crops exacerbating the pressure of use on existing forest areas. Secondly, in this region, and particularly in Central America, it is essential that bioenergy strategies are flanked by food security policies and measures for preventing displacement of food production by energy crops. In addition, large-scale agro-industrial cultivation of bioenergy plantations may be obstructed by the increasing frequency and severity of storm and flood events as a consequence of climate change (WBGU, 2007).

6.7.2
China and neighbouring countries

The scale of theoretical bioenergy potential in China including Hong Kong and the neighbouring Asian economies ranges from 4 EJ per year in the least favourable case (Scenario 1 non-irrigated) to 15 EJ per year (Scenario 4 irrigated), equivalent to around 12–13 per cent of global potential or 5–20 per cent of China's primary energy requirement of 72.9 EJ in 2005 (IEA, 2007d). The key country here is China, for which the modelling indicates potentials under the two above scenarios of respectively 4 EJ and 13 EJ per year, which in both cases is just under 90 per cent of the potential of the entire region.

Socio-economic and institutional conditions for the sustainable mobilization of bioenergy potential in China are favourable, provided that efforts are made to introduce and implement the guard rails and sustainability standards proposed by WBGU; in particular, competition with food and feed production would need to be considered (Box 5.2-2). The Failed State Index currently gives China a rating of 80.3, which indicates a risk of political instability. However, the Chinese government has proved to be competent in matters of economic policy and is becoming increasingly aware of environmental issues. The Global Competitiveness Index reflects the economic dynamics of the past three decades with a figure of 4.57, placing China on the same level in this regard as Tunisia (4.59), the Czech Republic (4.58) and Saudi Arabia (4.55). If the Chinese government and national and international companies decide in favour of systematic use of the country's bioenergy potential, there is in principle no reason why the necessary investment should not be made and the relevant capacities developed. In view of the rapid growth of energy demand in China, bioenergy would be a component of a more climate-friendly growth strategy, particularly if the expansion of bioenergy enabled the use of fossil energy carriers (particularly the combustion of coal) to be reduced.

6.7.3
Pacific Asia

The scale of theoretical bioenergy potential in Pacific South-East Asia ranges from 1 EJ per year in the least favourable case (Scenario 1 non-irrigated) to 11 EJ per year (Scenario 4 irrigated), which corresponds to around 2–9 per cent of the modelled global potential for energy crop cultivation. The range spanned by the scenarios is larger than in most of the other regions modelled on account of the large amount of land reserved for nature conservation in the more restrictive scenarios.

Some newly industrializing South-East Asian countries are growing bioenergy crops on a large scale. This is particularly apparent in Malaysia and Indonesia (Box 5.4-2), which have in recent years become the two leading producers of palm oil in the world; however, many of their plantations have replaced areas of tropical rainforest and cannot be considered sustainable (FWA, 2007). In WBGU's view, significant expansion of sustainable bioenergy potentials in the Pacific/Asian region to 11 EJ is only possible if the need for farmland is low and if lower standards of nature conservation are tolerated (Scenario 4). For the more restrictive scenarios the modelling identifies sustainable potentials ranging from only 1 EJ to 7 EJ (Scenario 3) per year.

The situation with regard to current political, institutional and socio-economic conditions in the region is mixed (Faust and Croissant, 2008). Some transition countries in the region whose political stability is questionable are classed as having relatively good conditions for investment and business. This applies, for example, to the island states of Indonesia (FSI index 83.3; GCI index 4.24) and the Philippines (83.4; 3.99) and also to Thailand (75.6; 4.70). The best conditions for further bioenergy expansion in the region are in Malaysia (FSI index 67.2); its authoritarian elective monarchy is capable of acting and stable and its global competitiveness, with a GCI index of 5.10, is rated nearly as highly as that of Australia (5.17). The Malaysian government is already actively promoting the development of the palm oil industry with the aim of consolidating its leading position in the world market and increasing the amount of value added in processing (FWA, 2007). Almost no consideration is given to the negative effects on climate change mitigation, species diversity or the price of palm fat, a staple food. Sustainable realization of bioenergy potentials in Malaysia, and also in Indonesia (Box 5.4-2) and other South-East Asian countries, would require major changes in policy, of which there is at present little prospect.

6.7.4
South Asia and India

The scale of theoretical bioenergy potential in South Asia ranges from 2 EJ per year in the least favourable case (Scenario 1 non-irrigated) to 4 EJ per year (Scenario 4 irrigated); this is equivalent to around 3–6 per cent of the modelled global potential. India produces by far the largest proportion of this, namely 2 EJ per year (Scenario 1 non-irrigated) or 3 EJ per year (Scenario 4 irrigated), which corresponds to 9–18 per cent of the Indian primary energy requirement of 22.5 EJ in 2008 (IEA, 2007d).

Solely India, with its total area of around 3.3 million km^2, has large areas – partly of marginal land – that are in principle suitable for sustainable bioenergy production (Box 6.7-2). The Indian states vary, sometimes widely, in their political and economic conditions. This means that in some parts of the country the feasibility of realizing these potentials is limited, while elsewhere rapid progress may be achievable. The prospect of progress is particularly high in areas where the production of energy crops is already being specifically promoted at state level (Box 10.8-1).

India has an FSI index of 72.9. With regard to global competitiveness it has a GCI index of 4.33, which puts it in roughly the same position as other newly industrializing countries such as South Africa (4.42) and Mexico (4.26). In principle, therefore, India is in a relatively favourable position to mobilize its bioenergy potential, although actual potentials may turn out to be more modest than envisaged in the modelling of technical bioenergy potential or the plans of the Indian government (Box 6.7-2). Whether a sustainable bioenergy strategy can succeed depends above all on the political will to observe sustainability standards and to pursue a path towards food security at the same time as developing a bioenergy strategy. In order to avoid land-use conflicts, it is also important to consider the rights of the poor population groups who live in many marginal areas.

6.7.5
Sub-Saharan Africa

The scale of theoretical sustainable bioenergy potential in sub-Saharan Africa ranges from 5 EJ per year in the least favourable case (Scenario 1 non-irrigated) to 14 EJ per year (Scenario 4 irrigated); this constitutes 12–15 per cent of the global potential. Within this macro-region there are great biogeophysical differences – one need only compare, for example, the humid tropical conditions of the Congo basin and the arid and semi-arid territory of the southern African states. Furthermore, the Republic of South Africa, being the dominant regional power and the largest national economy on the continent, occupies a special position. The largest potentials for bioenergy production are considered to lie in the Sahel belt, particularly in Nigeria, Mali and Sudan, and in parts of East and South-East Africa.

If these estimates of potential are considered alongside the political, institutional and socio-economic conditions in the region, it becomes clear that there is little hope of the theoretical potentials being even approximately realized in the medium term. Eleven of the twenty countries rated by Foreign Policy as particularly critical are in sub-Saharan Africa. They include Sudan (FSI index 113.0), Chad (110.9) and the Democratic Republic of Congo (106.7), which in terms of area are three of the largest countries in the region, with a combined area of more than 6 million km². The stability of a further fifteen countries south of the Sahara – including agriculturally important states such as Kenya (93.4) and Cameroon (91.2) – is considered to be acutely endangered. Prospects for political and economic consolidation are considered to be small (Grimm and Klingebiel, 2007). Subregions such as the Sahel zone, the Horn of Africa and southern Africa are likely to come under additional pressure in the near future on account of the consequences of climate change and persistently high food prices, limiting their scope for action (Bauer, 2007; WBGU, 2007).

In the light of the precarious political and institutional circumstances, business conditions and competitiveness in the region must also be considered critically (Altenburg and von Drachenfels, 2007). The ten lowest ratings on the Global Competitive Index include eight countries in sub-Saharan Africa with GCI indices between 3.29 (Zambia) and 2.78 (Chad). South Africa, with a GCI index of 4.42, is the only sub-Saharan country in the top third of the GCI ranking.

Furthermore, the Democratic Republic of Congo, Zambia and Nigeria are the countries with the fifth-, sixth- and seventh-highest deforestation rates in the world (Section 5.5.1.1; Figure 5.5-1). This means that when the sustainability guard rails for climate protection and biological diversity are considered, it becomes clear that the preconditions for expansion of sustainable bioenergy production are unlikely to be met. There is no prospect in the immediate future of any serious attempt to control logging in these countries or in the other timber-rich countries of western and central Africa. Nevertheless, the specific opportunities for mobilizing sustainable bioenergy production should continue to be examined in sub-Saharan countries that have comparatively favourable institutional structures, such as Mali, or that are pinning

great hopes on the cultivation of energy crops, such as Mozambique (Namburete, 2006).

6.7.6
Community of Independent States (CIS)

The scale of theoretical sustainable bioenergy potential on the territory of the former Soviet Union ranges from 2 EJ per year in the least favourable case (Scenario 1 non-irrigated) to 9 EJ per year (Scenario 4 irrigated). This is equivalent to 7–8 per cent of global potential.

This is another major region in which consideration of socio-economic conditions and political stability reveals a mixed picture. In Russia (FSI index 79.7; GCI index 4.19), Kazakhstan (72.4; 4.14) and especially in Ukraine (70.8; 3.98), socio-economic conditions are clearly conducive to the mobilization of bioenergy potentials. Ukraine is traditionally an important agricultural producer; with a GCI index comparable to that of Brazil it appears well placed to realize optimistic expectations with regard to bioenergy production. In all countries mentioned, though, it is questionable whether there exists the political will to combine such an option with sustainability criteria. Moreover, the region is vulnerable to political crises and conflicts and is overall relatively unstable (Grävingholt, 2007).

6.8
Interpretation and conclusions

The global modelling commissioned by WBGU of the technical sustainable bioenergy potential from the cultivation of energy crops, in the light of sustainability guard rails, indicates an annual potential of 34–120 EJ (Section 6.5) in 2050, depending on the scenario used. In view of the uncertainties described in Section 6.6 and the large differences in potential between the scenarios, this range will be given as 30–120 EJ in this report. This represents the gross energy contribution, i.e. not including losses that occur during conversion to final energy. To this potential from the cultivation of energy crops can be added a contribution from the utilization of residues from agriculture and forestry, which could amount to around 50 EJ per year (Section 6.1). The global sustainable potential for the use of bioenergy thus amounts to 80–170 EJ per year, which is less than 10 per cent of the anticipated primary energy requirement in 2050.

In evaluating this estimated potential it should be borne in mind that this is the technically sustainable potential: the potential that can actually be realized is significantly less. The figures should therefore be interpreted as an upper limit. Identifying the realizable economic potential requires information about geographical, political and socio-economic conditions in the various regions (Section 6.7), which is difficult to predict for 2050. As an initial rough estimate WBGU assumes that economic potential may equate to around half of the technical sustainable potential, but points out the need for considerable further research in this area.

Moreover, it is likely that the bioenergy potentials identified here overlap with the traditional use of biomass, which currently amounts to around 45 EJ per year (Section 6.1); this overlap is difficult to quantify. This means that the figures quoted do not necessarily represent additional potentials for bioenergy use, but simply a total potential. However, WBGU assumes that the level of traditional bioenergy use will decline by 2050 as a result of efficiency improvements and shifts to other forms of energy.

A further factor that could reduce the use of biomass for energy is the need to use biomass as an industrial feedstock – a need that will increase as oil resources dwindle (Section 5.3). It is estimated that the quantity of biomass required for feedstock use in 2050 will be around 100 EJ per year (Section 6.1). The present requirement of around 25 EJ per year (Hoogwijk et al., 2003) is produced on land that is not included in the estimate of potential in this chapter (e.g. existing cotton-growing land and forests used by the forestry industry), but for the additional requirement of around 75 EJ per year feedstock use would almost certainly be in direct competition with the bioenergy potential considered here. However, it should be borne in mind that at least some of the biomass initially used as material feedstock could subsequently, through cascade use, be used for energy.

Section 7.3 explains that the use of bioenergy from crop cultivation and from the deployment of wastes and residues without indirect land-use changes and with the best energy conversion methods through the substitution of fossil fuels, observing the promotion criteria, can save greenhouse gas emissions of 60t CO_2eq per TJ of raw biomass used. In particularly favourable cases this figure can be as high as 100t CO_2eq per TJ. This applies only if energy carriers with high specific emissions are replaced. If one assumes a maximum bioenergy potential of 80–170 EJ per year, of which around half will be realizable, this is equivalent to around 2–9 Gt CO_2eq or around 1–2 Gt C per year. This can be compared with the annual anthropogenic carbon dioxide emissions from the use of fossil fuels and cement production of 32 Gt CO_2eq (8.5 Gt C) in 2007, or those from land-use changes of 6 Gt CO_2eq (1.6 Gt C) (GCP, 2008).

Annual global emissions of all greenhouse gases amounted in 2004 to around 49 Gt CO_2eq or around 13 Gt C (IPCC, 2007c). Projections for 2050 assume annual greenhouse gas emissions of 50–100 Gt CO_2eq (13–26 Gt C; IPCC, 2000) for the various IPCC scenarios.

In the light of the figures for the global potential of bioenergy obtained here the importance of bioenergy should not be overestimated. At the same time, however, the expected scale is significant and should certainly not be neglected in the future development of energy systems.

In addition, the ability of plants to remove carbon dioxide from the atmosphere through photosynthesis could open up a valuable option for climate change mitigation. The use of bioenergy in combination with carbon capture and storage could result in a slowing of the increase of the CO_2 concentration in the atmosphere or contribute to a reduction in the atmospheric CO_2 concentration once fossil fuels have ceased to be used. Realistically, however, the atmospheric CO_2 concentration can only be reduced by around 0.2 ppm CO_2 per year by this means (Box 6.8-1), while the mean rise in the atmospheric CO_2 concentration in recent years has been around 2 ppm per year (GCP, 2008). For comparison: this technically realizable sequestration rate is significantly below the net rates at which the oceans (2.3 Gt C or 1.1 ppm CO_2 per year) and the terrestrial vegetation (3.0 Gt C or 1.4 ppm CO_2 per year) currently remove carbon dioxide from the atmosphere (GCP, 2008) as a result of the elevated CO_2 concentration.

In view of the major challenge presented by the need to prevent warming exceeding 2°C from pre-industrial levels (Section 3.1.1) one must ask whether and how a larger sustainable bioenergy potential could be realized. As the discussion in Section 6.1 shows, other studies tend to arrive at significantly higher bioenergy potentials if they assume major yield increases on existing farmland and hence project that some of this land will be available for future energy crop cultivation. In addition, the majority of other studies define less strict criteria for nature conservation than WBGU does (Section 3.1.2). A significantly higher bioenergy potential is therefore only sustainably realizable if, as a result of efficiency increases or less land-intensive dietary habits, land previously used for food production becomes usable for energy crop cultivation. This would have to take place in a manner that does not jeopardize the food security of the growing world population and does not infringe the guard rails for soil and biosphere protection.

The land requirement for the future food security of humankind is very uncertain. It depends not only on population growth but also on the development of

Box 6.8-1

Potential for reducing the atmospheric CO$_2$ concentration by deploying bioenergy with carbon capture

Combining bioenergy use with carbon capture and storage technology is increasingly viewed as an option for removing CO$_2$ from the atmosphere. We therefore estimate how big a reduction in the atmospheric CO$_2$ concentration could be achieved by such measures.

The modelling identifies a sustainable bioenergy potential from energy crops of around 30–120 EJ per year. This would involve 1–3 Gt C being fixated annually in the biomass that could be used for energy generation. This is, however, a theoretical potential (Box 6.1-1): WBGU assumes that, for socio-economic and political reasons, only half of this potential can actually be realized (Section 6.8). If ones adds to this potential a further assumed technical sustainable residue potential of around 50 EJ per year, of which it can again be assumed that half is realizable (i.e. an additional 0.6 Gt C), biomass with a stored carbon content of 1–2 Gt C would be available each year.

How much of this carbon can ultimately be sequestered depends on the deployment pathway. If the biomass is converted into biomethane by gasification or fermentation, around 40 per cent of the carbon must in any case be separated in the form of CO$_2$ and is thus directly available for storage. Higher proportions can be separated if the biomass is converted into hydrogen or used to generate electricity. Rhodes and Keith (2005) describe a model system for generating energy from biomass in which up to 55 per cent of the carbon can be sequestered. WBGU assumes that, if bioenergy deployment were to focus on achieving as high a sequestration rate as possible, across-the-board capture of up to 50 per cent of the stored carbon might be possible. This would mean that around 0.5–1.0 Gt C would be available annually for sequestration.

WHAT EFFECT WOULD SUCH A SEQUESTRATION RATE HAVE ON THE ATMOSPHERE?
In the atmosphere 1 ppm CO$_2$ corresponds to 2.123 Gt C. However, if 2 Gt C is removed from the atmosphere, that does not mean that the atmospheric CO$_2$ concentration falls by 1 ppm. The atmosphere is involved in a process of constant exchange with the oceans and the biosphere. The resulting effect can be roughly calculated as follows. At present, half of the CO$_2$ emissions from fossil energy carriers are relatively quickly absorbed, two-thirds of them by the oceans and one-third by the terrestrial biosphere, so that only around 50 per cent of the emitted CO$_2$ remains in the atmosphere. Both sink processes are driven by the CO$_2$ increase itself. The sink function of the oceans is a response to the anthropogenic rise in the atmospheric CO$_2$ concentration, since the oceans only continue to absorb CO$_2$ until the partial pressures between the oceans and the atmosphere are equalized (WBGU, 2006). The sink function of the terrestrial biosphere is at least partially caused by the CO$_2$ fertilization effect as a result of the CO$_2$ increase in the atmosphere (House et al., 2002). If one accordingly assumes by way of an initial approximation that these feedback effects also operate in reverse and ignores non-linearities, around 4 Gt C would have to be removed from the atmosphere in order to reduce the quantity of CO$_2$ in the atmosphere by 2 Gt C or reduce the concentration by 1 ppm.

Similar results are arrived at by using a model function that describes the atmospheric response to an emissions pulse, taking various feedback effects into account (Joos, 2002). According to this model, at a constant sequestration rate over a period of 100 years around 40 per cent of the sequestered CO$_2$ would be effectively removed from the atmosphere. At a sequestration rate of 0.5–1.0 Gt C per year over a 100-year period, this would result in the atmospheric CO$_2$ concentration at the end of the period being between 9 ppm and 18 ppm lower than the concentration without this sequestration. However, this estimate takes no account of non-linear effects such as saturation of the ocean or biosphere sink. Such non-linear effects could cause sequestration to have a stronger effect, and the reduction in the CO$_2$ concentration could amount to 14–28 ppm over a 100-year period. For comparison: the current annual increase in the CO$_2$ concentration of Earth's atmosphere is around 2 ppm per year and is therefore about ten times greater (IPCC, 2007a).

dietary habits, technological progress and the degree of intensification of agricultural production (Section 5.2). It should be borne in mind that intensification of agricultural production is accompanied by increased greenhouse gas emissions (for example, from nitrogen fertilizer use and the use of agricultural machinery) and can therefore have a negative climate impact. Moreover, it is plausible that securing the food supply of a growing world population may in fact require the use of additional farmland, as the FAO, among others, anticipates (FAO, 2003a).

A significant increase in global sustainable bioenergy potential could therefore result from major efficiency improvements (which would have to be sustainable and environmentally friendly) in the production of food and feed, but it is more likely to be brought about by a shift to a diet involving fewer milk and meat products. At present 69 per cent of all farmland is pasture. If one includes land used for growing feed, around 80 per cent of farmland is used for livestock management (Section 5.2). A less land-intensive nutrition style could therefore free up farmland not included in the model used here – land that could be used for the sustainable cultivation of energy crops. However, dietary trends are at present tending in the opposite direction (Section 5.2.3).

In total, the potentials for the use of biomass for energy described here can meet a small but nevertheless significant part of the world's future energy requirement. Moreover, bioenergy is a useful strategic option in a more climate-friendly energy system, particularly because it can be stored and used as control energy. In combination with carbon storage it is even possible to achieve negative CO$_2$ emissions, in other words net sequestration. This can only occur, however, if cultivation meets all the sustainability criteria relating to climate, nature and soil protection and food security (Chapter 3). On account of the extensive competition for land and land-use (Chapter 5), the use of biomass for energy should always involve the most efficient technical processes available; this will ensure that the greatest possible climate change mitigation effect is obtained from the limited land or the limited quantity of suitable biomass available. A more detailed description and assessment of different cultivation systems and technical conversion pathways is contained in Chapter 7.

Cultivation systems for biomass production as energy resource

The cultivation and management methods chosen for energy crops not only influence production volume, but also determine the wider impacts on ecological resources and the atmosphere. Nitrous oxide (N_2O) and methane (CH_4) emissions resulting from agriculture rose by 17 per cent over the period from 1990 to 2005, primarily from burning biomass, livestock farming and soil emissions (Smith et al., 2007a). Annual emissions of greenhouse gases (GHGs) other than carbon dioxide (CO_2) from agriculture are estimated at 5.1–6.1 Gt CO_2eq, which corresponds to around 10–12 per cent of all anthropogenic GHG emissions (Smith et al., 2007). An additional 35–60 per cent increase in N_2O emissions is expected by the year 2035 from increased use of nitrogen and farmyard manure alone (FAO, 2003a).

This section focuses on the effects of large-scale bioenergy production on key ecosystem services (Fig. 7.1-1). The advantages and disadvantages of various cultivation systems will be discussed and illustrated with examples of energy crops commonly grown.

There are considerably more plants that supply biomass for energy (e.g. Sudan grass, giant reed, reed canarygrass, Jerusalem artichoke, *Pongamia*, Acacias, etc.) than can be discussed in the scope of this report. Algae as suppliers of bioenergy are discussed in Box 7.1-9. The selection in this report is limited primarily to promising energy crop species on which there is sufficient qualitative and quantitative production data available to produce an energy and GHG balance (Section 7.3). The competition for available land for energy crop production versus other land uses has already been discussed in Chapter 5.

7.1.1
Energy crop cultivation in monoculture

Many energy crop species such as sugar cane, maize and soya are grown in large monocultures around the world. Although this enables efficient management and high yields in a short period of time, it also results in high levels of greenhouse gas emissions (especially N_2O, CH_4 and CO_2) through soil tillage and fertilization. Over the long term, monocultures have an adverse impact on soil fertility and biological diversity (Matson et al., 1997; Table 7.1-1).

There are suitable agricultural management practices for reducing these negative impacts or preventing them in the first place. An example is crop rotation, which is widely practiced in Europe and also used in energy crop production. In this cultivation system, different crops are grown in succession (in rotation). As a general rule, the nutrient supply is more balanced, there is less pest and disease pressure, and soil structure as well as humus content remain more stable compared with monocultures. Three-year crop rotations to conserve organic matter and soil structure have been required in Germany since the European agricultural reform of 2005 (BMELV, 2006).

Depending on the planting time as well as the temperature and growing season needs of the crops, organic matter and nutrients can be added to the soil by sowing and then ploughing under catch crops (e.g. legumes, grass-clover mixes, nectar/pollen sources for bees, etc.), or by mulching or bi-cropping (e.g. clover in wheat fields). This is especially well-suited to energy crop cultivation, in which the entire aboveground biomass is harvested, for example, in the use of the whole plant for second generation biofuels or for generating biomethane by gasification. Suitable seed mixtures for fallows sown in wildflowers and herbs also make an important contribution to maintaining biological diversity in agro-ecosystems. In semi-arid regions, however, permanent cultivation rather than crop rotation with fallow land has proven to be more favourable in terms of quality and carbon content in the soil (Antle et al, 2003; Manley et al., 2005). As is the case with standard crop production, energy crop production also needs to be adapted and modified to suit each region.

With annual crops, annual tillage leads to reduction of carbon (C) in the soil. In order to minimize the loss of organic matter in the soil and protect the

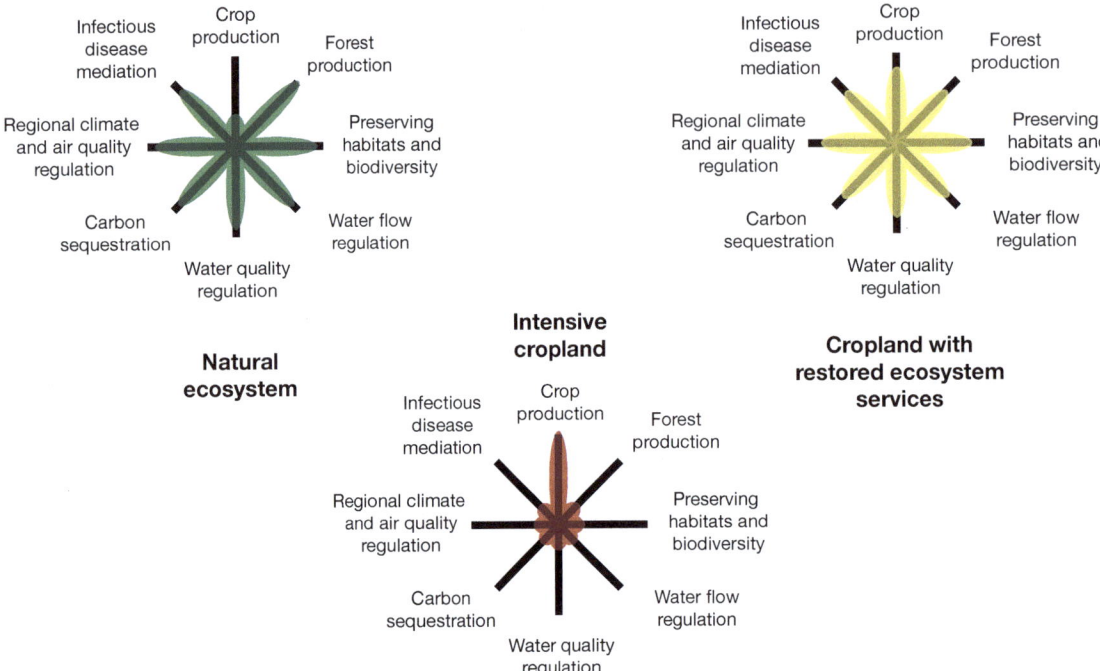

Figure 7.1-1
Schematic illustration of different land-use methods and their effects on ecosystem services. The provision of a wide variety of different ecosystem services with different land uses can be illustrated with simple 'branch diagrams', in which the status of each ecosystem service is shown on the corresponding axis. In this qualitative presentation the axes are not labeled with units. For the sake of illustration, three hypothetical landscapes are compared : a natural ecosystem (left), an intensive cropland (centre), and a cropland with restored ecosystem services (right). In natural ecosystems, many ecosystem services besides food production are present at very high levels. In contrast, the intensively managed cropland is capable of producing a surplus of food (at least in the short term), but at the expense of other ecosystem services. A middle way – i.e. a management method that supports other ecosystem services – promotes a more diverse portfolio of ecosystem services.
Source: Foley et al., 2005

soil surface from erosion and consequent degradation, alternatives to conventional tilling such as conservation tillage or no-till, and leaving or shallow tilling of crop stubble have become popular, as among other advantages they provide an opportunity to sequester carbon in the soil (Batjes, 1998; Paustian et al., 2000). If conventional tillage were dispensed with on a worldwide basis, an estimated 12–25 Gt of carbon emissions could be prevented by the middle of this century (Pacala and Socolow, 2004). The average sequestration potential of no-till management systems is 160 kg of C per hectare per year (Freibauer et al., 2004). For no-till (incl. savings in fuel and increase of organic carbon in the soil), Smith et al. (2000) estimate the overall annual C emission reduction potential for Europe at more than 40 million t of C. However, these cultivation techniques – particularly no-till – are not suitable for all crops. In terms of C sequestration, profitability varies from region to region, as under certain circumstances reduced tillage is concomitant with a decrease in production (Manley et al., 2005). Furthermore, the sequestration effect can be nullified by increased N_2O emissions (Six et al., 2002).

The breakdown of inorganic nitrogen fertilizers can result in the release of N_2O. A fertilization programme tailored to nutrient needs and the growth stage of the crops helps in preventing nitrogen losses (Crews and Peoples, 2005). There is indeed considerable room for improvement in nitrogen use efficiency, not only in Europe but also worldwide: on average, only approx. 50 per cent of the nitrogen from fertilizer is taken up by the plants. The remainder either escapes from the soil in gaseous form or leaches deeper into the soil beyond the reach of roots (FAO, 2001b). In intensive bioenergy cultivation systems, it is therefore essential to focus on a suitable fertilization programme (e.g. by nitrogen balances, precision agriculture, etc.).

7.1.1.1
Perennial crops in the tropics

SUGAR CANE
Sugar cane is a common tropical crop that is being grown with increasing frequency for ethanol production (Box 7.1-1). It can be grown as either an annual

Table 7.1-1
Advantages and disadvantages of energy crop cultivation in monocultures.
Source: WBGU

Advantages			
Economic efficiency	**Soil quality**	**GHG balance**	**Ecosystem services**
Specialized machinery, not as many different machines required	Protection from erosion and soil improvement are possible with perennial crops on marginal land (depending on management intensity)	Emissions can be reduced with machine syndicates	Production of food and/or raw materials as feedstocks for manufacturing processes and as an energy resource
Price discounts for seed, fertilizer, agrochemicals purchased in bulk quantities			Nutrient recycling, air quality regulation, water flow regulation
Easy planning and calculation			
Better marketing opportunities because of large production volumes			

Disadvantages			
Economic efficiency	**Soil quality**	**GHG balance**	**Ecosystem services**
Heavy dependency on demand for raw materials and political influences (duties, taxes, subsidies)	One-sided nutrient depletion	High C losses through intensive soil tillage plus CO_2 emissions from farm machinery	Favourable conditions for plant diseases and specialized pests
Heavy pesticide use due to high risk of plant diseases which may result in total loss	Mechanical stress from heavy machinery leads to soil compaction and severe surface run-off	With heavy N applications: N_2O losses through inefficient N use	Loss of biological diversity (above and below ground) due to large populations of a single plant species and generally heavy pesticide use
	High chemical fertilizer requirement poses a threat to soil microfauna		Less protection from abiotic environmental factors (wind, heavy rain, hail)
	High erosion hazard		Contamination of the regional water supply (negative impact on drinking water supply and biotope protection from large plantations), competition with land for food production

or a perennial crop. Currently, nearly 1600 million tonnes of sugar cane are grown per year in some 120 tropical countries. With 514 million t, Brazil was by far the largest producer in 2007, followed by India (356 million t) and China (106 million t). The average worldwide biomass yield for 2007 was 70.9 t per hectare (FAOSTAT, 2007).

Various adverse effects on the environment are linked with sugar cane cultivation. Hardly any other crop leads to biological diversity losses of comparable magnitude through alteration of the primary vegetation in croplands. In the Brazilian state of Alagoas, only 3 per cent of the virgin forest still exists, the remainder having been clearcut for sugar cane cultivation (WWF, 2005a). Wetlands have also been and are still being lost to sugar cane production, primarily

through drainage of their nutrient-rich soils. In Australia (Queensland), 60-80 per cent of the freshwater coastal wetlands have already fallen victim to sugar cane cultivation (WWF, 2005b). The heavy harvesting machinery compacts the soil. Cultivation on steep slopes and artificial irrigation lead to water erosion and salt accumulation in the soil. In many countries sugar cane fields are burned off in order to facilitate harvesting. This practice not only leads to emission of greenhouse gases and soil degradation and subsequent declines in future production, but also has an adverse impact on human health (Ribeiro, 2008). The processing of the harvested biomass into ethanol produces potassium-rich, acidic vinasses (fermented molasses) as a by-product, which in some cases is

Box 7.1-1

Sugar cane (*Saccharum officinarum* L.)

Sugar cane is a perennial grass that grows as tall as 7 m. It is originally from tropical South-East Asia and was introduced to America by European settlers. As a tropical plant, sugar cane cannot tolerate frost and needs a minimum annual mean temperature of 18°C and, for rain-fed agriculture without irrigation, at least 1000 mm of annual rainfall. The 2–5 cm thick stalk (cane) of the plant contains a soft, sugar-storing pith. Depending on the growing region, the crop is ready for harvest after 10–24 months. If the stubble is not ploughed under after harvest, the regrowth can be harvested up to 4–8 times with suitable fertilization. Yields average 10–120 tonnes of biomass per hectare.

At harvest the canes are cut and the leaves removed. After chopping, the pieces of cane are crushed and pressed several times in order to extract the sugar. The sugar cane syrup is clarified and then crystallized by boiling. The crystallized raw sugar is refined to 99.8 per cent pure sucrose. The remaining sugar syrup (molasses) is fermented to produce alcohol or is used as feed or for yeast culture. The

fermented molasses (vinasses) is often applied back on the fields as fertilizer (Lieberei et al., 2007).
Photo: Hannes Grobe, AWI

drained into water bodies, thus endangering their aquatic ecosystems (Rosebala et al., 2007).

Various means are available to improve sugar cane production. Using efficient irrigation systems (trickle irrigation) and mulch helps to conserve water. In order to prevent water erosion, the slope of land for sugar cane production should not exceed 3 per cent (WWF, 2005a). Cutting off the leaves prior to harvest and using them as mulch rather than burning them off increases the organic matter content of the soil, reduces the rate of evaporation, and prevents soil erosion (WWF, 2005b).

OIL PALM

The oil palm is one of the traditional oil and energy crops (Box 7.1-2). The product, palm oil, is used mainly in the food and cosmetics industries. The main producers and exporters of palm oil are Malaysia and Indonesia (world production in 2007: 39.3 million t; FAOSTAT, 2007). Both countries are striving to export 40 per cent of their palm oil production as fuel. Although the worldwide land area devoted to oil palm cultivation is a mere 10 per cent of that devoted to soya cultivation, worldwide production is comparable for both crops. According to FAO estimates, by the year 2030 palm oil production will be double that of 1999/2001 (FAO, 2006c).

Oil palm cultivation leads to substantial ecological damage, particularly in Indonesia. Whereas in Malaysia new oil palm plantations may only be established on existing cropland or fallow land, in Indonesia virgin forests on bog soils are often sacrificed to oil palm plantations (Stone, 2007; Box 5.4-2). More than 25 per cent of Indonesian oil palm plantation allotments

Box 7.1-2

Oil palm (*Elaeis guineensis* Jacq.)

The oil palm is originally from Africa and is nowadays grown in tropical America and South-East Asia. As a humid tropical plant, the oil palm requires 100 mm of precipitation monthly and an average temperature of 24–28°C (minimum temperature 15°C, drought period no longer than three months). The perennial plant grows as tall as 30 m and bears fruit clusters weighing up to 50 kg with several thousand fruits. The first harvest is possible after five years, and full harvest potential is reached after 12–15 years. The plants can live as long as 80 years. After harvest the rapidly perishable fruits are immediately treated with steam in order to destroy a fat-splitting enzyme. Palm oil is extracted from the orange-coloured pulp and palm kernel oil is extracted from the kernels. The yield of palm oil from the pulp averages 2.5–5 tonnes per hectare per year (Lieberei et al., 2007).
Photo: Frank Krämer, GTZ

have been issued for areas with bog soils. The production of one tonne of palm oil results in 10–30t of CO_2 emissions, which are generated from oxidation of the drained organic soils (fires not included in the calculation; Hooijer et al., 2006).

Furthermore, destruction of virgin forest leads to habitat loss for numerous plant and animal species (Section 5.4). In Sumatra, three-quarters of the native species of bats have disappeared, and less than 10 per cent of all birds and mammals of the virgin forests are able to find new habitats in the plantations (Stone, 2007). In addition, the processing of palm fruits in oil mills produces effluent, which with traditional treatment methods leads to further environmental contamination. The effluent is ultra-rich in nutrients and is drained into large ponds, in which anaerobic decomposition of the organic compounds leads to high methane emissions if the methane is not used in some way, such as in biogas plants. The production of 1 tonne of palm oil in this manner results in an additional 756 kg CO_2eq of GHG emissions (Schuchardt, 2007).

In order to prevent these enormous CO_2 emissions and conserve biological diversity, it is essential that no more swamp forests on bog soils be destroyed (Hooijer et al., 2006). Where possible, bog soils should be restored to their natural state. In terms of CO_2 reduction, oil palm cultivation is most efficient on marginal land (Section 7.3). With sustainable plantation management (e.g. better water management) and improved palm oil processing methods, considerable savings in energy and considerable reductions of GHG emissions could be achieved (WWF, 2007). The nutrient-rich waste water from the processing of oil palm fruit along with the empty and ground-up fruit residue can be used in biogas plants (Schuchardt, 2007). This ensures that the nutrients are not lost, and hardly any methane and nitrous oxide are released.

Jatropha

Jatropha curcas (physic nut or Barbados nut), an oleiferous member of the spurge family, is frequently mentioned as the new 'miracle plant' for biodiesel production. This tropical plant is relatively easy to grow. It grows in arid and semi-arid regions as well as in areas with more precipitation (200–1500 mm per year), and also grows on nutrient-poor soils (Box 7.1-3). The high oil content of *Jatropha* and the erosion protection provided by its deep roots make the plant an interesting possibility for biodiesel production on marginal lands in tropical regions (Openshaw, 2000; Augustus et al., 2002; Wiesenhütter, 2003; Sirisomboon et al., 2007).

Although the plant has a long history of diverse uses, e.g. hedgerows, erosion prevention, traditional human and veterinary medicine, soap manufacture

Box 7.1-3

Jatropha (*Jatropha curcas* L.)

The succulent spurge Jatropha curcas or physic nut is originally from South America and has since spread throughout the tropics. It grows on various soils, even nutrient-poor ones, and under various climatic conditions. Within three years the plant grows as tall as 3–5 m and can live as long as 50 years. Yields range from 0.5 t to 12 t per hectare per year, depending on the site and water availability. The seeds have an oil content of approx. 30 per cent (Openshaw, 2000). The press cake from the seeds contains approx. 6 per cent nitrogen. The nitrogen requirements of Jatropha have still not been clearly determined. Openshaw (2000) recommends growing Jatropha as a companion crop to nitrogen-fixing trees.
Photo: Meinhard Schulz-Baldes, WBGU

and fertilizer, research on it is still in the early stages. Unlike other crops, *Jatropha* is not a cultivated plant. The plant used today is a wild form, which has only recently been selectively bred (Rosegrant and Cavalieri, 2008). For profitable production, it is first necessary to develop new cultivars, as the yields of the wild form are highly variable and difficult to predict (Fairless, 2007). *Jatropha* is relatively resistant to pests and diseases, and livestock (including goats) find it unpalatable because of its milky, poisonous sap (Augustus et al., 2002; Wiesenhütter, 2003). The press cake from oil production is likewise unsuitable as feed, and is used as fertilizer or in organic pest control. Should an economical means of detoxifying the residue be found, it could be used for feed.

Planting trials are currently being conducted in India in order to identify the site requirements and productivity of various landraces. Preliminary intercropping trials by ICRISAT indicate that other crops can be interplanted with *Jatropha*. At this point, however, it is still too early for an ecological and economic evaluation of such cultivation systems. Although the plant can grow on marginal land, it requires fertile soils and adequate water (up to 750 mm of water/year according to ICRISAT) to produce high per-hectare yields, and therefore competes with food production on sites that meet these conditions. Nevertheless, *Jatropha* is seen as a source of hope for biodiesel production and large areas of land are already under cultivation (Boxes 6.7-2 and 10.8-1). There may already be as many as 500,000 to 600,000 hectares of *Jatropha* plantations in India, and even two million hectares in China (Fairless, 2007). It is still not possible, however, to produce biodiesel from *Jatropha* economically – that is, without subsidies (Openshaw, 2000; Wiesenhütter, 2003).

7.1.1.2
Rotational crops in temperate latitudes

Maize

Whereas maize is primarily grown as livestock feed (silage maize) in Europe and in North America, in many developing and newly industrializing countries it is one of the most important staple foods (grain maize; Box 7.1-4). World grain maize production was 785 million tonnes of grain in 2007. The largest producers by far were the USA (332 million t or 42 per cent of world production) and China (nearly 152 million t or 19 per cent of world production; FAOSTAT, 2007). In 2006, Germany produced nearly 3.4 million t of grain maize, representing a mere 0.5 per cent of worldwide production, but nearly 47 million t of silage maize (Destatis, 2006).

In 2007, the area planted with genetically modified maize (GM maize) cultivars amounted to 35.2 million hectares, which represents 24 per cent of the worldwide maize production area. Eighty per cent of the maize grown in the United States now comes from GM maize cultivars (ISAAA, 2008).

Modern maize monocultures have a number of adverse effects on the environment. Groundwater is contaminated by nitrate leaching associated with the use of nitrogen fertilizers and by herbicides. The soil is endangered by prolonged fallow periods, and is compacted and eroded by heavy farm machinery. In addition, conventional maize cultivation practices require a high energy input (ITADA, 2006). In addition, maize monocultures are low biological diversity agro-ecosystems. According to Searchinger et al. (2008), if changes in land use to produce ethanol from maize are figured into the GHG balance, the GHG emissions over 30 years are twice those associated with fossil fuel use.

Various measures can be employed to mitigate ecological impacts. The agro-environment can be upgraded by increasing the number of different crops within the crop rotation and planting suita-

Box 7.1-4

Maize (*Zea mays* L.)

Maize is a 'true grass' (i.e. a member of the family Poaceae = Gramineae) and comes originally from Mexico, where it was already being grown as a crop between 5000 and 3400 B.C. The plant grows as tall as 2.5 m and has a pith-filled stalk up to 5 cm in diameter. After pollination by wind, the ears grow out of the leaf axils. Depending on the variety, the kernels may be yellow-gold, white, red or dark purple in colour. Maize kernels are composed of approx. 70 per cent starch, from which ethanol can be produced. Approximately 2.5 t of maize are needed to produce 1 m³ of ethanol.

As a tropical-subtropical plant, maize is not frost resistant. The optimum temperature for growth is 30°C, although some varieties also grow in temperate latitudes. The plant is relatively drought-tolerant and will also grow on poorer soils (Farack, 2007). The nitrogen content of grain maize (grain and stalk, 86 per cent dry matter) is 2.41 kg N per tonne of fresh weight, and in Germany the average grain yield is 90 tonnes per hectare. (LfL Bayern, 2008).
Photo: ©gabriele.moser

ble catch crops. Mulch seeding, better ground cover, and reduced tillage contribute to protecting the soil from erosion and compacting. A well-balanced fertilization programme and reduced tillage help reduce energy inputs (ITADA, 2006).

Rape

In 2007, 49 million tonnes of rape were produced worldwide (FAOSTAT, 2007; Box 7.1-5). After China, Canada and India, Germany (5.3 million t, almost 11 per cent) ranked fourth in world production. Genetically modified cultivars account for 20 per cent or 5.5 million hectares of the worldwide land area under rape cultivation (ISAAA, 2008).

As a consequence of the market for biofuels, the production volume has increased substantially in recent years. In 2006, 12 per cent of available arable land in Germany was devoted to rape production

Box 7.1-5

Rape (*Brassica napus* ssp. *oleifera* L.)

Rape is originally native to the Mediterranean area of south-eastern European and, like its close relative turnip rape (*Brassica rapa* L.), it is grown primarily for the production of high-quality oil. Depending on the variety, the plant grows as tall as 160 cm and produces clusters of bright yellow flowers. The small round seeds are borne in seed-pods known as siliques and have an oil content of 40–50 per cent. Rapeseed oil formerly had a high content of erucic acid, which was responsible for the bitter taste of the oil. Through selective breeding in the 1980s, however, it was possible to replace nearly all of the erucic acid content with more digestible oleic acid (known as Canola = Canadian oil, low acid).

Winter rape is sown in late summer, and annual summer rape is sown in the spring. Rape grows well everywhere that wheat can be grown. Without a protective snow cover, however, winter rape does not readily tolerate temperatures below -15°C that last for more than a few days. A further requirement of the species is that the autumn climate should enable the seedlings to become established and form leaf rosettes. Poorly-drained soils as well as those that dry out severely are less suitable for rape growing.

To extract rapeseed oil, the seeds are rolled and pressed. The press residue, known as press cake, is a livestock feed

rich in protein. Rapeseed oil is a high-quality culinary oil and its many uses include the production of margerine. It is also used as an industrial lubricant. Nowadays the main use of rape is the production of biodiesel (RME = rapeseed oil methylester; Lieberei et al., 2007). The nitrogen content of rape (seed and straw, 91 per cent dry matter) is 4.54 kg N per t fresh weight, and in Germany the average grain yield is 30t per hectare (LfL Bayern, 2008).
Photo: Meinhard Schulz-Baldes, WBGU

(incl. turnip rape; Destatis, 2006), albeit with considerable regional variation. In 2007, almost 23 per cent of the arable land in Mecklenburg-Western Pomerania was devoted to oilseed rape production (Grunert, 2007). Grunert estimates the tolerable threshold for the proportion of land under rape cultivation to be 25 per cent, which corresponds to a rotation sequence with a three-year recess interval between rape crops.

The ever-shorter recess intervals and the increasing proximity of different rape fields increase disease and pest pressure, which in turn requires increased pesticide usage and favours the development of resistance, are issues of concern (Grunert, 2007). In addition, rape requires large amounts of nitrogen fertilizer, which is a major cause of environmental contamination (from high energy consumption to manufacture the fertilizer and nitrate contamination of the groundwater due to improper application techniques). The following rotational crop, however, benefits from a good soil structure and a high nutrient content in the soil (Grunert, 2007). However, nutrient leaching must be prevented by selecting a suitable rotational crop. Although conservation tillage methods such as mulch seeding and leaving the stubble on the field are beneficial for the soil, they also favour the survival of pathogens and pests (Alpmann, 2005).

CEREALS

Due to rising crude oil prices, cereals (wheat, oats, barley, rye) for bioenergy production have attracted increasing interest from an economic standpoint in recent years (Tuck et al., 2006; Box 7.1-6). Ethanol can be produced from the grain, which contains starch, and the straw is suitable as a solid fuel. The wheat-rye hybrid triticale, previously used mainly as livestock feed, has attracted the attention of bioenergy producers in recent years due to its high biomass yields and good combustion properties (Jorgensen et al., 2007). In 2007, a total of 12.6 million t of triticale was produced worldwide in 36 countries. Poland (4.2 million t), Germany (nearly 2.2 million t) and France (1.5 million t; FAOSTAT, 2007) were the three largest producers.

Since 2001, however, cereal cultivation for energy production in the EU has been permitted only on set-aside land, and then only if the energy produced from it is used on the same farm (EU Regulation No. 587/2001).

The environmental contamination from intensive cereal cultivation is the same as that from maize and rape production (nitrate and herbicide contamination of soil and groundwater, high energy consumption). Even the use of residual biomass such as straw for bioenergy production is not without controversy, as doing so removes additional carbon from the soil (Reijnders, 2008; Saffih-Hdadi and Mary, 2008).

Box 7.1-6

Triticale (*Triticum aestivum* L. x *Secale cereale* L.)

Triticale has been selectively bred since the 1930s. The hybrid grain triticale is a cross breed between wheat and rye and combines the characteristics of both cereal species. Rye is less demanding than wheat as far as site conditions are concerned, and also grows in harsher climates and poorer soils. Wheat by comparison produces higher yields and a flour with good baking properties. Triticale cultivation of biomass for energy is increasing. The grain is especially suitable for this purpose, as it has firmly-adhering kernels that do not drop-off, even with a late harvest when the plant has a low moisture content (Lewandowski and Schmidt, 2006). Like rye, triticale produces more dry matter than wheat with less nitrogen fertilization and is therefore better suited for the production of biomass for energy (Jorgensen et al., 2007). The nitrogen content of triticale (grain and straw, 86 per cent DM) is 2.1 kg N per t of fresh weight, and in

Germany the average grain yield is 6.0 t per hectare (LfL Bayern, 2008).
Photo: Klaus Münchhoff, Gut Drerenburg

7.1.1.3
Perennial crops in temperate latitudes

MISCANTHUS GRASS
Miscanthus grass is a perennial grass originally native to South-East Asia and a prolific producer of biomass and fibre (Box 7.1-7). A danger associated with Miscanthus grass cultivation is that it may become invasive if, for instance, the roots are not completely dug up when removing the plant (Box 5.4-3). In the eastern USA, the plant is on the list of invasive species that threaten to displace native plant species (Swearingen et al., 2002).

The sequestration potential of Miscanthus grass plantations ranges from 5.2 to 7.2 C per hectare per year (Clifton-Brown et al., 2007). In a comparative study of nitrogen, energy and land-use efficiency of the three energy crops triticale (x *Triticosecale*), reed canarygrass (*Phalaris arundinacea L.)* and Miscanthus grass, Miscanthus grass clearly outperformed the other two (Lewandowski and Schmidt, 2006). In order to ensure optimum resource management, the authors of this study recommend Miscanthus grass for biomass production, and in regions where the climate is too cool for it, they recommend short-rotation plantations with woody plants.

SWITCHGRASS
Switchgrass is a prairie grass that is becoming increasingly important not only in terms of its traditional use as a high quality forage but also for bioenergy production (Box 7.1-8). The positive environmental effects of switchgrass cultivation are erosion mitigation through wind screening, year-round ground cover and good root penetration of the soil. Switch-

grass stands in the USA also provide shelter and food for various birds and small animals (USDA, 2001). In its report on climate protection and biomass, however, SRU (2007) rated switchgrass monocultures as a medium risk to biological diversity.

In a life-cycle analysis by Adler et al. (2007) on GHG fluxes in energy crop cultivation, the GHG reduction of switchgrass compared with fossil fuels over a 30-year period averaged 115 per cent. The value greater than 100 per cent can be explained by carbon sequestration in the soil, which results in a net sink. Compared with a maize-soya crop rotation system, switchgrass has greater carbon sequestration potential (Al-Kaisi and Grote, 2007).

7.1.2
Short-rotation plantations (SRPs)

The main purpose of a short-rotation plantation is to produce timber as an energy resource and as a feedstock for manufacturing processes. To this end, fast-growing species such as poplar and willow are grown in temperate zones and eucalyptus or *Pinus radiata* is grown in tropical and subtropical zones. Nutrient loss and contamination of local water sources in tropical soils are considerable, even after short rotation periods. In addition, the litter from some eucalyptus species can be phytotoxic, thus suppressing erosion-preventing undergrowth (Poore and Fries, 1985). The general advantages and disadvantages of SRPs are summarized in Table 7.1-2. Because the plantations are on agricultural land and in Europe are thus subject to agricultural rather than forestry legislation,

Box 7.1-7

Miscanthus grass (*Miscanthus sinensis Anderss.*)

Miscanthus grass grows to a height of over 4 m after only three years. Like maize and sugar cane, this grass owes its rapid growth rate to the C4 photosynthesis mechanism, which when more heat and light are available enables a more rapid rate of photosynthesis and uses water and CO_2 more efficiently than the C3 mechanism. The interspecific hybrid *Miscanthus x giganteus* is grown for biomass production. Miscanthus grass is a heat-loving plant that is capable of producing biomass for more than 20 years under ideal conditions. In its growth stage, however, the plant is very sensitive to cold. Late frosts can lead to a total yield loss. In Central Europe the plant does not produce germinable seeds, but instead reproduces by rhizomes (Stolzenburg, 2007). The nitrogen content of *Miscanthus* (whole plant, 80 per cent dry matter) is 0.15 kg N per t of fresh weight, and in Germany the average yield is 22.0 t per hectare (LfL Bayern, 2008).

Photo: www.hpc-group.com

Box 7.1-8

Switchgrass (*Panicum virgatum* L.)

Switchgrass is native to the North American prairies. The prairie grass is a perennial, rhizomatous plant (a rhizome is perennial stem that grows underground or just below the surface) that grows as tall as 3 m and has an equally deep root system. It is relatively winter hardy. Like Miscanthus grass, switchgrass is in the group of plants with a C4 photosynthesis mechanism. The useful life of the plants is 15–20 years, and it is cut once a year (twice a year with suitable fertilization). Full yield potential is reached approximately after the third year. The annual nitrogen removal with harvest varies between 48–276 kg N per hectare, depending on the site and frequency of harvest. Maximum yields are as high as 36.7 t dry matter per hectare per year (Parrish and Fike, 2005). Maximum yields of 10–17 t dry matter per hectare per year are reported for Germany (TFZ, 2008).

Photo: Michael Hassler, Bruchsal

there are no legal restrictions as regards clearcutting and clearing.

SRPs are suited for non-cultivated land (fallow land, contaminated soils). In Germany the development of non-cultivated land for the production of renewable biomass is subsidized. The timber can be used both as a feedstock for manufacturing processes and as an energy resource.

If intensively used grassland is ploughed up to produce energy crops, short-rotation poplar plantations for producing wood chips are more sustainable in Germany than maize cultivation for ethanol production, according to a study by the Karlsruhe Research Centre (Rösch et al., 2007). However, the ploughing up of grassland to grow biomass for energy production should be avoided (EU Commission, 2005d). In its report on biomass and climate change mitigation, SRU calls for a general ban on the conversion of grassland to biomass production (SRU, 2007).

As a general rule, biological diversity is greater in natural forests than in tree plantations (Raison, 2005). Nevertheless, planting SRPs as spatial and temporal mixed cultures can help create more diverse landscape structures. Placing green manure or mulch between the tree rows not only serve as nutrient sources and replace chemical fertilizers, but also provide niches for micro-organisms.

7.1.3
Agroforestry

Agroforestry is a combination of agriculture and forestry on the same land (e.g. trees, fruit or nut-bearing hedgerows, etc.). Annual crops are interplanted with perennial trees or shrubs. The term agrosilvopastoral system is used if livestock are also raised on the same land. Agroforestry systems are most common

Box 7.1-9

Algae as bioenergy sources

ALGAE FOR HYDROGEN PRODUCTION

When under stress, the unicellular algae species *Chlamydomonas reinhardtii* has the unique characteristic of generating hydrogen as a photosynthetic by-product (hydrogenesis). This algae species is suitable for climate-friendly energy production, as no greenhouse gases are released. In laboratory experiments it was found that the algae produce hydrogen particularly when they are temporarily deprived of the trace element sulphur. Furthermore, it was possible to increase hydrogen production significantly through genetic engineering, thus increasing the efficiency of the conversion of sunlight to hydrogen from 0.1 to 2.5 per cent.

In spite of this progress, however, there are still some limiting factors that need to be overcome: the algae need to be maintained under stress conditions for a prolonged period in order to generate hydrogen. Prolonged periods of stress, however, can damage the organisms considerably, which is why periodic regeneration phases in which no hydrogen is produced are necessary. Furthermore, the algae require a considerable amount of sunlight, and therefore hydrogen is only produced in a very thin surface layer of an aqueous solution populated with *Chlamydomonas*. Even if it were possible to grow the algae on light-conducting fibres, very large areas would be needed for industrial use.

Other research approaches are focusing on isolating the hydrogenesis enzymes from the algae species in order to be able to generate hydrogen independently of living cell cultures. This approach, however, is in the developmental phase and still a long way away from pilot applications (Melis and Happe, 2001).

BIOMASS AND DIESEL PRODUCTION FROM WASTE GASES

In principle it is possible to convert dried algae biomass into biodiesel, bioethanol or biogas. The oil content of different algae species is as high as 40–50 per cent by weight, thus making these species potent biodiesel producers (FAO, 2007c). In the recently launched pilot project in Hamburg 'Technologies for Exploiting the Resource Microalgae' (Technologien zur Erschließung der Ressource Mikroalgen, TERM), exhaust emissions from a small-scale combined heat and power (CHP) unit are fed into a tank with algae cultures. The algae are able to take up part of the CO_2 and convert it into biomass. According to the project report, it is possible to sequester as much as 450 t CO_2 per hectare per year. This corresponds to around 150 tonnes of biomass, which in turn would correspond to a tenfold increase in productivity per unit land area compared with agricultural biomass production. The pilot project plans to convert the algae into biodiesel and use it as an energy source. While this would not result in a direct sink effect, the emitted CO_2 would be divided between two energy products in the entire system, thus lowering the greenhouse gas intensity per unit energy.

In terms of a large-scale industrial use, however, the strategy appears to be of limited applicability: sequestering the CO_2 emissions of a hard-coal-fired power plant with 800 MW capacity would require an algae bed with an area of 100 km² in the vicinity of the plant. Nevertheless the system could have advantages on sealed and contaminated sites (industrial brownfield sites), which have few other possible uses. Coupling with smaller facilities seems quite promising in this case. So far, however, there is no evidence that large-scale algae production with the necessary complex infrastructure and associated processing of the fuel has an overall positive CO_2 balance (Ullrich, 2008). The production of biodiesel from algae is also still quite cost-intensive (Ackermann, 2007; FAO, 2007c).

---

among small farmers in the tropics. The primary purpose of such systems is food production. It is therefore difficult to estimate how much overall biomass for energy use could be produced in these tropical systems. Energy crops (e.g. oil palms) grown in mixed cultures have potential as a sustainability strategy for small farms.

Agroforestry offers certain advantages over pure agriculture (Table 7.1-3). The mixed culture with trees provides different ecological niches in a small space and enhances the diversity of the cultivated landscape. Thanks to the year-round ground cover and the dense and deeper root penetration, the soil is better protected from erosion, nutrient leaching is reduced, and nutrient availability is improved. The comparatively small quantities of products, however, may result in a marketing disadvantage for the producers. Furthermore, agroforestry systems require longer-term planning and only become profitable after a certain period of time.

In the European SAFE project (Silvoarable Forestry for Europe, 2001–2005), the benefits of agroforestry were studied in various trials and in several European countries. The researchers were able

to demonstrate that the available nutrients were used more efficiently in mixed cultures than in monocultures (Dupraz et al., 2005). According to Reisner et al. (2007), the European tree species poplar, holm oak, stone pine, walnut, and cherry could be grown profitably in combination with standard field crops on 56 per cent of the arable land in Europe. Doing so could also reduce nitrate leaching, protect the soil from erosion, and enhance biological diversity on approx. 40 per cent of the arable land in Europe (Reisner et al., 2007).

Agroforestry systems are still not widespread in Germany. As part of the Agroforestry Research Cooperative Project (Agroforst-Forschungsverbund), the German Federal Ministry of Education and Research commissioned the Jülich Research Centre to study agroforestry management strategies in terms of economic, ecological and social criteria, particularly in terms of a potential alternative to the conventional spatial separation of agriculture and forestry and promotion of sustainable development. The project will run from 2005 to 2008, and the research focuses on the production of timber biomass as well as other topics.

Box 7.1-10

Short-rotation plantations (SRPs)

The term short-rotation plantation (SRP, also known as short-rotation coppice or short-rotation forestry) refers to the cultivation of fast-growing tree species on agricultural land to produce biomass. The origin of SRPs goes back to coppicing, which in the past was used to produce firewood. The turnover period describes the growth period until the trees are cut and it depends on the use of the wood. For pulpwood or for chip production, the trees are harvested after 3–5 years, for industrial timber after around 20 years, and after as many as 30 years for standing timber.

In addition to rapid growth, suitable tree species for SRPs must also have other characteristics, such as tolerance to extreme planting densities, narrow (columnar) canopy, favourable ratio between water and nutrient consumption and wood production, ability to regrow after harvest, and resistance to biotic and abiotic damage.

Frequently planted tree species in temperate latitudes include willows (*Salix* sp.) and poplars (*Populus* sp.) and their hybrids, as they grow quite rapidly under conditions of average nutrient availability and a good water supply in the soil. Aspens (Populus tremula) are relatively drought-tolerant and are also suited to exposed sites (e.g. knolls), whereas willows grow best on moist sites. Birches (*Betula* sp.) can be planted on very dry, nutrient-poor soils and

alders (*Alnus* sp.) are suited to wet areas (Röhricht and Ruscher, 2004; LWF, 2005). Eucalyptus plantations are common in subtropical and tropical regions.

Nitrogen removals by SRPs at a medium yield level average around 64 kg N per hectare with a yield of 10t of dry matter per hectare per year for poplars and 32 kg N per hectare with yields of 7 t of dry matter per hectare per year for willows (KTBL, 2006).

Photo: CLAAS Harsewinkel

7.1.4
Permanent grassland and pastures

As a result of decreasing livestock populations (due to structural change in dairy cattle husbandry, etc.), the area of excess grassland in Germany is increasing. In a study by the Karlsruhe Research Centre, researchers therefore investigated ways of using grassland in Baden-Württemberg as a bioenergy resource, as alternatives to dairy cattle husbandry or pure grassland husbandry (Rösch et al., 2007). Using grass silage to produce biogas (fermentation) and using hay to produce heat have become commonplace. The gasification of grass cuttings is still in the research and developmental stages (Section 7.2). With extensive grassland management, the use of hay as a fuel is more sustainable than pure grassland husbandry in the categories of energy conservation, climate change mitigation, income and employment, according to the study by Rösch et al. (2007). The drawback is that thermal use produces emissions that are harmful to human health and the environment, although it may in future be possible to reduce such additional contamination through advanced technology. On the other hand, the production of electricity from grass silage has the adverse consequence of releasing methane, whereas the effects on ecological and socio-economic indicators are positive (Rösch et al., 2007).

Another interesting aspect of extensive grassland use is the conservation or enhancement of biological diversity. Results from the Jena Experiment, a large biological diversity experiment in Germany, convincingly show that grasslands with greater biological diversity are able to achieve higher ecosystem services (e.g. productivity, carbon sequestration, nutrient use, etc.) than species-poor systems (Oelmann et al., 2007; Weigelt et al., 2008). Even forage quality and calorific values increase with increasing biological diversity (Dr. Michael Scherer-Lorenzen, ETH Zürich, and Prof. Dr. Michael Wachendorf, University of Kassel, personal communication). Results for the North American prairie are similar. According to a study by Tilman et al. (2006), the prairie with high biological diversity produced even more bioenergy per unit land area than a maize cultivation system for ethanol or a soya cultivation system for biodiesel, and with fewer GHG emissions and less soil contamination from agricultural chemicals. According to this study, there are 5 x 10[8] hectares of eroded, marginal land worldwide that would be suitable for such low intensity/high diversity grassland cultivation. Even though the actual amount of marginal land available worldwide and its productivity is still under discussion (Russelle et al, 2007; Tilman et al, 2007), these new approaches to sustainable biomass production nevertheless show great potentials for ecologically upgrading degraded areas. Furthermore, with new vegetation cover and the resultant input of organic

Table 7.1-2
Advantages and disadvantages of short-rotation plantations.
Source: WBGU

Advantages

Economic efficiency	Soil quality	CO_2 balance	Ecosystem services
High biomass production on small land areas	Year-round vegetation cover prevents erosion	'Climate-neutral' resource (except for nutrient removal and emissions at harvesting)	Improvement of the local climate (windbreaks, mitigation of temperature extremes)
Relatively rapid amortization of invested capital	Dense root systems improve top soil structure	Carbon sequestration in the soil. However: dependent on the time period	Habitats for animals (nesting sites, food, etc.), mixed culture SRPs as bridging elements in habitat networks
Simple planning and calculation	Undisturbed development of soil fauna		Flood protection due to improved sequestration effect
Securing of raw material supplies	No soil compaction due to infrequent driving on the land		Green areas make aesthetic landscapes
Uniform availability of raw materials	Depending on the tree species: detoxification of soils contaminated with heavy metals		Air and water filtration, O_2 production

Disadvantages

Economic efficiency	Soil quality	CO_2 balance	Ecosystem services
Fertilization/irrigation required, depending on the site and tree species	The shorter the rotation period, the greater the adverse ecological impacts of soil contamination, nutrient needs, pesticide use, etc.	GHG emissions dependent on overall eco-balance (cultivation scheme, length of rotation, etc.)	Competition with land use for food crop production
Market price for timber biomass becomes a risk			Contamination of the regional water supply (negative impact on drinking water supply and biotope protection with large plantations)
Additional machinery for forestry			Large-scale plantations threaten agro-ecosystems

matter into the soil, it is possible to restore such degraded land areas and thus make them available for long-term food production or for the production of industrial feedstocks.

The effects of resource, pollutant and nutrient management in agricultural biomass production on GHG emissions and ecological sustainability were assessed in the scope of a life-cycle assessment (LCA) study conducted by the Swiss Federal Offices of Energy, Environment and Agriculture (Kägi et al., 2007). According to this study, extensive meadows produce on average 2.7 tonnes of dry matter (DM) per hectare, organically managed permanent meadows 9.9 t DM per hectare, and permanent meadows in integrated production systems 11.7 t DM per hectare. Integrated production (IP) is a method of farming with minimal environmental impacts, but with less strict requirements than for certified organic production. According to this study, for bioenergy produc-

tion from grass the more intensive IP management system is preferable, partly because this results in 10–15 per cent more yield than organic farming, but also because only minor differences in environmental impacts are ascertainable. The ethanol yield conversion factor for grass (starch equivalents) is given as 0.24 kg per kg DM. Extensively managed meadows result in less overall environmental contamination per kg DM than intensively managed ones. However, both intensively and extensively managed meadows achieve better results per production unit in terms of environmental contamination and yield loss than medium management intensity (Kägi et al., 2007). According to the EMPA life-cycle assessment study on energy products (Zah et al., 2007), greenhouse gas emissions are less in extensive grass cultivation systems than in intensively managed ones, but biomass production and ethanol yield also decrease.

Table 7.1-3
Advantages and disadvantages of agroforestry.
Source: WBGU

Advantages			
Economic efficiency	**Soil quality**	**CO_2 balance**	**Ecosystem services**
Diversification	Year-round vegetation cover prevents erosion	With ground cover there are fewer C emissions than with monocultures	Mixed cultures more resistant to diseases and pests
Self-sufficient, not dependent on large markets and agriculture industry	Mixed culture prevents one-sided nutrient depletion	Dense and deep root systems of the trees sequester C	Greater habitat diversity than with monoculture
			Creation of a suitable microclimate (shade, windbreaks, water storage) for certain field crops

Disadvantages			
Economic efficiency	**Soil quality**	**CO_2 balance**	**Ecosystem services**
Long-term planning, only profitable after a certain period of time			Depending on root characteristics, the trees may compete directly with the interplanted crop for nutrients and water
Small harvest quantities make market access more difficult			
More labour intensive, as large farm machinery cannot be used			

As a consequence, the study makes no clear recommendations in favour of either cultivation method.

As part of a long-term experiment conducted by the Swiss Research Institute of Organic Agriculture (FiBL) known as the DOK Trial, various cultivation systems (organic, biodynamic, conventional in accordance with integrated production) and various fertilizer types (farmyard manure, farmyard manure and chemical fertilizer, chemical fertilizer) and rates were compared (FiBL, 2001; Maeder et al., 2002). The yields of grass leys (that is, meadows rotated with cropland) were only 11–13 per cent lower in the first two rotation sequences (in each case 7 years) with organic management as compared to conventional production. The yield differences increased slightly in the third rotation sequence (FiBL, 2001). As a general rule, a yield reduction of approx. 20 per cent can be expected in organic farming without chemical fertilizers and without chemical or synthetic pesticides (FiBL, 2001; Maeder et al., 2002). Compared with conventional production, however, fertilizer and energy inputs were reduced by 34 per cent and 53 per cent respectively, and pesticide use was reduced by 97 per cent (Maeder et al. 2002).

Carbon uptake in temperate grassland can be influenced by nitrogen fertilization (Soussana et al., 2004). A moderate nitrogen application promotes carbon uptake in the soil, whereas excessive nitrogen fertilization stimulates mineralization of organic carbon (Soussana et al., 2004). In order to stock as much carbon as possible in the soil, the authors recommend reduction of nutrient input on heavily fertilized grasslands and moderate fertilization of extensively managed grasslands (Soussana et al., 2004). This recommendation does not apply to alpine and wet meadows, which already have naturally large carbon stocks.

7.1.5
Forests as biomass producers

7.1.5.1
Biomass use in tropical forests

According to FAO estimates, tropical forests make up 42 per cent of the world's forested land (Hakkila und Parikka, 2002). A total of approximately 2800 million m³ of timber from tropical forests was used worldwide in 2005 (FAO, 2006c). The percentage of timber used directly as fuel worldwide (mostly by developing countries) is estimated at approx. 40 per cent (Africa 88 per cent, North and Central America 13 per cent), with the remainder being industrially processed (FAO, 2006c). Illegal timber harvesting and the collection of firewood by individuals only feature in FAO forest statistics when such figures are reported by individual countries (FAO, 2006c).

The expansion of road systems in tropical forests almost inevitably leads to deforestation in the medium term (Asner et al., 2006; Fearnside, 2008). In the Brazilian Amazon region, rainforest within a 25km radius of a road is at a fourfold greater risk of being cut down than forested land outside this radius (Asner et al., 2006). Improper selective timber harvesting, illegal cutting, and fires started by people have additional negative impacts on the carbon balance of tropical forests and pose a threat to biological diversity (e.g. WBGU, 1998; Cochrane, 2003; Nepstad et al., 2008; Fearnside, 2008). After cutting, not only is the species diversity of the regrowth poorer compared to natural, undisturbed forests, but invasive species also spread much faster (Baret et al., 2007). Furthermore, a statistical study on bird populations in exploited versus undisturbed Bolivian forests showed that 40 per cent of the bird species inhabiting undisturbed forests need protection. In contrast, most of the bird populations in exploited forests consist of species that can withstand human disturbances (Felton et al., 2008).

In contrast to clearcutting, which is primarily carried out to obtain more land for farming and grazing, selective cutting is frequently used for harvesting timber. But even though 'only' individual trees are selectively harvested and a large part of the forest is left – albeit in a heavily damaged state – selective timber cutting causes genetic depletion of the population, depending on the tree species (Farwig et al., 2007; de Lacerda et al., 2008), and gives impetus to deforestation (Asner et al., 2006; Table 7.1-4). Almost a third of selectively harvested forests eventually fall victim to clearcutting after an average of four years (Asner et al., 2006). Furthermore, the sparser forest canopy resulting from selective timber harvesting leads to weakening of the ecosystem (Alongi and de Carvalho, 2008). Pereira et al. (2002) measured the gaps in the forest canopy in the Brazilian Amazon region. Whereas in undisturbed forests gaps comprised 3.1 per cent of the canopy, with selective timber harvesting they comprised more than 20 per cent of the canopy, and even with more sustainable reduced-impact logging (RIL), they still comprised approx. 10 per cent, which affects the microclimate and the soil (Pereira et al., 2002). A sparse forest canopy leads to greater water stress during droughts and, together with the dry wood residue, poses a greater risk of forest fires (Cochrane, 2003; Asner et al., 2006). Rainforest ecosystems in the humid tropics are considerably more sensitive to fires than arid forest ecosystems, as the plants are not adapted to fire events (Cochrane, 2003; Nepstad et al., 2008). A selectively

Table 7.1-4
Advantages and disadvantages of reduced-impact logging in tropical rainforests.
Source: WBGU

Advantages			
Economic efficiency	**Soil quality**	**CO$_2$ balance**	**Ecosystem services**
Long-term economic use possible due to sustainability	Minimal soil compaction due to carefully planned harvesting operations	Biomass-neutral, as only forest growth is harvested	No-go areas conserve biological diversity
			Intensive monitoring for early detection of ecosystem changes

Disadvantages			
Economic efficiency	**Soil quality**	**CO$_2$ balance**	**Ecosystem services**
Greater management input (land-use planning, tree inventories)	Soil moisture affected by sparser canopy		Decrease in drought resistance, forest fire hazard increased
			Risk of deforestation increases from development
			Signs of genetic depletion in harvested trees, long-term effects on biological diversity unknown
			Subtle change in the local microclimate

logged forest may remain sensitive to fire for decades (Cochrane, 2003).

In timber harvesting by the RIL method, various measures can be taken to promote maximum sustainable management: these include conservation of the forest substance (only regrowth is harvested), taking tree inventories, careful layout of access roads in order to damage as little of the surrounding vegetation as possible, a network of defined no-go areas to serve as biological corridors, and the conservation of mature trees as seed producers. In spite of these measures, RIL can lead to alteration of the gene pool and the spatial genetic structure of the population of a tree species, thus increasing the risk of inbreeding (de Lacerda et al., 2008). However, not all tree species appear to be equally sensitive from a genetic standpoint to economic exploitation of their population (Borges Silva et al., 2008). Castro-Arellano et al. (2007) measured the impacts of RIL on bat populations in the Amazon region. Although the changes in biological diversity appear to be minor in short-term measurements, long-term studies on changes in biological diversity are lacking.

In principle, according to the present state of knowledge it seems that sustainable use of tropical virgin forests is nearly impossible, as on the one hand the ecosystem reacts with extreme sensitivity to disturbances, and on the other hand interventions involving even a small area of land, such as the building of a road, lead to deforestation within a few years due to socio-economic factors. Nowadays there are various certification systems in the area of sustainable forestry (Section 10.3.2.1).

7.1.5.2
Biomass use in temperate forests

Temperate forests are found predominantly in the northern hemisphere, in Europe, East Asia, and in eastern North America. They make up 25 per cent of the world's forests (Fischlin et al., 2007). Around one-third of the entire land area in the EU is covered in forest, of which 12 per cent are under protection. Around two-thirds of European forests are privately owned (EU Commission, 2005d). An estimated 65 million m^3 of timber is used annually in Germany (BMELV, 2008).

Along with use as a feedstock for industrial production (in the construction, paper and pulp industries), the use of wood residues for energy production is gaining in importance in developed countries. The focus here is on the slash left over from logging, unmanaged growth and individual trees unsuited for industrial use. Different forest management schemes result in varying amounts and types of wood residues.

Unmanaged, 'overripe' forests that are clearcut contain more wood residues from dead and diseased trees, whereas wood residues from managed forests consist mainly of topwood and wood from the thinning of undersized trees (Hakkila and Parikka, 2002).

Along with the socio-economic aspects, criteria for sustainable forest management (SFM) also include forest health, production capacity, biological diversity, water balance, soil quality, and carbon balance (Raison, 2005). According to the sustainability principles of the Forest Stewardship Council (FSC, 1996), unique and sensitive ecosystems and landscapes must be conserved in spite of use, and the ecological functions and the integrity of the forest must be assured (BUWAL, 1999).

According to a metastudy by Guo and Gifford (2002), broadleaf tree plantations have no adverse impacts on carbon stocks in the soil compared to virgin temperate forests. In pine plantations, however, the carbon stocks in the soil decline when annual rainfall exceeds 1500 mm compared to virgin forests (Guo und Gifford, 2002). A chronosequence study of a commercially exploited beech forest shows that the quantity of sequestered carbon in the soil does not change significantly within the rotation cycles (Hedde et al., 2008). When the impact of wood residue use on the carbon balance of a forest is modeled, however, the result is highly variable depending on the model used (Palosuo et al., 2008).

Future wood residue use will have strong impacts on forest flora and fauna, as the removal of biomass alters the microclimate, soil properties and nutrient proportions, which in turn affects interspecies interactions in the ecosystem (EEA, 2007a). Resource use in European forests varies greatly by region. Whereas the rate of biomass extraction from forests is very high in certain countries, particularly Finland, the Baltic states and Belgium, in countries such as France it is 56 per cent and in Italy a mere 47 per cent of the resource potential (EEA, 2007a). In order to assure a sustainable use of wood residues, the European Environment Agency has established a maximum extraction rate of 60 percent of the resource potential in European forests on the basis of model data; parameters such as hillside slope, water balance and soil fertility are to be taken into account (EEA, 2007a).

7.1.5.3
Biomass use in boreal forests

Boreal forests form the northernmost biome with tree growth and span Eurasia and North America approximately between 50° north latitude and the

Box 7.1-11

Potentials and risks of green genetic engineering

The discussion of the use of biomass for energy production raises questions about the contribution genetic engineering might make to increasing the potential of bioenergy use and the risks associated with doing so. Against the background of increasing scarcity of high-yielding agricultural lands due to various reasons, as early as the 1990s there was frequent mention of the relevance of future developments in green genetic engineering (WBGU, 1998). Green genetic engineering refers to the application of genetic engineering techniques to plant breeding. The modified plants are also known as transgenic plants. Particularly in regard to the question of using marginal sites for bioenergy production, the possibility of growing genetically engineered crops with greater stress tolerance is repeatedly raised.

The following potential applications of green genetic engineering for improving the potentials of bioenergy use are under discussion:

- Yield increase in the narrower sense: genetically engineered increase of biomass production in the plant;
- Yield increase in the broader sense: yield assurance by minimization of yield losses (e.g. through genetically engineered resistance to insect pests);
- Alteration of plant contents in terms of a higher biofuel yield (e.g. through genetically engineered higher starch content or genetically engineered production of enzymes in the crop that facilitate the conversion of starch to ethanol);
- Development of stress-tolerant crops (e.g. salt or drought-tolerant crops) for cultivation on extreme sites that were previously unsuitable for agricultural production.

The options for yield increases are a highly controversial subject in both public and scientific discussions. A realistic, quantitative assessment of the actual improvement potential of these options, however, is impossible at the present time, as the data base for such an assessment is still insufficient.

The options proposed must be considered critically from various points of view. Transgenic plants with altered characteristics of agronomic interest (especially herbicide and insect resistance; '1st generation') have been grown commercially for at least a decade. The growing of these transgenic plants does not lead to yield increases in the narrower sense. What is involved instead are new approaches to pest and weed control, and therefore anticipated yield increases in the broader sense through minimization of yield losses. So far, however, sound data on large-scale and long-term effects versus actual savings in pesticide use are lacking (IAASTD, 2008; Levidow and Paul, 2008). All of the other above-mentioned transgenic approaches or developments which, if implemented, could offer additional optimization potentials in bioenergy use are still in the research and development phase.

An increase in the biomass yield of the entire above-ground crop biomass, i.e. a yield increase for bioenergy production in the narrower sense, is a substantial objective in terms of bioenergy production. Efforts to achieve such an increase, either by conventional or biotechnology methods, have so far been unsuccessful (Levidow and Paul 2008). This raises the question as to whether it is even possible to increase the total energy content of a plant without auxiliary inputs such as nutrients, water, etc., or whether as a general rule it is only possible to redistribute the available energy. Specific breeding successes, e.g. a higher grain yield, frequently go hand in hand with a reduction, such as in stalk weight. Attempts to improve the photosynthetic mechanisms of plants have so far been unsuccessful (Rosegrant and Cavalieri, 2008).

The use of sites unsuitable for agricultural production to grow energy crops is considered an option in terms of a sustainable biomass use strategy. Only stress-tolerant plants, however, can grow on such sites. The development of stress-tolerant, transgenic cultivars has been actively pursued since the 1990s (Schmitz and Schütte, 2000). The comparatively complex physiological and biochemical mechanisms responsible for tolerance to abiotic stress factors (drought, heat, salt, flooding) in plants, however, are still poorly understood. Characteristics corresponding to such tolerance are frequently based on a number of genes and complex regulatory mechanisms. Most of the genetic engineering research approaches so far have focused on specific components of these complex characteristics, although nearly all previous findings in this area indicate that only the simultaneous transmission of the genetic bases for several stress responses in a plant can lead to the creation of stress-tolerant plants (Holmberg and Bülow, 1998). Knowledge of the regulator genes coordinating the complex gene responses to abiotic stress in plants is considered a key factor (Datta, 2002). Teufel (2005) summarizes the approaches to the development of transgenic, stress-tolerant crops currently being pursued (2005).

As a general rule, results on stress-tolerant transgenic plants published to date are based on experiments conducted under laboratory conditions or in greenhouses. A realistic assessment as to when and if genetically engineered, stress-tolerant crops will be commercially available is not possible at the present time.

Another group of genetically engineered crops is characterized by altered use characteristics known as 'output traits'. These are crops that have been improved or selectively bred for bioenergy production. For example, there are research efforts to develop high starch yielding maize cultivars for ethanol production and oilseed crops with a higher oil content. Although intensive effort has been devoted to such strategies for many years, and although the first two genetically engineered crops (lauric acid-producing rape and a high oleic acid soybean) have been approved in the USA, but have yet to be marketed successfully, use of such genetically modified organisms (GMOs) does not appear to be achievable in the near future (TAB, 2005). A 2005 report by the German Parliamentary Office of Technology Assessment (Büro für Technikfolgen-Abschätzung, TAB) on 2nd and 3rd generation transgenic plants suggests a possible reason for this: namely that in some cases the expectations, particularly in terms of achievable product yields, have yet to be fulfilled in spite of years of research and development. In many cases efforts to maximize the content of certain ingredients have been/are concomitant with undesired side effects, which in turn lead to yield losses. It has also been found that using genetic engineering techniques to target lipid metabolism in plants is a considerably more complex process than was previously thought, although the lipid metabolism in plants on a molecular level is very well researched and understood (TAB, 2005).

ECOLOGICAL AND SOCIO-ECONOMIC RISKS ASSOCIATED WITH THE USE OF GMOS FOR BIOENERGY PRODUCTION
The ecological risks linked to the use of GMOs for bioenergy production are in general similar to those discussed in

connection with the cultivation of 1st generation transgenic crops for food and feed production. These include the general effects of uncontrolled and undesired spreading of GMOs and their transgenes by proliferation in the wild (by vegetative propagation, for example), outcrossing and horizontal gene transfer, as well as specific effects the transgenic traits may have on non-target organisms or on the entire ecosystem (TAB, 2000, 2005). The damage potential of such an uncontrolled and undesired proliferation depends mainly on the type of transgene and/or the traits it imparts. One example is the outcrossing of rape cultivars that have been genetically engineered for herbicide resistance with closely related wild species in Canada, which as a consequence have become problem weeds in agriculture.

As a general rule it must be assumed that with increased stress tolerance, a plant's fitness and ability to compete likewise increases. The potential for proliferation of such plants is thus increased. The same applies to closely related wild species, which may hybridize with such plants. If the given stress factor to which the plant is tolerant is a limiting factor for proliferation, there is a real danger that stress-tolerant plants could spread widely and turn into problem weeds with economic consequences for agriculture (Schmitz and Schütte, 2000).

Transgenic trees are also being discussed as bioenergy resources. Due to the long life of trees as well as the fact that it usually takes a long time before they are able to reproduce, it has so far not been possible to make any substantiated statements about the stability of the introduced traits and about their effects on the environment. Authoritative conclusions with regard to this effect would have to be based on data acquired over several decades (Pickardt and de Kathen, 2002; Farnum et al., 2007; Schmidt, 2008). At the 9th meeting of the Conference of the Parties to the Convention on Biological Diversity, it was therefore agreed that the release of transgenic trees may only take place on the basis of the precautionary approach and that such release should otherwise be prohibited on the grounds of insufficient data on biological safety. This agreement also stipulates that long-term monitoring of the risks of release should be carried out, and that statements concerning potential environmental effects must be substantiated with authoritative, experimentally obtained data. In addition to the ecological effects on the environment, the potential

socio-economic effects on local and indigenous communities also need to be studied. Specific risk assessment criteria are to be developed for transgenic trees (CBD, 2008a).

The cultivation of genetically modified energy crops poses still another risk if the plant species is also grown for food production. The danger here is potential contamination of the food chain with potential risks to human health. In addition to these environmental and health-related effects there are also economic risks, which play an important role in developing countries in particular. Cultivar protection and patents can promote concentration of agricultural property and considerably impede the exchange of plant genetic material. Alternatives to genetically engineered plants could thus be displaced in the long run (IAASTD, 2008).

CONCLUSIONS
Based on the current state of knowledge, it cannot be assumed that genetic engineering will make a substantial contribution to expanding the potential of bioenergy in the next ten years. In light of the potential ecological and economic risks, the approval of a genetically modified plant for bioenergy production is indeed controversial and should only be granted after thorough environmental risk assessment studies, which should be carried out step by step. This means that only after thorough laboratory and greenhouse studies may a decision be made as to whether additional field experiments in support of the approval are justifiable. The use of GMOs for bioenergy production can only be considered in a responsible manner on the basis of comprehensive, accurate, and carefully collected data. This involves considerable research effort and expense.

WBGU therefore recommends the promotion and performance of additional research projects on the uses and applications of biotechnology methods, particularly in the areas of marker-assisted breeding and the use of 'white' biotechnology in closed systems, e.g. to improve the exploitation of available biomass resources. These and other recommendations are also given in the International Assessment of Agricultural Knowledge Report (IAASTD, 2008). To prevent undesired effects of GMOs arising when energy crops are cultivated, the issues surrounding GMOs need to be taken into account when formulating sustainability standards (Section 10.3).

Arctic Circle. One-third of the world's forests are in the boreal zone (Fischlin et al., 2007). Boreal forests sequester 26 per cent of terrestrial carbon stocks, which is equivalent to the carbon stocks of tropical and temperate forests combined (UNEP, 2002). In the 1990s, forest fires, insect pest infestations and unfavourable climatic events along with clearcutting led to tremendous losses in Russian forests (UNEP, 2002).

Forest timber accounts for 10 per cent of the total biomass used in the northern European nations (Lunnan et al., 2008). Finland and Sweden have the largest biomass reserves (Röser et al., 2008). In Finland 20 per cent of the energy demand is met by wood products – mostly wood residues and slash (Lunnan et al., 2008). Röser et al. (2008) estimate the annual potential of wood residue use for energy production at 58

million m³ per year in Baltic and northern European forests, which corresponds to around 116TWh. Commercial use of wood residues deprives organisms that live on dead wood of their food source. Jonsell et al. (2007) found more than 160 species of saprophilic beetles on slash, 22 of which are on the red list of threatened species. Whereas a large diversity of beetles was found on the dead wood of broadleaf trees such as alders and oaks, the diversity on dead spruce wood was low (Jonsell et al., 2007). It is essential to take the species composition of the tree population into account in considering sustainable wood residue use.

Table 7.1-5

Summary and qualitative rating of the productivity and impact on biological diversity and carbon sequestration in the soil of the proposed cultivation systems. A yellow rating indicates that different impacts on these factors are possible, depending on previous use and cultivation methods. The colour for the overall rating corresponds to the colour rating of the majority of the factors studied.

Sources: Productivity: see caption; biological diversity and carbon sequestration in the soil: WBGU qualitative estimate

Cultivation systems	Productivity [t of dry matter/ha/year]	Sources	Productivity rating	Biodiversity rating	Soil carbon rating	Overall rating
Tropical monocultures						
Annual						
Sugar cane	10–120; 80; 70 (a)	1, 3, 5	positive	negative	negative	negative
Perennial						
Sugar cane	10–120; 80; 70 (a)	1, 3, 5	positive	negative	unclear	unclear
Oil palm	30; 13,8 (b)	3, 5	positive	positive	unclear	unclear
Jatropha	0,2–8; 0,5–12 (b)	3, 4	unclear	unclear	unclear	unclear
Temperate Monokulturen						
Annual						
Maize	8–14*; 9; 9 (d)	2, 5, 6	positive	negative	negative	negative
Rape	4,1; 3,4; 3 (c)	1, 5, 6	unclear	negative	negative	negative
Triticale	3,5–9**; 5,6; 6 (d)	2, 5, 6	positive	negative	negative	negative
Perennial						
Miscanthus grass	up to 30; 10–27,5; 11–40 (a)	1, 2, 3	positive	unclear	unclear	unclear
Switch grass	12–17; 5,2–11,1 (a)	7, 8	positive	unclear	unclear	unclear
SRPs: poplar/willow	4–16/2–14; 12–15/5–20 (a)	2, 3	positive	unclear	unclear	unclear
Grassland						
Tropical grassland, pasture	not specified		unclear	positive	positive	positive
Temperate grassland, grassland systems						
Grass leys	7–15 (e)	2	positive	unclear	positive	positive
Permanent grassland	7–12 (f)	2	positive	positive	positive	positive
Forests						
Agroforestry	not specified		unclear	positive	positive	positive
Wood residues from forests						
Tropical	not specified		negative	negative	negative	negative
Temperate	not specified		negative	positive	positive	positive
Boreal	not specified		negative	positive	positive	positive

Impacts

- negative (red)
- unclear (yellow)
- positive (green)

Sources

1: Lieberei et al., 2007
2: KTBL, 2006
3: El Bassam, 1998
4: Openshaw, 2000
5: FAOSTAT, 2007
6: LfL Bayern, 2008
7: TFZ, 2008
8: Schmer et al., 2008

(a) Worldwide
(b) Seed yield, worldwide
(c) Seed yield, Germany
(d) Grain yield, Germany
(e) In rotation, Germany
(f) Permanent grassland

* DM content = 70 %
** DM content = 86 %

7.1.6
Summary evaluation of currently predominant cultivation systems

As a general rule, crops are less beneficial for biological diversity and carbon stocking in the soil than forests, grasslands or pastures. Perennial crops such as *Jatropha*, oil palm and SRPs rate better on these factors than one- to three-year crops such as rape, cereals or maize. Sugar cane, which can be grown either as an annual or a perennial crop, likewise has a less favourable balance as an annual crop than as a perennial crop (Table 7.1-5).

7.2
Technical and economic analysis and appraisal of bioenergy pathways

7.2.1
Overview of energy conversion options

The possibilities for providing energy from biomass are numerous (Fig. 7.2-1, Box 7.2-1). Biomass has one decisive advantage over other renewable energy sources: it occurs mainly in the form of an energy store or can be stored without additional technical input. A further advantage is the fact that it can be used universally for power, heat and transport. Since large parts of heat and transportation demand can be met using electricity generated from other renewable energies, however, its storability may be considered the most significant advantage of bioenergy over other renewable sources.

A pathway or supply chain encompasses all processes, beginning with the cultivation of energy crops or supply of biogenic residues or wastes and extending to the provision of final energy (electricity, heat or fuels). These feedstock life cycles may be split into several sub-processes. Distinctions are made between
- Biomass production,
- Biomass supply,
- Biomass conversion processes,
- Energy use,
- Recovery and disposal of residue and/or waste arisings.

The sub-processes are mutually interdependent, and thus the entire process chain must be examined for an assessment to be made.

Each sub-process is the sum of many individual steps (FNR, 2005). The principal technical processes are briefly discussed below.

7.2.2
Energy conversion technologies

7.2.2.1
Combustion and thermochemical processes

Combustion is the oxidation of a fuel in the presence of oxygen with an accompanying release of energy. In chemical terms carbon (C) or hydrogen (H) is oxidized to carbon dioxide (CO_2) or water (H_2O) with the aid of oxygen (O_2). The reaction is exothermic, i.e. energy is released. For example, the maximum energy released in the combustion of carbon to CO_2 is 394 kJ per mol. In the combustion of hydrogen to water vapour the maximum energy released is 242 kJ per mol.

More oxidation agent (oxygen) is usually supplied to the process than is necessary for complete oxidation. Combustion of a solid fuel consists of the sub-processes of pyrolytic decomposition (see pyrolysis), gasification and the subsequent oxidation of the decomposition products. If the combustion is complete, it is termed stoichiometric combustion (Kaltschmitt and Hartmann, 2003).

The direct combustion of solid biomass in combustion systems (large-scale furnaces, room fireplaces, pellet stoves, etc.) or boilers is the most common energy conversion process for biogenic solid fuels such as wood or straw. Combustion devices produce heat that can be used as secondary energy (e.g. as steam for the mechanical operation of turbines that convert the energy into electrical energy in a generator), as final energy (e.g. district heat) or as useful energy (e.g. radiated heat from a tiled stove).

It is possible to deploy a combination of solid biofuels and fossil energy carriers in what is known as co-combustion, and this is state of the art in many combustion plants. An example of this is a small-scale local district heating system in which biomass is used to meet the baseload requirement while demand peaks are covered using light heating oil or natural gas. Large coal-fired power stations use biomass in the form of pellets in co-combustion systems to provide electricity and heat.

In developing countries the burning of solid biofuels represents the most widely used source of heat, especially for cooking (Section 4.1).

Biomass is converted in a thermochemical conversion process (carbonization, gasification, pyrolysis or liquefaction), primarily under the influence of heat, to energy carriers in gas, liquid or solid form (Fig. 7.2-1; FNR, 2005).

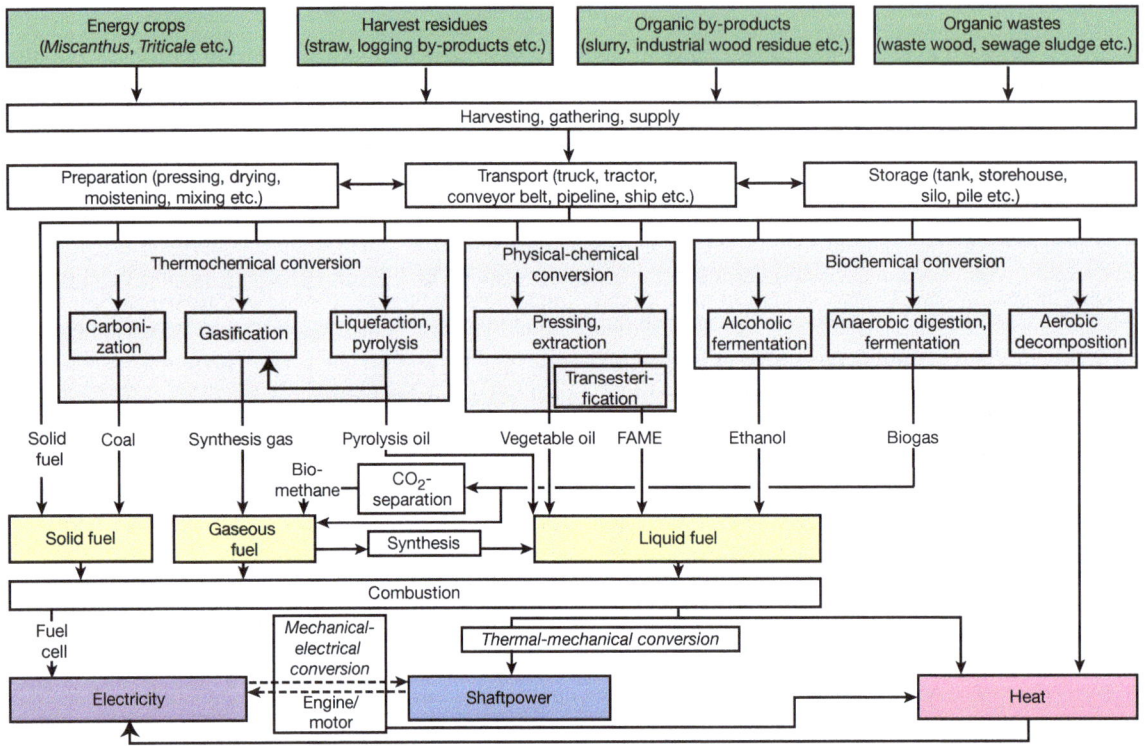

Figure 7.2-1
Simplified representation of typical feedstock life cycles for final or useful energy provision from biomass.
Source: WBGU, adapted after Kaltschmitt and Hartmann, 2003

CARBONIZATION

The carbonization of biomass is the processing of biomass to obtain a high yield of solid fuel (charcoal) with specific characteristics (low weight, high energy content). In this process biomass is broken down under the action of heat. The process energy necessary for this is often obtained by combustion of a part of the feedstock. The charcoal thus obtained can then be used in the appropriate plant to provide heat or electricity. It can also be used as a material for other purposes (e.g. as activated carbon). The charcoal production process is well established. In industrialized countries charcoal is used almost exclusively as a material feedstock in the chemical industry and elsewhere (FNR, 2005). In developing and newly industrializing countries, carbonization takes place locally and the charcoal is used to supply heat (in particular for cooking). The intermediate step of carbonization, however, results in a conversion ratio of only approx. 15–40 per cent of wood to charcoal and compared to direct combustion is neither technically nor economically attractive for energy applications. For this reason it is not used for energy conversion in industrialized countries (FNR, 2005). It is, however, widely used in developing and newly industrializing countries since the specific energy content of char-

coal is much greater than that of firewood and it is thus easier to transport. The efficiency of charcoal use in these countries is even more reduced by the poor efficiency of the charcoal furnaces employed. An alternative use that has been little discussed so far is the emplacement of charcoal in soils to increase their fertility and as a means of carbon sequestration. CO_2 can thus be removed from the atmosphere and the carbon (C) can remain stored for an extended period (Section 5.5; Box 5.5-2).

GASIFICATION

In gasification, prepared biomass in solid or liquid form is converted under high process temperatures as completely as possible into a high-calorific biogenic gas (raw gas, lean gas). For the gasification process less oxygen is fed to the biomass via a gasification agent than would be required for complete combustion (partial or substoichiometric oxidation). A supply of process heat is required since the reactions occurring in the gasification are predominantly endothermic, i.e. they require an external input of energy. Air, oxygen, water vapour or carbon dioxide can be used as gasification agents. A disadvantage of air is the high proportion of inert gas obtained in the

Box 7.2-1

Bioenergy: Definitions

Biomass is diverse – both in its origins and in its options for conversion and technical application. Biomass, bioenergy and their possible uses are defined below.

BIOMASS

Biomass stores solar energy. In photosynthesis, CO_2 and water are converted with the aid of solar radiation into organic matter. Some of the energy thus absorbed is released again when biomass is burnt and thus becomes available for use. Biomass consists primarily of the elements carbon (C), oxygen (O) and hydrogen (H) and may be described by the empirical formula $C_nH_mO_p$. More generally, according to Kaltschmitt and Hartmann (2003) the term 'biomass' encompasses all material of organic origin:

- all living phytomass and zoomass (plants and animals),
- the residues that are formed from these (e.g. animal excrements),
- dead (but not yet fossilized) phytomass and zoomass,
- in the wider sense all (waste) substances that have been formed through a technical conversion process and/or through the use of biogenic resources for production processes (e.g. black liquor, cellulose and pulp, residues from animal carcass disposal and from the waste management industry, etc.).

Biomass is classified into primary and secondary products. Primary products are formed directly through photosynthesis, and thus include all plant biomass (energy crops and vegetable by-products from farming and forestry operations). Secondary products arise indirectly through the transformation of primary products, i.e. they are created by the decomposition or conversion of organic matter in heterotrophic organisms (e.g. animals or bacteria). These include all zoomass, its excrements and sewage sludge.

Biomass is either produced deliberately by cultivating farmed feedstocks or arises as organic residue in other production processes. Farmed feedstocks include energy crops (e.g. cereals, *Miscanthus* grasses, harvested timber), while organic residues include harvest residues (e.g. straw, logging residues) and organic wastes and by-products (e.g. slurry, household organic wastes, animal fats, green cuttings: Kaltschmitt and Hartmann, 2003; FNR, 2005).

BIOENERGY

Bioenergy is the final or useful energy that can be released and made available from biomass.

BIOFUELS

The term biofuels refers to fuels in liquid or gaseous form of biogenic origin that are used primarily as transport fuels but also have application in electricity and heat generation, e.g. in small-scale combined heat and power (CHP) units. A distinction is made between 1st-generation and 2nd-generation biofuels. The 1st generation includes vegetable oil, biodiesel and bioethanol, obtained through established physical and chemical (pressing, extraction, esterification) or biochemical (alcoholic fermentation) processes. The 2nd generation includes synthetic biofuels such as BtL (biomass-to-liquid, Fischer-Tropsch diesel), biomethane (or bio-SNG, synthetic natural gas) and biohydrogen, produced using thermochemical processes (gasification, pyrolysis). Almost without exception these technologies remain at present at the laboratory or demonstration stage. Biomethane produced by fermentation can also be included in the 2nd generation. The division of biofuels into the 1st and 2nd generations is not very strict and is based on different parameters depending on the literature consulted, such as whether parts of the plant (1st generation) or the entire above-ground plant (2nd generation) is used, or even whether or not the fuels in question are already established in the marketplace. WBGU therefore designates liquid and gaseous fuels as 2nd generation in the case of biomethane or where the fuels have been obtained by thermochemical processes.

BIOGAS

Biogas is a gas mixture of approx. two-thirds methane (CH_4) and approx. one-third carbon dioxide (CO_2). It also contains small quantities of hydrogen, hydrogen sulphide, ammonia and other trace gases. Biogas is formed by the anaerobic fermentation of organic matter. The component of the gas with a usable energy content is the methane (FNR, 2006a).

BIOMETHANE

The gases with no usable energy content such as CO_2 and other noxious components (e.g. hydrogen sulphide) can be removed from biogas, resulting in a fuel of the quality of natural gas. Known as biomethane, this can be fed into the existing natural-gas network and used in all final energy sectors (electricity, heat, shaftpower – for electrical, thermal and mechanical energy). Through the gasification of solid and liquid biomass it is also possible to produce a raw gas that, after cleaning (clean gas) and conditioning (synthesis gas), is converted, via methane synthesis, into biomethane (bio-SNG; bio-synthetic natural gas) (IE, 2007a).

---

raw gas. For synthetic fuel manufacture pure oxygen or water vapour is therefore mostly used.

The gasification process can be divided broadly into four separate stages. The fuel is first heated and dried at temperatures of up to 200°C. This is followed by pyrolytic decomposition: at approx. 200–500°C in the absence of oxygen, gaseous hydrocarbon compounds, pyrolysis oils and pyrolysis coke are formed. The next stage is oxidation, in which at temperatures of approx. 2,000°C the coke and some of the higher hydrocarbons split into smaller gaseous molecules (CO, H_2, CO_2, CH_4 and water vapour). The majority of the combustible components of the raw gas are formed in the subsequent reduction of carbon

dioxide and water. The raw gas consists principally of carbon monoxide and hydrogen, carbon dioxide, methane, higher hydrocarbons and water vapour and in some cases nitrogen. The composition of the gas depends on the type of gasification, the gasification agent (type and quantity) and the conditions surrounding the reaction (temperature and pressure; Sterner, 2007; Kaltschmitt and Hartmann, 2003). In addition to these main components (CO, CO_2, H_2, CH_4, C_2 carbon compounds, water vapour, N_2), the gas resulting from gasification also contains various noxious components (tars, particles, alkalis and sulphur, halogen and nitrogen compounds), which must be removed from the raw gas before it is used further. Some aspects of this cleaning of the gas are complex and form the bottleneck in the development and market introduction of gasifiers for distributed use (Kaltschmitt and Hartmann, 2003; Knoeff, 2005).

The uncleaned gas can be used directly in burners to provide heat. For electricity generation it must, however, be cleaned, since if used directly the raw gas would severely contaminate the engine and render it inoperable. Various possibilities exist for converting the clean gas to electricity: it can be burnt as a diesel substitute component in a diesel engine, or used in its pure form in gas engines or gas turbines coupled to generators. It is also possible to use the clean gas in fuel cells to provide electricity. The most advanced developments in this area are those based on solid-oxide fuel cells (SOFC) (Aravind et al., 2006; IISc, 2006). There are many further possible uses that are not discussed here. Alternatively, following further preparation (in particular in order to achieve the required H_2/CO ratio) the cleaned gas can be fed into a synthesis process again and converted into liquid energy carriers (Fischer-Tropsch diesel, ethanol, methanol) or useful biogenic gases (biomethane, dimethyl ether, hydrogen) (Vogel, 2006).

In India and China wood gasifiers have been in successful use for electricity and heat generation for some time. In the industrialized countries the environmental regulations (air and water emissions) for the plant are generally more stringent, and for this reason there have until now been only a small number of commercial-scale gasification plants for electricity and heat generation (Vogel, 2007). This is, however, only a technical problem concerning the cleaning of the gas, one that is solvable and mainly only a question of cost. Biomass gasification plants are the core process in the production of synthetic biofuels, also known as 2nd-generation biofuels. In some cases the biomass feedstock must be shredded and dried before gasification. The raw gas formed in the gasification process is cleaned and conditioned, i.e. impurities such as tars, particles or sulphur compounds are removed (clean gas) and the composition of the gas is adjusted to suit the fuel synthesis. The synthesis gas thus obtained is converted in a synthesis process into the desired fuel. There are various types of synthesis: Fischer-Tropsch synthesis converts synthesis gas into a raw Fischer-Tropsch product that is further refined by additional processes to usable Fischer-Tropsch diesel fuel, often known as BtL (biomass-to-liquid) diesel. Methanation (methane synthesis) converts synthesis gas into methane and carbon dioxide. The CO_2 must be separated off and can be stored (Box 7.2-2). This process chain is energy-intensive and involves major losses. The production of Fischer-Tropsch diesel requires that the aggregate state is changed twice, as predominantly solid biomass is converted into first

Box 7.2-2

Biomethane: A highly promising bioenergy carrier

For a number of reasons biomethane production is a particularly interesting pathway. The feed material is either biogas, formed in biogas plant from the fermentation of wet biomass, or synthesis gas from the gasification of predominantly solid biomass.

Biogas consists of methane (CH_4) and carbon dioxide (CO_2) as its main components, plus very small quantities of the compounds H_2O, H_2S, NH_3, N_2 and O_2. The unwanted components can be removed by various processes, such as pressure-swing absorption and amine scrubbing. Membrane separation processes promise increased efficiency, but are at present only at the development stage. There are today approx. 80 biogas processing facilities in operation in Europe that feed into natural gas networks or provide a natural gas substitute. Depending on the process used, purities in the range of 96–99 per cent methane are achieved.

Synthesis gas is made up of other components. Depending on the gasification process, carbon monoxide (CO) and hydrogen (H_2) form as principal components, while CH_4 and water vapour occur only in small concentrations. This mixture is converted to CH_4 and CO_2 in synthesis reactors and the CO_2 separated off. The process of converting synthesis gas to CH_4 and that of removing the CO_2 are managed on a large scale and both processes have been successfully applied for over 20 years (IPCC, 2005). In contrast to the direct deployment of the biogas or synthesis gas in decentral electricity generation plant, in which waste heat utilization is only partly feasible, the feeding of biomethane into natural gas networks allows more flexible utilization. Biomethane can then be supplied via the network to those users (CHP plant) that can optimally utilize waste heat. Alternatively the gas network can serve a collecting function, supplying the biomethane product from multiple plant for maximum-efficiency deployment in large combined-cycle power plant.

When using biomethane it is generally possible to make use of the entire infrastructure developed for natural gas (distribution networks, combined-cycle power plant, gas engines, gas turbines, natural-gas vehicles).

The CO_2 inevitably produced in the gas processing can, at least in larger plant, be captured and stored. This improves the climate change mitigation effect of bioenergy use by approx. 20 per cent. The technology for CO_2 storage is at present being developed. WBGU has expressed its views on the sustainability requirements for CO_2 sequestration elsewhere (WBGU, 2006).

The emissions of CH_4 caused during the gas processing should be considered critical. While they are of the order of only a few percentage points, they cause a distinct reduction in the climate change mitigation effect of biomethane production and utilization owing to the strong global warming potential of CH_4. The goal of future development must be to drastically reduce these leakages.

a gaseous and then a liquid energy carrier, with the result that the final product of BtL diesel contains only approximately half the original energy (Sterner, 2007). Biomethane can be more efficiently produced since the final product is a gas. In many cases the starting product is woody biomass such as wood chips that nevertheless can be better deployed in direct combustion for electricity and heat provision than in synthetic fuel production by gasification. The latter is advantageous only where the feedstock used is a biomass that is difficult to process.

The development of biomass gasification plant for synthetic fuel production in the countries in which it is being pioneered (Germany, Sweden and Austria) is still at the pilot and demonstration stage. The first commercial facilities are now being constructed, but a significant macro-economic contribution from this technology cannot be expected before 2020 at the earliest. Particularly promising as regards energy yield are highly integrated processes in which a technically mature fluidized bed gasification process is used for the polygeneration of electricity, heat and fuel (synthetic biomethane) (Choren, 2007; Chemrec, 2007; TU Vienna, 2005).

PYROLYSIS (LIQUEFACTION)
Pyrolysis involves the thermal decomposition of solid biomass in the absence of oxygen, and is thus also known as degasification. The objective of the process is to obtain the largest possible proportion of liquid components (pyrolysis oil). Solid and gaseous by-products are also produced (pyrolysis coke, pyrolysis gas), some of which can be used within the process. In principle cleaned and prepared pyrolysis oil can be used as a heating or transport fuel in combustion plant or combustion engines, gas turbines or small-scale combined heat and power (CHP) units. Technical problems in the production and use of pyrolysis oil and the low economic viability of its preparation have, however, prevented a breakthrough in this technology thus far. The significant advantage of the process is the high energy density of pyrolysis oil. Pyrolysis oil is an outstanding medium for energy transport between locally arising feedstocks with lower energy density (e.g. straw) and large-scale central systems (gasification for fuel production). This concept is being examined at the Forschungszentrum Karlsruhe in particular (FZK, 2007).

7.2.2.2
Physical-chemical processes

In physical-chemical conversion processes oils and fats with usable energy content are pressed or extracted from certain bioenergy carriers (e.g. rape-seed or Jatropha plants). They can be efficiently used directly in small-scale CHP units for electricity and heat provision or as a fuel for modified engines in road transport. Vegetable oils can also be converted by transesterification into a biodiesel whose characteristics closely match those of conventional diesel fuel. Engine modification is then unnecessary and the application range of the biogenic transport fuel is increased (Fig. 7.2-1; FNR, 2005).

PRESSING AND EXTRACTION
Simple mechanical pressing of parts of plants that contain oil (e.g. the seed) separates the liquid oil from the solid component, known as press cake. The press cake is mostly used as animal feed, except where toxic plant residues are present, such as those of *Jatropha*. The pressing technology is available both on a small scale (e.g. in farming) and on a large scale (e.g. oil presses). The same technology is used to obtain vegetable oil in the food sector. Following a stage of purification the oil can be used in special vegetable oil engines or for stationary use in small-scale CHP.

Alternatively or additionally to pressing, oil can be extracted from bioenergy carriers through the use of a solvent (e.g. hexane). Oil and solvent are then separated by distillation. The solvent is reused. This process is also large-scale in application. Vegetable oil can be obtained from a large number of oleiferous plants; examples include *Jatropha* and oil palms.

ESTERIFICATION
To broaden the application range for biogenic fuels and to avoid the need for engine modification for use with vegetable oil, vegetable oil can be converted to fatty acid methyl ester (FAME, commonly known as biodiesel) with losses of approx. 5–10 per cent (TUM, 2000). The esterification process is large-scale in application. FAME can be used pure in adapted and all newer diesel engines and in a blend of up to 5 per cent in all conventional diesel engines. Apart from its use in the transport sector, biodiesel can technically be used in small-scale CHP. This is not an efficient application, however, as vegetable oil can also be burnt directly in CHP without further treatment (FNR, 2005; 2006b).

7.2.2.3
Biochemical conversion

The conversion of biomass into final energy by biochemical processes involves the use of microorganisms (Fig. 7.2-1; FNR, 2005).

ANAEROBIC DIGESTION – FERMENTATION

The process of anaerobic digestion involves the decomposition of organic matter in the absence of oxygen through the activity of certain bacteria. The final product of this process is a water-vapour saturated, combustible mixed gas (biogas) that consists substantially of methane (50–70 per cent) and carbon dioxide (25–40 per cent) (BayLfU, 2004; FNR, 2006a). The process of biogas formation has four stages: (1) hydrolysis, (2) acidification, (3) acetic acid formation and (4) methanogenesis.

In hydrolysis the complex compounds of the starting substrate (e.g. carbohydrates, proteins, fats) are broken down into simpler compounds (e.g. amino acids, sugars, fatty acids). The bacteria involved in this process use enzymes that break the material down by biochemical means. In the acidification phase (acidogenesis) the intermediate products formed in hydrolysis are broken down further by acid-forming bacteria to lower fatty acids (acetic, propanoic and butanoic acids), carbon dioxide and hydrogen. Other products, in small quantities, are alcohols and lactic acid. In the next stage – acetogenesis, or the formation of acetic acid – these products are converted by bacteria to precursor substances of the biogas (acetic acid, hydrogen and carbon dioxide). In the final methanogenesis the methane is formed from the products of the acetogenesis (FNR, 2006a).

These processes take place in one or more fermenters. For the process to have a high yield, the temperature in the fermenter must be maintained at approx. 35–37°C. The heat required for this usually comes directly from waste heat from the combustion engine that is powered by the biogas. Fermenters form the core of conventional biogas plant. Contaminated material, for example waste from the food industry, must be pre-treated in a hygienization stage. On removal from the fermenter, the digestate is stored in closed post-fermenters, in which the biogas is used, or in open digestate containers, and usually deployed as liquid fertilizer on cultivated land (IE, 2007a). This fertilizer is lower in odour and richer in nutrients than unfermented slurry. Methane and nitrous oxide emissions from the post-fermentation have a negative effect on the greenhouse gas balance of the biogas plant (methane slip). Covering the post-fermenters prevents these emissions from escaping. Covering is obligatory in Germany but not yet required in other countries (Zah et al., 2007). The collected biogas can be used as a heat and light source by direct burning (mainly in developing countries) or in small-scale CHP through combustion in piston engines or micro gas turbines for electricity and heat provision (mainly in industrialized countries).

An elegant process step is the preparation of biomethane of natural gas quality from biogas. Here the CO_2 is separated off from the gas stream and the remaining methane cleaned. It can now be fed decentrally into the existing natural gas network and used for the provision of electricity and heat both in decentral micro-CHP units (combustion engine, fuel cell, etc.) and in large-scale combined-cycle power plant (Box 7.2-2). In Germany and other European countries many large-scale biogas facilities have been built in recent years that use not only animal excrements as the substrate but also farmed biomass such as maize (IE, 2007b).

In large areas of Asia small biogas plant are widely distributed (FAO, 1992). This technology has been successfully used for decades on a local basis. In particular the low maintenance requirements (no moving mechanical parts, little wear and tear) and uncomplicated operation have assured this technology success. The addition of external heat for the fermenter is not required in these latitudes owing to high average annual temperatures. A combination of tillage and livestock farming is necessary since pure arable farming does not produce dung, while nomadic livestock breeders without tillage cannot use the resultant digestate. Separation of arable and livestock farming is common in many parts of sub-Saharan Africa. In Asia however the commonest practice is combined arable and stock farming, which allows a very good integration of biogas in its rural regions (SNV, 2008).

AEROBIC DECOMPOSITION

In aerobic decomposition the biomass is oxidized by bacteria in the presence of oxygen. Unlike anaerobic digestion, in this process heat is released that can be exploited with heat pumps to provide low-temperature heat. Owing to the low demand for heat in composting systems and poor availability of system technology, there has been no penetration of this method so far (FNR, 2005).

ALCOHOLIC FERMENTATION

Alcohol can be produced from various organic substances containing sugar, starch or cellulose by the use of yeasts or bacteria and purified by distillation or rectification. If substances containing starch or cellulose are used, these must first be saccharified. The processes are state of the art and have been long established for the production of potable alcohol. Bioethanol can be used as an engine fuel (automotive, CHP, etc.) and thus finds application in all areas of energy provision (electricity, heat, transport). 'Flexible-fuel vehicles' with E85-optimized engines that can burn fuel mixtures with up to 85 per cent ethanol are becoming ever more popular and are being mass-produced by more and more car makers (FNR, 2005). Synthetic ethanol production from matter containing lignocellulose by a microbiological fermentation process has not so far gone beyond the pilot scale (Igelspacher et al., 2006). Ethanol production based on lignocellulose is theoretically also possible by a gasification process (Abengoa, 2006; Sterner, 2007).

7.2.3
Efficiencies of various modern conversion processes

In the following section the processes described above, embedded in selected bioenergy pathways, are assessed in terms of both technical and economic parameters. The most robust and important technical criterion for the general assessment of a technology and for the appraisal of a pathway is efficiency. For the present WBGU report, expert reports were commissioned from the German Biomass Research Centre (Deutsches Biomasseforschungszentrum Leipzig; Müller-Langer et al., 2008) and the Institute for Applied Ecology (Öko-Institut; Fritsche and Wiegmann, 2008) in which data was compiled for selected conversion pathways. Efficiencies are calculated on the basis of this data in accordance with the VDI 4661 standard. An overview of the bioenergy pathways investigated is given below.

7.2.3.1
Overview of the bioenergy pathways investigated

The selection of bioenergy pathways for investigation included those of current market relevance and additionally such pathways that WBGU considers to be of particular future significance in environmental and technical terms, from cultivation or production through to final energy use. An important selection criterion was the availability of data. From more than 120 pathways, 66 were selected and analysed under these criteria, comprising 25 in the transport sector (biofuels and electromobility), four heat pathways and 37 pathways in the electricity and heat generation sector (predominantly CHP). In all pathways the reference point is bioenergy use in Germany, and this also applies for the recovery of residues and the cultivation of temperate energy crops. The countries of origin selected for tropical crops were Brazil (sugar cane), Indonesia (oil palm) and India (Jatropha) (Table 7.2-1).

THE CULTIVATION SYSTEMS AND RESIDUES INVESTIGATED
Various cultivation systems were selected as the upstream chain for the technical conversion pro-

cesses. These include widely-used systems such as maize, rape and sugar cane cultivation, but also the cultivation of perennial C4 grasses (switchgrass) that provide long-term ground cover, and *Jatropha*, which can be cultivated on degraded land. The deployment of residue and waste feedstocks was also included. For the temperate energy crops, cultivation was considered both on land previously used as cropland and on former grassland (Fritsche and Wiegmann, 2008). For clarity, only the results for cultivation on cropland are shown below. The differences resulting from grassland conversion are discussed in Section 7.3. The cultivation systems and residues analysed in Sections 7.2 and 7.3 are listed in Table 7.2-1.

THE TECHNICAL CONVERSION PROCESSES INVESTIGATED
In the first conversion step (biomass to bioenergy carrier), in addition to conventional processes, pathways that involve the processes of fermentation or gasification were also selected, since these are considered by WBGU to be a highly promising option for the future (Box 7.2-2).

The next conversion step (bioenergy carrier to energy service, product conversion) considered not only transport options based on the conventional drive concept using a combustion engine but also pathways involving electromobility. Owing to their higher fuel utilization and overall energy efficiency, the emphasis in electricity generation was placed on combined heat and power systems – both central and decentral, with existing technologies such as small-scale CHP units, large combined-cycle power plant and solid-oxide fuel cells (SOFC). Some of the conversion pathways examined reflect the current state of the art (2005); others are referenced to the year 2030, i.e. new and ongoing technological developments are expected. All the conversion pathways are listed in Table 7.2-2.

THE EFFECT OF TECHNOLOGICAL SCALE ON EFFICIENCIES, PRODUCTION COSTS, GHG ABATEMENT COSTS AND GHG ABATEMENT EFFECTIVENESS
For the analysis, representative plant with predefined output capacities were used in each case. The size of the facility has a decisive influence on its efficiency figures. As a rule, the larger the plant, the more efficient it is. This is particularly the case for the mobility pathways, where plant size varies by a factor of 300: the smallest plant (biogas for electromobility) has a size of 1.6 MW (approx. 0.4 t biomass input per hour); the largest (Fischer-Tropsch diesel) is 535 MW (approx. 140 t biomass input per hour). The size of the plant thus exerts an influence on the potential for greenhouse gas abatement and on production costs,

Table 7.2-1
Selection of the different cultivation systems examined by WBGU.
Source: WBGU

Name in pathway	Previous land use and origin of feedstock	Cultivation system, feedstock
Tropical monocultures		
Sugar cane (degraded)	Degraded land (Brazil)*	Sugar cane
Sugar cane	Cropland (Brazil)	
Oil palm (rainforest)	Tropical rainforest (Indonesia)*	Oil palm
Oil palm (degraded)	Degraded land (Indonesia)*	
Jatropha	Cropland (India)	*Jatropha*
Jatropha (marginal)	Marginal land (India)	
Temperate monocultures		
Maize silage	Cropland (Germany)	Maize
Maize grain	Cropland (Germany)	Maize
Cereal, wheat	Cropland (Germany)	Cereal
Rapeseed	Cropland (Germany)	Rapeseed
Switchgrass	Cropland (Germany)	Millet, switchgrass
Short-rotation plantations (fast-growing tree species)		
SRP	Cropland (Germany)	Poplar, willow
Temperate grassland		
Grass silage	Meadow (Germany)*	Grass
Residues and wastes		
Harvest residues/slurry	-	Harvest residues/slurry
Slurry	-	Slurry
Wood residue	-	Wood residue
Straw	-	Straw
Waste fats	-	Cooking fat, animal fat, waste fats
Organic waste	-	Organic waste
Mixture of energy crops and residues		
Grass silage/slurry	Meadow (Germany)*	Grass silage (70%), slurry (30%)

* No GHG emissions from indirect land-use changes are included in these cultivation systems. Regarding meadow (Germany), owing to changing milk quotas and increasing use of high-quality crop feeds for cattle there is currently grassland available for which a use is being sought. For this reason using grass and slurry as a substrate in biogas plant is presently expedient. This is, however, a special case arising out of particular current market conditions in Germany and/or Europe that cannot be transposed to other regions, so that changes in grassland use elsewhere may cause indirect land-use change.

and thus also on the greenhouse gas abatement costs (Sections 7.3 and 7.4). Small plant tend to perform less well in this context and large plant better.

7.2.3.2
Efficiencies

The declared efficiencies of different bioenergy pathways are set out separately for transport, electricity and heat. Efficiencies capture the ratio of outgoing target energy flows (electricity, heat, shaftpower) to the energy flows expended in a plant (biomass feedstock, auxiliary energy). This technical parameter is calculated by WBGU using the declared efficiency method (Nennwirkungsgradmethode) in accordance with the VDI 4661 standard, which is described in more detail in Box 7.2-3. An advantage of this method is that it is established and recognized, thus allowing direct comparisons with other studies. Moreover, an exergetic assessment in the heat calculations allows the distinction to be drawn between exergy and energy, thus enabling the three sectors – transport, electricity and heat – to be cross-compared on the basis of the mechanical or electrical equivalent of the thermal energy. In many publications the efficiency for providing one thermal energy unit of heat is equated with one chemical energy unit of fuel

Table 7.2-2
List of technical conversion processes examined by WBGU.
Source: WBGU and Müller-Langer et al., 2008

Name in pathway	Conversion steps	Product or product conversion	Capacity of installation Biomass input of thermal combustion capacity [MW]
Transport			
Ethanol passenger car	Alcoholic fermentation, dehydration	Bioethanol (1st gen. apart from straw: 2nd gen.) Otto engine / flexible-fuel vehicle	Maize – 192 MW Sugar cane – 319 MW Cereal – 229 MW Straw – 378 MW
Biodiesel passenger car	Extraction, transesterification	Biodiesel (1st gen.)	Rapeseed – 175 MW Oil palm – 298 MW Jatropha – 291 MW Waste fats – 61 MW
Vegetable oil passenger car	Extraction	Vegetable oil (1st gen.)	Rapeseed – 2.9 MW
Biomethane passenger car	Anaerobic digestion, gas processing	Biomethane (1st gen.) Gas Otto engine	Maize – 3.2 MW Slurry / harvest residue – 5.0 MW Grass silage/slurry – 3.8 MW Organic waste – 3.9 MW
Biomethane passenger car	Gasification, methanation	Biomethane (2nd gen.) Gas Otto engine	SRP – 39 MW Wood residue – 39 MW
Fischer-Tropsch diesel BtL	Fuel gasification, Fischer-Tropsch synthesis (biomass-to-liquid, BtL) Upgrading	BtL diesel (2nd gen.)	SRP – 518 MW Wood residue – 518 MW Straw – 536 MW
Hydrogen fuel cell (PEM)	Gasification, gas scrubbing	Biohydrogen (2nd gen.) Fuel cell (H2) (proton-exchange membrane, PEM)	Wood residue – 250 MW
Small-scale biogas-CHP electric passenger car	Anaerobic digestion, combustion in small-scale CHP for electricity and heat	Bio-electricity Electric motor	Millet – 1.6 MW Slurry / harvest residue – 2.5 MW
Heat			
Pellet heating system	Pelleting	Pellets Mini-combustion plant (pellet heating system)	Millet – 0.017 MW SRP – 0.015 MW Wood residue – 0.016 MW Straw – 0.019 MW
Electricity and heat – combined heat and power (electricity only in (hard) coal-fired power stations)			
Small-scale biogas CHP	Anaerobic digestion	Biogas Decentral CHP unit Gas Otto engine	Maize – 1.6 MW Millet – 1.6 MW Slurry/harvest residue – 2.5 MW Grass silage/slurry – 1.9 MW Organic waste – 3.9 MW
Small-scale biomethane CHP	Anaerobic digestion, gas processing	Biomethane Decentral CHP unit Gas Otto engine	Maize – 3.2 MW Millet – 3.1 MW Slurry/harvest residue – 5.0 MW Grass silage/slurry – 3.8 MW Organic waste – 3.9 MW
Biogas fuel cell (SOFC)	Anaerobic digestion	Biogas Solid-oxide fuel cell (SOFC)	Maize – 1.6 MW Millet – 1.6 MW Slurry/harvest residue – 2.5 MW Grass silage/slurry – 1.9 MW Organic waste – 3.9 MW

▶

Table 7.2-2 (continued)
List of technical conversion processes examined by WBGU.

Name in pathway	Conversion steps	Product or product conversion	Capacity of installation Biomass input of thermal combustion capacity [MW]
Biomethane combined-cycle	Anaerobic digestion, gas processing	Biomethane Central combined-cycle power plant	Maize – 3.2 MW Millet – 3.1 MW SRP – 39 MW Wood residue – 39 MW Slurry/harvest residue – 5,0 MW Grass silage/slurry – 3,8 MW Organic waste – 3,9 MW
Small-scale vegetable oil CHP	Extraction, (transesterification)	Vegetable oil Decentral CHP (diesel engine)	Rapeseed – 2.9 MW Oil palm – 3.9 MW Jatropha – 3.7 MW
Pellet-coal power plant	Pelleting, co-combustion	Pellets Central hard coal-fired power station	SRP – 100 MW Wood residue – 103 MW Straw – 144 MW
Central woodchip CHP – steam turbine	Combustion	Wood chips Central CHP Steam turbine	SRP – 22 MW Wood residue – 22 MW Straw – 22 MW
Raw gas – gas turbine	Fluidized bed gasification	Raw gas Gas turbine	SRP – 90 MW Wood residue – 90 MW
Raw gas – fuel cell (SOFC)	Fluidized bed gasification	Raw gas Solid-oxide fuel cell (SOFC)	SRP – 18 MW Wood residue – 18 MW

or one electrical energy unit of electricity. However, the three quantities each have different energetic natures and can only be meaningfully compared to one another and evaluated by consideration of the Second Law of Thermodynamics, as shown in Box 7.2-3. If, however, comparisons are drawn within the heat sector, the thermal energy is the relevant quantity of useful energy.

EFFICIENCY COMPARISON FOR COMBINED ELECTRICITY AND HEAT PRODUCTION FROM BIOENERGY

Thirty-seven pathways were investigated from the area of CHP, and additionally one pathway for pure electricity generation from the co-combustion of biomass in a hard coal-fired power station. In energy terms the biggest yield from biomass, and woody biomass in particular, is obtained through the direct combustion of woody biomass in a central CHP plant or its gasification and direct use in gas turbines. Equally energy efficient is the use of rapeseed oil in small-scale CHP or of maize and switchgrass in biogas plant. From an exergetic perspective the differences are less significant; exergetic efficiency averages approx. 30 per cent, and the most efficient processes are biomass co-combustion in hard coal-fired power stations, the use of fuel cells or the use of rapeseed oil in small-scale CHP. Solid-oxide fuel cells

demonstrate very high electrical efficiencies overall, which is an advantage of this technology.

A comparison of the deployment of different feedstocks in the same technology, e.g. in the biomethane combined-cycle pathway, shows that woody biomass such as SRP and wood residue and grasses such as switchgrass can be more efficiently converted than, for instance, organic waste or harvest residues. Efficiency comparison for heat production from bioenergy

The use of different feedstocks for the production of heat was analysed on the basis of a pellet heating system with a capacity of 15 kW. The exergetic efficiency for the heat pathways examined lies in the range of 15–20 per cent and is thus above most of the mobility pathways examined. It can be seen from Figure 7.2-3 that the use of SRP and wood residue is more efficient than that of switchgrass and straw. The reason for this is that the relative energy input for the pelleting of stalky biomass is higher than that for woody biomass on account of its lower energy density. However, in exergetic efficiency terms heat provision via CHP is superior to the pathways examined here for heat alone.

EFFICIENCY COMPARISON FOR TRANSPORT PATHWAYS

Figure 7.2-3 compares the efficiencies and other parameters of 25 bioenergy pathways for the transport sector that differ in terms of the fuel supply and vehicle technology employed (cf. characteristics of vehicle types in Table 7.2-3). Using the mechanical drive of the vehicle as the target energy reference, 1st and 2nd-generation biofuels can be compared with biomass-derived electromobility.

In the application of biomass in mobility, biogenic electricity used for electromobility demonstrates particular advantages over biofuels in conventional combustion engines: the exergetic efficiency of the pathway is almost double that of conventional biofuels in combustion engines. The reasons for this are, firstly, the efficient production of electricity and its deployment in modern electric cars, and secondly, the efficient use of the biomass feedstock, since in stationary energy conversion processes the waste heat of the combustion process can be utilized, whereas in mobile applications it can not (Section 8.1). The comparison further shows that 2nd-generation biofuels are not generally more efficient in use than those of the 1st generation. The production and use of ethanol with a maximum efficiency of 11 per cent is inefficient. Higher efficiencies are achieved by some biodiesel pathways that do not, however, use the whole aboveground plant but, for example, only parts of the plant such as rape seeds or *Jatropha* nuts. An exception is waste fats as a feedstock, which can be converted very efficiently into biodiesel since waste fat is already a converted vegetable product and this conversion step does not need to be included in the efficiency calculation.

In the production of Fischer-Tropsch diesel a small quantity of electricity can also be generated. Although the three theoretical BtL routes examined assume a very large plant size with a combustion capacity of approx. 500 MW, the efficiency of BtL production and utilization is in the range of 14 to 16 per cent and thus comparable with other mobility pathways based on much smaller-scale facilities. In the biomethane sector SRP and wood residue yield the highest efficiencies since the polygeneration concept can be applied in the gasification of these feedstocks to provide fuel, electricity and heat (Fürnsinn, 2007). The gasification of wood residue to hydrogen in a plant with a combustion capacity of 250 MW and its deployment in modern fuel-cell vehicles is hardly more efficient than conventional biofuels in combustion engines.

In the mobility pathways examined, the significance of plant size for efficiencies was particularly clear. For electricity generation for electromobility, for example, the 1.6 MW unit examined represented a relatively small class in terms of capacity. If biomass is converted to electricity in larger power stations and the waste heat put to effective use, the efficiency of the electromobility pathway will be still higher than given here.

COMPARATIVE ASSESSMENT

Overall it can be seen that although the highest energy efficiencies are obtained from pure heat provision, the exergy – that is the mechanical or electrical equivalent of this heat – is considerably lower than that for CHP or even that for pure electricity generation in a power station. From an exergetic perspective it can thus be seen that power generation and CHP are more efficient than the provision of heat alone.

The lowest exergetic weighting is obtained in the use of biofuels in vehicle combustion engines, corresponding in most cases to only about half the equivalent for combustion for pure heat production or even a third of the values achievable in CHP plant or pure electricity generation. A more efficient deployment than in transport can be achieved by using biofuels for combined heat and power generation, for example bioethanol in combined-cycle power plant or vegetable oil in small-scale CHP. Electromobility represents an exception: not only is the conversion to the drive energy more efficient here than in a combustion engine, but an exergetic component also arises from the waste heat in the CHP. It should be noted, however, that electromobility is an application of the supplied electricity and can thus be compared only within the mobility pathways, and not with the electricity pathways.

A complete, cross-sector comparison can only be made by taking account of exergy. Use of bioenergy is, however, also dependent on the demand in the various sectors and on the energy system concerned, which varies from country to country. As regards technical efficiency and exergetic evaluation, CHP and pure electricity generation are clearly the preferable utilization options. If, however, the electricity requirement is covered in future by another renewable energy, for example, then bioenergy deployment in specific instances for heat-only applications such as stand-alone pellet heating systems or central woodchip-fired heat plants is also conceivable.

In comparing pathways that have similar technical conversion processes but differ in the biomass used, no systematic efficiency differences are indicated between residues and energy crops. As a rule, however, the values for residues are more widely scattered, as is illustrated by comparison of the pathways based on wood residue, waste fats, harvest residues/slurry and organic waste.

Box 7.2-3

Declared efficiencies: Methodology, inventory boundaries and calculation

The method of calculating the declared efficiency of bioenergy pathways as set out in VDI standard 4661 is described below. In the view of WBGU this method is particularly appropriate as a means of comparing the uses of biomass for transport, heat and electricity across sectors.

The efficiency of a system is defined as the quotient of the useful work output (benefit) and work input (expenditure; VDI, 2000, 2003). The definition of the benefit and expenditure is determined in each case by the particular application and supply task in question (Stiens, 2000). The target energy is the desired form of energy of a conversion process. WBGU defines mechanical energy, electrical energy and thermal energy (as useful heat) as target energies. The deployment of energy provided from biomass is then viewed in terms of electricity, heat or vehicle drivepower. In all bioenergy conversion processes the core process is the combustion of the energy carrier. All processes are therefore inventorized inclusive of this core process and compared with one another on the basis of the target energy.

Some conversion processes provide not only the target energy but also other forms of energy. In CHP the generation of shaftpower or electricity is coupled to the generation of heat. In polygeneration, combined heat, power and fuel or combined cooling, power and fuel, the biomass feedstock is converted into three target energies: electricity/shaftpower, fuel and heat/cold. All three target energies are included as benefits in the efficiency calculation (Equation 7.2-1).

Alongside these forms of energy are by-products that do not take the form of target energy but have an energy value. These then do not have to be created by other processes since they arise as by-products and can therefore reduce expenditure or alternatively substitute for products being produced. They are therefore deducted, in energy terms, from the expenditure or the biomass input. Such by-products include glycerin, naphtha, press cake, dried distillers' grains with solubles (DDGS) from ethanol production and rapeseed meal after rapeseed oil extraction. Some such by-products require further preparation (e.g. drying) before they can be deployed as energy sources. Therefore not all residues are weighted at 100 per cent in energy terms. Certain by-products are used almost exclusively in their capacity as materials (predominantly as animal feedstuffs or fertilizers), for example fermentation residues, rapeseed meal and DDGS; these are therefore given a reduction factor of expenditure for bioenergy provision in the present method of only 50 per cent. Some other by-products such as glycerin and naphtha are liquid energy carriers that, while also used predominantly as materials, can nevertheless be 100 per cent thermally utilized without processing and are

therefore also weighted at 100 per cent. Glycerin can be fermented as a co-substrate very successfully in biogas plant, giving a high yield of methane. An exception is the bagasse produced during ethanol production from sugar cane. Normally this is burned for electricity generation after drying. Bagasse is therefore given an 80 per cent expenditure reduction weighting.

Auxiliary energy, applied at any point in the feedstock life cycle from cultivation through to target energy, is normally added as expenditure. Examples include the use of machinery for sowing, the energy used in transporting the biomass and that required in the conversion processes in the form of electricity and heat. Thermal and electrical auxiliary energy can also be deducted from the benefit in an energy conversion process, but only when the converted energy is in the same form (VDI, 2000). Thus the electricity requirement of the conversion plant is deducted from the electricity produced, i.e. from the benefit, to give the net electricity production. For economic reasons this is not always the practice in Germany, however, since under the Renewable Energy Sources Act (EEG) regenerative electricity can be sold at a higher price than that for which conventional electricity is obtained. Auxiliary thermal energy can only be deducted from the heat generated when it is integral in the energy conversion process. In many cases the temperature of the output heat is considerably lower than that of the necessary process heat. In thermal power plant engineering it is therefore conventional practice to add the auxiliary thermal energy to the expenditure (Bolhár-Nordenkampf, 2004; Strauss, 2006). In the method used here, all forms of auxiliary energy are counted as expenditure with the exception of auxiliary electrical energy where electricity is produced in the facility itself. Where it does not exceed the electricity production of the facility, this is then deducted from the 'electricity' benefit. If it exceeds the latter, it is included in the expenditure.

Definition of the inventory boundaries and the efficiency calculation

The inventory boundaries (Fig. 7.2-2) of the selected bioenergy feedstock life cycles (pathways) encompass the feedstock cultivation or supplying of residues, feedstock transport, their conversion via various conversion steps into a 'fuel' product (biodiesel, bioethanol, bioelectricity), product transport and conversion of the product into the target or useful energy – heat, electricity or shaftpower (mechanical energy to the wheel of a vehicle). Co-products and auxiliary energies are integrated as described above.

The efficiency as per VDI 4661 (VDI, 2003) used to evaluate the bioenergy pathways is defined as follows:

For the present report and in Müller-Langer et al., 2008 energy expenditure for feedstock transportation was based on a distance of 50 km and expenditure for product distribution (predominantly biofuels) on a distance of 300 km. Capacity utilization of plant was assumed at 7000 hours per year, but with co-combustion of biomass in hard coal power stations assumed at 5000 hours per year.

$$\eta_{ex} = \frac{P_{out} \text{ (benefit)}}{P_{in} \text{ (expenditure)}} = \frac{P_{elec\_net} + P_{shaftpower} + P_{heat} \cdot \eta_{Carnot}}{P_{biomass} + P_{AE\,cultivation} + \Sigma\,P_{AE\,trans} + P_{AE\,therm.} + P_{AE\,elec.} - P_{by\text{-}products}}$$

Equation 7.2-1
Calculation of the exergetic efficiency η_{ex}. For the exergetic efficiency the heat is evaluated by means of the Carnot efficiency η_{Carnot}. Calculation of the energetic efficiency η_{en} is analogous, whereby the factor η_{Carnot} no longer applies. AE = auxiliary energy.

Figure 7.2-2
Inventory boundaries for calculation of efficiency. All energy flows that occur along the bioenergy supply chain are inventorized. This encompasses the main energy flows of biomass feedstock as input and electrical, mechanical and thermal energy as output plus all auxiliary energies in cultivation, transport and conversion of the biomass or energy carrier and any co-products such as naphtha, press cake, bagasse, glycerin, etc.
Source: WBGU

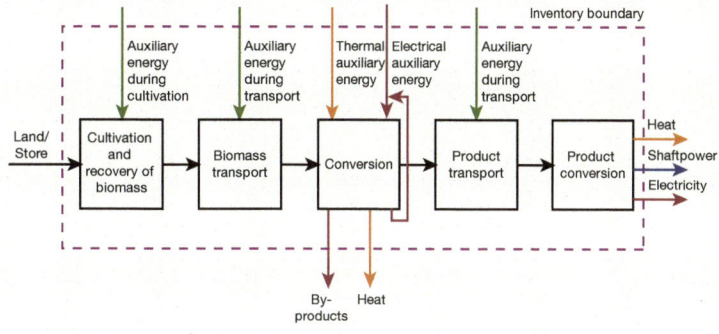

The data basis for all efficiency calculations was Müller-Langer et al., 2008.

EXERGETIC APPRAISAL OF HEAT
A direct comparison of thermal energy with electrical or mechanical energy can lead to erroneous conclusions, since these forms of energy cannot be converted from one to the other on a unit-for-unit basis. Thus even in a loss-free and reversible conversion, thermal energy can only be partially converted into mechanical or electrical energy. The proportion that can be converted is dependent on the temperature of the available thermal energy and is termed exergy. It may be represented via the thermal efficiency of the Carnot cycle as expressed in Equation 7.2-2:

$$E_{ex} = E_{th} \cdot \eta_{Carnot} = E_{th} \cdot \left(1 - \frac{T_U}{T_H}\right)$$

Equation 7.2-2
Calculation of the exergy of heat.

Here the exergetic fraction of the heat is described by E_{ex}, the thermal energy by E_{th}, the temperature of the heat in Kelvin by T_H and the ambient heat by T_U. The entire thermal energy can, however, be supplied in a conversion in the opposite direction, from mechanical/electrical energy of the same value, e.g. via a heat pump process.

The exergy is thus the mechanical or electrical equivalent of the heat energy and is shown in Figure 7.2-3 for the various conversion procedures.

For the efficiency calculation and weighting of the heat, an ambient temperature T_u of 293 K (20°C) is assumed and a temperature of 373 K (100°C) is used for the extracted heat T_o. The heat quantities in the efficiency calculation are then multiplied by a factor of 0.214, obtained from the Carnot efficiency. This proportion is the exergy of the heat. In the following illustrations the efficiency of the heat provided is shown in its exergy and anergy components. Energy always consists of exergy and anergy (Baehr, 1965; Baehr and Kabelac, 2006). This distinction is common in thermodynamics, whereas outside technical discourse it is often not made, with the result that technical data may be wrongly interpreted.

For the calculation of the mobility pathways in the report, average values from the medium vehicle category were used; these are given in Table 7.2-3. All values are based on the New European Driving Cycle (NEDC), which defines uniformly the conditions and speed patterns under which a vehicle is driven in determining energy or fuel consumption and the resultant greenhouse gas emissions.

Table 7.2-3
Characteristic values for the vehicle types used in the mobility pathways, as per the New European Driving Cycle. The MJ quantity related to input describes the energy carrier in the vehicle, i.e. one MJ fuel or one MJ electricity.
Source: Müller-Langer et al., 2008

Vehicle type – drive system	Time horizon	Mileage related to input [km/MJ]	Efficiency (mechanical drive energy related to input)
Otto combustion engine for petrol and gas (methane)	2005	0.37	0.26
	2030	0.48	0.29
Diesel combustion engine	2005	0.43	0.29
	2030	0.53	0.32
Electric motor	2030	1.11	0.78
PEM fuel-cell passenger car with electric motor	2030	0.71	0.39

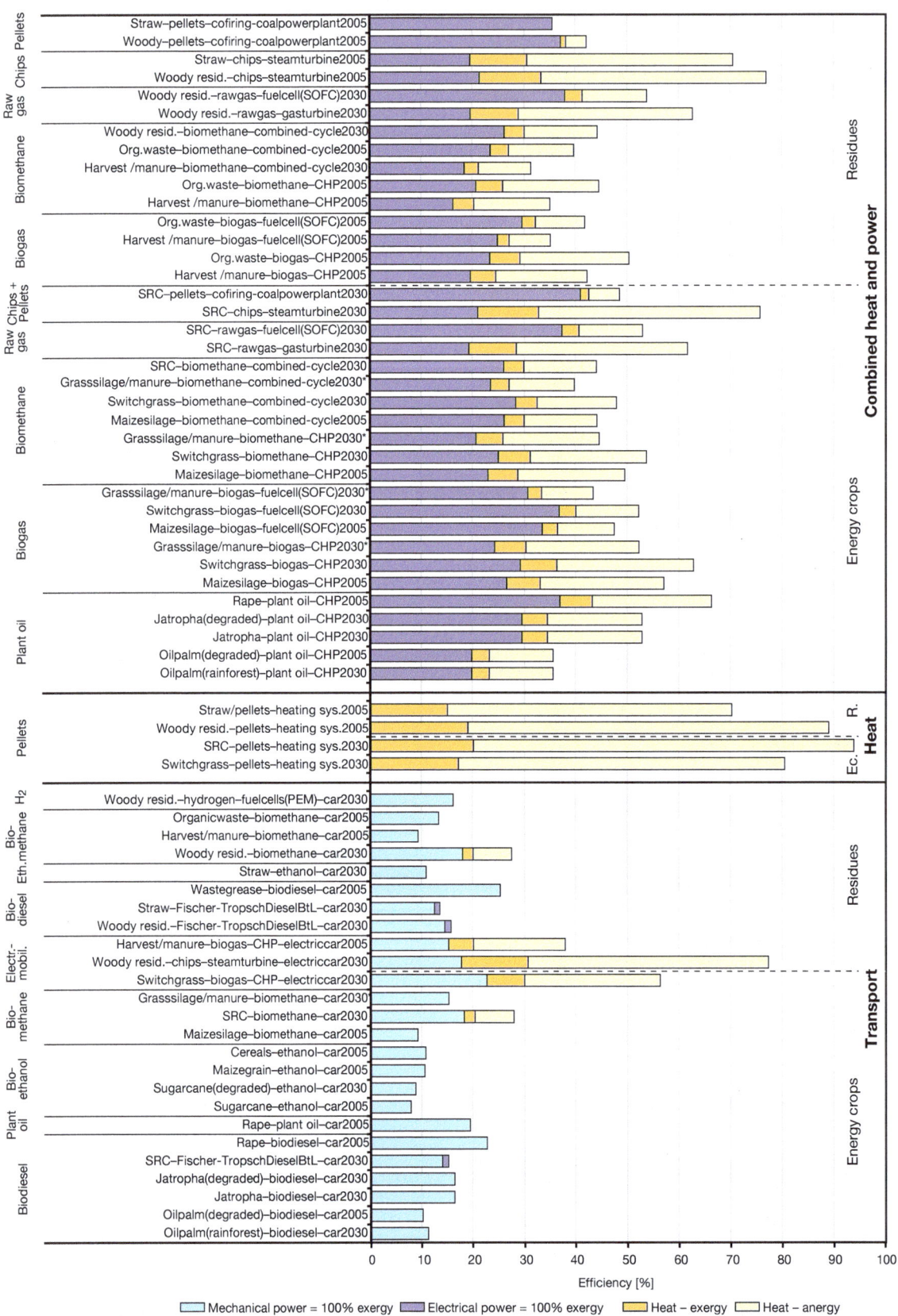

Figure 7.2-3
Overview of exergetic and energetic efficiencies (with and without light yellow bars respectively) of the bioenergy pathways examined. Efficiency of biomass pathways in per cent. The names of the pathways relate to the cultivation systems and conversion processes listed in Tables 7.2-1 and 7.2-2.
Source: WBGU

7.2.4
Efficiencies of various traditional conversion processes

The traditional use of biomass takes place predominantly in developing countries. For transport and electricity generation very similar efficiencies are obtained in these countries to those in industrialized countries; for this reason they are not specifically discussed here. However, of interest for biomass utilization in the developing countries in terms of efficiency is the area of heat applications, in particular the provision of heat for cooking.

Wood stoves
Globally the largest use of biomass is in developing countries for cooking, heating and lighting. Cooking is traditionally done using gathered firewood in a three-stone hearth that, owing to poor combustion and heat utilization, has an efficiency of only 5–15 per cent (Mande and Kishore, 2007). Also widespread is the production and use of charcoal; its overall efficiency is only 4 per cent but it is widely used, particularly in the growing urban centres, owing to its high energy density and ease of transportation. The low overall efficiency arises from, firstly, the low efficiency of approx. 18 per cent in charcoal production and, secondly, a combustion efficiency of 23 per cent in the stove. Wood and charcoal stoves are the most common source of heat in developing countries. Kerosene and electric stoves are used mainly by the numerically small wealthier population groups (FAO-RWEDP, 2008).

The efficiency of wood stoves can be greatly increased, simply and at low cost, by a change in their construction. In southern India, for example, simple clay ovens have been developed that enable a more efficient combustion. The heat is shared between two hotplates and the hot exhaust gases entering the chimney are passed through a metal heat exchanger in which water is heated (IISc, 2006). The efficiency achieved here is 40 per cent, which represents a quadrupling of the efficiency, i.e. for the same utility biomass consumption can be reduced to one quarter of that necessary with a three-stone hearth. Even if the heat from the exhaust is not utilized, the efficiency of this stove is still in the region of 25–30 per cent, reducing firewood consumption to between a half and a third of that previously required (Jagadish, 2004). The efficiency of clay or metal charcoal-burning stoves can also be improved by means of an appropriate design. Here again efficiencies of up to 40 per cent are possible, and charcoal production can also be made more efficient (with efficiencies of up to 20 per cent). These technological improvements can double overall efficiency to 8 per cent (Kumar et al., 1990).

Micro biogas devices
Fuelwood can be replaced by biogas. By feeding animal excrements mixed with water into a micro biogas plant, up to 80 per cent of the energy of the residue can be converted into methane for cooking and lighting. The methane can be converted to thermal energy in a simple biogas stove with an overall efficiency of approx. 40–60 per cent or to electricity via a generator, with an overall efficiency of 15–25 per cent (FAO-RWEDP, 2008).

Gasification plant
Gasification of wood for electricity generation has been the state of the art in India for several years. The gasification route allows waste and residues such as coconut shells or waste wood to be utilized. These residues are converted to raw gas with an efficiency of approx. 80 per cent. This raw gas can be converted to electricity in a diesel generator operating in dual-fuel mode with 80 per cent raw gas and 20 per cent diesel with an efficiency of 20–25 per cent. If gas engines are used, efficiency is increased to 25–30 per cent. This allows an overall efficiency of approx. 15–25 per cent to be reached (Dasappa et al., 2003). Gasification plant can also be used to provide pure heat (e.g. for drying) and can achieve overall efficiencies of 30–45 per cent depending on the biomass feedstock, thus roughly three or four times the value for traditional heat provision (Mande and Kishore, 2007). One challenge in the use of this technique is the purification of the contaminated effluent that forms in the gas scrubbing and also the reduction of the noxious substances in emissions to air (Dasappa et al., 2003; Mande and Kishore, 2007).

Vegetable oil engines, generating sets and small-scale CHP
Combustion engines are often also used for a variety of stationary purposes, e.g. to grind foodstuffs (maize, cereal) or to power water pumps. They are often coupled to a generator to form a generating set and used to supply electricity (e.g. for public buildings such as hospitals and schools or for mini-grids). These generating sets can be powered with unrefined vegetable

oil (from *Jatropha* or palm oil), offer efficiencies of 20–25 per cent and have large potential for rural off-grid electrification since they are low-maintenance and relatively easy to handle (FAO-RWEDP, 2008). Waste heat from the sets can be used e.g. for drying agricultural products. If the waste heat is thus utilized, the sets are termed small-scale CHP units.

7.2.5
Economic analysis and assessment of conversion processes

7.2.5.1
Production costs of modern conversion processes

The production costs of traditional conversion processes such as the simple wood stove are difficult to establish. Neither the investment cost nor the cost of fuel can be established clearly, as can be readily seen in the case of the simple three-stone hearth: no investments apply for such a stove, and the fuel costs are characterized solely by labour. Owing to the difficulty of data acquisition for the costs of traditional conversion processes, only the costs of modern pathways listed in Table 7.2-2 are discussed here.

The costs of providing bioenergy carriers (production costs) determine the economic potential for substituting biogenic energy carriers for fossil energy carriers. These costs, for the pathways examined by WBGU, were determined by Müller-Langer et al. (2008) using a calculation model based on the annuity factor method as set out in VDI 2067 and VDI 6025. Here, and likewise for the greenhouse gas balance discussed in Section 7.3, the 'allocation method' was used (Box 7.2-4). For the bioenergy pathways for electricity and heat the cost analysis extends to include the combustion engine or heating system (small-scale CHP), but for transport it extends 'only' up to the biofuels or (transport) bioelectricity, excluding the combustion engine or vehicle. For transport, therefore, the extra costs incurred by a natural gas or electric vehicle are added to the production costs of the fuels. The assumptions for mathematical financial boundary conditions, capital and operating costs and revenues are set out in Müller-Langer et al. (2008). The year 2005 was selected as the reference year for cost calculations on the basis of data availability, even for those bioenergy pathways that are technically construed on the 2030 time horizon.

Figures 7.2-4a, b and c show the production costs for one energy unit of electricity or heat or for one vehicle kilometre in euro cents.

For electricity generation the co-combustion of biomass in coal-fired power stations is particularly

cost-effective and economically efficient, with electricity production costs of approx. €ct 4–5 per kWh, since almost no technology costs are incurred. For this form of electricity generation no dedicated bioenergy plant need to be built; existing facilities can instead be converted or expanded. All technologies that use fuel cells have very high production costs. Also expensive are all pathways involving biomass gasification or the deployment of organic waste. Relatively inexpensive, however, is the established technology of biogas plant for biogas or biomethane production, whose production costs are in the region of €ct 10 per kWh, their specific level depending on the feedstock used.

Heat generation in the pathways examined exhibits production costs of approx. €ct 15 per kWh. This relatively high value results primarily from the high investment costs for a pellet boiler, estimated at € 14,000 for a 15 kW unit and thus forming approx. 50 per cent of the production costs.

The deployment of bioenergy in transport in the form of biofuels is currently considerably more cost-effective than the electromobility route. The additional costs of the latter are due to the high investment costs and the still unsatisfactory service life of the storage batteries or fuel cells. The production costs of 2nd-generation biofuels per kilometre travelled are approximately double those of the now technically advanced 1st generation that are based on maize, sugar cane, *Jatropha*, oil palms and waste fats.

7.2.5.2
Discussion of future developments of bioenergy pathway costs

For all pathways, the costs of biomass feedstock had risen considerably by 2008 in comparison to the selected reference year of 2005. With few exceptions the feedstock costs have been the dominating factor in the production costs of bioenergy pathways (Müller-Langer et al., 2008). However, the costs of fossil energy provision have also increased significantly in this period owing to rising fuel prices. In estimating the future trends of production costs, two parameters in particular require discussion – the feedstock costs and the technology (capital) costs.

For biogenic feedstocks, further cost increases can be expected in the long term, predominantly caused by increasing scarcity of land. It may be predicted that fossil fuel costs for crude oil, natural gas and coal will likewise rise in the long term. No accurate forecast of the relationship between the costs of fossil and biogenic fuels in 2030 can be made, however, and for this reason the following discussion is based

Box 7.2-4

The allocation method: Its application for determining specific energy expenditure

In order that by-products (co-products) are also included in determining the specific energy expenditure, a proportion of the expended energy is assigned to these in what is known as allocation. Allocation is done on the basis of allocation factors along the inventory boundaries. These factors determine what fractions are allocated to the main product and what to the co-product. In CHP, electricity is considered a main product and heat a co-product.

The allocation factors are determined by the 'heat value' method, which expresses the energy content of the main and co-products (e.g. main product: rapeseed oil; co-product: press cake) as fractions of the sum of both products. These factors are used to allot the energy expenditures. In some bioenergy pathways electricity or heat occur as co-products that are considered as electricity or heat equivalents respectively. The electricity equivalent is obtained assuming a simplified generating mix of 50 per cent of each of natural gas combined-cycle plant (η_{el} of 60 per cent) and hard coal power stations (η_{el} of 44 per cent) is assumed. For

the heat equivalent, a natural gas condensing boiler (η_{th} of 95 per cent) is used (Equation 7.2-3).

The allocation of electricity and heat from CHP was based on a 'heat value' for electrical energy of 2.5 kWh / kWh$_{el}$ (Fritsche and Wiegmann, 2008). The allocation factors for CHP were obtained as the ratios of the individual efficiencies for electricity and heat, taking into account the energy weighting ('heat value') of electricity (2.5) and of heat (1), to the sum of both efficiencies including their weightings (Equation 7.2-4).

$$AF_{el} = \frac{2{,}5 \cdot \eta_{el}}{(2{,}5 \cdot \eta_{el} + \eta_{th})}$$

Equation 7.2-4
AF allocation factor; η_{th} = individual efficiency, thermal; η_{el} individual efficiency, electrical

$$AF = \frac{m_{MP} \cdot H_{u,MP}}{m_{MP} \cdot H_{u,MP} + \Sigma\,(m_{CP,n} \cdot H_{u,CP,n}) + W_{el} \cdot F_{el\text{-}equ} + W_{th} \cdot F_{th\text{-}equ}}$$

Equation 7.2-3
AF – allocation factor; m_{MP} – mass of main product; $H_{u,MP}$ – lower heat value of main product; $m_{CP,n}$ – mass of co-product(s); $H_{u,CP,n}$ – lower heat value of co-product(s); W_{el} – electricity as co-product; $F_{el\text{-}eq}$ – electricity equivalent; W_{th} – heat as co-product; $F_{th\text{-}eq}$ – heat equivalent

Table 7.2-4
Efficiencies and allocation factors for the bioenergy pathways with CHP analysed in the report.
Source: Müller-Langer et al, 2008

Technology	Electrical efficiency η_{el} [%]	Thermal efficiency η_{th} [%]	Allocation factor for electricity as main product
Small-scale CHP unit	38	44	0.68
Fuel cell (SOFC)	48	23	0.84
Steam turbine	23	60	0.49
Gas turbine	25	55	0.53
Hard coal-fired power plant	45		1.0
Combined-cycle power plant	43	30	0.78

on the simplified assumption of a constant cost relation.

As concerns technology costs, two groups may be distinguished: the relatively young technologies whose investment costs may fall substantially and that present steep learning curves, and those technologies that are already established or semi-established in the market, whose costs are also falling, but by a smaller proportion. The fraction of the production costs made up of capital or technology costs is

particularly high for the young technologies, as can be clearly seen in Figures 7.2-4a, b and c.

The young technologies include lithium-ion batteries, fuel cells and all biomass gasification systems that convert woody biomass to biomethane, raw gas or Fischer-Tropsch diesel. A high potential for cost reduction can be expected for these technologies. The learning curve for the technology costs can thus be estimated at 80 per cent, i.e. these costs will fall to 80 per cent with a doubling of installed capacity. This

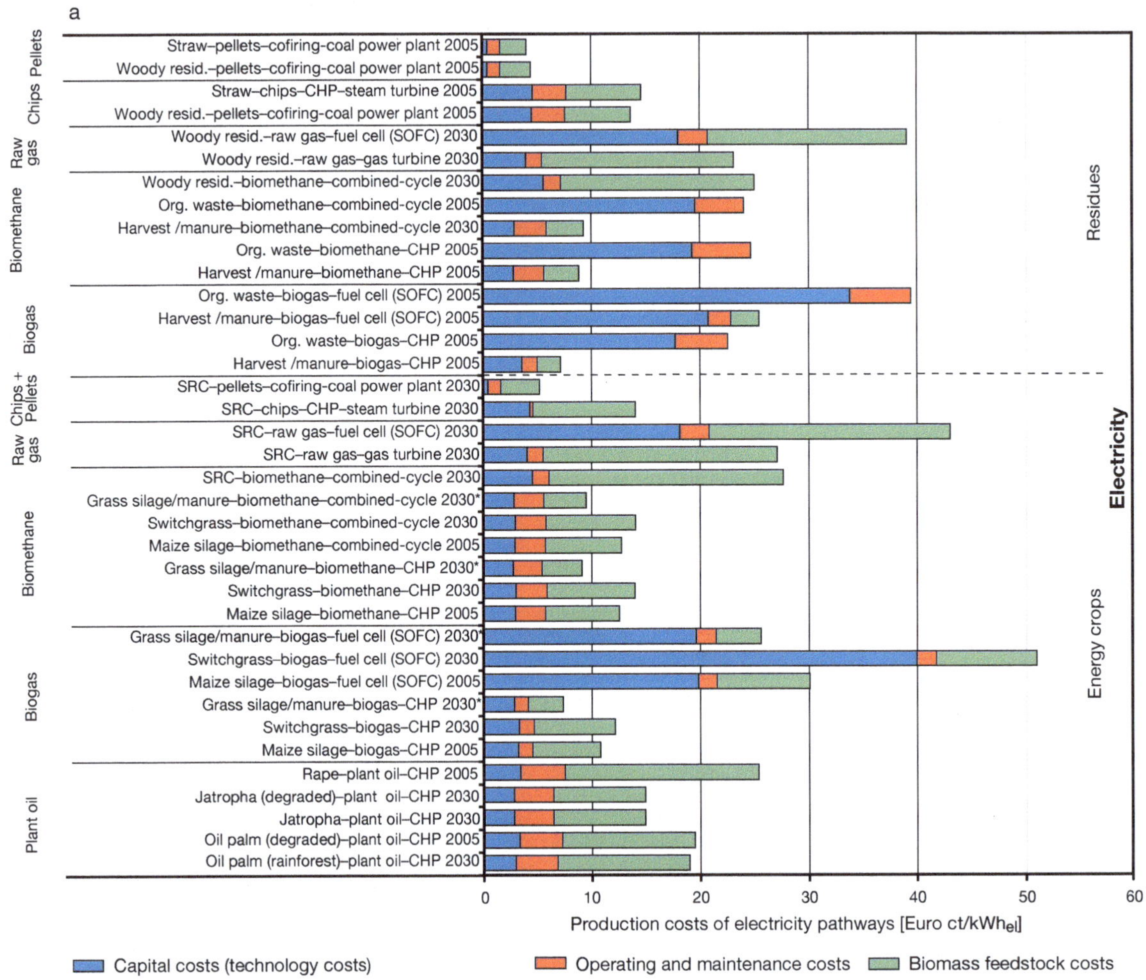

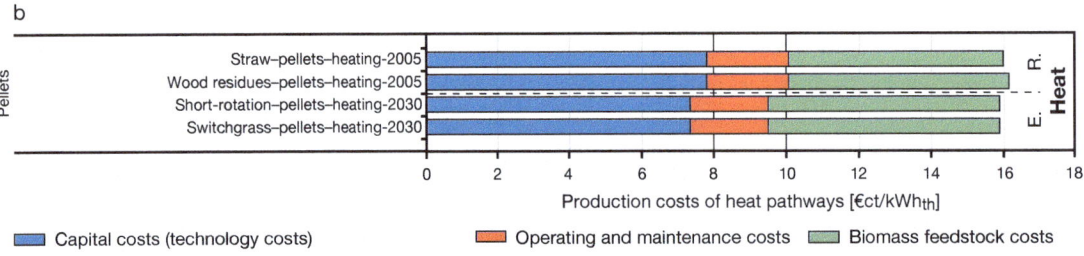

a

Production costs of electricity pathways [Euro ct/kWh_el]

Capital costs (technology costs)   Operating and maintenance costs   Biomass feedstock costs

Figure 7.2-4a
Production costs of bioenergy pathways for electricity generation. The proportions of capital/technology costs, operating costs and feedstock costs are shown in each case. * For these pathways, a mixture of 70% grass and 30% manure was assumed. The names of the pathways relate to the cultivation systems and conversion processes listed in Tables 7.2-1 and 7.2-2. Source: WBGU and Müller-Langer et al., 2008

b

Production costs of heat pathways [€ct/kWh_th]

Capital costs (technology costs)   Operating and maintenance costs   Biomass feedstock costs

Figure 7.2-4b
Production costs of bioenergy pathways for heat production. The proportions of capital/technology, operating and feedstock costs are shown in each case. The names of the pathways relate to the cultivation systems and conversion processes listed in Tables 7.2-1 and 7.2-2. R. = residue pathways, E. = energy crop pathways. Source: WBGU and Müller-Langer et al., 2008

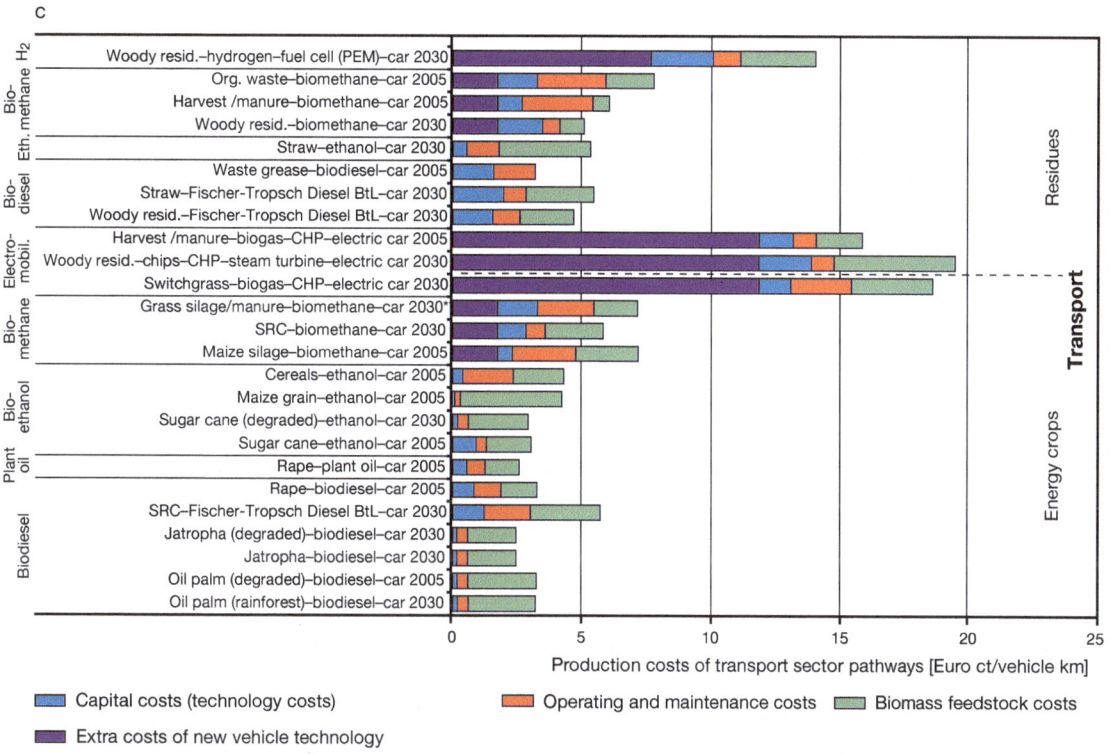

Figure 7.2-4c
Production costs of bioenergy pathways in the transport sector. The proportions of capital/technology costs, operating costs and feedstock costs are shown in each case. In addition, the extra vehicle costs are shown. * For these pathways, a mixture of 70% grass and 30% manure was assumed. The names of the pathways relate to the cultivation systems and conversion processes listed in Tables 7.2-1 and 7.2-2.
Source: WBGU and Müller-Langer et al., 2008

value is based on the learning curve for photovoltaic systems, which in recent years have demonstrated a similar cost-reduction potential (Staffhorst, 2006).

The (semi-)established technologies include those already on the market such as biogas plant, biofuel plant for bioethanol production, biodiesel (1st generation), small-scale vegetable oil CHP, pellet heating systems, central CHP plant and the co-combustion of biomass in hard-coal-fired power stations. For these technologies only a moderate cost reduction potential should be assumed, with a learning curve of 90 per cent. This value is based on the learning curve for wind power systems, which is of the same order of magnitude (Durstewitz et al., 2008).

Considering the period from 2005 to 2030 and assuming a global growth in installed capacity for all technologies of 20 per cent per year, the costs of the young technologies at the end of the period will be only about one quarter of the original costs at 2005 levels cited in the report; costs for the (semi-) established technologies will fall to about one half. In most cases, under these conditions these technologies, which today are still very expensive, will be fully competitive and thus will permit a transformation of energy systems to highly efficient, low-emissions technologies. Large-scale introduction of electromobility in road transport, use of CHP for electricity and heat provision and the increasing direct generation of electricity from wind, water and solar energy will also lead to a shift in the applications for bioenergy. The reduction in costs will make the production of some bioenergy carriers such as biomethane competitive with fossil energy provision. In countries with developed natural gas networks the use of biomethane for electricity generation in decentral CHP and combined-cycle power plant will be of interest. In countries without developed natural gas networks liquid bioenergy carriers such as vegetable oil or bioethanol can be used for stationary, combined heat and power provision as an addition to the direct electricity generation from wind, water and solar energy, and also to supply control and balancing energy for fluctuations in the amount of power fed to the grid.

7.3
Greenhouse gas balances

7.3.1
Life-cycle assessment methodology

The methodology of life-cycle assessment (LCA) is set out in detail in the ISO 14040 ff. series of standards, and has been tried and tested in numerous case studies. LCA covers, across the entire product life cycle, inputs such as metals or fossil/renewable energy carriers, and outputs such as emissions of substances hazardous to the environment or to human health. These are aggregated in a number of impact categories. For instance, substances with radiative forcing potential or their emissions are aggregated in the global warming potential (GWP) impact category. The individual gases (CO_2, CH_4, N_2O, etc.) are weighted with specific GWP values reflecting their contribution to global climate change.

As is the case for all accounting methods, within the prescribed LCA standard there is also scope for interpretation and, above all, the possibility of applying different system definitions and boundaries, all of which can influence the outcome. This applies, for example, to the question of how to take account of co-products if a process involves the simultaneous production of rapeseed oil for energy and rapeseed meal for animal feed, or the simultaneous production of electricity and heat in cogeneration systems. Similar problems of attribution also arise in economic accounting in companies. A further point is that LCA outcomes can differ according to the region to which they apply: For instance, one tonne of aluminium is produced with less environmental impact in Norway than in Germany, because in the former country the high electricity requirement is met almost exclusively from hydropower sources.

LCA studies generally differentiate inputs of energy and the associated primary energy (CER; cumulated energy requirement) according to fossil and renewable sources and, where appropriate, nuclear power. In greenhouse gas (GHG) balances biogenic CO_2 is generally not specified, but is assumed to generate zero emissions. Accounting methods seeking to establish climate change mitigation potential often only inventorize energy and GHG emissions, dispensing with the impact assessment step and the evaluation (of different types of environmental impact) otherwise required in LCA studies. The accounting of product-related GHG emissions has recently come to be termed 'carbon footprinting', and proposals have been made for a specific interpretation convention (PAS 2050, second draft, British Standard), which are presently being explored and refined in a German research project. It also must be noted that there are further climate-related accounting methods whose methodologies and system boundaries differ: Corporate GHG emissions accounting without inclusion of upstream chains (e.g. Carbon Disclosure Project, Greenhouse Gas Protocol), GHG emissions inventories of countries and selected systems (under the Kyoto Protocol, and in connection with emissions trading) and accounting of the GHG emissions reductions of offset projects (CDM projects).

Furthermore, EU directive are due to stipulate new accounting modalities in statute, e.g. in the guise of the EU Fuel Quality Directive (GHG reporting commitments from 2010 and a phased reduction of the CO_2 emissions of fossil fuels by 2020) or the EU Renewable Energies Directive with its 'Guarantee of Origin' for green electricity and allocation rules for biofuel accounting. A major need for harmonization arises here. Activities undertaken by the EEA in the EU, and by the GBEP GHG Task Force at global level, are seeking to meet this need.

It is therefore essential when comparing different GHG balances to consider which methodology and which system boundaries and demarcations (e.g. for co-products) were used. If system boundaries differ, outcomes can diverge substantially. When engaging in bioenergy accounting, the following six determinations are key:

- Which system and which product was inventorized, and for which functional unit? For instance, petrol can be inventorized either prior to its combustion in the car or inclusive of its combustion in the car.
- To which period does the accounting apply? For example, this is important when considering land-use changes.
- Were only greenhouse gases inventorized and evaluated, or were other resource extractions and environmental pressures (e.g. primary energy requirement, water consumption, fertilizer application, land requirement or particulate emissions) addressed? The findings presented in Section 7.3.2 below show that analyses and evaluations can differ depending upon the parameter in question (e.g. energy consumption or land requirement).
- Were direct land-use changes included?
- Were indirect land-use changes included?
- How were environmental aspects such as land requirement, energy consumption or greenhouse gas emissions allocated to co-products (Box 7.3-1)?

CHARACTERISTIC VALUES AND OPTIMIZATION
The GHG balances presented in Section 7.3.2 of the cultivation and industrial processing of biomass for

Box 7.3-1

Handling co-products – The allocation method

Greenhouse gas balances and life-cycle assessments are performed for selected processes or products. However, in agriculture and in technical processes it is often the case that one or several by-products (co-products) are generated in addition to the main product. The accounting procedure must then divide resource inputs such as energy, land or water, as well as emissions, among the products generated.

The LCA standard ISO 14040 ff. proposes a number of different approaches for this, in the following order of priority: (1) avoiding allocation by expanding the system boundary, (2) introducing credits (specific to each co-product) or (3) allocation according to criteria such as energy content, mass or market price.

When conducting a comparison between numerous different options (as, in the present case, several dozen bioenergy pathways) the approaches that normally have priority – system boundary expansion or comparative effective credit generation – are scarcely practicable. Hence usually allocations are performed. For bioenergy pathways it is expedient to perform the allocation on the basis of energy content. This has also been the approach taken in the course of the statutory design of sustainability standards for biofuels, and in work performed for the German Federal Environment Agency UBA on the question of allocation in combined heat and power (CHP) production. The German Sustainability Ordinance to the Biofuels Quota Act and comparable proposals made by the European Commission stipulate an allocation method that apportions environmental burdens among main products and co-products on the basis of their calorific values (IFEU, 2007).

When performing an interpretation of findings, it is important to take account of the allocation previously conducted. For instance, an average required area of cropland is set for the cultivation of a defined quantity of rapeseed. As, however, cultivation and processing give rise to co-products, e.g. rapeseed oil and rapeseed mean (which is used as animal feed), the actual cropland area is apportioned among these products and a part of that area assigned to each of the two. If, then, the cropland area associated with a certain process (e.g. electricity from rapeseed oil) is stated (Fig. 7.3-3), because of the allocation performed only the associated part of the area is stated. The actual rape cropland used is larger. The allocation or, in this case, the "deduction" of the part of the area for rapeseed meal is also justified in substance, as otherwise cropland elsewhere would be needed to product the animal feed.

---

several dozen bioenergy pathways, together with the respective GHG abatement costs, were based on average or characteristic values. The values can fluctuate widely depending upon the input parameter. Per-hectare yields can differ depending upon climate, soil, fertilization, irrigation and cultivation type, and along the conversion route there are different processes and, above all, different facility sizes. All of this can influence the outcome.

This implies conversely that targeted selection of cultivation areas and systems, together with process selection and optimization, can substantially reduce greenhouse gas emissions from the characteristic values ascertained.

For instance, the production process of biogenic methane gives rise to major greenhouse gas emissions. This is due to methane and nitrous oxide emissions that occur in the post-fermentation of fermentation residue. The Swiss EMPA study (Zah et al., 2007) shows that targeted measures such as covering the post-fermentation container can reduce the bulk of these emissions, improving CO_2 savings from approx. 35 per cent to more than 90 per cent. Such covers were already state of the art in 2007. They are stipulated by law in Germany, but in the developing world they are scarcely used.

7.3.2
Greenhouse gas balances of selected bioenergy pathways

A number of studies have produced greenhouse gas (GHG) balances of bioenergy uses; these, however, have mostly focussed on liquid fuels. For the present report, GHG and energy balances were generated not only for the transport sector, but also for other applications such as heat and power. To that end, a set of representative pathways was defined (Section 7.2.3.1). The balances integrate emissions from direct and indirect land-use changes. For that purpose WBGU commissioned external reports (Fritsche and Wiegmann, 2008; Müller-Langer et al., 2008). The GHG balances take the status quo as their baseline, and inventory the changes induced by the cultivation of biomass. Regardless of biomass use, every type of land use generates continuous greenhouse gas fluxes. Cropland is generally a source, grassland a sink of greenhouse gases. The GHG balances of the various bioenergy pathways presented here do not integrate the GHG fluxes of unchanged land use, but only those of changed land use. Box 7.3-2 presents the methodology used to identify emissions from indirect land-use changes.

Figure 7.3-1 first provides an overview of different cultivation systems, giving only those emissions that arise due to direct and indirect land-use changes. The emissions relate to the gross energy content of the biomass cultivated or harvested, and are averaged across a 20-year period. The emission level of

Box 7.3-2

Quantifying emissions from direct and indirect land-use change

Direct land-use change (dLUC) arises when a plot of land was characterized by another use (e.g. forest, grassland or cropland for food) or was unused prior to cultivation of energy crops. The emissions associated with direct land-use changes are taken into account in life-cycle analyses and default values are available for such emissions that can be used in the analyses (Gnansounou et al., 2008). Table 7.3-1 shows the default values of annual emissions discounted over 20 years for various types of land-use change. These figures are based on IPCC (2006), and were used in the study commissioned by WBGU (Fritsche and Wiegmann, 2008). Depending on the specific management and tillage methods applied, these values can diverge substantially from the default values on individual plots of land.

In addition to dLUC, indirect land-use change (iLUC; the associated effects are often also termed leakage) can also arise if a different use – such as food or feed produc-

Table 7.3-1
Default values for per-hectare GHG emissions induced by direct land-use change for various species utilizable as energy crops, in kg CO_2 per ha and year. The per-hectare emissions are discounted over a 20-year period following land-use change and do not include the emissions from the further processing stages across the life cycle of a given bioenergy pathway,
Source: Fritsche and Wiegmann, 2008

Crop	Previous use	GHG emissions [kg CO_2/(ha · a)]
Wheat	Grassland	2,630
	Cropland	0
Maize	Grassland	2,630
	Cropland	0
Poplar (SRP)	Grassland	1,255
	Cropland	-1,375
Sugar cane	Savannah	14,428
	Degraded land	-3,722
	Cropland	-55
Rapeseed	Grassland	2,630
	Cropland	0
Oil palm	Tropical rainforest	28,417
	Degraded land	-13,750
Jatropha	Cropland	-458
	Degraded land	-4,125
Switchgrass	Grassland	1,897
	Cropland	-733

* light red = C release
  light green = carbon sequestration
  white = CO_2 neutral

tion – took place on land used for energy crop cultivation and is thereby displaced. To the extent that demand remains for the food or feed previously produced on this land, its production is likely to shift elsewhere. The result may be that production on existing cropland is intensified, or further land is developed as cropland or pastureland. This can generate substantial CO_2 emissions, especially if additional land is converted for this purpose that previously had a large carbon stock, such as forests, wetlands or peat-land (Section 4.2.3). While these emissions arise elsewhere, they were essentially caused by energy crop cultivation and are therefore attributable to it. Indirect land-use change can also be induced in situations where there is no direct land-use change for energy crop cultivation, but the use of harvested products changes (for instance, if maize is used for biogas instead of feed).

Greenhouse gas emissions from indirect land-use change cannot be identified and quantified directly, but only modelled. To generate model-based statements, various models have been proposed or are currently being developed. If the specific plots of land affected by displacement were known, emissions could be determined without major effort, as they correspond to those of direct land-use change. However, due to global trade displacement effects can also arise beyond a region or country, and hence it is not possible to attribute them unequivocally to biomass production on certain areas. In the medium term, a reliable global register could at least record at an aggregated level year-on-year changes in land use – the problems of attribution to the diverse causes and allocation to any intensified bioenergy cultivation taking place would remain.

QUANTITATIVE MODELLING OUTCOMES
Searchinger et al. (2008) use an econometric equilibrium model, which involves a simulation of global trade in order to estimate the land requirement induced by displacement effects and the resulting CO_2 emissions. This analysis refers to the market situation and dynamics in the USA and is relevant above all to ethanol from maize. One of the criticisms levelled against the method is that the model does not capture production increases achieved by increasing agricultural yields or avoiding logistic losses and market distortions such as taxes (Fritsche and Wiegmann, 2008).

The balances presented here use the 'iLUC factor' approach developed by the Institute for Applied Ecology (Öko-Institut) in Darmstadt, Germany (Fritsche and Wiegmann, 2008): the indirectly induced land-use changes are derived for the 2005 reference year from the globally traded agricultural products that could have theoretically been displaced by energy crop cultivation. For the purposes of the approach, these are simplified as maize, wheat, rapeseed, soya and palm oil. The shares in trade of these products of the key countries in this field – the EU, USA, Brazil and Indonesia – and the respective yields are used to derive a weighted global 'land appropriation' that would result due to displaced food and feed. The resulting land-use changes assumed in the model for the above countries and the EU are as follows: in the EU and USA grassland (pastureland, grassland) is converted into additional cropland for the displaced land, in Brazil savannah is converted and in Indonesia tropical rainforest is converted.

The theoretical GHG emissions potential induced by indirect land-use change is characterized by the quantity of carbon stored per unit land in both the soil and in the aboveground vegetation. As this quantity varies depending upon climate zone and soil, the shares of the corresponding land areas are relevant. Proceeding from the aboveground and underground carbon inventories for these regions, a globally weighted theoretical emissions potential amounting to 400 t CO_2 per ha was calculated. Discounted over 20 years, this gives a theoretical per-hectare CO_2 emissions potential of 20 t CO_2 per ha and year.

In reality, the modelled theoretical iLUC potential will not come fully into effect, at least currently and in the next few years, as displaced food and feed production can induce not only additional demand for land, but also increased yields on existing cropland and (re)activation of presently unused land. The maximum emissions potential is therefore estimated to be 75 per cent of the theoretical potential. The medium value is taken to be 50 per cent of the theoretical value, and a low value is assumed to be 25 per cent. Using these figures and taking account of the respective per-hectare yields of bioenergy cultivation, an energy-related emission factor can then be determined for indirect land-use effects – the iLUC factor. If there is increasing cultivation of energy crops and rising demand for food, the iLUC factor will rise over the coming decades and must then be adjusted accordingly. That adjustment could increasingly be based on real inventory values.

Table 7.3-2 gives an overview of CO_2 emissions from direct land-use change and of emissions from indirect land-use change determined by means of the iLUC factor (50 per cent) for various types of land conversion. The values relate to the energy content of the biomass produced. The impact of indirect land-use change therefore tends to be higher the smaller the per-hectare energy yield of the energy crop in question is. This leads to the relatively high iLUC values for rapeseed, wheat and Jatropha.

For the use of residues and wastes, WBGU assumes that the iLUC factor is zero. It is albeit possible for certain residues and wastes for which there are already applications today (such as residue use as livestock feed) that use for energy causes resource competition resulting in increased cultivation of plant feedstocks and thus also indirect land-use changes with the corresponding emissions. WBGU expects these effects to be slight, however.

WBGU considers emissions from indirect land-use change to be an indispensable part of any appraisal of the climate change mitigation effect of bioenergy use. Although research on the quantification of such emissions has only just started, it is necessary to produce quantitative estimates of these effects even today. WBGU therefore proposes using the iLUC factor (50 per cent) set out above for standard-setting (Section 10.3), while adjusting it in future in line with new scientific findings. Dispensing with application of an iLUC factor because of inescapable uncertainties in modelling would mean that indirect land-use change is not considered at all, although it does have a very major impact on the GHG balances of bioenergy.

Table 7.3-2
GHG emissions per unit energy induced by direct (dLUC) and indirect (iLUC) land-use change for different cultivation systems and different previous uses. Emissions relate to the gross energy content of the biomass feedstock. Negative values mean that energy crop cultivation results in carbon storage. The figures do not include the emissions from the further processing stages across the life cycle of a given bioenergy pathway.
Source: Fritsche and Wiegmann, 2008

Crop	Previous use	dLUC [t CO_2/TJ]	iLUC 50% [t CO_2/TJ]	Total LUC [t CO_2/TJ]
Wheat-meadow	Grassland	26	100	126
Wheat-cropland	Cropland	0	100	100
Maize-meadow	Grassland	17	63	80
Maize-cropland	Cropland	0	63	63
SRP-meadow	Grassland	9	74	83
SRP-cropland	Cropland	-10	74	64
Sugar cane-savannah	Savannah	21	0	21
Sugar cane-degraded	Degraded land	-5	0	-5
Sugar cane-cropland	Cropland	-0,1	15	15
Rapeseed-meadow	Grassland	31	119	150
Rapeseed-cropland	Cropland	0	119	119
Oil palm-trop. rainforest	Trop. rainforest	172	0	172
Oil palm-degraded	Degraded land	-83	0	-83
Jatropha-cropland	Cropland	-4	88	84
Jatropha-marginal	Marginal land	-76	0	-76
Switchgrass-meadow	Grassland	9	50	59
Switchgrass-cropland	Cropland	-4	50	46

fossil fuels is given by way of comparison. The figure illustrates that, in relation to the energy content of the biomass produced, some types of land-use change and their indirect effects generate or can generate emissions that are already comparable or even greater than those of fossil fuels – without the emissions and conversion losses associated with cultivation and processing along the further bioenergy utilization chain having yet been taken into account.

In the view of WBGU, this analysis already makes certain types of land-use change for energy crop cultivation non-tolerable. The largest emissions result if tropical rainforests are converted to oil palm cultivation, whereby the emissions result exclusively from direct land-use change. Switching to energy crops on cropland does not generally lead to emissions from direct land-use change; if energy crop production takes the form of short-rotation plantations, carbon is even sequestered in the soil. But the previous use is displaced from the land – this must be expected to lead to substantial emissions from indirect land-use changes. The conversion of grassland causes emissions from both direct and indirect land-use changes. Cultivation of perennial crops on degraded land has the best outcome, as carbon can be stored in the soil and no indirect land-use changes are to be expected

(apart from possibly displaced grazing). In these cases, which include the cultivation of oil palm or Jatropha on degraded land, this cultivation alone can deliver a climate change mitigation effect, without yet taking account of the substitution of fossil energy carriers.

For the following analysis, a range of forms of biomass cultivation and use, as characterized in Section 7.2, was selected for transport, electricity and heat (biomass pathways), the GHG emissions of which were inventoried across the entire life cycle (Fritsche and Wiegmann, 2008). Emissions from energy crop cultivation are based on conditions in Germany (Table 7.2-1), with the exception of tropical energy crops. The following analyses place these life-cycle emissions in relation to the emissions arising in a reference system, in order to produce statements concerning the GHG abatement potential of the various bioenergy pathways. The choice of reference systems has a major effect on the results. If, for instance, it is assumed that bioenergy use substitutes the use of natural gas, the GHG abatement that results is much smaller than if a reference system is based on coal. Which energy carrier is displaced in the real world by bioenergy use depends not only on the present energy mix, but also on present and

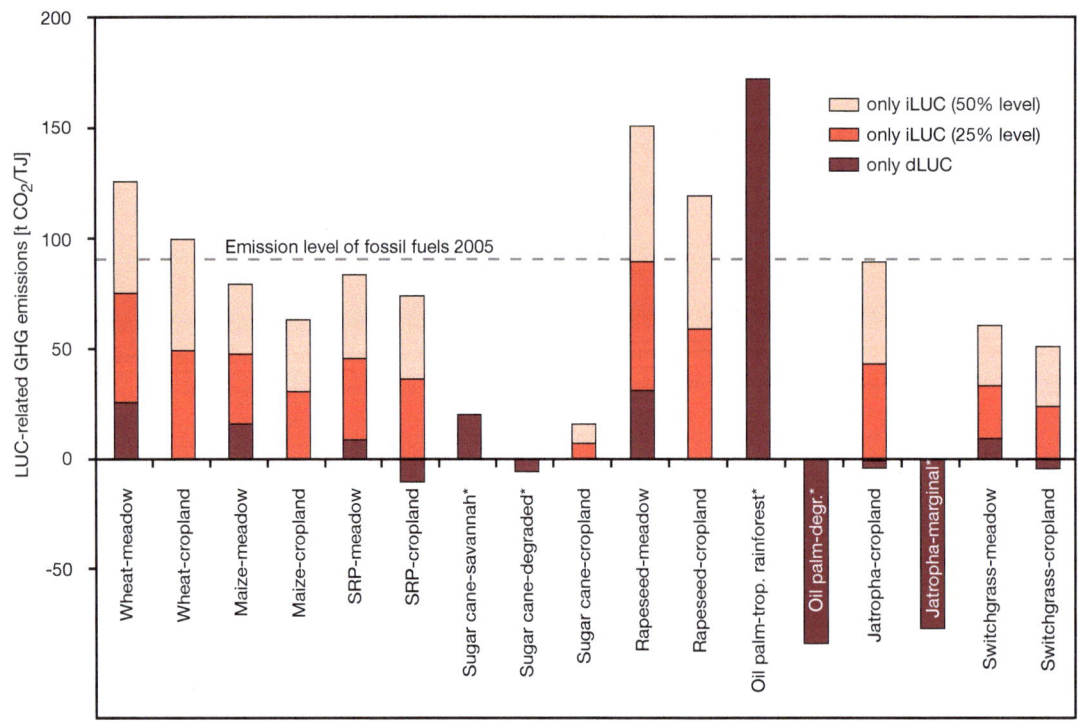

Figure 7.3-1
GHG emissions from direct (dLUC) and indirect (iLUC) land-use change for different energy crops and previous land uses, in relation to the gross energy content of the biomass utilized in t CO_2eq per TJ biomass. The values are discounted over 20 years (Box 7.3-2). No indirect land uses arise for systems marked with *, as it is assumed that no previous use is displaced. The cultivation systems are characterized in Table 7.2-1.
Source: Fritsche and Wiegmann, 2008

Table 7.3-3
Emissions of the fossil reference systems used by WBGU to derive the GHG abatement potentials of the individual bioenergy pathways. Veh-km = vehicle-kilometre.
Source: WBGU based on data from Fritsche and Wiegmann, 2008 and Müller-Langer et al., 2008

	Fossil reference system	Emissions per unit fuel [g CO_2eq / kWh_{th}]	Emissions per unit final or useful energy	Share of mix [%]	Reference value
Electricity	Hard coal power plant	411	1.085 g CO_2eq / kWh_{el}	80	953 g CO_2eq / kWh_{el}
	Natural gas combined-cycle	234	425 g CO_2eq / kWh_{el}	20	
Heat	Oil heating	321	376 g CO_2eq / kWh_{th}	40	327 g CO_2eq / kWh_{th}
	Natural gas heating	252	295 g CO_2eq / kWh_{th}	60	
Transport	Petrol car	328	250 g CO_2eq / Veh-km	60	230 g CO_2eq / Veh-km
	Diesel car	316	201 g CO_2eq / Veh-km	40	

future political and economic conditions. WBGU has selected as reference system a mix of fossil energy carriers oriented to the fossil energy mix in Germany in 2005; the specific mix is defined separately for the electricity, heat and transport sectors (Table 7.3-3). In the transport and heat sectors, fossil energy carriers are used almost exclusively today, with the result that the selected fossil reference systems can be defined clearly. In the electricity sector, emissions in 2005 in Germany averaged 648 g per kWh_{el} (Fritsche and Wiegmann, 2008). This figure relates to the entire electricity mix and thus also comprises renewables and nuclear power. The fossil contribution to electricity generation was more than 60 per cent in 2005; of this, approx. 80 per cent was based on hard coal and lignite, and approx. 20 per cent on natural gas. WBGU has selected a mix of 80 per cent hard coal and 20 per cent natural gas as reference system. At 953 g per kWh_{el}, the emissions of the selected reference system for electricity are above those of the overall electricity mix. A reference system based on natural-gas-fired combined-cycle power plants would have lower emissions, namely 425 g per kWh_{el}.

In the view of WBGU, the political and economic conditions need to be shaped such that the use of bioenergy primarily substitutes fossil energy carriers and mainly coal. Only in this case will the level of GHG abatement presented in the following analyses be achievable. This applies equally if the share of renewable energies is very high in the future. In the same vein, these assumptions are not generally transferable to other countries. The reference value used in the present study is far above the present emissions of power generation in Norway, a country with a large share of hydropower and thus a very small

contribution of fossil sources to electricity generation. In a country where the share of fossil electricity is very high, such as China, which mainly uses coal-fired power plants, the actual emissions of electricity generation are higher than the reference value used here. An additional sensitivity analysis explores the effect of different reference systems (Fig. 7.3-5).

The emissions and costs attributed to the specific reference systems relate to the technology status of 2005 in Germany (Nitsch, 2007; Fritsche and Wiegmann, 2008; Müller-Langer et al., 2008; BMWi, 2008). The bioenergy pathways analysed relate either to the technology status of 2005 or to the anticipated technology status of 2030, and are labelled accordingly. Because of the uncertainties attaching to the composition of future energy systems and the major difficulties in performing cost appraisals that result, as well as to improve comparability, the reference systems for the year 2005 (Table 7.3-3) were used for all pathways – with regard to both costs and emission values. It is evident that the specific emissions of fossil energy supply will drop by 2030, as further development will make these technologies more efficient. It can therefore be expected that the GHG abatement levels of those bioenergy pathways that relate to 2030 will in fact be lower than the levels presented here.

The following discussion presents three parameters that characterize the climate change mitigation effects of the various bioenergy pathways. The suitability of each parameter as a basis for comparing the climate change mitigation effect of different bioenergy pathways is discussed; their suitability for developing standards (Section 10.3) is one of the aspects explored. WBGU then derives its proposals on that basis. Improvements to the efficiency of tra-

Box 7.3-3

GHG mitigation through efficiency improvements in traditional biomass use

Bhattacharya and Salam (2002) show that if traditional wood stoves are replaced by efficient wood stoves, this can reduce GHG emissions by approx. 60 per cent while delivering the same quantity of useful energy; if they are replaced by biogas stoves, emissions are even reduced by 95 per cent. CO_2 emissions are not the only issue here: generally (and in the GHG balances presented above) the CO_2 emissions

arising in biomass use are not counted as emissions, as no more CO_2 is released than was taken up by the plant in its growth. This applies equally to traditional and modern biomass use. A further aspect, however, is that because of incomplete combustion processes traditional wood stoves emit larger quantities of other greenhouse gases such as CH_4 and N_2O. These emissions are reduced if efficient wood stoves are introduced. Moreover, if the wood is harvested in a non-sustainable manner, i.e. causes a decline of the carbon stock in the biosphere, as is the case if harvesting leads to deforestation, this, too, can be reduced, which delivers a net emissions reduction.

ditional bioenergy use can also deliver GHG emissions reductions (Box 7.3-3). However, WBGU has not performed any calculations of this.

PERCENTAGE GHG REDUCTION IN RELATION TO FINAL ENERGY

Figure 7.3-2 gives an overview of the relative GHG reduction potentials of different bioenergy pathways in relation to final or useful energy. The assumption underlying this parameter is that a certain energy service (i.e. one kWh electricity, one kWh heat or one vehicle-kilometre) that was previously delivered by utilizing fossil energy carriers is now delivered on the basis of biomass. The GHG reduction is stated as a percentage reduction for a constant level of energy service. While this parameter is often chosen in the bioenergy debate, its informative value is in fact limited. It can albeit be used to 'knock out' particularly poor options, but, as discussed below, the parameter is not suited for comparisons between different application sectors of bioenergy (electricity, heat, transport). Nor does this parameter permit conclusions concerning the quantity of biomass deployed in a specific case or the land area required to cultivate it.

As it is assumed for residue use that this does not lead to land-use changes and associated emissions, the climate change mitigation effect of such use is positive in all cases. The relative mitigation effect of energy crop use, in contrast, depends greatly upon

the emissions arising from direct and indirect land-use changes. If indirect land-use changes are to be expected for a pathway, taking the associated emissions into account generally causes the mitigation potential to be halved at least. Because it is essential to take account of indirect land-use changes in all cases, energy crops cannot generally be assumed to have a satisfactory mitigation effect. In unfavourable circumstances, if indirect land-use changes are taken into account, some pathways can even exhibit negative values in the balance, i.e. higher emissions than the reference system. Land-use changes can have both a positive and a negative impact upon the GHG balance. This is exemplified by oil palm cultivation. If tropical rainforest is cleared for such cultivation, greenhouse gas emissions can be up to four times higher than in the fossil reference system (Hooijer et al., 2006). If, in contrast, oil palm is cultivated on marginal land that was previously scarcely used, a particularly great mitigation effect can be achieved. The emissions reduction compared to the fossil reference system can then reach 200 per cent and more – the pathway is then a real carbon sink. Regardless of which technological pathway is chosen, particularly high relative emissions reductions can be achieved by cultivating tropical, perennial crops (oil palm, *Jatropha*, sugar cane) on marginal land.

Emissions reductions can reach more than 100 per cent. This occurs when, as a result of the cultivation of energy crops, so much carbon is absorbed per

Figure 7.3-2

Percentage reduction of GHG emissions by the substitution of fossil fuels relative to a fossil reference system, in relation to final or useful energy for selected bioenergy pathways. The chosen reference systems are as follows: for the electricity pathways a mix of 80% hard coal and 20% natural gas; for the heat pathways 60% natural gas and 40% mineral oil; and for the transport pathways 60% petrol and 40% diesel (Table 7.3-3). The yellow bars contain the life-cycle emissions inclusive of emissions from direct land-use change (dLUC). The green bars further take into account emissions from indirect land-use change (iLUC 50%; Box 7.3-2). If not otherwise indicated, it is assumed for the energy crop pathways that cultivation is on former cropland. For pathways involving residue use only one bar is shown, as no emissions from indirect land-use change are anticipated. Negative values indicate an increase in emissions relative to the reference system. * It is assumed for pathways which use grass silage or slurry as substrate that in Germany grass silage generates no emissions from land-use change; this special case, however, is not globally transferable. The pathway designations refer to the cultivation systems and conversion processes listed in Tables 7.2-1 and 7.2-2.

Source: WBGU based on data from Fritsche and Wiegmann, 2008 and Müller-Langer et al., 2008

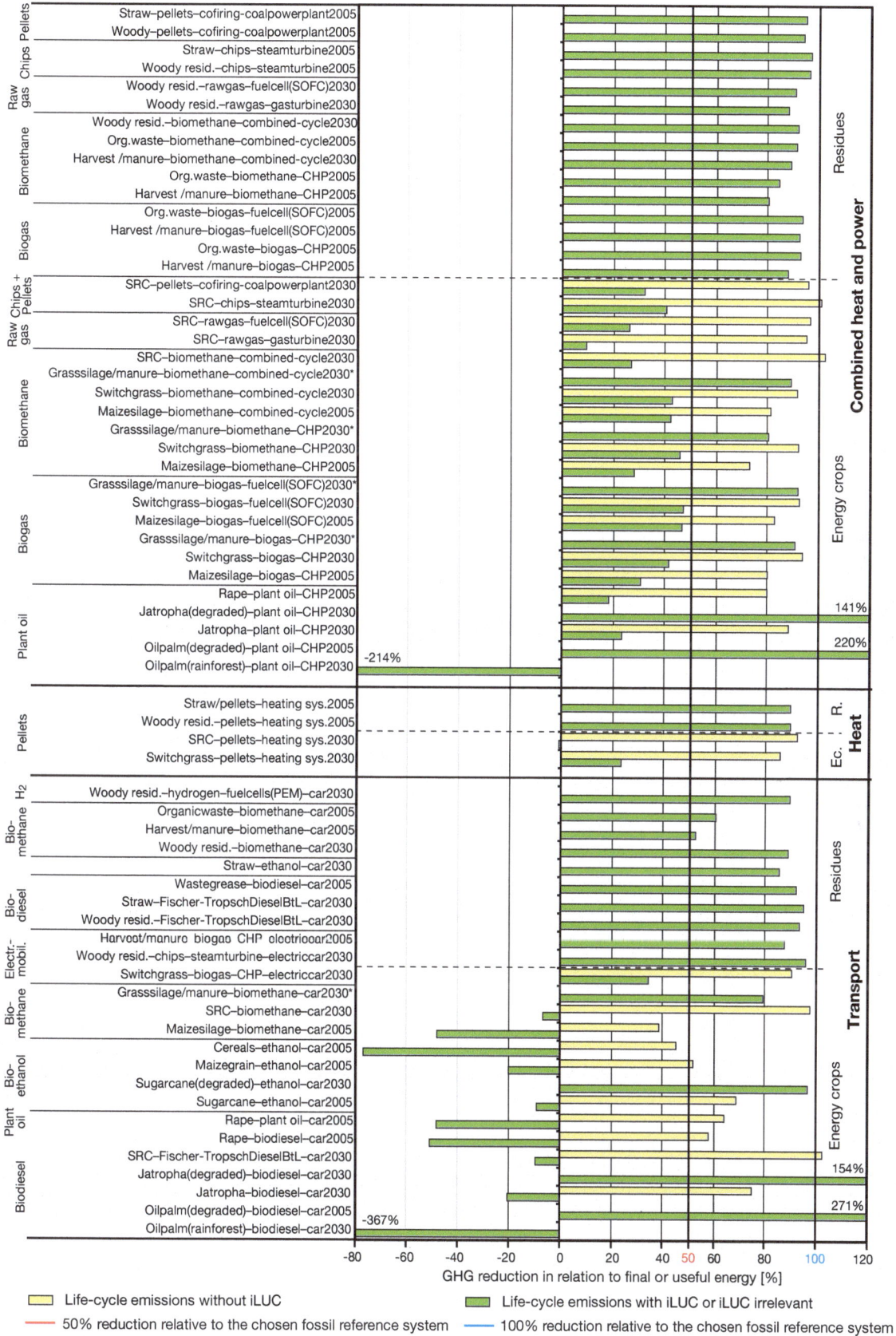

GHG reduction in relation to final or useful energy [%]

☐ Life-cycle emissions without iLUC

☐ Life-cycle emissions with iLUC or iLUC irrelevant

— 50% reduction relative to the chosen fossil reference system

— 100% reduction relative to the chosen fossil reference system

unit land (generally in the soil) that the greenhouse gas emissions arising in biomass cultivation and use are over-compensated. With sufficiently good management practices, this can be achieved especially on marginal land. A number of pathways involving short-rotation plantations (SRPs) also exhibit emissions reductions in excess of 100 per cent if indirect land-use changes are not taken into account. This is because the pathways shown here presuppose that SRPs are established on arable land, which leads to an accumulation of carbon in the soil. However, such conversion must be expected to lead to indirect land-use changes, as discussed. If grassland is converted to establish SRPs, a poorer GHG balance must be expected, as then emissions are approx. 20 per cent higher than in the case of direct cultivation on arable land that does not involve ploughing up grassland (Fig. 7.3-1; Fritsche and Wiegmann, 2008).

In the transport sector, the GHG abatement percentages shown in Figure 7.3-2 correspond to the strategic parameter targeted by the German Biofuels Sustainability Ordinance and by the draft directive of the European Commission on biofuels, although in the case of the latter other reference systems are applied (BMU, 2007b). It is proposed there that biofuels must deliver at least 35 or 50 per cent GHG abatement compared to the equivalent quantity of fossil fuel in order to meet the standard (Section 10.3). Very high GHG abatement levels of more than 50 per cent result for almost all pathways examined as long as, as in the case of residues, no indirect land-use changes affect the balance (Fig. 7.3-2). An abatement level of 50 per cent is generally not achieved if indirect land-use changes must be taken into account, i.e. if cropland or grassland is converted to energy crop cultivation. In all pathways for liquid fuels in the transport sector, the analysis shows that if energy crops are deployed whose cultivation leads to indirect land-use changes the emissions balance is even negative, i.e. emissions are higher than they would be if fossil fuels were used.

WBGU has analysed this parameter because it is the subject of the present debate on standards in the field of biofuels. For this specific field – the comparison of the climate change mitigation effect of different biofuel pathways – this parameter is indeed purposeful and applicable, as the efficiencies of the fuel pathways are comparable. In WBGU's view, however, a broader analysis is necessary that allows purposeful comparison of all energy pathways and not just the fuel pathways. This can reveal in which field of application the greatest absolute climate change mitigation effect can be achieved within the limits of the potential set by the sustainably available quantity of biomass. The 'application-specific GHG abatement percentage' parameter does not reveal, for instance, when examining residue pathways, any systematic difference between the electricity, heat and fuel routes. Quite evidently the contribution that bioenergy can make to climate change mitigation is not limited by the quantity of fossil fuel or fossil-generated energy that may potentially be substituted, but rather by the land area available for sustainable cultivation of energy crops or by the quantity of sustainably available biomass that can be deployed to substitute fossil energy carriers. The 'application-specific GHG abatement percentage' parameter is therefore of only limited value in answering the questions posed by WBGU in the present report.

ABSOLUTE ANNUAL GHG ABATEMENT PER UNIT LAND

In order to identify those bioenergy pathways that perform best in terms of climate change mitigation, the pathways are evaluated using the 'absolute area-specific GHG abatement' parameter. This gives the annual GHG reduction that can be achieved by the energy crops cultivated on a specific area of land (expressed as CO_2eq per ha and year). In addition, the annual GHG reduction is stated that can be achieved with a unit of biomass feedstock or primary energy (expressed as t CO_2eq per TJ biomass feedstock).

Figure 7.3-3 has two parts: (a) shows the absolute GHG emissions reduction achieved by fossil fuel substitution for a range of bioenergy pathways in relation to the land required to cultivate energy crops in temperate climatic zones; (b) does the same for tropical energy crops. No pathways that exclusively involve the utilization of residues and wastes are listed here, as it would either be impossible or inexpedient to place them in relation to crop cultivation area.

The purpose of this analysis is to provide an answer to the following question: Which bioenergy pathway delivers the maximum GHG reduction in view of the limited amount of land available for sustainable energy crop cultivation? Before continuing the discussion, it must be noted once more that the outcome is influenced by the allocation method chosen (Box 7.3-1). Land areas do not correspond to the entire real cultivation area for biomass, but only to those parts of the area allocated to the co-products. It is because of this allocation that the analysis cannot be used to extrapolate the globally required real land area needed to achieve certain reduction goals.

It is assumed for all pathways for temperate energy crops that these are cultivated on land that was already previously cropland. If grassland were converted instead, the emissions generated by energy crop production would be approx. 20 per cent higher and the GHG abatement performance would be poorer. The absolute mitigation potential values in relation to the allocated cultivation area scatter much more widely than the relative mitigation potential values in relation to final energy shown in Figure 7.3-2. This spread is due in part to the different efficiencies of energy conversion; the values also depend greatly upon the per-hectare yields of the different cultivation systems. These depend in turn greatly upon the climatic zone, which is why temperate and tropical cultivation systems are presented separately here. Due to the all-year-round vegetation period, the higher temperatures and the higher solar irradiance, substantially greater yields are in principle possible in the tropics than in temperate regions, insofar as soil characteristics are favourable and water supply is assured.

But cultivation methods and soil quality also cause major differences. For example, the yield of sugar cane can range between 5 and 120 t dry matter per hectare and year (Section 7.1). Characteristic per-hectare yields were used for the present calculations; these are listed in Table 7.3-4, together with the range found in the literature.

It is apparent that, relative to cultivation area, the absolute GHG reduction achievable is generally higher in electricity generation and combined heat and power production than in pure heat production or in transport.

To provide a benchmark for GHG abatement performance per unit area, the level of carbon sequestration achievable by afforestation on the same land can be taken as reference. According to Righelato and Spracklen (2007), for instance, 12 t CO_2 (corresponding to 3.2 t C) can be stored on average per ha and year over a 30-year period by means of pine afforestation of temperate cropland. On the other hand, it may be expected that such conversion would trigger the same indirect land-use changes as would energy crop cultivation on the same land, and thus emissions from indirect land-use changes amounting to 10 t CO_2 per hectare and year should be integrated in the calculation (Box 7.3-2). It is apparent that, with the exception of electromobility, in the transport sector the GHG abatement performance of the temperate pathways analysed is poorer in all cases than the value for afforestation stated here, while in the electricity and CHP sectors comparable or greater emissions reductions can be achieved. In WBGU's view the cultivation of energy crops is only then an expedient climate change mitigation option if it is ensured that the associated emissions reductions are greater than those achievable – taking indirect land-use changes into account – on the same land by means of other measures such as afforestation.

It needs to be taken into consideration in this context that the carbon storage achievable by means of afforestation can differ from region to region. The carbon storage capacity in the biomass of afforested land depends greatly upon the location, stand age and tree species (Nabuurs et al., 2007). This is illustrated by the following examples from the tropics: a tropical afforestation planted with 13 tree species on former pastureland in the 1930s hosted 57 tree species a little less than 60 years later, and stored an average of 5.1 t CO_2 (or 1.4 t C) per hectare and year (Silver et al., 2004). On former barren land in India, a tropical tree plantation achieved net storage of 14.3 t CO_2 (3.9 t C) per ha within five years following afforestation with *Gmelina arborea*, which translates into an average of around 3 t CO_2 (0.8 t C) per ha and year (Swami and Purim, 2005). Righelato and Spracklen (2007) report CO_2 storage of 15–29 t CO_2 (4–8 t C) per year and hectare for natural succession on abandoned tropical cropland.

ABSOLUTE ANNUAL GHG ABATEMENT PER UNIT BIOMASS UTILIZED

The land-referenced parameter discussed above is suited for comparing energy crop pathways, but cannot be used to assess the use of residues and wastes. A suitable approach for that purpose is to reference emissions to the primary energy content of the biomass feedstock. This permits a comprehensive comparison of bioenergy pathways.

Figure 7.3-4 shows the absolute reduction of GHG emissions provided by substituting fossil fuels, for a range of bioenergy pathways and in relation to the gross energy content of the biomass utilized. The reference quantity is only that proportion of energy content which is allocated to the respective final use. The other proportions are allocated to co-products (Box 7.3-1).

This analysis pursues the question of relevance to WBGU: Which technology pathways can deliver the greatest greenhouse gas reduction with a given quantity of biomass? In the view of WBGU, this energy-referenced parameter is the suitable one for a comprehensive assessment of the GHG abatement performance of bioenergy pathways, and should be taken as the basis for standard-setting.

Comparison of Figure and 7.3-4 and 7.3-2 shows that while in the transport sector similar percentage emissions reductions can be achieved in relation to final energy, in stationary applications substantially greater absolute GHG abatement can be achieved for the same quantity of biomass resource utilized. Electromobility, with a performance comparable to electricity generation and to combined heat and power production, is an exception in the transport sector. The use of biodiesel produced from oil palm cultivated on degraded land is a further exception. This exhibits a very high abatement performance which can, however, in turn be further exceeded

a

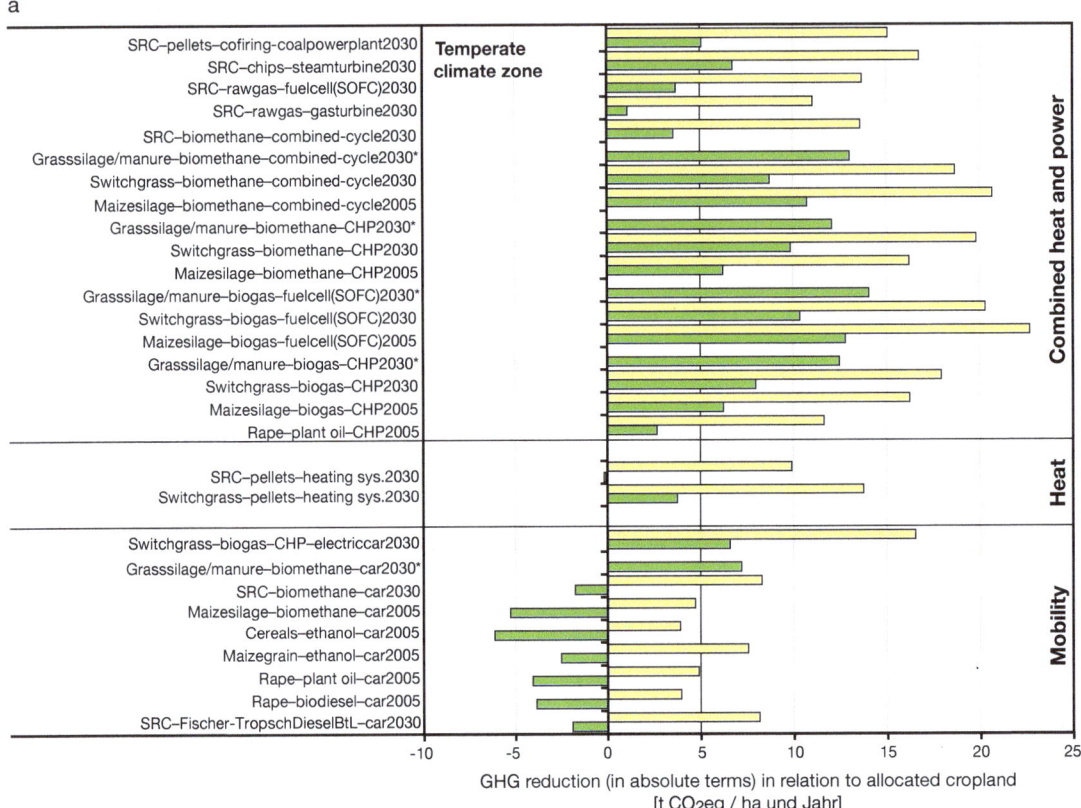

b

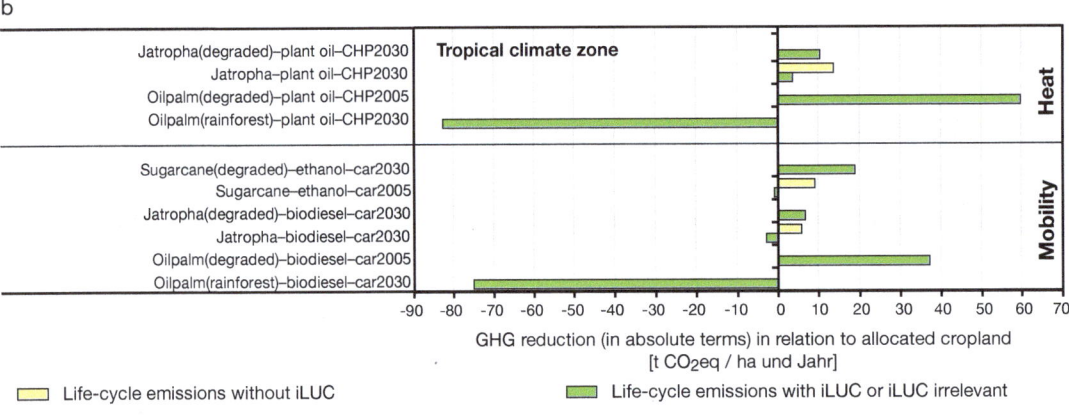

Life-cycle emissions without iLUC   Life-cycle emissions with iLUC or iLUC irrelevant

Figure 7.3-3
Absolute GHG emissions reduction through the substitution of fossil fuels for different energy crops in (a) the temperate climate zone and (b) the tropical climate zone, in relation to the allocated cropping area (Box 7.3-1) in t CO_2eq per ha and year. The chosen reference systems are as follows: for the electricity pathways a mix of 80% hard coal and 20% natural gas and for the transport pathways 60% petrol and 40% diesel (Table 7.3-3). The yellow bars contain the life-cycle emissions inclusive of emissions from direct land-use change (dLUC). The green bars further take into account emissions from indirect land-use change (iLUC 50%; Box 7.3-2). If not otherwise indicated, it is assumed for the energy crop pathways that cultivation is on former cropland. Negative values indicate an increase in emissions relative to the reference system. * It is assumed for pathways which use grass silage or slurry as substrate that in Germany no emissions result from land-use change; this special case, however, is not globally transferable. The pathway designations refer to the cultivation systems and conversion processes listed in Tables 7.2-1 and 7.2-2.
Source: WBGU based on data from Fritsche and Wiegmann, 2008 and Müller-Langer et al., 2008

Table 7.3-4
Gross energy yields per hectare used to calculate GHG emissions in the individual bioenergy pathways, and range calculated from the various per-hectare yields cited in the literature. In the case of palm oil, the literature values refer solely to the oil fruit, while the value used here refers to the entire harvested biomass. This is made necessary by the allocation methodology (Box 7.3-1). The values are therefore not comparable.
Source: Fritsche and Wiegmann, 2008 and WBGU

| Crop/product | Climate zone | Gross energy yield [GJ/(ha · a)] | | Range calculated from the literature cited in Section 7.1 |
| | | Used for the calculations (Fritsche and Wiegmann, 2008) | | |
		2005	2030	2005
Palm oil	Tropical	500	660	220–480
Palm oil (degraded land)	Tropical	350	462	110–240
Jatropha (cropland)	Tropical	–	113	5–310
Jatropha (marginal land)	Tropical	–	54	5–155
Sugar cane	Tropical	650	700	160–1,960
Maize silage	Temperate	211	250	
Maize grain	Temperate	159	–	120–210
Rapeseed	Temperate	84	–	75–105
Triticale	Temperate	100	–	50–105
Switchgrass	Temperate	–	200	90–300
Poplar (SRP)	Temperate	–	135	35–350
Grass silage	Temperate	100	–	100–210

if the same feedstock is utilized in stationary applications. Figures 7.3-2 and 7.3-3 clearly illustrate the major impact of land-use changes on abatement performance.

In the realm of applications involving residue use, it is apparent that the different electricity generation technologies have no great impact on abatement potential. Biomethane production can achieve an abatement potential that is increased by a further approx. 20 per cent if the CO_2 which is to be separated is stored permanently (Box 7.2-2). With the exception of electromobility, pathways involving the use of residues in the transport sector achieve only around half the abatement performance of pathways involving conversion to electricity. High abatement performance further exceeding that of residue use can be achieved by cultivation systems such as cropping Jatropha and palm oil on marginal land and processing it to liquid fuels. But even for these bioenergy carriers it is apparent that greater performance can be achieved in the electricity sector than in the transport sector. The abatement performance of pathways using energy crops cultivated on land that was already cropland is constrained greatly if indirect land-use changes are taken into account.

Comparison of bioenergy pathways using this parameter reveals that a reduction by a certain percentage across all fields of application (electricity, heat and transport) involves substantially greater absolute emissions reductions in the case of electricity generation. It follows that if the goal is to promote bioenergy pathways that deliver maximum climate change mitigation impact, it is not purposeful to establish a standard requiring a percentage GHG reduction referenced to final or useful energy. WBGU therefore proposes the development of a standard guided by the 'absolute annual GHG abatement per unit biomass utilized' parameter. The numerical values for such a standard need to match the chosen reference system, as this has a major impact on the outcome (Fig. 7.3-5). When applied as a precondition for the promotion of bioenergy in industrialized countries (Sections 10.3 and 10.7), the standard should be in line with the best available systems. For example, in relation to the reference system chosen by WBGU, a minimum abatement of 60 t CO_2eq per TJ raw biomass utilized could be stipulated. This would need to take account of emissions from both direct and indirect land-use changes. As a minimum standard (Section 10.3), WBGU considers abatement by 30 t CO_2eq per TJ to be suitable. In the field of biofuels, such a stipulation would translate approximately into the requirement to cut emissions compared to the reference system by 50 per cent in relation to final energy.

A sensitivity analysis of the 'SRP–biomethane–combined-cycle 2030' pathway shows how greatly the abatement performance of bioenergy pathways depends upon the chosen reference system or fossil energy carriers substituted (Fig. 7.3-5). If, in the ideal case, electricity generated from lignite is substituted by electricity generated from SRP biomethane, then, if indirect land-use changes are not taken into account, approx. 150 t CO_2eq per TJ biomass could be saved. If, in contrast, electricity generated from natural gas is substituted, savings are only approx. 50 t CO_2eq. These values deteriorate or even become negative if emissions from indirect land-use changes are taken into account.

This underscores that bioenergy delivers the greatest climate change mitigation effect when it substitutes coal. The abatement performance of biomethane is therefore not particularly high when it substitutes natural gas – an outcome that appears particularly probable when biomethane is used in heat applications. If, however, biomethane is used explicitly to generate electricity and especially to displace coal-fired electricity, it delivers a far greater climate change mitigation effect.

It is therefore essential to set appropriate standards and create suitable political and economic settings that make it probable that bioenergy will be used to substitute those fossil energy carriers that incur high emissions.

GHG ABATEMENT COSTS

When bioenergy substitutes fossil energy carriers, extra costs may be incurred; they are the difference between the specific production costs of a bioenergy pathway and those of the reference pathway. The cost per unit of greenhouse gas abated is the ratio of these extra costs to the GHG emissions reduction achieved (Equation 7.3-1; Müller-Langer et al., 2008).

Table 7.3-5 lists the production costs of fossil reference systems used for the calculations. Figure 7.3-6 shows the extra costs per tonne GHG emission reduced that are incurred by the use of bioenergy as compared to the reference system. In the heat and electricity sector the comparison is referenced to final energy, in the transport sector to the vehicle-kilometre. Other publications concerned with the use of biofuels in the transport sector often only produce accounts up to the energy of the fuel, as the vehicle used usually has the same characteristics as the reference vehicle. In the present study, however, accounting extends up to the vehicle-kilometre in order to be able to include electromobility and fuel-cell vehicles in the analysis. In this manner, account is taken of both the extra costs and the improved efficiencies of the corresponding electric and fuel-cell vehicles. This approach permits a full comparison of all mobility pathways examined.

Where a negative GHG abatement cost results, supplying energy from biomass is more cost-effective than the fossil reference case. Where bioenergy pathways deliver no emissions reduction compared to the reference case or even result in increased emissions, no GHG abatement costs can be defined. These pathways are labelled with 'no reduction'.

To place these costs in relation, it is useful to initially examine characteristic abatement costs of other climate change mitigation options. The cost of emissions reduction by means of afforestation is stated today at approx. US$ 22 per t CO_2 (Section 5.5). Other renewable energies are also suitable mitigation options and have substantially lower abatement costs in some cases. Wind power can be supplied in Germany at an average of €ct 8 per kWh, the figure falling to €ct 5 per kWh at favourable locations, and on average generates emissions of only 25 g CO_2eq per kWh (Durstewitz et al., 2008; Wagner et al., 2008). Compared to the reference system (€ct 6.16 per kWh and 935 g CO_2eq per kWh), this results in abatement costs of wind power amounting to € 22 per t CO_2eq on average, and even an economic gain, i.e. negative abatement costs of -€ 11 per t CO_2eq at excellent locations. Photovoltaic electricity can be supplied in Germany at an average cost of €ct 42 per kWh and in southern, sun-rich countries for approx. €ct 25 per kWh, with emissions amounting to 75 g CO_2eq per kWh electricity (Wagner et al., 2008; EPIA, 2008).

Figure 7.3-4
Absolute GHG emissions reduction through the substitution of fossil fuels for different bioenergy pathways, in relation to the gross energy content of the biomass utilized. The chosen reference systems are as follows: for the electricity pathways a mix of 80% hard coal and 20% natural gas; for the heat pathways 60% natural gas and 40% mineral oil; and for the transport pathways 60% petrol and 40% diesel (Table 7.3-3). The yellow bars contain the life-cycle emissions inclusive of emissions from direct land-use change (dLUC). The green bars further take into account emissions from indirect land-use change (iLUC 50%; Box 7.3-2). If not otherwise indicated, it is assumed for the energy crop pathways that cultivation is on former cropland. For pathways involving residue use only one bar is shown, as no emissions from indirect land-use change are anticipated. Negative values indicate an increase in emissions relative to the reference system. * It is assumed for pathways which use grass silage or slurry as substrate that in Germany grass silage generates no emissions from land-use change; this special case, however, is not globally transferable. The pathway designations refer to the cultivation systems and conversion processes listed in Tables 7.2-1 and 7.2-2. The vertical red lines mark the value proposed by WBGU for a minimum standard (30 t CO_2eq per TJ) and the value that should be achieved as a precondition for the promotion of bioenergy pathways (60 t CO_2eq per TJ; Section 10.3).
Source: WBGU based on data from Fritsche and Wiegmann, 2008 and Müller-Langer et al., 2008

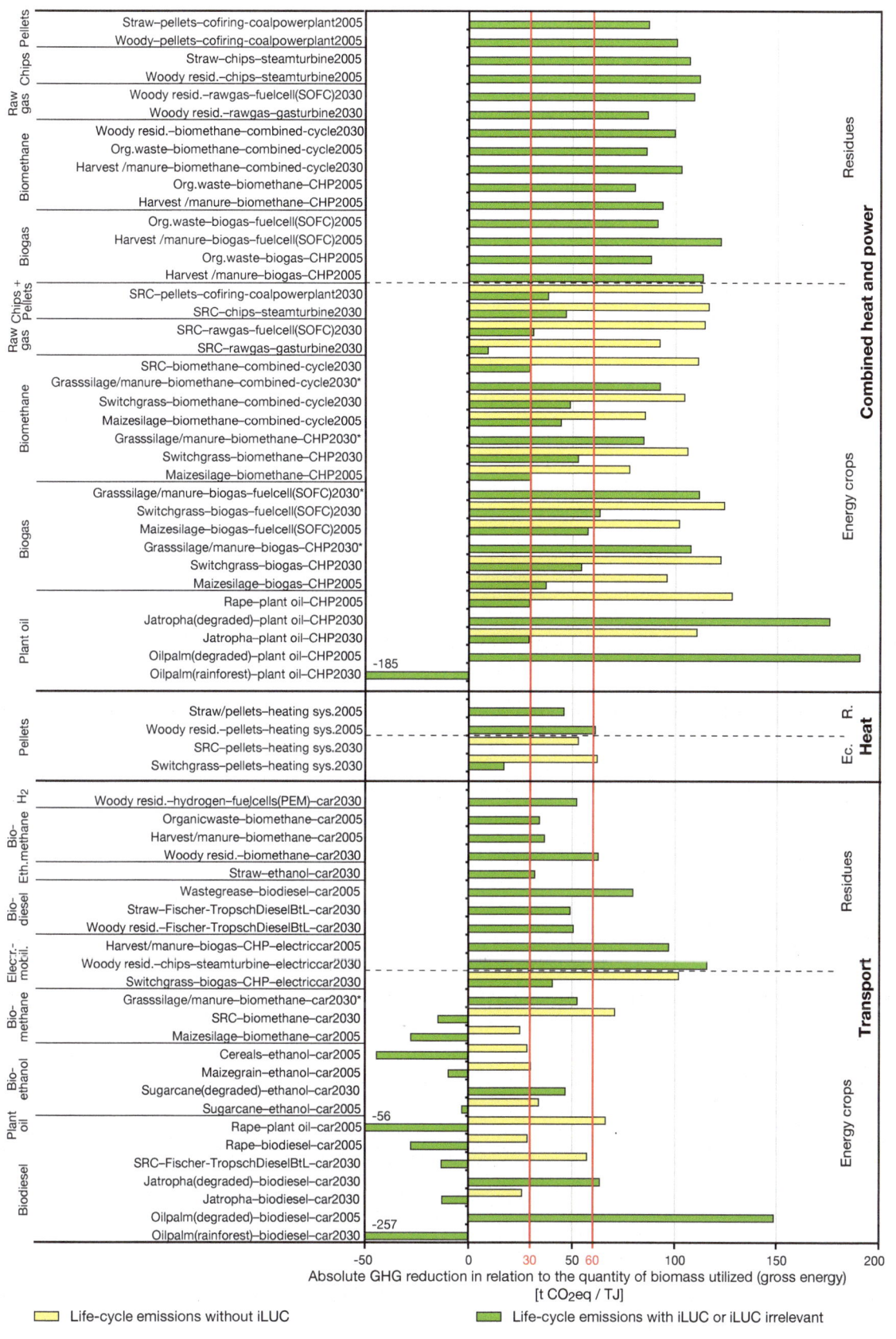

Absolute GHG reduction in relation to the quantity of biomass utilized (gross energy)
[t CO$_2$eq / TJ]

☐ Life-cycle emissions without iLUC   ☐ Life-cycle emissions with iLUC or iLUC irrelevant

The abatement cost of photovoltaics is thus approx. € 420 per t CO_2eq or € 220 per t CO_2eq, respectively. The production cost of hydropower is approx. €ct 6 per kWh in Germany; globally, the cost of large-scale hydropower is approx. €ct 4 per kWh (Fichtner, 2003). The emissions of hydropower are similar to those of wind power, at approx. 25 g CO_2eq per kWh (Wagner et al., 2007). Hydropower thus saves costs, i.e. even has negative abatement costs of up to -€ 24 per t CO_2eq. However, especially in the case of systems using storage reservoirs, the GHG balance of hydropower is subject to major uncertainties because of potential methane emissions; substantially different values can therefore result from case to case.

For a basic evaluation of the abatement costs presented here, WBGU makes use of studies that have determined the marginal damage costs of climate change, i.e. the aggregated macroeconomic net costs incurred worldwide by damage induced by climatic changes (the social cost of carbon), as well as studies that have identified the marginal abatement costs at which emissions reductions can be realized which are globally sufficient to stabilize the CO_2 concentration in the atmosphere at 400 ppm. Based on figures reported in studies on marginal damage costs (Clarkson and Deyes, 2002; Pearce, 2003; UBA, 2007) and on the marginal costs of various abatement options (Enkvist et al., 2007) pathways are classed in the following in three categories: those with abatement costs above € 60 per t CO_2eq are classed as presently economically inefficient, i.e. too expensive under current conditions, those with abatement costs of € 40–60 per t CO_2eq are classed as economically acceptable, and those with less than € 40 per t CO_2eq are, in the view of WBGU, in the presently cost-efficient range. Because of uncertainties resulting from the modelling that underpins the figures in the studies used, the class boundaries stated here are provisional; they do, however, mark out plausible evaluation corridors.

Taking present costs as a basis, the following conclusions result: cost-effective climate change mitigation can be achieved today primarily by means of those pathways involving cultivation of tropical energy crops on marginal or degraded land. This applies equally for applications in the transport sector. By using biodiesel from Jatropha cultivated on marginal land, climate change mitigation can even avoid costs. Furthermore, those pathways are favourable that are based on relatively simple technologies, such as the use of unrefined vegetable oil, or the simple co-firing of straw or wood residue pellets in coal-fired power plants. Established technologies such as co-combustion or the fermentation of residues in biogas facilities to produce biogas or biomethane are favourable and are initially to be preferred over other technologies whose core process is biomass gasification (e.g. Fischer-Tropsch diesel) or that make use of expensive generating sets (e.g. fuel cells). It is only when their costs have been reduced in the course of technological progress that their broad-scale deployment for climate change mitigation becomes recommendable.

As the analysis must take account of emissions from indirect land-use change (green bars), under current conditions only a small number of residue pathways are attractive, together with those energy crop pathways that use degraded or marginal land as a part of the cropping system.

This assessment matrix is not, however, tantamount to a recommendation not to pursue the technologies whose abatement costs are rated as too expensive today. Here it is rather essential to carry the analysis forward with a differentiated consideration of the reasons for the high costs and the anticipated cost reductions.

It further needs to be considered when examining abatement costs that these depend greatly upon the production costs of the reference system. For exam-

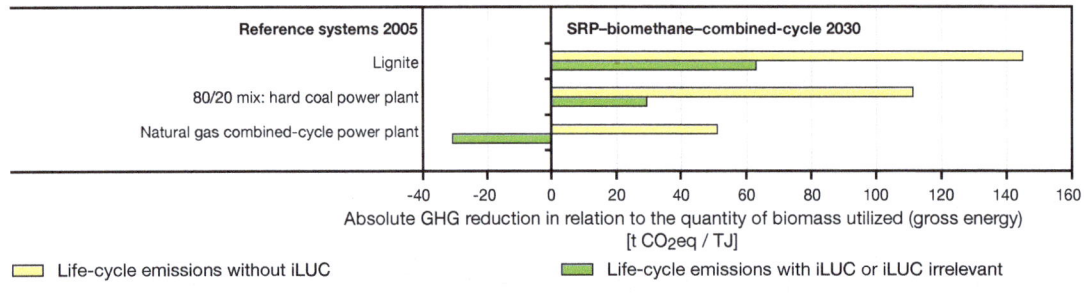

Figure 7.3-5
Sensitivity of absolute GHG reduction in relation to the quantity of biomass utilized relative to the reference system, for the example of the conversion of wood from short-rotation plantations to biomethane for a combined-cycle power plant. The abatement performance is shown for three different reference systems, without (yellow) and with (green) consideration of emissions from indirect land-use change. The greatest abatement performance is achieved if lignite is substituted. The lowest abatement performance – or even an increase in emissions – results if natural gas is substituted.
Source: WBGU based on data from Fritsche and Wiegmann, 2008 and Müller-Langer et al., 2008

Table 7.3-5
Production costs of fossil reference systems and reference values for specific emissions used by WBGU to derive the GHG abatement costs of the individual bioenergy pathways. Veh-km = vehicle kilometre.
Source: Müller-Langer et al., 2008

	Fossil reference system	Production costs per unit final or useful energy		Share of mix [%]	Reference value for specific emissions
Electricity	Hard coal power plant	5.20	€ct / kWh$_{el}$	80	
	Natural gas combined-cycle	10.0	€ct / kWh$_{el}$	20	6.2 g CO_2eq / kWh$_{el}$
Heat	Heating oil heating	10.8	€ct / kWh$_{th}$	40	
	Natural gas heating	11.6	€ct / kWh$_{th}$	60	11.3 g CO_2eq / kWh$_{th}$
Transport	Petrol car	2.68	€ct / Veh-km	60	
	Diesel car	2.58	€ct / Veh-km	40	2.6 g CO_2eq / Veh-km

ple, these are relatively low for the case of electricity generation from coal at €ct 5.2 per kWh$_{el}$, while for electricity generation from natural gas they are almost double at €ct 10.0 per kWh$_{el}$ (Table 7.3-4). It must further be kept in mind that all costs of reference systems refer to the 2005 baseline year and that changes are to be expected in future. In that connection a further distinction needs to be made for both the bioenergy pathways and the reference system between technology costs and fuel costs.

The technology costs can be expected to decrease in both cases. As the relatively old conventional fossil technology is technologically advanced, the cost reduction potential will be smaller than is the case for bioenergy technologies, some of which are relatively recent.

The situation can be different for fuel costs. From the 2005 baseline year, the fuel price for imported hard coal already rose over the three years to 2008 by approx. 50 per cent in Germany, from € 65 to € 95 per tonne (Müller-Langer et al, 2008; Statis, 2008). On the other hand, the fuel prices for biomass also rose in that period, partly because they correlate positively with those of fossil energy carriers (Section 5.2.5.2). Whether rising fossil energy prices really cause the abatement costs of bioenergy pathways to drop will

$$K_{GHG,S} = \frac{K_S - K_{REF}}{e_{REF} - e_S} = \frac{\Delta K}{\Delta e_{GHG}} \quad \text{with } \Delta e_{GHG} > 0$$

Equation 7.3-1
$K_{GHG,S}$ specific GHG abatement cost of a pathway [€/t$_{GHG}$]; K_S specific production cost of a pathway [€/GJ$_{EE}$]; K_{REF} specific cost of the reference pathway (taking account of the marginal cost of fossil energy carriers) [€/GJ$_{EE}$]; ΔK differential cost of the pathway compared to a reference pathway [€/GJ$_{EE}$]; e_S specific GHG emissions of a pathway [kg GHG/GJ$_{EE}$]; e_{REF} specific GHG emissions of the reference pathway [kg GHG/GJ$_{EE}$]; Δe_{GHG} specific GHG abatement of the pathway compared to a reference pathway [kg GHG/GJ$_{EE}$].

depend upon how strongly technology and fuel costs change and on the specific contributions of the two to the production costs per unit energy. In the case of energy crops, increasing competition for land as well as rising costs of inputs (agricultural machinery, fertilizers, water, etc.) may cause fuel costs to rise, while these factors will be less effective in driving prices of residues upwards. This is a point in which bioenergy also differs from other renewables: while learning curves can be expected to deliver further cost reductions for wind and solar energy, this only applies to technology in the case of bioenergy but not to energy

Figure 7.3-6
GHG abatement costs incurred by the use of different bioenergy pathways, calculated in accordance with Equation 7.3-1. The chosen reference systems are as follows: for the electricity pathways a mix of 80% hard coal and 20% natural gas; for the heat pathways 60% natural gas and 40% mineral oil; and for the transport pathways 60% petrol and 40% diesel (Table 7.3-3). The yellow bars contain the life-cycle emissions inclusive of emissions from direct land-use change (dLUC). The green bars further take into account emissions from indirect land-use change (iLUC 50%; Box 7.3-2). If not otherwise indicated, it is assumed for the energy crop pathways that cultivation is on former cropland. For pathways involving residue use only one bar is shown, as no emissions from indirect land-use change are anticipated. Negative values indicate an increase in emissions relative to the reference system. * It is assumed for pathways which use grass silage or slurry as substrate that in Germany grass silage generates no emissions from land-use change; this special case, however, is not globally transferable. The pathway designations refer to the cultivation systems and conversion processes listed in Tables 7.2-1 and 7.2-2.
Source: WBGU based on data from Fritsche and Wiegmann, 2008 and Müller-Langer et al., 2008

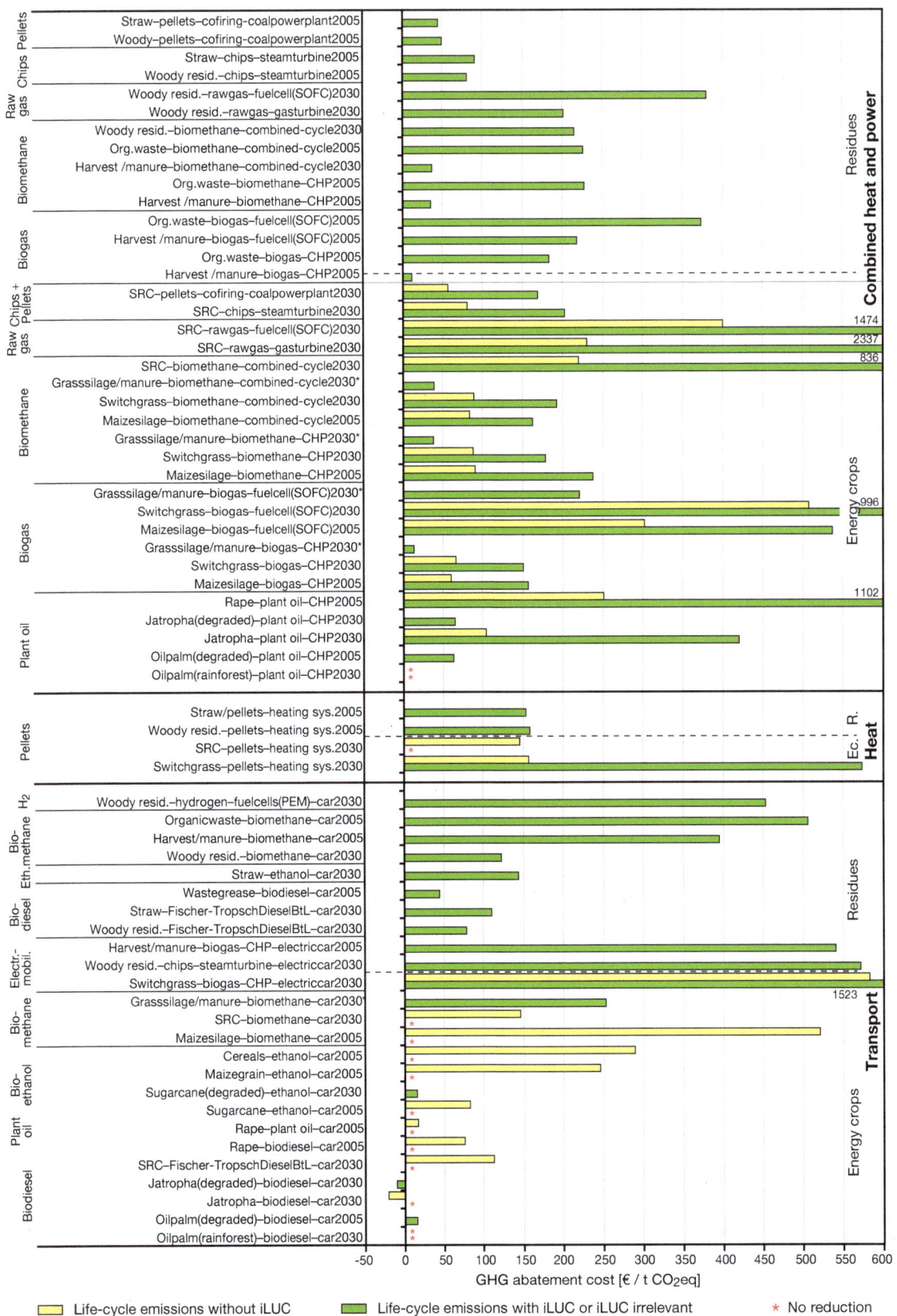

production itself, as here the biomass fuel costs are a further and major factor determining production costs per unit energy.

In the view of WBGU substantial cost reductions can be expected in future in electromobility and in gasification technologies for biomethane production (Section 7.2); these will result in a marked reduction of the GHG abatement costs of these technologies. WBGU therefore views these as highly promising climate change mitigation technologies in addition to the abatement options that are already cost-effective today.

Optimizing bioenergy integration and deployment in energy systems

8

Bioenergy was the first source of energy utilized by humankind. Fossil energy, in contrast, has only been utilized for two to three centuries. The scarcity of fossil energy carriers is not the only reason why a transformation of energy systems is essential. The other is that their use causes the bulk of anthropogenic greenhouse gas emissions, which lead in turn to dangerous climate change (IPCC, 2007d). Transforming energy systems towards sustainability has top political priority in order both to mitigate climate change and overcome energy poverty (Chapter 2).

The energy supply structures of industrialized and developing countries differ sharply. In newly industrializing countries such as China and India, both biomass-based traditional energy use and fossil-based energy uses are widespread. In more than 75 countries bioenergy is the principal energy source, while in more than 50 countries its contribution to energy supply is even greater than 90 per cent (IEA, 2006b). These are almost all developing countries where biomass is used with traditional techniques. By deploying relatively minor technological and financial resources, it would be possible to greatly improve the efficiency of biomass use and also to substantially curb greenhouse gas emissions (Section 8.2). In industrialized countries, too, bioenergy can contribute to mitigating climate change and ensuring the technical security of supply (as a source of control energy) of future energy systems based on renewable resources (Section 8.1).

8.1
Bioenergy as a part of sustainable energy supply in industrialized countries

8.1.1
Transforming energy systems for improved energy efficiency and climate change mitigation

In order to comply with the 2°C guard rail (Chapter 3), it will be necessary to stabilize the greenhouse gas concentration in the atmosphere at a level below 450 ppm CO_2eq. In 2004, the use of fossil energy carriers was the largest source of global greenhouse gas emissions, accounting for 56.6 per cent or 28 Gt. To stabilize the atmospheric greenhouse gas concentration between 445 and 490 ppm CO_2eq, global greenhouse gas emissions will need to be reduced by 2050 by 50–85 per cent from their level in 2000. Studies of the distribution of emissions reduction commitments among states show that stabilization at 450 ppm CO_2eq is feasible if by 2020 the emission rights of industrialized states are 25–40 per cent below the emissions of 1990 and, in tandem, emissions from newly industrializing countries drop substantially below present projections. By the year 2050, the emission rights of industrialized countries will need to be 80–95 per cent below the emissions of 1990, and in all other regions emissions will need to drop substantially compared to projections (IPCC, 2007c). These targets can be achieved in industrialized countries and in the industrialized regions of emerging economies by means of energy-saving measures and with the help of renewable energies such as biomass. This will require a targeted transformation of energy systems.

8.1.1.1
Transformation components

The transformation proposed by WBGU is based on expanding the use of renewable energies in tandem with combined heat and power production (CHP), preventing waste heat in the transport sector, utilizing ambient heat for heat supply, and engaging in energy-saving measures across all sectors.

EFFICIENCY GAINS THROUGH INCREASED DIRECT ELECTRICITY GENERATION FROM SOLAR, HYDRO AND WIND SOURCES
Electricity is presently generated largely from fossil energy carriers. The associated conversion generates large amounts of CO_2. In power plants, most of which are large-scale, only approx. one-third of the energy contained in the fuel can be converted into electricity,

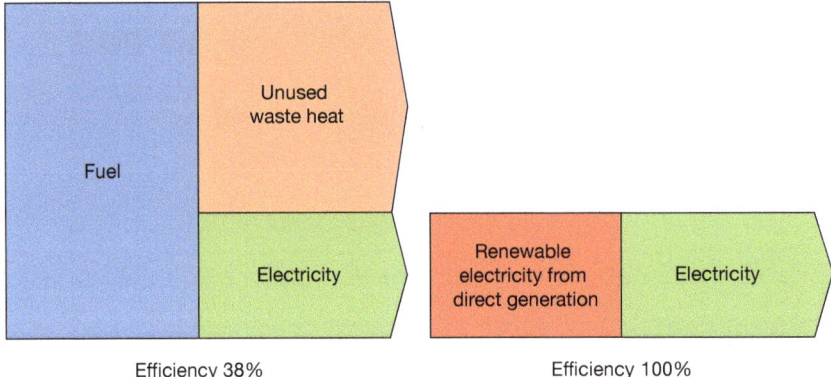

Efficiency 38%   Efficiency 100%

Figure 8.1-1
Efficiency gain through the transition to renewable energies involving the direct generation of electricity from solar, hydro and wind sources. In conventional electricity generation processes without heat extraction that make use of fossil sources, on average worldwide only a good third of the primary energy is converted into electricity, two-thirds being lost as waste heat. A switch to direct generation from renewables therefore slashes the primary energy requirement and the energy-related CO_2 emissions while delivering the same level of electricity generation.
Source: WBGU

while the rest is lost as waste heat insofar as no heat is extracted (BP, 2008). In contrast, electricity generated directly from hydro, solar and wind sources avoids the waste heat losses of thermal energy conversion and thus contributes decisively to improving energy efficiency (Fig. 8.1-1).

With increasing renewable direct generation, the fossil primary energy requirement for electricity production is reduced, and the associated GHG emissions drop in step (Section 4.1; Box 4.1-1).

EFFICIENCY GAINS THROUGH EXPANDED COGENERATION
Cogeneration (CHP) helps to improve the utilization of fossil and biogenic fuels and thus to reduce greenhouse gas emissions. The use of waste heat, transported via local or district heat networks, for space or process heat saves energy carriers and thus reduces the primary energy requirement in the heat sector. The share of CHP in energy systems can be increased by tapping the major potential for industrial cogeneration, by carefully planning and siting new cogeneration plants, and vigorously expanding heat networks (Fig. 8.1-2).

EFFICIENCY GAINS THROUGH SWITCHING TO ELECTROMOBILITY
Present mobility systems and the associated transport infrastructure have major inefficiencies: on average, an internal combustion engine only converts 20 per cent of the fossil energy into shaftpower (determined in accordance with the New European Driving Cycle, NEDC). Apart from a small proportion used to heat the interior of the vehicle, the remainder of the energy is lost as ambient waste heat. Drives using electromotors are far more efficient, as these make

approx. 80 per cent of the energy stored (in the form of electricity) utilizable as mechanical shaftpower.

The 80 per cent efficiency of electric drives (from socket to wheel) arises as follows: in the electric vehicle, electricity is stored in a modern lithium battery by means of a rectifier with an efficiency of up to 95 per cent; when travelling, this is converted back into alternating current via an inverter with the same efficiency, driving an electromotor that has an efficiency of approx. 95 per cent. This makes electric drives four times more efficient than conventional drive systems using internal combustion engines. This factor would still be around 3, if, under optimistic assumptions, internal combustion engines would reach efficiencies of 25 per cent in future and those of electric drives were assumed to be 75 per cent. Even if the higher vehicle weight attributable to the heavy lithium batteries is taken into account, the efficiency improvement factor is still around 2–2.5.

These benefits are not harnessed as yet, however, because the origin of the electricity crucially determines overall efficiency. For instance if this is fossil electricity generated with low efficiency, this negates the energetic benefit of the electric drive. It is only from a certain efficiency of electricity conversion onwards that the use of electromobility becomes technically more efficient than conventional drive systems (Fig. 8.1-3). This is illustrated by combining the efficiency of conventional electricity generation of 38 per cent as shown in Figure 8.1-1 with the efficiency of an electric vehicle of 80 per cent: the overall efficiency is then merely a good 30 per cent, which is only slightly above that of conventional drive systems. If, however, the heat is utilized by means of CHP, delivering a power generation efficiency of 80 per cent, the fuel efficiency of electromobility

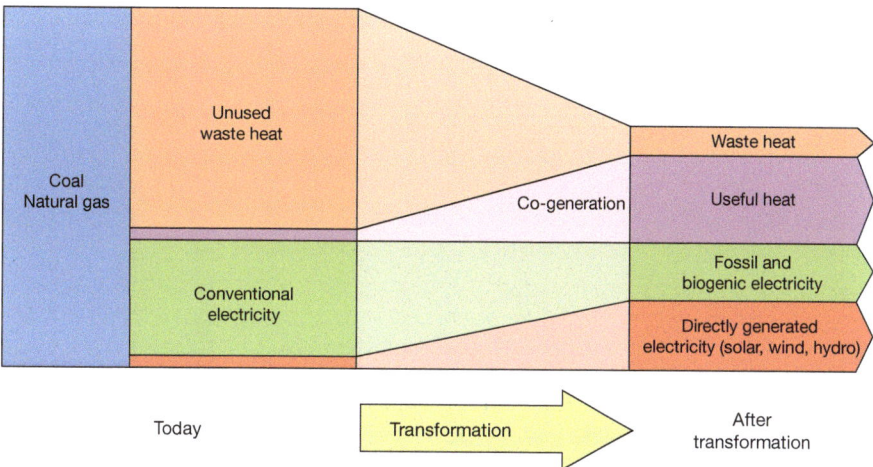

Figure 8.1-2
Electricity sector transformation. Two key components deliver efficiency gains in the sector: the expansion of direct generation from renewable sources and the increased use of combined heat and power (CHP). CHP processes make it possible to harness the bulk of the waste heat from thermal electricity generation. The share of directly generated electricity from renewable sources grows in tandem, substituting a part of the electricity that would otherwise need to be supplied by combusting fossil or biogenic energy carriers.
Source: WBGU

becomes greater than that of internal combustion vehicles in all configurations. Thus, only a combination of electric drives with directly generated renewable electricity from solar, hydro and wind sources fully taps the efficiency potential of electromobility (Fig. 8.1-4).

Electromobility delivers further benefits compared to conventional compulsion systems: the thermal conversion process does not take place in the vehicle, but in a stationary system. This makes it possible not only to utilize waste heat, but also to sequester CO_2. Moreover, it resolves particulates issues and mitigates noise pollution. Electromobility also provides potential benefits for energy generators and transmission system operators. It represents an energy store that is available for 90 per cent of the day (non-driving times of vehicles), which, by means of suitable information and communication technologies, can be integrated and used to balance fluctuating feed-in from renewable sources. The prevention of conversion losses in internal combustion motors and the deployment of directly generated electricity in electric vehicles thus present a major efficiency potential in the transport sector. Electromobility systems in which the electricity utilized comes from renewable sources are therefore a key component of the transformation of energy systems towards sustainability.

The above findings apply mainly to road transport, which has the largest share of energy consumption in the transport sector. Broad-scale deployment of electric vehicles, however, can only be realized over longer periods (Fig. 8.1-5). With the exception of

hybrid cars, electric drives for series vehicles are still at the development stage. The battery, which must store large amounts of energy yet also be light and have a long service life, is a neuralgic point. Nonetheless, electric vehicles with ranges of 100–200 km are already being manufactured today, and many carmakers plan to include electric and hybrid vehicles in their fleets (Engel, 2007). In the medium term, it can even be expected that electric drives will be used in heavy goods vehicles.

In aviation, however, there is presently no alternative to liquid, carbon-based energy carriers. While the situation is similar in shipping, new propulsion systems such as controllable kites have the potential to reduce the fuel consumption of a ship by 10–50 per cent. Prototypes are already in use (Skysails, 2008). Rail transport already runs mainly on electricity in numerous countries and uses a dedicated electric grid (Oeding and Oswald, 2004; DB, 2008). In Austria, for instance, the share of hydropower in railway electricity supply already amounted to 89 per cent in 2007 (ÖBB, 2008). If it should become possible in future to no longer consume mineral oil in passenger cars thanks to electric drives, then the oil can be deployed in long-transport, in aviation and in shipping until alternatives are found.

EFFICIENCY GAINS THROUGH USING ELECTRIC
HEAT PUMPS FOR HEAT SUPPLY
Conventional oil- and gas-fired heating systems have efficiencies of 70–110 per cent based on the net calorific value (condensing boiler technology; BHD, 2008; DIN, 1990). Through direct combustion in oil-

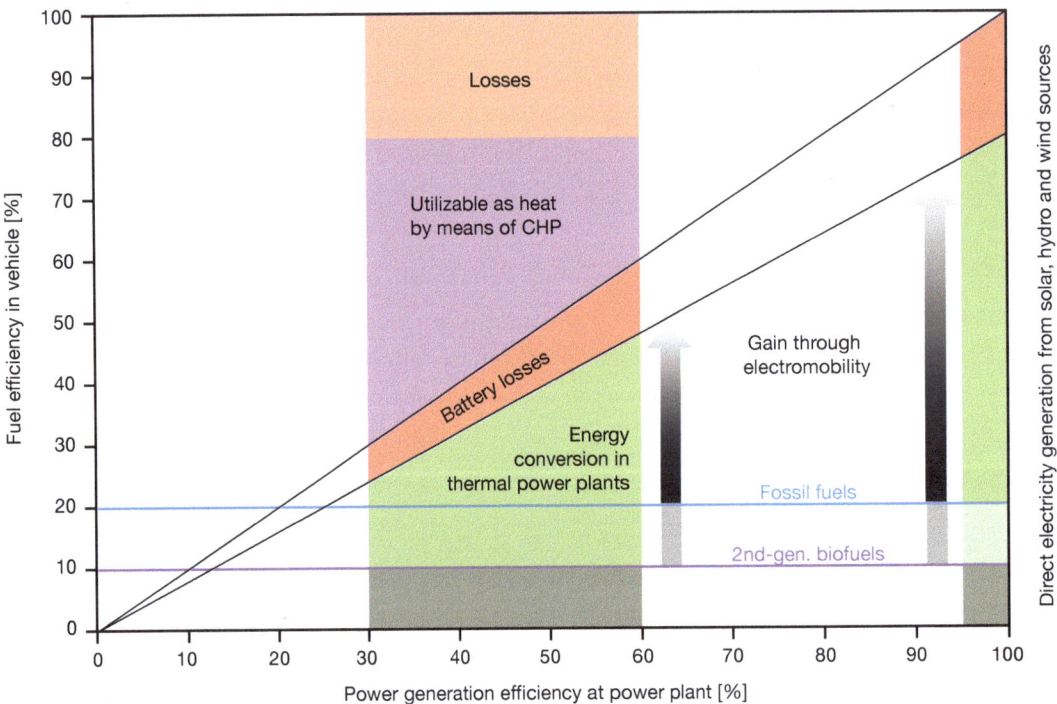

Figure 8.1-3
Comparison of the efficiencies of fossil or biogenic fuel use in vehicles with internal combustion motors and in electric vehicles. In vehicles with internal combustion motors, approx. one-fifth of the chemical energy filled in the tank is converted into mechanical shaftpower. In vehicles using electric drives, up to 80% of the 'filled' electrical energy can be converted into mechanical propulsion. It is the type of electricity generation, however, which crucially determines the overall efficiency gain. Electromobility delivers an efficiency gain with power plant efficiencies of 25% and more – the Figure indicates the typical range of power plants, which is 30–60%. Electromobility is worthwhile in all circumstances if the waste heat arising at the power plant is utilized. As this is not possible in conventional vehicles with internal combustion motors, they can only utilize 20% of the fossil fuel as mechanical shaftpower (horizontal line marked 'fossil fuels'). If biofuels are used, this value is reduced to 10%, as in the best case (2nd-generation biofuels) only half of the bioenergy can be converted from biomass to fuel (horizontal line marked '2nd-gen. biofuels'). Electromobility achieves its maximum efficiency potential when fed with directly generated electricity from hydro, solar or wind sources (column on right hand of Figure). This avoids the substantial thermal conversion losses.
Source: WBGU

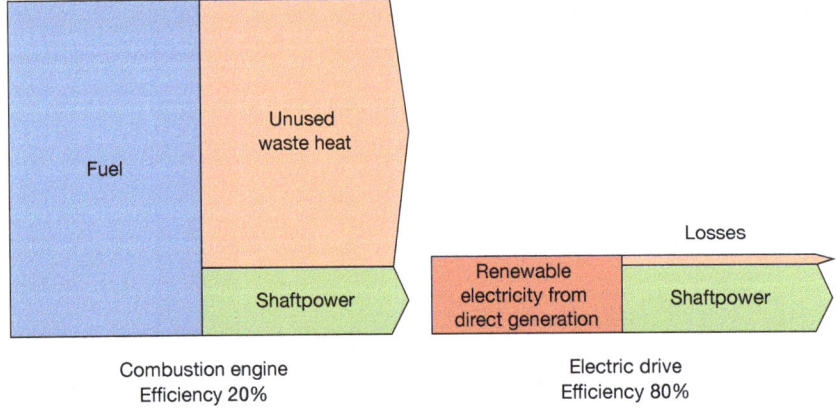

Figure 8.1-4
Efficiency gain in the transport sector: energy input and efficiency of a conventional drive system using fossil and biogenic fuels compared to those of an electric drive using renewable, directly generated electricity from hydro, solar and wind sources. The efficiency with which the same quantity of mechanical shaftpower is delivered is several times greater if renewable electromobility is used than if fuel is used in a conventional internal combustion engine.
Source: WBGU

Figure 8.1-5
Transport sector transformation: key component renewable electromobility. By gradually increasing the use of renewable electricity sources, of which a large proportion comes from direct generation using wind, hydro and solar energy, the energy requirement and GHG emissions of the transport sector can be reduced substantially – ideally to one-quarter of present energy consumption levels.
Source: WBGU

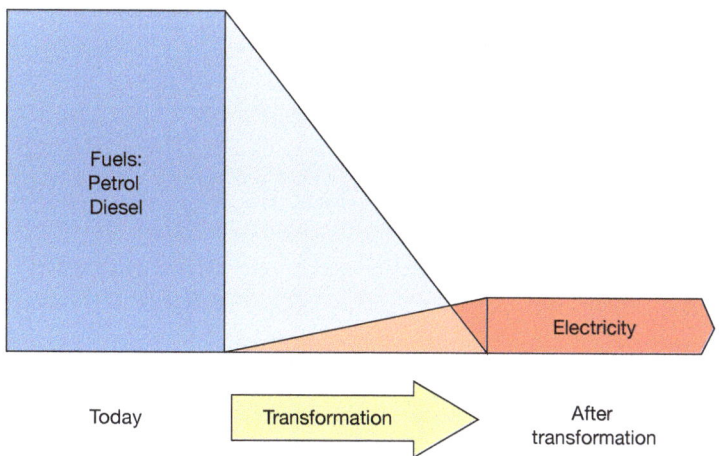

and gas-fired heating systems, the energy stored can be converted to 100 per cent into heat and almost to 100 per cent into useful heat (hot water, space heat, etc.). If electricity is used in an electric heat pump, that raises the available ambient heat to a suitable level and renders it utilizable, substantially more heat can be supplied. The quotient between the utilizable heat output and the electrical energy consumed in the compressor is termed the coefficient of performance, and is determined according to defined conditions in various standards such as EN 14511 (DIN, 2008a, b; VDI, 2008). Assuming that electric heat pumps have an average performance coefficient of 3.5, an input of 1 kWh electricity can deliver 3.5 kWh heat, of which 2.5 kWh come from the ambient heat (Baumann et al., 2006). This value is especially likely to be achieved if heat pumps are linked to geothermal systems. Under favourable conditions heat pumps can then achieve high annual performance factors. As in the field of electromobility, the origin of the electricity or the efficiency of power generation is decisive. Only from a certain power plant efficiency onwards is the use of electric heat pumps more favourable

than the direct thermal use of fuel, as illustrated by the following example.

The present generation mix, with its efficiency of 30–35 per cent, can generate approx. 0.30 kWh electricity from 1 kWh fossil or biogenic energy. With the performance coefficient assumed here, the electric heat pump can use this electricity to deliver at most 3.5 times more heat, i.e. in our example approx. 1 kWh heat. Under such circumstances the use of electric heat pumps is pointless, as the direct combustion of fossil or biogenic fuel would deliver the same utility. However, the useful heat ratio rises to approx. 200 per cent if the electricity is generated in combined-cycle power plants with efficiencies around 60 per cent, and even climbs to 350 per cent if the electricity was generated directly from solar, hydro or wind sources. It follows that the energy efficiency of electric heat pumps is best harnessed by using directly generated electricity (Figs. 8.1-6 and 8.1-7). Thermal heat storage systems can decouple electricity demand from heat demand. This allows efficient load management; for instance, in periods of high wind power generation, the surplus electricity can be stored in this manner. The observed trend towards

Figure 8.1-6
Efficiency gain through using ambient heat by means of heat pumps running on renewable electricity.
Source: WBGU

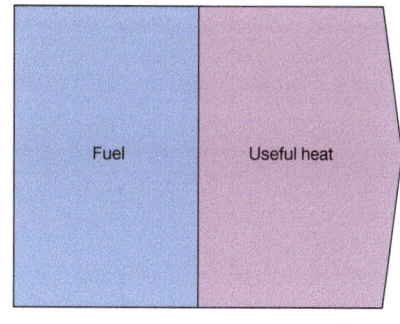

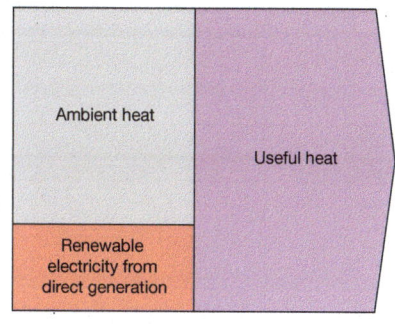

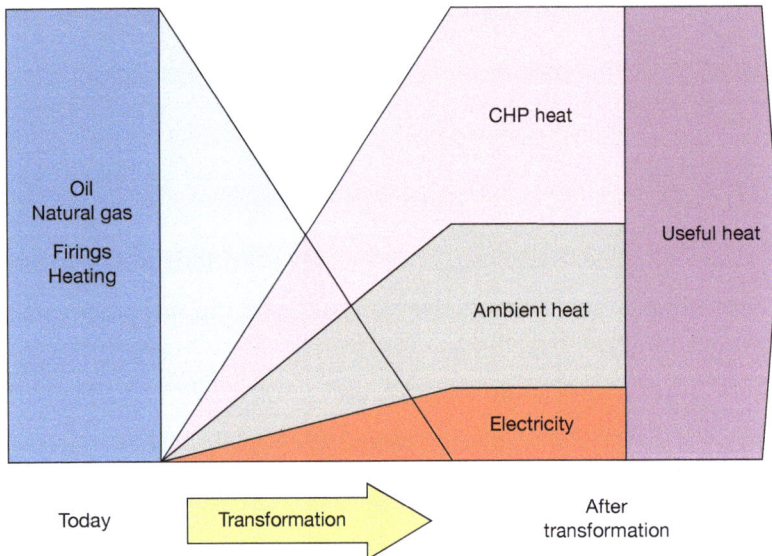

Figure 8.1-7
Heat sector transformation: through CHP expansion and the greater use of electric heat pumps, process and space heat demand can be met entirely in future. To fully tap the efficiency potential of electric heat pumps, the electricity powering them should come predominantly from renewable, directly generated sources (hydro, solar or wind).
Source: WBGU

an increasingly larger proportion of direct generation from renewable sources will in future substantially improve the overall energetic efficiency of electric heat pumps.

Broad-scale introduction of heat pumps greatly reduces the consumption of fossil and biogenic fuels in the heat sector. In combination with waste heat from CHP, the fossil primary energy requirement and thus GHG emissions in the heat sector can be reduced greatly; in the ideal case, directly combusted energy carriers can be substituted (Fig. 8.1-7). Space heat, hot water heating and a part of process heat can all be supplied in this manner. A further part of process heat demand can be met by renewable electricity.

EFFICIENCY GAINS THROUGH ENERGY
CONSERVATION MEASURES
There are many ways to make the use of energy more efficient – i.e. to reduce energy requirement while delivering the same level of utility. Such options are available in all energy sectors. This is highly evident in the heat sector: thermal insulation, meeting the 'passive house' standard in the ideal case, can greatly reduce the energy required for space heating. Improved space heating systems, cooking stoves and hot water production systems are further examples.

In the electricity sector, too, there are many options. For example, in industrial processes compressed air is often used highly inefficiently. Installing improved sealings and exchanging leaky components can save electricity consumed in compressors. Savings in lighting are a further example. An industrialized country such as Germany only consumes 5 per cent of electricity consumption for lighting; however, 95 per cent of this is dissipated as waste heat

when used in conventional incandescent lamps. If in each of the 39 million German households 75 W standard incandescent lamps were replaced by 15 W compact fluorescent lamps, which have the same light output, 2.3 GW generating capacity, which translates into two large-scale power plants, would theoretically no longer be needed during the period of lighting demand. All household appliances (refrigerators and freezers, stoves, etc.), lighting devices (LED technology) and industrial processes (electric drives, power electronics, etc.) could be designed so as to be more efficient, meeting the same purpose with less electricity consumption. Switching off stand-by circuits by installing switchable sockets, or even banning stand-by, is a further measure that would promote energy efficiency.

In the transport sector, the energy consumption of all aircraft and vehicles can be reduced by improving their aerodynamics and reducing their weight or rolling resistance. This goal can also be furthered by means of socio-economic and organizational measures such as improved local public transport systems, better capacity utilization of buses, trains and aircraft, or improved traffic flow organization. The period considered decisively determines all quantitative savings potentials.

8.1.1.2
Transforming energy systems by combining the components

If the five components set out above are combined, the fossil and nuclear primary energy requirement of an industrialized country can be reduced by more than 80 per cent, and energy-related GHG emissions curbed accordingly. This corresponds to the reduction obligations that may result from a newly negotiated Kyoto Protocol for Annex I countries such as Germany. In a transformed energy system, GHG emissions of quantitative relevance only arise from fossil and biogenic electricity generation in CHP systems and in highly efficient combined-cycle power plants. The very low emissions from wind, hydro and solar energy are negligible in comparison.

The five components described in Section 8.1.1.1 boost efficiency in an energy system that need not be a distant vision, but is in fact a route that can be taken with present technology (Fig. 8.1-8).

8.1.2
The role of bioenergy in the sustainable energy supply of industrialized countries

The sustainably available bioenergy potential is limited (Chapter 6). It is plain that it will not suffice to meet a major proportion of today's global energy requirement. This makes it important to make optimum use of the strategic attributes of the limited biomass resource (Chapter 2). In industrialized countries, it is recommendable to deploy this resource where its specific attributes cannot be substituted by other energy carriers and its benefits in terms of energy efficiency and climate change mitigation are greatest, without jeopardizing other sustain-

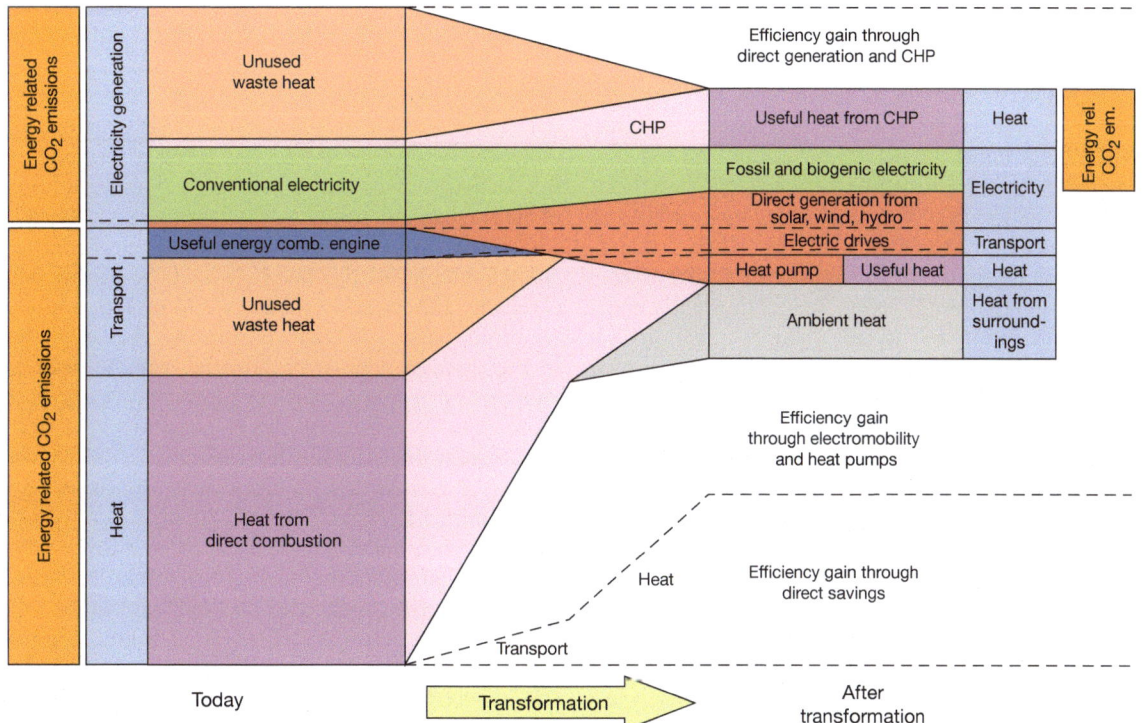

Figure 8.1-8
Energy system transformation – the example of Germany, an industrialized country: five key components can deliver both energy and climate efficiency. The primary energy requirement in Germany in 2005 – excluding the part going to non-energy uses such as mineral oil for the chemical industry – amounted to 13.4 EJ, of which 34% went to the electricity sector, 44% to the heat sector and 23% to the transport sector. The ratios are different when examined in terms of final energy: the share of electricity is then smaller (18%) and that of the other two sectors correspondingly larger (heat 54%, transport 28%). Electricity is also used in the transport sector, but at present this only supplies 2% of that sector's demand and is therefore not taken into account in the Figure. In future, direct combustion for heat uses is to be replaced by heat from CHP and electric heat pumps. The proportion of heat produced from electricity, including heat from CHP, is contained under 'electricity generation' in the diagram. More than 70% of electricity is to be generated by means of direct generation using solar, hydro and wind sources. Load management of intermittent energy sources is to be facilitated by various means: a massively expanded electricity transmission and distribution network; the integration of storage power plants (pumped-hydropower, compressed-air); integration of the transport sector (electric vehicles); and heat pumps to make use of miscellaneous heat sources – all of these are to be integrated by means of an equally expanded information network (smart grid). Overall, the heat and transport sector are to account for 25% of electricity supply. Such a transformation permits energy supply systems in industrialized countries that are environmentally sustainable and economically viable, and is feasible by 2050.
Source: WBGU using data from BMWi, 2008

ability criteria such as maintaining biological diversity and ensuring food security. The following sections explore the three main potential applications: as a fuel in the transport sector, for heat supply, and for electricity generation.

8.1.2.1
Bioenergy for transport: Bio-electricity versus biofuels

RESOURCE CHALLENGES

The transport sector is almost entirely dependent today upon fossil mineral oil. In Europe, for instance, approx. 98 per cent of its demand is met by mineral oil, and in many other regions the level of dependence is similarly high (Boerrigter and van der Drift, 2004). Fossil fuels are becoming increasingly scarce and are tending to become more expensive. Biofuels are hoped to provide a way out of this dilemma; their relevant potential, however, is tied to the available land area (Chapters 6 and 7). Synthetic 2nd-generation biofuels promise a higher energy yield than those of the 1st generation. Their aggregate environmental impact, however, is not substantially better than 1st-generation biofuels, and – as is the case for all assessment results for energy crops – depends greatly upon direct and indirect land-use changes (Section 7.3; Jungbluth et al., 2008).

Not only the land resource for energy crops is limited; so, too, is the quantity of residues available for recovery. Moreover, the use of biomass as industrial feedstock will continue to grow from mid-century, when mineral oil becomes uneconomic for the chemical industry and the industry must turn to other hydrocarbons (Section 5.3). Biofuels cannot solve the climate problem and can only perform a bridging function in terms of providing security of supply. Food and feed production as well as certain branches of industry (chemicals, construction, textiles, etc.) need carbon compounds. Energy for mobility, in contrast, can be supplied through other avenues.

EFFICIENT USE OF BIOENERGY IN THE TRANSPORT SECTOR

Beside resource availability, conversion efficiency is the key determinant of the energy contribution of biomass in the transport sector. Conversion of biomass to biofuels delivers substantially lower overall efficiencies compared to conversion to electricity for use in electric vehicles (Sections 7.2 and 7.3). Therefore, it makes more sense in terms of energy yield to deploy biomass in CHP, rather than converting it into fuel. This is illustrated by the example of woody biomass: The state of matter of the solid fuel is changed twice in processes involving energy losses of

up to 60 per cent, and the fuel thus derived used in a combustion engine, with the result that only approx. 10 per cent of the energy is available as useful energy at the wheel (Fig. 8.1-9). In contrast, the biomass can be converted efficiently to electricity (stationary conversion process, operating at full load), the waste heat can be utilized and the electricity deployed in electric vehicles with low particle emissions and little noise. Figure 8.1-9 compares four representative pathways for vehicle drives. Starting from 100 per cent primary energy (the energy content of the whole plant at harvest), if biofuels are used substantially less energy is available for propulsion at the wheel than would be possible on the electromobility pathway. The Figure shows by way of comparison the highly efficient direct electricity generation from hydro, wind and solar sources. The superiority of electric drives is also highly apparent when comparing vehicle mileage per unit primary energy (kWh) (Fig. 8.1-10).

WBGU does not consider it purposeful to entrench the existing, inefficient transport infrastructure by adapting bioenergy use in the form of biofuels to this structure. Various stakeholders pursue the goal of deploying bioenergy in the transport sector as a source of innovation for technology and climate change mitigation; this leverage needs to be targeted carefully. Innovations should be directed towards electromobility – a form permitting the efficiency of the drive system to be boosted from its present level of approx. 20 per cent in combustion engines to more than 70 per cent if the electricity is generated from renewable sources. Hybridization (the combination of combustion engines and electric drives) has an important role to play in the process of transition from combustion engines to electric drives. The IPCC has identified the further development of electric and hybrid drives with stronger and more reliable (i.e. more durable) batteries as a key technology for emissions reduction in the transport sector (IPCC, 2007c). WBGU therefore recommends, for reasons of both technical efficiency and environmental benefit, that in industrialized countries biomass is deployed via conversion to electricity to drive electric road and rail vehicles, rather than entrenching an inefficient technology via biofuels.

8.1.2.2
Bioenergy for central and decentral heat supply

The use of biomass for heat supply is often recommended. The climate change mitigation effect of such a use follows from the substitution of the displaced conventional fuels. In an energetic perspective, heat used for space heating has a substantially lower value than the same quantity of energy in the form of elec-

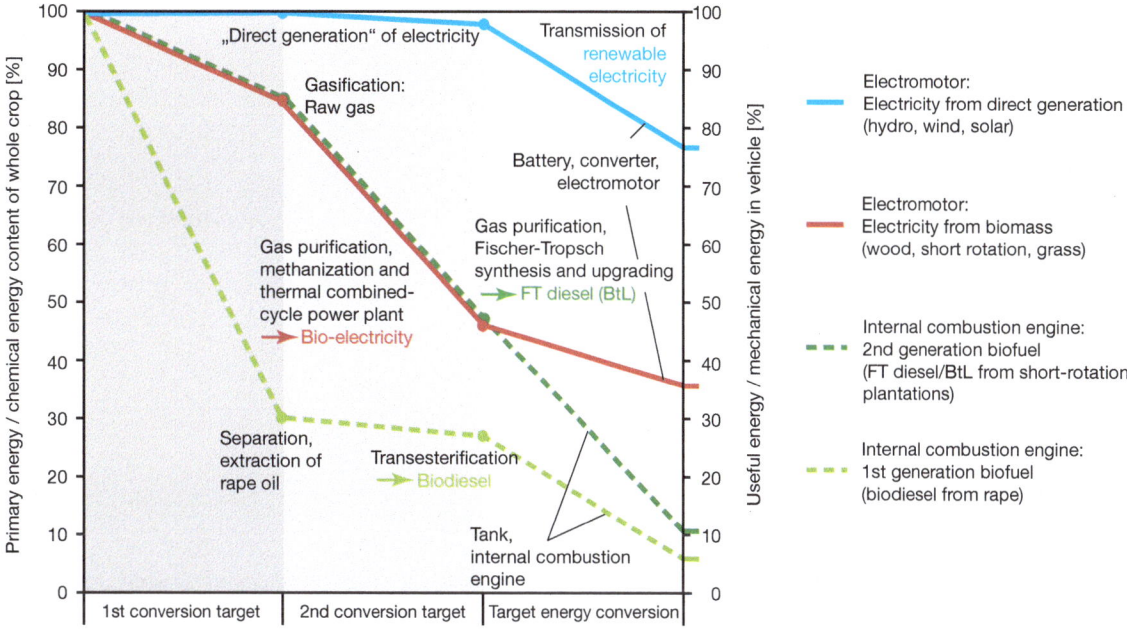

Figure 8.1-9
Comparison of different conversion pathways in the transport sector in terms of the mechanical energy utilizable at the wheel. The diagram only considers the main energy flows. It does not take account of by-products such as the useful heat extracted from CHP processes, nor of material products such as fertilizer or animal feed. In the best case, approx. 75% of the primary energy from hydro, wind and solar sources can be utilized as shaftpower in the electric vehicle. In the worst case, only a good 5% of the primary energy contained in the rape crop is utilized in the combustion engine. When biodiesel is produced from rape, the rapeseed is separated from the straw, meaning that approx. 50% of the chemical energy contained in the crop is lost to fuel production. Approx. 60% of the energy contained in the seed is extracted as rapeseed oil, and 90% of the oil is esterified to biodiesel. Overall, approx. 25% of the energy contained in the crop is filled as biodiesel fuel, of which in turn only 20% is converted by the combustion engine to shaftpower. Ultimately, therefore, only a good 5% of the original energy contained in the crop is utilized as shaftpower. The balance is not substantially better for 2nd-generation biofuels. Here the whole crop can be used and approx. 85% of it converted to raw gas, but major conversion losses occur in gas processing and synthesis, with the result that only approx. 45% of the energy originally contained in the crop is filled as Fischer-Tropsch diesel and thus only approx. 10% of the energy in the crop can be utilized as shaftpower. If the raw gas is converted on the same initial pathway to biomethane which is then used to generate electricity in efficient combined-cycle power plants, a good 45% of the energy contained in the crop can be charged as electricity in the electric vehicle; with 80% efficiency of conversion in the vehicle, overall a good 35% of the energy originally contained in the crop can be utilized as shaftpower.
Source: WBGU, using calculations from Ahmann, 2000; Dreier and Tzscheutschler, 2000; Kaltschmitt and Hartmann, 2003; EAA, 2007; Engel, 2007; Sterner, 2007

tricity or mechanical energy. It follows that if the requisite heat can be supplied from the waste heat extracted from CHP systems or can be provided by means of heat pumps, greater exergetic value can be delivered from the biomass than through pure combustion (Fig. 7.2-3). Using biomass at lower temperatures below 100°C to produce hot water and for space heating is to be seen as a transitional application in industrialized countries (heating systems firing wood, woodchips and pellets). Heat is supplied most efficiently by means of CHP and heat pumps (Section 8.1.1). Decentral CHP based on biomass is relatively problematic. Woody and moist biomass can only be made available for distributed use via the route of gasification or fermentation to biomethane (Section 7.2). WBGU therefore recommends utilizing the waste heat from CHP facilities, deploying biomethane in decentral CHP systems and, over the

long term, deploying heat pumps running on renewably generated electricity.

8.1.2.3
Bioenergy for electricity generation: Control energy and cogeneration

Biomass is most purposefully used in electricity generation. This is because, firstly, it can be deployed to produce control energy in the same way as fossil energy carriers, and, secondly, its use in combined heat and power production (cogeneration) or in combined-cycle power plants provides the highest fuel utilization factor and the greatest GHG abatement performance per unit biomass consumed. Moreover, some pathways, such as co-firing pellets in coal-fired power plants or utilizing residues and

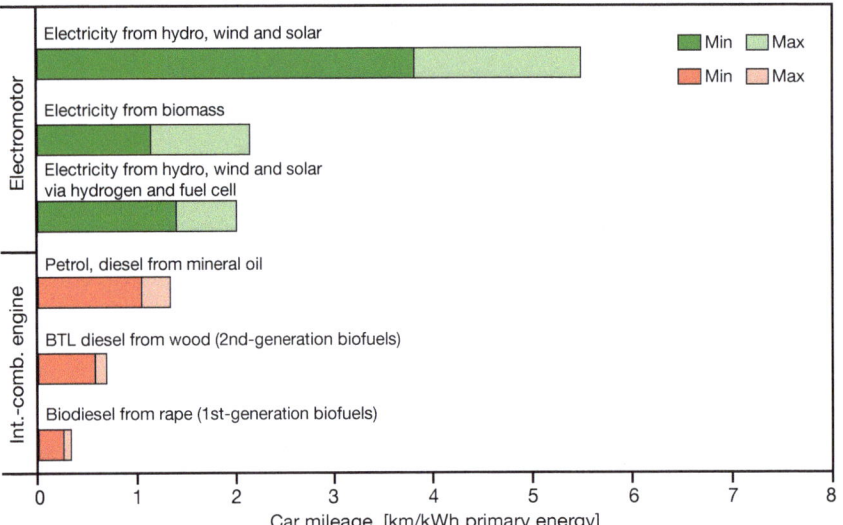

Figure 8.1-10
Car mileage per unit of primary energy. Thanks to the substantially greater efficiencies of electromotors compared to combustion engines, electromobility generally results in greater mileages per unit primary energy input. The calculations assume the consumption of an electric vehicle to be 15–22 kWh_{el} and that of a conventional vehicle to be 60–80 kWh_{fuel} (corresponding to 6–8 litres diesel) per 100 km. The conversion rates of wood to Fischer-Tropsch diesel (BtL) are assumed to be 40–50% and those of rape to biodiesel 20–30%. The thermochemical conversion of biomass to biomethane with subsequent electricity generation was calculated for an efficiency range of 30–60%. Direct electricity generation using hydro, wind and solar sources has a conversion rate of 96–100% and the use of hydrogen-based electricity has a range of 35–42%, calculated from the conversion rate of direct generation and the efficiencies of electrolysis (70%) and a polymer electrolyte fuel cell (50–60%). It follows that bioenergy can be utilized much more efficiently in the transport sector via the electromobility pathway than via the path of fuels in combustion engines. The greatest mileages are achieved with directly generated renewable electricity.
Source: WBGU, using calculations from Ahmann, 2000; Dreier and Tzscheutschler, 2000; Kaltschmitt and Hartmann, 2003; EAA, 2007; Engel, 2007; Sterner, 2007

wastes in biogas facilities, have very low abatement costs (Sections 7.2 and 7.3). In an electricity supply system based entirely on renewables, control energy is required to balance out the intermittent feed-in of electricity from wind farms and solar power plants in electricity grids. Deployed in conjunction with massively expanded electric grids that have large transmission capacities, together with electricity storage systems such as pumped-storage hydropower or compressed-air systems and the integration of electric vehicles, control energy generated from biomass sources can serve to level the intermittent feed-in from other renewable sources and the fluctuating electricity demand of consumers. This is where the special attribute of biomass as a medium that stores energy comes to the fore. Biomethane, in particular, is a suitable energy carrier, because it can be produced from a great number of wastes and residues as well as from the most varied energy crops, and can also be stored well in gas networks. Taken from there, it can be used in existing gas-fired power plants designed to generate control energy in the same way as natural gas. The losses arising in the process of conversion from biomass to biomethane are offset by the great versatility provided in conjunction with the natural gas network (storage functions, decentral CHP, etc.).

Because of its high fuel utilization, CHP makes optimum use of both fossil and biogenic fuels in energy terms. For decentral applications, the conversion of gaseous and liquid bioenergy carriers in small-scale CHP units is an appropriate route. Beside biogas, fuels such as vegetable oil, biodiesel or bioethanol can also be used in CHP systems, where, due to the use of waste heat and the greater conversion efficiency, they deliver a much greater climate change mitigation effect than when used in the transport sector (Section 7.3; Fig. 7.3-4). Solid bioenergy carriers do not yet play any dominant role in decentral CHP. This is because the technology (gasifiers, Stirling engines) are partly not yet mature or are still uneconomic (FNR, 2007b). In contrast, the use of small cogeneration systems in municipalities is a route that is already mature and can be pursued expediently in distributed applications, generating electricity and making use of waste heat to supply heat to public buildings such as swimming pools or schools. Such systems procure the biomass locally, for example in the form of woodchips or wood pellets. In industrial structures, biogas facilities can be integrated beside cogeneration systems in such a way that heat can be usefully extracted (e.g. for drying systems, vegetable production, or manufacturing; FNR, 2006d; Roy, 2008). A further prospect for the future is the efficient use of decentral mini-CHP units in households.

As concerns large-scale power plants various options arise to make use of bioenergy: these include co-firing in conventional power plants; use in central cogeneration plants or in power plants with integrated fuel gasification, or the supply of biogas (bio-methane) from biogas production facilities (fermentation process) and large gasifiers that convert wood and plastics wastes. Both systems involving fermentation and gasification facilities generate CO_2 at high concentrations, which needs to be captured before the gas is fed into natural gas networks. At this point, already available technologies permit a first step towards decarbonization, making it possible to halve the CO_2 emissions of these conversion processes if the CO_2 captured is then consigned to storage. Numerous further CHP technologies that can use biomass, such as Organic Rankine Cycle systems, are available and technologically mature. The precondition to broad-scale use of the heat extracted is proximity to a large heat consumer or, alternatively, the intensified expansion of heat networks.

8.1.2.4
Overall assessment of bioenergy in industrialized countries

The conclusion for the optimum integration of biomass in future energy supply systems is that it should not be used as fuel to drive vehicles, and should only be deployed in a transitional period in direct combustion for heat production. In the energy systems of the future, suitable biomass uses are rather as follows: to generate electricity in CHP facilities such as central cogeneration plants and small-scale CHP units, to be co-fired in coal-fired power plants, and to be used in combined-cycle power plants with maximum efficiency. By consistently implementing the systems presented above and making use of electric energy in the transport sector, efficiencies can be improved by more than a factor of two. Moreover, wastes from bioenergy use, such as ash or fermentation residue, can be recycled as mineral fertilizer. This is an important criterion of the sustainability of bioenergy use.

Biomass, and particularly residues, can thus be deployed in an optimum manner via the fermentation (biogas) and gasification (raw gas) pathways (Fig. 8.1-11). Gasification processes allowing conversion of almost all forms of biomass are presently still at the development stage, but promise a broad range of applications in future. Biomethane is in any case the main product of the fermentation processes established today, which are already highly flexi-ble. These two pathways allow production of biomethane, which can be fed decentrally into existing natural gas networks as synthetic natural gas and can in turn be converted in decentral processes for power and heat supply. Leakage of methane, a potent greenhouse gas, must be prevented as far as possible at all points in the chain. This pathway presents the opportunity of universal use of biomethane in regions with well developed natural gas networks, especially in distributed CHP with small unit capacities and high fuel utilization factors. Compared to other forms of bioenergy use, the production and use of biomethane achieves a substantially greater climate change mitigation effect (on the scale of 20%; Table 7.3-2), if the CO_2 which needs to be separated in any case during the conversion process is not released to the atmosphere but is stored securely. WBGU has discussed the risks of and criteria for sustainable CO_2 storage elsewhere (WBGU, 2006). A further aspect is that residues such as those from forestry operations cannot be utilized in decentral CHP in households at present. Only the biomethane pathway makes it possible to consign numerous residues to more efficient use with high climate change mitigation effect.

Figure 8.1-11 summarizes the post-transformation elements of energy systems and the specific role of bioenergy. A particularly attractive feature is that the two classic energy carriers – natural gas and coal – can be integrated within this future system. In particular, the use of coal via the gasification pathway with subsequent conversion to synthetic methane of the synthesis gas allows, due to the separation of CO_2 which is requisite in this process in any case, a use of coal that generates much less GHG emissions than would be possible when it is used in conventional coal-fired power plants. Through carbon capture and storage (CCS), the emissions levels of coal use approach those of natural gas use. Coal gasification technology is still a niche technology at present, used successfully in a small number of facilities in the USA (IPCC, 2005). As does CCS technology, it needs to be further developed and improved before it can be deployed on a broad scale.

8.1.2.5
Stages en route to sustainable bioenergy use in industrialized countries

WBGU views climate change mitigation as the principal goal of bioenergy use in industrialized countries. It follows that residue use and cascading use have top priority, as these scarcely cause any land-use changes and the associated emissions. In the view of WBGU, energy crops should only be used if their climate change mitigation effect inclusive of emissions

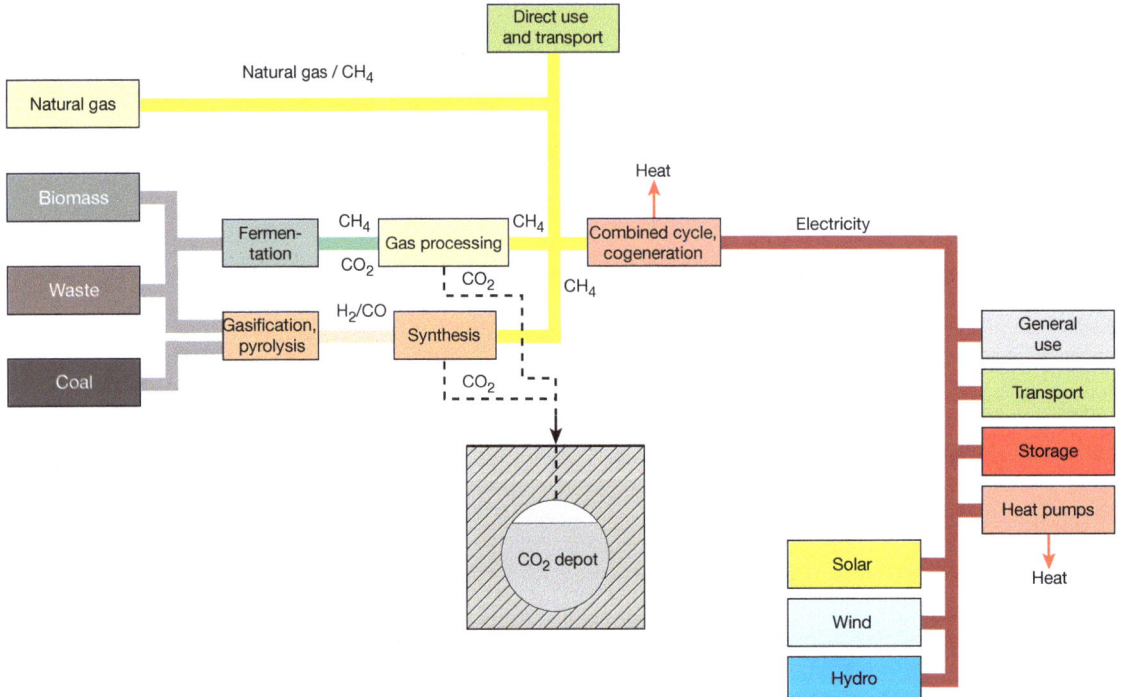

Figure 8.1-11
Future, sustainable energy supply structures in industrialized countries. Building upon an integrated gas and electricity network and the capture and storage of the CO_2 arising in the use of fossil and biogenic fuels, energy supply is largely via electricity. Useful heat is supplied by cogeneration processes and by heat pumps. In biomethane production, CO_2 capture is essential and its storage feasible.
Source: WBGU

from land-use change has been proven to be particularly positive and the sustainability criteria of the minimum standard (Section 10.3) are complied with. The use or import of biomass derived from energy crop cultivation that does not meet these criteria should not be permitted.

Initially those applications appear most attractive in which energy carriers with high CO_2 emissions are displaced. These are above all hard coal and lignite, but also mineral oil. This is the case if, for instance, wood is co-fired in coal-fired power plants or if it is used to heat buildings. While using wood to fuel power plants reduces demand for coal accordingly, its use to heat buildings and provide process heat frees up corresponding quantities of oil and gas that can then be deployed for a transitional period in the transport sector. Due to conversion performance in the heat sector, more mineral oil can be substituted there than in the transport sector (Fig. 8.1-12). This is the case if, for instance, direct combustion of wood in a heating system substitutes mineral oil. If the same quantity of wood is converted to Fischer-Tropsch diesel (BtL), it can only substitute half as much mineral oil, because the other half of the energy is required for the conversion process itself.

With increasing efficiency of global energy systems – in which, beside improved thermal insulation, building heating and process heat are supplied by means of cogeneration and heat pumps, and electromobility is well-established in the transport sector – the requirement for direct combustion and fuels drops. Demand for electric energy rises in parallel. In this situation, which could emerge in several decades, most of the electricity is generated directly from hydro, solar and wind sources. Their output, however, is subject to substantial fluctuations over time – the biogenic and fossil energy carriers then increasingly have the task of balancing these fluctuations in output. This can be done either by means of decentral cogeneration or with rapidly dispatchable combined-cycle power plants. Today, both power plant types operate mainly on natural gas (i.e. methane). This makes it expedient to utilize biogenic methane (biomethane), which can be conveyed by the natural gas network infrastructure already in place today. In the ideal case, the use of biomass proceeds in two stages: initially, biomass is co-fired in coal-fired power plants or is deployed in heating systems (Fig. 8.1-12). Later on, the biomethane produced by fermentation and gasification is fed increasingly into natural gas networks and is used in cogeneration processes to pro-

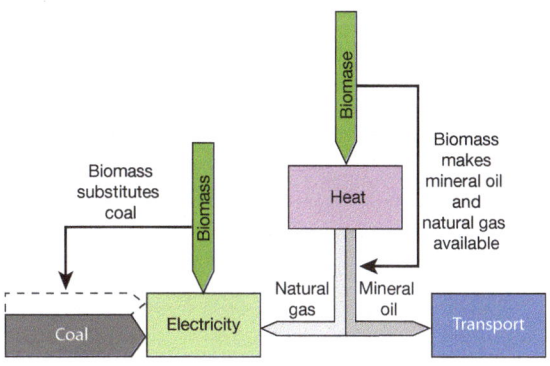

Figure 8.1-12
First stage of sustainable bioenergy use in industrialized countries. Biomass is deployed primarily where it can directly substitute fossil energy carriers that have high emission levels without incurring major conversion losses or costs – thus primarily to displace coal through co-firing in coal-fired power plants. In the heat sector, natural gas and mineral oil are freed up, which become available for use in a transitional period in, for instance, the transport sector.
Source: WBGU

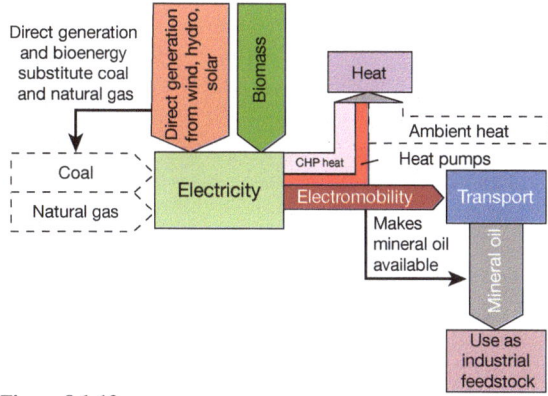

Figure 8.1-13
Second stage of sustainable bioenergy use in industrialized countries. In future, sustainable and integrated energy supply systems, direct generation and bioenergy replace fossil energy carriers in electricity generation. By utilizing the waste heat from cogeneration processes and making use of ambient heat, heat supply can be ensured on the basis of renewable electricity. Electromobility makes mineral oil available that can be used as industrial feedstock.
Source: WBGU

duce electricity and heat. The electricity generated in this manner is also used in electromobility and in heat pumps (Section 8.1.1; Fig. 8.1-13). Electromobility in turn frees up mineral oil, making it available for use as industrial feedstock to produce plastics, pharmaceuticals and other carbon-based products.

8.2
Bioenergy as a part of sustainable energy supply in developing countries

In developing countries, as in industrialized ones, bioenergy will play a complementary role in conjunction with other renewables. Increasingly, electricity generation in rural areas is involving photovoltaic units, small windpower systems and mini-hydropower systems that are used in combination with biofuel generators to provide a secure supply. Replacing classic diesel generators with biogenically powered CHP systems enables locally produced biofuels to be used while at the same time providing useful heat. This section specifically examines the role of biomass in developing countries; other energy sources that might be used to overcome energy poverty are not the subject of consideration here. Biomass is the fuel that currently forms the principal source of energy in most rural areas, and techniques of handling and using it are well established. This contrasts with the finding, based on experience, that the spread of new technologies such as photovoltaics is slow because the systems often cannot be maintained by the users

themselves. Since firewood is a familiar fuel, technologies based on it are better able to establish a foothold in real-life applications.

8.2.1
A revolution in traditional biomass use

The majority of traditional biomass use takes place in the rural areas of developing countries where poverty and a low level of technological development are the norm. New, more efficient technologies for using bioenergy must therefore be simple to install, use and maintain, as well as being affordable for the population. If these conditions are met, modern bioenergy use provides opportunities for creating jobs in rural areas, increasing the degree of comprehensive energy coverage, reducing import dependency and – through the use of more efficient technologies – reducing deforestation and energy-linked greenhouse gas emissions. Replacing traditional stoves with efficient wood stoves can reduce greenhouse gas emissions by about 60 per cent for the same level of use; in case of replacement with micro biogas devices this figure rises to 95 per cent. These reductions are the result of more efficient combustion that releases less methane and nitrous oxide, or of avoided deforestation (Section 7.3). To promote the spread of modern bioenergy use, two strategies need to be pursued in parallel: firstly, efficient low-cost technologies must be made available and, secondly, the sociocultural constraints – which are sometimes considerable – that

Box 8.2-1

Health-related and ecological impacts of traditional biomass use

The modern use of bioenergy has great potential for reducing inefficient traditional biomass use that has adverse impacts on health and the environment. This can substantially improve living conditions in private households and micro-businesses, particularly in the rural areas of developing countries (WBGU, 2004a). In many developing countries biomass is still used mainly in traditional, inefficient ways for cooking and heating. Some 2500 million people (52 per cent of the population of the developing countries) depend on biomass as their primary source of energy and have no access to efficient technologies. Fuelwood, charcoal, agricultural waste and animal dung are burnt to provide heat. In many countries of Africa and Asia these energy carriers are used to meet up to 95 per cent of domestic demand for heat (Table 8.2-1; IEA, 2006b).

The International Energy Agency (IEA) forecasts that unless appropriate action is taken the number of people dependent on traditional biomass use and on inefficient ways of using it will rise to 2600 million by 2015 and to 2700 million by 2030 (IEA, 2006b). Traditionally, fuels are usually burnt highly inefficiently in three-stone hearths or other simple stoves under poor stoichiometric conditions; this results in severe indoor pollution from soot particles and other harmful substances and leads to the formation of poisonous carbon monoxide. The damage to health that this causes is considerable. Each year more than 1.5 million people – two-thirds of them in south-east Asia and Africa – die from the consequences of indoor pollution (WHO, 2006). More than 1.3 million of these deaths arise from the use of biomass; the rest are attributable to the use of coal (IEA, 2006b). This means that traditional biomass use causes

Table 8.2-1
People who are dependent on biomass as the primary source of energy for cooking.
Source: adapted from, 2006b

	Land		Stadt	
	[%]	Mio.	[%]	Mio.
Southern Africa	93	413	58	162
North Africa	6	4	0,2	0,2
India	87	663	25	77
China	55	428	10	52
Indonesia	95	110	45	46
Rest of Asia	93	455	35	92
Brazil	53	16	5	8
Rest of Latin America	62	59	9	25
Total	**83**	**2,147**	**23**	**461**

more deaths than malaria (which kills around 1.2 million people per year).

An additional consideration is the large distances that must frequently be covered in order to collect fuel. The task falls usually to girls and women, exposing them to risk and occupying time that might otherwise be available for education or economically profitable activity. In addition, through increasing clearance of forests and destruction of the steppes this type of biomass use contributes to the degradation of natural ecosystems and to climate change and in the long term reduces the development prospects of the regions in which it is used (WBGU, 2004a).

impede the use of these technologies must be overcome.

8.2.2
Supplying energy in rural areas with the aid of modern biomass use

IMPROVED WOOD STOVES
It is more efficient to use wood directly in improved wood stoves than to carbonize it to make charcoal that is then burnt in charcoal stoves (Section 7.2). Improved wood stoves range in their construction from very simple clay stoves that cost almost nothing to stoves of metal. They are therefore available in principle to all population groups, provided that sociocultural barriers do not stand in the way of their introduction (Section 10.8). Since they enable fuelwood consumption to be reduced – depending on the type of stove used – to a half or a quarter of the previous level, they represent a simple and cost-effective means of using bioenergy more efficiently and hence

in a more environmentally friendly manner (Sections 7.2 and 9.2.2; Box 8.2-2; Kumar et al., 1990).

MICRO BIOGAS UNITS
Micro biogas units can also help to reduce fuelwood consumption. Biogas units have additional advantages over improved wood stoves in that they are associated with better indoor air quality, a greater reduction in greenhouse gas emissions (Section 7.3; Boxes 7.3-3 and 8.2-1) and better fertilizer for agricultural purposes. Studies from Nepal show that women can save up to three hours' work a day when they cook with biogas, since they do not need to spend time gathering fuel and the cooking itself is easier (ter Heegde, 2005). In addition, the improved air quality reduces expenditure on medication and increases life expectancy. Biogas can also be used to provide lighting. This technology is not the most efficient, but it represents a good compromise between ease of handling, installation and maintenance, the energy benefit and the costs of the energy service. In Nepal and Vietnam more than 200,000 micro biogas units are already in use. In India more than 4 million such units

Box 8.2-2

Country study: Uganda – Tackling traditional bioenergy use through active bioenergy policy

Uganda is a small, relatively densely populated landlocked country in east Africa; its energy system, in which traditional biomass use plays a large part, is typical of countries in sub-Saharan Africa. 93 per cent of the country's primary energy requirement is met through traditional biomass use. Fuelwood accounts for 82.4 per cent of energy use, followed by charcoal, which accounts for 5.8 per cent, and biogenic residues (mainly sugar cane bagasse), which accounts for 5.0 per cent. Almost 80 per cent of the biomass used is deployed in the home for cooking and water heating. Biomass is the most important energy carrier in Uganda and will remain so for the foreseeable future (Turyareeba and Drichi, 2001).

In rural areas, where almost 85 per cent of the population lives, three-stone hearths are used almost exclusively. In larger settlements and towns, however, charcoal stoves made of metal, which are almost twice as efficient as the three-stone hearths, are also used (Pesambili et al., 2003). In large areas of the country there are no energy supply structures, such as an electricity grid (World Bank, 2008f). Small quantities of agricultural residues such as bagasse and fuelwood are used in industry to meet the need for power and heat. Electricity in Uganda is generated mainly from hydropower; excess capacity is exported to areas such as western Kenya. Bioelectricity, by contrast, is generated almost entirely in CHP plants and used within the country.

The proportion of petroleum-based fuels such as diesel, petrol and kerosene in the national energy budget is very low at 6.0 percent; electricity generation, at 0.8 per cent, is even lower. All fossil fuels are imported, since Uganda has no oil refinery (MEMD, 2004). The government has recently approved a new Oil and Gas Policy which regulates exploitation of the country's crude oil resources. From 2009 the policy envisages production of 4000 barrels per day, which will cover more than a third of the national requirement. The petroleum produced will be used in the transport sector and for electricity generation (Olaki, 2008).

Access to sustainable modern energy services forms part of the Ugandan government's poverty reduction strategy. The issue is addressed in two strategy papers, the Uganda

Energy Policy (2002) and the country's Renewable Energy Policy (MEMD, 2007). The main aim of the Uganda Energy Policy is to meet the energy requirement needed for social and economic development in an environmentally friendly way (MEMD, 2002). One of the aims of the Renewable Energy Policy is to improve the socio-economic situation of, in particular, women and the poor (MEMD, 2007). As in other developing countries, the objectives of development, supply security and independence of supply are in Uganda regarded as in principle more important than climate change mitigation. Alongside its role in rural households, biomass would be of particular interest for the fuel sector, because electricity can also be generated from other energy carriers and there is little requirement for heat for uses other than cooking. Since one-third of the fuel requirement is due to be met from the country's own sources of crude oil, a biofuel programme is not a high priority for the state planners.

The government has in recent years attempted to improve energy efficiency and has implemented a programme for the introduction of improved wood stoves. The Energy Advisory Project (EAP) of the Ugandan Ministry for Energy and Minerals aims to replace three-stone hearths with improved wood stoves. The wider use of these wood stoves is linked with the training of craftsmen and the provision of instruction to users. At the start of the century approx. 125,000 households – 2.7 per cent of all the households in the country – were equipped with improved technology (Turyareeba and Drichi, 2001). Through the EAP, which is supported by GTZ, and despite rising population numbers this figure has now risen to 357,500 households or 8 per cent of the total (GTZ-EAP, 2007). In 2007 Uganda set itself the goal of increasing the number of improved stoves to 4 million by 2017 (REN21, 2008).

The EU Energy Initiative for Poverty Eradication and Sustainable Development and GTZ regard Uganda's endeavours in the area of bioenergy very positively: a coherent and trans-sectoral biomass strategy has been developed and significant progress has been made in implementing it (Teplitz-Sembitzky, 2006). However, on account of the governance problems in relation to nature conservation, it remains questionable whether the potential interest in the production of biofuels can be implemented by the Ugandan government in a sustainable manner (Biryetega, 2006; NFA, 2006; ABN, 2007).

have been installed since 1980; more than 70 per cent of these are still in use. Households in Nepal have on average avoided 5 t CO_2eq greenhouse gas emissions annually by using biogas units instead of traditional biomass (ter Heegde, 2005). This emissions reduction results from the avoided methane emissions of dung used in the unit and from the substitution of fertilizers and of fuelwod harvested in non-sustainable ways. Micro biogas units are particularly suitable for households, schools and public institutions; their use reduces the community's dependence on fuelwood.

Biomass gasification

Biomass gasifiers can convert residues and wastes such as end-of-life wood, coconut shells, coffee chaff and rice chaff into raw gas. The raw gas can be used, for example, to generate process heat in drying plants

or bakeries, or it can be converted to electricity in a generator. Wood gasifiers of this type can drive generators in rural areas and help to advance rural electrification. According to TERI (2008) they are economically efficient, environmentally friendly and appropriate for rural communities, because local waste can be utilized and the technology can be maintained without outside help. For households small gasifiers are also available. If residues are used in this way, less soot is produced and indoor air pollution is reduced. Particularly exemplary and successful programmes are being conducted in India, where up to 1 million micro and small business are benefiting from this technology (Mande and Kishore, 2007).

BIOMASS COMBUSTION

In many branches of industry in developing countries large quantities of residues arise. Residues and wastes such as bagasse from sugar manufacture or end-of-life wood can be used directly in CHP to generate power and provide heat for drying. Other areas of industry in which biogenic residues arise are distilleries, textile and paper factories and food processing plants. Small CHP systems can convert wastes such as maize cobs, peanut shells, rice chaff, coffee chaff and sawdust into power and heat. In India and other countries systems of this type are used, for example, in sugar factories; skilful integration enables them to generate all the electricity they need for their own needs (MEMD, 2007). If the efficiency of these systems were to be improved and a small electrical grid set up, surrounding households could also be supplied with electricity.

VEGETABLE OIL FOR LOCAL PROVISION OF
ELECTRICITY AND MECHANICAL ENERGY

Diesel generators are in widespread use in developing countries. These generators can fairly simply be converted to run on locally produced vegetable oil. Oil plants such as Jatropha can be grown on marginal land for this purpose, with the fruits being pressed manually. Such a pathway is simple to implement, because mechanical presses or a motor are easier to repair than a photovoltaic unit or a wood gasifier. The converted plant-oil motors can then provide mechanical energy for water pumps or grain mills, or they can be connected to an electricity generator. Portable batteries can then be charged at village charging stations and used to light houses and huts or to provide distributed power for small devices such as mobile telephones and radios.

8.2.3
The role of bioenergy in the sustainable and integrated energy supply of developing countries

8.2.3.1
Bioenergy for transport

Many developing countries are highly dependent on petroleum imports. They are now hoping that biofuels will reduce this dependency and create an additional source of income in rural areas (Box 8.2-3). First-generation biofuels such as bioethanol or biodiesel are straightforward to produce in developing countries. Cultivation systems based on oil palms or sugar cane are state-of-the-art and an option for fuel manufacture. Vehicles can be converted to run on vegetable oil. Alternatively, as is happening in Brazil, it is

possible to build 'flexible fuel' vehicles that can run on both bioethanol and petrol.

With the help of oil plants such as *Jatropha* and *Pongamia*, which can be grown on marginal land that is unsuitable for food production, the problem of competition between energy crops and food for land can be largely avoided. In developing countries the use of degraded land for biofuel production is appropriate as a transition measure until electric transport has become established and liquid fuels are therefore no longer required. Electrically powered two-wheelers represent a first stage in the introduction of electric transport. Several million electric bicycles are already in use in China. Another step in the right direction would be an expansion of public transport in newly industrializing countries to make greater use of trams and trains powered by green electricity since this, too, would reduce liquid fuel consumption.

Biofuels can in general be used more efficiently in the electricity sector. For example, bioethanol can be burned in the turbine of combined-cycle power plants, and biodiesel or vegetable oil can be used in small-scale CHP units. The decision on whether to use biofuels in transport or in the electricity sector should be based on the share of fossil energies in electricity generation. Where fossil electricity has a large share in the electricity mix, as for example in China, biofuels can achieve their greatest climate change mitigation effect by being converted into electricity (Sections 7.3 and 9.2). In countries such as Uganda that meet their electricity requirements almost exclusively from renewable energies such as hydropower, biofuels can achieve a greater climate change mitigation effect by being used to replace fossil petroleum in the transport sector.

8.2.3.2
Bioenergy for heat and light

The most important energy service in developing countries is the provision of heat for food preparation and heating. Modernization of traditional biomass use is therefore a key element in achieving a sustainable energy supply. The production and maintenance of improved wood stoves and of micro biogas units and the associated biogas stoves requires no more than simple craft skills; in South-East Asia this approach has enjoyed success for a number years, with many advantages for all participants (ter Heegde, 2005; ADB, 2008).

Industrial process heat can be produced directly by wood gasifiers, preferably from residues (Dasappa et al., 2003). This represents an alternative to the previous form of heat generation, involving, for example, firing with natural gas or mineral diesel. In the longer

Box 8.2-3

Development opportunities presented by bioenergy production for supra-regional internal markets and exports

The extent to which the large-scale deployment of biogenic energy carriers can bring major benefit for the economy of the country in which they are produced and prove sustainable is hotly debated (Peskett et al., 2007). The advantages of a shift to modern bioenergy on a scale significant for the economy as a whole include the diversification of energy sources and technologies and greater independence of price fluctuations on the international oil market. This is particularly important for developing countries that import oil (Kojima and Johnson, 2005). Many emerging and developing countries have therefore set targets for the production and use of bioenergy or plan to do so. These countries include China, India, South Africa and many developing countries in south-east Asia, west and east Africa and South America (Section 4.1.2). Brazil has succeeded – partly through the expansion of bioethanol production – in becoming largely independent of oil imports (Box 8.2-4; Luhnow and Samor, 2006).

Many countries also hope that the export of biogenic energy carriers will give rise to development opportunities (UNCTAD, 2006b). By comparison with industrialized countries, many developing countries have comparative cost advantages in the production of agricultural goods, including energy crops. Exporting these goods brings in foreign currency, creates income and generates direct or indirect state revenue, thereby promoting economic growth. The export of biogenic energy carriers can only succeed if exporters have access to the markets of potential importing countries; in other words, potential importing countries must not impose major restrictions on imports and must not attempt to nullify or even reverse the comparative cost disadvantages of their producers through state (agricultural) subsidies (Worldwatch Institute, 2006). If such subsidies in the industrialized countries were removed and if the demand for 1st-generation biofuels were to rise, a general rise in agricultural prices could be expected. At national and international level this would increase production incentives and incomes for farmers in the developing countries (Section 5.2.5.2), with beneficial consequences for developing countries in both cases.

However, experience to date has shown that biofuels are rarely competitive, even in developing countries. Most biofuel programmes require high levels of subsidy (Section 4.1.2). Brazil is so far the only country to have developed a viable ethanol industry, and this required 20 years of state support (Box 8.2-4; Kojima and Johnson, 2005). It is fundamentally questionable whether the use of public funds for large-scale bioenergy production is economically justifiable and whether it might not be better to apply these funds to other purposes such as education, health, poverty reduction or infrastructure development. Another factor to be borne in mind is that the beneficiaries of biofuel production subsidized by the taxpayer are mainly large agricultural businesses; this makes little contribution to the reduction of poverty among the rural population. However, if it is assumed that technology will advance, that the price of oil will tend to rise and that blending quotas will be maintained and implemented on a wider scale, then the production of biofuels, including biofuels on the basis of raw materials other than sugar cane, is very likely to become economi-

cally worth while in other countries (de La Torre Ugarte, 2006).

Because agriculture in developing countries is still very labour-intensive, the production, transport and processing of plant raw materials create jobs – although some may only be seasonal – and generate income (Kojima and Johnson, 2005). The infrastructure required for production, transport and processing can give rise to further positive development effects. In order to ensure that these positive effects reach rural areas, it is often argued by NGOs and others that promotion must in particular target small-scale and cooperative production, in order to avoid the disadvantages of the large-scale production of cash crops.

It is, however, evident that the blanket objection that large-scale cash crop production inevitably leads to very poor working conditions and exploitation of the workers is untenable. For example, in the Brazilian province of São Paulo sugar-cane cutters were receiving a wage of around US$ 140 per month as far back as the early 1990s. This meant that their wages were higher than those of 86 per cent of all agricultural workers and 46 per cent of all industrial workers (UNCTAD, 2006b).

There are advantages to be gained from mass production in both the production and the processing of agricultural goods. In consequence, it can be assumed that both domestic and foreign investment will tend to flow into large-scale projects. There are fears that this will give rise to further concentration processes in agriculture and in land distribution, increasing the hold of elite groups within the country and large transnational companies and encouraging the spread of non-sustainable monocultures (ABN, 2007; Biofuelwatch et al., 2007). These concentration processes seem to be particularly vulnerable to corruption and may be accompanied by the exploitation of small farmers who are contract producers, displacement of farmers in situations in which land rights are unsecured, violence, rising land prices and environmental damage as a result of large-scale clearance (Kojima and Johnson, 2005; Misereor, 2007). Measures such as cooperatives and specific promotion programmes can help to ensure that smallholders and small businesses are able to benefit from the bioenergy boom. For example, Brazil has instituted a biodiesel programme directed specifically at small farmers: in order to qualify for a 'social seal' and the associated tax concessions, some of the raw materials used by the biodiesel producers must come from family-run farms. The programme has not yet been quite as successful as was hoped; this is because biodiesel is produced mainly from soya, the cultivation of which is controlled by large-scale producers. Nevertheless, the programme's supporters regard the difficulties as being merely startup problems (Fatheuer, 2007).

Appropriate bioenergy policy is not in itself enough to ensure that large-scale bioenergy production is sustainable and promotes development for broad sectors of the population; well-functioning public institutions and good governance are also necessary. The required factors include in particular effective administrative and legal structures, the assurance of legal certainty and the avoidance of corruption, fair and secure distribution of disposal rights and – in particular – land rights, fair opportunities for participation including economic rights and effective environmental and employee protection legislation.

Box 8.2-4

Country study: Brazil – a newly industrializing country with a long-standing bioenergy policy

Brazil consumes 40 per cent of the energy used in South America and is thus the largest energy consumer on the subcontinent. The primary energy supply is met mainly from petroleum (42.1 per cent in 2006). Other important sources of energy are hydropower (14.2 per cent), energy from sugar cane (16.6 per cent) and traditional biomass use. Natural gas and coal, which account respectively for 8.3 per cent and 1 per cent, play a minor role. Nuclear energy accounts for 1.1 per cent of the energy mix, renewables other than bioenergy account for 3.2 per cent (World Bank, 2007; MME and EPE, 2007). The electricity sector is dominated by hydropower (ca. 77 per cent of electricity generation in 2006); the rest is generated from biomass, gas, coal and nuclear energy or covered by imports. Brazil's energy supply is thus highly diversified and includes a relatively large proportion of renewable energies (ca. 45 per cent of the primary energy supply in 2006; IEA, 2006a; GTZ, 2007a; MME and EPE, 2007).

Favourable climatic conditions and the availability of much potentially usable land make the cultivation of energy crops, especially sugar cane, particularly attractive in Brazil. Biomass can therefore make a significant contribution towards meeting Brazil's increasing energy requirements. Brazil already produces ca. 2 EJ of bioenergy per year, making it the world's fourth-largest user after China, India and the USA (GBEP, 2008). In 2006 4.1 per cent of electricity from biomass, most of it from sugar cane bagasse, was generated by industry for its own use. In that year the country's 320 sugar and ethanol mills processed 430 million tonnes of sugar cane for sugar and ethanol production (MME and EPE, 2007; GTZ, 2007a; GBEP, 2008). Around 50 per cent of Brazil's petrol requirement in the transport sector is currently met by the ethanol that is produced; a further 3.5 million litres of ethanol are exported (WI, 2007; GBEP, 2008). In 2007 Brazil produced around 19,000 million litres of ethanol, making it the world's second-largest producer after the USA (26,500 billion litres; quoted in OECD, 2008).

The fact that so much ethanol is produced can be attributed to the long-term targeted promotion of ethanol production by the Brazilian government that has been taking place since the 1970s. The oil price rise of 1973 and the fall in sugar prices led the government to subsidize ethanol production through the ProAlcoól programme. Brazilian ethanol production is now competitive without the need for direct subsidies (GBEP, 2008). In addition, at the end of 2004 the Brazilian government launched a wide-ranging programme intended to promote the development of a competitive biodiesel sector. The programme is targeted at the poorest regions of the country in the north-east and the Amazonas area. The 'social seal' is intended to guarantee that a certain proportion of the raw materials for biodiesel come from small family-run businesses in poor regions (GBEP, 2008).

Under the Brazilian Agroenergy Plan 2006–2011 the Brazilian government plans to make greater use not only of biofuels but also of electricity generation from biomass. CHP from sugar cane bagasse has great potential in this area. Timber and paper residues, rice husks and coconut and cashew nut shells are also available for conversion to energy. Expansion of the modern use of residues from agriculture and forestry is also planned (Ministry of Agriculture Livestock and Food Supply, 2006). In addition, biomass is playing an important part in rural electrification through the programme Luz para Todos ('Light for All'). Through the PROINFA programme, initiated in 2002, the contribution of wind, biomass and mini-hydropower to power production is due to be increased to 3300 MW (GTZ, 2007a; GBEP, 2008).

Nevertheless, biofuels continue to be a priority area of Brazil's bioenergy strategy. The country intends to expand ethanol production on a large scale and also increase the volume of exports (Ministry of Agriculture Livestock and Food Supply, 2006). Its aim is to become the world market leader in biofuel manufacture. To this end Brazil has already concluded partnership agreements on the development of a global biofuel market with some Latin American, African and Asian developing countries in which climatic production conditions are comparable (Stecher, 2007; Biopact, 2007a,b).

Many of these other countries see Brazil as a model of a successfully implemented bioenergy policy that has brought with it positive socio-economic developments. It is thought that the sugar industry has created around 1 million jobs, which has led indirectly to the creation of a further 6 million jobs (GBEP, 2008). However, the country is facing large-scale ecological and social problems that are directly or – more often – indirectly linked to bioenergy production and use. For example, large areas of the Amazonas forest are being taken over for agricultural use, largely on account of the expansion of soya production and cattle rearing. If sugar cane production is increased further, indirect displacement effects may also occur (Fatheuer, 2007; Bringezu and Schütz, 2008). Many NGOs are critical of the cultivation of sugar and soya in monoculture, the increased use of pesticides and herbicides, the health problems caused by the continuing practice of burning sugar cane fields before harvest, the working conditions of some plantation workers which the Brazilian government is continually seeking to improve, migration of people associated with the expansion of plantations and the non-observance of traditional land rights (Fritz 2007; Stecher, 2007; GBEP, 2008).

The Brazilian government must address these issues before it permits the bioenergy sector to exapnd further. Biofuel production is in principle an effective strategy for increasing export-driven growth and decoupling the transport sector from fossil energy consumption. Nevertheless, in Brazil's case it would be advisable for the expansion of bioenergy use to be pursued more rigorously than hitherto within strict sustainability guard rails, as has already been specified formally in the Brazilian Agroenergy Plan 2006– 2011. In the long term Brazil also aims to pursue a shift towards electromobility, particular in large urban areas, on account of its greater energy efficiency (Section 8.1.). As ethanol, sugar cane could then be used profitably for electricity generation in combined-cycle power plants, with the ethanol being combusted in the turbine, or in small-scale CHP units. From an energy perspective the conversion of biomass to biomethane and the flexible use of this methane, for example in CHP, is also an efficient pathway.

term it will be possible to consider providing heat – in developing countries and elsewhere – through the use of biomethane, which can be distributed and used via gas grids (Section 8.1).

Kerosene lamps are frequently used in developing countries as a source of light. However, rising oil prices render them prohibitively expensive for many people. One solution is to produce vegetable oil locally and use it in lamps. Such lamps are not as efficient as electric lamps and they create soot; however, the plant oil can be produced locally by very simple means. The use of green electricity in electric lamps would have advantages for indoor air quality but would be more expensive.

8.2.3.3
Bioenergy for central and decentral electricity generation

For rural regions in developing countries, small-scale technologies such as vegetable-oil motors and generators can be used for local distributed electricity generation. Since the efficiency of a stationary motor is higher than that of a mobile one, biofuels can displace a greater quantity of fossil fuel and thus avoid more greenhouse gases when used in electricity generation rather than in a mobile way in the transport sector. Generating sets that combine a biofuel-driven combustion engine and an electricity generator can be used both to produce electricity (small networks for public buildings, schools, hospitals, settlements) and to provide flexible mechanical energy as required for purposes such as grinding foodstuffs (maize, cereals) or driving water pumps. In combination with other renewable energy technologies such as photovoltaics or mini hydropower such bioenergy generating sets, as hybrid systems, can secure the electricity supply in rural regions.

In urban, industrialized regions where power requirements are high, the conversion to energy of wastes and residues and of sustainably cultivated energy crops can be effected by means of the same electricity generation systems as are used in industrialized countries (Sections 7.2 and 8.1). These will comprise on the one hand systems for the fermentation and gasification of biomass to methane that is then converted to electricity and on the other systems for the direct combustion of biomass, such as large-scale CHP plants or co-combustion in fossil-fired power plants. In central CHP plants the waste heat can be used for industrial processes, such as drying harvested products. Just as in industrialized countries, the use of wastes and residues to generate electricity is particularly to be recommended. In predominantly agricultural countries there arise considerable quantities of residues (e.g. from aquaculture, sawmills, tea and coffee plantations) that can be used for energy. Agro-industrial biogas plants and large-scale CHP plants are particularly suited to the utilization of residues; ideally the waste heat will be used in the manufacturing process of their products.

8.2.3.4
Overall assessment of bioenergy in developing countries

It will be possible for significant progress in overcoming energy poverty in developing countries to be made through bioenergy, particularly through major increases in the efficiency of traditional biomass use. Replacing old and inefficient wood stoves by modern ones saves large quantities of fuelwood, avoids greenhouse gas emissions and prevents deforestation. Within the context of local cycles the use of bioenergy to overcome energy poverty is to be recommended (Section 9.2.2). Subject to correct management and the appropriate choice of cultivation system, growing energy crops on degraded land can improve soil quality (Section 7.1), and the trade in bioenergy carriers generates income. The use of residues from industrial food production in biogas and gasification systems or in large-scale CHP plants represents a climate-friendly way of generating energy that should be promoted in developing countries. However, large-scale cultivation of energy crops in developing countries is not sustainable unless social labour standards are complied with, land use is regulated and food security is given precedence (Box 8.2-3). Energy crop use is most efficient in CHP plants; in addition, the greatest climate change mitigation effect is achieved via this route if fossil electricity generation is replaced or avoided (Section 9.2.1). In this way electricity generation from biomass combined with other renewable energy technologies can make an important contribution to rural electrification and the supply of electricity to urban centres in developing and newly industrializing countries.

8.2.3.5
Technological stages en route to sustainable bioenergy use in developing countries

Various technological stages can contribute to the overcoming of energy poverty in developing countries. Some technologies are particularly cheap and simple to implement – among them the use of improved wood stoves, micro biogas units and vegetable oil that is produced and used locally for lighting, electricity and power generation and transport. In introducing technology it is important at all stages to factor in and take account of the cultural features of the population. In development cooperation it is necessary to ensure that the technologies are accepted and that they can be maintained by the user (Section 9.2). To a certain extent the use of biofuels for transport is also justified; however, it is technically more efficient and therefore preferable to use it in the combined generation of power and heat in combined-cycle power plants or small-scale CHP units. In producing and using biofuels the priority of food production and other sustainability criteria and standards must always be borne in mind. As in industrialized countries, the use of wastes and residues and cascade use should be emphasized and should be preferred to the cultivation of energy crops. Some developing countries will change their status to that of newly industrializing and industrialized countries. In those countries, pathways to sustainable bioenergy use similar to those adopted in industrialized countries should be pursued (Section 8.1).

Sustainable biomass production and bioenergy deployment: A synthesis

9

Any intensified use of bioenergy must be measured against the extent to which it promotes a shift towards sustainable energy systems (Section 2.2). Specifically, it must be evaluated in terms of its contribution to climate change mitigation (Section 9.2.1) and its role in overcoming energy poverty (Section 9.2.2). The following synthesis is based on the analysis presented in Chapters 6, 7 and 8; it specifies WBGU's vision for sustainable bioenergy use.

9.1
Sustainable production of biomass as an energy resource: The key considerations

9.1.1
Biogenic wastes and residues

In the production of biomass for conversion to energy a fundamental distinction must be made between wastes and residues on the one hand and energy crops on the other. Using biogenic wastes and residues for energy production has the advantage that no additional land is required and in consequence competition with existing land uses is unlikely to arise. Greenhouse gas emissions from land-use change and crop cultivation are almost entirely avoided, so that the climate change mitigation effect is determined primarily by the emissions from the conversion and use of the bioenergy carriers and the emissions saved through the replacement of fossil energy carriers. In many cases the use of biogenic wastes for energy also reduces other greenhouse gas emissions, such as methane emissions from slurry or landfills. However, the removal of residues from agricultural or forestry ecosystems for conversion to energy must be limited, in order to prevent too much organic matter being removed from the soil (Section 6.1.2). The use of residues must not jeopardize soil protection – and hence climate change mitigation. Waste and residue use should not give rise to pollutant emissions. In principle WBGU accords the conversion of biogenic wastes and residues to energy (including cascade use;

Section 5.3.3) a higher priority than the use of energy crops (Table 9.2-1).

9.1.2
Land-use changes

Where crops are specially cultivated for energy purposes, land use is also a factor. Land-use changes caused by cultivation of the energy crops must be taken into account, because they have a crucial impact on the greenhouse gas balance of the different bioenergy pathways. Directly triggered land-use changes can be monitored, because they can be directly attributed to the specific case of energy crop cultivation. But when farmland is converted for energy crop cultivation, the previous agricultural production is very likely to be wholly or partly displaced onto other land. The indirect land-use changes thus brought about – together with the direct land-use changes – are the deciding factor in the assessment of greenhouse gas balances over the entire value chain. They often cause around half the climate change mitigation effect to be lost and may even result in a net release of greenhouse gases. The estimation of greenhouse gas emissions from indirect land-use change is, however, a task beset by considerable uncertainties. These emissions cannot be quantified without understanding of the complex relationships involved; at present, calculations are based on models only. The long-term goal must be a global land-use strategy that aims to prevent greenhouse gas emissions from land-use change in general, seeks to conserve terrestrial carbon stocks, and ensures the conservation and sustainable use of biological diversity (Sections 10.2 and 10.5).

WBGU entirely rejects the direct or indirect conversion of forests and wetlands into cropland for energy crop cultivation, since such conversion is usually associated with non-compensatable greenhouse gas emissions and has a negative impact on biological diversity and on soil carbon stocks (Section 4.2). The conversion of grassland to cropland also reduces the climate change mitigation effect.

In WBGU's view, conversion of land for energy crop cultivation is only expedient if the emissions resulting from direct and indirect land-use change, including the lost sink capacity of the land in question, do not exceed the amount of CO_2 that can be re-sequestered through energy crop cultivation on that land (i.e. in the soil and vegetation and in the harvested products) within ten years. The calculation should also include emissions that are likely to arise in the course of cultivation (such as N_2O emissions from fertilizer use). As a rule, the maximum greenhouse gas emissions that can be saved by replacing fossil energy carriers with bioenergy correspond to the amount of carbon stored in the biomass (Section 6.4.3.3).

For reasons of food security, energy crop cultivation should be restricted to land whose conversion for bioenergy production is unlikely to compete with food production; this ensures that the risk of indirect land-use changes is kept to a minimum. The cultivation of energy crops on marginal land (land with limited productive or regulatory function; Box 4.21) is to be preferred. This must not take place, however, without consideration of the interests of the local community, who may use the marginal land for grazing or other purposes. Before the land is used, an assessment of its nature conservation value must also be carried out. Particularly suitable for climate change mitigation is marginal land – especially degraded land – where conversion for energy crop cultivation may result in enrichment of soil carbon. The cultivation of oil palms or *Jatropha* on degraded land in the tropics is an example of this. Particularly cost-effective greenhouse gas savings can be achieved in such cases.

9.1.3
Cultivation systems

The criteria used by WBGU for the sustainability of cultivation systems are the effects on biological diversity and soil carbon storage. Bioenergy can only be classed as sustainable if, on a long-term basis, biomass regrowth keeps pace with the amount of biomass that is harvested from the same land – in other words, only if long-term soil fertility can be assured. Only under these conditions can it also be assumed that the carbon that is removed from the atmosphere and stored by the energy crop, and that is released again in the form of CO_2 when the crop is used for energy, does not result in an increase in the atmospheric concentration of CO_2 and therefore does not need to be regarded as an emission. In order to minimize the likelihood of bioenergy triggering competition for

land use, the different specific crop yields per unit area must also be taken into account.

According to these criteria, perennial crops such as *Jatropha*, oil palms, short-rotation plantations (fast-growing woody plants) and energy grasses score better than annual crops such as rape, cereals or maize. In WBGU's view, perennial crops should therefore in principle be preferred (Table 9.2-1). Wherever possible, plant mixtures rather than monocultures should be used for biomass production. This is particularly important in view of evidence that grassland with greater biological diversity can also provide more ecosystem services (Section 7.1.4). If appropriate cultivation systems are selected, organic carbon can also be incorporated into the soil, benefiting both the greenhouse gas balance and soil fertility. At the same time, inappropriate application of nitrogen fertilizers must be avoided, as this results in N_2O emissions, among other effects. In view of the rising demand for wood products, WBGU regards additional forest growth as having little bioenergy potential (Section 6.1.2).

9.2
Conversion, application and integration of bioenergy

Each of WBGU's goals for sustainable bioenergy use (Section 2.2) results in a different perspective on bioenergy. In relation to the climate change mitigation goal (Section 9.2.1), there are several key factors. These include, in addition to the manner in which the biomass is supplied, both the way in which the biomass is converted into usable products (such as gases, vegetable oils, biofuels and wood pellets) and the type of application, for example in mobility, heating, or combined heat and power generation (CHP). The impact of these factors, however, is usually less significant than the effects triggered by direct or indirect land-use changes as a result of the production of bioenergy carriers. In connection with use in energy systems, the key issue is the nature of the energy carriers that the biomass replaces and the magnitude of the losses in the conversion path. Careful integration of bioenergy into existing energy systems is therefore extremely important, as is the specific contribution of bioenergy to the sustainable transformation of energy systems. Maximizing the climate change mitigation effect should be a priority, particularly in industrialized countries but also in the rapidly developing urban and industrialized regions of emerging and developing countries; the same should apply, with appropriate reservations, elsewhere in developing countries. Developing and newly industrializing countries have not as yet made any quantified international commitment to limit their greenhouse gas emissions. Nonetheless, pursuing the development of technologies that are as modern, energy-efficient and cost-effective as possible – and that therefore promote climate change mitigation – should be an important guiding principle in these countries as elsewhere.

In relation to the goal of overcoming energy poverty (Section 9.2.2), the key issues are to modernize traditional bioenergy use and provide access to modern forms of energy such as electricity and gas. These are the key challenges in the rural regions of developing countries. Yet in such settings, too, bioenergy can have a positive impact on climate change mitigation.

9.2.1
Climate change mitigation

9.2.1.1
Reducing greenhouse gases through bioenergy use: Measurement and standard-setting

The contribution made by bioenergy to climate change mitigation is frequently measured as a percentage reduction in greenhouse gases by comparison with a reference system in relation to final or useful energy consumption. For example, this is one of the sustainability criteria for liquid biofuels proposed by the Council of the European Union as part of the planned EU directive on the promotion of renewable energies (Box 10.3-2). Such a parameter captures the climate change mitigation effect that can be achieved through the production of a particular quantity of energy from biomass, without asking what quantity of biomass is needed to generate this energy. However, the factor that limits the mitigation effect of biomass is not the level of energy demand which can potentially be replaced by bioenergy, but the quantity of sustainably available biomass.

For comparing the climate change mitigation effect of different biomass deployment options WBGU therefore considers the key parameter to be the absolute greenhouse gas reduction potential either in relation to the cultivation area or in relation to the quantity of biomass consumed (Figs. 7.3-3a, b and 7.3-4). These two parameters are also a good basis for standard-setting (Section 7.3.2). In specific terms WBGU recommends that bioenergy should only be used if over the whole life cycle, including emissions from direct and indirect land-use changes, a greenhouse gas reduction of at least 30t CO_2eq per TJ of raw biomass used can be achieved (Section 10.3.1.1; Table 9.2-1). For biofuels this corresponds roughly

to a requirement to reduce emissions, in relation to final energy, by 50 per cent by comparison with the fossil reference system. WBGU proposes that a condition of state promotion should be attainment of twice this value – this is, at least 60t CO_2eq per TJ of raw biomass consumed (Section 10.3.1.2). This figure is somewhat more than half of the climate change mitigation effect that can be achieved using present technology (Section 7.3.2). WBGU emphasizes that a standard that prescribes a particular mitigation effect to be achieved from bioenergy use should be regarded as an interim solution. Apart from anything else, the figure defined in a quantitative specification of this sort is inevitably somewhat arbitrary. In principle, therefore, the aim should be to work towards a global system of mandatory limits on greenhouse gas emissions that covers all relevant sources of emissions, including those from land use and land-use change (Section 10.2).

9.2.1.2
Taking account of indirect land-use change

WBGU considers it essential for emissions from indirect land-use change to be included in any evaluation of the climate change mitigation effects of bioenergy. Quantifying these effects is a science that is still in its infancy; in consequence, there is as yet no recognized method that is based on a scientific consensus. Nevertheless, emissions from indirect land-use change need to be included even now in greenhouse gas balances used in standard-setting.

Every indirect land-use change is of course also a direct land-use change somewhere else; this means that any land-use change is in principle measurable and the associated emissions are quantifiable. However, the causal link to energy crop cultivation is not directly verifiable. WBGU therefore proposes that calculation of emissions from indirect land-use changes should for the time being be based on the iLUC factor method developed by the Öko-Institut (Fritsche and Wiegmann, 2008; Box 7.32). This method allows for an initial, albeit rough, estimate of such emissions to be made. Further development of this parameter is an important research task (Chapter 11).

Since indirect effects must be taken into account, the conversion of cropland for the cultivation of annual energy crops in temperate regions leads to emissions that are so high that they cannot be compensated within 20 years in the biofuel pathways common in the transport sector today (Section 7.3.2). In comparison with these pathways, continuing the use of fossil fuels would be the better climate protection option in the transport sector. When biomass is used for power generation or for combined heat and power (CHP), it continues to have a climate change mitigation effect even when indirect land-use changes are taken into account; however, the effect is only around half that achieved when indirect effects are not considered (Figure 7.34).

The use of wastes and residues gives rise to very few emissions from land-use changes; these emissions can therefore usually be ignored when calculating greenhouse gas balances.

9.2.1.3
Replacing fossil energy carriers

From a climate protection perspective, the energy applications that are most attractive are those in which fossil energy carriers with high CO_2 emissions are displaced. Among fossil energy carriers the displacement of coal thus saves the largest quantity of greenhouse gas emissions, while the displacement of petroleum products saves less and the displacement of natural gas saves least of all. Competition with or even displacement of other renewable energy sources through bioenergy should definitely be avoided.

9.2.1.4
Climate change mitigation effect of different technical applications/pathways

The level of the climate change mitigation effect that can be achieved through bioenergy depends not only on the land-use changes associated with energy crops but also on the technical field of application in which the bioenergy is used. This section considers the technical applications studied by WBGU from the point of view of their mitigation effect and greenhouse gas abatement costs (Table 9.2-1).

CO-COMBUSTION OF BIOMASS IN POWER PLANTS
The production of wood chips or pellets from lignocellulose biomass (e.g. timber wastes, or timber from short-rotation plantations) results in only small conversion losses. If these products are then used as a fuel alongside coal in large-scale power plants (co-combustion), a very favourable mitigation effect is achieved for moderate CO_2 abatement costs; this is the case both for the combined generation of power and heat (combined heat and power, CHP) and for conventional coal-fired power plants without heat extraction. The use of biogenic wastes and residues from agriculture and forestry is particularly advantageous in this context, since virtually no emissions from land-use change are incurred. However, the co-combustion of biomass in coal-fired power plants

should not result in the use of conventional coal-fired power plants coming to be regarded as a viable option for the future or in such use being continued for longer than necessary; this would create lock-in effects with undesirable implications for system sustainability. Co-combustion should therefore be promoted only in particularly climate-friendly power plants with heat extraction.

ELECTRICITY GENERATION AND CHP
The greenhouse gas abatement potential in relation to the quantity of biomass used is greatest when biomass is used for power generation or to generate combined heat and power (CHP), or when it is used in high-efficiency large-scale power plants such as combined-cycle plants. Most of the pathways studied by WBGU in connection with pure heat generation or use as biofuel in the transport sector achieve only around half of the greenhouse gas savings achievable in the power sector. On account of the high energy efficiency that results from the utilization of waste heat, CHP technology is in principle to be preferred to pure power generation, provided that appropriate uses exist for the heat. For regions with a significant requirement for cold or cooling, CHP can also be used to generate cold, for example through absorption chilling processes. A very good mitigation effect with low abatement costs can be achieved through the use of wood (preferably from residues) in direct combustion with CHP.

Biogas systems that ferment wastes and residues represent an efficient method of generating power and heat that has a high mitigation effect and very low abatement costs. High-productivity energy crops (such as grasses) are also a suitable substrate for biogas systems, provided that associated emissions from land-use changes are low. The biogas thus obtained can be processed into biomethane; it can be transported via the natural gas grid and used for power generation or in CHP systems.

Combined-cycle power plants are at present the most efficient established technology for generating electricity from natural gas or biomethane. The use of fuel cells is expected to result in further efficiency increases in the future. All these pathways have comparatively high greenhouse gas abatement potentials, although the associated abatement costs vary considerably. The production of biomethane in biomass fermentation systems is not yet competitive, but further technological developments are likely to result in significant cost reductions (Section 7.2.5.2).

BIOMASS HEATING
In the field of pure heat production WBGU has considered pellet boilers that utilize residues and energy crops (short-rotation plantations). The greenhouse

gas reduction potential of these pathways is significantly less than the reductions achievable in the power sector. One of the reasons for this is that the use of bioenergy in the heat sector replaces petroleum and natural gas, which have lower energy emissions per unit final energy than coal. If wood is used from short-rotation plantations established on cropland, the mitigation effect of the representative pathway considered in this study may even be completely negated by the anticipated emissions from indirect land-use changes. When used exclusively for heat, the technologies considered have relatively high abatement costs and achieve on average only around half the absolute reduction in greenhouse gases achievable from use for combined power and heat through CHP. Larger systems such as wood chip boilers or large-scale heating facilities tend to have lower heat production costs and hence lower abatement costs. Nevertheless, the mitigation effect per unit of raw biomass is usually higher in combined power and heat generation than in pure heat generation; CHP pathways are therefore to be preferred to pure heat pathways.

TRANSPORT FUELS

Where residues are used (e.g. timber waste, slurry, straw), the mitigation performance of the use of biofuels in the transport sector is similar to that of biomass use for heat generation: it is relatively poor. In most cases, greenhouse gas savings in relation to the quantity of biomass used are at least 50 per cent less than those achievable in the power sector.

The production and use of biofuels in conventional combustion engines is a highly inefficient use of resources. Only approx. 5–10 per cent of the energy stored in the crop or biomass can be utilized as vehicular propulsive energy (Section 8.1.2.1). Moreover, the use of biofuels in the transport sector actually results in higher emissions than the use of fossil fuels (Figure 7.34) if biomass from temperate, annual energy crops (e.g. maize, rape) is used as a substrate (Section 9.1.3) and the cultivation of these crops causes high emissions from indirect land-use changes (Section 9.2.1.2). The energy balance of second-generation biofuel use is not in general better than that of first generation biofuels. Second-generation biofuels utilize the entire above-ground plant, but during the process of conversion into biofuel around half the original energy content of the biomass is lost (Figure 8.1-9).

For the future of road transport, WBGU considers electricity generation from renewables combined with the use of electric vehicles to be the most appropriate solution. A vehicle operating on bioenergy achieves the greatest range for the same biomass input by using biomass-derived electricity in an electric engine, not by using biofuels in a combustion engine (Figure 8.110). For this pathway, however, the climate change mitigation costs are at present still very high. Nevertheless, if electric vehicles were introduced on a large scale, costs – particularly the costs of storage batteries, which are at present very expensive – could be significantly reduced within 15–20 years, resulting in a concomitant fall in greenhouse gas abatement costs (Section 7.2).

However, the use of electromobility only holds a medium-term promise of environmental benefit. Only beyond a certain efficiency level of electricity conversion is electromobility technologically more efficient than conventional drives. Other than for hybrid vehicles, electric engines for mass-produced vehicles are still at the development stage. A weak point is the need for batteries that combine high energy-storage capacity with lightness of weight and long service life. The bottom line is that the efficiency potential of electromobility is utilized to the full only by the combination of electric propulsion and directly generated, renewable electricity from solar, wind or hydropower sources.

Through electromobility bioenergy use achieves a significantly higher climate change mitigation effect than it does through the blending of biofuels with fossil fuels used in the transport sector. For this reason pathways that generate power and heat from bioenergy should be given precedence over the use of biofuels for transport. WBGU therefore considers the production of biofuels for road transport in industrialized countries an essentially inappropriate mitigation option. WBGU recommends that promotion of biofuels for the transport sector be phased out speedily. The quotas for blending biofuels with fossil fuels should be frozen and then withdrawn completely within the next 3–4 years.

By contrast, biofuel pathways involving tropical, perennial energy crops such as Jatropha, oil palms or sugar cane, when grown on marginal land, have an appreciable mitigation effect at moderate cost. The additional storage of carbon in the soil as a result of their cultivation and the avoidance of indirect land-use changes have a positive impact on the mitigation effect. However, if the same crops are grown on cropland and thus prompt indirect land-use change, or if forest is directly cleared for energy crop cultivation, their use usually gives rise to significantly more emissions than the use of fossil fuels would have done. Biofuel use can therefore quickly tip from major benefit to major harm (Section 7.3.2). Since GHG abatement costs are in some cases very low, some of these pathways are also candidates for promotion through international climate protection instruments.

Since there are as yet no established sustainability standards for biofuels from tropical production,

Table 9.2-1
Synthesis of the evaluation of bioenergy pathways, broken down according to cultivation systems, technical analysis and greenhouse gas balance. Pathways shaded grey are residue pathways. * For pathways that have grass silage/slurry as a substrate, it has been assumed that in Germany grass silage does not cause any emissions from land-use changes; this does not necessarily apply to the rest of the world. The labelling of the pathways refers to the cultivation systems and conversion processes listed in Tables 7.21 and 7.22.
Source: WBGU based on the data of Fritsche and Wiegmann, 2008 and Müller-Langer et al., 2008

Pathway	Section 7.1: Cultivation systems — Overall assessment	Section 7.2: Technical analysis — Energy efficiency [%]	Section 7.3: GHG balances — GHG reductions with iLUC per unit of raw biomass [t CO_2-eq/TJ]
	positive	over 30	over 60
	unclear	18–30	30–60
	negative	below 18	below 30
Switchgrass-pellets-heating-2030	unclear	17	17
Short rotation-pellets-heating-2030	unclear	20	-1
Wood residues-pellets-heating-2005	positive	19	61
Straw-pellets-heating 2005	positive	15	46
Oil palm (rainforest)-vegetable oil-small-scale CHP-2030	negative	23	-185
Oil palm (degraded)-vegetable oil-small-scale CHP-2005	unclear	23	190
Jatropha-vegetable oil-small-scale CHP-2030	unclear	34	27
Jatropha (degraded)-vegetable oil-small-scale CHP-2030	unclear	34	176
Rape-vegetable oil-small-scale CHP-2005	negative	43	29
Maize silage-biogas-small-scale CHP-2005	negative	33	37
Switchgrass-biogas-small-scale CHP-2030	unclear	36	54
Grass silage/slurry-biogas-small-scale CHP-2030*	positive	30	107
Maize silage-biogas-fuel cell (SOFC)-2005	negative	36	57
Switchgrass-biogas-fuel cell (SOFC)-2030	unclear	40	63
Grass silage/slurry-biogas-fuel cell(SOFC)-2030*	positive	33	112
Maize silage-biomethane-small-scale CHP-2005	negative	29	30
Switchgrass-biomethane-small-scale CHP-2030	unclear	31	53
Grass silage/slurry-biomethane-small-scale CHP-2030*	positive	26	84
Maize silage-biomethane-combined-cycle power plant-2005	negative	30	44
Switchgrass-biomethane-combined-cycle power plant-2030	unclear	32	49
Grass silage/slurry-biomethane-combined-cycle power plant-2030*	positive	27	93
Short rotation-biomethane-combined-cycle power plant-2030	unclear	30	29
Short rotation-raw gas-gas turbine-2030	unclear	28	9
Short rotation-raw gas-fuel cell (SOFC)-2030	unclear	41	31
Short rotation-wood chips-central CHP-steam turbine-2030	unclear	33	47
Short rotation-pellets-coal-fired power plant-2030	unclear	43	38
Harvest residues/slurry-biogas-small-scale CHP-2005	positive	24	113
Organic wastes-biogas-small-scale CHP-2005	positive	29	88
Harvest residues/slurry-biogas-fuel cell (SOFC)-2005	positive	27	122
Organic wastes-biogas-fuel cell (SOFC)-2005	positive	32	91

	Section 7.1: Cultivation systems	Section 7.2: Technical analysis	Section 7.3: GHG balances
	Overall assessment	Energy efficiency [%]	GHG reductions with iLUC per unit of raw biomass [t CO_2-eq/TJ]
Harvest residues/slurry-biomethane-small-scale CHP-2005		20	94
Organic wastes-biomethane-small-scale CHP-2005		26	80
Harvest residues/slurry-biomethane-combined-cycle power plant-2030		21	103
Organic wastes-biomethane-combined-cycle power plant-2005		27	86
Wood residues-biomethane-combined-cycle power plant-2030		30	100
Wood residues-raw gas-gas turbine-2030		29	86
Wood residues-raw gas-fuel cell (SOFC)-2030		41	109
Wood residues-wood chips-central CHP-steam turbine-2005		33	112
Straw-wood chips-central CHP-steam turbine-2005		30	107
Wood residues-pellets-coal-fired power plant-2005		38	101
Straw-pellets-coal-fired power plant-2005		35	87
Oil palms (rainforest)-biodiesel-car-2030		11	-257
Oil palms (degraded)-biodiesel-car-2005		10	149
Jatropha-biodiesel-car-2030		16	-13
Jatropha (degraded)-biodiesel-car-2030		16	63
Short rotation-Fischer-Tropsch diesel BtL-car-2030		15	-13
Rape-biodiesel-car-2005		23	-28
Rape-vegetable oil-car-2005		19	-56
Sugar cane-ethanol-car-2005		8	-3
Sugar cane (degraded)-ethanol-car-2030		9	47
Maize grain-ethanol-car-2005		11	-10
Cereals-ethanol-car-2005		11	-45
Maize silage-biomethane-car-2005		9	-28
Short rotation-biomethane-car-2030		20	-15
Grass silage/slurry-biomethane-car-2030*		15	53
Switchgrass biogas small scale CHP-electric car-2030		30	40
Wood residues-wood chips-central CHP-steam turbine-electric car-2030		31	116
Harvest residues/slurry-biogas-small-scale CHP-electric car-2005		20	97
Wood residues-Fischer-Tropsch diesel BtL-car-2030		16	51
Straw-Fischer-Tropsch diesel BtL-car-2030		14	49
Waste fat-biodiesel-car-2005		25	80
Straw-ethanol-car-2030		11	32
Wood residues-biomethane-car-2030		20	63
Harvest residues/slurry-biomethane-car-2005		9	36
Organic wastes-biomethane-car-2005		13	34
Wood residues-hydrogen-fuel cell (PEM)-car-2030		16	52

the import and use of these biofuels presents problems. Once suitable standards and certification systems have been introduced (Section 10.3), it may be appropriate to promote the import of vegetable oils and bioethanol if their compliance with the eligibility criteria can be verified (Section 10.3.1.2). In order to achieve the greatest possible climate change mitigation effect, use in CHP systems or for power generation – if coal is thereby replaced – is still to be preferred to transport sector use, even if evidence of the sustainability of these biofuels is provided. For example, in Brazil bioethanol from sustainable sugar cane cultivation can also be used in efficient combined-cycle power plants for the combined generation of power and heat.

BIOMETHANE

Biomethane can be regarded as a very promising option for the future (Box 7.22). Biomethane produced from the fermentation of wet biomass is already a highly cost-effective mitigation option, for example when used to replace coal in small-scale CHP units. Systems for producing biomethane from solid biomass by gasification are still relatively expensive, but WBGU anticipates significant cost reductions in this area in the future. Both biomethane production pathways, when used for electricity generation, can achieve high absolute greenhouse gas reductions in relation to the quantity of biomass used; these reductions are comparable to those achieved by other electricity pathways (such as co-combustion in coal-fired power plants or the use of wood chips in large-scale CHP plants). Moreover, in both processes of biomethane production it is necessary to separate CO_2 from the biogas or product gas. Should sustainable storage of this CO_2 become possible in future, this would reduce the specific emissions of the biomethane pathway, thus further increasing the climate change mitigation effect. WBGU has elsewhere submitted its recommendations relating to the criteria for sustainable sequestration (WBGU, 2006). Biomethane can be readily transported via the natural gas grid and can thus be made available to users who are able to make optimum use of heat extracted in CHP systems; conversely, the biomethane can be collected from distributed systems and made available for maximum-efficiency deployment in large-scale combined-cycle power plants.

9.2.2
Energy poverty

Overcoming energy poverty is key to tackling poverty in general, particularly in the rural regions of developing countries but also in urban areas. Overcoming energy poverty involves providing a choice of options for access to affordable, reliable, high-quality, secure, safe and environmentally sound energy services to meet basic needs, especially through access to electricity and gas (WBGU, 2004a; Section 2.2.2). In rural areas, small to medium-scale off-grid technologies are particularly suitable for heat and electricity generation. They provide an important lever for significantly improving the quality of life of many hundreds of millions of people at low cost and within a short period of time. Traditional biomass use accounts for 90 per cent of the bioenergy currently used worldwide. Modernization of this use should be pursued above all other measures, since a major contribution to tackling energy poverty can be made through, for example, efficiency improvement. In urban areas the opportunities for overcoming energy poverty are more diverse – in relation to access to energy services, however, distributional issues present serious problems.

The production and use of modern bioenergy on a larger scale, which can also contribute to overcoming energy poverty in developing countries, should be assessed in terms of its climate change mitigation effect (Section 9.2.1). Where the greenhouse gas abatement costs associated with a bioenergy pathway are low, new financing sources can be accessed via international climate policy instruments. The aim should therefore be to identify conversion pathways for bioenergy use that perform well in terms of abatement per unit of biomass consumed and also have low abatement costs.

WOOD AND CHARCOAL STOVES

WBGU recommends that the complete phase-out by 2030 of forms of traditional biomass use that are harmful to health should be made an international target. Some technologies that would help to achieve this can already be implemented quickly and cost-effectively. The use of improved cooking stoves can reduce fuel consumption to a half or a quarter for the same level of utility, while at the same time dramatically reducing the health risks involved. This frees up the time of women and girls in particular, since in developing countries collecting fuelwood is an important task for females. More time is then available to earn an income or pursue an education (Box 8.2-1). This applies both to simple wood stoves and to the simple charcoal stoves that are particularly widely used in urban areas.

MICRO BIOGAS SYSTEMS

Micro biogas systems enable biogenic residues such as animal excrement to be converted into methane and used for cooking and lighting. Use of these systems can also save fuelwood and improve indoor air quality. This technology is not the most efficient, but it represents an acceptable compromise between simple installation, maintenance and use, energy benefits and the costs of the energy service. This technology is particularly suitable for households, schools and public institutions (Section 8.2.2)

BIOMASS GASIFICATION SYSTEMS

Biomass gasification systems that utilize wastes and residues such as coconut shells, waste wood, coffee chaff and rice chaff can be used for power generation or for heat production. Depending on the size of the system, this technology is appropriate both for households and for rural communities. For example, the raw gas can be used directly to generate process heat for drying systems or bakeries. Wood gasifiers use a broad and flexible range of feedstocks and can also contribute via generators to rural electrification (Sections 7.2.4 and 8.2.2).

VEGETABLE OIL ENGINES, GENERATING SETS AND SMALL-SCALE CHP UNITS

Oil plants (e.g. *Jatropha*, oil palms) can be processed locally using simple mechanical presses to produce unrefined vegetable oil. This can be used in combustion engines to drive a range of stationary machines such as corn mills or water pumps. Combustion engines can also be coupled to a generator to form a generating set and used to generate electricity (e.g. for public buildings, hospitals, schools, mini-grids). This technology has great potential in connection with rural, off-grid electrification, since it requires little maintenance and is relatively simple to operate (Section 8.2.2). The waste heat from generating sets can be used for processes such as drying agricultural products. Larger small-scale CHP units that operate on sustainably produced vegetable oil can also be used for electrification in urban, industrial regions; if a large number of them are established they can even replace large fossil-fuel power plants or render the building of new ones unnecessary.

9.2.3
Bioenergy as a bridging technology

As the world moves towards energy systems based on renewable sources, the sustainable use of bioenergy from energy crops can be expected to fulfil an important function as a bridging technology until around the middle of the century. There are two reasons for the limit on its extensive use.

Firstly, after the middle of the century it is likely that most renewable energy will be generated directly – by wind and hydropower, and in due course also by solar energy on a large scale (WBGU, 2004a). Once these energy carriers are sufficiently available in the energy system and the electricity grid is well developed, energy crops as energy carriers will largely have fulfilled their function of bridging the way to sustainable energy provision. However, the renewable sources will mostly be used to generate electricity directly and output will be subject to major temporal fluctuations. Biogenic wastes and residues and the residual use of fossil energy carriers will then be increasingly required to level these fluctuations in output (balancing power). The sustainable biomass potential identified by WBGU can in future make a significant contribution to meeting the need for balancing power; it thus also underpins the technical security of supply and the stability of electrical grids in a sustainable and integrated energy supply system involving a large element of wind and solar power. Through the use of smart electricity grids, electromobility can also contribute to balancing power (Section 8.1). If the use of the sustainable bioenergy potential is combined with the capture and secure storage of CO_2, it may even be possible to generate 'negative' CO_2 emissions (Box 6.8-1).

Secondly, as a result of dynamic trends, demands on global land use will increase dramatically in the coming decades. These trends include a growing world population with increasingly land-intensive dietary patterns, increasing soil degradation and water scarcity. Furthermore, for reasons that include climate change mitigation, more and more petrochemical products will in future be made from biomass. This non-substitutable land-use requirement for the manufacture of textiles, chemical products, plastics, etc. may constitute around 10 per cent of world agricultural land, although at the end of their service life it will be possible to use some of these biomass-based products in the form of biogenic waste for conversion to energy (cascade use; Section 5.3.3). By contrast, the supply of energy is not tied to carbon; it can also be generated directly. All this takes place against the backdrop of increasingly manifest anthropogenic climate change, which will affect future harvest yields. In consequence, energy crop cultivation is likely to decline in the second half of the century. The base supply of bioenergy from biogenic wastes and residues will not be affected by this, since it has very little connection with land use; the part that this energy plays in the provision of balancing power in the electricity supply system is therefore assured for the long term.

Global bioenergy policy

Introduction

WBGU considers that bioenergy policy should be geared primarily towards climate change mitigation and the elimination of energy poverty (Section 2.2). The modelling presented in Chapter 6 shows that there is substantial global sustainable potential for bioenergy. However, the bioenergy strategies that can currently be observed worldwide are not specifically geared towards the exploitation of this sustainable bioenergy potential, but often promote non-sustainable bioenergy production. Moreover, the current focus on biofuels ignores the potentials afforded by efficiency increases in the globally more relevant traditional use of bioenergy, as well as the potential for utilization of wastes and residues.

Any meaningful contribution by bioenergy to the sustainable transformation of energy systems must comply with the relevant guard rails (Chapter 3). Measured against this benchmark, it is clear that by no means every bioenergy pathway is appropriate (Chapters 5 and 7). The task of policy-makers is therefore to establish framework conditions for channelling bioenergy use into sustainable pathways. WBGU recommends minimum standards which should be the prerequisite for the use of all types of bioenergy products. As these minimum standards can only achieve limited validity, non-sustainable bioenergy use should in the long term be restricted worldwide by a comprehensive and effective global regulatory system, e.g. through revised incentives within the United Nations Framework Convention on Climate Change (UNFCCC) and the development of better protection mechanisms under the Convention on Biological Diversity (CBD). Only those applications which offer the potential to generate positive ecological and socio-economic impacts, improve the efficiency of traditional biomass use and are based on the utilization of residues and wastes should be promoted (promotion criteria).

The key task, then, is to devise a regulatory framework which puts in place the conditions to maximize the use of potentials while minimizing risks. Diffi-culties with regulation arise at various levels, however. First and foremost, bioenergy is a cross-cutting theme which impinges on a wide range of policy areas and interests. 'Bioenergy policy' does not only encompass energy, agricultural and climate policy; transport policy and foreign trade policy as well as environmental, development and security policy all play an important role in this emerging policy field. The complex dynamic of the markets is another relevant factor. Energy and agricultural markets are becoming increasingly interlinked via bioenergy, and energy markets in particular are strongly influenced by countries' strategic interests.

In other words, there are complex political issues which must be resolved and which transcend the boundaries of established policy arenas, requiring cooperation among actors who, in the past, have shared little common ground. Given that policy-making is largely structured along sectoral lines, this poses major challenges in terms of governance and integration capacities. Bioenergy policy also transcends the framework of an international system that is based on nation-states. For example, a blending quota for biofuels in Europe can contribute to an increase in deforestation in other parts of the world. Bioenergy is thus an example of a complex global issue in which the actions of state and non-state actors at national or local level may have unintended consequences of a transregional or even a global nature. Bioenergy policy therefore requires a multi-level policy approach. The situation is further exacerbated by the fact that policy-makers are required to take action on the basis of great uncertainty, as the scientific fundamentals and correlations have not yet been adequately elucidated. And finally, bioenergy also involves aspects of global equity which are neatly summed up by the 'food/fuel' nexus.

Bioenergy policy is thus a highly volatile policy field with economic, technological, scientific, environmental and social dimensions. It spans the local, regional and global levels and – in view of the highly dynamic developments in this field – requires swift policy decisions and accountability for them. In short,

'Biofuels, while seemingly simple, are incredibly hard to do right' (Greene, cited in Conniff, 2007).

Because the cultivation of energy crops is proceeding at a rapid pace, the task now is to develop instruments and adopt measures which promote sustainable developments in the short and long term. At present, no international organization or treaty exists which would be specifically responsible for the issue of bioenergy. Instead, a plethora of private forums, UN activities and intergovernmental processes have developed over recent years at national, regional and multilateral level which address the issue of bioenergy with a variety of partners and with differing objectives. The result is a fragmented institutional picture which lacks clarity, although more intensive efforts are now being made to create coherence between the individual initiatives and processes. There is an ongoing struggle to achieve viable and binding standards, the best possible promotion strategies and new institutional arrangements, or at least ensure that existing steering mechanisms are utilized appropriately. Within this debate, and in light of previous analyses and existing regulatory endeavours, WBGU seeks to chart a viable course towards the further development of a sustainable global bioenergy policy for the future. The following sections reflect the logic of this new policy approach.

To ensure that the expansion of bioenergy use contributes to climate change mitigation, the right conditions must be in place. Section 10.2 therefore looks at incentives and commitments under the UN climate protection regime. As adaptation of the rules and accounting modalities cannot be accomplished in the short term or guarantee that other sustainability criteria (e.g. food security, conservation of biological diversity, etc.) will be met, work on drawing up and applying bioenergy standards must be undertaken simultaneously. The issues of standard-setting and initiatives aimed at achieving more comprehensive instruments for global land and area management for all types of biomass/land use are therefore dealt with in Section 10.3. The aim of achieving not only sustainable bioenergy use but also a reduction in land-use competition cannot be attained through appropriate standards alone, however. More comprehensive flanking measures to safeguard global food production, biological diversity, and the protection of water resources and soil are therefore required. The existing UN institutions can make an important contribution here. Section 10.4 describes relevant initiatives and the role of the Food and Agriculture Organization (FAO) in safeguarding global food security. Section 10.5 looks at the options afforded by the CBD for better conservation of biodiversity, while Section 10.6 outlines measures for the protection of water resources and soil. Finally, the sections on state promotion policy (Section 10.7) and development cooperation (Section 10.8) consider which forms of bioenergy use should be given explicit support, and what form a national and international promotion policy systematically geared towards the objectives of climate change mitigation and overcoming energy poverty might take.

10.2
International climate policy

10.2.1
The UNFCCC as an actor in global bioenergy policy

As described in Chapter 2, climate protection is not the only reason for the growing interest in bioenergy use, so it would be erroneous to view global climate change mitigation and the UNFCCC as the key drivers of increasing bioenergy use. Indeed, the UNFCCC offers only limited opportunities for steering bioenergy policy, as other objectives besides climate change mitigation play a very significant role in bioenergy use. Nonetheless, the climate policy oriented scope for steering bioenergy policy that the UNFCCC affords should be fully utilized. As the minimum, the international climate regime should not create incentives to engage in a bioenergy policy that is counterproductive for climate change mitigation, and in an ideal scenario international climate policy should be shaped in such a way that bioenergy use is consistent with the requirement to avoid dangerous climate change (Chapter 3). This entails an integrated approach towards activities in the bioenergy sector, and especially its implications for greenhouse gas emissions in the energy and land-use sectors. Bioenergy use invariably affects both these sectors, which is why they should not be viewed in isolation from each other. In WBGU's view, there are two key requirements in this context:

Firstly, the UNFCCC is the most important global reference institution for data on emissions and emissions reductions. This data, which is broken down by country and sector, facilitates not only the evaluation of countries' individual contributions to global emissions but also their contributions to climate change mitigation. It is therefore particularly important that data on the greenhouse gas emissions produced through bioenergy use are collected in a way that they provide an accurate picture of its actual contribution to climate change mitigation. The same applies to the processes established by the Kyoto Protocol for the accounting of emissions and their allocation to the countries which have adopted binding emission reduction targets.

Secondly, it is anticipated that with the adoption of ever more ambitious reduction targets the steering effect of the UNFCCC/Kyoto Protocol and its successor regime on bioenergy use will increase. The direct and indirect incentives which the regime creates for bioenergy use should be coordinated to ensure that they contribute to the attainment of the maximum possible global emissions reductions. This applies to the flexible mechanisms in general and the Clean Development Mechanism (CDM) in particular.

Furthermore, it should also be ensured that the UNFCCC's steering effect on bioenergy use does not conflict with other sustainability criteria (Chapter 3).

The following section looks at the impact of existing rules on the bioenergy sector, provides an overview of current negotiation processes and, on that basis, evaluates and discusses how the rules within the UNFCCC framework should be developed further in order to achieve maximum compatibility between bioenergy use and climate change mitigation.

10.2.2
Evaluation, attribution and accounting of emissions

10.2.2.1
The current rules and associated problems

The parties to the UNFCCC are obliged to produce, publish and regularly update inventories of their national greenhouse gas emissions in accordance with the Guidelines for National Greenhouse Gas Inventories of the Intergovernmental Panel on Climate Change (IPCC, 2006). These inventories provide the core data for the global monitoring of anthropogenic emissions. The accounting of emissions within the framework of the Kyoto Protocol is distinct from inventorization, firstly because this accounting applies only to those countries which have committed to binding emission reduction targets (Annex I countries), and secondly because only a part of the inventorized emissions is included in the accounting.

INVENTORIES: CARBON NEUTRALITY AND ATTRIBUTION OF EMISSIONS
Under the IPCC's current guidelines, the use of biomass for energy generation or transport is treated generally as carbon-neutral. While the CH_4 and N_2O emissions produced by bioenergy use are counted in the inventories, the CO_2 emissions from biomass are reported as zero in the energy sector (IPCC, 2006). For biomass with a short lifespan (e.g. annual energy

crops), this treatment is justified on the grounds that the carbon released was in the recent past captured in the plant by photosynthesis and was thus removed from the atmosphere. In this way, two processes which may lie at some spatial and temporal distance from one another are summed up to zero from the outset (Table 10.2-1). CO_2 emissions from energy recovery from wood are also reported as zero in the energy sector. However, CO_2 emissions are calculated as soon as the wood is felled; the harvest of wood is treated as if the carbon stored in the wood is released to the atmosphere immediately (Table 10.2-1, Part B). This does not take account of the possible use of wood products as an industrial feedstock, which under some circumstances would capture the carbon for several decades (Section 5.3 and 5.5). When preparing the inventories under the UNFCCC, countries can opt to use alternative accounting methods (Box 10.2-1), but for the purposes of the Kyoto Protocol the method described here is mandatory.

INCOMPLETE ACCOUNTING IN ANNEX I COUNTRIES
The attribution of emissions described above gives rise to specific consequences for the accounting of emissions from bioenergy use under the Kyoto Protocol. Whereas Annex I countries must as a general principle include emissions produced in the energy, industrial, waste and agricultural sectors in their accounting, this is only partly the case in the land use, land-use change and forestry sector (LULUCF). Under Article 3.3 of the Kyoto Protocol, only the net changes in carbon stocks resulting from afforestation, reforestation and deforestation since 1990 have to be accounted mandatorily towards the commitments of each Annex I party. Other activities, whose inclusion is optional, are listed in Article 3.4. These include changes in greenhouse gas emissions by sources and removals by sinks in the agricultural soils and the land-use change and forestry categories. Prior to the start of the commitment period, each Annex I country must state whether and which of these activities are to be used to meet its commitments (UBA, 2003b). As the CO_2 emissions from land use are covered by Article 3.4, this in practice leaves only the non-CO_2 emissions to be accounted for on a mandatory basis in the agricultural sector (Benndorf, personal communication). When it comes to the harvesting of timber, there is no overall regulation defining exactly what constitutes the deforestation that must be accounted for under Article 3.3, or whether the forestry activity in question is the type of activity which only needs to be accounted for if the country concerned has opted for its inclusion under Article 3.4. It is up to each individual country to propose an appropriate arrangement here (Höhne et al., 2007).

Table 10.2-1
Inventory and accounting practices employed to date in the first commitment period of the Kyoto Protocol for the greenhouse gas balance chain associated with the use of bioenergy. A: short-lived biomass (energy crops). B: wood. The columns show the different sub-processes associated with the use of bioenergy. These phases can occur separately at different times and in different locations. Most notably, some parts of this chain can be implemented in non-Annex I countries, and some in Annex I countries. For phases taking place in non-Annex I countries, neither emissions nor removals are accounted for. The colours show, in each case, whether and in which sector the emissions or changes in emissions in Annex I countries are accounted for. The scheme for wood is consistent with the accounting system laid down in the Kyoto Protocol (Section 10.2.2.2).
Source: WBGU

A: Scheme for energy crops	Land-use change (direct conversion and indirect effects)	Cultivation (agriculture)	Processing and transport	Use (combustion)
Sources	CO_2 (biogenic)	CO_2 (biogenic): e.g. higher emissions through changes in cultivation	CO_2 (biogenic)	CO_2 (biogenic)
		CO_2 (fossil): use of machinery	CO_2 (fossil)	
		N_2O, CH_4		N_2O and CH_4
Sinks		CO_2 uptake of crop		
	CO_2 uptake: e.g. carbon reservoir in soil increases	CO_2 uptake: increased uptake of the soil through changes in cultivation		

B: Scheme for wood		Cultivation and harvest (forestry)	Processing/ transport	Use (combustion)
Sources		CO_2 (biogenic): Carbon reservoir decreases (wood harvesting counted as emission)	CO_2 (biogenic)	CO_2 (biogenic)
		CO_2 (fossil): use of machinery	CO_2 (fossil)	
				N_2O and CH_4
Sinks		CO_2 uptake: carbon reservoir in forest increases		

☐ Automatically added to make zero in the inventories and thus not accounted anywhere

☐ Reported as zero in the inventories, not accounted for in the energy sector

☐ Energy sector: accounted for in Annex I countries

☐ Agricultural sector: accounted for in Annex I countries

☐ LULUCF sector: accounting only mandatory in Annex I countries if deforestation or afforestation is involved. Otherwise optional (Article 3.4)

☐ LULUCF sector: can be accounted for in Annex I countries (optional) (Article 3.4)

In sum, this plethora of provisions means that in Annex I countries not all emissions resulting from the cultivation and harvesting of energy crops or the direct or indirect land-use changes associated with its cultivation have to be included in the accounting for the purpose of the reduction commitments (Table 10.2-1). At the same time, the CO_2 released through energy recovery is reported as zero, so the substitution of other emissions through bioenergy use can thus be accounted for as fully avoided emissions. It can therefore be assumed that the eligible emissions reduction resulting from bioenergy use is usually greater than the reduction actually achieved.

LACK OF INCENTIVES
The practice of reporting CO_2 emissions from bioenergy use as zero means that various activities which would contribute to genuine emissions reductions

Box 10.2-1

Harvested wood products

The possible methods of accounting for emissions associated with the use of wood products are discussed in a technical paper by the UNFCCC secretariat in 2003 (UNFCCC, 2003). The uses discussed explicitly include the use of wood as a fuel. The following accounting methods are presented as options:

THE IPCC DEFAULT APPROACH
The assumption in this approach, which is the one currently practised, is that there is no change in the size of the wood products pool, so only (positive or negative) CO_2 emissions resulting from the deforestation or reforestation/afforestation of woodland are counted. Emissions from deforestation (or harvesting) are attributed to the year of deforestation and to the country in which this took place. In the greenhouse gas inventories, deforestation or forest harvesting is therefore treated as an immediate emission, regardless of whether the resulting products (wood) are initially stored, exported, etc. To avoid double counting, the CO_2 emissions resulting from the combustion of wood products (for energy production, waste disposal) are not in principle counted as emissions. This has repercussions, particularly when wood products from a country that has no Kyoto commitments are exported for energy purposes to a country that has such commitments. This method offers especially high incentives to use wood for energy production, since the wood is counted as an emission when it is harvested (and there are no incentives to avoid these emissions for a country with no commitments), while the importing country receives an emissions-free source of energy in line with the Kyoto provisions (although not in the real world).

THE STOCK-CHANGE APPROACH
An alternative to the IPCC default approach is the stock-change approach: this counts changes in both the forest stock and the wood products pool, with the former being allocated to the country in which the wood is grown (producing country) and the latter being attributed to the country in which the wood products are used (consuming country). In both cases the emissions associated with these changes are then allocated to the country in which they occur and at the time at which they are produced. Com-

pared to the IPCC default approach, the use of wood for energy purposes appears to be less attractive in this case. A country that has Kyoto commitments can, for example, import wood products and offset this as a CO_2 sink against its own account. Only when the wood products are used as energy, or when they decay as waste products, will the resulting CO_2 be counted as an emission again.

THE PRODUCTION APPROACH
Like the stock-change approach, both the changes in the forest stock and those in the wood products pool are counted at the point in time at which they occur; however, both are attributed to the producing country. This approach differs from the IPCC default method only in the time at which the emissions resulting from the use of wood products are accounted for. With this method of counting, as with the IPCC default approach, the incentive to use imported biomass as energy is particularly high for countries with commitments as the resulting emissions are not attributed to them. By contrast, the exporting country should be more interested in the export of products with a long lifespan, as the emissions will not be attributed to it until a later stage.

THE ATMOSPHERIC FLOW APPROACH
This approach counts emissions at the place and time at which CO_2 is released into the atmosphere. Only the emissions that enter the atmosphere directly as a result of harvesting are attributed to the producing country. The emissions from the use of wood products, or those caused when they decay, are attributed to the consuming country. In this case, biomass imported for energy use is accounted for in exactly the same way as the use of fossil energy carriers for the consuming country, and therefore cannot be used to reduce its emissions. This means that there would be no incentive to use bioenergy in countries with commitments. However, this process, in the form described by the UNFCCC, does create incentives for the export of bioenergy carriers to developing countries, which is equally questionable in terms of climate protection policy: a country with commitments could use the sink represented by the growth of a forest as an offset, and export the harvested wood to a country with no commitments which is not required to offset the emissions. WBGU therefore presents a modified proposal relating to trade between countries with and without commitments (Section 10.2.2.2).

cannot be included in the accounting. The first to be mentioned here is the use of wood products as an industrial feedstock (Box 10.2-1). As the carbon contained in wood is considered to have been emitted at the time of felling, delayed CO_2 release through long-term use of wood products is not rewarded under the current rules, even though it is desirable from a climate perspective (Sections 5.3 and 5.5). A further example is CO_2 capture and storage in energy generation from biomass. This type of technology would offer the opportunity for net removal of CO_2 from the atmosphere (Chapter 6; Box 6.8-1). However, as the CO_2 is treated in the accounting as if it had never been emitted at all, even without storage, the current accounting modalities in the first commitment period of the Kyoto Protocol offer no incentives for

the sequestration of biogenic CO_2 (Grönkvist et al., 2006).

TRADE IN BIOMASS FOR ENERGY USE
If biomass is produced in a non-Annex I country and is then used for bioenergy in an Annex I country, the problem of incomplete accounting is exacerbated. In this case, the emissions arising during production are not accounted for at all under the Kyoto regime, with the result that the eligible emissions reduction is almost always greater than the reduction actually achieved. The conversion of tropical forests into bioenergy plantations, in particular, can produce very high emissions and thus give rise to an extremely negative Greenhouse Gas (GHG) balance for bioenergy use (Section 7.3). So without additional meas-

ures, even the use of bioenergy which actually results in a net increase of emissions could be recorded as an emissions reduction by an Annex I state. The current accounting system thus encourages the use of biomass imported from developing countries for energy production, regardless of whether it reduces or increases emissions.

CONFLICT WITH OTHER CLIMATE PROTECTION MEASURES IN THE LAND-USE SECTOR

The scenarios described above show that the direct and indirect incentives created by the UNFCCC can steer bioenergy use into channels which are unfavourable from a climate perspective. Moreover, further indirect effects can be anticipated as a result of increased bioenergy use which put other climate protection efforts at risk (Section 5.5). Of particular significance here is deforestation in developing countries, which currently accounts for more than 20 per cent of global anthropogenic CO_2 emissions (IPCC, 2007c), but the conversion of grassland for the cultivation of energy crops (Section 4.2.3.3) should also be mentioned. Even if no forested areas are cleared specifically for this purpose, or grassland converted, the indirect conversion of such areas is likely to increase as the production of energy crops squeezes out other types of usage, with the result that the cultivation of other crops or pasturage is displaced to other hitherto relatively untouched or less intensively used land (Searchinger et al., 2008; Section 5.5). At present, the UNFCCC does not offer developing countries any direct incentives to reduce the conversion of forested areas or grasslands into cropland. The emerging bioenergy boom therefore makes it even more imperative to account for all emissions from land use, land-use change and forestry (LULUCF) in the UNFCCC and the (post-) Kyoto mechanism and to implement adequate incentives for developing countries to protect their terrestrial carbon stocks and combat deforestation.

10.2.2.2
Criteria and opportunities for the further development of the rules

The problems described above demonstrate the need for bioenergy production and use to be put on the agenda of the UNFCCC bodies as a matter of urgency, for reform of the accounting modalities for the Kyoto commitments, and for modification of the procedures for the attribution of emissions in the greenhouse gas inventories.

CAP AND TRADE FOR ALL COUNTRIES AND ALL EMISSIONS

In theory, there is an elegant solution to many of the problems described above: namely to agree emissions limitations ('caps') for all countries, sectors and emissions, including LULUCF. If a system of 'cap and trade for all countries and emissions (and sectors)' were established, the incentive to cut emissions would be created via the price of emissions allowances in a global carbon market. Provided that all emissions are included as close to the time of their occurrence as possible, this would be an effective and economically efficient approach, regardless, initially, of which actor the emissions are assigned to. It is unlikely, however, that developing countries and newly industrializing economies will agree to emission limitations of their own in the foreseeable future. Moreover, there are considerable practical problems associated with the implementation of a purely market-based solution, especially for the LULUCF sector. So although emission limitations for all countries and all greenhouse gas emissions are a desirable long-term goal for current policy-makers, and market-based solutions – as far as practicable and effective – should serve as a model in the choice of instruments, the initial task must be to identify options for a transitional regime. Apart from closing the loopholes described above, removing other deficits and remedying the inadequate treatment of LULUCF, a further task is to identify accounting modalities which make it easier for developing countries to sign up to commitments which may even include emission limitations.

FROM CRADLE TO GRAVE

One way of avoiding any incentives to displace emissions through the import of bioenergy carriers would be to follow the approach of various certification initiatives and attribute the emissions from the entire production chain to the end user of bioenergy, namely the emissions from land-use change, cultivation, processing, transport and use. However, there are considerable problems with this method: as the production chain associated with this end product, i.e. the bioenergy carrier, is not transparent, there must be full and complete accounting of all steps. Furthermore, a decision must be taken on the allocation of emissions to co-products and the point in time at which land-use emissions should be allocated. In order to avoid double counting, the emissions allocated to the end product must be deducted from the original sectors in the inventories. This means, for example, that the emissions resulting from the use of machinery during the production of biomass would have to be deducted from the emissions produced in the energy sector. This would not only entail a considerable amount of bureaucracy and monitoring, but would ultimately require the complete restructuring of inventories, whereas at present – and for good reason – the emissions are generally assigned to the country on whose territory they arise and which can best control them, at the point in time at which they are released. Furthermore, it would be hard to justify why this established principle should not then be broken in respect of other products as well; indeed this would be imperative with co-products. If this 'Pandora's box' is opened, the inventorization and accounting procedures would ultimately become unmanageable.

ATTRIBUTION OF EMISSIONS: THE ATMOSPHERIC FLOW APPROACH

WBGU takes the view that the emissions should, as a matter of principle, be attributed to the state on whose territory they are produced, and calls for the existing exemption for bioenergy to be abolished. The accounting of emissions from the use of wood and wood products should also be based on the time and place that actual emissions are produced. The atmospheric flow approach (Box 10.2-1) seems generally suitable for this purpose and should be extended to bioenergy production as a whole. This would mean that during harvesting, only the CO_2 emissions which are actually produced would be counted; the same applies, accordingly, to the CO_2 emissions produced during the use of the harvested product. However, only some of these emissions can be calculated reliably. This includes those from bioenergy, waste incineration, landfill gas, domestic heating and the natural decomposition of building materials made from timber (UNFCCC, 2003). For that reason, a combined arrangement may be sensible: harvest products which constitute or have been converted into a tradable energy carrier (wood for energy recovery, biofuels) are treated in accordance with the atmospheric flow approach, as their use and the associated CO_2 emissions take place at a defined point in time and are easily measurable. These emissions are also already reported in the national inventories, although they are not counted when summing up national emissions, as CO_2 emissions from bioenergy are treated as zero. This can create incentives for a greater focus on technical efficiency in bioenergy use, rather than treating efficiency improvements and fuel substitution as being of equal value. Emissions from wood used as an industrial feedstock, on the other hand, should be counted using a hypothetical annual emissions rate based on a previously determined country-specific lifespan. Although a flat-rate and very low lifespan should be the basis at the start, countries should have the option of extending this period on production of appropriate evidence. For the other harvest products (food, residues), where a very short

lifespan can be assumed, the IPCC default method should continue to be used. To safeguard consistency, the real sources must be offset by real sinks. In the case of wood products, this is already guaranteed in the sense that the increase in the carbon reservoir through the growth of the forest is counted as a sink, i.e. carbon uptake. In order to transfer the atmospheric flow scheme to other biogenic energy carriers, carbon uptake during growth should be counted as a CO_2 sink. Full accounting of LULUCF emissions in Annex I states

Even if emissions are attributed on the basis of the atmospheric flow approach, the problem of incomplete accounting in Annex I countries is not yet resolved. Therefore in addition to the sectors which hitherto have been included, the emissions arising in LULUCF in Annex I countries must also be set against emissions caps in the Annex I countries on a full and binding basis (or, if appropriate, against emissions caps in the LULUCF sector). This entails mandatory accounting of the activities currently covered by Articles 3.3 and 3.4 of the Kyoto Protocol and its extension to all anthropogenic emissions in the LULUCF sector.

This arrangement, which combines the atmospheric flow approach with the full accounting of anthropogenic greenhouse gas flows in the LULUCF sector, offers incentives for Annex I countries to substitute the use of fossil energy carriers with biomass produced in their own countries, if this results in overall emissions reductions. However, it does not offer any direct incentive to use imported biomass for energy generation, as the use of imported biomass for energy would thus be equivalent to the use of fossil fuels.

ACCOUNTING IN THE TRADE BETWEEN ANNEX I AND NON-ANNEX I COUNTRIES

The accounting modalities described would automatically reduce the problem of emissions displacement to developing countries that arises in the context of imported biomass. As in the context of biomass imports for energy use or wood as an industrial feedstock, it is the importing country which is responsible for the emissions released during use, there is initially no incentive to import biomass on climate protection grounds. A specific mechanism could be established to create opportunities for the recognition, as a climate protection measure, of biomass produced in non-Annex I countries but used in Annex I countries, and whose cultivation-related emissions are low. A project-based approach similar to the CDM could be adopted here. In 'projects for the export of bioenergy carriers/wood products to Annex I countries', the cultivation-related emissions and CO_2 uptake could be accounted for in the products destined for export, and 'carbon fixing certificates' could be issued. The issue of a certificate would be conditional on the biomass produced genuinely being exported to an Annex I country. These certificates would be recognized as greenhouse gas reductions in Annex I countries and treated like a Certified Emission Reduction (CER) from the CDM (Section 10.2.3). In this context it must be ensured that double counting, e.g. in the context of forestry projects, is avoided. Only in cases in which the GHG balance looks promising will project developers be found for this type of 'carbon fixing project'. In all other cases, exports from non-Annex I countries should not be promoted through the climate regime. This approach would help to ensure that land-use changes e.g. deforestation, that are the direct result of bioenergy exports are avoided in non-Annex I countries. However, additional instruments are required to prevent substantial carbon emissions from occurring indirectly if, for example, the previous land use is displaced to other carbon-rich sites. One method of undertaking a quantitative estimate of the associated emissions is discussed in Section 7.3 (Box 7.3-2).

Logically, this system would mean that the import of bioenergy carriers from Annex I countries by non-Annex I countries would result in the CO_2 contained in the products immediately being accounted for as emissions of the Annex I country, as the products would have left the accounting area.

LAND-USE CHANGES IN NON-ANNEX I COUNTRIES/
AVOIDED DEFORESTATION
The accounting system described above could defuse most of the problems outlined in Section 10.2.2.1. However, it does not resolve the negative effect that increasing bioenergy production in non-Annex I countries intensifies the pressure on natural ecosystems and therefore also on forests. While the direct conversion of forest into plantations for the cultivation of energy crops or the 'forest yield' could largely be controlled through a bioenergy-based scheme (Section 10.3), the indirect effects, namely the displacement of other forms of crops to previously forested areas, ultimately require a more comprehensive regulatory approach.

Options for reducing the increasing deforestation in developing countries are currently being negotiated within the UNFCCC framework (Box 10.2-2).

In the event that an effective regime for reducing the emissions from deforestation and forest degradation is established within the REDD process (Reducing Emissions from Deforestation and Degradation), membership of this REDD regime should be a prerequisite for developing countries' participation in 'carbon fixing projects'. WBGU considers that an appropriate REDD regime should comprise at least the following elements:

- It should create effective incentives for the swift achievement of real emissions reductions by reducing deforestation. These emissions reductions should be achieved at national level in order to avoid leakage.
- Beyond the direct reduction of emissions, incentives should also be created to permanently protect the natural carbon reservoirs, such as tropical primary forests, from deforestation and degradation.
- Incentives should also be established to limit emissions from grassland conversion.
- Perverse incentives, e.g. incentives which encourage more destruction in order to generate particularly high rewards for ceasing this destruction, must be excluded.
- The regime must mobilize adequate international financial transfers.

In order to fulfil these requirements, the regime should consist of a combination of national targets to limit emissions and project-based procedures: for example, participating developing countries could commit to limit their future national emissions from land-use changes. The emissions limit could be based on past average annual emissions, while the ceiling could be made dependent on the amount of emissions and the country's economic performance. In addition, financial transfers should be provided by Annex I countries if emissions reductions go beyond the level agreed. This would create an incentive to reduce the current high emissions from land-use changes quickly, while at the same time encouraging the developing countries to take some responsibility. In addition, it could be considered to promote the permanent protection of terrestrial carbon reservoirs by giving participating developing countries financial support if they place designated areas under protection. The REDD regime would ideally form part of a comprehensive agreement to preserve carbon stocks of terrestrial ecosystems within the UNFCCC (Section 10.2.4), with financial transfers also being regulated within this framework.

Measures to protect sinks in developing countries are currently being financed through various government/private sector mechanisms (Box 10.2-3). In some cases, these mechanisms go beyond sink protection as defined in the UNFCCC and also finance measures for carbon stock protection, including the protection of tropical forests. These mechanisms could in future supplement a UNFCCC regime, but cannot replace a comprehensive solution within the UNFCCC (Box 10.2-2).

Box 10.2-2

Reducing emissions from deforestation and degradation (REDD) in the UNFCCC

At the suggestion of Papua New Guinea and Costa Rica, a new topic was placed on the agenda for the 11th session of the Conference of Parties to the United Nations Framework Convention on Climate Change in Montreal in 2005: the inclusion of emissions from deforestation in developing countries within the climate protection regime. These emissions account for more than 20 per cent of anthropogenic CO_2 emissions. While there is an incentive for the industrialized countries to abandon their deforestation activities during the first Kyoto commitment period from 2008 to 2012, as the emissions associated with this are counted in their permitted emissions as a matter of principle, there are no such inducements for developing countries and emerging economies. Yet all the signatory states of the UN Framework Convention on Climate Change have undertaken to promote the conservation and enhancement of sinks, including forests. (Article 4.1d UNFCCC).

The topic was addressed in the Bali Action Plan in 2007 which governs the negotiation of a new climate regime for the post-2012 period. The regime is expected to be adopted at the 15th Conference of Parties in December 2009 in Copenhagen. In specific terms, policies and incentives should be created to reduce emissions from deforestation and forest degradation in developing countries. The role of

forest conservation, the sustainable management of forests, the enhancement of forest carbon stocks and afforestation are also to be discussed. The choice of policy instruments to be used is as yet undecided and the various financing opportunities, together with the type of activities to be included, are a matter for the negotiations. In the interim, the countries are required to step up their ongoing measures to avoid these emissions, and adopt measures to prepare for a future regime, e.g. collecting data, conducting pilot projects, etc.

The costs of reducing a relevant proportion of the emissions from deforestation are estimated at several billion US dollars per year, with opportunity costs (foregone income from deforestation) accounting for the majority of these expenses (Grieg-Gran, 2006; Nabuurs et al., 2007). Many developing countries take the view that the industrialized countries should raise these funds by tightening their emission reduction goals and then acquiring REDD certificates on the international carbon markets (UNFCCC, 2007c). However, one of the countries that does not share this view is Brazil: Brazil is interested in a voluntary agreement whereby the industrialized countries pay into a fund from which the emission reductions achieved in the developing countries are subsequently rewarded (UNFCCC, 2007d). Whereas the industrialized countries would like to use national emission or deforestation rates as benchmarks in order to avoid displacement effects, some developing countries (e.g. Colombia, Paraguay, Peru) are interested in project-based or sub-national solutions.

Box 10.2-3

International payments to conserve carbon stocks and sinks

Payments for the conservation of carbon sinks and stocks in developing countries and emerging economies are made within the framework of various, generally project-based mechanisms. A distinction can be made between mechanisms which in principle permit offsetting against reduction commitments within the framework of the Kyoto Protocol, and those that may not be offset under Kyoto. The flexible mechanisms form the core of the first group, i.e. the Clean Development Mechanism (CDM; Section 10.2.3) and Joint Implementation. On the other hand, payments are also made which are not directly induced by the Kyoto Protocol; these include voluntary payments by private individuals, companies and public sector actors wishing to demonstrate their commitment to climate change mitigation or offset their own greenhouse gas emissions (e.g. climate-neutral flying). The granting of credits related to conserving carbon stocks and sinks is, depending on the organization, linked to criteria of varying degrees of stringency which aim to achieve a lasting mitigation effect (GTZ, 2007b; Neef et al.,

2007). Payments also come from various state-backed environmental funds which pool financial resources and specifically invest in climate protection projects abroad. These mainly comprise projects on land use, land-use change and forestry. The initiators of the funds range from actors at federal state level (including the Oregon Climate Trust) to those at multilateral level (including the BioCarbonFund of the Carbon Finance Unit or the Forest Carbon Partnership Facility, both of which come under the umbrella of the World Bank; World Bank 2006a, b, 2007; Neef et al., 2007; UNFCCC, 2007a). In principle, international compensation payments for conserving carbon stocks or sinks could come from national emissions trading systems: companies that are committed to acquiring emission certificates within their own countries could fulfil this obligation by making a corresponding investment in the LULUCF sector for eligible carbon credits (from their own country or from abroad). Due, among other things, to the difficulty of safeguarding a lasting mitigation effect, this opportunity has scarcely featured in national trading systems to date. The international payments for such land uses are therefore hardly significant in this context and, outside the CDM, sink projects abroad, including forests, are only reported for the Chicago Climate Exchange (Neef et al., 2007).

10.2.3
Bioenergy and the Clean Development Mechanism

The purpose of the CDM is defined in Article 12 (2) of the Kyoto Protocol: it is to assist developing countries in achieving sustainable development and in contributing to the ultimate objective of the

UNFCCC. It is also intended to assist industrialized countries in achieving compliance with their quantified emission limitation and reduction commitments cost-effectively. The CDM offers incentives to investors to finance or implement greenhouse gas reduction measures in developing countries. In return, the investors receive tradable certified emissions reductions (CERs). Annex I countries can offset their CERs against their own emissions reductions targets.

10.2.3.1
Existing rules on bioenergy and its evaluation

In the CDM, bioenergy is mainly dealt with via projects for bioenergy use in developing countries. In certain cases, aspects of land use and land-use change and their treatment may also be relevant in the CDM.

Bioenergy use in developing countries can achieve emissions reductions through the substitution of fossil fuels by biogenic fuels (Section 7.3). In addition, under certain circumstances, efficiency increases in existing traditional forms of bioenergy use can also contribute to emissions reductions. This is the case if, for example, less fuelwood is used as a result of increased efficiency, so that existing terrestrial carbon sinks are not degraded or long-lasting wood products are not converted. Reductions of non-CO_2 emissions can also be achieved through more efficient end-energy use (i.e. combustion; Jürgens et al., 2006). Finally, the cultivation of energy crops can be shaped in such a way that a (temporary) CO_2 sink arises. This means that through targeted crop cultivation for energy recovery, CO_2 is temporarily removed from the atmosphere and captured in biomass (Schlamadinger et al., 2001).

Under the current CDM rules, however, only some of the possible measures to reduce emissions and create sinks are promoted. The granting of CERs and the implementation of CDM projects take place in line with specific modalities and procedures. In order to guarantee the effectiveness of a climate protection project, the anticipated emissions reductions must be determined in advance with reference to a theoretical baseline. For the production of the baseline and for monitoring purposes, methods approved by the CDM Executive Board must be deployed. For new project proposals, recourse can be made to methods which have already been approved or a new methodology can be submitted for approval (Sterk and Arens, 2006).

CDM ELIGIBILITY OF BIOENERGY PROJECTS
The baseline covers, in principle, only those emissions which are listed in Annex A of the Kyoto Protocol, i.e. mainly emissions from fossil energy carriers and non-CO_2 emissions (Jürgens et al., 2006). Emission reductions achieved through the substitution of fossil fuels by biomass are therefore eligible for CDM on principle; examples are the use of biogenic residues and renewable organic biomass.

What is less clear, however, is the CDM eligibility of emissions reductions which are achieved in private households or small businesses, for example, through more energy-efficient forms of traditional bioenergy use. In the case of energy use, the CO_2 emissions produced by combustion are not allocated to the energy sector (Section 10.2.2), but instead are included indirectly in the emissions in the land-use sector which – with the exception of afforestation and reforestation measures – are not eligible for CDM (Höhne et al., 2007). A reduction in CO_2 emissions from traditional bioenergy use – unlike the reduction of non-CO_2 emissions or the substitution of fossil energy carriers – is therefore not eligible for CDM, at least initially.

And yet CDM eligibility of projects for the distribution of energy-efficient stoves, cookers, etc. in households can certainly be justified: if inefficient traditional forms of biomass use result in short- or long-term damage to natural resources, e.g. through deforestation or soil degradation, thus forcing a switch to fossil fuels, then a more efficient thermal use of biomass at present actually is a substitute for the fossil fuels whose use would otherwise be necessary in the long term. The health and socio-economic benefits and the opportunity to carry out projects to a greater extent in poorer developing countries and their rural regions in the future, which are clearly underrepresented in the CDM project portfolio at present, are other arguments in favour of extending CDM eligibility to this type of efficiency-enhancing project (Schneider, 2007; JIKO, 2007; UNFCCC, 2008b).

Current developments in the UNFCCC are pointing towards eligibility for small-scale CDM projects which facilitate the transition to sustainable use of biomass energy, e.g. via biogas stoves or highly efficient cookers which run on biomass (Schneider, 2007; Schlamadinger et al., 2007; JIKO, 2007; UNFCCC, 2008b).

CDM ELIGIBILITY OF SINKS COMBINED WITH BIOMASS PRODUCTION FOR ENERGY
Biomass which is produced specifically for energy recovery has at least a temporary CO_2 sink effect which, under certain circumstances, could be rewarded in the CDM and thus create incentives for certain forms of energy crops, e.g. short-rotation plantations (Schlamadinger et al., 2001; Dutschke et al., 2006). For projects which envisage a carbon uptake (sink function), the current rules on land-use changes and forestry apply; these are set out in the Marrakesh Accords that flesh out the Kyoto Protocol. Under the current procedural rules, only those methods of land-use are permissible which lead to afforestation and reforestation (Höhne et al., 2007; Box 10.2-2). Measures for the direct conservation of

Box 10.2-4

The Global Environment Facility and bioenergy

The conversion of biomass to energy as an option for reducing greenhouse gas emissions is being addressed by the Global Environment Facility (GEF) within its focal area of climate change. The GEF is, however, not intended to take over any tasks that are already being financed through the flexible mechanisms, especially the CDM. Accordingly, the GEF's strategy is not to directly promote specific technologies for renewable energies, but primarily to create a suitable market environment and remove barriers to (market) development. The promotion of sustainable energy from biomass is one of the issues being addressed in the climate change window of the current GEF-4 strategy. Overall, stronger support is to be given not only to the extraction

of bioenergy from residues but also to the production of biomass specifically for energy purposes. The focus will be placed specifically on efficient technologies as well as on eliminating the negative effects of bioenergy use, such as deforestation or soil degradation, and risks to food security. Similarly, a scheme to certify sustainable biomass is to be developed for the programme area (GEF, 2007a, c). Demonstration projects on biofuels have already been financed within the framework of the Small Grants Programme (GEF and UNDP, 2006), and in the current GEF-4 period (2006–2010) funds are also to be deployed for research and development on the sustainable use of biofuels (GEF, 2007b). This programme area is part of the mitigation subprogramme of the GEF's climate change window, which has been allocated funds of just under US$ 1000 million for four years (GEF, 2006).

existing terrestrial carbon stocks or for the greening of degraded land are not rewarded with CDM credits (Jürgens et al., 2006).

The first reforestation and afforestation project to receive registration under the Clean Development Mechanism (China, project duration: 2006–2036) does not relate to bioenergy (UNFCCC, 2008a). Other reforestation and afforestation projects are currently being assessed, along with the relevant methodologies, and methods to couple reforestation/afforestation and subsequent (commercial) biomass use (UNEP-Risoe, 2008).

OVERVIEW OF CDM PROJECTS ON BIOENERGY

Biogenic energy carriers are the subject of the CDM in various project categories. Bioenergy projects (biomass energy) relate to solid fuels; gaseous energy carriers are covered by biogas and landfill gas projects, if they are destined to be used in the generation of electricity (IGES, 2008). By October 2008, a total of 609 bioenergy projects had been submitted (224 had been registered), along with 243 (67) on biogas use in energy recovery. A further 299 (103) are landfill gas projects, of which 101 (37) include electricity generation. The anticipated emissions reductions from all 609 bioenergy projects amount to 36.6 Mt CO_2eq p.a., i.e. 6.7 per cent of the annual emissions reductions expected from all 3,967 CDM projects submitted (546 Mt CO_2eq). Biogas projects are trailing behind (11.5 Mt CO_2eq; 2.1 per cent). For landfill gas projects for electricity generation, an annual reduction of 20.8 Mt CO_2eq (3.8 per cent) is anticipated. Compared with other market segments, the bioenergy sector thus ranks midfield (UNEP-Risoe, 2008; IGES, 2008). In 2006, CERs amounting to 475 million tonnes were traded worldwide, with a value of more than US$ 5000 million (Worldwatch Institute, 2008).

The bioenergy projects registered to date have almost all involved biogenic residues arising in agriculture (mainly bagasse and rice bran), the forestry sector and wood-processing industry (e.g. sawdust, black liquor) (UNEP-Risoe, 2008; IGES, 2008). In 2006, a small-scale project on forest biomass was registered for the first time; this replaces liquid gas in a combustion process in Brazil with wood from eucalyptus forests planted for energy recovery purposes on degraded land (UNFCCC, 2008b). In 2008, methods were approved for the first time that focus on the production of biofuels on the basis of recycled fats and oils and on vegetable oil. Other biofuel-related methods are currently in the assessment phase (JIKO, 2008). Based on the two approved methods, seven biodiesel projects have so far been submitted for assessment (UNEP-Risoe, 2008). Some of these projects involve the use of energy crops, but only for the producer's own use of the fuels in agriculture or in public transport (UNFCCC, 2008c, d).

In a regional breakdown of the 609 bioenergy projects submitted, India (306) and Brazil (100) currently predominate, followed by China (52) and Malaysia (33). Of the biogas projects (totalling 242), the majority have been developed in Thailand (46), Mexico (34), India (32) and the Philippines (31). China (32) and Mexico (11) also account for most of the landfill gas projects for energy recovery (101). As in other CDM sectors, it is apparent that projects for bioenergy use tend to be implemented in Asia and Latin America and in the emerging economies/middle-income countries. International transfers within the CDM framework are complemented by payments via the GEF climate protection window (Box 10.2-4).

EVALUATION

Even allowing for the expected future increase in the number of projects, it is clear that the lion's share of global bioenergy production is taking place outside the CDM. The CDM is at best no more than a flanking mechanism which helps to ensure that bioenergy production and use are directed into a sustainable pathway. Nonetheless, it is important to ensure that incentives for sustainability are created and, above all, that the CDM rules do not create incentives for increasingly non-sustainable bioenergy production and use.

The CDM has been successful in that it has established an international market for reduction activities. On the other hand, the question which arises is to what extent the CDM has achieved its goal, namely to promote sustainable development in developing countries, and whether it genuinely supports ecologically sustainable climate change mitigation (Schneider, 2007). The scope afforded by bioenergy projects illustrates these fundamental problems besetting the CDM:

In terms of the goal of promoting sustainable development, it is notable that for bioenergy, as for other sectors, poorer countries in which traditional forms of bioenergy predominate (Sections 4.1 and 10.8) have rarely been the target of project activities to date (Jürgens et al., 2006; JIKO, 2007). And yet it is in precisely these countries that the CDM could make a contribution to sustainable development: through adapted bioenergy use, local outdoor and indoor air pollution can be reduced, rural energy supply improved and poverty-related deforestation and soil degradation minimized. WBGU therefore recommends that in the debate in the CDM bodies on non-renewable biomass, a greater focus be placed on projects for the substitution of inefficient traditional forms of biomass use. In this context, stringent but attainable benchmarks should be established to measure the success and especially the permanence of efficiency increases and modern biomass use. Otherwise, one of the key reasons for including biomass in the CDM is undermined, namely that modern biomass use can reduce fossil fuel use in the long term. These criteria apply to an even greater extent to projects which involve the substitution of fossil fuels by biofuels. In other respects, the outcome of the REDD process should be awaited (Box 10.2-2). If an appropriate regime for rewarding avoided deforestation and degradation is established, this could create scope for the promotion of emissions reduction projects for traditional biomass use.

However, CDM projects which involve the substitution of fossil energy with bioenergy – and indeed the increasing use of bioenergy overall – also pose some risks to the climate. Many bioenergy projects – with the possible exception of those involving the use of residues – are associated with an increased, rather than reduced, requirement for biomass. This directly increases the withdrawal of biomass from the natural environment, and intensifies the pressure to convert forests and grasslands to create space for the cultivation of energy crops. There is also a risk that in adjacent regions, in particular, the conversion of largely intact forest or grassland for other agricultural usages (e.g. pasturage, food production) will increase as an indirect consequence of more intensive cultivation of energy crops and the displacement of these other forms of use to other areas. As a result, greenhouse gas emissions are thus caused by land-use changes outside the CDM projects, but these emissions are not deducted – or are only deducted to a limited extent – from the certified emissions reductions. The issue, then, is whether leakage effects resulting from CDM promotion of bioenergy use can be mitigated by combining CDM projects on bioenergy with afforestation measures (Dutschke et al., 2006; Schlamadinger et al., 2006).

10.2.3.2
Options for further development of the rules

The development of CDM projects currently has only a very limited influence on bioenergy use in developing countries. Any expansion of CDM projects that includes the cultivation of energy crops should be viewed with scepticism unless it can be ensured that the use of land for this purpose will not give rise to leakage effects and result in terrestrially stored carbon being released elsewhere. As in practice, it is impossible to undertake comprehensive GHG balances for such projects and quantify all side- and remote effects, including leakage effects, as precisely as necessary, ancillary criteria could be applied. For example, every certified and credited emissions reduction could be mitigated by an average or country-specific leakage deduction.

The alternative approach – to couple projects for more intensive bioenergy use to afforestation/reforestation measures – may not mitigate leakage effects; indeed, in extreme cases it could actually increase them, as it could result in the displacement of activities from the areas to be reforested to other regions. Ultimately, however, it is the treatment of land use and land-use changes, rather than bioenergy policy in particular, which plays a key role in this context. For example, adaptations of the CDM to the new rules governing the LULUCF sector in the framework of a post-2012 regime must be anticipated. It is possible that the parties to the UNFCCC will adopt a raft of new rules on the treatment of the LULUCF sector, based on experience in this first commitment period. It must then be determined which role CDM should play in this land-use regime. In particular, the problems of leakage and permanence are factors in favour of using the ancillary criteria mentioned above, and CDM should not be given a key role in reducing emissions in the LULUCF sector.

On the other hand, the potential scope afforded by CDM for contributing to the substitution of fossil fuels through bioenergy projects should be expanded. However, it is essential to review more stringently and consistently whether, in the overall balance, the bioenergy used genuinely reduces greenhouse gas emissions compared with the use of fossil fuels (Section 7.3). CERs should only be issued for these reductions. In cases of doubt, project activities should be rejected rather than be promoted by CERs; this should be safeguarded through the application of stringent criteria. The current rules on leakage effects (UNFCCC, 2006) form a good basis for stringent criteria, or should be developed in this direction.

As regards bioenergy projects to reduce inefficient traditional forms of biomass use, the CDM should be utilized in order to contribute to the dissemination of appropriate technologies in poorer developing countries and in rural regions in general, thus preventing the increased use of fossil fuels. Current developments within the UNFCCC framework are moving in the direction of allowing simplified methods for small-scale CDM projects which facilitate the transition to sustainable biomass energy use, e.g. via biogas stoves or highly efficient biomass cookers (UNFCCC, 2008b; Chapter 8). Other amendments to the rules could create additional scope for projects geared towards more efficient energy use of biomass at the level of individual households and small enterprises, and simplify procedures for project developers in the interests of transparency. However, clear incentive effects must be created through the CDM. At the same time, flat-rate deductions on reduction credits could play a role here, e.g. to take into account the fact that emissions reductions through improved efficiency are partly offset by the more frequent cooking or more intensive heating that accompanies greater efficiency.

A very different approach to CDM would arise if in the Kyoto Protocol the automatic assumption of zero emissions from the burning of bioenergy carriers were abandoned and a different approach were adopted to the attribution and accounting of emissions associated with bioenergy (Section 10.2.2). If CO_2 uptake in plant growth is offset against the emissions released during cultivation and harvesting and the emissions associated with use were counted at the place of use, projects to increase the efficiency of traditional biomass use could be directly eligible for CDM. The quantitative potential of CERs which could be issued for the substitution of fossil fuels by bioenergy would noticeably decrease, however.

10.2.4
Approaches to an integrated post-2012 solution

From WBGU's perspective, it is extremely important that the decisions on a post-2012 regime to be taken in Copenhagen in 2009 are shaped in a way which leads to a swift reversal of trends in global greenhouse gas emissions. In order to comply with WBGU's climate guard rail, it is important, on the one hand, for this reversal of trends to be achieved as early as possible, i.e. within a few years of the adoption of the regime, and on the other, for the foundations for a long-term and continuous decrease in global emissions beyond the mid 21st century to be put in place now (WBGU, 2008). To this end, it is essential to agree appropriate incentive schemes to reduce emissions in all the relevant sectors, each encompassing as many countries as possible. It is not necessary for all the various sectors to be subject to the same incentive scheme, however. However, the absorption and release of CO_2 by the terrestrial biosphere differs from the emissions of fossil energy sources in a number of fundamental respects, including measurability, reversibility, long-term controllability and interannual fluctuations (Section 5.5). WBGU has therefore proposed, in its Special Report 2003 (WBGU, 2003), that a separate protocol on the protection of terrestrial carbon reservoirs be adopted within the UNFCCC process. This proposal is discussed in more detail below.

Annex A of the Kyoto Protocol lists those sectors to which the reduction commitments of the Annex I countries apply. It does not include emissions from land-use changes and forestry. Nonetheless, under the current rules Annex I countries may opt for selected emission reduction measures in these sectors to be offset against their reduction commitments.

As discussed above, WBGU considers that CO_2 emissions arising from land use, land-use change and forestry (LULUCF) should be fully and systematically included in the post-2012 regime in order to ensure that the incentive given to bioenergy use by the UNFCCC is based on the actual contribution to climate change mitigation made by this use. This correction would be facilitated even more if bioenergy was treated with the principle used elsewhere of always allocating emissions to the place and time of their release (Section 10.2.2).

WBGU recommends that in future, rather than setting national limitations for all emissions (sectors listed in Annex A and the LULUCF sector), separate commitments should be envisaged for LULUCF. There are various arguments in favour of this approach:

1. It avoids a situation in which the substantial inaccuracies and measuring difficulties, interannual fluctuations and relatively poor amenability to

planning, etc. in the LULUCF sector adversely affect the appropriateness of measures adopted in the Annex A sectors.

2. It avoids the emergence of lock-in effects in emissions-intensive technologies in the Annex A sectors (e.g. energy generation, transport infrastructure, etc.) such that as a result of short-term successes in the LULUCF sector, investments in other sectors are dispensed with and technological developments are delayed (WBGU, 2007). Ultimately, as LULUCF measures are limited in scope, it is fossil energy use which will determine whether the 2° guard rail can be complied with or not (Section 5.5).

3. Since the different sectors also have very different characteristics in terms of time-related dynamics and amenability to planning, it would seem more appropriate, from the point of view of remaining within the 2°C guard rail, to define separate reduction targets rather than one overarching target where the allocation of the target across the various sectors takes place via the market according to the marginal abatement costs for short-term emissions reductions.

4. For emissions from the LULUCF sector, a different approach for the allocation of commitments than that applied in the Annex A sector would seem appropriate. Whereas for the Annex A sectors, an overall allocation key for commitments is a sensible option (e.g. emission rights based on harmonization of absolute per capita emissions), harmonization of per capita emissions from the LULUCF sector would not appear to offer any advantages, especially as the general objective is to increase its sink function.

5. In light of the debate about REDD in the LULUCF sector, the opportunity arises to reach an agreement with emissions reduction targets for a larger group of countries than the Annex I countries.

WBGU therefore recommends that a comprehensive separate agreement on the conservation of the carbon stocks of terrestrial ecosystems be negotiated. This agreement should (1) take up the debate on REDD, (2) replace the existing rules on offsetting reduction commitments in the sectors listed in Annex A to the Kyoto Protocol against sinks (including through CDM activities) and (3) fully include all CO_2 emissions from LULUCF. The allocation of commitments should take place in line with the principle of common but differentiated responsibilities, whereby, besides historical and current emissions, a country's stock of forests and other terrestrial carbon sinks, as well as its economic capacity, should undoubtedly play a role. At the same time, the targets and international compensation mechanisms should be designed in a way which creates incentives for the participation of as many non-Annex I countries as possible (Section 10.2.2.2).

Despite the above-mentioned arguments in favour of separate target agreements, WBGU considers it appropriate from the point of view of economic efficiency to aim, in a second step, for a certain level of fungibility of emission rights from the regime to reduce emissions in the Annex A sectors and the regime for the LULUCF sector. However, on account of measurement problems and other uncertainties attaching to LULUCF emissions, this fungibility should be clearly demarcated and associated with deductions. WBGU identifies a considerable need for research on the precise form that such an arrangement should take.

10.2.5
Conclusions

The UNFCCC, including the Kyoto Protocol and the post-2012 regime, plays a key role in international bioenergy policy, firstly in the evaluation of the impacts of bioenergy use on climate change mitigation, and secondly in terms of steering the incentives for forms of bioenergy use which are at least sustainable from a climate perspective. The existing rules and allocation modalities within the Convention and the Protocol fall far short of these objectives. This is partly due to the fact that bioenergy production and use have an impact on emissions from energy and agriculture, on the one hand, and on emissions from land use and land-use change, on the other. These sectors are subject to divergent requirements in relation to the rules on emissions and emissions reductions. Existing rules which initially served to simplify complex emissions data, or which arose as a result of political compromises that were designed to resolve conflicts of interest, currently impede a consistent approach to the treatment of bioenergy in the climate regime and are creating the wrong incentives.

These insufficiencies in the climate regime must be corrected so that the regime – including the CDM – promotes the use of bioenergy in a way which genuinely makes the greatest possible contribution to preventing dangerous climate change. It is clear, however, that even a highly effective and efficient climate regime which consistently evaluates and monitors all the processes taking place in the bioenergy sector cannot safeguard compliance with other sustainability dimensions (e.g. biosphere conservation, global minimum food production). For that reason, and because it is unlikely that all the relevant countries will commit to emissions ceilings for the time being, other institutions must be involved and other mechanisms deployed (Section 10.3 ff.).

10.3
Standards for the production of bioenergy carriers

In view of the risks identified by WBGU in relation to bioenergy production, notably its potential impacts on the climate and biosphere, as well as the social problems with which it is associated, defining and introducing comprehensive sustainability standards for bioenergy carriers is essential as a regulatory measure. This is the only way to ensure that the international community and individual countries can exert influence over bioenergy production methods (cultivation of feedstocks, conversion) and guarantee their sustainability.

10.3.1
WBGU's criteria for a bioenergy standard

A number of European countries – including Germany, the United Kingdom, the Netherlands and Switzerland – have pioneered the development of sustainability standards for bioenergy carriers, especially liquid biofuels. In Switzerland and the United Kingdom, legislation is already in force which makes the promotion or import of biofuels conditional on compliance with appropriate sustainability criteria. In 2008, in its Proposal for a new directive on the promotion of the use of energy from renewable sources, the European Commission also defined sustainability criteria that would apply to liquid biofuels produced in the EU or imported from other countries and calculated as part of a national blending quota or covered by other promotion measures such as tax reductions. However, the European Commission's approach and the unilateral schemes in Switzerland and the United Kingdom do not impose a general ban on the import and use of bioenergy products which do not meet these standards.

WBGU, by contrast, recommends the introduction of minimum standards as a basic prerequisite for the use of all bioenergy products. These minimum standards could be introduced at national level initially but should be established at international level in the long term, not least in view of trade rules. Moreover, the minimum standard should apply not only to biofuels but to all bioenergy carriers produced from renewable feedstocks and crop residues. This includes end products such as biomethane, biofuels, electricity from biomass and wood pellets, and also bioenergy feedstocks, i.e. energy crops, wood products, vegetable oils and crop residues such as straw and forest residues used in energy generation. Cultivation and/or supply of renewable energy feedstocks (energy crops, organic residues) for energy recovery should only be promoted if they make a specific contribution to sustainable land use (promotion criteria for biomass production: Section 10.3.1.2).

In principle, it would be desirable, from an ecological perspective, for the production of all biomass products, including food and animal feed and biomass for feedstock use, to meet the same minimum standards. The only exception should be stipulations on greenhouse gas reduction potential: these should not be applied as stringently to the production of vital goods, such as foodstuffs, compared with bioenergy carriers. Nonetheless, the aim should still be to minimize greenhouse gas emissions from food and animal feed cultivation. The fact remains, however, that introducing a general biomass standard is a complex task and is difficult to enforce politically in the short term, so the minimum standard for bioenergy carriers, as called for by WBGU, must be viewed as the first step towards a global land-use standard.

In implementing the minimum standard, it must be borne in mind that some bioenergy feedstocks such as rapeseed and palm oil, soya or grain can be used both for conversion to energy and for food and animal feed. A minimum standard for these products would automatically affect food and feed producers as well. However, WBGU only recommends this over the longer term. The duty to furnish proof that the standard has been adhered to should therefore lie in the first instance with the entity marketing the end product (e.g. biomethane, electricity from biomass, biofuels, wood pellets). For feedstocks and inputs, this would initially only create an indirect obligation to comply with the required minimum standard, which would only apply once it became clear that the biomass is destined for use as an energy source. The entity marketing the bioenergy end product would thus be required to provide evidence that the feedstocks it has purchased meet the minimum standard.

A different situation arises in relation to the promotion of feedstock cultivation. Here, verification of compliance with the promotion criteria (Section 10.3.1.2) should take place directly with the feedstock producers concerned (generally farmers and forest managers). In both cases, regulations are therefore required to establish a methodology for the assessment of feedstocks and semi-finished bioenergy products. For GHG emissions from biomass inputs, WBGU recommends criteria for land use, e.g. a maximum limit for emissions from direct and indirect land-use changes and from cultivation, as emissions from feedstock cultivation crucially influence a bioenergy carrier's energy balance throughout its life cycle. Failure to comply with this maximum limit would make feedstocks ineligible for promotion schemes for cultivation, or for further processing into bioenergy products. Definition of this land-use standard (Section 10.3.1.1) would prepare the way for the subsequent extension of binding standards to all biomass products, including food, animal feed and biomass for feedstock use. If the conversion of biomass is promoted in addition to feedstock cultivation – e.g. within the framework of development cooperation – technical standards should also be defined, to be applied to the various conversion methods, in order to limit GHG emissions during the conversion process.

10.3.1.1
A minimum standard for bioenergy carriers

WBGU's recommendation for a minimum standard for bioenergy carriers is based on the criteria, set out in Chapter 3, for the sustainability of bioenergy (ecological and socio-economic guard rails and other sustainability criteria). The minimum standard is initially formulated in general terms without specific reference to subsequent implementation (whether this be national/regional/international or voluntary/statutory). The options for implementing the standard are described in Section 10.3.2. The minimum standard takes account of climate protection principles, on the one hand, and, on the other, principles pertaining to biosphere and soil protection, sustainable water resources management and land use as well as the need to safeguard decent working conditions in the production of bioenergy carriers. WBGU confines itself in this context to a small number of key principles in order to safeguard compliance with international trade rules and increase the likelihood of the minimum standard being implemented swiftly. In WBGU's view, the aspect of food security cannot be adequately encompassed in a minimum standard for bioenergy at individual producer level, so food security – although defined as a guard rail and sustainability factor in Chapter 3 – is not included in the following catalogue of principles.

REDUCING GREENHOUSE GAS EMISSIONS THROUGH THE USE OF BIOENERGY CARRIERS
WBGU considers that it would be suitable to introduce a rule which, taking direct and indirect land-use changes into account, ensures that the use of bioenergy carriers reduces greenhouse gas emissions by at least 30 t CO_2eq per TJ of raw biomass used in comparison with fossil fuels. For biofuels, this is equivalent to a GHG reduction of around 50 per cent compared to fossil fuels. If an additional GHG reduction from co-products is demonstrated, these products can also be factored in. The methodology for calculating GHG emissions is explained in more detail in Section 7.3.

For the cultivation of biomass feedstocks, the greenhouse gas emissions produced from direct and indirect land-use changes from a specific reference date onwards, including the loss of the sink effect, should not exceed the amount of CO_2 that can be fixed on the same site (i.e. on the land itself and in harvest products) by the energy crop within ten years (land-use standard). The analysis should also take account of the predicted emissions from cultivation, e.g. N_2O emissions from the use of fertilizers.

AVOIDING INDIRECT LAND-USE CHANGE
Indirect land-use change – i.e. the displacement of productive forms of land use (e.g. cultivation of food and animal feed, pasturage) – by the cultivation of energy crops in areas that are valuable for biodiversity and climate protection should be avoided. GHG emissions from indirect land-use change should therefore be factored into the GHG life-cycle analysis of a bioenergy carrier. WBGU recommends that initially, accounting for GHG emissions from indirect land-use change should take place using the 'iLUC factor' proposed by the Öko-Institut (Institute for Applied Ecology) (Fritsche and Wiegmann, 2008; Box 7.3-2), using 50 per cent of the theoretical value. Box 10.3-1 presents this and other methods of accounting for indirect land-use change in the context of a bioenergy standard.

PRESERVING PROTECTED AREAS, ECOSYSTEMS AND AREAS WITH A HIGH NATURE CONSERVATION VALUE
In order to safeguard biodiversity and ecosystem services, energy crops should not be cultivated in existing protected areas or in any elements of protected area systems (e.g. corridors). Similarly, energy crops should not be cultivated in areas which, on a specific reference date (e.g. 1.1.2008), have been identified as areas of high conservation value, particularly natural ecosystems such as primary forests or wetlands, species-rich grasslands or savannahs. These exclusion zones must be identified prior to cultivation.

Cultivation systems for energy crops should be embedded in the landscape (linkage with protected area systems, preservation of landscape diversity and agrobiodiversity, identified of unused sub-areas). When cultivating energy crops, it is therefore essential to establish adequate buffer zones adjacent to protected areas/areas with a high nature conservation value. In some cases, the use of biomass from protected areas or areas with a high nature conservation value may be compatible with protection (Section 5.4.1). Prior to any use of potentially invasive non-local species, an assessment of the ecological risks must be undertaken (Box 5.4-2). To facilitate the effective monitoring of land use and land-

Box 10.3-1

Ways of accounting for indirect land-use changes in a bioenergy standard

Indirect land-use changes (iLUC, also referred to as leakage) are difficult to account for in a bioenergy standard using current methodologies. Nevertheless, some methods for dealing with this problem have been proposed: for example, a criterion could be formulated that limits the cultivation of energy crops to marginal land, e.g. fallow land and land with low productivity in its previous use. This largely prevents competition among uses but also means that, in some cases, the sustainable potential of bioenergy is not fully exploited. In addition, the existence of regulations on the planning of land use and protection areas in the producer country could be a prerequisite for certification (Fehrenbach et al., 2008). Such a criterion would be feasible but inadequate; on the one hand, legislation in many developing countries and emerging economies is not implemented effectively, and on the other, it would only address local and national displacement effects, not international ones.

Alternatively, the life-cycle analysis could include an additional GHG factor which factors in the risk of a potential, indirect land-use change and puts the additional GHG emissions onto the GHG balance of the bioenergy car-

rier. The Öko-Institut, for example, has proposed such a factor, which has already been used in model calculations (iLUC-Faktor; Fritsche and Wiegmann, 2008; Section 7.3). WBGU supports the further development of this factor. It should be borne in mind that such a factor would also have to take account of the impacts of indirect land-use change on biodiversity and food security, in addition to the GHG emissions. In the opinion of WBGU, this could take place in a separate assessment model that can estimate which type of land in which region is more likely to be converted to replace the displaced use; this allows conclusions to be drawn about the consequences for biodiversity and food security. On the basis of these results, a bioenergy carrier can be awarded a bonus or a penalty in the overall assessment. Further research is required, as specific causality relationships can only be represented realistically by means of complex models (Sections 7.3 and 11.1.2).

Ultimately, the problem of indirect land-use change in connection with the cultivation of energy crops can only be completely resolved if all the countries and all types of biomass are included within a uniform standard, or if binding international agreements are concluded on national land-use planning criteria (including systems of protected areas). These must be demonstrably implemented by all the relevant biomass-producing countries (Box 10.3-5).

use change, a global satellite-based land-use register should be created (Section 12.6).

MAINTAINING SOIL QUALITY

The only bioenergy carriers which should be used are those whose feedstock cultivation is proven not to impair soil functions or soil fertility in the long term, i.e. over a period of 300–500 years (e.g. through erosion, salinization, compaction or nutrient depletion; Section 3.1.3). The renewable energy feedstocks used must comply with the provisions of national or regional (e.g. EU) law for the agricultural, forestry and fisheries sector (e.g. correct use of fertilizers, restrictions on pesticide use, avoidance of sediment input into neighbouring ecosystems). When using agricultural residues for energy recovery, it must be demonstrated that an adequate proportion of the residues is left in the fields for the maintenance of nutrient cycles and for humus formation.

SUSTAINABLE USE OF FOREST RESIDUES

Forest residues which are used for energy recovery must come from a sustainably managed forest. It must also be shown that during their production, an adequate proportion of dead wood remains on the forest floor for the maintenance of nutrient cycles, and that the biological diversity of the forest ecosystem is preserved.

ENSURING THE SUSTAINABLE MANAGEMENT OF WATER RESOURCES

When cultivating energy feedstocks, it must be ensured that water quality and the hydrological regime are not significantly impaired and that there is no overuse of groundwater resources. The provisions of national or regional (e.g. EU) law concerning the protection of water resources in the agricultural, forestry and fisheries sector must be complied with.

AVOIDING THE UNWANTED EFFECTS OF GENETICALLY MODIFIED ORGANISMS

Genetically modified organisms (GMOs) should only be used if introgression from genetically modified plants into wild plants can be prevented, contamination of, or inputs into, the food and animal feed chain can be ruled out, and there are demonstrable benefits backed by reliable statistical data (e.g. improved productivity, reduced environmental impacts). GMOs must, as a matter of principle, comply with national and international biosafety standards, for which the Cartagena Protocol on Biosafety, adopted within the framework of the CBD, is the recognized basis in international law.

OBSERVING BASIC SOCIAL STANDARDS

The minimum standards for bioenergy carriers should also encompass a number of the core labour standards of the International Labour Organization (ILO), particularly a ban on forced labour and child labour (in line with ILO Conventions 29, 105, 138 und 182) and standards for adequate protection of health and safety at work. This is intended to ensure that as well as complying with sustainability criteria, which are mainly motivated by ecological concerns, the production of biomass does not violate basic social standards.

Monitoring of compliance with labour standards is often quite complex and entails a reversal of the burden of proof, so WBGU – for pragmatic reasons – limits its proposal for a minimum standard to a small number of core issues. More far-reaching criteria relating to working conditions, fair business practices and respect for land rights should, however, be met in case of any explicit promotion of biomass feedstock production (Section 10.3.1.2).

As already mentioned, food security is not dealt with specifically in WBGU's minimum standard, as the impacts of bioenergy production on the availability of foodstuffs cannot be determined at individual-producer level. However, the competition between bioenergy cultivation and food security (i.e. the food/fuel nexus) is taken into account in the minimum standard through the evaluation of indirect land-use change (Box 10.3-1). A monitoring and reporting duty for food security would certainly be useful for the evaluation of bioenergy programmes and projects at country level, however.

It is important to ensure that as the next step the recommended principles for the minimum standard are translated into criteria which are as clear and verifiable as possible. In their further elaboration, cropping systems' feedstock- and country-specific characteristics may have to be taken into account. Some degree of flexibility in the national interpretation of the criteria would also improve compliance with the rules of the World Trade Organization (WTO) (Fehrenbach et al., 2008).

In Germany and the European Union, the criteria must be harmonized with standards of good professional practice and Cross Compliance rules in particular. The sustainability standards for bioenergy carriers should be adapted to any legislative amendments relating to agricultural standards in Germany and the EU. Conversely, the good professional practice and Cross Compliance rules should also be reviewed as part of the debate about sustainability criteria for biofuels and other bioenergy carriers and, if necessary, more stringent criteria should be adopted. In particular, they should be enhanced with climate protection aspects as part of the review of the EU's common agricultural policy (CAP) in 2008 and 2009 (Section 12.1). This would ensure that bioenergy products from Europe automatically meet high environmental and climate standards.

10.3.1.2
Promotion criteria for biomass production

Compliance with the minimum standard should be a fundamental prerequisite for the production of bioenergy carriers. If the minimum standard is implemented effectively, bioenergy carriers which are classed as particularly unfavourable in terms of their ecological and social implications would be excluded from the market. Bioenergy usage should only be actively promoted if it enables particularly high climate protection impacts to be achieved. To be eligible for promotion, the use of the bioenergy, taking account of direct and indirect land-use change, should result in a life-cycle greenhouse gas reduction amounting to at least 60 t CO_2eq per TJ of raw biomass used, as compared with fossil fuels (Section 10.7.2). The promotion of biomass feedstock production based on other additional criteria could also be considered. Promotion should take place if the cultivation of energy crops and the provision of other biomass feedstocks help to achieve demonstrable improvements, e.g. in the form of reduced energy poverty or increased climate, biodiversity or soil protection. The latter, in particular, can be achieved if direct and indirect land-use change can be avoided. For that reason, the use of biogenic waste and residues in particular and the cultivation of energy crops at sites which induce little or no displacement of previous forms of use (e.g. food/animal feed cultivation), especially marginal land, are particularly worth promoting.

CRITERIA FOR THE PROMOTION OF BIOGENIC WASTE AND RESIDUES USE
Energy recovery from biogenic waste (including cascade use) and residues should generally be promoted. In the case of residues from agriculture or forestry, sustainability – i.e. the maintenance of soil fertility – should be the criterion determining eligibility for promotion schemes. For the promotion of energy recovery from waste, an analysis of current use should ensure that no displacement effects occur, e.g. to ensure that if biogenic waste has hitherto been utilized in the substance cycle, its diversion into energy recovery does not trigger an unwanted demand for alternative resources.

CRITERIA FOR THE PROMOTION OF ENERGY CROP CULTIVATION
Beyond the minimum standard, the cultivation of energy crops should meet all the following criteria in order to be eligible for promotion:
- *Increase of carbon uptake on the site as a result of cultivation:* Cultivation systems based on perennial energy crops – in which the entire above-ground biomass (grass, wood) or the fruits of per-

ennial oil plants are used as feedstocks – should be promoted. As the below-ground parts of these energy crops are not harvested as biomass, carbon uptake in the soil occurs, thus enhancing soil fertility.

- *Minimization of greenhouse gas emissions in the cultivation process:* Turning and regular tillage of the soil should largely be dispensed with, and there should be an emphasis on year-round ground cover. There should be a demonstrable reduction in the use of primary energy (and therefore greenhouse gas emissions) in the cropping system, with the food or feed production typical of the region serving as the baseline for comparison. Biogenic fertilizers (e.g. farmyard manure, slurry and mulch, but also green manuring through catch crops or bi-cropping, or ash spreading) should be used in preference to synthetic (especially nitrogen) fertilizers. Leaching of nutrients from the land should not occur.

- *Integrated plant protection:* Pesticide use should be substituted as far as possible with integrated plant protection. Among other things, this entails the preferential use of biological and mechanical protection measures and the selection of resistant species.

- *Sustainable use of water resources:* If irrigation is used, this should be based on an effective integrated water resources management plan, to be implemented over a period of at least 15–20 years. Salinization and waterlogging should be avoided.

- *Preservation of biodiversity:* Cropping systems with maximum possible diversity (of varieties, species, cropping sequences, landscapes) should be promoted. Cultivation of potentially invasive species must be avoided. An assessment of the nature conservation value should be undertaken prior to any utilization of marginal land.

- *Decent working conditions:* Only those feedstocks should be promoted for which the producer can provide evidence of the adoption of measures to improve working conditions in the production process, beyond compliance with the ILO's core labour standards. In particular, this must include the payment of a living wage and an agreement on fair working conditions, and it must be demonstrated that other measures to improve health and safety at work have also been adopted.

- *Fair terms of trade for feedstocks, and respect for the local communities' land rights and interests:* If biomass feedstocks are being produced within the framework of contract farming, the purchaser must also provide evidence that all feedstocks have been acquired at the usual prices for the sector, and that he maintains reliable and transparent business relations with local feedstock producers.

In developing countries in particular, the interests of local and indigenous communities and landless persons should be respected. It must be demonstrated that land tenure and ownership rights are also being respected and that cultivation areas have been acquired lawfully.

It is assumed that these criteria are most likely to be met in the cultivation of energy crops on marginal land, which is why WBGU considers this to be particularly worth promoting.

The modernization of traditional bioenergy use can make a valuable contribution to overcoming energy poverty, especially in rural regions in developing countries. Here, WBGU considers that promotion of bioenergy-based projects is justified even if climate protection and promotion criteria are not being met in full.

10.3.2
Schemes for the implementation of standards for bioenergy carriers

The implementation of the minimum standard for bioenergy carriers, described in Section 10.3.1, can in principle take place in various ways. On the one hand, it can be introduced unilaterally by a single actor (private organizations, the state, the European Union) on a unilateral basis; on the other, it can be negotiated between various parties on a bi- and multilateral level. It can also be introduced with varying levels of binding legal force: possible options include, for example, a binding legal standard at national and international level, or the adoption of voluntary guidelines or general principles. A minimum standard can also be introduced in the financing of bioenergy projects by multilateral development banks and international financing institutions.

In light of the high level of uncertainty and risks associated with the production and use of bioenergy, WBGU recommends that the European countries initially introduce the binding standard on a unilateral basis at the level of the European Union, but then ideally extend it via bi- and multilateral agreements between key bioenergy producer and consumer countries. To some extent, existing voluntary standards can serve as a frame of reference in this context, and existing voluntary certification schemes can be used in the verification of binding minimum standards. Rules based solely on voluntary standards are inadequate in the case of bioenergy production, in WBGU's view, as they cannot safeguard sustainability on a broad basis with the requisite level of impact.

Box 10.3-2

EU sustainability criteria for liquid biofuels

Within the framework of the planned EU directive on the promotion of the use of energy from renewable sources (as at: September 2008), the Council of the European Union has proposed the following sustainability criteria for liquid biofuels:

1. GHG REDUCTION
The reduction in greenhouse gases resulting from the use of biofuels compared to fossil fuels should be at least 35 per cent and should increase to at least 50 per cent from 1 January 2017. Biofuels from feedstocks originating from restored and degraded land should be awarded a bonus and the entitlement to higher GHG output.

2. CONSERVING BIODIVERSITY
Biofuels should not be produced from feedstocks that have been extracted from high-value areas in terms of biodiversity, especially those that were primary forests, protected areas or biodiverse grasslands prior to 1 January 2008.

3. PRESERVING CARBON RESERVOIRS
Biofuels should not be produced from feedstocks that have been extracted from areas with large carbon reservoirs, especially those that were wetlands or woodland areas with an area of more than 1 ha (tree height > 5 m; canopy cover > 30 per cent) prior to 1 January 2008.

4. CRITERIA FOR AGRICULTURAL CULTIVATION
Feedstocks produced within the European Union must satisfy the same requirements that have to be met in order to be eligible for direct payment within the framework of the EU agricultural policy (cross compliance). The Commission will submit a report to the Parliament and the Council every two years, starting in 2012, on national compliance with the sustainability criteria in 1.–3. above and the cultivation conditions for biomass feedstocks in member states and third countries (including water and soil conservation, air pollution, and the use of agrochemicals). The Commission will also investigate the possibilities of introducing mandatory criteria for the protection of soil, water and the air by 2015.

5. SOCIAL CRITERIA
The reporting requirement of the Commission mentioned in 4. above should also be extended to the national implementation of the ILO core labour standards – in both the member states and the countries producing biomass feedstocks – as well as to the socio-economic impacts of biomass use in general. In particular, the Commission's reports should comment on the development of food prices and the potential risks to food security. The observance of land rights and other development issues should also be discussed.

The above-mentioned sustainability criteria will initially relate to all liquid biofuels, and the Commission is to investigate whether they can be extended to include other uses of biomass by 2010. In order to establish comprehensive and more stringent sustainability criteria, the development of bi- and multilateral agreements with producer countries, as well as voluntary standards, is to be encouraged (Council of the European Union, 2008).

As regards certification, the EU intends to recognize various schemes, provided that their testing criteria ensure compliance with the requirements of the directive. This procedure, also known as a meta standard approach, enables proof to be furnished of compliance with a standard by means of evidence of adherence to several sub-standards, the individual criteria of which, when taken together, fully cover those of the meta standard.

10.3.2.1

Standards established by private, state and supranational organizations

LEGALLY BINDING MINIMUM STANDARDS AND THEIR IMPLEMENTATION

The introduction of a statutory minimum standard is the most binding form of standard-setting. A minimum standard defines limit or threshold values and criteria which must be met in the production process, with non-compliance resulting in special treatment of the product in question, e.g. a ban on import or use. A general statutory obligation requiring all types of bioenergy carrier to comply with sustainability criteria is currently not envisaged in any country, as far as can be ascertained. Relevant plans by the European Union and individual European countries mainly entail the identification of liquid biofuels which qualify for promotion. The European Union, for example, as part of its strategy on climate change, aims to substitute an increasing share of fossil fuels with biofuels and make compliance with binding minimum standards a prerequisite for promotion in the Member States via national blending quotas or tax reductions

(Section 4.1.2). The requisite sustainability criteria were defined in the Commission's Proposal for a directive on the promotion of the use of energy from renewable sources (Box 10.3-2).

The proposed criteria described in the Box (Council of the European Union, 2008) therefore cover many of the areas which are also addressed in WBGU's minimum criteria for a standard for bioenergy carriers (Section 10.3.1). The proposal can thus be regarded overall as a step in the right direction. However, as the criteria on local environmental impacts and working conditions lack binding legal force, the planned EU directive falls short of the principles which WBGU considers essential. In order to be effective, the proposal should contain binding criteria concerning the impacts of feedstock cultivation on soils and water resources, and also in respect of individual ILO core labour standards (especially forced and child labour). The proposal should also be expanded to include the methodologies for the calculation of indirect land-use changes, as recommended by WBGU (Section 7.3). The use of genetically modified organisms (GMOs) should also be subject to specific criteria. This is the only way to safeguard the ecological and social sustainability of bioenergy with any degree of credibility. The EU's plans to give preference to biofuels from biogenic waste and residues and the planned bonus for the rehabilitation of degraded areas are to be welcomed.

From WBGU's perspective, however, the European Union should go further and introduce a requirement for all biomass products destined for use in energy production in the EU to comply with the minimum standards. Explicit promotion of energy crop cultivation and the supply of other biomass feedstocks should take place only if more stringent promotion criteria (Section 10.3.1.2) are fulfilled and if the cultivation or supply contribute to sustainable land use or a reduction in energy poverty. The European Commission launched a public consultation process in July 2008 to look at the expansion of the sustainability standard to all types of bioenergy; one issue under discussion is the introduction of minimum standards as a general prerequisite for the marketing of bioenergy products. This is WGBU's preferred option.

If legally binding minimum standards for bioenergy carriers are established as envisaged by WBGU, an appropriate internationally applicable certification scheme must be developed within the European Union so that compliance by enterprises within the EU and abroad can be properly documented. The certification scheme should be structured so as to allow the inclusion of other forms of biomass use in the scheme in the medium to long term (Section 10.3.1). This option is provided for, for example, in the International Sustainability and Carbon Certification (ISCC) scheme, which was developed by Meó Consulting Team and is supported by the German Federal Ministry of Food, Agriculture and Consumer Protection (BMELV). The ISCC covers all types of energy carrier and is designed in such a way that all types of biomass can ultimately be covered by the scheme. Depending on the crops and regions, various minimum standards and GHG balances (default values) have been developed as a basis for certification. The ISCC project is based on a system of meta-standards, so existing certification schemes, e.g. for wood or food, are recognized in the validation of compliance with the criteria (Meó Corporate Development, 2008). WBGU supports the ISCC approach. WBGU also recommends that, as envisaged in the EU proposal, products be certified on the basis of the mass balance system. In contrast to the book-and-claim system, which is also under discussion, the mass balance system allows traceability of product flows, which makes the system less susceptible to fraud.

For the implementation of this type of certification scheme, independent certification agencies would have to be created, along with supervisory bodies to monitor the market for bioenergy certification and enforce compliance with the standards both nationally and internationally, and which could also impose penalties in the event of non-compliance. Whereas standard-setting and monitoring are best performed by an individual state or the EU, the creation of appropriate certification systems and the certification process itself can take place in cooperation with market actors. General outsourcing of certification to private accredited certification agencies would reduce the costs of such a scheme to the public purse. However, compliance with the standards must nonetheless be safeguarded by government agencies through random checks. Developing countries and emerging economies should receive financial and technical support for the establishment of their national supervisory bodies.

VOLUNTARY STANDARDS
Various voluntary standards and certification schemes for biomass and energy products have already been established, and in some instances their substantive focus, global applicability and global acceptance mean that they could serve as a reference system for a standard for bioenergy carriers and perhaps be used in the certification process to demonstrate compliance with certification criteria. In general, voluntary standards and certification schemes have an advantage over a binding minimum standard: they can include more stringent criteria as the voluntary nature of the scheme results in a higher level of acceptance among the market actors concerned, and compliance with international trade rules is also more likely to be favourable (Section 10.3.4).

Voluntary standards are attractive to producers if the additional spending on compliance with the standards and the certification process is offset by higher revenue from the sale of the certified products. This only occurs, however, if consumers are willing to pay more for bioenergy carriers from certified sources, allowing the producers to charge a price premium. The extent to which consumers are willing to pay more determines the level of demand for certified products and therefore also the market share of certified bioenergy carriers. Raising consumer awareness of possible negative environmental and social impacts in the production of bioenergy carriers could create a demand pull for sustainably produced bioenergy carriers along the entire supply chain.

In assessing the effectiveness of voluntary certification schemes in the bioenergy sector, it is helpful to cast a glance at existing certification schemes for wood, food and energy products (Table 10.3-1). It is clear that state regulation and incentive schemes play a far more important role in the market for certified renewable energies than in the market for eco-

Table 10.3-1
Selected examples of existing standards and certification systems, and those in the development phase, for biomass products by sector.
Sources: van Dam et al., 2008; Lewandowski and Faaij, 2006; Zarrilli, 2006; Fritsche et al., 2006; Fehrenbach, 2007; Paul, 2008; SEKAB, 2008; Nordic Ecolabel, 2008; MDA, 2008

	Voluntary standards and certification systems at international level	Voluntary standards and certification systems at national and EU level
Forestry	Forest Stewardship Council (FSC) Principles and Criteria, Programme for the Endorsement of Forest Certification Schemes (PEFC), International Tropical Timber Organization (ITTO) guidelines	Malaysia: Malaysian Timber Certification Council (MTCC) certification; Indonesia: Indonesian Ecolabelling Institute (LEI) certification
Agriculture	International Federation of Organic Agriculture Movements (IFOAM), Good Agricultural Practices of the Euro-Retailer Produce Working Group (GlobalGAP, formerly EurepGAP), Sustainable Agriculture Network/Rainforest Alliance (SAN/RA), Generic Fairtrade Standards for Small Farmers' Organizations of Fairtrade Labelling Organizations International (FLO)	Germany: VDLUFA/USL certificate; EU: CAP Cross Compliance
Food	Fairtrade Labelling Organizations International (FLO) standards, UTZ Certified Codes of Conduct (coffee), Max Havelaar	Germany: Fairtrade mark, Bioland label; Switzerland: Bio Suisse label, Max Havelaar label; USA: Organic label; EU: EU organic label
Energy products		EU: European Green Electricity Network standard (EUGENE standard); USA: Green-e; Netherlands: Green Gold Label; Benelux: Electrabel label; Germany: the ok- Power label, Grüner-Strom (green electricity) label; Finland: Ecoenergia label; Switzerland: Naturemade label
General industry	SA 8000 Social Accountability International (SAI) standard, Ethical Trading Initiative (ETI) Base Code; International Labour Organization (ILO) standards; Fairtrade Labelling Organizations International (FLO) standards	EU: European eco-label; USA/EU: Energy Star
Bioenergy carriers *(some in development)*	Roundtable on Sustainable Palm Oil (RSPO), Round Table on Responsible Soy (RTRS), Better Management Practices of the Better Sugarcane Initiative (BSI), Roundtable on Sustainable Biofuels (RSB) at the Swiss EPFL (École Polytechnique Fédérale de Lausanne) in Lausanne	Brazil: Social Fuel seal; Germany: International Sustainability and Carbon Certification (ISCC); Sweden: Verified Sustainable Ethanol Initiative of the company SEKAB; Nordic Ecolabel (The Swan) for biofuels; EU: European standardization process CEN/TC/383 Sustainability criteria for biomass

certified foods, for example. That being the case, experience with the certification of food cannot be applied directly to the bioenergy sector. In contrast to eco-certified foods, which also appeal to consumers' health awareness, voluntary certification of energy products can only achieve a significant market share if there is a generally high level of environmental awareness among consumers. In reality, however, the comparable product, i.e. eco-certified electricity, only accounts for a small share of the mar-

ket in most countries (Willstedt and Bürger, 2006), which gives some indication of consumers' probable level of willingness to pay for certified bioenergy. In WBGU's view, there is a need for further research on consumer behaviour in relation to voluntary standards and certification in the bioenergy sector (acceptance, information needs) (Section 11.4.3).

Nor can it be expected that energy and fuel producers will show sufficient willingness to submit voluntarily to a stringent sustainability standard. In the forestry sector, it has been observed that as the criteria become more stringent, the market share of a voluntary certification scheme tends to decrease (Fehrenbach et al., 2008). Less stringent voluntary schemes such as the Programme for the Endorsement of Forest Certification (PEFC) in the forestry sector or the Global Partnership for Good Agricultural Practice (GlobalGAP) for foods may well be backed by the industry, but in a dynamically expanding bioenergy market they cannot achieve the deep impact that WBGU regards as essential.

It must therefore be assumed that voluntary standards and certification schemes alone cannot guarantee the sustainability of bioenergy use. Nonetheless – as will be demonstrated below – voluntary schemes do have a role to play to some extent as reference systems within the framework of binding minimum standards. As the market for bioenergy carriers is closely linked with other forms of biomass use, e.g. biomass as food or animal feed and as industrial inputs, for which voluntary national and international certification schemes already exist to some extent (Table 10.3-1), potential synergies between certification schemes for bioenergy carriers and existing certification schemes for other forms of biomass use should be exploited.

Table 10.3-1 lists selected examples of existing standards and certification schemes for various product categories and, specifically, standards currently being developed for bioenergy carriers:

Various biomass standards and certification schemes have already been established for biomass for non-energy use, and in some instances their substantive focus, global applicability and global acceptance mean that they could serve as a reference system for a standard for bioenergy carriers and could perhaps also be used in the certification process to demonstrate compliance with certification criteria. In the forestry sector, the standard adopted by the Forest Stewardship Council (FSC) is the benchmark against which a bioenergy standard should be measured. The FSC standard, alongside the less stringent PEFC standard, is currently the most important international standard for sustainable forest management (Kaiser, 2008). It is also recognized as a reference standard for many other standards (e.g. the Nature-Made, ok-Power, and Green Gold labels). In the agricultural sector, the Sustainable Agriculture Standard adopted by the Rainforest Alliance's Sustainable Agriculture Network (SAN/RA) could serve as a model for the elaboration and implementation of the minimum standard recommended by WBGU. The Generic Fairtrade Standards for Small Farmers' Organizations developed by Fairtrade Labelling Organizations International (FLO) specifically for agricultural smallholdings could serve as a model

for the certification of small producers in developing countries. In relation to social standards, the internationally recognized core labour standards defined by the International Labour Organization (ILO) can be used as a basis, although additional elements relating to adequate health and safety at work should be included. More comprehensive social criteria, to be complied with if biomass feedstock production is to be eligible for promotion, are defined in the Social Accountability 8000 standard (SA 8000).

Some European countries already have labelling and certification schemes for green electricity, such as the Green Gold, Electrabel, Ecoenergia, Green Electricity and ok-Power labels (Zarrilli, 2006; Table 10.3-1). In most cases, however, these focus primarily on the origin of the renewable energy and additionality along the entire electricity supply chain. These certification schemes only rarely make specific reference to sustainability criteria. The German ok-Power label and the Netherlands' Green Gold label are exceptions in that they require biomass from agriculture and forestry to achieve compliance with certification schemes in line with the FSC scheme for wood and the 'organic' schemes for agricultural products, or the GlobalGAP standard.

In the past, the European Union has established eco-standards for various product categories such as the EU organic label for foods and the EU's Eco-Label for other products. Standards of good agricultural practice are incorporated into the rules on Cross Compliance within the common agricultural policy (CAP) and already set basic standards for land use in the EU.

Countries which are key exporters of bioenergy carriers, such as Indonesia and Malaysia, have also developed standards and certification schemes for sustainable agriculture and forestry. For example, the Malaysian Timber Certification Council and the Indonesian Ecolabelling Institute run voluntary certification schemes for sustainable forestry.

It should be noted in this context, however, that while the standards and certification schemes mentioned here include criteria for the protection of natural resources such as soils, water resources and biological diversity as well as social criteria, none of the existing schemes includes criteria relating to GHG emissions reductions. This criterion is relevant only in the specific case of bioenergy and must therefore be newly defined, and methodologies discussed, in an appropriate standard or certification scheme.

Due to the increasing production and use of bioenergy, some schemes have now been developed specifically for the bioenergy sector that also contain criteria relating to GHG emissions reductions. A first private certification scheme for bioethanol from Brazil was recently unveiled by Swedish ethanol manufacturer

SEKAB (SEKAB, 2008). At the same time, the Nordic Ecolabel (the Swan) has developed certification criteria for biofuels so that in future the logo – which until now has applied primarily to food and household products – can also be displayed on sustainably produced biofuels (Nordic Ecolabel, 2008). In Germany, the International Sustainability and Carbon Certification (ISCC) scheme has been developed for the certification of bioenergy (Section 10.3.2.1). However, it is currently in the pilot phase and is not yet fully operational. With its Social Fuel Seal, which is tied to the new National Biodiesel Program, Brazil now has a certification scheme for biodiesel which aims to integrate small family farms into the biodiesel production process to a greater extent, although the scheme is at present confined to this particular social dimension of biofuel production (MDA, 2008; GBEP, 2008). In parallel to the development of sustainability criteria for biofuels under the EU directive on the promotion of the use of energy from renewable sources (Box 10.3-5), the European Committee for Standardization (CEN) has initiated the standardization procedure CEN/TC/383 for sustainably produced biomass for energy applications.

As shown in Table 10.3-1, various voluntary sustainability standards are now also being developed specifically for bioenergy carriers at international level. For example, the Roundtable on Sustainable Palm Oil (RSPO), the Round Table on Responsible Soy (RTRS) and the Better Sugar Cane Initiative (BSI) have developed feedstock-specific standards (Doornbosch and Steenblik, 2007), although with the exception of the RSPO these standards are not

yet operational; in other words, standards have been developed, but no related certification schemes have been established. The involvement of feedstock producers, banks and investors, wholesalers and the consumer goods industry in the decision-making bodies is one reason why these multi-stakeholder initiatives sometimes arouse controversy (Fritz, 2007). The Roundtable on Sustainable Biofuels (RSB) was also launched in 2007 as an initiative of the Swiss EPFL (École Polytechnique Fédérale de Lausanne), with a specific focus on standard-setting for the biofuels sector. The RSB is committed to a highly participatory and transparent standard-setting process, with participation being in principle open to anyone (Maier, 2008; Box 10.3-3).

Finally, in response to the growing demand for bioenergy, many international organizations and multilateral initiatives have also looked at the issue of appropriate standards for bioenergy carriers. They include, for example, the UNCTAD BioFuels Initiative, the FAO's International Bioenergy Platform (IBEP), UNEP initiatives on the sustainability of bioenergy, and the Global Bioenergy Partnership (GBEP; Box 10.3-4). Significant research on the sustainability of bioenergy use is also being undertaken by IEA Bioenergy Tasks 31 and 40 (van Dam et al., 2008). At present, however, there is very little international coordination of these numerous projects. More bundling and coordination of activities should therefore be the goal.

Box 10.3-3

Roundtable on Sustainable Biofuels

The Roundtable on Sustainable Biofuels (RSB) is a multilateral forum which has its origins in an initiative of the Swiss EPFL (École Polytechnique Fédérale de Lausanne) in Lausanne. The forum brings together more than 300 different actors, including international companies (e.g. Shell, Bunge), NGOs and associations (e.g. World Wide Fund for Nature, WWF), the Brazilian Sugarcane Industry Association (UNICA), organizations of small-scale farmers, international institutions (e.g. the International Energy Agency, the Forest Stewardship Council, FSC), and UN organizations (UNEP, FAO, UNIDO). Its aim is to work together with civil society groups and experts to produce both global standards and a certification scheme for biofuels. The FSC seal, the standards of SAN/RA, BSI, RSPO, ILO (Table 10.3-1) and current standard-setting processes for bioenergy, including those of the Netherlands and UK governments (RSB, 2008a, b) will serve as the main references. The RSB follows the ISEAL Code of Good Practice for Setting Social and Environmental Standards.

Since the launch of the initiative in 2007, the RSB has been working actively in four open working groups on envi-

ronmental and social standards, the methodology of GHG accounting and the implementation of standards. Principles are formulated via virtual networks, websites (Bioenergy-Wiki), telephone conferences and workshops on various continents. The proposals for standards on sustainable biofuel production, and the relevant discussions, may be viewed in the Internet at any time, and there is also a facility to submit comments. A coordinated list of criteria was published for comments in August 2008, and a revised version is expected to be available in April 2009 (RSB, 2008b).

The RSB therefore occupies an important position in the process of formulating global bioenergy standards. Through its broad and open participation, its work enjoys a high degree of credibility. The Inter-American Development Bank, for example, announced a partnership with the RSB in April 2008, with a view to integrating the sustainability criteria to be formulated into its lending practice (IADB, 2008). It has not yet been decided how the standards, once developed, will be applied. A voluntary certification scheme could be created; however, the proposals of the RSB could also form the basis for a more formal standard-setting process which is linked to political decision-makers and international financial institutions.

The Global Bioenergy Partnership (GBEP) is an important forum at government level. It originated at the suggestion of the UK at the 2005 G8 Summit in Gleneagles, with the aim of advancing the development of a biomass and biofuel market as part of the promotion of renewable energies. In 2007 the partnership was reaffirmed in Heiligendamm with a clear mandate. In addition to the G8 and outreach countries (China, Indian, Mexico, Brazil, South Africa), several UN organizations are involved such as the FAO, UNEP and UNDP. Institutionally, the GBEP is located within the FAO in Rome.

The GBEP is a high-level, intergovernmental discussion platform which engages industrialized countries and emerging economies in a dialogue. Its main tasks are to summarize developments and policies in the individual countries and identify best practice projects (GBEP, 2008). Methods relating to global carbon balancing and sustainability standards are also to be formulated. One working group has been dealing with GHG balances since 2007, and the establishment of a working group on global sustainability standards began in mid-2008 (GBEP, 2008). To take advantage of the momentum created by the global sustainability debates, it intends to present the results as early as 2009 at the G8 summit in Italy.

STANDARDS IN PROJECT FINANCING

The bioenergy standards developed within the framework of the various initiatives can, and should, also be used by international financial organizations and development banks as standards for project financing. For example, the Global Environment Facility (GEF) is planning to adopt sustainability guidelines for its bioenergy projects and, in project implementation, to develop a certification scheme for sustainably produced biomass (Box 10.2-4; GEF, 2007a, b). The Inter-American Development Bank (IADB, 2008) has announced a partnership with the RSB for the development of standards for bioenergy projects (Box 10.3-3). The sustainability standards for the production of bioenergy carriers mentioned here go beyond the safeguard policies customarily applied in project financing.

10.3.2.2
Bilateral agreements

The schemes for the implementation of standards discussed above are based on a unilateral approach by standard-setting agencies. Unilateral approaches reach their limits, however, where wider issues relating to land-use planning, nature conservation and food security are concerned. These issues require central coordination at national (and ideally international) level and can only be dealt with to a very limited extent via unilateral standards and related certification schemes.

In general, therefore, establishing bioenergy standards in bi- and multilateral agreements is preferable to a unilateral approach. This can also be justified on economic grounds, in that a larger share of the global market will then be covered by the standards and the likelihood of leakage occurring can be reduced. This applies especially if – as with the EU's unilateral bioenergy standards – producers have recourse to large alternative markets for bioenergy products

(such as the US, China, Japan as importing countries) where no sustainability standards are in force. Furthermore, the acceptance of bioenergy standards among trade partners is likely to be greater if they are involved in the standard-setting process. And finally, bi- and multilateral schemes are also preferable to a unilateral approach from the perspective of the WTO rules (Section 10.3.4).

Bilateral agreements merely create obligations for the contracting parties; third countries are not directly affected by their provisions. This means that in relation to import goods, the inclusion of social criteria is more likely to be possible here than at the level of unilateral statutory minimum standards. In order to address the problem of leakage, which is difficult to resolve through certification, the producer countries could pledge to stop or at least substantially reduce the conversion of natural ecosystems at home, or ensure that the cultivation of energy crops does not displace other forms of use. In return, the trade partners should be granted free market access for bioenergy carriers if they comply with the minimum standard.

Agreements on bilateral cooperation in the bioenergy and especially the biofuels sector already exist. Brazil in particular is working hard at present to establish an international biofuels market and is therefore offering developing countries in Latin America, Africa and Asia (including Venezuela, Colombia, Paraguay, Ecuador, Senegal, Angola and Indonesia) technical assistance and knowledge transfer for the development of their biofuels sectors. In many cases, industrialized countries such as Sweden, the United Kingdom and Italy are envisaged as financial partners in the Brazilian projects. However, these 'South-North-South' projects are mainly geared towards market development (Biopact, 2007a, b). The introduction of sustainability standards is therefore likely to play a subordinate role for the time being.

The agreements between the industrialized countries and the emerging economies, too, have so far only included superficial statements on sustainability standards. The agreement signed between Germany and Brazil in May 2008 on comprehensive cooperation in the field of renewable energies and energy efficiency does not contain any specific goals relating to ecological and social criteria for the sustainable production of bioenergy. It merely contains declarations of intent with a view to initiating a dialogue on issues of sustainable production. At best, this is only a preliminary step towards more substantive commitments under a further treaty. The agreements between Brazil and the Netherlands/Sweden on cooperation in the bioenergy sector are very similar. Here too, there is only a fleeting reference to planned cooperation in the field of sustainable production and use of biofuels. It can thus be assumed that specific environmental and social standards for the production of biofuels are unlikely to be included in bilateral agreements between countries very soon. In the case of the agreement between Brazil and Sweden, it is the private sector which has taken on the task of implementing the sustainability criteria via a certification scheme (Section 10.3.2; SEKAB, 2008).

10.3.2.3
Multilateral approaches

Unilateral measures by individual states, as well as bilateral agreements, only have limited geographical impact. What's more, a plethora of unilateral and bilateral standards would fragment the global market and greatly impede trade between countries with different bioenergy standards. By contrast, the range and compatibility of bioenergy standards would be greatest if a uniform multilateral approach were pursued at international level. This would also reduce the information and certification costs for producers. The establishment of standards in a multilateral treaty-based regime would also be beneficial in terms of compliance with world trade rules (Section 10.3.4).

Existing multilateral environmental conventions (e.g. the CBD, Section 10.5; UNFCCC, Section 10.2; UNCCD, Section 10.6) could be used as an initial starting point. Within the framework of these conventions, specific thematic contributions could be developed for standard-setting in the field of bioenergy, primarily in the form of optional protocols. Within the framework of the CBD, negotiations on bioenergy have already been agreed, which should be expanded in this direction (Section 10.5.3). As the individual multilateral environmental agreements have only limited thematic objectives which are confined to their specific field, they are not suitable as a framework for the development of comprehensive bioenergy standards.

The negotiation of a separate multilateral 'Bioenergy Convention' has one potential disadvantage, however: it would take a considerable length of time. Furthermore, in view of stakeholders' divergent interests, it would probably only be possible to establish relatively general minimum standards in a multilateral convention of this kind. Nonetheless, in the interests of achieving sustainability goals over the longer term, despite the initial problems just mentioned, it is essential to pursue a multilateral approach in parallel towards standards for sustainable bioenergy production.

There are already various global initiatives which could be used as a starting point in this context (Section 10.3.2.1). In particular, the Global Bioenergy Partnership (GBEP; Box 10.3-4), which brings together the main producer and buyer countries, would in principle provide an appropriate institutional forum for effective policy debate. As a relatively transparent international forum, GBEP could channel the formal and informal processes involved in drawing up global sustainability standards and accelerate the formulation of standards. The proposals of WBGU, which has taken up important ideas put forward by the Roundtable on Sustainable Biofuels (Box 10.3-3), could be incorporated in the work of the G8+5 Task Force on Sustainability. With political support from the G8, it would be possible to ensure that the decisions are channelled into policy-relevant forums, institutions and processes. However, efforts should be made to ensure that relevant civil society stakeholders have greater involvement in the dialogue so that all interested parties (countries, industry, trade, NGOs, etc.) participate in the discourse.

The World Commission on Dams (WCD) could provide ideas for the institutional expansion of this body and its activities. The Commission, consisting of government representatives, international organizations, NGOs, industry, etc., was established in the late 1990s in order to develop global standards and criteria for an equally controversial sector, namely dams. Here too, despite a wide range of political, environmental and economic interests, there was a consensus that an international agreement for this sector was essential. The principles developed by the Commission do not have binding legal force, but as 'soft law', often perform an important reference function by providing ecological and social guidelines as an important basis for many decision-making processes (Thürer, 2000). Ideally, a catalogue of sustainability principles for bioenergy production would have binding legal force and would be transposed into appropriate legislation by the participating countries. Whereas the World Bank and the IUCN were initiators of the WCD, the FAO would need to play a key role in any similar initiative for bioenergy.

Box 10.3-5

Vision of a Global Commission for Sustainable Land Use

In the medium term, all forestry and agricultural practices should be shaped in a sustainable manner and be subject to uniform principles. However, this aspiration goes far beyond the formulation of standards. There is considerable potential for conflict in the issue of competing land-use claims, as the WBGU analysis shows. Critical trends in world food security, or as a result of the loss of ecosystems, are becoming apparent even now, and this pressure on land use and social systems will continue to increase. In the future, the growing world population, increasingly land-intensive food consumption patterns – not only in the OECD countries but also in the rapidly growing emerging economies, especially China and India – and a rising demand for biomass as an industrial feedstock must be reconciled with increased soil degradation, growing water scarcity and climatic stress. This is a global challenge on a scale and of a complexity that is still little understood, only partial aspects of which have been addressed in international governance processes (e.g. within the framework of the FAO and the CBD).

In view of this challenge to land-use management, WBGU considers it necessary to intensify the debate on the issue of global land use at international level, and also embed the topic in an institutional context. To this end, a new Global Commission for Sustainable Land Use should be set up. The commission's task should be to identify the key challenges arising from global land use and pool current scientific knowledge. On this basis, the commission should then elaborate the principles, mechanisms and guidelines required for global land-use management. This opens up a complex new field of global governance, in which food, energy, development and environmental policy issues interact. The commission will deal with complex issues that involve conflicting interests. The starting point will be to improve the international data on current global land use and soil conditions, and to harmonize it on an international basis. A common understanding of the underlying problems must be created and the diverse interests surrounding land use must be identified, from the local level right up to the global level. Ultimately, this will also promote an extensive change in established perceptions and practices: it is no longer possible to see land use solely as an issue for action at national level. On account of the diverse global interactions and linkages involved, developments that impact on land use should no longer be understood and addressed at this level only. This is illustrated by the example of the indirect land-use changes associated with the expansion of bioenergy, as well as by the issue of equitable per-capita land use. For this reason, transboundary cooperation is required to debate these complex issues of sustainable and equitable land use in an open process, to address the problems and develop long-term regulatory proposals.

Such an institution could, for example, be modelled along the lines of the Financial Action Task Force on Money Laundering (FATF). The FATF, convened by the G7, was launched in 1989 as a small, dynamic group of experts who were appointed to combat money laundering, a new policy field at international level. The task force has since expanded and has formulated guidelines and standards which are recognized by many countries and organizations. To this end, a global peer review and monitoring procedure has been established (Reinicke and Reinicke, 1998; Sharman, 2008).

The new Global Commission for Sustainable Land Use, whose mandate – as outlined above – would extend far beyond land-use issues in the narrower sense, could be located within UNEP and work closely with other UN organizations such as the FAO. The findings of this process should regularly feature on the international agenda, for example within the framework of the agenda of the UNEP Global Ministerial Environment Forum. It would also be important to address and discuss the issues within the strategically important G8+5 gatherings of heads of state and government. This would significantly increase the probability of the findings and decisions reaching a broader public and being fed into relevant political processes.

The goal of sustainable bioenergy use cannot be achieved solely by means of relevant standards in the bioenergy sector, however. Standards for energy recovery from biomass can only ever cover sub sectors of agricultural production and do not adequately resolve the key issue, namely displacement effects. Bioenergy must be embedded in a sustainable cross-sectoral land-use management system which complies with the guard rails and sustainability rules developed by WBGU (Chapter 3). This type of global, cross-sectoral land-use management could be supported by a Global Commission for Sustainable Land Use, which has yet to be established (Box 10.3-5).

10.3.3
Implications of the adoption of standards for trade in bioenergy carriers

As soon as a legally binding bioenergy standard is introduced in the EU, overseas producers will also be forced to seek certification of their products to demonstrate compliance with the standard if they wish to sell them in the EU. Bioenergy standards could thus become *de facto* barriers to trade if the required standards do not apply in the producer countries. In return for compliance with the standards, however, the EU could grant its trade partners preferential conditions for imports into the EU.

10.3.3.1
Standards as a barrier to trade

An individual country or supranational organization such as the EU can deploy binding standard-setting deliberately as a tool to influence the way in which bioenergy carriers are produced in the exporting countries. However, there is a risk that the exporting countries will regard the trade restriction resulting from standard-setting as a protectionist measure, with the result that this type of unilateral scheme will fail to find acceptance among trade partners. The fact is that the unilateral adoption of binding sustainability standards effectively operates as a ban on the import of non-sustainably produced bioenergy products, which would inevitably prompt a review of its compliance with WTO rules (Section 10.3.4).

In order to increase acceptance of the minimum standard, tangible measures to facilitate market access (e.g. clear reductions in tariffs) should be offered to trade partners which comply with the minimum standards. At the least, tariffs and export subsidies in the agricultural sector should be further reduced.

A far more stringent trade restriction would be the introduction of a general import ban, or a time-limited moratorium on imports, for bioenergy carriers. An alternative which could also be considered is only allowing biomass/bioenergy imports from specific producer countries, in line with the current practice applied in part to food and product safety. Ultimately these general import restrictions, which are a less differentiated mechanism than minimum standards, would reduce the export prospects for producers in the countries of origin to a far greater extent and would result in a far lower level of acceptance by trade partners.

10.3.3.2
Implications for trade relations with developing countries and emerging economies

A specific situation arises in respect of the developing countries and emerging economies. Since 1971, the European Community has granted these countries preferential access to the EU market for their products under the Generalised System of Preferences; this includes feedstocks which can be used to generate bioenergy. The GSP countries include key bioenergy producer countries such as Brazil, Argentina, Indonesia and Malaysia. However, a number of least developed countries (LDCs) which enjoy special preferential arrangements under the 'Everything But Arms' (EBA) initiative as part of the GPS also have significant bioenergy production potential, including Malawi, Mozambique and Zambia (Johnson et al., 2006). The EU continues to grant the ACP countries duty-free access to the EU for their exports. In the EU Strategy for Biofuels, the European Commission emphasises that it will maintain a comparable level of preferential access for ACP countries (EU Commission, 2006b). Likewise, negotiations are currently under way with the MERCOSUR countries, i.e. Argentina, Brazil, Paraguay and Uruguay, about preferential conditions for their imports into the EU (Dufey, 2006).

A goal conflict thus arises between the promotion of trade with the developing countries and emerging economies on the one hand, and the need for sustainability standards on the other. For example, agriculture and agro-processing account for 30-60 per cent of GDP in two-thirds of the LDCs. What's more, agricultural exports from LDC countries are greatly dependent on the European market, with the EU absorbing 70 per cent of LDC agricultural exports to Japan, the US, Canada and the EU combined (EU Commission, 2008b). The enforcement of sustainability standards without a corresponding offer of technical and financial cooperation would constitute a disproportionately heavy burden for these countries (Grote, 2002) and would conflict with the main reason for these countries' duty-free access to the EU markets, which is to promote their development.

This goal conflict could be resolved by offering small and medium enterprises in developing countries and emerging economies support with the implementation of standards. The LDCs in particular, to which the EU currently grants duty-free or reduced duties on goods imports under the 'Everything But Arms' initiative, must be given financial and technical support with the implementation of the required standards. Furthermore, during an initial phase, simplified terms for the verification of certification criteria could also be offered to the developing countries to reduce their transaction costs (e.g. application of default values). Similarly, group certification is a good way of keeping down certification costs for agricultural smallholdings, primarily in the developing countries, but also in industrialized countries and the emerging economies.

10.3.3.3
Preferential treatment of bioenergy carriers through qualification as environmental goods and services

Preferential treatment of bioenergy feedstocks in the multilateral framework could also result from the liberalization efforts which the WTO has pursued for some time in the trade with environmental goods and services (EGS) (Chaytor, 2002; Iturregui and Dutschke, 2005; Singh, 2005; Dufey, 2006; Sell, 2006; Sugathan, 2006; Yu, 2007).

Within the framework of the Doha Round, under way since 2001, liberalization of trade in EGS has been defined as an objective under various mandates to develop WTO law in relation to measures to protect the environment (Doha WTO Ministerial 2001, Ministerial Declaration, para. 31 [iii]). The aim is to improve market access for selected products or services which are regarded as particularly environmentally friendly, by abolishing tariffs and other trade barriers for the goods and services defined as EGS.

The question of which goods should qualify as EGS is a contentious one among the WTO members. However, it can be assumed that goods from the renewable energy and especially the bioenergy sector would qualify for these liberalization measures (Singh, 2005). An initiative launched by Brazil is relevant in this context: in November 2007, it applied to the WTO Committee on Trade and Environment for biofuels to be classified generally as environmental goods.

In advance of the United Nations Climate Change Conference in Bali in December 2007, the EU and the US also submitted a joint proposal to the WTO to give preferential treatment to technologies which can combat climate change and classify them as EGS. In line with this proposal, in a first phase, a total of 43 key 'climate-friendly' goods and services' (including technologies such as solar panels and wind turbines) would be liberalized, i.e. exempt from tariffs and similar trade restrictions, with immediate effect. Second, an Environmental Goods and Services Agreement (EGSA) is to be negotiated by WTO members within the framework of the Doha process, which would foresee further binding commitments to eliminate tariffs and non-tariff barriers in trade in green technologies. However, the EU and the US are opposed to the inclusion of biofuels in the list of key 'climate-friendly' technologies, a move which has been particularly criticized by Brazil. The EU also argued in favour of listing non-agricultural products only.

The Doha Round has become deadlocked on numerous occasions over recent years. After the breakdown of the talks at the ministerial meeting in Geneva in July 2008, and in view of the ongoing political differences, it is unclear how the negotiations will proceed, and in which time frame. As with many other aspects of the current WTO Round, the outcomes that can be achieved are unclear, also in relation to EGS. The main points of contention here include the issue to be addressed by the Committee on Trade and Environment concerning the conditions under which the requisite environmental compatibility of a product or service is deemed to exist. The desired liberalization only makes sense from an environmental perspective if, through classification as an EGS, the goal of environmental protection is not undermined by a generalized rather than individual life-cycle evaluation of the goods or services declared to be EGS. Overall, this means that in the debate about EGS in the WTO framework, sustainability criteria (as defined by WBGU in Section 10.3.1) must also be taken into account, which must also relate to the manufacturing processes. However, it is debateable whether, in the selection of EGS, the application of specific standards, which WBGU regards as essential, is politically enforceable. Germany and the EU should therefore lobby for environmental protection goals to be taken into appropriate consideration at the EGS negotiations.

If an evaluation system based on the individual goods' life cycle balance cannot be achieved, WBGU argues against the inclusion of biofuels in the EGS list, as a generalized evaluation of their social and ecological sustainability is not viable. However, one option which could be considered is to classify as EGS specific bioenergy carriers which come from selected bioenergy pathways and which merit promotion.

10.3.4
WTO compliance of standards for bioenergy carriers

In relation to the statutory minimum standard called for by WBGU, it is important to consider the issue of compliance with the relevant provisions of WTO law. The sustainability principles for the minimum standard recommended by WBGU relate to production processes in the country of origin and therefore should not be regarded as product-specific measures (on the distinction between product-specific and non-product-specific measures, see Droege, 2001; Puth, 2003; Hilf and Oeter, 2005). Whereas with product-specific environmentally motivated trade restrictions – which relate to the properties of the product itself – compliance with the provisions of WTO law is usually a given, provided that the measures in question genuinely serve environmental policy goals (Hilf and Oeter, 2005), substantial deviations arise in relation to the WTO compliance of non-product-specific measures.

10.3.4.1
Relevance of WTO law in standard-setting

In evaluating the WTO compliance of standard-setting which relates to the production process, a distinction must first be made between government and purely private measures. In both mandatory certification through statutory standard-setting and in purely voluntary certification, albeit sponsored by the state, there is always the potential for a trade-restricting effect to arise. By contrast, in terms of GATT's provisions, which impose rights and obligations only on states, purely private measures are non-problematical from the outset (Droege, 2001; Blüthner, 2004; Hilf and Oeter, 2005).

In terms of the legal implications of standard-setting, Article III GATT (National Treatment on Internal Taxation and Regulation) and Article XI GATT (General Elimination of Quantitative Restrictions) are the main provisions of relevance. The former applies to national treatment, while the latter applies to restrictions and prohibitions at the border (Puth, 2005). If the trade restrictions introduced solely affect products from specific countries, Article I GATT (General Most-Favoured-Nation Treatment) would also be of relevance.

Under Article III:4 GATT, the principle of national treatment of imported products is deemed to be violated if, firstly, the measure in question is based on a domestic (national) legal provision; secondly, the imported and domestic goods concerned are 'like products', and thirdly, the imported goods concerned enjoy less favourable treatment than 'like' domestic products (Puth, 2003).

A potential conflict between sustainability standards and WTO rules can arise, in particular, as a result of the second criterion concerning the 'likeness' of products if foreign goods are treated less favourably than domestic (national) products. In practice, this may arise if there is *de facto* unequal treatment on the grounds that certain national products are more likely to meet specific standards. The decisions taken by the WTO's conflict resolution bodies have tended to assume that identical products should be classified as 'like' products even if they are produced by different methods (e.g. WTO Panel in the case of 'Japan – Taxes on Alcoholic Beverages', 1996). Differences in production measures therefore may not be taken into account if the production processes result in 'like' products (Droege, 2001). This criterion is generally fulfilled in relation to the measures concerning bioenergy, as the standards and certification processes are based on the type and method of production.

It must therefore be assumed that products which are physically identical, and differ only in the non-product-specific methods used in their manufacture, meet the requirement of 'likeness' for the purpose of Article III:4 GATT (Hilf and Oeter, 2005). The state-supported setting of environmental and/or social standards and control of their implementation through certification schemes therefore conflict with the principle of equal treatment. The same applies to other unilateral measures which subject specific products to less favourable treatment due to their methods of production, as would be the case within the framework of a promotion policy designed along these lines.

If the import of products, such as biofuels, is prohibited or restricted because these products do not comply with specific standards, a conflict with Article XI GATT also arises. Non-compliance with Article I GATT would occur if, in addition, bioenergy imports were only allowed from specific producer countries, e.g. because they do not recognize general standards of production. In sum, then, an infringement of the relevant provisions of WTO rules could arise on three counts in relation to state-supported measures to guarantee sustainable production of bioenergy, depending on the specific form such measures take. Nonetheless, there are good grounds in all cases for justifying the adoption of sustainability standards.

10.3.4.2
Justifying discriminatory measures

The various agreements contain exception clauses which may be applied specifically to environmental protection and could potentially justify a contravention of the GATT rules (as well as the supplementary WTO agreements, where applicable). The main question which arises is whether an assumed contravention of Article III:4 GATT through the adoption of sustainability standards (and their implementation through certification and labelling schemes) or a contravention of Article XI GATT may be justified on the basis of the clauses contained in Article XX GATT.

APPROACHES TO JUSTIFYING ENVIRONMENTALLY MOTIVATED MEASURES
Article XX GATT contains exceptions for measures to protect human, animal or plant life or health (clause b) and for those relating to the conservation of exhaustible natural resources (clause g). Admittedly, the environment as such is not contained in the list of legally protected assets, but the scope to include it is certainly provided for in the above-mentioned clauses b and g (Epiney, 2000; Epiney and Scheyli, 2000; Hilf and Oeter, 2005). On this basis, it is now widely recognized that 'general' concerns about protecting the environment can also justify exceptions in accordance with these provisions.

Even though the precise interpretation and application of Article XX GATT is still disputed, the actual rulings of the WTO dispute settlement bodies suggest that purely process-specific trade measures – and thus the setting of standards and monitoring of implementation through certification schemes – in principle also fall within the scope of the protection afforded by Article XX GATT (Hilf and Oeter, 2005; Droege, 2001 also points in this direction). A particular example of this is the 'US–Shrimp' case (1998), which concerned the dangers posed to sea turtles by shrimp harvesting. In this case, the Appellate Body clarified that the production method may indeed be admissible as a determining factor if a certain species of animal is endangered by this mode of production, and stated that there was no obvious reason why non-product-related trade measures should be regarded as in principle incompatible with the WTO rules (Althammer et al., 2001; Droege, 2001; Puth, 2003). This case also clarified that, in principle, measures affecting protected goods outside the territory of the states pursuing the environmental measure may also be justified (extraterritorial application of Article XX GATT).

One potential difficulty relating to a possible dispute about the WTO compliance of product-related environmental standards for imported products is that the state seeking justification for contravening the principles of world trade is required to furnish the proof that exception criteria exist. To defend discrimination under Article III:4 GATT on the basis of ecologically motivated sustainability standards, the state concerned would therefore have to be in a position to prove that there is no other, less radical way of achieving its environmental objective (principle of proportionality). The presence of objective risks must also be proven (on the relevant assessment process, see, most recently, the WTO 2007 Panel Report: 'Brazil – Measures Affecting Imports of Retreaded Tyres'; Qin, 2007).

The most persuasive means of justifying a violation of the relevant principles of WTO law on the basis of Article XX GATT is to demonstrate an international consensus on the indispensability of a specific environmental good, and the necessity of protecting it, with reference to a multilateral agreement. It may be assumed that such a consensus exists in relation to climate change mitigation and biological diversity conservation in particular, for which specific protection goals have been established within the framework of the various treaty regimes (the United Nations Framework Convention on Climate Change and the Convention on Biological Diversity, together with their additional protocols). These objectives are covered by the provisions of Article XX (b) and (g) GATT, and they are also inherently consistent with the chapeau clause of Article XX GATT, provided they are applied in a way that does not result in any arbitrary or unjustified discrimination.

Overall, it can thus be assumed that unilateral measures which have trade-restricting effects due to the definition of product-related requirements can be justified and are thus consistent with WTO rules. However, the decisions of the WTO dispute settlement bodies are not yet sufficiently well-established in doctrinal terms and can vary in some cases. It therefore cannot be predicted with certainty which decision the relevant dispute settlement bodies of the WTO would make if the applicable unilateral measures were contested by a member state. In view of this uncertainty, environmental policy measures to safeguard the sustainable production of bioenergy should be developed at multilateral level if possible (Section 10.3.2.3).

APPROACHES TO JUSTIFYING SOCIO-POLITICALLY MOTIVATED MEASURES

Separate issues arise in relation to measures, especially standards, which are socio-politically motivated. Here too, the associated trade restrictions are a source of potential conflict with the above-mentioned prohibition of discrimination contained in the WTO rules, particularly the principle of national treatment and the elimination of quantitative restrictions on imports or exports.

The exceptions contained in Article XX GATT contain no protection goals that relate explicitly to social or human rights issues. However, in this context too, the view is held that the corresponding legally protected assets are covered by the broadly formulated clauses that are akin to general clauses. These include the classification of socio-politically motivated action as a measure necessary to protect public morals (clause a) and as a measure to protect human, animal or plant life or health (clause b). The relevant literature also advances the opinion that the arguments (described above) advanced by the WTO dispute settlement bodies to justify environmental policy measures may also – at least potentially – be applied to socio-political goals (López-Hurtado, 2002). It is further argued that justification should be possible in those cases where socio-political measures aim to help enforce principles that are particularly well-established in multilateral international agreements and are thus supported by a far-reaching consensus. These include, most notably, the core labour standards laid down in the main ILO conventions (on the prohibition of child or forced labour, for example).

In addition to the uncertainties already mentioned in relation to environmental policy measures, this area has its own particular difficulties. The inclusion of specific social standards in the list of exceptions in Article XX has been specifically discussed within the framework of the WTO. However, even though they were to be limited to the generally accepted ILO core labour standards, this debate has foundered in the face of opposition from most of the developing countries. Accordingly, there is uncertainty as to whether a majority of the parties would endorse the argument that the existing exception clauses under Article XX (a) and (b) GATT also embrace both socio-political aspects and those relating to human rights in general. As things stand at the moment, it would be very doubtful whether recourse to the grounds for justification contained in Article XX GATT would be successful in the event of a potential dispute on the WTO compliance of social standards for imported products.

10.3.4.3
Legal assessment of the sustainability standards recommended by WBGU

As long as the WTO agreements – apart from the exception clauses mentioned above – contain no explicit objectives of an ecological and social nature, the question of whether sustainability-oriented measures comply with trade rules ultimately depends on the view of the WTO dispute settlement bodies. No relevant practice has been established to date, particularly on the WTO compliance of unilateral sustainability standards. If the EU, within the framework of its legislation, makes the use of bioenergy carriers conditional on compliance with minimum standards on a unilateral basis, as proposed by WBGU, it is quite within the realms of possibility for third countries to view such a course of action as an illegal trade restriction and plead a violation of the discrimination prohibitions before the WTO bodies.

The requirement to prove that a certain sustainability standard is indispensable and thus fulfils the principle of proportionality, and the question of the degree to which extraterritorial standard-setting is deemed to be permissible, create particular uncertainty in terms of a future ruling by the WTO dispute settlement bodies. Nonetheless, on this basis, it still appears possible to assess the likelihood of various categories of unilateral sustainability standards being accepted by the WTO bodies on the basis of Article XX GATT (BTG, 2008).

This kind of preliminary assessment produces particularly positive results for criteria aimed at improving the global GHG balance. On the one hand, this climate policy objective is clearly recognized within the framework of the international climate regime (UNFCCC, the Kyoto Protocol), and the indispensability of the climate policy criterion can be clearly demonstrated. Since climate change has a global impact, it is likely that suitable measures would be

regarded as justified even if the permissibility of an extraterritorial application of Article XX GATT were challenged.

A similar situation exists in relation to the criterion of biodiversity conservation. Here too there is a clear consensus based on the establishment of corresponding objectives in the Convention on Biological Diversity (CBD). Although the global impact of a loss of species diversity is less easy to prove than the impacts of climate change mitigation, it must not be dismissed out of hand. If the WTO dispute settlement bodies continue to take the view that measures relating to targets of protection located outside the territory of the state enacting the regulation can also be justified, then criteria aiming to conserve biodiversity would also be permissible.

On the other hand, it is more difficult to assess criteria that affect purely local ecological protection goals in the producer countries. The question which arises here is to what extent consensus on indispensability and adherence to the principle of proportionality can be proved. As illustrated by the ruling in the 'US–Shrimp' case, however, the possibility of a positive assessment in a dispute should not be ruled out.

By contrast, the question of whether socio-politically motivated standards are permissible from the perspective of WTO law has, by and large, still to be clarified. In view of the opposition from a large number of states to the introduction of even basic social standards, it is assumed that such criteria would be classified as not WTO-compliant in any dispute. However, this should not be seen as an obstacle to promoting adherence to key social standards, but more as an incentive to work towards the appropriate amendment of the WTO rules.

10.3.5
Interim conclusion

Bioenergy standards are essential in order to steer the production of bioenergy products along sustainable trajectories. The more bioenergy carriers are covered by a standard, the more effective this standard will be. Including as many countries as possible in the development of a global bioenergy strategy can achieve a broad impact. From the perspective of the WTO rules, too, a coordinated multilateral approach is always preferable to unilateral measures. However, due to the differing interests of the individual countries at international level, it is likely that, at least in the short term, it will only be possible to negotiate relatively weak and ineffective global minimum standards.

A broader impact would be achieved by a more stringent standard; however, this could only be enforced as a voluntary or unilaterally binding standard. Since voluntary certification schemes – as shown to date by the experience with voluntary schemes in the areas of forestry and green electricity – are likely to occupy only a niche position in a market for bioenergy carriers, unilateral minimum EU standards appear to be the most effective option in the short term. However, the question of whether in a dispute the WTO bodies would class such unilateral standards as compatible with international trade rules cannot be answered conclusively. In terms of the most important sustainability criteria, it is possible to pursue this line of argument based on the existing rulings (Section 10.3.4). Compliance with WTO rules should therefore not be viewed as a fundamental impediment.

Besides introducing a bioenergy standard on a unilateral basis, bi- and multilateral negotiations should be held to convince non-EU countries which produce or use bioenergy carriers on a large scale (e.g. the US, Brazil, India and Japan) of the importance of sustainable bioenergy production. If the EU acts alone, large parts of the world market for bioenergy carriers will remain unregulated. Bi- and multilateral agreements containing specific criteria for the sustainable production and use of bioenergy carriers must therefore also make a contribution to a sustainable global biomass strategy.

In WBGU's view, these considerations will produce a very promising approach in the form of a phased process, i.e. the combination of unilaterally binding minimum standards within the EU with the integration of sustainability standards in bi- and multilateral agreements between major producer and buyer countries of bioenergy products. The GBEP could be an important body at multilateral level as it could shorten international negotiation processes and accelerate the formulation of bi- and multilateral policies on global standards (Section 10.3.2.2). With political support from the G8, the decisions could be introduced into relevant political forums, institutions and processes, thus ensuring their implementation. In the longer term, the standards promoted by WBGU should apply to all types of biomass (Section 10.3.1; Box 10.3-5).

However, as long as there is no uniform global standard for all types of biomass, the more realistic scenario is that a standards or certification scheme for bioenergy carriers may well reduce the sustainability problems associated with bioenergy extraction from energy crops, but will not be able to eradicate them completely, for although indirect effects may be taken into account, they cannot be ruled out entirely. It is also expected that even if bioenergy standards are implemented on a global basis, they will not be able to guarantee sustainability in

bioenergy production across the board, due to gaps in the monitoring system and weak institutions in some developing countries and emerging economies. Yet standards and certification in the bioenergy sector are important tools in preparing the way for the mainstreaming of sustainability in all agricultural and forestry practices worldwide. Unilateral standards can also perform this function as they will inevitably trigger debate about production methods in both the producer countries and the agricultural and forestry sector.

As yet, no established sustainability standards for biofuels or bioenergy carriers from renewable feedstocks are being applied across the board. Even if the EU's political processes in relation to bioenergy continue to make the dynamic progress witnessed to date, an EU-wide minimum standard for bioenergy, as promoted by WBGU, is unlikely to be introduced before 2012. Until then, there could be three to four years in which bioenergy carriers and especially biofuels could continue to be imported into the EU on an unregulated basis. In order to limit the non-sustainable production and use of bioenergy, bioenergy carriers that do not satisfy the desired minimum standards should not be promoted (Section 10.7).

In terms of introducing a statutory minimum standard for bioenergy carriers within the EU, a pragmatic first step would be to draw on existing certification schemes for biomass used for non-energy purposes, as well as on the certification schemes in the bioenergy sector currently being developed (Table 10.3-1), and make these count towards bioenergy certification in the sense of a meta standard. This is planned by the EU and is also provided for in the certification system developed in Germany. The recognition of national and voluntary certification schemes will also reduce the work involved in the certification process for foreign producers and increase the acceptance of a unilateral approach. Accompanying bi- and multilateral agreements on the internationally coordinated, sustainable promotion of bioenergy (Sections 10.7 and 10.8), the establishment of protected areas and networks of protected areas (Section 10.5), safeguarding global food security, and agreements on agricultural land and land use (Section 10.4) can further improve the effectiveness of certification.

10.4
Options for securing the world food supply in the context of a sustainable bioenergy policy

10.4.1
New challenges arising from bioenergy use

The growing importance of bioenergy use worldwide brings with it new challenges for the security of the world food supply. Efficiency improvements in the use of traditional biomass for energy purposes impact positively on the food situation, since they reduce health risks and energy poverty. At the same time, however, the cultivation of energy crops competes directly with food and animal feed production for land; in consequence, unchecked expansion of energy crop cultivation can displace food production and pose a major risk to food security. This displacement problem can to some extent be regulated by setting up standards and certification systems, including monitoring processes (Section 10.3). In the light of changing patterns of diet – particularly in the major newly industrializing countries – and of the emerging impacts of climate change, such steps need to be accompanied by reconsideration and revision of present policy on agricultural policy and trade in agricultural goods. These policy measures must be based on the FAO guidelines on the right to food (Eide, 2008).

There is general agreement that food prices are likely to remain high in the long term (Section 5.2.5.2). For people living in poverty and spending the majority of their income on food, this poses a threat to life and livelihood. This is particularly the case in Low Income Food Deficit Countries (LIFDCs). Increasing demand for energy crops is one of the causes of rising food prices. If the bioenergy sector continues to expand, the price effects thereby induced will have an increasing impact on food security. Figure 10.4-1 shows that many of the regions with bioenergy potential identified by WBGU are located in LIFDCs, where for reasons of food security the cultivation of energy crops must be approached with particular caution.

In view of the significant price rises on the world agricultural markets during 2008 and the ever closer coupling of the agricultural and energy markets, many actors in development cooperation and in the UN system have drawn up strategies and action plans aimed at limiting the emerging risks to food security. There is broad consensus on the key elements of such strategies. In the first place, conditions for food production in regions at risk must be directly and imme-

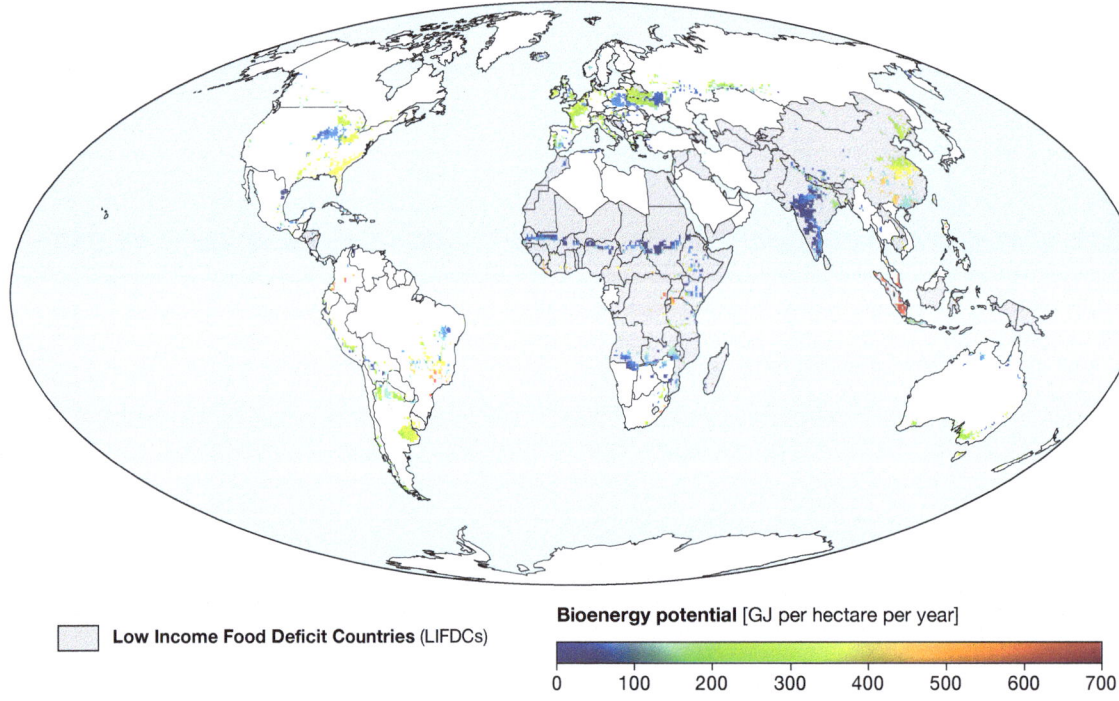

Low Income Food Deficit Countries (LIFDCs)

Bioenergy potential [GJ per hectare per year]

0   100   200   300   400   500   600   700

Figure 10.4-1
Potential regions for bioenergy and countries classified as LIFDCs. The map shows the regional distribution of possible land on which energy crops could be grown in 2050 for a WBGU scenario with low farmland need and high biodiversity conservation in non-irrigated cultivation (Scenario 3; Section 6.5).
Source: WBGU using data from Beringer and Lucht, 2008

diately improved (e.g. through provision of seed for the next harvest). Secondly, attention must focus on improving conditions for food security and food production in the medium to long term (e.g. through conversion to more productive farming systems). This would help prevent escalation of competition for land use. Thirdly, these activities must be coordinated with each other and must at the same time be consistently integrated into other areas of policy such as climate change mitigation (Section 10.2) and biodiversity conservation (Section 10.5). Fourthly, coordination of the individual policy areas must be governed by an overarching vision that also guides energy crop cultivation. The requirement for food and feed production to take precedence over energy crop cultivation is in WBGU's view a key component of this vision.

10.4.2
Short-term coping measures

To achieve rapid success in coping with crisis, steps must be taken that will in the short term increase the availability of food in crisis areas and/or improve access to food for those in need. These actions must be accompanied by measures to increase production – measures that will have a rapid impact and that can be implemented swiftly and without extensive preparation at national and international level (e.g. distribution of seed to secure the next harvest).

10.4.2.1
Safety nets and other fiscal measures

Safety nets involve quasi-monetary transfers, e.g. in the form of food stamps or coupons, and direct income transfers for people in need. If such programmes are to be effectively applied, the need of individuals must be verified; this entails extensive organization and is often costly, making the system difficult to apply in countries in which the state's administrative capacity is weak. One solution may be to use a cruder measure of the needy person's circumstances, for example by basing transfers on the place of residence or some other easily verified criterion. Other transfers are made through state employment and training programmes in which food is provided in exchange for work (food-for-work programmes). School feeding programmes operate in a similar manner.

In principle, transfer programmes allow benefits to be directed more precisely at needy target groups than general fiscal measures such as tax reductions and direct subsidies. Nevertheless, measures of the latter type are used by many affected countries to cushion the effect of food price rises. A study of the World Bank found that between 2007 and 2008 more than 40 per cent of 58 developing countries surveyed cut taxes or customs duties, while more than 30 per cent introduced price subsidies (World Bank, 2008c). Price subsidies are a burden on the public finances of the countries concerned and have only a very limited impact on the level of need among consumers. In the worst case such a policy collapses as higher agricultural prices drive up state expenditure still further; social unrest and political crisis may then follow. Because of this, international organizations advise against fiscal measures of this type. Responsibility for the use and expansion of transfer programmes lies ultimately with the affected states themselves. In the context of development cooperation, support can be provided through financial assistance and advice on good governance.

10.4.2.2
Administrative price ceilings

In addition to fiscal instruments and expenditure programmes, some affected countries also use price ceilings as a regulatory measure. A maximum price is set for the sale of food or grain by producers. In the short term, this can ensure that food remains affordable for much of the population. However, price ceilings reduce profitability for individual producers. Depending on the level at which prices are set and the return available from other types of farming or land use, they may provide a false incentive. In other words, energy and other non-food crops may be produced instead of food if a higher market price can be realized from these non-food products. By contrast, demand-oriented measures such as transfer programmes induce market effects that will ideally be transformed into production incentives and thus influence land use competition in favour of food security. Thus income or quasi-monetary transfers can increase demand for food and so indirectly generate production incentives.

10.4.2.3
Short-term aid for smallholders

Subsistence farmers and smallholders, who produce food primarily for their own use, benefit very little or not at all from rising food prices. Instead they are often adversely affected by the consequences of higher energy prices, since agricultural inputs (in particular fertilizers) that are linked to the price of energy become considerably more expensive. To secure the forthcoming harvests, steps must be taken

Box 10.4-1

The role of the FAO in global bioenergy policy

The Food and Agriculture Organization of the United Nations (FAO) started looking at the issues surrounding bioenergy 20 years ago. Its focus has traditionally been on fuelwood use and energy crops, with the primary aim of improving the energy supply in rural and remote areas in developing countries. The FAO produces and distributes information on the production, trading and use of bioenergy. In addition, the FAO supports member states at local and national level on technical issues (e.g. in connection with the development of bioenergy programmes). The Natural Resources Management and Environment Department was set up in 2007; one of the department's divisions, the Environment, Climate Change and Bioenergy Division, deals specifically with the links between bioenergy and climate change. In order to improve the coherence of its bioenergy policy, the FAO has also set up an Inter-Departmental Working Group on Bioenergy. The Working Group's task is to strengthen the FAO's profile on bioenergy issues and specify priorities for action. In 2006, the FAO set up the International Bioenergy Platform (IBEP) to coordinate

policy outside its own organizational boundaries; the IBEP is intended to promote inter-disciplinary and trans-regional cooperation between relevant actors in the fields of policy-making, business and science who are involved in issues of sustainable energy, agriculture and the environment. One of the platform's aims is to carry out studies that will help policy-makers to take decisions on the sustainable production and use of bioenergy; the Millennium Development Goals provide a frame of reference for this work. Another organization working in the same field is the Global Bioenergy Partnership (GBEP), which was set up in 2006 by the UN Commission on Sustainable Development; its secretariat is located at the FAO in Rome. Conflicts between bioenergy policy and global food security are also addressed by the FAO through its Bioenergy and Food Security Project (BEFS), which was set up in 2007. The BEFS project, which is currently funded by Germany, is charged with exploring the opportunities and risks of bioenergy use and in particular identifying possible impacts on food security. The FAO is currently in crisis; whether it can establish itself as a leading institution in the field of bioenergy depends in part on the outcome of the internal reform process that is currently under way (Windfuhr, 2008).

---

before the next season to ensure that small-scale producers have better access to the resources they need (Ressortarbeitsgruppe Welternährungslage, 2008; UN, 2008). This involves providing prompt assistance with the provision of loans, seed, fertilizers and technology. The aim must be to improve the productivity of small-scale farming in the worst affected regions before the next sowing season. In addition there is a need for ad-hoc measures to improve rural and agricultural infrastructure such as irrigation systems, roads and marketing infrastructure. For landless people, food-for-work programmes are particularly effective. Finally, immediate steps should also be taken to reduce harvest losses by improving crop and animal health and to prevent post-harvest losses by improving storage conditions (UN, 2008). It should, however, be borne in mind that low-cost loans and other subsidies for the provision of agricultural inputs are only feasible if functioning distribution and advice networks are in place.

10.4.2.4
Export restrictions on agricultural products

In order to check the rise in food prices and prevent an internal supply crisis, many countries resort to trade policy measures. Depending on the country's importance on the world market, such measures may have international repercussions. In the wake of the drastic rise in food prices of 2007/2008, at least 20 per cent of the newly industrializing and developing countries surveyed in a World Bank study introduced export

restrictions for cereals and other agricultural products (World Bank, 2008c). While these export restrictions may in the short term boost domestic supply, they also penalize domestic agricultural producers and thus in the medium term impede development of the country's agricultural sector. Moreover, export restrictions are not a target-group-oriented measure, since any benefits resulting from domestic price reductions are not confined to those on low incomes. And, finally, such measures cause prices on the world markets to rise even higher; this has a particularly detrimental effect for low-income countries that are dependent on food imports (Rudloff, 2008; World Bank, 2008c). Export restrictions on food and agricultural goods may conform with WTO rules (Art. XX GATT; Section 10.3.4.4) but at present there is no more detailed provision for limiting them to countries in need or setting a time limit on them. In order that WTO-conformity can be monitored and disproportional side-effects prevented, it is relevant to consider application criteria, such as trigger levels that would take account of impacts on food security in other countries (Rudloff, 2008).

10.4.2.5
Removal of distortions of trade in world agricultural markets

Agricultural subsidies, minimum prices and import restrictions are most often applied in industrialized and newly industrializing countries, usually with the aim of securing farmers' incomes. In 2006, financial assistance to agricultural producers in the OECD countries totalled US$ 268,000 million. Agricultural subsidies account for around 1.1 per cent of GDP in OECD countries (OECD, 2007a). The situation is similar in many newly industrializing countries: China spent more than 2.4 per cent of GDP on agricultural subsidies in 2005, while Russia spent around 1 per cent, Brazil around 0.8 per cent and South Africa around 0.7 per cent (OECD, 2007b). Many poor developing countries, on the other hand, have little capacity for promotion of the agricultural sector, because their governance capacities are too weak and their funds too limited. In many cases this sector is instead subjected to disproportionate taxation (World Bank, 2008c).

These trade-distorting measures come under criticism because they protect domestic suppliers and thus obstruct access to the market for more competitive providers, including those in developing countries. However, subsidies – unlike import restrictions – can have a positive impact on global food security, because they tend to keep world agricultural prices down. At the same time, they have adverse consequences for farmers who compete with the subsidized agricultural goods; ultimately, therefore, they also have an adverse effect on the food situation in the country where those farmers are located. Coordinated multilateral dismantling of these distorting measures would open up growth and export opportunities for the agricultural sector in developing countries. The associated production increases would contribute to supply security. The Doha Round of WTO negotiations continues to provide an opportunity for adjusting and developing the rules of world trade. In the short term, the aim must be to reduce distortions by dismantling the obstacles that have been described. In the long term, the functional capability of the agricultural markets should be underpinned by further liberalization of world agricultural trade. Such a process should, however, take account of the differing conditions and various needs in developing countries.

10.4.2.6
Financial assistance, emergency aid and reform of the Food Aid Convention

FINANCIAL ASSISTANCE
Rising food prices present a major problem for LIFDCs who have very limited funds available to finance more expensive imports. The rapid rise in cereal prices in the recent past has jeopardized food security in the developing regions, particularly in the LIFDCs of Africa and Asia (Figure 10.4-1). According to the FAO, prices of imported grain rose on aver-

age by 56 per cent in 2007/2008; this contrasts with a rise of 37 per cent in 2006/2007 and relatively stable prices between 2000 and 2005 (FAO, 2008a). The World Bank and the IMF have already responded rapidly to the crisis and have offered financial assistance to countries that are in difficulty with their balance of payments on account of the high food prices. These programmes need to be as flexible as possible so that they can adapt swiftly to dynamic developments.

EMERGENCY AID
In critical situations of acute food shortage, internationally coordinated emergency aid measures are implemented (World Bank, 2008c). The most important international actor is the United Nations World Food Programme (WFP), which has a budget of US$ 2,800 million. On account of rising food prices and a growing number of people in need, the programme is likely to require additional financial resources in the foreseeable future if it is to properly perform its task. The World Bank and the IMF put the additional short-term need at US$ 500 million (joint spring meeting of the World Bank and IMF, 2008). To ensure adequate long-term funding of the WFP, it may be appropriate to set up independent sources of funding for it. In view of the growing global competition between different forms of land use, one option – based on the ëpolluter paysí principle – would be to consider a levy on forms of land use that serve neither food production nor biodiversity conservation. Energy crop cultivation would be one of the uses subject to such a levy.

REFORM OF THE FOOD AID CONVENTION
These measures must be accompanied by reform of the international Food Aid Convention (FAC), which was adopted in 1999. Within the framework of the Convention there is, in particular, a need to improve steering options, introduce a system of needs analysis and integrate emergency food aid into food security strategies. In this connection the cross-party motion in the German Bundestag (March 2008) for a re-negotiation of the FAC is a step in the right direction. Until now the most important target of the FAC has been to ensure that each of the 23 participating industrialized nations makes a certain quota of food available each year to countries where there are acute shortages. The re-negotiated convention should include measures that will prevent subsidized production surpluses from the industrialized countries being dumped on developing countries, damaging or displacing local food production. In cases of acute food shortage, various instruments must in WBGU's view be brought to bear; they include monetary payments if sufficient food is available on local

markets. In addition, it is essential that emergency aid is replaced by long-term food security. Food aid should, in principle, be confined to acute emergencies. It is particularly important to involve the FAO, IFAD and the WFP in these measures (Ressortarbeitsgruppe Welternährungslage, 2008).

10.4.3
Medium-term and long-term measures

The factors that have contributed to food price rises (including population growth, changing dietary styles, energy prices, increasing competition for resources) will continue to operate in the long term. In consequence, steps must be taken now to improve the global food situation not only immediately but also in the medium and long term. Attention should focus on instruments for boosting agricultural production potential, reform of world agricultural trade and increased promotion of agricultural research.

10.4.3.1
Bioenergy strategies to avoid land-use competition

The global growth in energy crop cultivation can exacerbate competition for land use, in particular competition with food production (Section 5.2). At the same time, however, it provides many countries with an opportunity to reduce their expenditure on imports of fossil energy carriers. Many newly industrializing and developing countries have for this reason already agreed national strategies for increasing their use of bioenergy, especially biofuels (Section 4.1.2). Some of these countries are classed as LIFDCs (e.g. Senegal, Mali, Ghana, Nigeria, Burkina Faso, Kenya, Tanzania, Malawi, Mozambique, Zimbabwe). Whether this has an adverse impact on food security depends ultimately on the type and extent of bioenergy use and on the opportunities for sustainable biomass production. In particular, the cultivation of energy crops on cropland is viewed as undesirable. On account of the shortage of land reserves and the need to increase global food production by around 50 per cent by 2030, cultivation of energy crops needs to be restricted. At the same time, however, it is also important to increase food production, in particular by increasing productivity per unit area. In WBGU's view, energy crop cultivation should be governed by a bioenergy and food security strategy that gives priority to food security.

There is undeniably a place for energy crop cultivation in the context of an off-grid rural energy supply, provided that the crops are grown mainly on marginal or degraded land, in agroforestry systems or in a mixed culture (Chapter 9). The final decision on where energy crops can be sustainably grown can only be made on a regional and context-specific basis and with the involvement of all the relevant actors. However, proposals for growing energy crops in poor developing countries, especially LIFDCs, should be evaluated particularly carefully. Appropriate national bioenergy strategies and compliance with corresponding standards for sustainable bioenergy use are essential (Section 10.3). The regional development banks can play an important role in this area.

Controlled expansion of bioenergy must in WBGU's view be accompanied by worldwide efforts to strengthen agriculture in the developing countries. Otherwise, and if farming in developing countries continues to be neglected, it is likely that competition for land use will escalate and that food security will be jeopardized.

10.4.3.2
Promotion of the small-scale agricultural sector in developing countries

Agriculture in the rural regions of developing countries can play an important part in preventing food crises (Section 5.2). It has, however, long been neglected in international cooperation and this has contributed to the global rise in food prices. In particular, small farmers often lack secure access to capital and agricultural resources, such as seed, fertilizer or loans. The International Assessment of Agricultural Knowledge, Science and Technology for Development commissioned by the World Bank and the FAO (IAASTD, 2008; Box 10.4-2) calls in this connection for changes in agricultural policy in order to better reflect the complexity of agricultural systems in their differing social and ecological contexts (Kiers et al., 2008; Butler, 2008). According to IAATSD, a global agricultural development strategy, if it is to be successful, must be based on the conditions faced by the 400 million small farms that cover less than 2 hectares. A support policy involving tied subsidization of agricultural inputs or the development of broad-based systems of agricultural loans could then be purposeful, provided that sustainability criteria are met (Sachs, 2008). This can only succeed if decision-makers in the developing countries place a correspondingly high priority on rural development. Multilateral development cooperation, which has in the past decade also neglected rural development, should support a radical re-orientation. In the medium to long term, the following measures for promotion of the small-scale farming sector should be pursued (Ressortarbeitsgruppe Welternährungslage, 2008; UN, 2008):

Box 10.4-2

The World Agricultural Council as a new stakeholder in global agricultural policy

The International Assessment of Agricultural Knowledge, Science and Technology for Development (IAASTD or World Agricultural Council) is an initiative of the World Bank and the FAO; it was set up in 2004 by the World Bank, FAO, GEF, UNDP, UNEP, UNESCO and WHO. Its structure and methods are similar to those of the Intergovernmental Panel on Climate Change (IPCC) and the Millennium Ecosystem Assessment (MA). The Millennium Development Goals that are of particular relevance to agriculture (tackling hunger and poverty, improving rural living conditions and health) provide a general normative frame of reference. IAASTD's first global report, drawn up by some 400 agriculture experts, was published in 2008. Fifty-seven countries have approved the synthesis report and the summary. In conclusion, the World Agricultural Council calls for radical changes in global agricultural production to take account of the needs of the poor and hungry. So far, however, the World Agricultural Council appears to lack political weight. It was, for example, noticeable that the findings and recommendations of IAASTD were largely ignored at the High-Level Conference on World Food Security in 2008. IAASTD was also unable to resolve the major disputes of agricultural policy, such as the relationship between policies on small-scale farming and the promotion of large-scale agriculture, the role of green genetic technology, or the question of whether LIFDCs should pursue policies of self-sufficiency or place their trust in international trade. In WBGU's view further work must be done on these contentious issues; questions of weighting and role distribution also need to be considered in this context.

- Cautious expansion of bioenergy, in order to avoid competing use. To this end mandatory sustainability standards and certification systems should be agreed and effective monitoring put in place as soon as possible (Section 10.3).
- Intensification of agricultural research along the entire agricultural chain, including supplier industries, in order to increase yields (Section 11.4).
- Boosting of agricultural productivity through increased investment, particularly in rural development, sustainable small-scale farming and plant breeding.
- Improvement of institutional and legal conditions in developing countries through radical structural changes on a wider scale than just the agricultural sector. Such changes must include improvements in the rule of law, the creation of instruments ensuring market and price transparency and other pro-poor (socio-)political measures.

10.4.3.3
More extensive and more differentiated liberalization of world agricultural markets

The objective of the stalled Doha Round of WTO negotiations is to bring about a prompt and significant reduction in agricultural subsidies in the industrialized countries and to lower other trade barriers. Export subsidies for agricultural goods and other forms of agricultural aid in industrialized countries distort competition and discriminate against providers from developing countries, increasing the dependence of these countries on agricultural imports. In addition, more extensive liberalization of world agricultural trade must be achieved in the medium to long term, in order to improve the efficiency of the agricultural markets. This would improve sales opportunities and production incentives for many developing countries that have comparative advantages in the agricultural sector.

In WBGU's opinion subsidies for industrial-scale agricultural production, such as those provided in Europe, are also questionable because sustainability criteria are often not met. For example, agroindustrial production methods often cause ecological damage (FOES, 2008; OECD, 2005). When food supply shortfalls occur, it is usually inefficient to subsidize local agricultural and food production on a large scale if sufficient food can be procured from surrounding regions. Furthermore, when subsidizing the production of agricultural goods, it is impossible to be certain that the products will be used for food or energy purposes; additional monitoring would have to be put in place to ensure that the goods are used directly for food production. If promotion of bioenergy leads to displacement of food production, it is more efficient to cut back on bioenergy promotion rather than initiate competition between food production and energy crop cultivation for subsidies.

PARTICULAR CONDITIONS AND NEEDS OF DEVELOPING COUNTRIES

Developing countries are affected in different ways by the dismantling of agricultural subsidies, depending on their existing agricultural and food policies. Removal of subsidies will initially cause world market prices to rise. From the point of view of food security, this is advantageous, since it will create production incentives and, if the price rises reach the farmers, cause agricultural production to increase in most developing countries. Developing countries that are net agricultural exporters thus profit directly from the removal of subsidies, particularly if import barriers worldwide are lifted at the same time. Net importers of food, on the other hand, lose out in the short to medium term as a result of the removal of subsidies. The situation is particularly serious for LIFDCs. To cushion the adverse effects, international support and compensation measures would be required immediately to enable LIFDCs to feed their population adequately (WBGU, 2005). Pledges to this effect were made by the industrialized countries in the Marrakech Declaration at the conclusion of the Uruguay Round of GATT negotiations. While further progress in the worldwide dismantling of agricultural subsidies is in principle desirable, it would place LIFDCs under severe strain; were it to happen, therefore, it would be crucial to ensure that sufficient compensatory funds were available.

However, greater food security in LIFDCs is not the only outcome of multilateral import liberalization and subsidy reduction. For many developing countries that impose import restrictions on agricultural goods and pay agricultural subsidies themselves, further liberalization pledges raise problems. Greater liberalization could inflict considerable damage on domestic food production in these countries; small farmers, in particular, would be unable to compete with cheap imports (since imports are no longer restricted) while at the same time foregoing subsidies. A solution would be to allow for exceptions to the general liberalization for a small number of poor developing countries.

Such exceptions were again discussed at the 7th WTO Ministerial Conference in July 2008. Talks centred on the possibility of enabling developing countries to partially exempt 'special products' that are particularly important for food security from tariff reductions. In addition, through the 'special protection mechanism', developing countries would be able to levy additional customs duties on agricultural imports in response to temporary increases in imports or to price decline (Deutscher Bundestag, 2008). Both instruments can therefore serve to support national strategies for promotion of the small-scale farming sector and sustainable food production. Subsidies for agricultural materials for small-scale producers could also be initially exempted from the liberalization process. Rulings on exemptions are discussed in the context of a 'development box' under the WTO Agricultural Agreement (Murphy and Suppan, 2003). In view of the current lack of consensus on the wording of the exemption rules in the WTO negotiations, WBGU calls for appropriate leeway for securing strategies for promotion of the small-scale farming sector. However, the right to make a claim should be restricted to the poorest developing countries (Rudloff, 2008).

INTERNATIONAL AGREEMENTS ON THE USE OF AGRICULTURAL LAND FOR FOOD PRODUCTION

In principle there could be international agreements under which states agree to bear responsibility for adequate food production. In contrast to the Food Aid Convention (Section 10.4.2.6), such agreements do not involve quotas for food aid. Instead they operate more broadly, entailing quotas for a minimum level of food production or pledges as to the amount of land that will be retained nationally for food production. The required coordination could take place under the umbrella of the United Nations, possibly through the FAO.

A useful starting point for negotiation of a food convention that seeks to safeguard production land would be for the international community of nations to specify a minimum area for food production. At regular intervals (e.g. every 10 years) this figure should be adjusted to meet food policy requirements, depending on the development of land productivity and the world population. If the amount of land actually in use for food production falls below this critical threshold, mechanisms should be triggered to ensure that the minimum level of food-related land usage is achieved again as quickly as possible. Such an approach would at present be difficult to implement politically, especially as extensive administration would be involved; however, the concept could become more attractive in future and it should therefore be debated in international policy forums at an early stage.

10.4.3.4
Promoting awareness of the consequences of different dietary habits

The change in dietary habits that is observable in many parts of the world (Section 5.2.3) serves to exacerbate global land-use competition. Rising consumption in industrialized countries and, increasingly, in newly industrializing countries of meat and milk products with a large footprint is the main driver of this development. It is assumed that these altered dietary habits will absorb around 30 per cent of the food production increases required by 2030 (Section 5.2). It is also likely that the problems of food security in the poorest countries will become more severe if this large-footprint dietary style becomes established and extends further as a result of the development of rapidly growing economies such as China and India.

There are also repercussions on the use of biomass for energy. If, as is desirable, land use for food production is given priority, land-intensive dietary styles reduce the global sustainable potential for bioenergy. Economically stronger countries would be able to meet their food needs even in the face of rising prices, but poorer countries run the risk of being increasingly exposed to food crises. There are initiatives, mainly in industrialized countries, that seek to inform consumers about the effects of diet on personal health. However, there are few attempts to raise public awareness of the links between individual eating habits on the one hand and global land use and food security on the other. Where such information is provided, it is often overlaid by consideration of health issues and thus has little conscious impact on the consumer. Growing environmental awareness, however, has brought with it a demand for information on the environmental consequences of food production. This development is supported by state schemes for the certification of organically produced products. Both trends could form the basis of strategies for raising consumer awareness of the consequences of different dietary styles. If this succeeds in creating awareness that ultimately leads to behaviour change, it will help to counteract growing land-use competition.

Activities in this context can be organized at national or local level by private actors, such as retail chains or NGOs, and public bodies. Initiatives can also be instigated at international level, for example in the arenas in which United Nations bodies operate. However, awareness campaigns alone are unlikely to prompt the introduction of sustainable dietary patterns on a worldwide scale. In the medium to long term more forceful instruments will have to be considered by the international community. These could include standards and internalization measures such as levies on food with a large footprint.

Land take for per-capita food consumption
Initiatives for influencing dietary habits can be supported by international agreements on land take for the per-capita consumption of food; indirectly, this could play an important part in reducing global competition for land. In a process coordinated at inter-country level it would first be necessary for each state to assess in quantitative terms the amount of land already in use at home and abroad for production of the food consumed within the country. The concept of the ecological footprint can form a starting point for operationalizing land take (Hails, 2006). However, factors to be borne in mind include the methodological critique of this concept and the fact that calculation of the ecological footprint includes not only land use for food production but other land and biomass uses as well (IMV, 2002; Venetoulis and Talberth, 2008).

This approach requires clear guidelines on accounting and inventorization methods so that the land use associated with international trade flows of food can be calculated reliably in a sustained process. It thus entails extensive assessment and monitoring work and a willingness to accept generalizations. Once appropriate data was available, countries with a significantly above-average per-capita land take could be identified, in much the same way as greenhouse gas emissions are calculated under the UNFCCC. The countries thus identified could be put under pressure to act on the principle that globally all human beings have the right to an equal quantity of natural resources to meet their needs (in terms of an adequate supply of food and energy) (von Koerber et al., 2008). Countries exceeding this level would have to describe in national strategy programmes the measures they intend to apply to reduce per-capita land take. Monitoring of programme outcomes and exchange of experience in programme components could take place in international forums. Implementation of such strategies would ultimately result in stabilization or progressive reduction of per-capita consumption of foods with a large footprint in the countries identified.

It is necessary to consider whether such a regime, initially designed only for food-related land use, could serve as a basis for or be expanded to include global land-use management covering additional biomass use, particularly bioenergy. A global regime of this type might also include mechanisms for increasing the flexibility of the corresponding implementation obligations; the flexible mechanisms of climate policy could provide a conceptual model for this approach. In theory, it would for example be possible for excessive land take in one biomass sector or in one country to be (temporarily) offset by 'reserves' not taken up in another sector or country.

10.4.3.5
Establishment of early warning and risk management systems

Improving early warning and monitoring capacities
In order to be better prepared for crises, an effective early warning system is needed. Existing monitoring capacities, e.g. those of the FAO and the World Food Programme (WFP) need to be better networked and their efficiency needs to be improved (Ressortarbeitsgruppe Welternährungslage, 2008). In addition, WBGU sees an increasing need for prompt identification of risks to food security arising from competition with energy crop cultivation for the use of resources.

CONSIDERING MEASURES FOR REDUCING FOOD PRICE VOLATILITY

The observable correlation between (bio)energy prices and food prices is likely to further increase food price volatility. Frequent sharp downward price fluctuations increase investment uncertainty for agricultural producers and have a particularly detrimental impact on the small-scale farming sector; producers in this sector are unlikely to have reserves and have no security against price risks. Sharp upward price fluctuations, on the other hand, jeopardize net food consumers with low incomes. There is a need to consider whether an internationally coordinated expansion of food reserves could represent a viable means of increasing supply on the world market in the event of a sudden significant price rise, or of generating demand if prices fall dramatically.

10.4.4
Conclusions

The international community has acknowledged that development of global agriculture and, in particular, worldwide food production faces enormous and previously underestimated challenges. Energy crop cultivation and the effect this has on land-use competition represents one of these challenges. In the first six months of 2008 policy-makers responded to the worldwide rise in agricultural prices and drew international attention to the issue at the High Level Conference on World Food Security. It there became clear that the international community of nations is still far removed from a global consensus on the sustainable regulation of bioenergy use. At the same time, it became evident that there is an urgent need for action, since without regulation the globally increasing demand for energy will add significantly to the pressure of use on fertile land.

The Task Force on the Global Food Crisis convened by UN Secretary General Ban Ki-moon has drafted a strategy paper on management of the food price crisis. The paper also lists issues relevant to the development of an international consensus on sustainable biofuels. Recommendations include the development of guidelines to minimize the adverse impacts of energy crop cultivation on global food security and the environment, preparation of a common frame of reference for the development of a certification process for sustainable bioenergy use, the promotion of private investment in agro-energy production in developing countries, re-assessment of subsidies for energy crop cultivation and a review of methodologies for measuring and monitoring biofuel impacts (UN, 2008). In response to current developments the World Bank has in addition set up the Global Food Crisis Response Facility.

In Germany, the Departmental Working Party of the Federal Government on World Food Affairs has submitted a report to the Cabinet (Ressortarbeitsgruppe Welternährungslage, 2008; Box 10.4-3). According to the authors, a response to the critical development of the world food situation requires a coordinated, comprehensive and long-term strategy that must be agreed between states and international institutions. The next major task is to consistently promote implementation of these measures.

Over and above these measures WBGU envisages four priority action areas that are key to securing the world food supply in the face of the worldwide bioenergy boom:
1. Energy crop cultivation must form part of integrated national bioenergy and food security strategies that aim to prevent competition with food production. This is particularly important for the group of LIFDCs. The cultivation of energy crops must be accompanied by increases in food production and in the productivity of land used for food purposes. These national endeavours should be

---

Box 10.4-3

Key recommendations of the Departmental Working Party of the German Federal Government on World Food Affairs

SHORT-TERM MEASURES
- Boost humanitarian on-the-spot aid, emergency aid and transition aid, especially food aid;
- Ensure that food and income transfers reach the weakest members of society;
- Improve access to agricultural operating resources;
- Lift export restrictions with immediate effect;

- Bring the Doha Round to a successful conclusion (in particular by reducing export subsidies and export promotion);
- Tackle budget and balance of payments imbalances.

MEDIUM-TERM AND LONG-TERM MEASURES
- Improve institutional and legal conditions in developing countries;
- Increase sustainable food production;
- Increase investment in sustainable agriculture;
- Boost agricultural research;
- Avoid competition for use through responsible development of bioenergy;
- Introduce mandatory sustainability standards and effective certification systems;
- Improve early warning systems.

underpinned at regional level, particularly by the regional development banks. This would result in a convergence of bioenergy, agricultural and development policies.

2. The greatly increased pressure on land use as a result of changed dietary habits in industrialized countries and the spread of these styles in dynamically growing newly industrializing countries represents a major challenge for the future; the extent of this problem is still widely underestimated and deserves fuller attention. At the same time, the spread of these dietary habits limits the potential for sustainable energy crop cultivation. Pressure on land use is increasing, particularly in LIFDCs, and is exacerbated by the growing demand for food and energy from other countries. In Africa, it is already possible to observe agricultural land being bought up on a large scale by foreign investors (country case studies; Section 8.2).

3. In order to be better prepared for crises, an effective early warning system is needed. Existing monitoring capacities, e.g. those of the FAO and the World Food Programme need to be better networked and their efficiency needs to be improved. In addition, WBGU sees an increasing need for prompt identification of risks to food security arising from competition with energy crop cultivation for the use of resources. Global monitoring and early warning systems are also important for prompt identification of risks arising from the growing pressure on global land use. For these reasons, among others, WGBU recommends the creation of a global commission for sustainable land use (Box 10.3-5). A particular lack is the absence of a global land register providing information on the status and dynamics of soil, water resources and land cover (especially deforestation). The rapid change in global land use is in WGBU's view an insufficiently addressed issue in research and policy-making.

4. The challenges involved in securing the world food supply must now be tackled against the backdrop of increasing pressure on global land use; they can no longer be surmounted by purely national efforts. In view of growing resource scarcity and the ever-closer links both between national and global markets and between food and energy markets, ways must be found to manage land use on a global basis in order to secure an adequate level of food production worldwide. A global agreement on securing a minimum area of land that would remain available for food production is a possible element of global land-use management.

10.5
International biodiversity policy and sustainable bioenergy

The objectives of the Convention on Biological Diversity (CBD) and specifically those referring to the conservation of biodiversity and the sustainable use of its components are directly relevant to the deployment of bioenergy resources. The CBD, by virtue of its 'conservation of biological diversity' objective, is the key international agreement promoting compliance with the WBGU guard rail for biosphere conservation, which is designed to prevent expanding land use from encroaching on ecologically valuable sites (Section 3.1.2). Equally, the 'sustainable use of biological diversity' objective backed by appropriate regulations offers a means of avoiding potential adverse impacts of bioenergy use on biodiversity (Section 5.4). Therefore, in the following section WBGU concentrates on the tasks and scope of the CBD.

The CBD first took up the issue of bioenergy only in 2007, when it was put on the agenda for a meeting of its scientific advisory body, and it became one of the key themes of the ninth meeting of the Conference of the Parties (COP-9) in Bonn. The dynamically expanding bioenergy sector is heightening the pressure on land use, and hence on biodiversity. Consequently, the ongoing efforts to conserve biological diversity and ensure the sustainable use of its components must be intensified. In addition, the CBD has a part to play in regulating the sustainable use of bioenergy. The issue of bioenergy poses the following specific challenges for biosphere conservation and the CBD:

First, enlargement and effective management of the global system of protected areas can help to limit the conversion of natural ecosystems caused, directly or indirectly, by increased utilization of bioenergy (Section 5.4). The Convention has at its disposal a vast reservoir of relevant experience, instruments and objectives integrated within its programme of work on protected areas (Section 10.5.1.1). The financing of the system of protected areas presents special challenges (Section 10.5.2).

Second, it is important to safeguard the conservation and sustainable use of biodiversity in the context of energy crop production or forest product use for bioenergy generation. The CBD can help to tackle this – after the next meeting of the COP in 2010, at the earliest – by developing biodiversity guidelines for sustainable production and deployment of bioenergy which can then be incorporated into relevant standards. The principles and programmes of work drawn up by the Convention form a good working basis for this task (Section 10.5.1).

In the following sections sustainable land use is discussed beyond the specific remit of bioenergy, since no clear distinction can be drawn between ensuring sustainable land use for bioenergy and sustainable land use in general.

10.5.1
Protected areas and protected area systems

Section 5.4 illustrates how the expansion of bioenergy use could come to compete with nature conservation. An effective system of protected areas is indispensable for avoiding such conflicts. At the same time, the significance of protected areas extends far beyond their implications for bioenergy issues. Protected areas are essential instruments for the conservation of biodiversity (MA, 2005b, c; CBD, 2004b). The CBD and the World Summit on Sustainable Development set themselves the target of significantly reducing the rate of biodiversity loss by 2010. In support of this target, WBGU proposed the following biosphere guard rail (Section 3.1.2): '10–20 per cent of the global area of terrestrial ecosystems (and 20–30 per cent of the area of marine ecosystems) should be designated as parts of a global, ecologically representative and effectively managed system of protected areas. In addition, approximately 10–20 per cent of river ecosystems including their catchment areas should be reserved for nature conservation.'

Within the system of protected areas it is by no means a requirement that conservation and use of biological diversity, including bioenergy use, should always be mutually exclusive (WBGU, 2001a). The World Conservation Union (IUCN) subdivides protected areas into categories ranked according to their conservation objective and intensity of use (categories I–VI; IUCN, 1994; Box 5.4-1). Only areas in categories I–IV, which denote conservation as the primary objective, should be counted towards compliance with area targets (Pistorius et al., 2008), since the emphasis of categories V and VI is on sustainable use and not on conservation. In UNESCO biosphere reserves, this principle of graded intensities of use has been applied as a zoning concept in which core zones are enclosed by buffer zones and buffer zones are surrounded by transition areas (UNESCO-MAB, 1995).

Parties to the CBD are bound by Art. 8(a) to establish a system of protected areas. They have thereby entered into a special form of land-use agreement: to conserve the global public good of biological diversity, they undertake that types of land use resulting in degradation will not be permitted in specific areas. Furthermore, the CBD calls for the use of biological resources to be regulated and managed even outside protected areas (Art. 8(c) CBD; Glowka, 1994). Thus, nature conservation cannot be 'delegated' solely to protected areas; instead, all types of land use must integrate aspects of conservation and sustainable use of biodiversity.

The amount of area designated under the current protected areas system is impressive: approx. 12 per cent of the terrestrial surface is already protected. Nevertheless, the target is far from having been met, since many of the areas fall short of fulfilling their conservation purpose and their protected status exists only on paper. Box 5.4-1 gives a survey of the situation and trends for protected areas.

10.5.1.1
CBD work programme on protected areas

It took 12 years from the signing of the Convention before the implementation of Article 8 was tackled with a special programme of work on protected areas (CBD, 2004b). This has since provided the framework for implementation by the Parties. The target is to establish an ecologically representative and effectively managed global network of national protected area systems by 2010 for terrestrial areas and by 2012 for the marine environment. The work programme reads like an ambitious action plan for nature conservation. It describes, in terms of outcome-based targets, what measures the Parties should implement by which point in time. However, a concrete, quantitative obligation to designate protected areas or to comply with global priorities, for example to ensure global representativeness, proved impossible to negotiate. But the programme does refer to the Global Strategy on Plant Conservation, which was likewise developed under the auspices of the CBD and specifies, among other targets, that 10 per cent of each of the world's ecoregions should be effectively conserved by 2010 (Section 3.1.2).

National implementation is carried out in the context of the Parties' own national priorities. The Parties undertook that they would report on the progress of implementation at every meeting of the Conference of the Parties (COP) until 2010 but very few countries are meeting this requirement: only 34 out of 190 Parties have reported on their progress (CBD, 2007). Many countries already have area protection targets or intend to define them: the figures are in the region of 5–30 per cent. China, for example, has made plans to designate 17 per cent of its terrestrial area as protected by 2010. As yet, the difficult task of networking the protected areas and integrating them with other land uses has met with very limited success. Equally, almost all reporting developing countries lack the capacity to develop and implement effective management plans for the existing protected areas. Implementation can be expected to be faltering even more in the 156 countries which have not delivered a report. Rapid progress can hardly be expected without substantially greater commitment in political and financial respects.

10.5.1.2
Further provisions of the CBD

In other sectors of the CBD it has been possible to agree specific area protection targets. In the Global Strategy for Plant Conservation the CBD defined the target of effectively conserving at least 10 per cent of all the world's ecoregions, 50 per cent of the most important centres of plant diversity and 60 per cent of threatened plant species by 2010 (CBD, 2002a; Section 3.1.2). In the context of the evaluation of the Convention's Strategic Plan (CBD, 2004c) the specific area target was likewise agreed of assuring the effective conservation of at least 10 per cent of all the world's ecoregions by 2010. Furthermore, the intention is to reduce the rate of loss and degradation of natural habitats. Within the provisional target framework of the Convention, the 10 per cent target is also being taken up and applied to the various thematic programmes of work: at least 10 per cent of marine and coastal ecosystems, inland waters, forest types, mountain ecosystems, arid regions and island ecosystems should be conserved and general protection afforded to all areas of particular importance for biodiversity (CBD, 2006b).

10.5.1.3
Options for further elaboration

Further elaborations of the protected areas policy must therefore do more to create enabling conditions to assure the establishment of a representative global network of protected areas and adequate financing to manage it effectively. Solutions worthy of consideration include additional agreements on area protection targets, their national implementation and international financing efforts.

THE LIFEWEB INITIATIVE
LifeWeb is a global initiative on protected areas instigated by Germany (BMU, 2008e). It builds on the programme of work on protected areas and is intended as a contribution to financing the implementation. To this end, interested developing countries should give notification of protected areas with underfinanced management capacities as well as 'candidates' for new protected areas for which they cannot

allocate the necessary resources. These candidates are listed in the UN database of protected areas at UNEP-WCMC (WDPA, 2008). Countries and public and private sector organizations interested in contributing to better financing can then cooperate on a bilateral basis with countries that have put forward candidates. This initiative, which is based exclusively on voluntary participation and creates no additional administrative structures, was welcomed at COP-9 in Bonn, where many developing countries proposed specific candidate protected areas amounting to a total area of approx. 460,000 km². The German federal government gave a commitment to contribute an additional € 500 million by 2012 to the LifeWeb Initiative (and the same sum annually thereafter). Unfortunately, at the conference neither EU partners nor G8 countries pledged offers of finance. Germany will therefore need to promote LifeWeb even more actively to its industrialized partner countries in order to guide the initiative towards success.

FURTHER ELABORATION OF CBD PROVISIONS TOWARDS A PROTOCOL ON PROTECTED AREAS
The 2010 target will be evaluated at COP-10. In all probability it will not have been attained, with the exception of a few minor successes. In addition, a fundamental evaluation of the programme of work on protected areas is planned. Here, too, the interim results indicate that most Parties will fail to meet the ambitious targets of the programme. The final version of the Sukhdev Report on the Economics of Ecosystems and Biodiversity (interim findings: Sukhdev, 2008) will be available in 2010, and will quantify the economic value of biodiversity conservation in much the same way as the Stern Review did for climate protection (Stern, 2006). Part of this value will be reflected in the REDD provisions under the UN Framework Convention on Climate Change, which propose to create incentives for forest conservation as a mechanism for climate protection (Box 10.5-3). If, in addition, negotiations on an international regime for access to genetic resources and equitable benefit-sharing can be concluded in 2010 to the satisfaction of developing countries, it will mark the emergence of a constellation of factors which creates a new policy framework. This could enable the further elaboration of the existing CBD provisions in the direction of a binding protocol on protected areas (WBGU, 2006). Such a protocol should encompass the full range of terrestrial, freshwater and marine protected areas and cover the spectrum from strict protection to conservation linked with sustainable use. The contracting parties to such a protocol would undertake to establish a system of protected areas according to quantitative targets. The specific reporting obligations should include notification of existing pro-

tected areas (geographical data, protected ecosystem types and species, protected area categories, details on management, financing, progress and constraints, etc.) and candidates for designation to the UNEP-WCMC global database. On this data basis, a scientific advisory board – emerging perhaps from the international discussion of a body equivalent to the IPCC to advise on policy in the sphere of biodiversity (Intergovernmental Panel on Biodiversity, IPBD: WBGU, 2001a) – could produce status reports on the system of protected areas at regular intervals, dealing with the issues of representativeness, effectiveness and financing and implementation constraints. Any associated interference with national sovereignty should be kept to a minimum. Success can only be achieved if the industrialized countries undertake, in return, to make available adequate financial resources and assistance with implementation.

10.5.2
Financing protected area systems through compensation payments

Areas that are especially rich in biodiversity are predominantly found in the territorial domains of developing and newly industrializing countries. Without additional incentives these countries will not be prepared to enter into conservation commitments on a sufficient scale to ensure the conservation of global biodiversity. They see such commitments as restraints on their sovereignty and their opportunities of economic development. Moreover, they would then be bearing a substantial proportion of the costs of conserving biological diversity, while the entire international community reaps the benefits. Additional financial incentives such as compensation payments are therefore necessary. They would primarily compensate for losses of income as a consequence of refraining from higher-revenue but unsustainable forms of land use (Endres, 1995; WBGU, 2001b, 2002). Seen in this way, compensation payments aim

to bring about the conservation of valuable ecosystem services which are threatened by intensifying pressures from competing uses of land for such activities as energy crop production (Section 5.4).

The level of finance required for compensation is primarily determined by the alternative uses. The opportunity costs of conservation in nature parks and landscape protection areas, for example, are often higher than in uninhabited, remote reserves where lucrative alternative uses such as energy crop production have not previously been feasible (James et al., 1999). Pearce (2007) calculates on the basis of James et al. (1999) opportunity costs of conservation in developing countries of under US$ 9 per ha per year. That would be significantly lower than the amount in excess of US$ 93 stated in an internal World Bank study (quoted in Pearce, 2007). Chomitz et al. (2004) assume that the opportunity costs are determined by the agricultural potential of the land; in developing countries these costs would be in the region of hundreds of US dollars per hectare if the alternative were extensive pasturage, and thousands of US dollars per hectare if it were possible to produce high-quality perennial crops. At national level there are already systems for the regulation of compensation payments (FAO, 2007b; Wunder, 2005). One example is the Pagos por Servicios Ambientales programme established in Costa Rica in 1997 (Box 10.5-1).

Companies offer compensation on the basis of national regulations to establish compensation areas. In several countries (including Australia, Brazil, Canada and the USA) companies intending to develop land intensively are obliged to provide compensating areas elsewhere. One way of doing so is by these companies compensating landowners for refraining from alternative, more profitable land uses in biodiversity-rich ecosystems. Under current national provisions, these obligations can only be fulfilled nationally by means of compensating areas (i. e. mitigation sites in the same country) (ten Kate et al., 2004; Carroll et al. 2007). In parallel, some corporate initiatives engage in voluntary payments for compensating

Box 10.5-1

Payments for ecosystem services in Costa Rica

Under the Pagos por Servicios Ambientales programme established in Costa Rica in 1997, landowners receive payments for the provision of forest ecosystem services. The state's role is complemented by private beneficiaries of ecosystem services, e.g. hydropower stations, which make voluntary payments. Potential participants in the system must initially submit a management plan covering such details as the characteristics of the forest parcel, the intended management method and the planned environmental measures, e.g.

to prevent degradation from forest fires or illegal logging. Over an initial five-year period, forest owners receive regular payments from the specially established fund, the Fondo Nacional de Financiamiento Forestal (Pagiola, 2002). In 1997, the first year of the programme, payments of approx. US$ 14 million were made to protect areas amounting to 79,000 ha (Pearce, 2004). In Costa Rica, as in Mexico and other countries, international donors like the World Bank and the Global Environment Facility (GEF) also contribute to financing. Costa Rica received a US$ 32.6 million loan from the World Bank, for example, and a grant of US$ 8 million from the GEF (World Bank, 2002, 2008c).

Box 10.5-2

Establishing an international market in certified conservation services

Establishing an international market in certified conservation services to finance a global protected areas network is a conceptual elaboration of the German federal government's LifeWeb Initiative (Section 10.5.1.3). The following proposal for such a market is based partly on the concept of tradable non-utilization commitment certificates (WBGU, 2002). The core idea is to secure the financing of protected areas by complementing the host countries' own contributions with a system of international compensation payments. To create and promote an efficiently structured system of compensation payments, it would be expedient to standardize the 'protected area' resource as an asset for which a global market can be established. Standards would need to specify the quality criteria that units of area should meet in order to gain certification, making them eligible to be offered on the market as candidate sites. It would be useful to work towards a globally recognized certificate for high-quality conservation of biodiversity, perhaps comparable to the CDM Gold Standard.

SUPPLY AND DEMAND
In principle, states and any other actors may be suppliers or demanders of certified conservation services in the marketplace. The greatest supply is likely to come from biodiversity-rich developing and newly industrializing countries. Supply-side activities should be voluntary; the incentive for supply should spring from the revenue-earning opportunities created by the market. To generate a minimum level of demand for certified conservation services, the contracting parties should enter into commitments to acquire conservation certificates. The scale of these commitments should depend essentially on the principle of economic capacity – hence industrialized countries would contribute significantly more to the financing of global biodiversity than developing countries. Secondly, the distribution key should also take account of the principles of subsidiarity and ecological capacity – the more biodiversity a state is fundamentally able to conserve on its own territory, the greater its obligation in comparison to other countries at an equivalent stage of economic development. Provisions of this kind would not affect existing obligations under the CBD, according to which every Party must establish and maintain its own national system of protected areas (Art. 8 CBD). There must be safeguards so that biodiversity-rich states do not merely offer conservation services on the international market but are always required to enter into obligations themselves as well. To fulfil these obligations, certificates for conservation services in their own country may be used. For the concrete elaboration of these obligations, one option is to define a quantity of conservation certificates which must be acquired whenever there is sufficient supply. Since it is difficult to predict, at least in the start-up phase, the level at which certificate prices will settle, the quantity of certifi-

cates can be coupled with a spending limit, above which a state need not acquire any further certified conservation obligations. This would be a way of avoiding fiscal overload on purchaser states. However, it must be borne in mind that any such capping impairs ecological effectiveness and could lead to some states acquiring conservation services from certain other countries, perhaps out of foreign policy considerations, at deliberately inflated prices.

HOW THE MARKET OPERATES
It is desirable to incorporate into this market system the kind of steering that ensures representativeness of the global protected areas network. Candidate protected areas nominated by supplier countries are assigned to ecological categories as part of the certification process. This process could be based on the LifeWeb Initiative. Suppliers nominate candidate sites, and these sites become certified. The quantitative basis is the area of the site. Demander states must acquire a certain quantity of certified conservation services and can offset the corresponding area against their registered obligations. They can make their own choice of the sites they wish to purchase, however. Depending on their preferences, e.g. their favoured ecosystem type (rainforest, biodiversity hotspot or conservation of flagship species, etc.), they purchase the conservation of specific areas on the 'first come, first served' principle. On the assumption that demander states are especially interested in sites of high ecological value – whether because this is in the interests of the population, which exerts pressure on state decision-makers accordingly, or because the selection is left to expert representatives – there would be an incentive to release the funding as quickly as possible to maximize both value for money and ecological value in the acquisition of conservation services. At the same time, supplier states would have an incentive to nominate areas of greater appeal from the perspective of biodiversity policy rather than just nominating candidate areas with low opportunity costs. Regular assessments by an Intergovernmental Panel on Biodiversity (IPBD; WBGU, 2001a, 2001a) could ensure transparency in the development of the protected area system, and designate underrepresented ecosystems (currently, for example, wetlands or marine ecosystems) or species groups which would then be allocated a valuation bonus in the marketplace to promote the purchase of conservation certificates from these segments.

Given the current political climate, such a market system for certified conservation services, e.g. in the framework of a CBD protocol on protected areas, can only be a medium-term if not long-term project. Nevertheless, now is the time to mark out the parameters. The LifeWeb Initiative should recruit as many partner countries, NGOs and companies as possible, and set the certification process in motion. Research projects (Chapter 11) should explore whether such a protocol can expediently be linked to the currently emerging REDD regime under the UNFCCC (Section 10.2).

areas in their own countries and abroad as a way of improving their image (ten Kate et al., 2004; Bishop, 2007). Compensation payments and the establishment of compensating areas carry high transaction costs from the companies' point of view. The market-based institutions of 'habitat banking' schemes can reduce these costs. Here the actors are companies

which create certified compensating areas and subsequently sell the certificates to companies wishing to develop greenfield sites. This approach was first established in the USA in the 1990s in the form of 'wetland banking' (Sulzman and Ruhl, 2002) and is now being recommended by the EU Commission for implementation (EU Commission, 2007a, b). Preliminary approaches along the same lines have also been developed in Malaysia (Hawn, 2008). According to data in Carroll et al. (2007), the 400 wetland mitigation banks in the USA alone generate transactions amounting to more than US$ 1000 million per year. Furthermore, private organizations (NGOs, foundations, etc.) make additional financial contributions to protected areas and the conservation of biodiversity in general. According to cautious estimates, payments of up to US$ 1000 million per year are made for international biodiversity conservation (OECD, 2003; Gutman and Davidson, 2007). Added to this are market payments for certain privatizable ecosystem services such as carbon certificates (Box 10.2-3), bioprospecting or ecotourism (WBGU, 2001a; 2005). Currently the volume of this segment amounts to US$ 1000–2000 million and is deemed to be highly dynamic (Gutman and Davidson, 2007). It remains to be seen whether these kinds of market payments increase in future and, above all, whether they are deployed as compensation mechanisms in the international context.

International compensation payments are justified because biological diversity found locally and nationally often yields benefits beyond national borders (WBGU, 2001a; MA, 2005a; Perrings and Gadgil, 2005). Property Rights over the relevant resources are assigned to individual nation states, effectively giving them control over the supply of biodiversity and ecosystem services resulting therefrom. Should these globally valuable services threaten to decline, other countries can offer conditional payments that alter the cost-benefit ratios of alternative land uses in favour of sustainable uses in the host country of the subject of protection. These conditions apply particularly in newly industrializing and developing countries in the tropics with globally significant biodiversity resources. Many of these countries are considering a major expansion of bioenergy production or have already taken steps in this direction (Section 4.1.2). This trend directly or indirectly increases the risk of resource-degrading land-use practices (Section 5.4), a risk which has implications for an effective deployment of international payments.

10.5.2.1
Financing the global network of protected areas through international payments

Estimates of the required funding for a global network of protected areas, including marine protected areas, vary according to area delimitations and assumptions about the degree of protection (Gutman and Davidson, 2007; Schmitt et al., 2007). Thus Balmford et al. (2002) assume annual costs of

$US\$_{2000}$ 45,000 million. James et al. (1999) put the figure at $US\$_{1996}$ 27,500 million. The total expenditure on protected areas is estimated at an annual US$ 6,000–10,000 million worldwide, of which developing countries contribute US$ 1300–2500 million (James et al., 1999; Molnar et al. 2004; Gutman and Davidson, 2007). Various financing mechanisms come into play here (Gutman and Davidson, 2007). Payments are stated inclusive of expenditure on protected area management, so compensation payments only represent a small share. The current level of international payments for the conservation of biodiversity as a whole is put at US$ 4000–5000 million per year of which 30–50 per cent relates to the protected area network. Some US$ 2000 million are made available in the scope of bilateral and multilateral development cooperation by the OECD countries (OECD, 2002; Gutman and Davidson, 2007), the dominant mechanism being bilateral development cooperation. From the German perspective, the principal component to mention is the tropical forest conservation programme for which spending of over € 100 million was budgeted in 2007 (BMZ, 2008a, b). In addition, there are debt-for-nature swaps, in which foreign debt is exchanged for nature conservation (WBGU, 2001a, 2002). Between 1987 and 2003, the equivalent of around US$ 2200 million was invested in nature conservation under this mechanism (Pearce, 2004). Swaps can be financed and implemented by the private as well as the public sector.

In terms of future enhancement of the financing of global biodiversity conservation, great expectations are attached to the global carbon trading market and, in particular, to the REDD process under the UNFCCC (Section 10.2; Box 10.2-2; Box 10.5-3; Gutman and Davidson, 2007; Huberman, 2007). At the same time, the possible different forms of a REDD regime must first be explored to assess their impact on the functionality of the carbon market and to determine their realizable financing potential (Box 10.5-2; Section 11.5.4).

Over and above the financing of protected areas, compensation payments are applicable in a general way to the conservation of biodiversity in the surrounding cultural landscape. Conservation measures in cultural landscapes are, according to very rough estimates by James et al. (1999), associated with annual costs in the low hundred thousand millions US$. These costs of biodiversity conservation normally fall to landowners or are partially passed on to the consumers of agricultural products. According to Gutman and Davidson (2007) a substantial portion of these costs can be offset in the long run by efficiency gains from more sustainable land use, but compensation payments have an important contribution to make as a mechanism to bridge financing gaps.

Box 10.5-3

Climate protection and biodiversity conservation within international climate policy

An operable regime on the theme of Reduced Emissions from Deforestation and Degradation (REDD) in developing countries could produce synergies in climate protection and biodiversity conservation. Often, projects and activities on land use, where the primary purpose is climate protection, also contribute directly to the conservation of biodiversity, as when near-natural ecosystems with high carbon content are protected from conversion and degradation. However, land use optimized for climate purposes (Section 5.5) can equally render biodiversity conservation all the more difficult. One example is climate protection by means of afforestation, which can have adverse impacts on biodiversity if it involves the conversion of natural areas. There is a similarly ambivalent relationship between biodiversity conservation and the production of energy crops motivated by climate concerns: the problems are discussed in Section 5.4. Goal conflicts can be expected, e.g. when energy crop production involves the conversion of grassland or savannah. Whilst it only results in moderate carbon release, which is balanced out comparatively quickly by the subsequent utilization of bioenergy, the conversion can entail losses of biodiversity.

COORDINATING CLIMATE PROTECTION AND BIODIVERSITY CONSERVATION IN LAND USE

The Marrakesh Accords to the Kyoto Protocol explicitly state that land use, land-use change and forestry (LULUCF) activities must contribute to the conservation of biodiversity and the sustainable use of natural resources (UNFCCC, 2002). The UNFCCC mechanisms for avoiding adverse effects of climate protection measures on biodiversity conservation differentiate between project-based measures and non-project-based climate protection. For project-based measures (e.g. the CDM), the available options are sustainability standards or environmental impact assessments incorporating biodiversity aspects. Evidence of indirect land-use changes caused by displacement effects, which have adverse impacts on biodiversity, is particularly difficult to capture, however. It is even more difficult to frame the non-project-based climate protection incentives under the UNFCCC involving national maximum emission targets for the signatory countries in such a way as to minimize adverse effects on biological diversity. Greater scope could be achieved in this area by ensuring that LULUCF obligations are not completely flexible and interchangeable with obligations in other sectors. Then the possibility would arise of considering further criteria in the LULUCF sector over and above the avoidance of greenhouse gas emissions, without jeopardizing the effectiveness of the regime as regards fossil emissions reduction.

Hence, given the dynamic progress of the REDD negotiations within the UNFCCC, some timely groundwork from the CBD is more important than ever. At CBD COP-9, it was resolved that a CBD expert group should contribute its expertise to the UNFCCC negotiations in order to harness synergies and avert risks. Better networking between the scientific bodies affiliated to both conventions is also conducive to these aims. At the implementation level, efforts should be made to achieve closer coordination within national ministries and agencies responsible for UNFCCC and CBD negotiation and implementation (Roe, 2006).

Overall, there is a need for international discussion on and broad research into ways to harness synergies between climate protection and biodiversity conservation through improved institutional linkages and ways to resolve conflicts (Chapter 11).

In sum, the current global volume of payments falls markedly short of the estimated requirement for the financing of protected areas. So far, there has been no coordinated, institutionalized system of international (compensation) payments. Payments are rendered principally on a decentralized basis as conditional project financing in the course of development cooperation. Unlike funding in other fields of development cooperation, however, biodiversity conservation calls for more than the start-up or 'seed' finance that is common in sectors like industry, since it is concerned with the lasting conservation of globally valuable natural resources. Hence the payments – unlike conventional project financing – should be effected periodically and reliably over the long term (Swanson, 1999). Only this will pave the way for setting effective incentives to refrain from unsustainable land uses in the long run. For this reason, additional financing mechanisms are necessary.

10.5.2.2
Options for further elaboration – criteria for an international compensation regime

Compensation payments can only set an effective incentive if adequate financial resources are made available for this purpose. According to the principle of joint but differentiated responsibility, it is advisable for every country initially to undertake its own efforts, based on its ability-to-pay and its biodiversity endowment. The choice of the appropriate instruments, i.e. national compensation payments or other methods, rests with the respective sovereign states.

Within the CBD framework there have long been endeavours to develop a strategy for the mobilization of additional financial resources. So far little has come of this process, however. Specifically for the implementation of the programme of work on protected areas, at COP-9 the industrialized countries demonstrated only muted willingness to make additional resources available. One notable exception is Germany's pledge of € 500 million for the LifeWeb Initiative (Section 10.5.1.3).

In the view of WBGU the OECD countries as well as emerging economies like Russia or the oil-rich states have a responsibility to do considerably more than they have in the past to finance protected areas. The question of how much the biodiversity-rich developing and newly industrializing countries should contribute themselves may be debatable; nevertheless, in view of biodiversity's enormous benefit to benefit other countries globally, it is clear that the resources mobilized by the international donor community are far from adequate. Depending on the level of the share deemed to be appropriate for the economically poorer countries of origin to contribute, the high-income countries should be prepared to mobilize around € 20–30 per capita per year to close the financing gap. The benefit of the nature conservation thereby achieved is surely many times greater than this amount (WBGU, 2005).

The financial resources allocated should be employed efficiently, and windfall gain should be avoided. Compensation payments should be geared towards conserving biodiversity threatened by the heightened pressure from competing forms of land use, which is an expected consequence of the increasing use of bioenergy. If rising demand for bioenergy increases the yield from degrading land use, then compensation payments must be raised to a level that (still) permits adequate protection. According to a study by IFPRI (2008), 30 per cent of the recently observed average rise in cereal prices can be attributed to the increased demand for biofuels. With progressive expansion of biofuel use, by 2020 prices could be 18 to 26 per cent higher (in real terms) than in a situation where biofuel production is held at the current level (Section 5.2.5.2). Compensation payments to ensure effective conservation would therefore need to rise on a comparable scale, differentiated by type of land area, just to keep pace with a biofuels boom, i.e. without yet taking account of the effect of increased demand for land for industrial feedstock purposes (such as wood production). This brings the close interrelationship between bioenergy policy and biodiversity conservation, including the financing issues, into sharp relief. Economic efficiency requires conservation objectives to be attained with the least possible expenditure. Essentially, efficient payments can be assumed where processes of searching for the most cost-effective conservation are organized via market institutions. On the national level, this has already taken place through such initiatives as habitat banking. In an international market, host countries could offer to supply the protection of high-biodiversity sites, and countries with a demand for conservation could pay for site protection. As a point of principle, however, a functional market for conservation services must meet a series of criteria not hitherto satisfied on the international level (WBGU, 2002; Kulessa and Ringel, 2003; EcoTrade, 2008). These include the definition of the tradable services, which can be highly complex to operationalize in the sphere of biodiversity conservation since the system must reflect not just area size but also each area's inventory of ecological resources. One possibility is to specify core ecological criteria for the quality of the sites and to leave the subsequent selection to the demand-side actors. To prevent freerider behaviour on a large scale, which would ultimately result in a situation where few remaining actors would be willing to pay, states should agree to co-finance the conservation of biodiversity through protected areas, e.g. by undertaking to purchase sufficient quotas of conservation certificates (Box 10.5-2).

10.5.3
Contributions of the CBD to bioenergy standards development

The task of conserving biodiversity in tandem with bioenergy deployment involves not only establishing effective systems of protected areas, but also ensuring the conservation and sustainable use of biodiversity in bioenergy production systems in the cultural landscape (Sections 5.4 and 7.1). Standards for sustainable bioenergy production are important instruments to this end (Section 10.3).

On account of its limited framework of goals and objectives, the CBD is not the appropriate forum for developing comprehensive sustainability standards for bioenergy, because it cannot cover the full spectrum of the necessary ecological and social dimensions. Within certain parameters, however, it can use its expertise to develop biodiversity guidelines and thus provide building blocks for the kind of standards required.

10.5.3.1
Provisions of the CBD as the basis for bioenergy standards

The provisions of the CBD establish a basis for policy. For example, the Parties undertake to regulate and manage biological resources within and outside protected areas with a view to ensuring their conservation and sustainable use (Art. 8(c) CBD), to conserve ecosystems, habitats and species in their natural surroundings (Art 8(d) CBD), and to adopt measures relating to the use of biological resources to avoid or minimize adverse impacts on biodiversity (Art 10 (b) CBD). The obligation to set up national systems of protected areas is discussed in Section

10.5.1. The Convention's provisional framework for goals and targets (CBD, 2004c) includes the general requirement that biodiversity-based products should be derived from sources and sites that are managed sustainably and in consistency with the conservation of biodiversity. The following themes in the CBD provide key principles which should be used for the detailed elaboration:

- *Ecosystem approach:* As a strategy for the integrated management of land, water and biological resources, the CBD developed the ecosystem approach which is intended as an overarching, guiding framework for action to promote the conservation and sustainable use of biodiversity (CBD, 2000). The approach consists of 12 principles which are very highly aggregated and worded in general terms, which is why supplementary guidelines for interpretation were provided (CBD, 2004a). The decision adopted at COP-9 reaffirms the ecosystem approach and encourages its broad application by means of better communication (CBD, 2008b).
- *Addis Ababa Principles:* The 14 principles and guidelines on the sustainable use of biodiversity adopted by the Convention in 2004 (CBD, 2004d) provide an additional framework for action for ensuring that the use of components of biodiversity does not lead to long-term biodiversity loss. COP-9 resolved to refine the development of the principles in cooperation with the FAO.
- *Global Strategy on Plant Conservation:* With this strategy the CBD has set itself specific objectives: By 2010 at least 30 per cent of land in use (primarily agricultural and silvicultural land) should be managed in a way that is consistent with the conservation of plant diversity, also avoiding adverse effects on adjacent ecosystems (Target 6; CBD, 2002a). Moreover, by 2010, 30 per cent of plant-based products should originate from sustainably used ecosystems, integrating social and environmental aspects (Target 12). The spread of the use of sustainability standards and the proportion of certified products are proposed as possible indicators for this target.
- *Work programmes on agriculture and forests:* In many countries the vast majority of biodiversity is found in the cultural landscape outside protected areas. For effective systems of protected areas which are integrated into the landscape, the conservation philosophy must be integrated into such land as well. The demand for sustainable land use on all land used for agricultural or silvicultural purposes must therefore incorporate the conservation of biodiversity, since the growing need for higher-intensity land use to raise land productivity is otherwise associated with considerable risks

for biodiversity (MA, 2005b; Phillips and Stolton, 2008). This applies equally to production systems for energy crops (Section 7.1). The links between agriculture, forestry and biodiversity are the subject of work programmes under the CBD, which equip the contracting Parties with objectives and instruments for improving the compatibility of management with the objectives of the CBD. Neither work programme specifies targets. The COP-9 decision on agriculture makes reference to bioenergy only insofar as general guidance is given on promoting the positive and minimizing the adverse impacts of biofuels upon biodiversity. The decision on forests makes no reference to bioenergy.

- *Genetically modified organisms:* Article 8(g) obliges the contracting Parties to the CBD to establish means to regulate, manage or control the risks associated with the use or release of genetically modified organisms (GMOs). These are elaborated in detail in the Cartagena Protocol on Biosafety (CPB) in respect of safe transboundary transfer, handling and use of GMOs. Under the terms of the Cartagena Protocol the precautionary approach must be applied, meaning that in the absence of an adequate data basis on potential detrimental environmental impacts of GMOs, approval can be refused in order to prevent such adverse impacts.

10.5.3.2
Routes towards implementation of biodiversity-relevant guidelines or standards on bioenergy

These CBD provisions are appropriate, in principle, as a policy foundation for the development of biodiversity guidelines or standards for bioenergy production as well as generally for all types of land use, but must be firmed up and elaborated on the basis of the precautionary approach (Section 10.3). The following are the key dimensions:

- *Conservation of biodiversity:* The production of bioenergy or biomass should ensure the conservation of biodiversity. In particular, protected areas and elements of protected area systems and ecosystems of high biodiversity value must be exempt from (both direct and indirect) conversion for bioenergy purposes (Sections 5.4 and 10.5.1). To this end, the implementation and further elaboration of the programme of work on protected areas is required for the establishment and extension of effective national systems of protected areas (Section 10.5.1.1). As a prerequisite for the necessary monitoring, the extension of the World Database on Protected Areas is recommended (WDPA, 2008). But beyond this, a further concern is the small-scale conservation of areas of especially high value for biological diversity in the cultural landscape (Section 5.4.2). These areas should be identified, encircled with buffer zones and, where reasonable, networked with corridors with a view to establishing habitat connectivity. Prior to any conversion of unused land (e.g. marginal, degraded or fallow land) for bioenergy, its ecological value must be investigated.

- *Sustainable use of biodiversity:* Where bioenergy production is based on the use of residues or energy crops, sustainability must be guaranteed. In agricultural and forest ecosystems, the accompanying flora and fauna and genetic diversity should be safeguarded and adverse impacts on other ecosystems should be avoided. This calls for regulations tailored to the production system and to local conditions, with due regard to such factors as observing crop rotations, use of water and agrochemicals, and avoiding the cultivation of potentially invasive species (Box 5.4-3). For the use of genetically modified organisms, Art. 8(g) of the CBD and the provisions of the Cartagena Protocol on Biosafety provide a framework for action, yet to be worked out in detail, in order to prevent such risks as the spread of modified genes into wild populations.

Whether the CBD should contribute in such a way to guidelines or standards is a highly controversial topic of discussion. The EU would like to use the CBD to develop guidelines which should help to minimize possible adverse effects of biofuels on biodiversity. Brazil, supported by Argentina and some other countries, is taking a very optimistic counterposition with regard to the potential and the environmental safety of biofuels. Proponents of those views would not endorse the CBD as a forum for dealing with the issue of biofuels, and particularly sustainability standards. Nevertheless, many non-governmental organizations operating within the CBD framework are issuing emphatic warnings about the adverse effects of energy crop production on biodiversity.

In the run-up to COP-9 in Bonn, views on the issue appeared to have reached a deadlock. COP-9 brought a partial success for the EU since a specific resolution on biofuels was adopted, which reaffirmed the principle of sustainability and acknowledged the role of the CBD in this area. However, the task of the CBD as envisaged by the EU, that of drawing up concrete biodiversity guidelines for the development of standards, was delayed. Until the time of COP-10 in Nagoya, Japan (autumn 2010), the CBD will only use regional workshops to explore ways in which the positive effects of biofuels for biodiversity can be promoted and the negative effects minimized. It will not be possible to assess definitively how the CBD

can contribute to the development of standards until after COP-10 (Loose and Korn, 2008).

More extensive demands upon the CBD over the long term to develop relevant guidelines, not only for bioenergy but for all forms of biomass production, are not yet a specific theme within the Convention, not least because some Parties perceive a conflict with WTO free trade rules and fear the possibility of trade barriers imposed by industrialized countries (Sections 10.3.3 and 10.3.4).

10.5.4
Conclusions

COP-10 in Nagoya will probably be forced to concede that it has not been entirely possible to achieve the CBD's target of significantly reducing the loss of biodiversity by 2010. In view of increasing competition between different forms of land use, which is only exacerbated by bioenergy, and the resultant pressure on the remaining natural ecosystems, it will become rather more difficult in future to achieve appropriate biodiversity targets. Consequently, it is all the more important to mobilize the necessary political will to advocate the conservation of biodiversity sincerely and effectively from now onwards.

The implementation of the protected areas programme of work to establish and develop effective national systems of protected areas is an important prerequisite to enable advancement towards the 2010 target. As the host of COP-9 in Bonn, the German federal government took an exemplary lead by pledging a substantial sum of additional financial resources for the financing of the global network of protected areas. Germany will need to devote even greater effort to promoting the LifeWeb Initiative among its industrialized partner countries in order to ensure the success of the initiative. At the same time, the effectiveness of existing financing mechanisms and spending must be improved. The aim must be to institute a dependable and coordinated international financing regime for protected areas and for the conservation of biodiversity in general. In the view of WBGU, the LifeWeb Initiative could be a first step in the direction of an international regime of conservation and payment obligations in the form of a market for certified conservation services. To this end, the further elaboration of the CBD provisions in the direction of a protocol on protected areas should be examined as a possible long-term option.

The drafting of biodiversity guidelines within the framework of the CBD as a contribution to sustainability standards should be supported and, as far as possible, accelerated. As an important parallel contribution, the expansion of the World Database on

Protected Areas (WDPA, 2008) should be promoted in order to build up the necessary monitoring capacities.

10.6
Water and soil conservation in the context of sustainable bioenergy policy

10.6.1
Soil conservation and desertification control: Potential and limitations of the Desertification Convention

Bioenergy is a newly emergent theme with regard to the implementation of the United Nations Convention to Combat Desertification (UNCCD) on which intensive consultations already took place at several sessions of the eighth meeting of the Conference of the Parties (COP-8) (IISD, 2007). Discussion is now focusing primarily on the potential of energy crop production for income generation and as a means of diversifying income sources, meeting rural energy needs, earning export revenues and combating desertification through erosion control and reforestation. What becomes clear is that the growth in bioenergy production, along with its unintended adverse effects, creates a need for policy in each country to monitor and manage activities in the sector. In this context the UNCCD provides a platform for setting out programmes and strategies supporting sustainable land use geared towards poverty reduction in the countries affected by drought and desertification, particularly the instrument of National Action Programmes to Combat Desertification (NAPs). Within the NAP framework it would also be possible to advance the use of standards for sustainable soil use, also making specific reference to the production of energy crops. Furthermore, the UNCCD ten-year strategic plan adopted in 2007 offers a host of possibilities for promoting awareness-raising, assessment of bioenergy, standard-setting with specific regard to desertification control, and policy-making for sustainable bioenergy use in general (Paquin, 2007;

Box 10.6-1

Policy implications of biomass use as industrial feedstock

Phasing out the use of fossil fuels and feedstocks presents novel requirements for the development of biogenic feedstock streams as well as the design of products for cascade use with end-of-life conversion into energy (Section 5.3). These requirements upon products are additional to existing sectoral requirements such as energy efficiency in production and use, hazardous substance reduction, resource efficiency (so far mainly in relation to mineral resources) and closed-loop management. Based on the above requirements, corresponding laws and programmes have been passed; in the European Union, for instance, the EU Directive on energy end-use efficiency and energy services (EU Parliament, 2006), the Action Plan for energy efficiency, the Ecodesign Directive, the Action Plan for sustainable consumption, the REACH chemicals regulation, and the Thematic Strategy on the sustainable use of natural resources. Comparable activities are under way in other OECD countries and, with a slight time lag, in newly industrializing countries. However, some undesirable global evasion strategies have also become established. For example, end-of-life vehicle recycling in the EU is being circumvented by massive exports to West Africa and Eastern Europe, while the disposal of electronic waste is similarly affected by high levels of exports for less-than-adequate processing in China or India. Another example is the very high consumption of paper in industrialized countries and in some cases environmentally destructive and climate-damaging extraction of cellulose (which – unlike biofuels – is not a major focus of the political debate).

Even if it may make perfect pragmatic sense to craft strategies and instruments focused on specific industrial sectors and environmental media, it is becoming increasingly clear that a new, integrated approach is necessary. Such an integrated global product strategy should be structured in such a way that it does not address sub-processes but focuses on products, takes account of and optimizes the entire product life cycle, carries out integrated assessment of the multidimensional requirements mentioned above giving due consideration to goal conflicts, and also takes account of global aspects. Elements of such a strategy would be:

- Scenarios for trends in material flows for global mass products, for use in determining the demand for strategic resources (energy sources as well as biomass for industrial feedstocks and selected mineral resources) and the total pressure on the environment associated with these products, globally and regionally, including pollutant loads and arisings of secondary resources and wastes,
- Investigation of competing demands for use of biomass for food purposes and for energy and material feedstocks,
- Identification of innovation potential and regulatory needs,
- Setting of international and, where relevant, region-specific innovation targets for products, and corresponding alignment of research and technology promotion and, where relevant, market incentive programmes,
- Setting of reduction targets and investigation of appropriate instruments with the focus on an integrated global product policy and dynamic product standards – with the twin objectives of stimulating and fostering innovations as well as regulating problematic products,
- New development of a waste management strategy which incorporates cascade use of biomass utilized as industrial feedstocks with subsequent reuse for energy generation.

Pilardeaux, 2008). Two of its strategic objectives are of special relevance here:

- To improve the living conditions of populations affected by drought and desertification: the production of energy crops can contribute to diversification of income sources and to improvement of rural energy supply.
- To improve the condition of degraded dryland ecosystems: bioenergy production can contribute to harnessing the value of marginal and overexploited land and can support desertification control by preventing erosion.

Under the auspices of the Convention, work on assessing the potential and consequences of various forms of bioenergy in dryland areas could be carried out by the newly organized Committee on Science and Technology (CST) (Bauer and Stringer, 2008). Monitoring and evaluation of the concrete impacts of local and export-oriented bioenergy production on food prices, food security and rural income structures would be a task for the Committee for the Review of the Implementation of the Convention (CRIC) and the CST. The promotion of standards for bioenergy production in dryland areas and appropriate labelling could be taken forward by these two subsidiary bodies of the Conference of the Parties (CST, CRIC) in cooperation with the Global Environment Facility (GEF). In the context of the global bioenergy boom, the UNCCD presents a range of opportunities for promoting development-oriented agricultural use of dryland areas and steering it along sustainable lines. In order to limit institutional fragmentation and redundancies and, instead, to promote synergies among them, any measures agreed in the context of the UNCCD should be coordinated in advance with the instruments for implementation of the Convention on Biological Diversity (CBD; Section 10.5). Particular attention should also be devoted to better coordination between the content of UNFCCC National Adaptation Programmes of Action (NAPA) and NAPs. The expertise located within the UNCCD should also be integrated into standard-setting processes outside multilateral institutions.

Fundamentally the UNCCD offers an appropriate framework for combining general international strategies for managing the challenges of climate change with the specific needs of people living in dryland regions. Whilst the Convention process does not provide for micromanagement at project level, it can help by providing support for policy design at local and national level to make meaningful links between climate change adaptation and desertification control – with the production of energy crops as one strategy among many. Turning the Convention to this purpose presupposes, however, that the State Parties support the initiated reform process and move swiftly to implement the ten-year strategy. As an institution whose field of competence is restricted largely to dryland areas, it can provide specialist backup to other international institutions for measures in the sphere of rural development, conservation of biodiversity and adaptation to climate change.

In relevant countries, implementing the NAPs and developing a bioenergy strategy in the context of desertification control should be made integral components of their overarching national development strategies, such as PRSPs.

10.6.2
Conservation and sustainable use of freshwater

One consequence of the growing global importance of bioenergy is greater demand for freshwater resources, especially for food production, increasing the pressure on such resources. The effects of climate change and soil degradation will further heighten this pressure (IPCC, 2007b; FAO, 2008d). In its 2008 strategy paper entitled 'Elements of a Comprehensive Framework for Action', the United Nations High Level Task Force on the Global Food Crisis points out the rising competition for use of freshwater resulting from the global bioenergy boom. Among its recommendations are the development of standards which incorporate stipulations on the sustainable use of freshwater, as well as prioritizing its use for food production (UN, 2008).

So far, global freshwater resources have remained a largely unregulated sphere of international policy. There are no agreements on the subject under international law, such as those which exist for climate protection, conservation of biodiversity or soil protection in drylands (Pilardeaux, 2004). At the same time, the trends observed to date highlight an urgent need for regulation on the conservation and sustainable use of freshwater resources, given the growing utilization pressure.

A significant international forum that deals with questions of water policy is the World Water Forum that convenes on a three-year cycle. This is a regular gathering of international water experts, decision-makers, scientists and representatives of international organizations, who come together to form a stakeholder forum. It is organized by the World Water Council, which was founded in 1996 as a platform for international water policy actors, including government representatives, parliamentarians, NGOs, representatives of the private sector, science, development cooperation and the United Nations. It has close links with UNESCO. The Council wishes to raise political awareness of the theme, support the debate on sustainable water policy, help to work out concrete courses of action and motivate countries to enter into binding agreements. So far the World Water Forum has not yet dealt with the consequences of energy crop production for water use, but it is on the agenda for the fifth World Water Forum, which will be held in Istanbul in 2009.

The United Nations is likewise engaging with the theme of conservation and sustainable use of freshwater in the framework of the Commission on Sustainable Development (CSD). Meetings are held on a two-year cycle, each cycle consisting of a policy session alternating with a review session. The last review session of the Commission on Sustainable Development (CSD 16) took place in New York in May 2008, and addressed the themes of agriculture, rural development, land, drought, desertification and Africa. Discussion during the CSD encompassed both integrated water resources management and access to water and sanitation. The outcomes will form the basis of negotiations on policy recommendations for CSD 17 in the year 2009, in order to intensify work to achieve the Millennium Development Goals in the water sector. A further review session on the theme of water and basic sanitation is scheduled for CSD 20 in 2012. Any regulatory effect on the freshwater sector from the activities of the CSD is unlikely, however, since it is a dialogue forum that has repeatedly been criticized for its lack of power to translate its decisions into constructive action (Maier, 2007).

At the 2002 World Summit on Sustainable Development in Johannesburg, the international community undertook to halve, by 2015, the proportion of the world's population without access to safe drinking water and basic sanitation. Consequently, the various organizations of the United Nations (e.g. WHO, FAO, UNDP) are active in the water sector, supporting the theme of conservation and sustainable use of freshwater and driving forward the process of providing better and more sustainable basic sanitation, in a concerted effort to achieve the Johannesburg target in the year 2015.

10.7
State promotion of bioenergy: Agricultural and industrial policies

State promotion policies in industrialized and developing countries alike have a substantial influence on the shape and scale of bioenergy use. Along the entire production and value chain, a variety of enabling regulatory and fiscal instruments are brought into play. The essential concern is to eliminate support for any non-sustainable production or use of bioenergy whilst promoting particularly sustainable pathways of cultivation and deployment. In this process, attention

should not be confined to direct support for the sustainable cultivation of energy crops, mobilization of organic wastes and residues and facilitation of market access for biogenic energy sources or particular conversion and end-use technologies. Rather, there is a need for critical scrutiny of any indirect incentives or even perverse incentives arising from other policy measures which have implications for bioenergy deployment. A large number of environmental, energy, agricultural and economic policy measures and frameworks need to be reviewed to determine what influence they exert on the choice of (bio) energy carriers and pathways. To demonstrate this approach, WBGU examines a few key areas, namely measures to internalize external climate effects, specifically emissions trading and the international climate policy framework (Section 10.7.1), and energy subsidies (Section 10.7.7.3).

10.7.1
Promoting bioenergy pathways through climate policy

With regard to the climate change mitigation effect of bioenergy (Section 7.3), the climate policy frameworks at international and national level are of special importance. So far, energy-intensive sectors of industry in industrialized and increasingly also newly industrializing countries have been the prime targets of national legislation on climate protection. To set effective incentives for the efficient avoidance of greenhouse gas emissions on a global and cross-sectoral scale, ideally any incentive schemes would need to cover all emissions, in all countries, and from all sectors. Furthermore all emissions would have to be capped within emissions trading systems or carry such a tax burden that the external climate costs of emissions would be fully internalized, i.e. each emitter would bear the entire costs of its own emissions by purchasing certificates or paying taxes. That would raise the competitiveness of bioenergy pathways with a significant mitigation effect. In principle, under such a scheme neither wider-ranging support measures for sustainable bioenergy pathways nor support for other sustainable renewable energies would be required. State support could then only be justified in well-founded special cases, such as technologies with very high greenhouse gas reduction potential and high learning curve effects coupled with high start-up costs which inhibit market entry. All in all, the state could refrain from any targeted support of climate-protecting energy pathways, including bioenergy pathways. Instead, energy-path choices would be substantially determined by prices in the (global) carbon market.

The ideal-typical implementation of a cross-sectoral international emissions trading system along the above lines is unlikely in the foreseeable future, for political reasons. And there are further barriers: the task of allocating all the emissions emanating from the bioenergy sector to individual emitters and, above all, integrating such emissions into a universal trading system in which emissions rights can be traded among individuals may prove virtually unworkable, or workable only with disproportionately high costs and excessive interventions. Nevertheless, in the long term, the aim must be to create a climate policy framework in which the external costs of energy pathways are fully internalized from energy-carrier cradle to grave. A first step in that direction in relation to bioenergy is the removal of any perverse incentives generated by the inventory and accounting methods in existing emissions regulation systems.

The current inventory and accounting rules under the UNFCCC, whereby emissions from the combustion of energy crops are not allocated to the site and sector of emission, indirectly favour uses of bioenergy that are harmful to the climate (Doornbosch and Steenblik, 2007). To reverse such instances of counterproductive support for bioenergy, the UNFCCC inventory and accounting procedures for biogenic CO_2 emissions need reform (Section 10.2). At national and European level, any such accounting loopholes should be swiftly closed. Until accounting procedures for biogenic greenhouse gas emissions, e.g. in emissions trading systems, are appropriately developed or adapted to achieve this, sustainability standards for bioenergy should be drafted and implemented which ensure that bioenergy use is prevented unless it makes a verifiable contribution to reducing global greenhouse gas emissions (Section 10.3).

10.7.2
Promotion and intervention approaches under sustainable bioenergy policy

It is the view of WBGU that bioenergy deployment should be evaluated principally in terms of its contribution to a global transition towards sustainable energy systems, and particularly in terms of the sustainability objectives of mitigating climate change and overcoming energy poverty (Section 2.2). This calls for a differentiated analysis of each individual pathway (Chapter 7). In light of the risks posed by bioenergy utilization (Chapter 5 and Section 7.3), blanket promotion of bioenergy is not advisable. Bioenergy carriers that fall short of certain minimum sustainability standards (Section 10.3.1) should not be promoted directly or indirectly. Rather, the utilization

of bioenergy sources that do not meet the minimum standard proposed by WBGU (Section 10.3.1.1) should be prohibited in the long term, and ideally worldwide, by means of a comprehensive and effective regulatory system (Chapter 10).

Where bioenergy carriers meet the minimum standard, their cultivation should fundamentally be left to the market. Generally there is no need for explicit state support of cultivation. Nevertheless, appropriate supplementary frameworks – governing aspects such as climate change mitigation (Section 10.2) or the internalization of external environmental costs, in the agriculture and food sectors (Section 10.4) and for nature conservation and soil protection (Sections 10.5 and 10.6) – should ensure that the market mechanism produces outcomes in line with the objectives of a globally sustainable bioenergy policy. In the view of WBGU, the only bioenergy pathways eligible for explicit promotion are those which are particularly conducive to climate change mitigation and other sustainability goals.

Firstly, this means that bioenergy deployment must not only comply with the minimum standard but must also permit a saving of at least 60 t CO_2eq per TJ of utilized biomass, taking total life-cycle emissions into account (Section 9.2.1). Strictly speaking, assessment by this criterion is only possible upon final use of the bioenergy. Of course, any such test would generate an enormous verification workload. That aside, there would be no certainty that promotion at the end of the value chain would actually carry through to the entire bioenergy pathway. A further consideration is that separate stages in the production and utilization process (cultivation, conversion and end-use systems) can be located in different countries. Hence WBGU deems it appropriate to employ promotion instruments at each stage of the process whilst, as a rule, applying default values for the emissions of the other stages involved.

Secondly, with specific reference to the promotion of energy crop cultivation, WBGU considers that additional compliance with ecological and social criteria is essential (Section 10.3.1.2). Equally, in the mobilization of biogenic residues, ecological limits need to be observed so as to preserve soil fertility (Section 10.7.4).

Thirdly, support for conversion and end-use systems should be structured to operate in line with the vision of a turnaround towards sustainable energy systems. Undesirable lock-in effects should be avoided and promising technologies such as electric vehicles supported (Chapter 9).

Apart from climate change mitigation, sustainability in energy systems also involves overcoming energy poverty and giving up forms of biomass use that are harmful to health (Chapter 3). The modernization of off-grid or traditional uses of bioenergy can make a valuable contribution to overcoming energy poverty and hazards to health, particularly in the rural regions of developing countries. In this context, WBGU finds support for bioenergy-based projects justifiable even in the absence of full compliance with climate change mitigation and other promotion criteria (Section 10.8).

10.7.3
Agricultural policy: Promoting biomass cultivation for energy production

Current policies promoting energy crop cultivation, particularly for conversion to liquid transport fuels, give scant attention to the different specific contributions made to greenhouse gas abatement, and hence to climate protection, depending on the cultivation method, conversion method and particular form of use (Chapter 7; SRU, 2007). The same can be said of other sustainability aspects (Chapter 9; Doornbosch and Steenblik, 2007). WBGU advocates that states should actively prohibit any clearly non-sustainable cultivation. In order to achieve this, minimum standards would need to be implemented worldwide. Because the international community is unlikely to agree on and implement rigorous cultivation standards in the short to medium term, however, WBGU recommends that at least within the EU a binding minimum standard for bioenergy sources be adopted as soon as possible (Section 10.3). The use of bioenergy sources for energy generation should not be permitted in the EU until such a standard has been met.

10.7.3.1
Favouring particular cultivation methods and ecosystem services

The EU currently supports the cultivation of energy crops within the framework of the Common Agricultural Policy (CAP) with an annual subsidy of € 45 per ha (EU, 2003). The establishment of a suitable climate policy framework, functioning internalization mechanisms and globally implemented minimum standards would, WBGU believes, negate any need for specific support of energy crop cultivation justified on the basis of climate policy. In place of this, market processes would ensure that bioenergy developed its potential sustainably, in both economic and climate-policy terms. Only cultivation methods meeting sustainability criteria over and above the requirements of climate protection, e.g. contributing substantially to erosion control, (energy) poverty reduc-

tion or biodiversity conservation, would then be eligible for support. Promotion criteria of this nature could be satisfied by producing appropriate perennial energy crops on degraded sites and in compliance with social standards.

Promotion can be effected by means of agricultural subsidies such as production premiums. A complementary or alternative option is to promote the spread of land management techniques or agricultural technologies which improve the economic viability of especially sustainable cultivation systems. For example, financial and technical promotion measures could boost the economic efficiency of small-scale agroforestry in tropical regions (Section 7.1.3) or support applications for the sustainable use of wood residues and forest products in temperate and boreal forests (Sections 7.1.5.2 and 7.1.5.3). In principle another conceivable approach by which to promote sustainable cultivation is to target the use of biomass, e.g. through trade or tax incentives for generating energy from certain crops (Section 10.3).

10.7.3.2
International initiatives

The reconfiguration of national support programmes for energy crop cultivation should be reinforced and accelerated by international coordination. Existing forms of support are frequently reaching non-sustainable biomass production, which reinforces the incentives for land degradation and biodiversity loss (OECD, 2008). Moreover, it distorts the price mechanism in the global energy and agricultural markets, which could have undesirable side-effects on food supply or the deployment of other renewable energies. Ultimately, the need to avoid an inefficient promotion and subsidy race is a point in favour of an international agreement on principles for agricultural subsidies. Without international regulation, a situation would arise similar to that established some time ago by the present system of agricultural subsidies: poorer developing countries would be robbed of the opportunities for economic development that could be opened up to them by energy crop cultivation and the export of biofuels.

Ideally, such forms of promotion, especially agricultural and bioenergy subsidies, need to be realigned within the context of a holistic approach. Where possible, energy, agricultural and environmental policy must be coordinated interdepartmentally. On the international level, coordination could take place along the same lines as other agricultural subsidies under the auspices of the WTO, but also involving other institutions such as UNEP, FAO, GBEP, CBD, UNCCD and, where relevant,

UNFCCC. Their involvement is essential in order to incorporate meaningful sustainability criteria in the agreements, determining the conditions under which cultivation support remains permissible. This may take the form of support to control land degradation and desertification (Section 10.6), to promote nature conservation and climate change mitigation (Section 10.5 and 10.2) or to eradicate energy poverty (Section 10.8).

10.7.4
Promoting the conversion of biogenic wastes and residues into energy

To promote energy generation from biomass wastes and residues which, along with energy crops, form the basic feedstocks for bioenergy use (Section 4.1), their special characteristics as commodities must be taken into account. Biogenic wastes and residues occur in different aggregate states, e.g. as solid or liquid fuels (timber waste, slurry, straw, household organic wastes). Biogenic wastes and residues arise in different sectors (agriculture and forestry, manufacturing, municipal enterprises and private households) and at different stages of the value chain (biomass production and harvesting, processing, consumption and disposal; EEA, 2006; Sims et al., 2007).

Mobilizing biogenic wastes and residues for energy generation
Energy sources derived from biogenic wastes and residues are more conducive to climate change mitigation than farmed energy crops (Section 7.3) and should therefore be prioritized for use in bioenergy production (Section 9.1.1). Availability of primary resources can be a constraining factor. There is a need for action, firstly, where sustainable potentials have not yet been fully utilized due to lack of the requisite infrastructure, particularly for waste and residue separation and (interim) storage. Secondly, action is necessary to address information deficits or, thirdly, where efficient bioenergy generation has previously been impossible because, for example, the costs of transporting wastes or residues to conversion facilities are unduly high. That said, the predicted rising cost of fossil fuels (oil, gas) should improve the economic efficiency or profitability of biogenic wastes and residues in the long term and thereby contribute to the mobilization of previously untapped potential. In view of the urgency of climate change mitigation, WBGU nevertheless recommends targeted promotion of the mobilization of biogenic wastes and residues in addition to internalization of the costs of fossil fuels, so as to expedite this substitution process.

Availability of appropriate conversion technologies
The necessary conversion technologies are available, as is the know-how to operate them: biogas facilities for harvest residues, green cuttings, slurry and food waste, biodiesel facilities for used oil and animal fat, and biomass-fired cogeneration systems or pellet heating systems for ligneous or straw-like wastes and residues. Therefore, at least in industrialized countries, these conversion technologies need no direct promotion or subsidies, state market introduction programmes or similar support. Moreover, direct promotion of such facilities would not increase demand just for biogenic wastes and residues but might also – depending on the particular technology – have unintended benefits for other energy sources such as energy crops (Section 10.7.3). A detailed rationale for steering biogenic waste and residue streams into electricity generation and coal substitution, which would bring about the greatest reduction in greenhouse gases (Sections 7.3 and 9.2.1), is set out in Section 10.7.5.

Promotion in order to harness unutilized sustainable potential
For biogenic wastes and residues, only a small number of global assessments of potential exist and, of these, many are subject to considerable uncertainty (Section 6.1.2). Studies indicate significantly higher potential from wastes and residues in the agricultural and forestry sector than in other sectors (Table 6.1-1). Studies for EU member states (EU-25) similarly indicate the general potential of these sectors, although there is considerable divergence in sectoral classification and, especially, sector sizes between the individual countries (EEA, 2006). Promotion should target areas where significant levels of unutilized sustainable potential are found. The extent of this exploitable potential varies depending on the economic importance of biomass-intensive sectors and the degree of organization in the waste management sector (functioning infrastructure, capacities for reprocessing, recycling, conversion for energy production, or landfilling) in different countries. For example, in Germany the potential in the forestry, timber and paper sector and in the industrial and waste management sector is almost fully tapped. Exploitation of potential in the agricultural sector, e.g. for straw, is limited by the fact that further removal of residues would impair soil fertility (humus content). To what extent straw should remain on the land essentially depends on the site, crop rotation and input of other organic fertilizers. Empirical values for straw requirements on arable land are between 67–80 per cent (Fritsche et al., 2004; Knappe et al., 2007; Vogt et al., 2008). Similar constraints apply to removal from for-

ests and should be respected without fail. Assessing the entire energy potential of timber residue in European forests, EEA (2007a) recommends that 40 per cent of the exploitable potential should be left in forests for sustainability reasons. If these restrictions are not adequately observed by farmers and foresters in the course of the – essentially desirable – increase in residue use, they should be mandated formally.

Previously untapped sustainable waste and residue potential can be made attractive for use in energy generation if state promotion measures give them (enhanced) economic value. In the forefront of policy on the (indirect) promotion of the use of biogenic wastes and residues as energy sources, most countries rely on the general promotion of renewable energies in electricity generation, sometimes combined with promotion measures in local heating (combined heat and power; DEFRA, 2007). For feed-in payments, but also for tendering and quota schemes, greater differentiation should be used to make waste and residues more attractive in comparison to energy crops (Vogt et al., 2008; Section 10.7.5). As an alternative to such promotion, direct incentives (e.g. subsidies for collection or transportation) or standards (e.g. a ban on landfilling) would be other possible means of mobilizing unutilized potential. Nevertheless, the effectiveness of such direct supply-side incentives for climate protection would be lower than promotion on the electricity generation side, since residues and wastes attracting these kinds of subsidies may also be used for purposes with less greenhouse gas abatement potential. Hence, measures which directly increase supply should never be considered as an alternative, but merely an addition, to measures that promote electricity generation from wastes and residues (Section 10.7.5).

Direct and indirect steering of biogenic material flows

In industrialized countries the availability of biogenic wastes and residues for use in energy generation can be supported by making further landfilling of such materials less attractive, or altogether unviable, by means of higher landfill fees or stringent standards. Strict regulations make sense in this context, primarily to reduce the uncontrolled escape of landfill gas containing the greenhouse gas methane. Incentives are set for the controlled capture of gas and use of landfill gas from existing untreated landfill wastes as an energy source, and for similar use of sewage treatment plant gas elsewhere, by promoting the use of renewable energies.

Greater promotion for biogenic wastes and residues for use as energy sources not only contributes to harnessing unutilized potential but also to redirecting material flows from existing uses, especially material reprocessing and reclamation. Examples of these uses are particle board production from waste wood (Bensmann, 2004), soil improvement with sewage sludge or composted garden waste and food residues (Knappe et al., 2007) and efforts to recover phosphorus, e.g. from slurry (UBA, 2004). Potential competition for waste resources can be reduced by cascade use steered by waste management laws, but will not necessarily be an issue, since lower quality waste resources are often unsuitable for industrial feedstock use, e.g. smallwood in the forestry sector (Fritsche et al., 2004). Nevertheless, any system for promotion of waste and residue use for energy generation must keep track of competing demands for resources.

Promotion policy for newly industrializing and developing countries

Existing uses of waste and residues in urban centres in newly industrializing and developing countries make special demands on any specific policy to promote their use for energy generation. Unseparated municipal waste, subject to varying levels of control, is often burnt for waste disposal purposes rather than for energy production. In some places, disposal is limited to landfilling waste without any separation at all. Although organic components still constitute the bulk of municipal waste, the emphasis of waste separation is on non-organic secondary resources (metals, glass, plastics) and their reprocessing or recycling. Finally, the informal sector (small entrepreneurs, poorer households) plays a major role in the decentralized collection and manual separation of wastes (Bogner et al., 2007; Brock, 2008; Weltsichten, 2008).

An opportune approach for greater promotion by national programmes and large-scale support through development cooperation may be the diffusion of efficient conversion technologies including the know-how required for their operation. The focus of such efforts could be (community) facilities for population groups affected by energy poverty (Section 10.8). Better mobilization of organic wastes could be achieved by improving infrastructures and – despite past problems – by means of transparent incentive systems for the collection and separation of these kinds of wastes, e.g. through a functioning state remuneration system for informal waste collectors. In addition, income-based levies could be charged for waste disposal (Brock, 2008; Weltsichten, 2008). A final element is the controlled capture of biogas from landfill sites and water treatment plants. The funding for the necessary investment can be mobilized by CDM projects, among other sources, as long as project activities deliver additional greenhouse gas emissions reductions (Bogner et al., 2007; Section 10.2.3).

In rural regions in developing countries, wastes and residues from the agricultural and forestry sector are used primarily for distributed heating – often with inefficient traditional biomass-burning technologies (Section 8.2.1). The goal of national promotion or promotion within the framework of development cooperation must be to disseminate knowledge about sustainability limits on the removal of residues from forests and fields, and to promote local institutions which ensure observance of sustainability standards. A further important approach is to promote the wider use of efficient energy technologies such as modern wood stoves and small-scale biogas systems, and the transition towards modern forms of energy (Section 8.2). This is an area where the CDM (Section 10.2.3) has a continuing role to play. Nevertheless, other actors in the field of development cooperation are called upon in equal measure (Section 10.8).

10.7.5
Technology policy and the promotion of selected conversion pathways

Incorporating bioenergy into energy systems in the best possible way and using it optimally (Chapters 7, 8 and 9) involves selecting conversion technologies that are particularly beneficial from a climate and energy policy perspective. However, widespread application of these technologies is hindered by various imperfections of the market. Barriers exist both in industrialized countries and in the urban centres of newly industrializing and developing countries, whose energy systems face some of the same challenges as are encountered in industrialized countries. State support measures therefore need to be used to promote promising technologies for selected bioenergy conversion pathways.

A significant problem is that innovative technologies often do not become competitive until medium- and long-term learning effects have been realized. Without state promotion their use is often unattractive to market participants. Another factor is that private households, in particular, often continue to use familiar but less efficient technologies as a matter of habit (perseverance tendencies). Targeted promotion measures can provide the impetus for a desired technology switch. Furthermore, the market entry and diffusion of beneficial technologies is hampered by the fact that in many countries and sectors the external costs of fossil energy carriers are not yet internalized – e.g. through taxes or pollution allowances or rights – or are internalized very inadequately. For this reason compensatory promotion measures are required for these technologies.

The advantages and disadvantages of the various conversion pathways are set out in detail in Chapters 7, 8 and 9. One of the conclusions is that a particularly strong climate change mitigation effect can be achieved by using sustainably produced biomass to replace coal in electricity generation. In this context biogas from fermentation processes, raw gas from biomass gasification plants and biomethane are important secondary energy carriers (Section 10.7.5.1). The direct combustion in coal-fired power plants or cogeneration plants of wood chips or pellets from wastes and residues also has an above-average mitigation effect. Irrespective of whether biomethane or biomass is used for conversion to energy, combined-cycle power plants and combined heat and power plants are efficient technologies (Section 10.7.5.2). Likewise it is more efficient to use bioenergy in the transport sector to power electric vehicles rather than perpetuate an inefficient vehicle technology by replacing fossil fuels in combustion engines (Section 10.7.6). Greater consideration must also be given to the use of biomass in the heat sector (Section 10.7.5.3). These conversion pathways should be promoted, but – in order not to narrow the discovery function of competition – only in ways that leave adequate scope for market developments.

10.7.5.1
Conversion of biomethane to energy

Biomethane (biogenic natural gas) can be produced through conversion of biomass in two different ways. One method involves the fermentation of biomass substrates to produce biogas (fermentation, biogas systems), which is then processed into biomethane. The other is based on the gasification of solid biomass to produce raw gas; after cleaning the synthesis gas is converted to biomethane in the methanization process (gasification, biomass gasification systems). In both cases it is necessary to separate the CO_2. Biogas systems are already established on a broad scale; by contrast, fully developed biomass gasification systems suitable for broad market use are not expected to be available until 2015 at the earliest (Thrän et al., 2007). Even though it will be some years before biomethane will be available in sufficient quantity to have an impact on energy policy in countries such as Germany (BMU, 2008b), it is forecast that under the right conditions – including accelerated development of the necessary infrastructure – the EU's entire present requirement of fossil natural gas could be met through biomethane by 2020 (Thrän et al., 2007).

In terms of the objectives of a shift towards sustainable energy systems, biomethane has a number of advantages which justify promotion; they are set out in more detail in Box 7.2-2 and in Section 9.2.1. There are in principle two points of leverage for supporting the use of biomethane: promotion can target either its production or its use. To ensure that promotion policy is transparent and to facilitate retrospective evaluation of its effectiveness, only one of these levers should be selected. WBGU considers promotion of use to be the most suitable approach. This prevents biomethane being produced without regard to demand for it. Importantly, it also enables preference to be given to the conversion technology that is likely to have the greatest climate change mitigation effect, namely the use of biomethane for electricity generation in countries in which fossil energy carriers, particularly coal, play a major part in electricity generation and would be replaced by biomethane. It follows that in these countries the emphasis should be on promoting the use of biomethane in electricity generation. The climate change mitigation effect could be heightened further if the CO_2, which must in any case be separated during the methane production process, is securely stored (Carbon Capture and Storage, CCS). Pursuing the development of CSS is thus an important issue for bioenergy policy as well as in other respects.

FEEDING BIOMETHANE INTO THE GAS GRID

In industrial energy systems, the use of biomethane as a gaseous energy carrier is contingent upon certain conditions: an adequate and functioning grid infrastructure must exist, the operators of biogas plants and biomass gasification systems must have unfettered access to the (supra-regional) gas grid, and power plants must likewise be connected to that grid. Experience in processing biogas and feeding it into (local) gas grids has been acquired in Switzerland, Sweden and the Netherlands (FNR; 2006e; van Burgel, 2006). In Germany there are around a dozen biogas plants that feed processed biomethane into the natural gas grid; a further 20 or so are currently planned. In Europe as a whole 80 biogas processing plants are in operation (Bensmann, 2008; ISET, 2008). Within the EU the legal framework is provided by the rules governing the internal market for natural gas, the provisions on non-discriminatory access to the gas grid that these rules contain, and the quality and safety standards of member states (van Burgel, 2006). In Germany feed-in promotion is effected through grid access rules, obligations on gas grid operators to accept and pay for gas fed in, and payments for avoided grid costs. The latter mean the distribution of cost savings that arise in connection with local feed-in because the transportation distance between production and consumption is reduced (Leuschner, 2008).

In general, expansion of gas grids would need to be promoted. At the same time, grid operators would have to be required to undertake the investment needed – e.g. in pumping stations – to enable operators of biomethane plants to access the grid. These pumping stations increase the gas pressure of the injected biomethane to that needed for feed-in and transport in gas grids.

PROMOTING THE FEED-IN OF ELECTRICITY FROM BIOMETHANE

Gas prices have risen significantly in recent times. This may help to make the feed-in of decentrally produced biomethane more attractive, and the fact that coal prices have risen even more sharply is likely to increase interest in biomethane among electricity generators. Nevertheless, because in most countries – and in the EU – the emissions of fossil energy carriers are not linked to their external costs (Section 10.7.1), explicit incentives for the generation of electricity from biomethane are required. Feed-in tariffs as a tool for promoting renewable energies have proven their usefulness relatively well (UBA, 2006a). Nevertheless, countries should not be committed to feed-in tariffs by the EU or obligations originating elsewhere, since other instruments also have advantages (WBGU, 2004a; Ringel, 2004; UBA, 2006b; Finon, 2007; Umsicht, 2007). In the case of energy from renewable feedstocks, however, quotas can cause problems. The obligation to use at least a certain quantity of energy crops or residues in generating electricity can significantly distort the market; where energy crops are involved this can have a significantly stronger impact on competition for land and in particular on food prices than is the case with other renewable energies.

Many countries promote electricity generation from biomass: in addition to a general payment per kWh for the feed-in to the grid of electricity from renewable sources (e.g. wind and solar) they pay a premium for electricity produced from cultivated biomass (termed in German: nachwachsende Rohstoffe, NaWaRo). In Germany this is done through the 'NaWaRo bonus'. This bonus can be supplemented by additional premiums for the use of heat from combined heat and power systems (Section 10.7.5.2), for the deployment of new technologies, or for the use of organic residues (FNR, 2006e). In Sweden promotion is effected through tradable certificates for electricity from renewables (SBGF et al., 2008). In WBGU's view such promotion is appropriate, provided that steps are taken to ensure that the bioenergy carriers at least meet the WBGU minimum standard for sustainability; in addition, WBGU recommends considering special incentives if the above-mentioned promotion criteria for energy crops are met (Section 10.3.1). In view of the important contribution that biomethane can make to mitigating climate change and its technical advantages in electricity generation (Box 7.2-2), particularly when it leads to the replacement of coal, WBGU also recommends additional promotion measures for electricity from biomethane, provided that the CO_2 – which must in any case be separated in the manufacturing process – is captured and stored in a secure repository.

SUSTAINABILITY STANDARDS

If biomethane is indeed to make a major contribution to climate change mitigation, its life-cycle greenhouse gas emissions must be kept low and the separated carbon dioxide must ideally be stored. Promotion policies must be flanked by corresponding standards relating, for example, to the use of energy crops as a biogas substrate. To reduce methane emissions from digestate stores that are not completely sealed or from gas combustion in small-scale CHP, promotion must be made contingent upon compliance with best-practice guidelines. Similar steps must be taken to limit ammonia emissions from the storage of digestate and digestate application on farmland. In addition, sustainability standards relating to biodiversity conservation must be considered in connection with the use of energy crops (BMU, 2008d). Regardless of the specific promotion policy, the production of electricity from biomass should only be promoted if minimum sustainability standards are complied with. German and European promotion policy should be developed with this in mind and international bioenergy cooperation should also be based on these principles.

10.7.5.2
Efficient system technology in electricity and heat production

Bioenergy has the potential to transform energy systems, making them more efficient and helping to mitigate climate change (Section 8.1.1), but this potential will only be realized if the propagation and use of efficient system technologies is accorded high priority.

COMBINED HEAT AND POWER

When power and heat are generated simultaneously in combined heat and power (CHP) systems, the technical efficiency of fuel use is particularly high (Section 8.1.2.3). This is true regardless of whether the fuel is from fossil or renewable sources, including biogenic energy carriers. It therefore presents an opportunity for replacing a considerable quantity of fossil fuel and/or for reducing overall emissions.

Despite their high efficiency, CHP systems are not sufficiently widely used in industrialized countries. In the European Union's member states the extent of CHP use is very varied. Factors conducive to the widespread use of CHP are the availability of a local or district heat grid, access to natural gas as the most frequently preferred fuel, a suitable transport infrastructure and sufficient heat demand. Market effects such as rising fuel prices, and in some cases falling electricity prices, also make CHP systems less competitive. Promotion policies in the member states are underpinned by the EU CHP Directive of 2004; these policies include measures such as subsidies for building new CHP plants, energy tax discounts for the energy carriers used, guaranteed feed-in tariffs for electricity generation and state infrastructure investment (EEA, 2007b). Germany, too, makes use of such instruments (such as in its amended CHP Act). In particular, the feed-in tariff for biogas under the Renewable Energy Sources Act (Erneuerbare-Energie-Gesetz, EEG) has led to a significant expansion in biogas-fired CHP plants (IEA, 2008a). The general framework set by energy and climate policy will help set future investment incentives and thus determine the competitiveness of efficient CHP systems (UBA, 2008b). In this context WBGU welcomes the EU Commission's proposal to reduce free allowance allocations under the Emission Trading Scheme (ETS). As proposed by the Commission, the efficiency of CHP should be recognized in the ETS through the allocation of free allowances and/or discounts on allowance obligations (IEA, 2008b). In sectors not covered by emissions trading, partial tax exemptions should be continued. In countries such as developing countries that have no emissions trading scheme or CO_2 emissions tax, investment grants or output subsidies linked to the use of efficient system technology could be used to promote CHP.

COMBINED-CYCLE POWER PLANTS IN ELECTRICITY GENERATION

Combined-cycle power plants have the highest fuel efficiency. In addition, they can provide system services, such as high-quality control energy and variable control of electricity generation, that stabilize the electricity grid and increase the technical security of supply (Section 8.1.2.3). Combined-cycle power plants are usually conceived on a large scale. The advantages of this system technology are that it is cost-effective and power generation is both highly efficient and controllable. In industrialized countries the capital needed for investment in power plants is available and there is a high level of investment security. In WBGU's view, therefore, no special promotion of this technology is required. Attention should focus on a consistent emissions trading or emissions taxation system; this would provide effective incentives for the use of efficient and low-emission system technologies, such as those used in combined-cycle power plants. In developing countries the situation is different: these countries have no effective incentive mechanisms for reducing emissions, they have little capital and potential investors are deterred by the lack of investment security.

10.7.5.3
Direct combustion of solid biomass to generate heat for private households

The direct combustion of biomass in coal-fired power plants and cogeneration plants and the use of biogas or raw gas for the combined generation of power and heat is already promoted in many countries. Provided that sustainability standards are complied with, this should be continued or introduced, especially in countries in which coal plays a major part in power generation. By contrast, the direct combustion of biomass to provide heat should only be promoted in certain circumstances. While the use of solid bioenergy carriers, such as pelleted energy crops or residues, can contribute to significant greenhouse gas reductions in the heat sector when they replace oil as a fossil energy carrier (Section 7.3), there is clear evidence that bioenergy – including solid biomass – has a greater climate change mitigation effect in electricity generation when it replaces coal. In WBGU's view the heat needed for domestic purposes can be provided most efficiently in combination with electricity generation by using heat extracted from CHP (combined heat and power; Section 10.7.5.2). In the long term another option is to use ambient heat and heat pumps powered by renewable electricity. However, on account of the extensive infrastructure investment needed, particularly in rural areas, it is likely to be some time before CHP can meet a large proportion of the heat requirement. It will take even longer for renewably driven heat pumps to become standard, in particular because the shift to extensive renewable electricity generation has not so far been straightforward. Thus the use of boilers powered by wood, wood chips and pellets should be seen as a second-best solution and one that is most appropriate in situations in which the expansion of local and district heat networks cannot be expected (Section 8.1.1.1 and 8.1.1.2).

In the residential sector and also in relation to public institutions it may be expedient to provide transition assistance, since it cannot be assumed that a rapid transfer from fossil to biogenic fuels will take place. Households often do not use the most efficient technologies; they thus fail to exploit either greenhouse gas abatement potentials or money-saving options. The reasons for this include particularly strong perseverance tendencies and resistance to change, even when confronted with economically beneficial technologies. Investments that are expensive, although cost-effective in the long term, are often unattractive to property owners. This attitude may be linked to financing difficulties or to a general tendency to prefer present savings to future ones; in the rented property sector a key consideration is that the investment benefits the tenant more than the landlord (Levine et al., 2007; Schleich and Gruber, 2008). It follows that in this area promotion in the form of subsidies may be appropriate for a transition period of ca. 10–15 years; promotion could take the form of cut-price loans or other investment allowances for households and businesses when they convert to biomass heating systems (Levine et al., 2007; BMU, 2008c). In view of the problem of perseverance, regulatory law may also need to be invoked. For example, owners of new buildings in Germany must by law meet a certain proportion of the heat requirement from renewable energies.

This type of promotion policy should focus on promoting heating with pellets. However, a disadvantage is that in promoting heating systems it is not possible to differentiate according to whether systems use energy crops or wastes and residues. Countries that have not yet imposed strict emissions limits should do so at the time they set up their promotion programmes, since otherwise emissions of pollutants and particulates may increase. For newly industrializing and developing countries promotion measures, including emissions regulations, are also appropriate, particularly for relatively affluent urban households. Nevertheless, these countries need to focus primarily on increasing the efficiency of traditional biomass use in rural areas and in relation to poorer households (Section 10.8.2).

10.7.6
Promoting bioenergy in final use

Alongside the promotion of cultivation systems and conversion technologies that has been described, certain forms of final energy use are already prescribed to a greater or lesser extent. In particular, quotas are used as a means of obliging energy consumers or providers to use a minimum proportion of biogenic energy carriers in their energy use or provision. Blending quotas for fuels, which are used in many industrialized and newly industrializing countries, are classic examples (Section 4.1.2). For example, Germany's draft law on blending quotas for diesel and petrol is intended to ensure that biofuels account for 5.25 per cent of the energy content of total fuel consumption in 2009 and 6.25 per cent from 2010 onwards (BMU, 2008f). This is based on the EU's planned target under which biofuels will supply 10 per cent of all road vehicle fuel by 2020 (EU Commission 2008a). However, debate is currently focused on calling instead for a 10 per cent renewables component in all energy consumption for road traffic. The European Parliament wants 40 per cent of this share to be met from sources other than first-generation liquid biofuels (e.g. hydrogen, electricity for automobility or second-generation biofuels from residues; EU Parliament, 2008; Box 4.1-3). Set against this are 'soft' target quotas in the form of expansion targets set by national governments. For example, target quotas of this type are used in connection with biomethane use (Section 10.7.5.1). Thus by promoting the feed-in of biomethane into the grid, the German government is pursuing its target of feeding 6000 million m^3 biomethane per year into the German natural gas grid by 2020, rising to 10,000 million m^3 by 2030. The latter figure corresponds to around 10 per cent of the country's current consumption of natural gas. Bioenergy-specific expansion targets of this type set the goalposts for market stakeholders without directly constraining their (investment) decisions through rigid usage quotas. In addition, bioenergy use is influenced by the general expansion targets that are being set or planned in many countries. For example, there are plans in the EU to meet 20 per cent of the final energy requirement from renewables by 2020 (EU Commission, 2008a; Box 4.1-3). Expansion targets of this sort allow greater flexibility in the selection of the renewable energy form and thus facilitate more efficient investment decisions than narrow sector-specific quotas for bioenergy use, such as those that apply to biofuels (Egenhofer, 2007).

Rigid quotas of this type channel investment more narrowly and result in consistently high demand for biogenic energy carriers from market stakeholders, with inadequate account being taken of relative scarcities on the biomass markets or the amplifying of competition for land. Additional inefficiencies arise from the fact that the way in which the bioenergy is used – i.e. through conversion into a liquid biofuel that is subsequently used in motor vehicles – is prescribed by law, regardless of whether other bioenergy pathways are more efficient and whether greenhouse gas emissions can be reduced at lower cost (Chapter 7). Moreover, blending quotas promote technological lock-in effects in the motor vehicle sector, since they cause the infrastructure associated with less efficient combustion engines to be perpetuated. This consolidates barriers that delay the shift to electromobility that must take place in the medium term. This example illustrates how relatively rigid and intrusive policies can unintentionally be less beneficial for the climate and the environment; they can promote inefficiencies and have undesirable long-distance effects, for example on food prices (Kulessa, 2007).

For these reasons WBGU believes that the promotion of liquid biofuels for the transport sector in industrialized countries cannot be justified on sustainability grounds. WBGU therefore recommends rapid abandonment of the promotion of biofuels in industrialized countries. In particular, current blending quotas in EU countries should not be increased further and should be removed entirely within the next three to four years. However, this should not mean that the motor industry in the EU is released from its emissions reductions obligations. If blending quotas are abandoned, the specific emissions reductions that have been agreed must be achieved by other means.

In developing countries the production and use of liquid biofuels can be more readily substantiated, for reasons that include the greater poverty of the rural population, the widespread energy poverty, significantly poorer starting conditions for a rapid shift to electromobility and in some cases a chronic shortage of foreign currency. In some developing and newly industrializing countries, therefore, the expansion of biofuel production for use within the region for a transition period can under certain conditions be justified. Here too, though, subsidies should be critically evaluated in terms of efficiency and environmental as well as social considerations.

PROMOTION OF ELECTROMOBILITY IN PLACE OF BIOFUEL QUOTAS

The electrification of motorized personal transport is a building block in the overall strategy for reformed energy systems (Section 8.1 and 9.2). Because many countries are focusing on liquid biofuels as a means of reducing their dependency on petroleum or cutting transport-related greenhouse gas emissions, there is now an indirect connection between the promotion of electromobility and bioenergy policy. If vehicles with combustion engines were replaced on a large scale by plug-in hybrid cars or battery-operated electric vehicles, the demand for biofuels from energy crops would fall and bioenergy potential could be more efficiently and more flexibly used. If this is to happen, conditions for the expansion of electric vehicles must be significantly improved: at present plug-in vehicles, which combine a combustion engine and an electric motor, have a very small share of the market and are used almost exclusively in industrialized countries. Both plug-in hybrid vehicles and pure electric vehicles can also – when connected to the grid – represent an energy store, which can provide balancing energy to stabilize the electrical grid. However, both technologies are currently still at the stage of inner-city demonstration projects. The greenhouse gas balance of electromobility is determined primarily by the composition of the electrical energy used in terms of energy sources: electromobility can make a significant contribution to climate change mitigation if the electrical energy used comes mainly from renewable sources rather than from fossil ones (Section 8.1; WI-IFEU, 2007). The environmental impacts associated with the batteries are also relevant (BMU, 2008a).

Active state promotion of electromobility can be justified on the grounds of previously unrealized learning effects in the development of battery and vehicle technology. For example, the German government is promoting an industrial consortium for the purpose of improving lithium-ion batteries. While the costs involved in expanding the opportunities for plugging in hybrid and electric vehicles to the electric grid may limit the market development of these technologies, this is a much less expensive alternative than other options for the future, such as hydrogen technology, since the basic structure of the energy supply – the electricity grid – already exists. In some industrialized countries the motor industry, sometimes in cooperation with energy companies, has for some time been conducting fleet trials of electric vehicles in conurbations (Haines and Skinner, 2005) – frequently with state participation. The German government is currently considering providing financial support for a fleet trial of plug-in hybrid vehicles to be conducted by a German industrial consortium. In the start-up phase, infrastructure subsidies can make the propagation of electric vehicles significantly easier. Demand for electric and hybrid vehicles can also be stimulated through the tax system, in particular through higher taxes on fossil fuels than on electricity but also through reductions in vehicle tax or toll charges. In addition, national development plans for electromobility, such as that announced by the German government, create transparency and planning certainty for industry and the consumer (BMU, 2008b)

However, in developing and newly industrializing countries, where purchasing power is low, prices must fall considerably if electric cars are to make a breakthrough. Since these countries lack the resources for adequate price subsidies, rapid diffusion of relevant technologies in industrialized countries is needed, in addition to pilot projects. In the medium to long term and as a result of learning effects and economies of scale, this will enable electric and hybrid vehicles to be sold at affordable prices in developing and newly industrializing countries.

10.7.7
International initiatives and institutions for the promotion of sustainable bioenergy

A global reconfiguration of energy systems requires sustainable expansion of renewable energies, including bioenergy, worldwide (Chapter 2; WBGU, 2004a). A broad spectrum of international institutions to promote renewables, whose mandates include specific aspects of promotion, are already in existence. These include, for example, the Renewable Energy and Energy Efficiency Partnership (REEEP), the Renewable Energy Policy Network (REN21) and United Nations organizations like UNEP and UNDP (Pfahl et al., 2005; WBGU, 2004b). However, as yet the existing institutions have not been able to provide a centralized and coordinated approach to the promotion of renewable energies by means of policy advice and technology transfer.

10.7.7.1
International Renewable Energy Agency

To alleviate this shortcoming, recent efforts have been made – substantially on the initiative of the German federal government – to establish a new and autonomous international organization to fulfil these functions effectively. The International Renewable Energy Agency (IRENA) will be founded at the beginning of 2009. This specialized agency is initially sponsored by a sizeable group of industrialized and developing countries. Membership is open to additional countries (FES, 2007). In its function as a 'centre of competence' IRENA offers the advantages that services to governments like policy advice, technology transfer and competence building can be delivered in a more targeted and cost-efficient way thanks to coordination. As a result, transparency is increased, consultation improved and double promotion avoided (Pfahl et al., 2005; IRENA, 2008).

Concentration and reinforcement of international energy policy institutions by a newly established organization is in line with the ideas of WBGU. The model developed by WBGU (2004a) of an international agency for sustainable energy emphasizes, however, that as well as promoting renewable energies, energy systems in their entirety must be included in the process. The energy demand and special needs of developing countries also need to be considered. Overall, IRENA should be put in a position to tackle energy, environmental and development issues in an integrated way. Ideally it should take a role in convening and running the International Conference on Sustainable Bioenergy proposed by WBGU (Section 10.7.7.2). The outcomes of the conference should, in turn, inform IRENA's advisory and transfer services on bioenergy. The energy development pathways (Chapter 8) and promotion principles for sustainable bioenergy (Sections 10.7.3 to 10.7.5) described in the present report offer a good orientation aid for bioenergy-specific advisory and promotion services.

10.7.7.2
International Conference on Sustainable Bioenergy

The current national policies for the promotion of biomass use as an energy source in industrialized, newly industrializing and developing countries (Section 4.1.2) demonstrate the growing interest in bioenergy. These are accompanied by exploratory processes and projects which have been initiated by partnerships like GBEP, by networks and by international organizations like FAO or IEA. These policy measures and processes to promote bioenergy are based on different, sometimes obviously incomplete assessments of the opportunities and risks of bioenergy. Consequently, there is as yet no adequate global consensus on appropriate standards for the production and use of different forms of bioenergy. Given the increasing biomass trade flows and growing demands upon land use, it is becoming essential to call heightened international attention to the interactions between bioenergy and sustainable development and to work towards consensus building among the actors.

To this end, on the model of 'renewables 2004', an International Conference on Sustainable Bioenergy (ICSB) could be initiated, which would serve as a forum for an international and cross-sectoral dialogue and possible cooperation among agricultural, energy and development policy actors. It should create space for exchanging information on best practices and for adopting recommendations on bioenergy deployment targets. Specific items on the agenda could include proposals for agreements on promotion measures (e.g. subsidies) as well as competition and sustainability standards. Further elements of the conference might be an action programme, within which governments could frame their voluntary measures at national or international level, and (framework) agreements on (bilateral) partnerships on bioenergy technologies (Section 10.8; WBGU, 2004a, b; Pfahl et al., 2005). If appropriate, the conference could be the starting point for a longer-term follow-up process on sustainable bioenergy use. In any event, considering the dynamic nature of bioenergy use and the enormous pressure for action, a well-attended ICSB should be convened at the earliest opportunity. Inter-

vention by the German federal government to this effect would be welcomed.

10.7.7.3
Multilateral Energy Subsidies Agreement

Subsidization of different forms of energy is an element of national energy strategies geared towards securing an adequate and reliable energy supply at reasonable prices whilst avoiding adverse effects on the environment and the climate (IEA, 2006b). A precise quantitative survey of global energy subsidies is difficult on account of the different classifications and accounting methods in use. Different estimates put subsidies at a level of US$ 240,000 to over US$ 300,000 million. Fossil fuels attract the dominant share of subsidies in comparison with either renewable energies or nuclear power (IEA, 2006b; Morgan, 2007). In the bioenergy realm, producers of biofuels benefit from relatively high subsidies in developed countries. According to Steenblik (quoted in OECD, 2008) the combined total of state promotion of this sector in the USA, the EU and Australia amounted to US$ 11,000 million in 2006 (Section 4.1.2).

Given the climate policy challenges and the increasingly evident economic efficiency of various renewable energies, it is ineffectual to persist in directing the majority of energy subsidies to the fossil energies sector. WBGU has previously pointed out the necessity to redirect or eliminate energy subsidies worldwide (WBGU, 2004a). In relation to the promotion of sustainable bioenergy, promoting fossil energy sources can militate against desirable bioenergy pathways, or actually force up promotion budgets to compensate for the advantages subsidies confer on fossil fuels. To avert this, subsidies for fossil energy sources and non-sustainable bioenergy pathways (Section 10.7.3) should be removed. To avoid distortions induced by international discrepancies in subsidy regimes, WBGU recommends initiating a Multilateral Energy Subsidies Agreement (MESA) which would provide for the abolition of environmentally harmful energy subsidies and establishes globally valid subsidy principles. It should provide for the phasing out of subsidies for fossil and nuclear energy and establish rules for subsidizing renewable energies and efficient energy technologies. To improve its prospects of realization, perhaps such an agreement could be geared towards plurilateral interests in the first instance by involving the most important energy producers and consumers, e.g. the oil-producing states and Russia as well as the OECD countries and the newly industrializing countries. Long-term efforts could be geared towards the establishment of a multilateral framework and incorporation into the WTO regime (WBGU, 2004a).

10.7.8
Conclusions

Promotion policies in the bioenergy sector – from the cultivation of energy crops to energy generation and use – must be operated with caution so that the existing potential of bioenergy for a sustainable energy system is better harnessed while avoiding the substantial risks of non-sustainable expansion. This means that, ideally, energy sources should only be used if they meet the WBGU minimum sustainability standard. A first essential step is to at least cease any promotion of bioenergy which contravenes this minimum standard. Within this framework, to ensure the success of the most advantageous pathways from the viewpoint of climate and environmental impacts, climate and environmental policy must be designed in such a way that all emissions are inventoried and attributed, and external costs are internalized as comprehensively as possible. Since this state is a long way from the reality both globally and in individual countries, additional promotion of bioenergy appears justified. Further justifications may be drawn from development policy aspects such as the eradication of energy poverty, and from market introduction difficulties caused by resistance to the up-take of new technologies. These arguments are joined by anticipated learning-curve effects, as well as positive external effects such as the cultivation of perennial energy crops on degraded land. Fundamental preference should be given to lean promotion policies: this means, firstly, concentrating on policy levers small in number yet with greatest individual effect, and, secondly, selecting instruments with minimal intensity of intervention. Promotion for biomethane feed-in without conditions on usage is one example, the abolition of mandatory fuel blending quotas another. On the international level this approach calls both for diverse technology cooperation ventures and for a coordinated reduction of energy subsidies, including at least those bioenergy subsidies that are contributing to non-sustainable developments.

10.8
Bioenergy and development cooperation

Many developing countries pin great hope on an international bioenergy boom. The production and use of modern forms of bioenergy do indeed provide developing countries with a range of opportunities. At the same time, however, bioenergy also entails significant risks – risks that are all too easily overlooked due to the high hopes and initial worldwide euphoria with which the emergence of bioenergy has been greeted. These risks are particularly evident in the production of biofuels for supra-regional use and for export; in this context short-sighted investment can create new dependencies and sustainability criteria may be met insufficiently or not at all (Chapter 8; Box 8.2.3). Moreover, the current focus on liquid biofuels for the transport sector distorts the view of the overall potential of the deployment of biomass for energy in developing countries. For example, there is potential for increasing the efficiency of traditional biomass use and for using bioenergy to generate power and heat. From a development perspective, therefore, the opportunities and risks of bioenergy – particularly those associated with the large-scale production of energy crops – must be carefully evaluated, taking account of specific circumstances in different countries and of the differing energy needs of developing and newly industrializing countries. Promotion strategies can then be tailored to specific circumstances as part of development cooperation activities.

The negative impacts of bioenergy may in particular include the jeopardizing of food security through the conversion of cropland, and increasing dependence of food prices on energy prices. Realizing the opportunities associated with bioenergy while preventing these negative impacts requires a high level of governance and regulatory competencies on the part of states, extending far beyond the narrow field of bioenergy policy itself (Section 10.2–10.6). At national level, the preconditions for a bioenergy policy that has a positive impact on development include effective state structures, good governance, participatory land-use planning, mechanisms for the fair distribution of land and assured food security. In many developing countries these preconditions are not met. However, where they are met or are likely to be met in the foreseeable future, potential for sustainable energy crop cultivation certainly exists (Section 6.7). With this in mind, country-specific bioenergy strategies should be drawn up to ensure that development potentials can be harnessed in such a way that the risks associated with bioenergy production are avoided or at last significantly reduced and the opportunities for overcoming energy poverty are used sustainably.

Bioenergy is not a completely new issue for global development policy and its multilateral and bilateral actors. For example, the traditional use of bioenergy and efficiency improvements in this area have for some years been discussed as a means of tackling energy poverty. Improved efficiency of traditional use sometimes features, too, in programmes for the promotion of renewable energies, but in this context it is usually assigned a lower priority. Biofuels, on the other hand, are accorded great importance in the current debate, although the absence of clearly defined positions and coherent operative strategies is evident. Crucially, there is a lack of cross-sector coordination of international cooperation in the fields of bioenergy, food security, rural development and climate change mitigation. Instead, a large number of initiatives and policy-making processes have been set in motion. This has resulted in an uncertain and in some cases contradictory scene, as the following survey of the relevant international institutions and actors shows (Section 10.8.1).

Alongside this overview of international development cooperation in the bioenergy sector WBGU presents what it regards as the key elements of sustainable bioenergy strategies for developing countries and the implications for an active promotion policy (Sections 10.8.2 and 10.8.3). In the process WBGU looks beyond the restriction of the international debate to biofuels and emphasizes the different forms, depending on the different pathways, in which bioenergy can be a suitable means of tackling energy poverty in a climate-friendly way through the provision of electricity, heat, and mobility.

10.8.1
Current bioenergy activities in international development cooperation

A coherent and sustainable global bioenergy policy requires intelligent links between international energy policy and other areas – particularly international agricultural, environmental and trade policy – at all levels. This in turn calls for an effective and well coordinated institutional architecture, which exists at present only in rudimentary form or not at all. This section outlines the activities of the multilateral actors that are most important from a development perspective and then summarizes the specific contributions of European and German development cooperation.

10.8.1.1
The World Bank Group and regional development banks

THE WORLD BANK GROUP
The promotion of bioenergy is an aspect of the projects and programmes of the World Bank (IBRD and IDA) in the area of renewable energies and energy efficiency. However, it features much less prominently than the promotion of solar energy, wind power, hydropower and efficiency improvements in the use of conventional forms of energy (IBRD, 2007). Bioenergy has until now been specifically promoted only in small-scale projects aimed

at tackling energy poverty in developing countries (e.g. the Burkina Faso Energy Access Project for sustainable firewood management). Another current project involves additional financing of US$ 35 million to support the Household Energy and Universal Access Project in Mali, which is helping to develop capacity building, the expansion of rural electrification, municipal forest management and energy efficiency initiatives concerned with the use of energy in households (World Bank, 2008a, b). Work should be undertaken to identify what the project can contribute with regard to a bioenergy component of a sustainable national energy system in Mali. In the 2008 financial year World Bank funding for renewable energies and energy efficiency, in part co-financed by the GEF, was almost twice what it had been in the preceding year. Expenditure totalled US$ 1400 million, of which US$ 476 million was for projects involving renewable energies (Zabarenko, 2008).

The World Bank Group's attitude to biofuels is cautious; this is on account of the current debate on environmental impacts, competition with food production and the associated uncertainties (World Bank, 2008e). The World Bank has not yet finalized a strategy for bioenergy promotion. This lack of certainty is reflected in the Development Committee (a joint ministerial committee of the World Bank and IMF): for example, while India is strongly critical of the shift in agricultural production from food to biofuels ('bad policy and worse economics'), Brazil sees in biofuels a major development opportunity. The German representative on the Development Committee, development minister Wiezoreck-Zeul, has called on the World Bank and IMF to impose a moratorium on biofuel blending until its impacts on food production have been better analysed. It is generally agreed that the specific role of biofuels in development processes needs to be examined critically. The World Bank plans to contribute to this by investing further in research (Siegel, 2008).

The International Finance Corporation (IFC), which forms part of the World Bank Group and does a great deal of work with the private sector in developing countries, is involved in many biofuel projects. The number and volume of such projects cannot be specified exactly, since agricultural projects have until now been recorded only as a whole, with no separate reporting according to system components. A breakdown by system components should be introduced as a matter of urgency, in order to enable better monitoring and analysis of developments in the biofuel sector. Although competition with food production needs to be taken seriously, the IFC views effects on the agricultural sector as being fundamentally positive. In particular it is anticipated that as a consequence of energy crop cultivation local agricultural practices will be improved, additional infrastructure will be created, farmers' incomes will rise and research and development relating to new breeds and practices will be pursued. The IFC does, however, admit that links with food production and the extent of competition with it are not yet sufficiently well understood. The IFC takes the view that it is better to be involved in shaping development of the biofuel sector than to avoid all involvement on account of the risks and thus forego any influence. This applies in particular to ethanol from sugar cane, the production of which is already globally competitive, and to projects that produce biodiesel from material with low value-added potential (e.g. waste oil and animal fats). Apart from the general IFC safeguards, which specify ecological and social project standards, no additional standards are required for biofuels. The IFC sees further research as being required in particular in relation to the production of biodiesel from *Jatropha* and other energy crops that do not compete directly with food crops (Hamad, personal communication).

The World Bank Group should as a matter of urgency draw up a comprehensive strategy detailing how it can contribute to restructuring of the world energy system. The strategy should in particular include the role of greenhouse gas abatement technologies, including bioenergy use.

THE REGIONAL DEVELOPMENT BANKS
The African Development Bank (AfDB) is currently revising its energy policy strategy of 1994. Its development strategy for the bioenergy sector is still in its early stages. It is likely that sustainable energy management and climate change mitigation criteria will be incorporated into the AfDB's energy strategy and that minimum targets for loans and subsidies in this area will be specified (AfDB, 2007). Africa is seen as having in principle major potential for energy crop cultivation, but on account of the region's low level of development realizing this potential presents a considerable challenge. There are question marks over the availability of suitable technologies, infrastructure and logistics as well as issues of market demand and trading capacity. Questions also need to be asked about the risk of competition between food and biofuel production and about the specific potential of the rural population and of small and medium-sized enterprises (SMEs; Sanchez Blanco, 2008). The next partnership meeting on rural development in west and central Africa takes place in October 2008 and the AfDB, the World Bank, FAO and IFAD will all be represented at it. The consequences of biofuel production for food and energy supply in the region are due to be discussed in detail at this meeting. The AfDB is currently considering promoting small-scale biogas projects along the lines of those supported in Asian projects (AfDB, 2007).

The Asian Development Bank (ADB) has produced a strategy paper on climate change that is very cautious about the potential contribution of biofuels. According to this paper, biofuels could only result in lower greenhouse gas emissions than conventional fuels if the associated raw materials and inputs are selected carefully and the use of fossil fuels in production is minimized. Food security and other issues affecting poorer countries would also have to be explicitly considered in biofuel production. The ADB aims to analyse these conflicts of use carefully and to promote only projects that do not result in soil degradation, the establishment of monocultures or sharp rises in the price of vegetable oils. According to the ADB, this means that *Jatropha*, sweet sorghum and bagasse (sugar cane residue) and second-generation biofuels (ADB, 2007a, b) could be candidates for the promotion of biofuel production. Projects for tackling energy poverty are one of the ADB's priorities. For example, it funds bioenergy projects in the Mekong delta in which biogas is produced from agricultural residues and used to provide heat energy to rural households (ADB, 2007b). Another major project that is rated as successful is the Efficient Utilization of Agricultural Wastes Project in China: in a scheme co-financed by the Global Environment Facility (GEF), US\$33.1 million has been invested in small-scale biogas systems, increasing the income and standard of living of 18,540 households (ADB, 2008).

The Inter-American Development Bank (IADB) takes a significantly more optimistic view of biofuels than the World Bank or the Asian Development Bank. This can be attributed to the fact that 40 per cent of global bioethanol production comes from Latin America and the Latin American biodiesel market is also growing strongly at present. The IADB is a partner to the US-Brazil Memorandum of Understanding to Advance Cooperation on Biofuels and supports the development of corresponding national action plans in El Salvador, Haiti and the Dominican Republic. The IADB also supports the Meso-American Biofuels Working Group. In 2006 the IADB launched a Sustainable Energy and Climate Change Initiative (SECCI), which is made up of four pillars. One of these pillars is biofuel development; the other three are renewable energy and energy efficiency, increasing access to carbon markets, and adaptation to climate change (IADB, 2008). In April 2008 the SECCI announced a partnership with the Roundtable on Sustainable Biofuels; the aim is to integrate the sustainability criteria that will be elaborated by the Roundtable into its own funding practice and to support the involvement of various Latin American interest groups in this global standard-setting process (IADB, 2008). The IADB is involved in some 50 bioenergy projects in both the private and the public sector. These projects also involve energy efficiency and carbon finance components; they have all in all a budget of more than US$ 1000 million (personal communication, G. Meerganz of Medeazza, IADB).

10.8.1.2
Programmes and specialized agencies of the United Nations

The promotion and use of sustainable energy has gained importance in the United Nations context, largely on account of the resolutions adopted at the Earth Summit at Rio de Janeiro that sought to strike a balance between the interests of developing and developed countries (United Nations Conference on Environment and Development, UNCED 1992). A similar approach is also explicitly reflected in the energy policy mandate of the Commission on Sustainable Development (CSD), which was set up by the General Assembly of the United Nations in 1992. However, the CSD has remained largely ineffective; in particular, the Energy Cycle scheduled for its 14th and 15th sessions (2006/2007) has failed (IISD, 2007; Mittler, 2008). While the CSD-16, which introduced the Agricultural Cycle in 2008, put the spotlight on the general risks of bioenergy, concrete political decisions are unlikely to materialize from it (IISD, 2008). Another umbrella institution, UN-Energy, was set up in 2002; it is intended to function within the United Nations as a hub for the energy policy activities of different UN organizations and its involvement with bioenergy issues is growing (UN-Energy, 2007a). In 2007 it published a report on the sustainable use of bioenergy entitled 'Sustainable Bioenergy: A Framework for Decision Makers' (UN-Energy, 2007b). The report provides a general discussion of the links between bioenergy and sustainable development and describes options for promoting bioenergy in developing and newly industrializing countries. It stipulates that standardization and certification of bioenergy products are necessary to ensure their sustainability. However, UN-Energy is not in a position to drive the associated processes forward or to actively require the UN organizations concerned to take action. It thus remains the case that the specialized UN organizations and programmes that are active in the bioenergy sector play a far more important part than UN-Energy in strategic decision and tangible promotion measures.

United Nations institutions that deal explicitly with the promotion of bioenergy and that are generally active in the field of development cooperation are described below; they include UNDP, UNEP, UNIDO and UNCTAD, but this list is not necessarily exhaustive. Other UN institutions whose work impinges on both these areas but within a different overall context are mentioned elsewhere – for example, the FAO is mentioned in the context of the world food supply (Section 10.4) and the CBD in connection with nature conservation (Section 10.5).

UNITED NATIONS DEVELOPMENT PROGRAMME
Improving the energy supply is a classic task of the United Nations Development Programme (UNDP); better energy provision is in great demand in developing countries and is seen as an important element in tackling poverty. However, there are as yet few signs that climate policy considerations are being systematically considered or that a strategic focus on renewable energies and bioenergy as a means of tackling energy poverty is being adopted.

It remains to be seen whether current attempts to develop a closer and more effective operative working relationship with the United Nations Environment Programme (UNEP) will result in corresponding adjustments to the UNDP (Bauer, 2008). This applies in particular to the Partnership for Climate put forward at the World Climate Conference at Nairobi in 2006 (UN, 2006). Within this partnership UNEP and UNDP intend, among other things, to support developing countries in the use of the CDM. Irrespective of any specific cooperation with UNEP, UNDP sees itself as having an explicit responsibility to provide developing countries with access to investment, such as that available through the Global Environment Facility (GEF) and the CDM, through the promotion of clean energy technologies.

UNITED NATIONS ENVIRONMENT PROGRAMME
UNEP is visibly endeavouring to play an active role in international bioenergy policy; it emphasizes the links between the production/use of bioenergy and the classic UNEP issues of greenhouse gas emissions, loss of species diversity, degradation of water resources and soil degradation. In connection with these issues UNEP is involved in various processes and partnerships, such as the climate protection initiative with UNDP mentioned above and the Global Bioenergy Partnership (GBEP, Box 10.3-4). Bioenergy could in the foreseeable future come to play a more significant part in implementation and development of the Convention on Biological Diversity (CBD), the secretariat of which is based at UNEP (Section 10.5).

Particularly important in connection with the provision of bioenergy expertise is the UNEP Risoe Centre on Energy, Climate and Sustainable Development; the centre's experts advise UNEP on renewable energy issues and produce studies on the development of renewable energies in developing and newly industrializing countries (e.g. Kejun et al., 2007; La Rovere et al., 2007). A number of smaller projects and public-private partnerships operate in the same field – for example, the International Panel for Sustainable Resource Management, which is hosted by UNEP and focuses on the sustainability aspects of biofuels, and a planned GEF project, which will work with FAO, UNIDO and the International Energy Agency to improve knowledge of the sustainability of biofuels and draw up appropriate guidelines (Fritsche and Hennenberg, 2008).

In all these activities the Environment Programme emphasizes the potential of bioenergy in line with the UNEP motto 'Environment for Development'. However, a comprehensive and coherent bioenergy strategy has not yet emerged. Instead, UNEP typically functions as a knowledge manager, synthesizing information and preparing it for decision-makers in order to underpin policy-making and place issues on the agenda. Overall, the fragmented activities of UNEP largely mirror the structural weaknesses with which it is confronted in the tangled institutional setup of the United Nations (Biermann and Bauer, 2004).

United Nations Industrial Development Organization

Energy and the environment is one of the three thematic priorities of the United Nations Industrial Development Organization (UNIDO), whose core mandate is the promotion of sustainable industrial development in developing countries. This UN specialized agency, which was upgraded from a UN programme in 1985, could thus in principle play an important part in promoting climate-friendly technologies in developing countries.

UNIDO's present energy policy focuses on improving industrial energy efficiency and expanding renewable energies. In the context of renewables UNIDO promotes mini hydropower, solar energy, and wind power and the use of biomass for energy. UNIDO's bioenergy projects have so far concentrated on the utilization of industrial and agricultural residues – such as bagasse (sugar cane residues), timber waste from forestry operations and cattle dung – in a range of different ways. The promotion of such small-scale projects is a positive step and should be systematically expanded. In parallel with this, UNIDO is currently drawing up its own biofuel strategy in cooperation with the GEF project of UNEP mentioned above.

United Nations Conference on Trade and Development

In June 2005 the United Nations Conference on Trade and Development (UNCTAD) set up a Biofuels Initiative with the aim of promoting research, analysis, technical cooperation and consensus building in relation to biofuels. As part of this process, an international Expert Group was convened in the same year to provide developing countries with country-specific advice on technical matters in connection with biofuel production and international trade in biofuels. The intention is to support developing countries in drawing up appropriate strategies for making better use of opportunities for producing, using and trading in sustainably produced biofuels. In this connection the various investment possibilities are also evaluated, including the option of using the CDM for biofuel projects. The UNCTAD Biofuels Initiative also aims to network existing initiatives within and outside the UN (UNCTAD, 2008a).

The final document of the 12th session of UNCTAD, which met in Accra (Ghana) in April 2008, expressed cautious reservation with regard to the sustainable development potential of biofuels (UNCTAD, 2008b): 'Countries should also exchange experiences and analysis, in order to further explore the sustainable use of the biofuels alternative in a way that would promote social, technological, agricultural and trade development, while being aware of countries' needs to ensure a proper balance between food security and energy concerns.' The work of the UNCTAD Biofuels Initiative should therefore be continued, in order to maximize trade and development gains for developing and transition countries while at the same time minimizing the potential ecological and social risks of biofuel production (UNCTAD, 2008b).

10.8.1.3
Development cooperation activities of the European Union and Germany

European Union initiatives

Since 2005 the European Union has defined access to energy as a core task of development cooperation; improved access to energy is seen as a means of combating poverty in newly industrializing and developing countries (EU Commission, 2005e). The European Commission and EU member states have initiated various programmes aimed at promoting and expanding renewable energies, including biomass, in developing countries. Most of this work is carried out by the EU Energy Initiative for Poverty Eradication and Sustainable Development through the COOPENER Programme (2003–2006: € 17 million co-financing), the ACP-EU Energy Facility (€ 220 million) and the Partnership Dialogue Facility. In addition the EU supports the Global Energy Efficiency and Renewable Energy Fund (€ 120–150 million) and other multilateral initiatives that involve the use of biomass for energy. Energy is also an important element of the EU-Africa Strategy (EU Commission, 2005f, 2008c).

The aim of these EU programmes is to combat poverty directly through access to energy services (e.g. EU Commission, 2004). Promoting the use of biomass for energy in this context is only *one* technical option for improving rural energy provision among many. Promotion focuses on the more efficient use of traditional biomass – at household level or in the context of small businesses. The use of biomass for electricity generation is also covered (Intelligent Energy, 2007; Europe Aid, 2007).

Promotion of liquid biofuels plays only a minor part in the energy programmes mentioned. For example, the Gota Verde programme in Honduras is the only liquid biofuel programme funded through COOPENER. However, small-scale biofuel production and use is sometimes promoted through other thematic priority areas, such as the environment/forest conservation project 'RE-Impact: Rural Energy Production from Bio-energy Projects: Providing regulatory and impact assessment frameworks, furthering sustainable biomass production policies and reducing associated risks'. In connection with sustainable land use the project 'COMPETE: Competence Platform on energy crop and agroforestry systems – Africa', funded under the EU's 6th Framework Programme for research and technological development, is worthy of note. This project, which is being conducted with international partners on a multidisciplinary basis, aims to promote the sustainable use of modern bioenergy in arid and semi-arid parts of Africa and to develop corresponding capacities (COMPETE, 2008).

Although the promotion of biofuels in developing countries as part of the EU's biofuel strategy has been announced, this objective is not yet reflected in specific promotion instruments of EU development cooperation. In a strategy paper of 2006, biofuels were described as an opportunity for developing countries, as they could contribute to diversification of production structures, reduction of the use of fossil energy carriers and economic growth (EU Commission, 2006b). By trading in raw materials and biofuels worldwide, countries such as Brazil could satisfy the demand for fuels. EU promotion measures that were initiated at that time included the sugar policy reform (to aid development of the ethanol sector), special support programmes (Biofuel Assistance Packages) and the establishment of local, national and regional bioenergy platforms. Despite what was said, direct and indirect promotion of biofuels focuses on domestic producers within the EU and the raw materials that they produce. Until now the agreed liberalization of trade in sugar and ethanol has also not been implemented (ODI, 2008). It is true that under the reform of European sugar policy, countries with sugar industries, such as Mauritius, Jamaica and Fiji, have been helped to develop bioethanol production capacities, but the help provided has consisted only of minimal grants which, moreover, are not subject to any environmental or social standards.

The ambitious European strategy for the expansion of biofuels is therefore criticized for reasons that include lacking coherence with other EU policies, for example in the areas of development, food security and trade (EU Coherence, 2008). It is noticeable that EU development cooperation, in particular, lacks an integrated and adequately funded strategy for exploring the potential of the use of biomass for energy in the context of poverty reduction, rural development and climate change mitigation. Such a strategy would need to cover all the ways in which bioenergy is used – for power, heat, and transport fuels. The pioneering role in global climate change mitigation policy adopted by the EU should be reflected in its approach to development cooperation.

INITIATIVES OF BMZ AND THE IMPLEMENTING ORGANIZATIONS OF GERMAN DEVELOPMENT COOPERATION

To clarify the German government's position on bioenergy in the context of development policy, the Ministry for Economic Cooperation and Development (Bundesministerium für wirtschaftliche Zusammenarbeit und Entwicklung, BMZ) has published a discussion paper on the subject of biofuels. The paper was drafted in the context of the joint hearing of the Bundestag committees on economic cooperation and development, on food, agriculture and consumer protection and on the environment, nature conservation and reactor safety on 20 February 2008 and the plenary debate on the amendment of the Federal Emission Control Act on 21 February 2008 (BMZ, 2008c). Discussion has so far focused mainly on the opportunities and risks of biofuels; there are as yet no signs of a more broadly based bioenergy strategy.

With regard to the relevance of biofuels to German development cooperation the government warns of the considerable social and ecological risks and of unjustified expectations; its policy is accordingly one of risk minimization. It takes a sceptical view of development opportunities based on export-oriented biofuel production in developing countries. Promotion should instead focus on off-grid supply and energy systems for direct use by the local population (BMZ, 2008c).

The intention is to minimize risk by ensuring that management of development policy in the bioenergy sector is governed by clear ecological and social sustainability standards. The fact that there is no immediate prospect of an appropriate international regulatory framework should in the short term be addressed by testing and developing certification systems and intensifying research into improved land-use systems and appropriate bioenergy sources. In the eyes of the German government another important task of bilateral German development cooperation is to advise partner countries on implementation of national food security strategies and on biomass strategies in a way that is appropriate 'to the country's national potential and is integrated into an overall rural development strategy' (BMZ, 2008c). In view of the unanswered questions relating to Germany's own bioenergy strategy, however, it remains unclear how such strategies are to be implemented in the partner countries.

In a position paper of its own, KfW Entwicklungsbank (KfW development bank) takes a somewhat less sceptical view than BMZ of the opportunities for tropical countries associated with the cultivation of energy crops and the possibility of exporting biofuels (KfW, 2008). The overall picture that emerges is one of uncertainty, as is also the case at all levels with other actors and institutions involved with bioenergy. This means that there is correspondingly large scope for influencing strategy decisions, the outcomes of which are still unclear. In this context the present WBGU report can serve as an aid to decision making and encourage those involved in shaping development policy to more strenuously direct their efforts towards the sustainable use of existing bioenergy potentials.

Below this general strategy level the implementing organizations already contribute significantly to utilization of the development potential of various forms of bioenergy use in the partner countries of German development cooperation. Mention should in particular be made of the activities of the Kompetenzzentrum Energie (Energy Competence Centre) of KfW Entwicklungsbank and the Bioenergy Sector Project of the Deutsche Gesellschaft für Technische Zusammenarbeit (GTZ – German Technical Cooperation). This project includes GTZ's HERA programme, which targets the household level, and the associated BEST Initiative for the promotion of national energy strategies based on biomass (Section 10.8.2.1).

10.8.1.4
The state of international development cooperation in the field of bioenergy

As a result of the discussion of biofuels, bioenergy is currently a 'hot' topic of development policy; almost all the actors in this field have published reports and position papers on it or set up working parties to address the issue. In some cases bioenergy is already a well established component of poverty reduction strategies and programmes for the promotion of renewable energies. Mention has already been made of the multilateral development banks and UN organizations that operate in this area. Alongside them, specific transnational partnerships and networks – such as the Renewable Energy and Energy Efficiency Partnership (REEEP) and the Renewable Energy Policy Network (REN21) – are involved with various aspects of the promotion of bioenergy.

It is, however, a noticeable feature of the present international discussion of bioenergy in the context of development that combating energy poverty in developing countries through improvements in the efficiency of traditional bioenergy use and through modern uses of biomass for energy is not a key objective of policy-makers. The narrow focus on biofuels reinforces this impression. Biofuels for the national transport sector represent only one specific aspect of the use of biomass for energy. In the present discussion there is little integrated consideration of the different deployment opportunities and needs in the areas of electricity, heat and mobility and of the associated potential for the economic development of the countries concerned.

Moreover, the energy policy portfolios of the international actors in development cooperation remain dominated by conventional energy systems and large-scale hydropower. The imponderable factors relating to the opportunities and risks of bioenergy in general and energy crop cultivation in particular appear to present a further obstacle to more active involvement in this field. While almost all international development cooperation actors are taking an increasing interest in the issue of bioenergy, they shy away from large-scale promotion programmes.

In this context the United Nations presents a particularly fragmented picture. Its institutions and initiatives intended to coordinate the environmental, development, agricultural and energy aspects of bioenergy-related policy activities have no large-scale impact. Some, such as the Commission for Sustainable Development, fail to meet even the most modest of expectations that, having regard to the prevailing structural weaknesses, can be placed on them. The diffuse picture arises in part from the differing expectations and demands placed on the various UN organizations and programmes by member states. For example, many developing countries look to UNDP and UNIDO mainly for technical and financial support for their conventional energy strategies and programmes. In addition, the UN actors involved are overwhelmed by the dynamics and complexity of the issue of bioenergy; many of them have only just begun to seek realistic solutions for the conflicts of use associated with bioenergy and therefore tend to adopt a 'wait and see' approach, as the example of the development banks shows. In view of this reluctance to act, there is still a great deal of scope for strategy development and policy management.

The reverse side of the coin is that, apart from such organizations as the Inter-American Development Bank, very few international organizations have a comprehensive bioenergy strategy. This is despite the fact that many developing countries are already taking concrete steps to promote biofuels, foreign investors are showing great interest in the farmland and markets of the South, and energy crop cultivation is booming in countries such as Brazil, Malaysia and Indonesia. Small-scale initiatives and pilot projects that focus on the sustainability of biofuels but ignore this larger picture of an emerging and dynamic world market for bioenergy are not enough, although they may be appropriate components of more comprehensive approaches. The decisive dynamic in this process is coming from the private sector. Yet the organizations involved in multilateral development cooperation, including in particular the financing institutions, are well placed to play a part in shaping this development. In WBGU's view they should apply their influence in line with sustainability guard rails (Chapter 3) and the corresponding minimum standards and promotion criteria (Section 10.3.1) in order to help create an enabling setting for the sustainable use of existing bioenergy potential in developing countries.

There is an absence of policies and strategies for tapping the various potentials of bioenergy in the electricity, heat and mobility sectors without undermining important sustainability targets relating to the world food supply, climate change mitigation and nature conservation. Elected representatives at national and international level need a sound basis for taking immediate decisions by means of which the bioenergy sector can be appropriately regulated and the scene set for its future development. It should also be borne in mind that well-funded private investors who are interested in long-term planning certainty are pressing for a clear and reliable framework within which they can operate.

In order to amalgamate the findings of the many forums, task forces, commissions and reports about the opportunities and risks of bioenergy that have emerged in the last few years and accelerate the international learning process in a way that produces tangible outcomes, WBGU recommends that an International Conference on Sustainable Bioenergy be convened very soon. Such a conference could serve, along the lines of the Renewables 2004 conference, as a forum for inter-governmental and cross-sectoral dialogue and possible cooperation in the fields of international agricultural, energy and development policy (Section 10.7.7.2); it could also contribute to the emergence of a global consensus on appropriate standards for the production and deployment of different forms of bioenergy. It should further lay the foundation for the promotion of sustainable bioenergy through specific multilateral partnerships, which could be based on existing technology agreements or trading partnerships between industrialized and newly industrializing countries.

10.8.2
Bioenergy strategies for developing countries

International development organizations should help newly industrializing and developing countries to use their bioenergy potentials in a sustainable way. Strategic issues of bioenergy policy need to be resolved in the light of existing biogeophysical, political, socio-economic and infrastructure-related conditions; this should occur mainly at national level and should involve as wide a range of affected interest and population groups as possible. At the same time, suitable conditions must be created at international level so that national strategies develop in a context that is conducive to sustainable bioenergy policy on a global scale.

At all levels it is important to clarify the primary objectives of bioenergy use. These objectives will in turn spawn further questions about feedstock sources (crops, residues) and bioenergy usages (heat, electricity or mobility). The question of which promotion policies would enable these objectives to be met sustainably must also be answered. Since bioenergy is always only one of several options for achieving these objectives, it is also necessary to explore whether more suitable alternatives might be available.

Energy supply structures in developing countries vary widely and are in some cases inadequate for the energy services that need to be provided (Chapter 8). Energy services range from the provision of ultra-simple technology for cooking to the large-scale supply of energy and power. These differing applications and needs must be integrated into an overall strategy. For the purpose of analysing future bioenergy potentials in developing countries and evaluating the extent to which realization of these potentials might be promoted through international development cooperation, WBGU considers four dimensions of the production and use of bioenergy:

- ecological and socio-economic sustainability (Chapters 3 and 5);
- energy potentials (Chapter 6);
- technological opportunities and options (Chapters 7–9);
- cost-effectiveness of different pathways (Chapters 7–9).

In WBGU's view measures are worth promoting if they make a significant contribution to climate change mitigation and other sustainability objectives, or if they help reduce energy poverty or health risks while also being compatible with the goal of turning energy systems towards sustainability. With this in mind the key elements of the development of sustainable bioenergy strategies in developing countries will now be discussed.

10.8.2.1
Combating energy poverty through off-grid rural energy provision

In WBGU's view, overcoming energy poverty is one of the major goals of the global energy turn-around towards sustainability (Chapter 2; WBGU, 2004a). Overcoming energy poverty involves ensuring that people have access to affordable, reliable, high-quality, safe, healthy and environmentally friendly energy services to meet their basic needs. Modern bioenergy use can play an important part in this. If energy poverty is to be overcome, WBGU considers it essential to improve the traditional inefficient use of bioenergy that is harmful to both human health and the environment or to replace it with other low-emission forms of energy. Small- to medium-scale solutions provide significant development opportunities for large sections of the population in developing countries; these opportunities should be utilized as rapidly as possible (Section 9.2.2). With solutions of this type, which are usually targeted at individual household level, extensive investigation of the risks associated with the use of bioenergy can usually be dispensed with. A range of biomass-based technologies can thus help to reduce energy poverty in both rural and urban areas and to mitigate climate change.

Some of these technologies are particularly simple and cost-efficient to implement. They include improved wood stoves, micro biogas systems, and vegetable oil that is produced and used locally. Whether conditions are appropriate for the use of these technologies must always be analysed locally. Energy-efficient wood ovens and stoves and micro biogas systems reduce the need for fuelwood, help to improve the quality of indoor air, reduce greenhouse gas emissions and cut the time needed for collecting firewood. In urban areas more efficient charcoal stoves can reduce charcoal consumption. Biomass gasification of wastes and residues can be used to drive gas or diesel generators and contribute to rural electrification. At local level vegetable oil, perhaps produced from *Jatropha* grown on marginal land, can be used anywhere to generate electricity in diesel generators, to provide mechanical energy (milling, water pumps) or for mobility purposes (Sections 8.2 and 9.2.2). In urban areas the conditions for overcoming energy poverty are different, since access to energy services in this context is often a problem of distribution.

SITUATION AND OBSTACLES

Recognition of the fact that improving the existing use of bioenergy plays an important part in reducing energy poverty and thus contributes to poverty reduction and human development is not new. This was highlighted at the World Summit for Sustainable Development (WSSD) in Johannesburg (EUEI and GTZ, 2006). For example, the use of improved biomass stoves can cut wood consumption by between 25 and 50 per cent (Section 7.2). This increases the security of supply for many people who are affected by energy poverty. The major adverse impacts on health are reduced at the same time.

Encouraging progress has been made in replacing inefficient stoves: 220 million energy-efficient stoves are now in use worldwide. A large number of public programmes, as well as the market, have played a part in this. Some 570 million households worldwide depend on the use of traditional biomass for cooking. In China there are now around 180 million improved stoves, representing 95 per cent of all households that are dependent on traditional biomass use for cooking. In India some 34 million improved stoves represent 25 per cent of such households. Success has also been achieved in other countries, although not yet at the same level. For example, Africa has only around 8 million energy-efficient wood and charcoal stoves whose use has been promoted through distribution and marketing strategies. With the support of international development cooperation, one-third of all African countries have introduced programmes for improved biomass stoves. For example, in 2007 Uganda set itself the target of increasing the number of improved stoves to 4 million by 2017 (Box 8.2-2; REN21, 2008).

Examples such as GTZ's Household Energy Programme to promote the use of energy-efficient stoves or the biogas programme of the Dutch development organization SNV illustrate that development cooperation is in principle able to make a very valuable contribution to improving rural energy efficiency in situations of poverty. For example, GTZ has developed markets for energy-saving and low-emission stoves; on behalf of the German and Dutch governments it has made such stoves available to 1 million households in the last four years (GTZ, 2007c). In the last five years around 25,000 families in Vietnam have been provided through SNV with micro biogas systems with integrated toilet facilities which use biogas to generate light and heat for cooking (DMFA, 2008; SNV, 2008).

These are important achievements that must be continued and expanded as a means of combating energy poverty, and hence poverty in general, and of contributing on a larger scale to resource conservation and emissions reduction. However, reducing energy poverty involves more than just providing energy-efficient stoves and micro biogas systems. Biomass gasification and vegetable oils also provide small-scale solutions for reducing energy poverty in the areas of electricity supply and mobility (Section 8.2). However, only token successes have so far been achieved in this field. For example, in India and China micro biogas systems are used to generate electricity. In India the total capacity of all such gasification systems is put at around 70 MW; ten national manufacturers provide the gasifiers together with the corresponding engines (Loy, 2007; REN21, 2008). In 2004/05, however, biomass contributed only 0.03 per cent to gross electricity generation in India (Loy, 2007).

Reasons for the sluggish growth of these simple and cost-efficient technologies can be identified on both the supply and the demand side. On the supply side the energy policy portfolios of actors on the international development cooperation scene continue to be dominated by conventional 'modern' energy systems. For donor organizations, governments in developing countries and private investors, large infrastructure projects are more attractive than small-scale endeavours in sparsely populated rural areas with weak market structures. Moreover, the requirement of households and businesses for useful heat and the opportunities for modern applications of the use of biomass for energy are often underestimated in national energy policies or their relevance for the energy supply is overlooked (EUEI and GTZ, 2006).

Yet rural development experts repeatedly point out that for the least developed countries, in particular, the simple technological options are extremely important (REN21, 2008). However, production and marketing of energy-efficient stoves or micro biogas systems appears unattractive to potential local providers, since demand is small. It is likely that there is also often a lack of information on the demand side:

many rural households in developing countries are insufficiently well informed about efficient alternatives to burning firewood or they are sceptical of new technologies and the follow-up costs involved, e.g. in purchasing new cooking utensils. In many places, too, there is little pressure to change for reasons of convenience since heating materials such as firewood can still be found and, as with the water supply, it is regarded as normal to spend a great deal of time on the collection process. It is mainly women and children who suffer as a result, since they use up time that could otherwise be used for economic gain or spent at school or in training.

Additional obstacles arise in connection with the development of an off-grid rural electricity supply based on biomass. These obstacles involve access to technology, for example because of the high costs of the initial investment, lack of familiarity with the technology, and the end users' lack of willingness or ability to pay. There are also problems in connection with the continuation of existing projects, e.g. on account of insufficient investment in maintenance and servicing (Valencia and Caspari, 2008).

PRIORITIZING THE REDUCTION OF ENERGY POVERTY
Comprehensive strategies for reducing energy poverty presuppose that priorities in development cooperation are adjusted in line with the requirements of these strategies. This means that the international community must set time limits for the abandonment of forms of traditional biomass use that are harmful to health. WBGU recommends that this target should be fully achieved by 2030. If such a strategy is to be successfully implemented, the obstacles that stand in its way must be better understood and overcome; this requires both cross-cutting multi-country evaluations and locally specific studies. However, even promotion of off-grid energy supply at national or local level requires a certain political and social framework. Possible sources of finance in addition to public development cooperation funds include microfinance systems and public-private partnerships (Section 12.5; WBGU, 2004a). Approval of small-scale CDM projects aimed at improving the efficiency of traditional bioenergy use is also justifiable and can therefore contribute to funding (Section 10.2.3.1). In WBGU's view the phasing out of traditional biomass use should have priority over all other development-related strategies for bioenergy use. This is particularly important in respect of the approximately 75 countries in which bioenergy constitutes more than 50 per cent of domestic energy use; it is even more urgent in some 50 countries in which this figure is more than 90 per cent.

10.8.2.2
Modernization of the energy sector and export production

Access to electricity is key to overcoming energy poverty; modernization of the energy sector in developing countries is therefore crucial. Advanced technologies associated with the use of biomass for energy are in general also suitable for decentralized power provision and heat generation in rural areas (Chapter 8). The use of biomass for energy can thus contribute to the modernization of the entire energy sector, including transport, and as well as improving the domestic supply it also offers opportunities for the export economy (Section 8.2.2; Box 8.2-3).

As the modelling results presented in Chapter 6 show, some developing and newly industrializing countries – predominantly in tropical and sub-tropical latitudes – have considerable potential for sustainable energy crop cultivation. Of the modelled global sustainable potential for energy crops, which corresponds to 6–25 per cent of present primary energy demand, approximately 22–24 per cent is in Central and South America, 12–15 per cent in sub-Saharan Africa, 12–13 per cent in China and neighbouring countries, 7–8 per cent in the CIS countries and 3–6 per cent in South Asia. The largest continuous areas are in the transition zone between the Sahel and savannah, in southern Africa and on the Indian subcontinent. However, data on present land use is imprecise. Realization of these sustainable potentials is dependent upon creation of appropriate institutional conditions and capacity development enabling actors in these parts of the world to meet sustainability requirements (Section 6.7). It is particularly important to avoid undesirable direct and indirect land-use changes that might impact adversely on food security, threaten biodiversity conservation or result in high greenhouse gas emissions. With this in mind, it is therefore essential both to create and improve an enabling international environment and to pursue national endeavours in this area.

International development cooperation should take appropriate steps to support these countries in the sustainable realization of these potentials. This assistance may extend to countries in which, although political and socio-economic conditions are currently unfavourable, bioenergy may in the medium to long term become an economically useful resource that could reinforce a positive development dynamic. In countries that already have land for energy crop cultivation, sustainable management of bioenergy production should be called for and supported. By contrast, where the sustainable potential for energy crop cultivation is already being fully exploited or where realization of this potential is unlikely even in the

long term, expansion of bioenergy production should not be promoted.

Bioenergy is usually produced and used within a national context. Nevertheless there are at international level both positive and negative incentives that can impact on the use of bioenergy in developing countries and thus influence opportunities and risks. In WBGU's view the following elements should be taken into account so that favourable international conditions for the sustainable production and use of bioenergy in developing and newly industrializing countries can emerge:

- The concept of guard rails put forward by WBGU highlights the limits within which expansion of the bioenergy sector is tolerable globally. A sustainable promotion policy should therefore take account of the guard rails for climate, biosphere and soil protection and access to food and energy as well as health and sustainability criteria. In the EU and the supervisory bodies of international development organizations the German government should call for these guard rails to be observed. It should also take steps to make the 2°C guard rail mandatory in the United Nations Framework Convention on Climate Change (UNFCCC), to extend nature conservation areas to cover 10–15 per cent of the world's land area and to secure basic energy services by aiming for a minimum provision worldwide of 700–1000 kWhel per person per year (Chapter 3).

- Operationalization of the guard rail concept for bioenergy production requires both additional supporting measures and the development and application of universally recognized sustainability standards and certification systems. Appropriate support should be provided to developing countries in setting up and implementing such certification systems. In awarding loans for bioenergy investment, multilateral development banks and international financing institutions should require compliance with sustainability standards. The minimum standard proposed by WBGU and the promotion criteria should serve as a basis for this (Section 10.3.1).

- Trade barriers should in general be removed to enable developing and newly industrializing countries to utilize their comparative advantages in the production of sustainable bioenergy carriers and the associated development opportunities (Section 10.3.3). Multilateral cooperation can be complemented by specific inter-country partnerships. For example, countries that supply bioenergy products can enter into trading partnerships with countries that require such products; within these partnerships free market access can be granted in return for assurances of sustainable production that meets the minimum standard, and in addition specialized technologies may perhaps be exchanged. On account of the potential in African and Latin American countries, bioenergy partnerships of this sort would be particularly appropriate with countries in these regions (Section 6.7).

- Since centres of biological diversity and areas of land with high carbon storage are often located in developing countries with major bioenergy potential, special steps must be taken to protect these ecologically valuable areas – in particular the tropical primary forests. An international system of compensation payments for foregone income in agriculture and forestry should be set up and binding agreements on reduced deforestation should be entered into (Sections 10.2 and 10.5.2).

- A controlled expansion of bioenergy should be accompanied by worldwide endeavours to strengthen agriculture, particularly in developing countries. The aim should be to ensure that the increase in land productivity in the food production sector that is necessary for food security can be achieved (Section 10.4.3). Because bioenergy production can endanger food security if it involves taking land away from food production, a global monitoring system for land use is required, as is appropriate risk management. Suitable precautions involving system-wide coherence in the fields of developing, humanitarian aid and the environment should be put in place under the umbrella of the United Nations (Section 10.4). In addition, a global commission on sustainable land use with appropriate political authority should be set up so that competition for land use can be better understood and regulated (Box 10.3-6).

10.8.2.3
Core elements of national bioenergy strategies for developing countries

To avoid the risks of bioenergy use, effective political management at national level is essential. These risks include a potentially negative greenhouse gas balance of crop cultivation, the loss of biological diversity (e.g. through forest clearance) and soil degradation. In addition, there is a danger that displacement effects will have an adverse effect on food production and that smallholder production will be displaced by large-scale buying up of land. The task of averting these risks and ensuring that bioenergy production is sustainable is primarily one of national policy.

Particularly when bioenergy is used to restructure national energy sectors and energy crops are grown on large areas of land, developing and newly industrializing countries need country-specific strategies

that address the associated opportunities and risks. Different policy goals related to bioenergy production can be achieved via different conversion pathways and value chains; the choices made can help to minimize conflicts between goals (e.g. in India: Box 10.8-1). The core elements of a bioenergy strategy must be specified by the government of the country concerned in the light of the country's circumstances and with the involvement of the actors concerned. Large-scale bioenergy investment in developing countries should therefore not be undertaken, or at least should not be supported through development cooperation, until clearly elaborated and viable strategies and utilization concepts are in place. Systematic examination of country-specific circumstances and objectives and elaboration of the conditions required for sustainable production – including evaluation of possible alternatives – are essential if national bioenergy strategies are to be developed in an ecologically and socially sustainable way (decision aid: Figure 10.8-1). Development policy can play an important part in this.

BIOMASS PRODUCTION
The use of biogenic wastes and residues from agriculture and forestry for energy has many advantages over the cultivation of energy crops. For example, competition with existing land use is largely avoided, as are possible emissions from land-use changes and cultivation. In many developing countries agriculture plays a major role and significant quantities of suitable residues arise (e.g. in the fishing industry, sawmills, tea and coffee plantations) and can be used (Chapters 7 and 8). If residues are removed from agricultural and forestry land, effective soil protection must be ensured. Since the size of the sustainable economic potentials and the way in which they can be utilized have not been definitively established there is a major need for research in this area. The mobilization and use of wastes and residues should be promoted through pilot projects and best practices identified.

When energy crops are grown, particularly in developing countries, direct and indirect land-use changes (e.g. through tropical deforestation, competition with food production) are a risk factor. Land-use changes have a major impact on the greenhouse gas balance, which is why WBGU rejects the direct and indirect conversion of forests and wetlands. In developing countries and elsewhere, therefore, energy crop cultivation should be restricted to land which can in the main be converted for bioenergy production without occasioning emissions from land-use change. Such emissions, if they arise, can reduce the climate change mitigation effect by around 50 per cent or even cause climate damages (Section 7.3). Energy crop cultiva-

Box 10.8-1

Country study: India – *Jatropha* cultivation as a development model

The *Jatropha* plant has for some years been cultivated for biofuel production and has attracted growing interest worldwide. Since 2003 India has also increased its support for the development of biofuels based on *Jatropha*. Faced with levels of crude oil imports that are high and still rising, the country's aim is to increase supply security and promote rural development. It also sees *Jatropha* cultivation as providing opportunities for restoring degraded land and mitigating climate change. The ad hoc committee for biodiesel development, set up in 2003, had ambitious expectations: it was assumed that a 20 per cent blending quota could be achieved by 2012. To this end, it was envisaged that *Jatropha* would be grown on some 11.2 million hectares of wasteland (Planning Commission, 2003). The wasteland is defined as land that is not effectively used for agriculture or is degraded or of poor quality. A national 'Biodiesel Mission' was planned for the purpose of testing and developing the production and use of biodiesel (TERI and GTZ, 2005). Some Indian states, such as Uttarakhand and Chattisgarh, announced ambitious programmes involving various promotion instruments with the aim of setting up the first cultivation areas and creating refinery capacity. The Indian government has now adopted a new bioenergy policy. The intention is that by 2017 20 per cent of the demand for diesel should be met from farmed biomass (Economist, 2008b).

Despite the major endeavours of recent years, development of the Indian biodiesel sector is still in its early stages. The main reason is that *Jatropha* cultivation has not yet been profitable. The fruits cannot be harvested until the plant is 3–5 years old, and without additional fertilizer and irrigation the yield potential of the wild plant is very low (Section 7.1). The subsidizing of fossil diesel hinders the profitability of biodiesel production. In addition, it is not yet clear what the socio-economic and ecological consequences of *Jatropha* cultivation will be. Some conclusions can nevertheless be drawn from experience to date:

According to Altenburg et al. (2008), the creation of development opportunities through biodiesel production depends to a large extent on the organization of the value chain. In India a number of different chains have been identified, which can be divided into three categories: government-organized cultivation on state land, cultivation by smallholders and cultivation initiated by large companies. The main respects in which the categories differ are their objectives, the distribution of cultivation risks, and land-use rights.

The objectives of the government-organized model are rural development and the bringing into use of degraded land. Cultivation takes place on state land and the workers receive a regular income (collectors model). In some cases there is cooperation with the private sector (PPP model). The state bears the risk, irrespective of any profit that is made in the market. This gives rise to additional employment and income-generating opportunities. By involving the village assemblies or panchayats in cultivation and in decisions on which land should be used, risks of displacement and of jeopardizing food cultivation are reduced.

Organization takes place mostly through state promotion programmes, such as the Joint Forest Management Programme. This model has become established in the state of Andhra Pradesh. Because of a high level of state intervention, however, it is questionable whether it could ever become a sustainable and economically independent development pathway.

In the second model, small-scale farmers and smallholders farm their own land. Since these farmers are dependent on immediate income returns, cultivation is at present carried out only as an add-on to their actual farming activities, for example in the form of hedge planting under the Fences for Fuel programme in Rajastan. Under this scheme the farmers use the vegetable oil directly for the operation of generators and pumps, thus contributing to the local energy supply. In most cases, though, the farmers have supply contracts with companies. In pilot projects in the state of Karnataka the state is supporting the founding of cooperatives. The formation of such farmers' consortiums does not guarantee a market, but such a system considerably dilutes production risks and promotes the self-organization and personal responsibility of the smallholders. In this way *Jatropha* cultivation can help to ensure that unused land owned by small farmers is brought back into productive use.

Ecological and social risks are particularly associated with the third model, involving cultivation initiated by companies. Companies aim for large-scale cultivation and benefits of scale. They have the capital and know-how to provide the necessary irrigation and nutrients. While this creates potential for significantly greater yields and for rural employment and income opportunities, it may result in monocultures and the displacement of food production. However, since market prices for energy crops are currently low, this is not a risk at present.

A key factor in large-scale production is low-cost access to land. Companies use predominantly private land. Alongside this, states such as Chattisgarh make marginal state-owned land for *Jatropha* cultivation available. However, this land is often used by cowherds to pasture their animals; there is a risk that the land-use interests of these herders will be ignored and that people will be displaced. In some Indian states signs of conflict over land issues are already apparent (Grain, 2008; Peoples Coalition, 2008; Box 6.7-2). In order to avoid the risks associated with large-scale production, it is essential that local processes for making decisions on the use and leasing of land operate on a participatory basis. This condition is not yet met in all parts of India (Altenburg et al., 2008).

Despite some sobering truths, the development potential of biodiesel and vegetable oil production in India is high. Nevertheless, expansion entails risk. This risk, however, can be significantly reduced through the selection of value-adding models, the government's objectives (local energy supply versus supplying of the national or global market) and appropriate participatory processes. India is a pioneer in the cultivation of *Jatropha* for biodiesel production. The available experience of cultivation and production under different promotion models provides an opportunity to explore the risks and potentials of oil plant cultivation and develop reliable models and methods that can also be of benefit to other countries.

tion should therefore be promoted mainly on marginal land. Since in developing countries even marginal land is frequently used, the interests of the local community should be considered and its participation ensured. Where energy crops are grown through a contract farming system, steps should be taken to

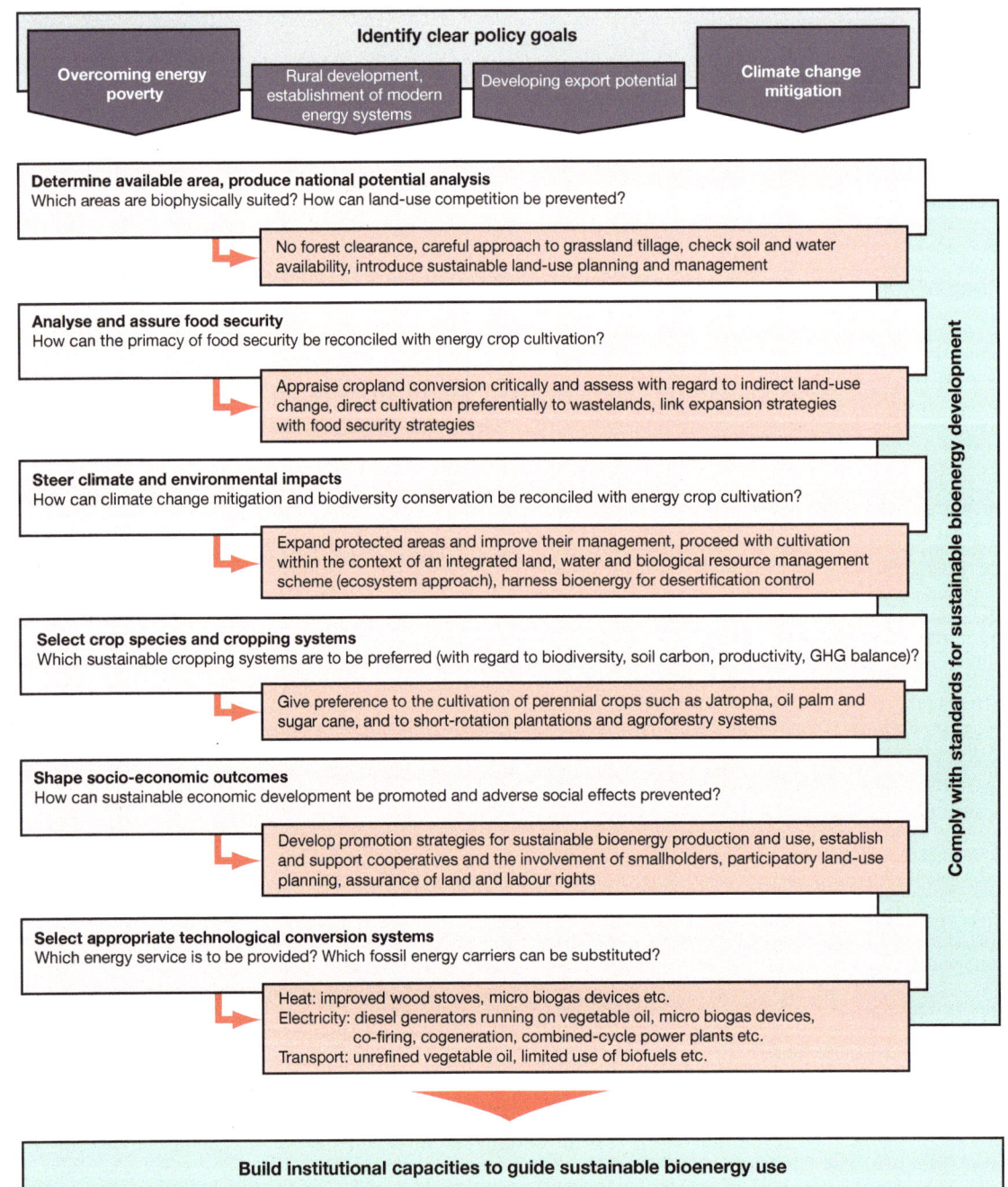

Figure 10.8-1
Decision tree for strategic national choices on biofuel development in developing and newly industrializing countries. The red arrows refer to WBGU recommendations.
Source: WBGU, building on Vermeulen et al., 2008

ensure that small-scale farmers are treated fairly (Box 8.2-3).

The choice of cultivation system also influences the sustainability of production. According to WBGU's sustainability criteria and taking account of land yields, perennial crops such as *Jatropha*, oil palms, sugar cane, short-rotation plantations of fast-grow-ing woody plants and energy grasses are in general to be preferred to annual crops such as rape, cereals and maize. Wherever possible, plant mixtures should be used rather than monocultures. Through suitable cultivation systems additional organic carbon can be incorporated into the soil, with beneficial effects for both the greenhouse gas balance and soil fertility (Section 7.1). This means that in developing countries, in particular, bioenergy production can be used as a means of tackling desertification and restoring degraded soils (Section 10.6). A participatory land-use planning system is an essential element of doing so. In view of these benefits, socially and ecologically sustainable cultivation systems of particular promise should be promoted through suitable pilot projects. The possibility of afforestation in combination with sustainable use should also be examined.

In summary, bioenergy from energy crops in developing countries should in WBGU's view only be promoted if two conditions are met. Firstly, the climate change mitigation effect, including emissions from direct and indirect land-use changes, must be shown to be unequivocally positive. Secondly, the other sustainability criteria – primacy of food security, conservation of biological diversity, protection of soil and water resources – must be met (Section 10.3.1).

Integrating bioenergy into the energy system and appropriate conversion pathways

WBGU describes how bioenergy can be used sustainably, depending on the objectives of developing countries and the existing structure of their energy systems. Bioenergy can help to improve the local energy supply at household and village level and in small businesses; it also has an important part to play in the modernization of the energy sector in an urban/industrial context. It is important to distinguish between on the one hand the use of bioenergy to provide power and heat and on the other its use in the transport sector. From the point of view of climate change mitigation, the most attractive areas of application – in developing countries and elsewhere – are those in which fossil energy carriers with high CO_2 emissions (particularly coal) are replaced (Section 9.2.1.3). This means that pathways that involve using bioenergy for power generation should generally be preferred to the use of biofuels for transport. What are the best applications in a given situation will, however, depend to a large extent on the energy supply structure and on costs. These costs include both initial costs and greenhouse gas abatement costs (Section 7.3).

Although developing and newly industrializing countries have not as yet entered into any international commitment to limit the quantity of their greenhouse gas emissions, they have committed to climate change mitigation. In economically weaker countries, however, the absolute reduction in greenhouse gas emissions is unlikely to be the deciding factor in the selection of conversion systems. Nevertheless, bioenergy pathways that represent particularly cost-effective climate protection options should also be given high priority in developing countries. Where greenhouse gas abatement costs are low, new sources

of finance can be developed through international climate protection instruments. The aim should therefore be to pursue conversion pathways for bioenergy use that combine a relatively high abatement outcome per unit of biomass used with low greenhouse gas abatement costs. The focus should be on technologies that facilitate a shift to a modern energy system with low greenhouse gas emissions. An inefficient infrastructure should therefore not be encouraged to stabilize.

In the light of these considerations the following applications are of particular interest, especially for developing and newly industrializing countries. In power and heat generation, pathways involving the co-combustion of solid biomass in coal-fired power plants result in major greenhouse gas reductions with low abatement costs. Particularly effective approaches in this context are the use of biogenic residues from agriculture and forestry (e.g. straw, wood chips, bagasse, etc.), cascade use and energy crop use in the form of wood or grass pellets. Co-combustion is particularly advantageous when coal plays a large part in the power supply, as for example in the growing newly industrializing countries India and China. The major advantage of stationary use is that the waste heat can be used. Because of its high level of energy efficiency, combined heat and power (CHP) generation should in general be preferred to power generation alone, provided that demand for the heat exists. For regions with a significant requirement for cold or cooling, CHP can also be used to generate cooling.

In countries such as Brazil ethanol from sugar cane could be used directly in combined-cycle power plants. The use of biofuels (such as vegetable oil and bioethanol) in small-scale CHP plants also has a greater impact on climate change mitigation than their use in transport. However, these bioenergy pathways have relatively high initial and greenhouse gas abatement costs if the infrastructure needs to be developed from scratch. As expensive climate change mitigation options they are therefore less attractive for possible financing through climate change mitigation instruments. They should therefore be supported mainly through development cooperation and technology cooperation schemes. In countries such as Uganda in which the power supply is largely based on hydropower, it is appropriate for biomass to be used as a liquid fuel in the transport sector if fossil fuels are thereby replaced. However, an even greater climate change mitigation effect could be achieved if the biomass or vegetable oil/bioethanol produced in tropical countries were used to generate power in small-scale CHP plants or were exported. In urban centres, large-scale programmes for the introduction of small-scale CHP plants could prevent the construction of large new coal-fired power plants. This provides biofuel producers with assured demand and (greater) investment security.

It should not be forgotten that biomethane is a flexible energy carrier that can be used almost universally. It is produced either from biogas arising from the fermentation of moist biomass in biogas systems, or from synthesis gas from the gasification of predominantly solid biomass. At present, however, gasification is still relatively costly (Section 7.2). In contrast to the direct use of biogas or synthesis gas in off-grid power generation systems, biomethane can be supplied via natural gas grids, enabling it to be used flexibly in power plants of different types. However, the biomethane option is only viable in newly industrializing or developing countries that already have a gas grid or are in the process of establishing one. Biomethane can, though, be readily transported or exported via pipelines or liquefaction (LNG) (Box 7.2-2, Sections 8.1.2.3 and 9.2.1.4).

In developing and newly industrializing countries in which bioethanol and biodiesel can be produced without difficulty, biofuels are an option in the transport sector. Provided that electromobility is not yet established and is able to replace the combustion engine, cultivation of tropical energy crops on degraded land in order to produce biofuels is a cost-effective climate change mitigation option. Biodiesel from *Jatropha* and bioethanol from sugar cane enable greenhouse gas reductions to be effected at low cost, if cultivation results in storing carbon in the soil and no indirect land-use changes are triggered. Projects involving these processes could be eligible for financing through climate protection instruments. However, larger greenhouse gas reductions could be achieved through power generation. To avoid perpetuating the old transport infrastructure in developing countries and elsewhere, specific promotion of conversion into electricity and electromobility should be undertaken. Provided that production of biofuels meets the minimum standard, opportunities arise for export in connection with the conversion of liquid fuels into electricity in mini-CHP plants, e.g. in industrialized countries (Sections 8.1.2.1 and 9.2.1.4).

It is also possible to identify a number of appropriate pathways for climate-friendly modernization of the energy sector in an urban/industrial context in developing and newly industrializing countries. Tropical countries have major potential in this area through the sustainable cultivation of energy crops; opportunities for rural development and the development of foreign markets are also opened up for them. However, as the example of oil palm cultivation quite clearly illustrates when direct and indirect land-use changes are taken into account, it is energy crop cultivation that harbours the greatest risks. If

the oil palms are grown on degraded land, there is potential for very significant greenhouse gas reduction. If tropical forest is cleared to make way for oil palm cultivation, the positive balance becomes a negative one. Issues relating to the cultivation site and to compliance with sustainable production methods are therefore crucial to evaluation of the benefits and prospects of promotion policy in relation to the sustainable expansion of the use of bioenergy.

In order to harness the existing sustainable potentials, development cooperation should support the partner countries in their strategy development and urge them to observe the minimum standards and the promotion criteria, particularly where relatively large-scale applications are concerned. To this end the necessary institutional capacities for establishment of sustainable bioenergy use should be strengthened, for example in areas such as land-use planning and certification. In particular, socially and ecologically sustainable cultivation systems should be actively promoted by development cooperation in pilot projects. To help promising technologies (CHP, mini-CHP, combined-cycle, co-combustion, etc.) to make a breakthrough and to facilitate effective technology transfer, selected bioenergy conversion pathways should be promoted through support measures (development cooperation, public-private partnerships, etc.). In addition, greater use could be made of CDM projects (Section 10.2.3).

10.8.3
Action under uncertainty: Consequences for active promotion policies

In WBGU's view the important opportunities that arise from bioenergy in relation to development are twofold. They involve, firstly the replacement of traditional forms of biomass use by more efficient technology and, secondly, the off-grid use of bioenergy for local production of heat and electricity. Both contribute to poverty reduction and attainment of the Millennium Development Goals. Importantly, more efficient use of combustible materials contributes to reduction of greenhouse gas emissions and hence to climate change mitigation. Reducing energy poverty in developing countries through modernized bioenergy use thus gives rise to clear win-win situations. Furthermore, in developing countries and elsewhere bioenergy can play an important part in the modernization of energy systems and hence also in transforming the world energy system into a climate-friendly one.

Development policy must now create the conditions in which the methodological and analytical insights acquired on the basis of current knowledge of bioenergy can be applied and operationalized in development cooperation. In parallel with this it should pursue the institutionalization of national and international regulatory frameworks that rule out unintended effects of bioenergy use and monitor the drivers of potential land-use conflicts, which will have their greatest impact on developing countries. In particular, competition between the requirements of a dynamic bioenergy system and those of food security cannot be properly addressed unless the corresponding sustainability standards and guard rails are duly considered in the development and implementation of a global bioenergy policy. The report shows that an expansion of bioenergy on a scale that is worthwhile in terms of climate and energy policy can only be justified if it is accompanied in the coming years by major investment in the agriculture of developing countries. This is necessary in order to improve food security worldwide. Theoretically existing bioenergy potentials must on no account be exploited at the expense of food security.

It is manifestly obvious that the findings presented here involve elements of uncertainty. It is clear that further information and research on the sustainability of future bioenergy use will be required (Chapter 11). Priorities and starting points for development-oriented research into the sustainable production and use of bioenergy are described in the next chapter (Box 11-1). Those involved in international development policy must take account of current knowledge gaps when developing their strategies and accept the uncertainties that prevail in the bioenergy sector. In consequence, their decisions and actions need to be based on the precautionary principle.

Bioenergy is a field in which there are large knowledge gaps and hence a major need for research. However, the considerable political pressure to act to reform the world's energy systems means that decisions must be taken now, despite the major uncertainties. WBGU therefore recommends, firstly, implementing robust win-win solutions without delay, while simultaneously scaling up research efforts, some of which will need to be coordinated internationally.

In this report, WBGU has already identified viable approaches to sustainable bioenergy use in some areas. Nevertheless, there remains a need for both single-discipline and integrating studies to be carried out in the coming years. WBGU has identified the six most important research fields (Box 11-1); specific recommendations for research in each area are presented below.

Germany is already intensively involved in research into global change, as is demonstrated by the numerous activities of the Federal Ministry of Education and Research (BMBF) in the areas of environment, sustainability and energy. For example, the German government's High-Tech Strategy includes the funding programme 'Bioenergy 2021 – Research for the use of biomass'. The 4th colloquium of the German National Committee on Global Change Research (NKGCF), which focused on 'Land Use in the Area of Conflict of Resource Conservation, Food and Energy', likewise highlighted the fact that global land use has been an established area of research for some years now. Internationally, these issues are addressed in such contexts as the International Geosphere-Biosphere Programme (IGBP) and the International Human Dimensions Programme on Global Environmental Change (IHDP), as well as in the European Research Framework Programme. For research into bioenergy the German government has recently set up the German Biomass Research Centre in Leipzig. In WBGU's view, a particular weak point in the assessment of bioenergy remains the need for evidence of the climate change mitigation effect of bioenergy and the necessary greenhouse gas balances of different bioenergy pathways including

emissions of land-use changes; these issues require further in-depth research.

In this chapter, WBGU has in particular attempted to highlight research issues that have arisen during the preparation of the present report. However, the research recommendations presented here in no way claim to constitute a systematic and complete portfolio for bioenergy research.

11.1
Bioenergy use and the greenhouse gas balance

11.1.1
Improving greenhouse gas balancing of energy crop cultivation

The conversion of biomass to energy is usually not CO_2-neutral; rather, the greenhouse gas balance of biomass use is extremely complex. It is true that when combusted biomass releases only the quantity of CO_2 that it previously absorbed from the atmosphere; however, emissions also arise during the production, supply and processing of biomass that is converted to energy. Depending on the soil, the cultivation system and the previous land cover, energy crop cultivation can lead to high CO_2 emissions as a result of land-use changes. Significant quantities of greenhouse gases may also be released during the cultivation of energy crops. Information on greenhouse gas emissions associated with different forms of cultivation is absent or inadequate; reliable data is rare. It is therefore essential that more in-depth studies of the main cultivation systems in Germany and other major producing countries are carried out. Greenhouse gas emissions also arise during the production and transport of the energy carrier, and these need to be taken into account too.

These considerations apply not only to bioenergy but also to food production. Better understanding of the greenhouse gas balance of food products can help farmers to adopt more climate-friendly agricultural

Box 11-1

Bioenergy and land use: The key research areas

1. *Improving the scientific basis of global land use:* To create the scientific basis for setting up a global GIS-supported land register, the state of global land use, land cover and soil and the dynamics of global land-use changes need to be observed and assessed more accurately than before (Bai et al., 2008). The work required includes collecting high-resolution data on vegetation cover, water levels and soil condition, agricultural use and ground sealing in the different world regions. To enable these dynamic processes to be observed and measured, it will be necessary to devise standardized indicator systems and develop methods of comparing and interpreting the data that is obtained (Section 11.4.1).

2. *Determining more accurate greenhouse gas balances for various bioenergy pathways:* As far as the climate is concerned, the greenhouse gas balance is the key variable that determines the benefit (or in many cases the harm) of a particular form of bioenergy. In the past it has not been possible to calculate this accurately, e.g. taking account of indirect effects such as the displacement elsewhere of previous land use (Section 11.1).

3. *Determining the potential, the greenhouse gas balances and the cost-effective pathways for utilizing residues.* Residues from agriculture and forestry as well as other sources have a potential for energy generation that has as yet barely been tapped; the future options for using them should be explored (Section 11.3.2).

4. *Analysing the role of bioenergy in an energy system of the future (nationally, regionally, worldwide):* The strategic importance of bioenergy in a particular energy system, and its incorporation into that system (e.g. as balancing power) should be explored in more detail. This is a key factor in selecting the preferred pathway, e.g. via the use of biomethane (Section 11.3.1).

5. *Clarifying the links between food security and bioenergy:* The complex local, national and global impact chain between bioenergy use and food security urgently needs to be explored from both socio-economic and geopolitical angles. A geopolitical question that arises is: Could the primacy of securing the energy supply of the western world and of other powerful political stakeholders in a world energy system of which bioenergy is an important component lead to worsening food problems in poor and politically less influential countries, and how could international cooperation agreements serve to prevent this (Section 11.4.4.)?

6. *Analysing international competition for land use and developing elements of a global land-use management system:* In the coming decade, as a result of various drivers (population growth, changing dietary patterns, use of biomass for energy generation, impacts of climate change) land will become a scarce resource worldwide. This means that land use will become a subject of global governance. Research should explore interest structures relating to worldwide use and contribute to the development of an effective global set of rules governing the management of land resources and to the avoidance of land-use conflicts (Section 11.5).

practices and enable consumers to choose food products that have a less harmful effect on climate.

11.1.2
Integrated assessment of climate change mitigation options in land and biomass use

The climate change mitigation effect of bioenergy can be only identified by considering the entire pathway. Biomass stores both energy and carbon. The deployment of bioenergy can therefore contribute to climate change mitigation by a number of routes. One option is to use the energy contained in the biomass, which can be done in various ways, thereby replacing various fossil energy carriers. An alternative is to make use of the carbon storage capacity of biomass (either by planting crops and thus increasing soil carbon stocks or in the form of harvested products); this involves not using (at least partially) its energy potential or using it as a material resource before conversion to energy. The associated climate change mitigation effects operate on differing time scales (Section 5.5.4).

There is therefore a need for integrated studies that compare the various deployment options over the entire life cycle and thus help to ensure that

the best possible climate change mitigation effect is achieved from the available limited raw biomass. These studies should consider not only relative savings in relation to the final energy but also the absolute values of greenhouse gas reductions in relation to the quantity of raw biomass used.

So far, life-cycle analyses of bioenergy carriers have often included only greenhouse gas emissions from direct land-use changes. But because the cultivation of raw materials for the production of bioenergy carriers competes with other forms of land use (including food and fodder production), a change of land use for bioenergy-related purposes can result in the conversion of land elsewhere. A new assessment methodology is needed to enable the potential additional emissions from such indirect land-use changes to be included in the assessment of energy crop use. In order to arrive at an indicator for the effects of indirect land-use change that is as realistic as possible, information on a number of factors needs to be coordinated. These factors include the previous use of the land in question, its productivity, the biogeochemical fluxes arising when land is converted and other economic determinants of land use (area productivity, involvement of the land in international trade and material flows, likely type of land use after conversion). The 'iLUC factor' (Box 7.3-2) is only a first and very rough approach.

11.1.3
Sequestration of CO_2 in depots and black carbon in soils

There are a number of ways in which biomass use can enable carbon dioxide to be permanently removed from the atmosphere. Important examples are the conversion of biomass to energy with carbon capture and storage in a depot, or the insertion of biochar into the soil. The capacity of plants to absorb carbon from the atmosphere and store it by means of photosynthesis opens up the possibility of sequestering the carbon thus obtained.

Energy crops and residues can be converted into biomethane by fermentation (biogas systems) or gasification (biomass gasification systems) (Box 7.2-2). During the production process, some of the carbon stored in the biomass arises as CO_2 and must be separated in any case. The separated CO_2 can be deposited in suitable storage sites. The technology associated with biogas systems is fairly mature; by contrast, biomass gasification systems still need further research. Both processes enable the use of biomass for two purposes simultaneously: for energy and for CO_2 sequestration. It is also possible to store black carbon in the soil in the form of biochar or charcoal. The gas that arises during carbonization or pyrolysis can be used for energy purposes, while the residual charcoal can be either completely removed from the biosphere (by being deposited in deep storage sites) or inserted into the soil, where it can act as a permanent soil conditioner for a long time (charcoal in the soil decomposes very slowly). Further research is required to determine how long carbon thus stored in the soil remains removed from the atmosphere, to calculate the associated greenhouse gas balance and to reveal any further ecological impacts that arise from the process of inserting charcoal into the soil. In addition, the global potential of this technology is at present unclear, as well as its cost-effectiveness (Box 5.5-2).

11.2
Sustainable bioenergy potential

11.2.1
Sustainable agriculture and energy crop cultivation

OPTIMIZING CROP MANAGEMENT
Increasing agricultural production while at the same time conserving ecological resources requires sustainable cultivation management involving practices such as efficient water management, an optimal fertilization regime and extensive soil management (e.g. no-tillage systems). There is still a major need for research into sustainable crop cultivation systems that enable biomass to be produced cost-effectively on degraded and marginal land. Research is also needed into various plants that are sources of biomass (e.g. pongamia – *Pongamia pinnata*, giant reed – *Arundo donax*, reed canarygass – *Phalaris arundinacea*) but for which insufficient qualitative and quantitative production data is available, in order to determine their suitability as energy crops.

PRIORITIES IN PLANT BREEDING
There are various plants, including *Jatropha*, switchgrass, miscanthus and willow (or other short-rotation coppice species), that have in recent years been found to be of interest for bioenergy production but that are not domesticated. In other words, they have not been genetically adapted and improved for biomass production by a process of breeding over centuries or millennia. There is great potential here for increasing biomass production; the agricultural industry is investing substantial sums of money in developing this potential, partly through genetic engineering research.

However, the financial resources available for research are sometimes limited, particularly in the case of tropical food and energy crops (cassava, oil palms, sugar cane), because the crops concerned are often grown on a small scale and are therefore not economically attractive for private-sector investors. Public funds must therefore be applied if research and development into sustainable bioenergy production in developing countries is to be advanced. In many parts of the world, the productivity of these crops can be increased considerably by conventional means (hybrid varieties, fertilizers, disease control, etc.). The highest production increases anywhere in the world could be achieved in the developing countries if research were intensified regionally. In developing genetically modified energy crops, it is essential that ecosystem impacts (outcrossing, introduction into the wild, toxicity, etc.) are identified and monitored. It is particularly important to consider socio-economic factors and to analyse the costs and benefits of research if the expected increase in yield of genetically modified energy crops or the stability of these yields is only slightly above that of adapted landraces.

INVASIVE POTENTIAL OF ENERGY CROPS

Many energy crops have high invasive potential (they are fast-growing, have many seeds and regenerate easily). Before large-scale cultivation commences, steps must be taken to identify for Germany, for Europe and on a global scale the potential problems associated with each crop species, both in the wider landscape and in agricultural land next to the energy crop.

AGROFORESTRY

Agroforestry – that is, a system in which forestry and agriculture are combined – has major potential for both food and energy crop cultivation. However, there is insufficient information on the 'best' mixtures and on the physiological and ecological processes that operate in such systems. Practices are often based on traditional knowledge which may lack any scientific foundation. In view of the climate-related challenges that the world will face in future, this knowledge cannot be relied on to ensure yield stability and quality. Work should be undertaken jointly with international organizations such as the FAO to set up research programmes that focus on food, fodder and energy crop cultivation in agroforestry.

IMPACT OF CLIMATE CHANGE ON ENERGY CROP CULTIVATION

Climate change will have a major impact on all cultivation systems. Previous discussion of energy crop cultivation has paid very little attention to this, although it is a factor that will play a key part in determining the success of such cultivation. As a matter of urgency, therefore, future research programmes on climate and agriculture should address the impact of climate change on energy crop cultivation.

11.2.2
International research programmes on sustainable and economic bioenergy potentials

Policies and promotion strategies for bioenergy are currently being developed in many developing and newly industrializing countries. At the same time, however, there is only limited knowledge of issues such as the potentials of the different forms of biomass use for energy generation, specific cases of competition for land use and the regulatory frame-

work required to prevent unintended ecological, social and economic damage. Drawing on the differentiated assessment of the opportunities and risks of bioenergy use drawn up by WBGU, the German government should exert its influence through international organizations such as UNEP, the World Bank and the regional development banks to encourage the establishment of regional research programmes to explore the sustainable potentials of bioenergy in the developing regions, with the aim of providing decision-makers with a more robust basis for bioenergy policies oriented towards the longer term. The research programmes should be networked so that trans-regional assessments can also be made. UNEP, as a leading international environmental organization, could take on a clearing-house role and provide a focal point for collection and assessment of the findings, giving decision-makers access to a comprehensive evaluation and to the current state of knowledge. The collated research findings will in addition assist the development of an international consensus on sustainability standards in the bioenergy industry. At the same time, development policy and/or international cooperation on research should be used as a vehicle for promoting national research programmes on sustainable bioenergy strategies, particularly in those countries that according to WBGU's analysis of potential could be significant providers of biomass from energy crops in future.

DETERMINING THE ECONOMICALLY SUSTAINABLE BIOENERGY POTENTIAL

Research needs to identify not only the size of the technically available and ecologically sustainable bioenergy potential, but also whether the potential is economically worthwhile and sustainable. The technically usable and ecologically sustainable potential of energy crops calculated in the present report needs to be validated in terms of the cost of tapping it. It is essential to distinguish between the potential of energy crops and that of residues and waste. In order to identify the potential of energy crops, existing (agro-)economic models must be developed further, and work on linking them to climate and vegetation models must be pursued (Section 11.4.2). For the potential of residues, global material flows in agriculture and forestry and in waste management must be analysed and assessed in terms of closed nutrient cycles and the cost-effectiveness of providing energy from residues. The economic models should be expanded to take account of technological change in order to yield a realistic picture of technological developments relating to land use and energy efficiency against the background of changing economic conditions.

Updating of the models must be accompanied by improved availability of economic land-related data. To this end, long-term data gathering (e.g. surveys) should be carried out and the data collected should be stored centrally. The data should be compatible with the spatial resolution of the scientific models; in addition to demographic and economic data, it should in particular include information on land-use decisions and lifestyles. WBGU estimates that the sustainable potential of residues and waste is of roughly the same order as that of energy crops. There is also a need for research into the institutional and structural conditions for exploitation of the economically and ecologically sustainable bioenergy potential. This will involve analysing individual regions with high economic energy crop potential in order to identify the conditions that are favourable for actual realization of this potential. The conditions in question are those relating to the distribution of land rights, institutions for enforcement of property rights and stakeholder structures in the agricultural sector (e.g. dominance of large companies vs. a large number of small suppliers).

In relation to developing and newly industrializing countries, it is in addition necessary to clarify the degree to which exploitation of the sustainable economic bioenergy potential is contingent upon the existence of infrastructures for the storage, processing, distribution, marketing and export of bioenergy carriers and what form these structures must take if the bioenergy potential is to be utilized in an economically and socially sustainable way (Rosegrant et al., 2008).

Costs and benefits of bioenergy use

In order to assess the socio-economic impacts and macroeconomic cost-effectiveness of bioenergy use within a region, studies of the regional (and in a subsequent step also the global) costs and benefits of bioenergy production and use must be carried out. Such studies should also explore the effects of energy crop production and use on raw material and food prices, food and energy security, the climate, economic growth, the balance of trade and dependence on oil imports. Distributional aspects should also be considered in this context, since costs and benefits may be very unevenly distributed among the stakeholders. It is important to ask who are the winners and losers under different bioenergy strategies (e.g. small-scale, decentral production and use vs. large-scale, export-oriented energy crop cultivation) and under what conditions (relating to production, marketing, distribution, etc.) different sections of the population (e.g. large landowners, smallholders, rural workers, urban dwellers) can benefit or lose from energy crop cultivation. Research should also pinpoint the sectors in

which residues can appropriately be used for energy purposes and products can be designed to be recycled for energy purposes at the end of their useful life (cascade use; Section 5.3.3). This information can be used to identify policy measures that will strike a balance between those who gain and those who lose from bioenergy use.

An initial tool, available for assessing the costs and benefits of energy crop use in different countries and regions, is the Bioenergy Assessment Tool produced by the FAO, the Copernicus Institute of Utrecht University and the German Öko-Institut. Use of this analytic framework, which is based on a combination of the FAO's partial-equilibrium economic model COSIMO and the QUICKSCAN model of the Copernicus Institute, provides a means of obtaining a more comprehensive assessment of the effects of energy crop use on macroeconomic variables, the incomes of individual households, price trends and food security. The methodology is currently being tested in Peru, Thailand and Tanzania (FAO, 2008e). Similar tools for comparing the costs and benefits of bioenergy use should be developed for residues. In both cases – for energy crops and for residues – the tools should be refined to deliver more robust results in future.

11.2.3
Social sustainability

The available case studies (Altenburg et al., 2008; van Eijck and Romijn, 2008; UNEP, 2007b) suggest that the economically sustainable and, in particular, socially sustainable development potential depends not only on biogeophysical conditions, the energy crop concerned and cultivation methods but also – especially where large-scale cultivation is concerned – on political and socio-economic conditions and on the opportunities that the local community has for participation.

The cultivation of energy crops (such as *Jatropha*) on marginal and degraded land in developing countries is at first sight an attractive option for reducing competition with food production. Often, however, this marginal land is already used (usually by sections of the population with no ownership rights, so that conflicts of use arise). Research is therefore needed to identify the measures and decision structures that will enable marginal or degraded land to be used for energy crop cultivation in such a way that the rural poor are not deprived of their means of livelihood or gain a stake in the value chain. These measures will involve both setting up incentive mechanisms and shaping institutional conditions (e.g. the division of responsibilities between different local and regional authorities, or land ownership issues).

11.3
Bioenergy and energy systems

Important conclusions can be drawn from the results of the assessment of different bioenergy pathways. From the point of view of climate change mitigation, the deployment of residues and waste, the use of combined heat and power (CHP) and the production of biomethane with storage of the separated CO_2 are attractive options that should be developed further. In addition, the strategic integration of bioenergy into future energy systems, particularly with regard to the stabilizing effect of biomass in electricity networks as balancing power, must be explored. Research can also help to make traditional biomass use significantly more efficient, particularly in developing countries.

11.3.1
Technologies of bioenergy use

BETTER INTEGRATION OF BIOENERGY INTO ENERGY SUPPLY STRUCTURES
As the share of fluctuating renewables such as wind and solar power rises, effective energy management to provide balancing power becomes ever more important. Bioenergy, like fossil energy, can be used when it is needed; this flexibility gives it a special position among the renewables. Ensuring the optimal integration of bioenergy systems into supply networks should therefore be a focus of future research efforts. A very promising approach is the development and dissemination of small biomethane-powered CHP units for households; such units can respond flexibly to the demand for power, and acting as stabilizing elements in a 'virtual power plant', they can operate highly dynamically to stabilize electricity supply. To enable bioenergy to fulfil its stabilizing role within the energy system in an optimal way, it is necessary to develop dynamic system analyses of the electricity system with various bioenergy variants to balance fluctuating feed-in. This will require forecasts of demand for bio-power and forecasts of the availability of bioenergy in each season of the year.

HIGHER CONVERSION EFFICIENCY THROUGH CHP APPLICATION OR POLYGENERATION
The highest fuel utilization rates are achieved with technical systems involving combined heat and power or combined heat/cooling, power and fuel (polygeneration). The simultaneous generation of power, heat and fuels by means of thermochemical or biochemical conversion processes is particularly promising and should be vigorously pursued.

A weakness in the present state of technology is the lack of small-capacity decentral CHP units for solid, woody biomass. In agriculture and forestry, in particular, large quantities of woody residues arise; in the past, it has been possible to use these in decentral applications only for heating. In future, however, the share of decentral facilities powered by renewables (e.g. wind turbines and photovoltaic systems) will rise, increasing the need for balancing power in the distribution networks. In addition to biogas systems, combustion plants combined with a steam power process (e.g. Organic Rankine Cycle plants) can be deployed as small CHP units to provide balancing power. So far, however, systems for converting biomass to electricity are not able to perform this regulating function. Targeted research programmes can help to overcome this.

INTEGRATING BIOMETHANE AS A UNIVERSAL, FLEXIBLE ENERGY CARRIER IN THE NATURAL GAS GRID
Biomethane can be deployed universally in all energy sectors to generate power and heat and can also be used as a transport fuel. It is a bioenergy carrier that is readily transportable; via the natural gas grid it can be easily distributed to decentral users or collected for use by central consumers such as combined-cycle power plants. Furthermore, on account of the generation processes used, biomethane has very low sulphur contents. A number of absorption and adsorption processes for processing biogas for small-scale, decentral plants are currently on the market.

There is a particular need for research into the specific energy consumption and the GHG emissions of biomethane processing. Much research is needed in connection with processes that have not yet become established for this type of application, including membrane processes and low-temperature processes. Processes of the latter type have the advantage that CO_2 that is both liquefied and very pure is produced as a 'waste product' of the technique, which simplifies the transport and further use of the CO_2. There is also a major need for research into technologies that will facilitate the secure storage of the CO_2 separated during processing.

The decentral systems must be integrated into existing gas networks and the gas quality must correspond to that of natural gas. Since the number of systems capable of feeding biomethane into the grid is rising rapidly, there is an urgent need for research into the integration of biomethane into gas networks. Research also needs to include the development of energy management approaches that model the full dynamics of gas, electricity and heat networks in a cross-sectoral way.

ADAPTING COMBUSTION AND GASIFICATION TECHNOLOGY TO BIOMASS

Biomass resources are highly heterogeneous and the combustion properties of biomass are not as homogeneous as those of fossil fuels. With this in mind, process technology must be developed in such a way as to ensure that pollutant emissions arising from biomass use are kept to a minimum.

Combustion and gasification processes must be developed so that a wide range of raw materials can be used without difficulty. Conversion of biomass into the secondary energy carrier synthesis gas enables the biomass to be used very flexibly, since fuels such as Fischer-Tropsch diesel, hydrogen and biomethane can be produced in addition to power and heat via polygeneration. Gasification processes for biomass starting materials of differing composition are therefore an important area for research.

FLEXIBLE BIOREFINERIES FOR ENERGY CARRIERS OR INDUSTRIAL FEEDSTOCKS

Crude oil, an important resource, will become a scarce commodity in the foreseeable future. Unlike energy services, which can be provided from other sources such as wind, hydro and solar energy, the chemical industry is dependent upon organic (carbon) feedstocks and will therefore need a substitute for crude oil (Section 5.3.4). Biorefineries that are currently being developed to produce biomethane or synthetic diesel should be designed in such a way that, once electromobility has become established, they can later provide feedstocks made from biomass for the chemical industry. Gas cleaning and processing should therefore be a focal area of research.

Like wastes and residues, algae are a form of biomass that does not compete with food. In contrast to energy crops, algae can be bred in closed systems with no condensation of water; they therefore need less water and can be produced even in arid regions. Initial experiments indicate that they have considerable potential for use as an energy resource and an industrial feedstock. This should therefore be a priority area for research too.

11.3.2
Potential for using residues and waste for energy

The technical potential of the use of bioenergy from residues and waste (including plant residues from agriculture and forestry, dung and organic waste) is in the region of 50 EJ per year (Chapter 6). However, it is still unclear what proportion of this techni-

cal potential can be utilized in a sustainable and cost-effective way. Dead wood and timber waste fulfil an important role in the ecology of natural or semi-natural forests. Plant residues are likewise highly important for the nutrient balance of the soil. Furthermore, for all sources of residues, it is unclear whether it is practical to use relatively small quantities of biomass that may often arise in spatially distributed locations. The importance of this issue in the context of energy has been underestimated and few studies have addressed it.

There is therefore a general need for research into the extent to which organic residues can be used sustainably and cost-effectively and in which manner this can be done. This includes ecological research into the significance of residues for biodiversity, soil conservation and climate change mitigation for forests and agricultural land located in various climate zones and used in various ways. It also covers research into the infrastructure required and into economically and technically appropriate ways of integrating this potential into global energy systems.

11.3.3
Modernizing traditional bioenergy use to overcome energy poverty

Bioenergy has major potential for reducing worldwide energy poverty, which affects 2.5 billion people, if the efficiency of traditional biomass use in developing countries can be significantly improved on a broad scale. This will also help to reduce GHG emissions. The technologies needed for this purpose (e.g. improved wood stoves, small biogas digesters based on residues) are available and many development measures in this area have already been implemented worldwide. However, the impact of these efforts remains limited. There has been no significant decrease in the number of 'energy poor' people, despite the fact that the technologies needed are relatively simple and inexpensive. Although evaluations of individual projects and programmes aimed to overcome energy poverty are available, there is a lack of reliable information that would explain why breakthroughs in this area have not been achieved yet. There is therefore a need for research to identify the cultural, institutional, property-right-related and economic factors that help or hinder improvement of the technical efficiency of traditional biomass use in developing countries through simple modernization of the technology involved.

11.3.4
Integrated technology development and assessment for bioenergy

So far, engineering research into the development of new bioenergy technologies has paid only little attention to socio-economic issues (economic cost-effectiveness of the new technology under various conditions, possible target users and their requirements, economic and social sustainability). Socio-economic conditions are, however, highly relevant factors influencing the successful adoption of new bioenergy technologies. In addition, an evaluation of technology by social scientists is needed in order to assess the impact of new bioenergy technologies on the competition between energy, food and fodder production and to analyse the role of the technology within an efficient, forward-looking energy mix. In current research and education in Germany, there is a clear lack of integrated technology development and assessment.

The funding activities of BMBF, such as 'BioEnergy 2021 – Research for the Use of Biomass', should therefore cover support for projects that provide for integrated development and assessment of new bioenergy technologies by engineers and social scientists. It would be important to launch research in which scientists link and integrate the approaches of the natural, engineering and social sciences from the very beginning. In addition, in interdisciplinary social science and engineering courses on bioenergy technologies, training should be specifically geared to the interface between technology development and assessment, as it already occurs in the master course in Bioenergy Technology at the Technical University of Lapeenranta in Finland (LUT, 2008) and the bachelor course in Renewable Resources and Bioenergy at Hohenheim University (Universität Hohenheim, 2008). Another pioneering course is the Concentrations in Environmental Sustainability (ConsEnSus) programme offered by the University of Michigan to engineering students. Under this programme, the engineers learn about legal and economic matters and about methods of assessing environmental impacts (University of Michigan, 2008). The Renewable and Appropriate Energy Laboratory (RAEL) of the University of Berkeley serves as an example of integrated research (UC Berkeley, 2008).

11.4
Bioenergy and global land-use management

11.4.1
Data on global land use and degradation

The principal source of information on current land use and cover changes is remote sensing data. But not all types of ground condition and land use can be identified from satellite data. For example, in many countries arable land is used post-harvest for grazing animals. Similarly, apparently unused fallow land may be used by the local community to pasture animals or collect fuelwood. On account of the food requirements of a growing world population and the increasing use of bioenergy, it is essential that land all over the world is used as efficiently as possible; the scientific basis must therefore be created for setting up a global GIS-supported land register enabling satellite data on land use to be aligned with and supplemented by local information.

Two factors essential to sustainable agricultural production are good water and soil management. Efficient water use requires a strategy that is adapted to local water availability and involves optimized production processes (e.g. re-use of (waste) water, efficient rainwater storage, etc.). This means that water resources must be mapped, local meteorological data must be made available and procedures must be in place for measuring and monitoring water quality by means of an indicator system.

At global level data on soil quality and soil degradation is outdated and not available for all countries. The only study that classifies worldwide soil degradation (GLASOD) dates from 1990–1992 (for the South Asia region: ASSOD study of 1995–1997). In addition to GLASOD there is the new initiative GLADA (Bai et al., 2008). While GLASOD is based on estimates and expert opinions, GLADA calculates soil degradation from net primary production (NPP) using remote sensing data. An urgent task in connection with implementation of a sustainable land-use strategy is therefore to measure soil types and soil condition on an up-to-date basis and by means of direct measurements. This should be done in cooperation with the FAO. Bearing in mind the shortage of available agricultural land it is particularly important to be aware of the condition and extent of degraded soils, since they have a great potential for sustainable production of biomass, which can be combined with soil improvement, carbon sequestration and ecological upgrading. In addition, research into marginal soils should be undertaken with the aim of identifying their extent, the reasons for their degradation, possible remedies and their potential yields.

11.4.2
Integrated scientific and economic land-use modelling

To identify the technically available, ecologically sustainable and economically viable bioenergy potential, improved integrated modelling of economic and scientific processes affecting land use is required. This gives rise to a variety of challenges, since existing scientific and (agro-)economic models are often incompatible in terms of their time scales, spatial resolution and subject matter. Nevertheless, it is essential to link the models in order to be able to properly depict feedback effects between the biological/physical and social systems. The task is therefore to pursue the linking of these models within a research consortium of natural and social scientists.

In addition, regional models should wherever possible be networked with each other so that interregional dependencies can be depicted. Further work on the dynamic aspects of the models should also be undertaken. The models should be based on standardized sets of scenarios, modelling assumptions and data sets since this will facilitate integration of different global and regional modellings and simplify comparison of modelling results. The expected conflicts of objectives relating to land use (e.g. between the provision of various agricultural products and various ecosystem services) could then also be assessed.

11.4.3
Agents and drivers

Existing information on the key agents and drivers of local, regional and global land-use changes must be specified in more detail, as must the factors that influence the decisions of individual stakeholders (microeconomic interests, social and cultural influences, political economics, social dynamics, etc.). As part of this process the role of differing lifestyles (e.g. consumption of animal products) and the effects of technology development (with regard to yield increases in resource production or greater efficiency in resource use) on land use should also be considered.

A necessary task is to identify the overall effect on land use of the sum of individual decisions. Methods such as Agent-based Modelling should be used for this purpose and developed further. Information on the effects of collective decision-making processes and social dynamics is an important requirement for the effective use of regulatory and control instruments.

Finally, information on the drivers and cause-and-effect relationships underlying land-use patterns should be used to manage land use in such a way that

WBGU's social and ecological guard rails (Chapter 3) are adhered to. In order to 'steer' land use it is necessary to identify appropriate institutions and mechanisms that can be used to coordinate individual decisions so that the desired land-use patterns are achieved. Related issues are the need for research into the steering effect of standards and certification systems (for agricultural and forestry products) and environmental economic instruments (such as payments for environmental services or emission trading schemes) on land use; the conditions under which these instruments have the greatest steering impact should also be identified.

11.4.4
Linkages between energy crop cultivation and food security

The causes of the rises in food prices that have recently been observed must be explored further. In particular, the role of the demand for bioenergy in these price rises needs to be clarified. Previous assessments of the influence of the demand for biofuels on food prices have arrived at very different conclusions. These variations can be attributed to the use of different models and assumptions; the various estimates cannot necessarily be compared directly with each other. For decision-makers this is a very awkward situation: which estimates should they rely on in, for example, in assessing whether biofuels should be subsidized? The first step should be to create transparency and to compare the various models and assumptions in metastudies. This will enable the relevance of the different models and assumptions to different contexts to be assessed; forecasts can then be drawn up for the future development of the agricultural markets and the future significance of biofuels and/or of bioenergy in general for food prices and food availability. In this way the necessary regulatory measures for biofuels and/or bioenergy can be identified.

11.4.5
Effects of changes in dietary patterns and lifestyles on climate and land use

Available data indicate that dietary habits (e.g. the amount of meat and milk products in the diet) have a major influence on the agricultural greenhouse gas balance and on the amount of land needed for food. This relationship should be explored in more detail in relation to different cultures and regions and in the light of observed trends over time (e.g. rising meat consumption in newly industrializing countries);

possible improvements should be identified (e.g. promotion of deliberate life-style choices as a win-win situation involving health benefits, an improved greenhouse gas balance and less need for land). Complete life-cycle assessments ('from farm to fork') have in the past been available for only a few food products. Studies should explore consumer behaviour with regard to the purchase of food and bioenergy with and without labels declaring their provenance or origin, covering aspects such as demand and price structure and consumers' need for information.

Indicators such as the 'ecological footprint per person per year' depict the impact of human consumption habits on the Earth's surface and the biosphere. Particularly significant is the demand for food products that are land-intensive in their production (meat and dairy products). This intensifies competition for land and resource scarcity. A shift by society as a whole towards lifestyles and consumption patterns compatible with sustainable land use would reduce the pressure of use on the biosphere. Further research is needed into appropriate instruments for changing dietary patterns in this direction. Such research should cover both 'soft' measures (education, awareness, labelling requirements) and the effects of more incisive instruments (e.g. subsidies, taxes on consumption). Research into the social implications and (ethical) limits of state influence on consumption patterns should also be intensified. To enable these issues to be assessed and discussed, work on a robust data basis and a reliable method of calculating individual and national land consumption must continue.

11.5
Shaping international bioenergy policy

11.5.1
Managing global land use

It is likely that competition for land will intensify on a global scale. Meeting the resulting challenges requires reliable information on the development of global land-use patterns in the light not only of climate trends but also of socio-economic and political developments. Interdisciplinary research on global and regional land-use models and scenarios should therefore be intensified (Section 11.4.2).

At present, global land use is determined to a significant extent by the market and by national policies. In view of the increasing competition for land, WBGU recommends that options for influencing land use should be explored in greater depth. On the one hand, research should cover the indirect impacts

on the allocation of land that arise from numerous international agreements and instruments. On the other, there is a need to consider whether local and national mechanisms and instruments of spatial planning can and should be developed further and transferred to the global level. International communication, reporting and monitoring commitments are at present important policy instruments. There are also some instances of intergovernmental land-use agreements, such as the international commitment to establish protected areas for nature conservation. A key task for research in this area is to identify the possible role of land-use agreements in other land-use sectors and to clarify how consistent overall coordination of sector-specific land-use agreements can be achieved. This will involve evaluating the effectiveness, efficiency, practicability and enforceability of international spatial planning instruments, assessing their economic distribution effects and analysing their impact on the structures of international division of labour. This research needs in addition to take account of development and geostrategic aspects.

11.5.2
Standard-setting and the WTO regime

FRAMEWORK FOR STANDARD-SETTING AND SUSTAINABILITY CRITERIA
From a legal perspective there is a need to identify the international law regime into which sustainability criteria relating to the production and use of bioenergy could best be incorporated. To clarify this matter, a comprehensive survey of the advantages and disadvantages of a range of possible procedures should be undertaken. The possibility of a separate multilateral agreement should be considered, as should the option of attaching the new provisions to an existing multilateral agreement (in particular the Convention on Biological Diversity or the United Nations Framework Convention on Climate Change; the Desertification Convention would be more difficult). In this context it is also relevant to enquire after the most appropriate methods for formulating sustainability criteria for bioenergy in whatever multilateral framework is selected. Development of a multilateral 'master agreement' relating to sustainability standards for bioenergy would make it possible to explore the potential for corresponding rules and regulations, the way in which these rules could be embedded in the existing body of law and their compatibility with the WTO regime.

CLARIFICATION AND SPECIFICATION OF THE WTO
REGIME

The WTO undergoes constant development (ongoing negotiating round; work of the dispute settlement bodies). The outcomes of negotiations should be monitored from political and legal perspectives with the aim of continuously analysing the impact of these outcomes on the realization of sustainable bioenergy production and promptly identifying any associated need for action at the various levels of legal regulation (international, supranational, national). Particular consideration should be given to the development of minimum requirements for recognizing bioenergy carriers as 'environmental goods'.

11.5.3
Bioenergy policy and security policy

Energy supply security has become a key subject of international policy. Bioenergy is an element of national, regional and international energy supply security. In future, competition for access to biomass will therefore increase, along the same lines as competition for access to oil and gas. The dynamics of the development of bioenergy management and the emerging patterns of international division of labour will therefore have geopolitical consequences and could give rise to new tensions in international politics. Security policy research should take account of these factors in its analysis of resource conflict risks.

11.5.4
Developing commitments under the UNFCCC and CBD

FUNGIBILITY OF EMISSION RIGHTS RELATING TO
SECTORAL TARGETS

For a post-2012 regime under the United Nations Framework Convention on Climate Change (UNFCCC) WBGU proposes sectoral sub-regimes with separate targets for emissions reductions – in other words, targets for the (global) land-use (LULUCF) sector and for the (global) non-LULUCF sector (WBGU, 2003). There could be a degree of fungibility of emission rights under the two sub-regimes; this would increase efficiency and promote liquidity in the global carbon market, thereby improving its functioning without compromising the ecological integrity of the overall commitment regime (Section 10.2). The form this fungibility should take needs to be explored: to what extent should the fungibility of emission rights be limited, and what ceiling should be imposed? To what incentive and distributional effects do alternative forms give rise? And

what are the consequences for political and administrative implementability?

COORDINATING THE FINANCING OF AVOIDED
DEFORESTATION AND OF PROTECTED AREA
NETWORKS

If attempts to avoid emissions from deforestation (REDD) in the context of the UNFCCC are to be successful, reliable agreements should be struck on the payments to be made to the affected countries (Section 10.2). The conservation of sinks and carbon stocks can have positive consequences for biodiversity conservation, and in particular for protected areas, and vice versa. Also required are rules of a more binding nature governing the financing of a global protected area network under the Convention on Biological Diversity (CBD) in order to ensure that adequate funds are mobilized for this purpose (Section 10.5). This opens up a range of research questions relating to the synergies between commitments under the two regimes. Research is needed, firstly, to define these synergies and, secondly, to explore options for regulating the way in which payments are credited between the two regimes (double payment or partial crediting of payments to commitments under both regimes, exclusive crediting of payments to commitments under one regime). The incentive and distributional effects of the individual criteria for assessing synergy effects and regulation options should be examined in detail. The political implementability and practicability of the various options should also be assessed, e.g. the development of a protected area protocol to the CBD and possible links between this and the emerging REDD regime under the UNFCCC (Section 10.5).

11.5.5
Methods of supporting decision-making under uncertainty

In the light of the fact that the bioenergy debate is beset by knowledge gaps, there is scope for exploring the decision-making methods on which individual stakeholders base their decisions, especially when decisions are taken in the face of uncertainty and considerable risk. New methods of supporting the process of decision-making under uncertain conditions should be developed for the different stakeholder groups (e.g. farmers and forestry managers, regional planners, political decision-makers), in order to help them make scientifically based land-use decisions in the future. There should be further development of computer-based decision support systems that can be used to simulate, contrast and systematically compare different scenarios.

The use of biomass as a renewable energy resource opens up opportunities but also involves risks. On the one hand, it raises hopes that dependence on imported oil and gas can be reduced, or that biofuels may mitigate the CO_2 emissions of road transport. On the other, it gives rise to new fears – for example, the possibility that the cultivation of energy crops may increase the likelihood of land-use conflicts arising from competition between food production, nature conservation and bioenergy. The issue is a dynamic and highly complex one, attended by much scientific uncertainty and involving a multiplicity of interests. For these reasons it has not been possible in the past to undertake an integrated assessment of the opportunities and risks of bioenergy in the light of the contribution it can make to sustainable development. WBGU sets out to show that bioenergy can be used in sustainable ways that exploit many of the opportunities while keeping many of the risks at a low level.

WBGU's core message is that potential for bioenergy exists in all parts of the world and should be tapped in any case. Bioenergy must, however, be used under conditions that ensure that sustainability is not jeopardized; in particular, risks to food security and to climate change mitigation and nature conservation targets must be precluded. Bioenergy strategies need to be differentiated for individual countries, taking account of their socio-economic and agro-ecological circumstances. The guiding principle governing the choice of policy must in WBGU's view be the strategic role of bioenergy as an element of the global shift towards sustainable energy systems. WBGU's guiding vision therefore incorporates two objectives:

- *Firstly,* bioenergy should contribute to climate change mitigation by substituting fossil energy carriers and thus helping to reduce greenhouse gas emissions in the world energy system. The fact that bioenergy carriers can be stored and used as control energy can play a strategically important part in stabilizing the electricity supply in the energy systems of industrialized, newly industrializing and developing countries when these systems involve a large proportion of wind and solar

energy. In the long term, bioenergy combined with carbon capture can even remove some of the emitted CO_2 from the atmosphere.
- *Secondly,* bioenergy use can help overcome energy poverty. The initial challenge here is to replace traditional forms of bioenergy use in developing countries that are harmful to health. Modernization of traditional bioenergy use can play an important part in reducing poverty, preventing health risks and reducing use-related pressures on natural ecosystems. Around 2500 million people currently have no access to safe and affordable forms of energy (e.g. electricity, gas) to meet their basic needs. Modern yet simple and cost-effective forms of bioenergy can play an important part in significantly reducing energy poverty in developing and newly industrializing countries.

The guiding vision makes clear that bioenergy is a cross-cutting issue. Bioenergy policy encompasses not only energy, agricultural and climate policy; transport policy and foreign trade policy as well as environmental, development and security policy all play an important role in this emerging policy field. The complex dynamic of the markets is another relevant factor. Energy and agricultural markets are becoming increasingly interlinked via bioenergy, and energy markets in particular are strongly influenced by countries' strategic interests. In other words, there are complex political issues that must be resolved and that transcend the boundaries of established policy arenas, requiring cooperation between actors who, in the past, have shared little common ground. This poses major challenges for the governance and integration capacities of the policy-making process, which is largely structured along sectoral lines.

However, bioenergy policy also transcends the framework of an international system that is based on nation states. For example, a blending quota for biofuels in Europe can contribute to an increase in deforestation in other parts of the world. Bioenergy is thus an example of a complex global issue in which the actions of government and non-government actors at national or local level may have unintended consequences of a transregional or even global nature. Bioenergy policy therefore requires a multi-level policy approach.

The following specific recommendations for action are grouped into six blocks. The inherent overlap between the blocks highlights the need for a multi-level policy. The starting point is the consistent inclusion of bioenergy in climate change mitigation policy, the first of the guiding vision's two objectives. Section 12.1 therefore deals with incentives and commitments under the UN climate protection regime. Measures in this area can, however, contribute simultaneously to the second of the guiding vision's objectives, the overcoming of energy poverty. But because the climate protection regime has little immediate impact and is unable to ensure conformity with other dimensions of sustainability (such as food security or the conservation of biological diversity), steps must be taken at the same time to set up standards and certification systems for bioenergy, initially at bi- and multi-lateral level. Section 12.2 contains a number of recommendations in this area. A key problem of bioenergy use is that the cultivation of energy crops can give rise to competition for land. Standards alone cannot solve this problem. There is therefore a need for more detailed accompanying measures to secure global food production, biological diversity and soil and water protection at all action levels; these measures are described in Section 12.3. The technical sustainable bioenergy potential, which WGBU estimates to be 84–166 EJ per year in the year 2050, should be tapped.

Promotion policy is an important instrument of state influence in this field. Its task must be to correct false incentives and promote developments that are in line with the guiding vision. WBGU's recommendations on this issue are presented in Section 12.4. Section 12.5 deals with the objective of overcoming energy poverty. In this area WBGU identifies both major action shortfalls and equally large development potentials. At present there is neither an international organization nor a convention with specific responsibility for bioenergy. Instead, a large number of private forums, UN activities and intergovernmental processes have arisen in recent years. Proposals for tackling this fragmented institutional situation round off the chapter in Section 12.6.

12.1
Making bioenergy a consistent part of international climate policy

REFORM ACCOUNTING MODALITIES FOR CO_2 EMISSIONS FROM BIOENERGY

The existing provisions in the United Nations Framework Convention on Climate Change (UNFCCC) and the Kyoto Protocol create false incentives for bioenergy production and use, and may even promote bioenergy use that is harmful to the climate (Section 10.2). In WBGU's view it is essential, therefore, that the UNFCCC bodies address the issue of bioenergy. The aim should be to reform the modalities for determining contributions to commitments under the Kyoto Protocol and its successor regime and adapt the allocation of emissions to greenhouse gas inventories in a way that avoids false incentives and does not distort the contribution made by bioenergy to climate change mitigation. WBGU favours a scheme that includes the following elements:

Firstly, the use of bioenergy must no longer be counted en bloc as free of CO_2 emissions ('zero emissions') in the energy sector. However, WBGU is not advocating replacing the presumed zero emissions with cumulated emissions from a life-cycle analysis of the bioenergy, since this would not be compatible with the other allocation modalities within the UNFCCC and would lead to double-counting (Section 10.2.2). Instead, within the energy sector, the actual CO_2 emissions arising from the combustion of the biomass should be counted and included. In return, the uptake of CO_2 from the atmosphere by energy crops in the land-use sector should also be counted. This correction would align the way in which bioenergy is treated with the principle used elsewhere of always allocating emissions to the place and time of their creation.

Secondly, the existing rules, under which only selected CO_2 emissions and absorptions from land use and land-use change are or can be set against the commitments made by states, should be replaced by full accounting of all emissions from these sectors. Ideally, this new arrangement should form part of a wider agreement on conservation of the carbon stocks of terrestrial ecosystems within the UNFCCC.

Thirdly, there need to be supplementary regulations governing trade between countries that have and countries that do not have binding commitments to limit emissions (Section 10.2.2.1).

DIFFERENTIATE CONSIDERATION OF BIOENERGY IN THE CDM

The Clean Development Mechanism (CDM) currently has only a limited influence on overall bioenergy use in newly industrializing and developing countries. Any expansion of CDM projects that include the cultivation of energy crops should be viewed with scepticism unless it can be ensured that the use of land for this purpose will not give rise to displacement effects (leakage) and result in terrestrially stored carbon being released elsewhere. Potential new rules governing the land use, land-use change and forestry sector (LULUCF) in the framework of a post-2012 regime are particularly important for bioenergy in the CDM context, and adaptations of the CDM will therefore be required. Leakage and the permanence problem (i.e. ensuring that climate change mitigation efforts have a lasting effect) in the CDM framework can be limited to some extent by lump-sum deductions on certified emissions reduction credits. Overall, however, these problems seem to be so important for ecological integrity and so difficult to control that the CDM should not be given a key role in reducing emissions in the LULUCF sector.

By contrast, CDM projects that aim to improve or replace inefficient traditional forms of biomass use should be utilized in order to further the diffusion of efficient technologies in poorer developing countries and rural regions. Current moves in the UNFCCC to permit simplified methods for small-scale CDM projects that facilitate the transition to sustainable biomass energy use – e.g. through highly efficient biomass stoves – are a step in the right direction. At the same time, however, the integrity of the CDM must be safeguarded; this means that only real and permanent emissions reductions from biomass use should be rewarded.

REDUCE EMISSIONS FROM DEFORESTATION IN DEVELOPING COUNTRIES

Since the current expansion of energy crop cultivation could result in an increase in tropical deforestation (Section 5.5), an effective regime for reducing the emissions from deforestation and forest degradation (REDD) within the UNFCCC framework is extremely important. An appropriate REDD regime should create effective incentives for the swift achievement of real emissions reductions by reducing deforestation. These emissions reductions should be effected at national level in order to avoid leakage. Beyond the direct reduction of emissions, incentives should also be created to permanently protect the natural carbon reservoirs, such as tropical primary forests, from deforestation and degradation, and to limit emissions from grassland conversion. The regime must mobilize adequate international financial transfers to establish effective incentives. In order to fulfil these requirements, the regime should consist of a combination of national targets to limit emissions and project-based procedures (Section 10.2.2). The REDD regime would ideally form part of a comprehensive agreement within the UNFCCC to preserve the carbon stocks of terrestrial ecosystems.

MOVE TOWARDS A COMPREHENSIVE AGREEMENT ON THE CONSERVATION OF TERRESTRIAL CARBON RESERVOIRS

CO_2 emissions arising from the LULUCF sector should be fully and systematically included in the post-2012 regime in order to ensure that the incentive given to bioenergy use by the UNFCCC is based on the actual contribution to climate change mitigation that it makes. However, the absorption and release of CO_2 by the biosphere differs from the emissions of fossil energy sources in a number of fundamental respects, including measurability, reversibility, long-term controllability and interannual fluctuations (Section 5.5).

Since the different sectors also have very different characteristics in terms of time-related dynamics and amenability to planning, it would seem more appropriate – from the point of view of remaining within the 2°C guard rail – to define separate reduction targets rather than one overarching target. WBGU therefore recommends that in future, rather than setting a national upper limit for all emissions, separate commitments should be envisaged for LULUCF. It proposes negotiating a comprehensive separate agreement on the conservation of the carbon stocks of terrestrial ecosystems. This agreement should (i) take up the debate on REDD, (ii) replace the existing rules on offsetting reduction commitments in the sectors listed in Annex A to the Kyoto Protocol against sinks (including CDM activities) and (iii) fully include all CO_2 emissions from the land use, land-use change and forestry sector (LULUCF). Despite separate target agreements, WBGU considers it appropriate from the point of view of economic efficiency to aim for a certain level of fungibility; however, on account of measurement problems and other uncertainties attaching to LULUCF emissions, this fungibility should be clearly demarcated and linked to deductions.

EUROPEAN CLIMATE POLICY: SET INCENTIVES FOR EMISSIONS REDUCTIONS IN THE LAND-USE SECTOR

The Annex I countries' commitments to reduce their GHG emissions under the Kyoto Protocol already cover emissions from agriculture and selected emissions from forestry (Section 10.2.2), but EU policymakers have yet to translate these commitments systematically into appropriate incentives for farmers and forest managers. In future this should be done. Within the framework of the EU's common agricultural policy (CAP), there is scope available in the context of direct payments (Pillar 1) (with binding cross-compliance commitments) and the promotion of rural development (Pillar 2) (e.g. agri-environmental measures). As also recommended by SRU

(2008), the German government should utilize the forthcoming CAP review in 2008 and 2009 to initiate comprehensive ecological reform of the CAP. If direct payments continue, cross-compliance should be systematically expanded to include climate protection aspects. Improving the focus of agri-environmental measures should be a further objective (SRU, 2008), and measures should also be geared towards mitigation of/adaptation to climate change.

12.2
Introducing standards and certification for bioenergy and sustainable land use

INTRODUCE A UNILATERAL MINIMUM STANDARD FOR BIOENERGY CARRIERS

WBGU recommends, as a first step, introducing a statutory minimum standard for bioenergy at EU level. The minimum standard should apply to all types of bioenergy carriers used for energy generation in the EU. Bioenergy carriers that do not satisfy the desired minimum standard should be excluded from the market. WBGU's proposal thus goes much further than the EU's current plans to make compliance with a sustainability standard a prerequisite for the promotion for liquid biofuels. For the minimum standard for bioenergy, WBGU recommends that the sustainability criteria for liquid biofuels set out in the EU directive on the promotion of energy from renewable sources be expanded to include the methodologies for the calculation of indirect land-use changes and specific criteria for restricting the use of genetically modified organisms (GMOs). Certain basic core labour standards of the International Labour Organization (ILO) should also be made mandatory as part of the minimum standard. With regard to greenhouse gas emissions, WBGU recommends a specific absolute emissions reduction in relation to the quantity of raw biomass consumed, rather than a relative emissions reduction based on the final energy or useful energy. The use of bioenergy carriers should reduce life-cycle greenhouse gas emissions by at least 30 t CO_2eq per TJ of raw biomass consumed in comparison with fossil fuels (Box 12.2-1).

Promotion of bioenergy should only take place if additional promotion criteria are fulfilled which help to reduce energy poverty or increase climate change mitigation or biodiversity or soil protection. As a general precondition for promotion, the use of the bioenergy should result in a life-cycle greenhouse gas reduction of at least 60 t CO_2eq per TJ of raw biomass used, as compared with fossil fuels. The use of biogenic wastes and residues in particular and the cultivation of energy crops on marginal land are particularly worth promoting if this achieves demonstra-

ble advantages for climate change mitigation, as well as for soil, water and biodiversity conservation, and if such cultivation also rates positively in terms of social criteria.

ESTABLISH CERTIFICATION SCHEMES FOR SUSTAINABLE BIOENERGY CARRIERS

If minimum standards for bioenergy carriers are introduced on a binding basis, appropriate certification systems must be created promptly in order to demonstrate compliance with these standards. Ideally, these should be internationally applicable certification schemes for all types of biomass for energy use as well as bioenergy end products. This makes it easier for the bioenergy standards to be extended at a later stage to all types of biomass. The International Sustainability and Carbon Certification (ISCC) scheme, which was developed for the German Federal Ministry of Food, Agriculture and Consumer Protection (BMELV) specifically for the certification of bioenergy carriers, is a first step in this direction. This, or a comparable scheme, should be put in place as soon as possible.

As the introduction of a general biomass standard can only be achieved in the medium term, the duty to furnish proof that the standards have been adhered to should lie in the first instance with the seller of the end product. This would require suppliers, in turn, to furnish proof that the production of their feedstocks also complies with the standards. This would not, at least initially, give rise to a duty to certify feedstocks that can be used for purposes other than conversion to energy, such as rapeseed, soya or palm oil. To ensure that feedstocks and raw materials can be certified, one option is to introduce a 'land-use standard' which imposes an upper limit on GHG emissions from the cultivation of feedstocks, including direct and indirect land-use changes. The introduction of this type of land-use standard for bioenergy feedstocks can prepare the way for later expansion to all types of biomass.

Whereas the creation of certification schemes and the certification process itself can be left to market actors, monitoring of compliance with the standards after certification is best performed by the state in order to enhance the legitimacy of the standards and prevent abuse. To this end, institutions must be created by the state to monitor implementation of the standards and impose sanctions in the event of non-compliance or inappropriate issuance or possession of certificates.

Box 12.2-1

WBGU's minimum standards for bioenergy production

- *Reducing greenhouse gas emissions by using bioenergy carriers:* The use of bioenergy carriers, taking account of direct and indirect land-use change, should reduce life-cycle greenhouse gas emissions by at least 30 t CO_2eq per TJ of raw biomass used in comparison with fossil fuels. For the cultivation of biomass feedstocks, the greenhouse gas emissions generated by direct and indirect land-use changes from a specific reference date onwards, including the loss of the sink effect, should not exceed the amount of CO_2 that can be re-sequestered on the same site by the energy crop within ten years (land-use standard).
- *Avoiding indirect land-use change:* This means avoiding the displacement of productive forms of land use by the cultivation of energy crops.
- *Preserving protected areas, natural ecosystems and areas of high conservation value:* No cultivation of energy crops in identified exclusion zones (protected areas or areas of high conservation value); establishment of adequate buffer zones; embedding of energy crop cultiva-

tion systems in the landscape; ecological risk assessment prior to any use of potentially invasive alien species.
- *Maintaining soil quality:* No long-term impairment of soil functions or soil fertility as a result of biomass cultivation for energy use; observance of good agricultural practice; where residues are removed, maintenance of nutrient cycles.
- *Ensuring the sustainability of logging by-product use:* Use of logging by-products must proceed in a manner that maintains nutrient cycles and preserves the biological diversity of forest ecosystems.
- *Managing water resources sustainably:* No significant impairment of water quality and the hydrological regime; no overuse of groundwater resources.
- *Controlling the effects of genetically modified organisms (GMOs):* Introgression from genetically modified plants into wild plants must be prevented; contamination of, or inputs into, the food and animal feed chain must be ruled out; and GMO use must comply with national and international biosafety standards.
- *Observing basic social standards:* Compliance with a number of ILO's core labour standards, particularly standards for adequate protection of health and safety at work.

SUPPORT DEVELOPING AND NEWLY INDUSTRIALIZING COUNTRIES IN ACHIEVING COMPLIANCE WITH THE MINIMUM STANDARDS

In order to achieve compliance with the guard rails and sustainability criteria, as called for by WBGU, the developing and newly industrializing countries, including the least developed countries (LDCs), must be required to meet minimum standards in the production of bioenergy carriers. It must be borne in mind, however, that compliance with sustainability criteria and the requisite certification procedures may impose major burdens on these countries. For that reason, the LDCs in particular should receive financial and technical support with the establishment of national certification schemes and supervisory bodies, as well as with the certification process in individual production facilities. Furthermore, during an initial phase, simplified terms for the verification of certification criteria could also be offered. Group certification is a good way of keeping down certification costs for agricultural smallholdings in the developing and the industrialized countries alike.

UTILIZE BILATERAL STANDARD-SETTING AS AN EFFECTIVE INSTRUMENT

Until a globally agreed land-use standard is created, the anchoring of bioenergy standards in bilateral agreements remains an effective instrument for increasing sustainability in the production of bioenergy carriers. The bilateral agreements concluded to date between bioenergy producer and importing countries have so far been limited in the extent to

which they have included environmental and social standards. An example is the agreement signed between Germany and Brazil in May 2008 on comprehensive cooperation in the field of renewable energies and energy efficiency. WBGU recommends that similar agreements concluded in future should contain specific and binding sustainability criteria for a minimum standard for the production of bioenergy. In return, the trade partners should be granted free market access for bioenergy carriers if they comply with the minimum standard. Wherever possible, existing bilateral treaties should be amended to that effect.

PURSUE MULTILATERAL APPROACHES TO THE GLOBALIZATION OF STANDARDS

The development and institutionalization of bioenergy standards in a multilateral framework is the preferred option in terms of WTO rules and should be pursued. A further argument in favour of this course of action is that if the EU acts alone, large parts of the world market for bioenergy carriers will remain unregulated. The German government should work to ensure that an international consensus on a minimum standard for sustainable bioenergy production and on a comprehensive international bioenergy standard is achieved as swiftly as possible. From WBGU's perspective, a promising approach is to develop existing international initiatives, especially the Global Bioenergy Partnership (GBEP), within the G8+5 negotiations. The GBEP is an important body as it could accelerate the formulation of multilateral policies. With political support from the G8, the decisions could be introduced into relevant political forums, institutions and processes, thus ensuring their implementation.

ACHIEVE WTO COMPLIANCE OF STANDARDS FOR BIOENERGY CARRIERS

The minimum standard, as called for by WBGU, and the promotion criteria for biomass production require bioenergy production to be compatible with environmental and social standards. If the EU were to introduce unilateral standards, producers from non-EU countries would also have to comply with these standards if they wished to export their products to the EU or access state aid for feedstock cultivation.

The question of whether environmental and social standards are compatible with WTO law has still to be clarified. The World Trade Organization (WTO) conformity of unilateral European bioenergy standards can be justified in law, particularly with regard to criteria for the reduction of greenhouse gas emissions and the conservation of global biodiversity, because the desirability of protecting climate and biodiversity is laid down in multilateral environmental agreements in international law (UNFCCC, CBD). Moreover, all the political options should be utilized in order to increase the acceptance of established environmental standards and the ILO's core labour standards in the existing WTO treaty regime and safeguard them in the long term. At present, major obstacles to the full recognition of environmental and social standards can be identified, notably in the opposition of certain groups of states, especially the majority of developing countries. It is therefore essential, within the framework of current and future WTO negotiations, to focus substantially on building trust in the relationship between the industrialized and the developing countries.

GEAR THE LIBERALIZATION OF TRADE IN ENVIRONMENTAL GOODS AND SERVICES TO ENVIRONMENTAL GOALS

Within the framework of the current WTO Round, one issue being considered is the abolition of tariffs and other trade barriers for goods and services which member states regard as particularly environmentally friendly. However, the intended liberalization of trade in relation to these 'environmental goods and services' (EGS) must not run counter to the goal of sustainable production and use of such goods and services. If the trade in bioenergy carriers undergoes generalized liberalization within the WTO framework, on the grounds that they qualify as environmental goods, without any consideration of the need for sustainability in the production pro-

cess, this would clearly conflict with efforts to influence the sustainability of production in the countries of origin through the adoption of a minimum standard. The German government should therefore work within the relevant negotiations to classify as EGS only those goods that satisfy the minimum standard called for by WBGU and come from selected bioenergy pathways.

EXTENSION OF THE MINIMUM BIOENERGY STANDARD TO ALL TYPES OF BIOMASS

There is a need in the medium term for a global land-use standard to regulate the production of all types of biomass for a wide range of uses (food and feed, use for energy and use as an industrial feedstock, etc.) across national borders and cross-sectorally. The EU member states should therefore prepare suitable provisions for extending the minimum standard to all types of biomass. At multilateral level, the establishment of a new Global Commission For Sustainable Land Use should be considered, whose task would be to elaborate the principles, mechanisms and guidelines required for global land-use management (Section 12.6).

12.3
Regulating competition between uses sustainably

Agriculture in the rural regions of developing countries can play an important part in preventing food crises but has for many years been a neglected area of development cooperation; this situation has contributed to the worldwide rise in food prices. WBGU therefore recommends that agricultural productivity should be raised through increased investment, targeted particularly at rural development, sustainable small-scale farming, and plant breeding. This can only succeed if decision-makers in the developing countries likewise accord a high priority to the development of rural regions. Multi-lateral development cooperation, which over the past decade has paid too little attention to rural development, should support a radical change of approach in this area.

12.3.1
Developing an integrated bioenergy and food security strategy

Over and above the measures specified by the departmental working party on world food affairs in its report to the German Cabinet (Box 10.4-3), WBGU recommends including the cultivation of energy crops in an integrated bioenergy and food security strategy in which food security has precedence. The integrated strategy must also cover sustainable water use and soil protection. This is particularly important for the low-income food-deficit countries (LIFDCs).

Energy crop cultivation must go hand in hand with increases in food production and in the productivity of land used for this purpose. Rural development and the agricultural sector have for many years been a neglected area in international cooperation; this has been a significant factor in the worldwide rise in food prices. This must change. A controlled expansion of bioenergy makes sense only if it is accompanied by worldwide efforts to strengthen agriculture.

Following the dramatic food price rises of 2008, there has recently been increased effort at international level to develop coordinated strategies for avoiding future food crises. In this discussion there is widespread consensus on the key strategy elements. What is required is, firstly, measures that will immediately and directly improve the food situation in the affected regions. Secondly, measures must at once be put in place to improve conditions for food security and food production in the long term; these measures must be consistently incorporated into other policy areas such as climate change mitigation and nature conservation. The World Agricultural Report commissioned by the World Bank and the FAO (IAASTD, 2008; Box 10.4-2) rightly calls for changes in agricultural policy in order to better reflect the complexity of agricultural systems in their differing social and ecological contexts. In WBGU's view, energy crop cultivation has a major part to play in this, particularly in strategies and projects involving off-grid, rural energy supply. Cultivation should, however, be promoted mainly on marginal or degraded land (Section 7.4). Finally, WBGU holds that the decision on where energy crops can be sustainably grown can only be made on a regional and context-specific basis and with the involvement of all the relevant actors.

In order to be better prepared for future food crises, an effective early warning system is needed. Existing monitoring capacities, e.g. those of the FAO and the World Food Programme (WFP) need to be improved and better networked (Ressortarbeitsgruppe Welternährungslage, 2008). In addition, WBGU sees an increasing need for identification of risks to food security arising from competition with energy crop cultivation for the use of resources. In order that the risks arising from the increasing pressure on global land use can be promptly recognized, global monitoring and early warning systems are also very important.

12.3.2
Taking greater account of the coupling of land use, food markets and energy markets

The challenges of securing the world food supply must today be dealt with against the backdrop of increasing pressure on global land use; they can no longer be addressed through national endeavours alone. In a globalized world, national and international policy-making must take much greater account of the ever closer links between land use, the development of agricultural prices and the energy markets. Measures in one sector must always be considered from the point of view of the (international) repercussions they may have on the other sectors. The task for the medium term is to design and set up regulatory mechanisms for reducing conflicts, for example if developments in the energy markets result in undesirable developments for food security.

TAKE ACCOUNT OF FOOD SECURITY IN THE LIBERALIZATION OF AGRICULTURAL TRADE

In the long term it is important for food security that signals from the world agricultural markets prompt production increases specifically in the poorer developing countries. To this end, import barriers for agricultural goods – particularly food – should be lowered further and export subsidies and other production promotion measures that do not comply with sustainability criteria should be reduced. This should happen worldwide, but particularly in the industrialized countries. In all probability prices on the world agricultural markets would then rise and in the medium to long term food production in developing countries could grow on account of the comparative production advantages that these countries enjoy. However, poorer developing countries, as net importers of agricultural commodities and food (LIFDCs), are often directly and adversely affected by such price rises on the world markets. Short-term international financial support and compensation measures would be needed to cushion these negative impacts. The smallholder sector in developing countries may be similarly affected if liberalization without further subsidies forces small farmers to compete with cheap agricultural imports. For the smallholder sector in poorer developing countries there should therefore be exemptions enabling these countries to levy (higher) import tariffs on certain agricultural commodities and pursue other promotion measures as a means of safeguarding long-term national strategies for regional development and sustainable food production. These regulations could be included as permissible Green Box subsidies in the WTO's Agriculture Agreement, or could be contained in a 'Development Box'.

Funding issues also arise in connection with emergency aid programmes – that is, the short-term direct provision of food to countries experiencing a crisis. The central institution here is the World Food Programme (WFP), which has a budget of US$ 2800 million. Following the global crisis in 2008 the Programme has announced that it needs an additional US$ 750 million. To ensure adequate long-term funding of the WFP, it may be appropriate to set up independent sources of funding for it. In view of the growing global competition between different forms of land use, WBGU recommends considering the possibility of international financing based on the polluter pays principle; this would involve a levy on forms of land use other than nature conservation and food production. For example, a contribution formula could be agreed in the donor community to take account not only of the donor state's ability to pay but also of the use it makes of biomass for conversion to energy.

CONSIDER MEASURES FOR REDUCING FOOD PRICE VOLATILITY

The closer the correlation between energy prices and food prices, the more volatile food prices are likely to be (Section 5.2). Frequent sharp downward price fluctuations increase investment uncertainty for agricultural producers and have a particularly detrimental impact on the small-scale farming sector; producers in this sector are unlikely to have reserves and have either wholly inadequate security against price risks or none at all. Sharp upward price fluctuations, on the other hand, jeopardize net food consumers with low incomes. There is a need to consider whether an internationally coordinated expansion of food reserves could represent a viable means of increasing supply on the world market in the event of a sudden significant price rise, or of generating demand if prices fall dramatically. Innovative measures for reducing volatility are highly desirable.

12.3.3
Taking greater account of increasing pressure on land use caused by changing dietary habits

The greatly increased pressure on land use as a result of changed dietary habits, namely the land-intensive dietary patterns in industrialized countries and the spread of these patterns to dynamically growing newly industrializing countries, increases global competition for land use. This is a challenge for the future that is still widely underestimated and deserves fuller attention. The spread of western dietary patterns with their high consumption of meat and milk products

also limits the potential for energy crop cultivation (Section 5.2). In LIFDCs, particularly in Africa, it is already possible to observe agricultural land being bought up on a large scale by foreign investors in order to meet the growing demand from other countries for animal feed.

The influence of dietary habits should not be underestimated: by 2030 they may account for around 30 per cent of the required increase in food production (Section 5.2). There is still far too little aware of these links – that is, of the importance of eating habits for land use and food security – and the facts therefore need to be brought home to consumers through education campaigns. The primary aim of such campaigns is to raise awareness of the issue, particularly in the industrialized countries, and to motivate people to change their behaviour. Initiatives should be instigated at international level, e.g. through the United Nations organizations, with the same aim.

These initiatives should be supported by international cooperation on land take for the per-capita consumption of food. At issue is the amount of land used at home and abroad for production of the food consumed within a particular country. Sustainability concepts such as that of the ecological footprint can highlight the fact that in global terms natural resources are currently being used at a rate that exceeds their regeneration limits. Although biomass use for food production cannot be considered separately from other biomass uses, and although issues relating to the methodology and operationalization of these analytical concepts still remain to be clarified, it is nevertheless incumbent upon industrialized countries that have a high per-capita consumption of land-intensive food to take the initiative. The initial aim should be to identify a precise basis for data relating to food-related land take. Countries could then undertake to reveal their per-capita land take for food. The next stage should be to develop strategies within the international community for reducing the per-capita land take. One aspect of this should be the very high secondary consumption of food (post-harvest losses, discards). These strategies could help to reduce the per-capita consumption of foods that are particularly land-intensive. In the long term countries could give a voluntary undertaking to reduce their per-capita land take. The rapidly growing newly industrializing countries should also be included in this collaborative endeavour.

12.3.4
Implementing biodiversity policy for sustainable energy crop cultivation

From the perspective of biosphere conservation the increasing use of bioenergy worldwide poses two key challenges to biodiversity policy. Firstly, within the global system of protected areas additional protected areas must be designated and the management of existing areas must be improved and made more effective. The purpose of this is to limit conversion of natural ecosystems triggered directly or indirectly by bioenergy use (Section 10.5.1). It is essential that this is underpinned by adequate financing of the protected area system from national and international sources (Section 10.5.2). Secondly, when energy crops are grown or forests are used as a source of bioenergy, the conservation and sustainable use of biological diversity must be ensured through appropriate bioenergy standards (Section 10.5.3). The Convention on Biological Diversity (CBD) can play a part in this through the development of biodiversity guidelines.

GET THE LIFEWEB INITIATIVE OFF THE GROUND
The LifeWeb initiative, which was introduced at Germany's instigation at the COP-9 of the CBD (BMU, 2008e), aims to promote implementation and financing of the CBD's protected areas programme through bilateral partnerships. The initiative, and in particular the German government's decision to make significant additional funds available to it, is very much to be welcomed. The challenge now is to persuade other donor countries to support LifeWeb. At the same time, concrete bilateral projects should be speedily progressed, using the funds that have already been approved. The success of this initiative will not only contribute to the 2010 target of the CBD but will also function as a nucleus for further development of the CBD's provisions relating to protected areas.

STABILIZE AGREEMENTS ON BIODIVERSITY CONSERVATION THROUGH COMPENSATION PAYMENTS
A strategy for ensuring globally sustainable land use should provide for worldwide expansion of international compensation payments. Without compensation for lost income from agriculture and forestry, developing countries in particular will be unable or unwilling to forego the opportunities for economic growth generated in these sectors, even if these opportunities are only of a short- to medium-term nature. This is particularly the case at times of rising agricultural and bioenergy prices. With this in mind, pilot projects should explore whether national habitat banking systems in industrialized countries can

be made accessible to providers of ecosystem services from developing countries.

MOBILIZE MORE FUNDS FOR PROTECTED AREAS IN THE FINANCING MECHANISMS
In view of the key importance of financing for the designation of new protected areas and the improved management of existing ones, there is a general need to make additional funds available. The intention here should be to mobilize € 20–30 per person per year in high-income countries in order to close the present funding gap (Section 10.5). Transition countries, newly industrializing countries and rich resource countries should be involved more strongly in the financing process as time goes on. The foundation should be laid now for a market-like mechanism in which assurances that previously certified land will be protected are traded for a compensation payment.

OPTIMIZE INTERACTION OF THE CBD AND UNFCCC
In view of the priority given to climate policy issues in global environment policy forums and the many interfaces between climate protection and biodiversity conservation, cooperation between the different political processes must take place in such a way that biodiversity conservation is promoted or at least not hindered. The substantive contribution to the UNFCCC process agreed at CBD COP-9 represents only a starting point. Better networking of the scientific bodies of the UNFCCC and CBD should also be established. At national level there should be better intra- and inter-ministry coordination of the responsible government bodies. Plans for using financing mechanisms related to climate change mitigation, such as proceeds from the auctioning of CO_2 certificates, or deploying the financing provided for avoided deforestation (REDD process, Box 10.2-2) to foster a global protected area network should be examined for suitability.

EXAMINE FURTHER DEVELOPMENT OF CBD PROVISIONS WITH A VIEW TO A PROTECTED AREA PROTOCOL
Evaluation of the 2010 target of the CDB and of implementation of the protected area programme will be undertaken at COP-10 in Japan. If previous trends continue, this target is likely to be missed. Whether this is so depends in part on the success of the initiatives agreed in Bonn. Recent studies highlight clearly the economic importance of conserving biological diversity. The REDD process in the UNFCCC, which is running in parallel, will underline the valuable contribution that conserving endangered natural ecosystems makes to climate change mitigation. Another

important component of development of the CBD is the setting up of a scientific advisory body for biological diversity, similar to the IPCC. The regular provision of reports by this body would raise awareness of the underlying issues. Under the changed conditions that may ensue, it could in the long term be possible to expand the CBD provisions to include a mandatory commitment linking the implementation of measures relating to protected areas to financing instruments. The content and political feasibility of a protected area protocol of this type and possibilities of linking it to the emerging REDD regime of the UNFCCC should be considered as an option now (Section 11.5.4).

PROMOTE DEVELOPMENT OF BIODIVERSITY
GUIDELINES FOR SUSTAINABILITY STANDARDS IN
THE CONTEXT OF THE CBD

In the light of the outcomes of COP-9, it cannot be assumed that the CBD will make rapid progress in the area of bioenergy. Nevertheless, the process of drawing up biodiversity guidelines under the CBD as a contribution towards sustainability standards should be promoted and where possible accelerated during the German CBD presidency. As an important element of this, development of the World Database on Protected Areas (UNEP-WCMC, 2008) should be promoted at the same time, in order to establish the necessary monitoring capacities. In the medium term the impetus for sustainability standards in the bioenergy sector should be used to drive forward the creation of general guidelines for all forms of biomass production.

12.3.5
Improving long-term water and soil protection through energy crop cultivation

The Comprehensive Assessment of Water Management in Agriculture (IWMI, 2007), the SIWI study (Lundqvist et al., 2008) and the GLASOD study (Oldeman et al., 1991; Oldeman, 1992) highlight the fact that present trends of water and soil use are pointing in the wrong direction (Section 5.6). Without a change of policy this will lead in many regions to more serious water crises and increased soil degradation.

MAKE PRIOR ANALYSIS OF REGIONAL WATER AND
SOIL AVAILABILITY A REQUIREMENT

In many regions, water and soil are precarious resources. Before bioenergy cultivation is promoted on a large scale, an integrated analysis of water and soil availability in the region should be carried out. Inappropriate cultivation systems (Section 7.1) and

an increasing hunger for energy worldwide can greatly increase the pressure of use on soil and water resources. This is not yet an acute issue on a global scale, but promotion of non-sustainable cultivation systems may cause it to become a problem in critical regions. In any case, energy crop cultivation should not result in a region being placed under intolerable water stress or in soil degradation being accelerated. In regions that are already experiencing significant water stress or soil degradation, energy crop cultivation should not exacerbate these adverse environmental impacts. Energy crop cultivation should therefore be integrated into a regional strategy for sustainable soil and water management.

USE ENERGY CROP CULTIVATION TO RESTORE
MARGINAL LAND

If the right cultivation system is used, growing energy crops on marginal and degraded land can actually help to improve the soil. In the long term it is likely that only the growing of energy crops on marginal and degraded land over several decades could form a strategic option because, in some cases at least, energy crop cultivation could in the long term make restored land available for food production. This could go some way towards reducing the growing pressure on land use.

EMBED BIOENERGY USE IN THE COMBATING OF
DESERTIFICATION

The United Nations Convention to Combat Desertification (UNCCD) provides a platform for supporting, programmatically and conceptually, sustainable and poverty-oriented land-use in countries experiencing droughts and desertification. It does this primarily through the instrument of National Action Programmes (NAPs) to combat desertification (Section 10.6). The use of standards for sustainable soil use, specifically including energy crop cultivation, could be promoted within the context of NAPs. In addition, the 10-year Strategic Plan of the UNCCD, adopted in 2007, provides numerous opportunities for promoting awareness-raising, bioenergy assessment, standard-setting in relation to combating desertification, and the development of policy for sustainable bioenergy use in general. In concrete terms, a sustainable bioenergy strategy should be developed and implemented within the context of the NAPs and steps should be taken to ensure that these NAPs are integrated into the overarching national development strategies (in most cases the PRSPs, Poverty Reduction Strategy Papers).

12.4
Targeting bioenergy promotion policies

The various bioenergy pathways have very different environmental impacts and contribute in varying degrees to climate change mitigation and biosphere conservation (Chapter 5). In the worst case they run counter to the German government's climate change mitigation and biosphere conservation targets. It follows that not all types of bioenergy are worthy of promotion. However, WBGU has identified some bioenergy cultivation systems and conversion pathways that make a sustainable contribution to mitigating climate change, overcoming energy poverty and securing the energy supply; these should therefore be promoted.

WBGU recommends that the use of non-sustainable energy carriers that do not meet its proposed minimum standard (Section 10.3.1) should cease entirely. Minimum standards can initially be agreed in the form of unilateral standards at EU level. Attempts should be made to encourage the adoption of these standards in other countries and at international level through bilateral and multilateral agreements (Section 12.3.1).

The direct promotion of bioenergy pathways must also be contingent upon compliance with more stringent criteria (Section 10.3.1.2). Promotion should only be accorded to those bioenergy pathways that contribute to climate change mitigation in a particularly sustainable way. WBGU defines such pathways as ones that not only meet the minimum standard but also permit a saving of at least 60 t CO_2eq per TJ of biomass consumed, taking total life-cycle emissions into account. However, since for practical reasons promotion needs to target different stages of the production process (cultivation, conversion and application systems), it is usually necessary, when targeting a specific stage, to make standardized assumptions about the other stages involved.

With regard to the promotion of energy crop cultivation, in particular, WBGU regards compliance with additional ecological and social criteria as essential. To conserve soil fertility, ecological limits should likewise be set on the use of biogenic residues. The promotion of conversion and application systems should be designed to be compatible with the overall vision of a shift towards sustainable energy systems. Undesirable lock-in effects should be avoided and forward-looking technologies such as electromobility promoted.

Sustainable energy systems involve not only attention to climate change mitigation but also the reduction of energy poverty. Modernization of distributed and traditional uses of bioenergy can make an important contribution to this latter aim, particularly in the rural areas of developing countries. In this context WBGU regards the promotion of bioenergy-based projects as justified even if the promotion criteria are not fully met.

12.4.1
Reforming agricultural subsidies

RESTRICT INCENTIVES FOR GROWING ENERGY CROPS TO EXCEPTIONAL CASES
Measures that specifically promote energy crop cultivation are often relatively inefficient and ineffective. They also tend to intensify competition for land use. Promotion measures should therefore always be examined for counterproductive effects. The existing promotion of non-sustainable biomass production that does not meet WBGU's minimum standards should be withdrawn. The promotion of sustainable biomass production for energy purposes can be recommended if the land use contributes to nature and soil conservation services or if it forms part of a project aimed at reducing energy poverty (promotion criteria for biomass cultivation). In such situations energy, agricultural and environmental policies need to be coordinated interdepartmentally.

TIGHTEN THE FOCUS OF SUBSIDIES FOR GROWING ENERGY CROPS
Production subsidies in the agricultural sector, especially in industrialized countries, distort the global agricultural market, create inefficient competition for subsidies between producer countries and impede the development of many poorer countries. This includes subsidies for energy crop cultivation, which have the additional disadvantage of intensifying competition with food production for land use. WBGU therefore recommends that policy-makers should work towards the widespread abolition of production subsidies in the agricultural sector. Nevertheless, certain subsidies that have major environmental and development benefits should be not only excluded from this abolition but actively supported by the international community. The subsidies in question are, firstly, those targeted at the sustainable cultivation of energy crops in developing countries in the context of projects that directly serve to reduce energy poverty and, secondly, subsidies for the cultivation of perennial energy crops on degraded and marginal land. WBGU therefore recommends that such subsidies should be explicitly classified as permissible subsidies for WTO purposes.

12.4.2
Advancing energy recovery from biogenic wastes and residues

The sustainable use of biogenic wastes and residues largely avoids the competition between different uses that energy crops provoke; this enables a consistently high climate change mitigation effect to be achieved. WBGU therefore recommends that the use of wastes and residues be given clear priority over the use of energy crops. Wastes and residues achieve their highest climate change mitigation effect in electricity generation, especially when they displace coal. Greater incentives need to be put in place so that residues are used for conversion to energy. The easiest way to mobilize the untapped potentials of biogenic wastes and residues is through targeted and differentiated promotion of renewable energies for generating electricity and heat, while ensuring that incentives for the use of biogenic wastes and residues are sufficiently large in relation to the promotion of other biogenic secondary energy carriers. At the same time regulations must if necessary be put in place to ensure that the removal of residues from forests and farmland meets ecological sustainability criteria. Additional incentives for mobilizing wastes and residues can be created through rules on the landfilling of waste and support for collection and transport infrastructures. Promotion of the use of wastes and residues as an energy resource must bear in mind the possibility of competition with the use of biogenic resources as an industrial feedstock; ideally, pathways involving cascade use should be promoted.

12.4.3
Reorienting technology policy

PHASE OUT PROMOTION OF LIQUID BIOFUELS AND PROMOTE ELECTROMOBILITY
For reasons of sustainability, WBGU takes the view that the promotion of liquid biofuels cannot be justified, particularly in industrialized countries. The arguments against promoting liquid biofuels include the high costs of abating greenhouse gas emissions, the low to negative abatement achieved per unit of land used, and the risk of lock-in effects which perpetuate an inefficient transport infrastructure based on the combustion engine. The expansion of biofuel production can at most be justified under certain conditions in some developing and newly industrializing countries, provided that the fuel is for regional consumption and is produced only for a transitional period; here too, though, subsidies must be evaluated critically from the point of view of efficiency and environmental and social considerations.

WBGU therefore recommends that the promotion of biofuels be rapidly phased out, at least in industrialized countries. In particular, blending quotas should be frozen with immediate effect and within the next few years the current quotas for blending biofuels with fossil fuels should be withdrawn completely.

In the transport sector bioenergy is most efficient when it is used to generate electricity to power electric vehicles (this is at least twice as efficient as using bioenergy in the form of biofuels in conventional combustion engines). Furthermore, the use of electric vehicles significantly reduces fine particle emissions and noise, and electric vehicle technology can be used to sequester CO_2 emissions in the transport sector. State promotion policies can help businesses develop energy storage and electric vehicle technology and expand opportunities for plugging in to the electricity grid. On the demand side, state promotion should in the main take the form of monetary incentives (e.g. taxes on motor vehicles and fuels) so that the process of technological innovation is not unnecessarily constrained (Section 10.7.6).

PROMOTE SELECTED BIOENERGY PATHWAYS FOR ELECTRICITY GENERATION
The greatest climate change mitigation effect is achieved when coal is substituted by bioenergy carriers for electricity generation. A range of conversion pathways – such as co-combustion in coal-fired power plants or large-scale CHP plants, the use of biogas from fermentation and raw gas from gasification in small-scale CHP units, or the use of biomethane in small-scale CHP units or combined-cycle power plants – all deliver roughly comparable reductions in greenhouse gas emissions. However, where biomethane is used a further enhanced climate change mitigation effect can be achieved if the CO_2, which must in any case be separated during the production process, can be securely stored. The conversion of biomass to electricity has the additional advantage that – unlike liquid biofuels – it facilitates the shift to electromobility for the transport sector.

WBGU therefore recommends that in countries in which coal plays an appreciable part in power generation, electricity from biogenic energy carriers should be promoted if a greenhouse gas reduction of at least 60 t CO_2 per TJ of raw biomass consumed (by comparison with the fossil reference system) can be achieved, and if policy conditions are such that it is likely that fossil energy carriers with high greenhouse gas emissions will be replaced. For the bioenergy pathways listed above this value is usually achieved if biogenic wastes and residues are used, whereas when energy crops are used it is not achieved for all pathways.

Feed-in tariffs are suitable promotion instruments, but other instruments including direct subsidies are also a viable approach. However, WBGU advises against quotas: quotas would generate an enforced demand for bioenergy which, in view of the significant competition for land associated with the cultivation of energy crops, could give rise to highly undesirable side effects – which the use of solar or wind energy would not do.

For reasons relating to the systems involved, the use of biomethane is particularly attractive. Not only is biomethane associated with relatively easy CO_2 sequestration: it can be collected and distributed via the infrastructure of natural gas grids that already exists in many industrialized countries and converted to electricity highly efficiently in small-scale CHP units or combined-cycle power plants near where it is needed. WBGU therefore recommends specific additional promotion of biomethane pathways, if the CO_2 arising during manufacture is safely stored and the biomethane is used for conversion to electricity.

In tandem with this, expansion of the gas grid should be moved forwards and grid operators should be obligated to set up pump stations to enable decentral biomethane to be fed into grids with higher pressure levels.

PROMOTE CHP AND COMBINED-CYCLE POWER PLANTS

If bioenergy is to be used efficiently in the energy system, the spread and use of technically efficient conversion systems – especially CHP and combined-cycle power generation – is particularly important. The competitiveness of these power plant technologies depends crucially on the framework of energy and climate policy. It is therefore important that allowances allocated under emissions trading systems are not issued free of charge; free allocations such as those under the European Emission Trading Scheme (ETS) should be phased out resolutely. In recognition of the high efficiency of CHP, in the ETS free allowances should continue to be issued or reductions in allowance obligations should be provided for. In sectors that are not covered by emissions trading, partial tax exemptions should be continued. In countries without an emissions trading system or without CO_2 emissions taxes, investment or output subsidies can be made contingent upon the utilization of efficient system technology.

PROMOTE HEATING OF BUILDINGS BY BIOMASS ONLY AS A TRANSITIONAL MEASURE

The heating of buildings is most efficient when it is combined with electricity generation via CHP or, as a longer term prospect, when it is provided by heat pumps operated by renewable electricity. On account of the extensive infrastructure investment that both local and central district heat networks require, it seems appropriate to deploy boilers fired by wood, wood chips and pellets as a transitional approach, especially in rural areas. However, resistance to change is strong, particularly among private households, and is encountered even in relation to technically efficient and economically worthwhile combustion technologies. In view of these perseverance tendencies, it may be expedient for regulatory sanctions to be supplemented by promotion through cut-price loans or investment bonuses for householders who switch to efficient biomass heating systems. If stringent conditions have not yet been imposed on fine particle emissions, these should be introduced and implemented in tandem with promotion.

ENGAGE IN STRATEGIC DEVELOPMENT OF THE USE OF BIOMASS AS AN INDUSTRIAL FEEDSTOCK

In order to draw up a strategy for the use of biomass from agriculture and forestry as an industrial feedstock, a material flow analysis should be carried out. The analysis should identify the present situation at national and global level (production, use, recycling and disposal, also including imports and exports) and inventorize the global land areas involved. A range of scenarios should then portray likely developments and options for action (the likelihood of increased competition between use for energy and use as an industrial feedstock and between the agricultural and forestry resource base; prospects for replacing petroleum-based products such as plastics, lubricants and bitumen; using long-lived wood products to bind carbon; cascade use of biogenically produced products that are finally used as an energy resource). For key material and product categories such as cellulose and paper products, sustainability standards for the cultivation and acquisition of the raw materials – analogous to those used for bioenergy – should be established and product standards with high recycling quotas should be developed. The present high consumption of resources and products should be drastically reduced by putting appropriate measures in place.

INITIATE AN INTERNATIONAL AGREEMENT ON (BIO) ENERGY SUBSIDIES

The subsidizing of bioenergy as a form of renewable energy should be seen in the context of other energy subsidies, particularly those for fossil energies. Since a shift towards sustainable energy systems is required, national policies on the production and use of energy should be revised, with sustainability criteria being accorded greater weight. This applies to the promotion policies of both industrialized and newly industrializing countries. In WBGU's view the requisite policy change can most readily be implemented without significant competition effects if countries jointly address the reduction of non-sustainable energy subsidies and coordinate their actions at international level. At the same time they should agree on principles for the admissibility of energy subsidies linked to sustainability criteria. This could take place in the context of negotiations on a Multilateral Energy Subsidies Agreement (MESA). Such an agreement, which would initially be plurilateral in nature and should cover at least the key energy-producing and energy-consuming countries, could in the long term be integrated within the WTO body of rules.

12.5
Harnessing sustainable bioenergy potential in developing and newly industrializing countries

Traditional biomass use forms the basis for cooking and heating for some 2500 million people worldwide. The use of low-cost, simple technologies can significantly increase the efficiency of traditional biomass use and improve the health and economic situation of very many people. In the context of international development cooperation many programmes have already been undertaken, but the problem is still far from being comprehensively solved. In addition, many developing countries are hoping to profit from growing energy crops. From its analysis of the situation WBGU derives recommendations for the deployment of bioenergy. These recommendations are intended to help mobilize the opportunities associated with bioenergy in developing countries identified in the report while at the same time minimizing the not insignificant risks, especially those of energy crop cultivation.

MAKE OVERCOMING ENERGY POVERTY A PRIORITY OF DEVELOPMENT POLICY
Access to energy is a crucial aspect of development. Local production and use of bioenergy therefore opens up major opportunities, particularly for overcoming poverty and for rural development. It therefore seems appropriate for the combating of energy poverty to be explicitly included in the Millennium Development Goals, as the UN Millennium Project (2005) proposes. This does not require a new Millennium Development Goal: access to energy should be included as a means for overcoming poverty and improving living conditions. In addition, WBGU recommends that complete abandonment of traditional forms of bioenergy use that are harmful to health by 2030 should be an international target (WBGU, 2004a). If this is to be achieved, overcoming energy poverty needs to be more strongly anchored in the energy policy portfolios of stakeholders involved in

international development cooperation. As a first step, overcoming energy poverty should be systematically included in Poverty Reduction Strategy Papers (PRSPs). The international community should particularly promote bioenergy projects that advance off-grid rural energy supply in developing countries. In this context WBGU welcomes the new Africa-EU Energy Partnership, which has set itself the task of improving electrification and capacity development in Africa (EU Commission, 2008c).

BASE STRATEGIES ON MULTI-COUNTRY CROSS-CUTTING EVALUATIONS AND LOCAL STUDIES
A development-policy strategy for overcoming energy poverty must start by considering alternative ways of providing energy services; obstacles need to be understood and overcome. The priorities of development cooperation should then be revised accordingly. Some programmes – such as those of the Asian Development Bank and of GTZ – are already adopting this approach. The development of appropriate strategies should be based on multi-country cross-cutting evaluations, which can identify the elements of best-practice approaches. However, in connection with the scaling-up of new technologies that could be based on biomass it is essential to carry out large-scale national, regional and local studies of the opportunities and obstacles. So far the only studies available are single-case analyses with a very narrow focus. Once this broader data basis has been created, stakeholders involved in international development cooperation should work with national institutions and decision-makers and affected groups to increase awareness and acceptance in this area, incorporate these factors into appropriate political strategies and actively pursue the replacement of traditional bioenergy use at local level. The wider use of improved wood and charcoal stoves is only a first step in this direction.

SUPPORT DEVELOPING COUNTRIES IN DRAWING UP NATIONAL BIOENERGY STRATEGIES
So that the opportunities and development potentials of bioenergy can be realistically assessed and any risks can be minimized, WBGU recommends that key strategic issues be discussed in the country context and with as broad a range of stakeholder groups and affected sections of the community as possible. This discussion should aim to clarify the principal goals that the promotion of bioenergy is intended to achieve. A national bioenergy strategy should examine all the forms in which bioenergy can be deployed and applied and evaluate their suitability in the context of the local situation. Since bioenergy represents only one of a number of options for achieving the required targets, other alternatives must also be

considered. To this end, developing and newly industrializing countries should draw up country-specific strategies that facilitate systematic examination of the targets, assessment of agro-ecological and socio-economic conditions and evaluation of institutional capacities (decision aid: Figure 10.8-1), in order to ensure sustainable cultivation. Development cooperation actors should support partner countries in developing these strategies and encourage compliance with minimum standards and promotion criteria. In order to achieve this, the necessary governance capacities needed to develop sustainable bioenergy use should be strengthened, particularly in areas such as land-use planning and certification. This will help to ensure that regions that have sustainable potential for energy crop cultivation, such as Latin America and sub-Saharan Africa, can realize their potential on a sustainable basis.

DOVETAIL FOOD SECURITY AND BIOENERGY STRATEGIES

Energy crop cultivation involves major risks – particularly for developing countries affected by a high level of food insecurity. A significant expansion of bioenergy policy on climate and energy-related grounds can only be justified if it is also linked to a food security strategy for the countries concerned.

TAKE ACCOUNT OF STANDARDS WHEN MAKING PROGRAMME- AND PROJECT-RELATED LOANS

In their programmes and projects multilateral development banks and international financing institutions have a lever for promoting sustainable development in their partner countries. In the context of energy crops this lever should be used by ensuring that minimum standards for bioenergy carriers are taken into account as a favourable factor when projects and loans are awarded, or that there is provision for project components that encourage compliance with WBGU's recommended minimum standards.

PROMOTE PILOT PROJECTS THAT INVOLVE PARTICULARLY SUSTAINABLE CULTIVATION SYSTEMS

While compliance with minimum standards should be a basic requirement for the production of bioenergy carriers, WBGU recommends that, particularly in the context of development cooperation, cultivation methods that are particularly sustainable and that help to combat erosion, conserve biodiversity, reduce (energy) poverty and advance rural development should be promoted in pilot projects (promotion criteria: Section 10.3). Pilot projects should also cover certifiable cultivation systems for the transregional market or for export. Such cultivation systems include, in particular, the socially acceptable

cultivation of multi-year energy crops on degraded land, or forest farming practices. This enables experience to be gained in new agricultural techniques and best practices to be identified and established. Pilot projects should form part of an overall strategy for improving agricultural practice; they should involve local participants in a joint decision-making process.

PROMOTE USE OF WASTES AND RESIDUES, PARTICULARLY IN ELECTRICITY GENERATION

In countries in which agriculture plays a major part, considerable quantities of residues arise (e.g. from the fishing industry, sawmills, and tea, coffee and sugar plantations); these can be converted to energy. Agro-industrial biogas and cogeneration facilities are particularly suitable for utilizing residues; ideally, the waste heat produced by electricity generation should be used in the manufacturing process that the plant supports (e.g. for drying products). WBGU recommends that these potentials should be examined on a country-specific basis when national bioenergy strategies are drawn up, and incorporated appropriately into these strategies. Technical and financial cooperation can provide valuable support in this area. Pilot projects should support the mobilization and use of residues and wastes and identify best practices.

IMPROVE ACCESS TO ENERGY SERVICES AND SUPPORT TRANSFORMATION OF ENERGY SYSTEMS

A range of instruments should be used to ensure across-the-board provision of energy services for the world's population and support developing countries in the transformation of energy systems. WBGU recommends that public and private-sector elements should be combined. On the demand side, the purchasing power of people affected by energy poverty needs to be improved. Stakeholders involved in development cooperation should therefore extend their financial support of microfinancing systems. Microfinancing can help improve energy provision both for private households and for small and micro-businesses, particularly in rural areas (WBGU, 2004a). Microfinancing systems can also be used to finance the technologies for replacing traditional biofuel use that WBGU classes as worth promoting. In order to generate additional funds and know-how, private capital should also be mobilized. To this end, cooperation between the private and public sectors should be promoted. There are also examples of successful cooperation in connection with the cultivation of energy crops such as *Jatropha* (Altenburg et al., 2008). Making projects that aim to improve the efficiency of traditional uses of bioenergy eligible as small-scale CDM activities is justifiable and can contribute to financing. In addition, greater use could be made of CDM projects for the large-scale substitution of fossil fuels (Section 10.2.3).

FORGE BIOENERGY PARTNERSHIPS
Multilateral cooperation in connection with sustainable bioenergy production and use can be supplemented by specific inter-country partnerships. In some cases these partnerships resemble earlier technology agreements or partnerships, such as those that already exist between industrialized and newly industrializing countries for purposes of technology transfer and joint technology development and expansion in other energy sectors (Gupta et al, 2007; Philibert, 2004). For example, technology-oriented partnerships between industrialized and newly industrializing countries can be considered in connection with the scaling-up of technologies for biomethane processing and use. Alternatively, pure technology partnerships in the bioenergy field can be linked to aspects of sustainable land-use policy. The same applies to trade partnerships: countries that supply bioenergy carriers and those that wish to buy them could enter into bilateral agreements under which the production of bioenergy carriers is linked to criteria for regionally sustainable land use. Supplier countries would undertake to ensure that production meets sustainability standards. In return, demand-side countries would provide technology or know-how relating to cultivation systems. The sustainably produced bioenergy carriers from the partner countries should be granted unhindered access to the market. In a positive move, the German government plans to enter into agreements that grant import concessions to developing countries that guarantee sustainable cultivation (BMU, 2008e).

PROMOTE RESTRUCTURING OF THE WORLD ENERGY SYSTEM
WBGU shows how bioenergy can be sustainably used in the energy systems of developing countries, taking account of differing goals and structures. For example, better use can be made of bioenergy in small-scale applications – for local energy supply, at household and village level and in small businesses. At the same time bioenergy can also make an important contribution to modernization of the energy sector in urban and industrial contexts. The recommended technologies, which range from improved stoves to micro biogas systems, small-scale CHP units, CHP, cogeneration plants, electromobility, etc. not only help to overcome energy poverty but also promote climate change mitigation. Restructuring of the world energy system should therefore also be coordinated and supported by the international community (Section 12.6; WBGU, 2004a).

In the context of development cooperation the World Bank Group, as an influential and financially powerful stakeholder, should draw up a comprehensive strategy detailing how it can contribute to restructuring of the world energy system along sustainable lines. WBGU recommends that the German government should place itself at the head of this process at European level and in the supervisory bodies of the relevant international organizations, in order to make full use of its pioneering role in climate change mitigation. A first step could be to convene an International Conference on Sustainable Bioenergy, in order to maintain and increase international awareness of the links between bioenergy and sustainable development (Section 10.7.7.2).

12.6
Building structures for sustainable global bioenergy policy

PROMOTE BIOENERGY THROUGH THE
INTERNATIONAL RENEWABLE ENERGY AGENCY
The necessary transformation of global energy systems requires coordinated action at global level and hence the streamlining of international institutions and stakeholders. WBGU therefore welcomes the founding of an International Renewable Energy Agency (IRENA) whose aim is to further the worldwide use of renewables through policy advice, technology transfer and promotion of competency acquisition and knowledge transfer. However, if the ultimate aim is to be a global reconfiguration of energy systems, promotion strategy must cover not only renewables but all aspects of energy systems as well as energy demand and development policy considerations. IRENA should address issues of energy, the environment and development in an integrated manner. The energy development pathways (Chapter 8) and sustainable bioenergy promotion principles (Section 10.7) that have been described provide effective guidance for bioenergy-related advisory and promotional activities.

SET UP A GLOBAL LAND-USE REGISTER
An important requirement for the monitoring of direct and indirect land-use changes in connection with the introduction of a standard and the required certification systems is that a global, GIS-supported land-use register should be set up. The register needs to incorporate high-resolution satellite data and land-use information. As an important element of this, rapid development of the UNEP-WCMC World Database of Protected Areas is recommended; this will make it possible to identify whether or not a particular product has been produced within a protected

area. The global land-use register must, however, go beyond this; for each imported bioenergy carrier it must be able to provide information on the land on which it was produced (geographical coordinates, type of cultivation, voluntary commitment to comply with all criteria, etc.). This information would enable plausibility checks to be carried out with the standardized assistance of the land-use register. The information provided would, however, need to be supplemented by random on-the-spot checks and evaluation of satellite data.

PURSUE MULTILATERAL APPROACHES TO THE GLOBALIZATION OF STANDARDS

From the point of view of trade rules, the option of developing and institutionalizing bioenergy standards in a multilateral framework is the preferred way forward and the one that should be pursued. The German government should take steps to ensure that international consensus on a minimum standard for sustainable bioenergy production and on a comprehensive international bioenergy strategy is achieved as soon as possible. In WBGU's view a promising approach would involve the expansion of existing international initiatives, particularly the Global Bioenergy Partnership (GBEP) in the context of the G8+5 negotiations. The GBEP is an important vehicle for accelerating multilateral policy formulation. It brings together key stakeholders and involves newly industrializing countries (especially Brazil and South Africa) whose interests are significantly affected. However, steps should be taken to achieve greater involvement of civil-society stakeholders than is currently the case, so that the discussion includes all affected parties. GBEP or the Task Force on Sustainability, as an intergovernmental forum, should be supported in streamlining the diverse formal and informal processes for drawing up global sustainability criteria and in working toward the development of global standards and guidelines. The proposals of the Roundtable on Sustainable Biofuels (RSB) could form a basis for this process of standard-setting with close links to political decision-makers; these proposals should be incorporated into the work of the Task Force. Under the political leadership of the G8 it could be possible to incorporate the decisions into policy-related forums, institutions and processes and to ensure that they are implemented.

SET UP A GLOBAL COMMISSION FOR SUSTAINABLE LAND USE

In view of the land-use management challenge, WBGU considers that the issues surrounding global land use should be addressed more vigorously at international level and institutionalized accordingly. A new Global Commission For Sustainable Land Use should be set up to organize this international 'search process'. The commission's task should be to identify the key challenges associated with global land-use issues and to bring together the most up-to-date knowledge available. The commission should then elaborate principles, mechanisms and guidelines for global land-use management. The new commission, whose responsibilities would extend far beyond narrowly-defined issues of land use, could be located within UNEP and work closely with other UN organizations such as FAO. The findings of the process should be regularly placed on the international agenda, for example in the context of the UNEP Global Ministerial Environment Forum. Even more importantly, the issues should be presented and discussed at the strategically important gatherings of heads of state and government (G8+5).

CONVENE AN INTERNATIONAL CONFERENCE ON SUSTAINABLE BIOENERGY

Sustainable bioenergy strategies can be better implemented in an increasingly globalized world if there is a shared international understanding of the opportunities and risks and consensus on appropriate norms governing the production and use of the different forms of bioenergy. Such understanding and consensus can only emerge from international cooperation in formulating goals and principles. Stakeholders from the agricultural, energy and development policy arenas must therefore be brought together. This could be achieved through an International Conference on Sustainable Bioenergy, which could be modelled along the lines of the renewables 2004 conference. It could be used to exchange ideas on best practice, formulate the general principles of promotion and conclude agreements on international bioenergy cooperation, such as pledges to promote bioenergy crop cultivation on degraded land or agreements on bioenergy partnerships. WBGU recommends that the German government supports steps to convene such as conference in the near future.

References

Abengoa (2006) *Greencell Abengoa Bioenergy: Description of 'Thermal Integration' and 'Maximum Production' Concepts on EF Ethanol Production. Internal Document.* Abengoa, Sevilla.

ABN (African Biodiversity Network) (2007) *Agrofuels in Africa – The Impacts on Land, Food and Forests. Case Studies from Benin, Tanzania, Uganda and Zambia.* ABN, Nairobi.

Ackermann, U (2007) Zukunftsworkshop Mikrotechniken für eine effiziente Bioenergieerzeugung, 15. October, Dresden. Konzeptpapier. VDI/VDE Innovation + Technik GmbH website, http://www.mstonline.de/mikrosystemtechnik/medien/konzeptpapier (viewed 12. September 2008).

Adams, R M, Fleming, R A, Chang, C-C, McCarl, B A and Rosenzweig, C (1995) A reassessment of the economic effects of global climate change on U.S. agriculture. *Climatic Change* **30**(2): pp147-67.

ADB (Asian Development Bank) (2007a) Climate Change – ADB Programms – Strengthening Mitigation and Adaptation in Asia and the Pacific, Manila. ADB website, http://www.adb.org/Documents/Brochures/Climate-Change/default.asp (viewed 07. May 2008).

ADB (Asian Development Bank) (2007b) Biofuel and Rural Renewable Energy Initiative in the Greater Mekong Subregion. ADB website, http://www.adb.org/documents/brochures/gms-biofuel/gms-biofuel-brochure.pdf (viewed 07. May 2008).

ADB (Asian Development Bank) (2008) Making Markets Work Better for the Poor – Market-Based Approaches to Low Carbon Development in Vietnam Domestic Biogas and CDM Financing – Possibilities and Problems. ADB website, http://www.markets4poor.org/?name=home&id=140 (viewed 18. September 2008).

Adelphi Consult and Wuppertal Institut (2007) *Die sicherheitspolitische Bedeutung erneuerbarer Energien. Im Auftrag des Bundesministeriums für Umwelt, Naturschutz und Reaktorsicherheit. Endbericht – FKZ 904 97 324.* Adelphi Consult, Berlin.

Adler, P R, Del Grosso, S J and Parton, W J (2007) Life-cycle assessment of net greenhouse-gas flux for bioenergy cropping systems. *Ecological Applications* **17**: pp675–91.

AfDB (African Development Bank) (2007) Renewable Energy Takes Centre Stage in Africas Development Efforts. AfDB website, http://www.afdb.org/portal/page_pageid=293,158705&_dad=portal&_schema=PORTAL&focus_item=22812297&focus_lang=us (viewed 08. May 2008).

Ahmann, M (2000) *Primary Energy Efficiency of Alternative Powertrains in Vehicles.* Department of Environmental and Energy System Studies, Lund University, Lund.

Al-Kaisi, M M and Grotte, J B (2007) Cropping systems effects on improving soil carbon stocks of exposed subsoil. *Soil Science Society of America Journal* **71**: pp1381–8.

Alongi, D M and de Carvalho, N A (2008) The effect of small-scale logging on stand characteristics and soil biogeochemistry in mangrove forests of Timor Leste. *Forest Ecology and Management* **155**: pp1359–66.

Alpmann, L (2005) Ist die Produktion von Raps in Deutschland noch steigerunsfähig? Rapool. Pressegespräch. Rapool website, http://www.rapool.de/index.cfm/startid/133/doc/292//cfid/1466506/cftoken/62397577.html (viewed 27. February 2008).

Altenburg, T, Dietz, H, Hahl, M, Nikolidakis, N, Rosendahl, C and Seelige, K (2008) *Biodiesel Policies for Rural Development in India. Preliminary Report.* German Development Institute (DIE), Bonn.

Altenburg, T and von Drachenfels, C (2007) *Creating an Enabling Environment for Private Sector Development in Sub-Saharan Africa. Paper Commissioned by BMZ/GTZ and UNIDO and Presented to the Donor Committee for Enterprise Development at the Africa Regional Consultative Conference on "Creating Better Business Environments for Enterprise Development – African and Global Lessons for More Effective Donor Practices" 05–07 November 2007, Accra, Ghana.* German Development Institute (DIE), Bonn.

Althammer, W, Biermann, F, Dröge, S and Kohlhaas, M (2001) *Ansätze zur Stärkung der umweltpolitischen Ziele in der Welthandelsordnung.* Analytica, Berlin.

Antle, J, Cabalbo, S, Mooney, S, Elliott, E and Paustian, K (2003) Spatial heterogeneity, contract design, and the efficiency of carbon sequestration policies for agriculture. *Journal of Environmental Economics and Management* **46**: pp231–50.

Aravind, P V, Ouweltjes, J P, de Heer, E, Woudstra, N and Rietveld, G (2006) *Impact of Biosyngas and its Components on SOFC Anodes: Proceedings of the 9th International Symposium on Solid Oxide Fuel Cells, May 15–20, 2005.* U.S. Department of Energy, Washington, DC.

Asner, G P, Broadbent, E N, Oliveira, P J C, Keller, M, Knapp, D E and Silva, J N M (2006) Condition and fate of logged forests in the Brazilian Amazon. *Proceedings of the National Academy of Sciences of the United States of America* **103**: pp12947–50.

Augustus, G D P S, Jayabalan, M and Seiler, G J (2002) Evaluation and bioinduction of energy components of Jatropha curcas. *Biomass and Bioenergy* **23**: pp161–4.

Badgley, C, Moghtader, J, Quintero, E, Zakem, E, Chappell, M J, Avilés-Vázquez, K, Samulon, A and Perfecto, I (2007) Organic agriculture and the global food supply. *Renewable Agriculture and Food Systems* **22**(2): pp86–108

Baehr, H D (1965) *Energie und Exergie: Die Anwendung des Exergiebegriffs in der Energietechnik.* Verein Deutscher Ingenieure, Beuth Verlag GmbH, Düsseldorf, Berlin.

Baehr, H D and Kabelac, S (2006) *Thermodynamik: Grundlagen und technische Anwendungen.* Springer, Berlin, Heidelberg, New York.

Bai, Z G, Dent, D L, Olsson, L and Schaepman, M E (2008) *Global Assessment of Land Degradation and Improvement 1. Identification by Remote Sensing*. Food and Agriculture Organization of the United Nations (FAO), Rome.

Baillie, J E M, Hilton–Taylor, C and Stuart, S N (2004) *2004 IUCN Red List of Threatened Species. A Global Species Assessment*. The IUCN Species Survival Commission, Gland.

Baldocchi, D (2008) Breathing of the terrestrial biosphere: lessons learned from a global network of carbon dioxide flux measurement systems. *Australian Journal of Botany* **56**: pp1–26.

Balmford, A, Bruner, A, Cooper, P, Costanza, R, Farber, S, Green, R E, Jenkins, M, Jefferiss, P, Jessamy, V, Madden, J, Munro, K, Myers, N, Naeem, S, Paavola, J, Rayment, M, Rosendo, S, Roughgarden, J, Trumper, K and Turner, R K (2002) Economic reasons for conserving wild nature. *Science* **297**: pp950–3.

Balmford, A, Green, R S and Scharlemann, J P W (2005) Sparing land for nature: exploring the potential impact of changes in agricultural yield on the area needed for crop production. *Global Change Biology* **11**(10): pp1594–605.

Baret, S, Le Bourgeois, T, Rivière, J-N, Pailler, T, Sarrailh, J-M and Strasberg, D (2007) Can species richness been maintained in logged endemic Acacia heterophylla forests (Réunion Island, Indian Ocean)? *Revue dEcologie Terre et Vie* **62**: pp273–84.

Batjes, N H (1998) Mitigation of atmospheric CO_2 concentrations by increased carbon sequestration in the soil. *Biology and Fertility of Soils* **27**: pp230–5.

Bauer, S (2007) Sicherheitsrisiko Klimawandel. *Entwicklung & ländlicher Raum* **41**(5): pp9–12.

Bauer, S (2008) "Admit that the waters around you have grown". Die Bedeutung des Klimawandels für die Vereinten Nationen. *Vereinte Nationen* **56**(1): pp3–9.

Bauer, S and Stringer, L C (2008) *Science and Policy in the Global Governance of Desertification: An Analysis of Institutional Interplay Under the United Nations Convention to Combat Desertification. Global Governance Working Paper 35*. The Global Governance Project, Amsterdam.

Baumann, M, Laue, H-J und Müller, P (2006) *Wärmepumpen – Heizen mit Umweltenergie, BINE Informationspaket*. Solarpraxis AG, Berlin.

Bayer CropScience (2006) *Volle Kraft vom Acker. Kurier: Das Bayer CropScience Magazin für moderne Landwirtschaft Nr. 3*. Bayer CropScience, Leverkusen.

BayLfU (Bayerisches Landesamt für Umweltschutz) (ed) (2004) *Biogashandbuch Bayern – Materialienband*. BayLfU, Augsburg.

Beese, F (2004) Ernährungssicherung als Produktions– bzw. Verteilungsproblem. Expertise for the WBGU Report "World in Transition: Fighting Poverty Through Environmental Policy". WBGU website, http://www.wbgu.de/www.wbgu_jg2004_ex01.pdf

Benitez, P C, McCallum, I, Obersteiner, M and Yamagata, Y (2005) Global potential for carbon sequestration: geographical distribution, country risk and policy implications. *Ecological Economics* **60**(3): pp572–83.

Bensmann, M (2004) Alles verfeuert. Altholz ist inzwischen ein beliebter Brennstoff. *Neue Energie* **9**: p52.

Bensmann, M (2008) Husch, husch ins Netz. Biogas ist ein universeller Energieträger, der in Erdgasqualität sogar ins Gasnetz eingespeist werden darf. *Neue Energie* **5**: pp64–9.

Beringer, T and Lucht, W (2008) Simulation nachhaltiger Bioenergiepotentiale – Übersicht. Expertise for the WBGU Report "World in Transition: Future Bioenergy and Sustainable Land Use". WBGU website, http://www.wbgu.de/wbgu_jg2008_ex01.pdf.

Berndes, G (2002) Bioenergy and water – the implications of large-scale bioenergy production for water use and supply. *Global Environmental Change* **12**: pp253–71.

Berndes, G (2008) Water Demand for Global Bioenergy Production: Trends, Risks and Opportunities. Expertise for the WBGU Report "World in Transition: Future Bioenergy and Sustainable Land Use". WBGU website, http://www.wbgu.de_wbgu_jg2008_ex02.pdf.

Berndes, G, Hoogwijk, M and van den Broek, R (2003) The contribution of biomass in the future global energy supply: a review of 17 studies. *Biomass & Bioenergy* **25**(1): pp1–28.

Bhattacharya, S C and Salam, P A (2002) Low greenhouse gas biomass options for cooking in the developing countries. *Biomass and Bioenergy* **22**: pp305–17.

BHD (Bundesindustrieverband Deutschland Haus-, Energie- und Umwelttechnik) (2008) Erdgas-Brennwerttechnik. BHD website, http://www.bdh-koeln.de/html/service_produkte/gas_brennwerttechnik.php?lng=de (viewed 18. September 2008).

Biermann, F and Bauer, S (2004) UNEP and UNDP. Expertise for the WBGU Report "World in Transition: Fighting Poverty Through Environmental Policy". WBGU website, http://www.wbgu.de/wbgu_jg2004_ex02.pdf.

Biofuelwatch, Carbon Trade Watch, Corporate Europe Observatory, Econexus, Ecoropa, Grupo de Reflexion Rural, Munlochy Vigil, NOAH (Friends of the Earth Denmark), Rettet den Regenwald and Watch Indonesia (eds) (2007) *Agrofuels. Towards a Reality Check in Nine Key Areas*. Transnational Institute (TNI), Amsterdam.

Biopact (2006) Indonesias Biofuels Program Seen as a Way to Alleviate Poverty. Article of 30. July 2006. Biopact website, http://biopact.com/2006_07_30_archive.html (viewed 27. November 2007).

Biopact (2007a) Senegal and Brazil Sign Biofuel Agreement to Make Africa a Major Supplier. Biopact website, http://www.biopact.com/2007/05/senegal-and-brazil-sign-biofuel.html (viewed 18. June 2008).

Biopact (2007b) Indonesia and Brazil Sign Agreement to Cooperate on Biofuels. Biopact website, http://www.biopact.com/2007/03/indonesia-and-brazil-sign-agreement-to.html (viewed 18. June 2008).

BirdLife International (2008) *Fuelling the Ecological Crisis – Six Examples of Habitat Destruction Driven by Biofuels*. BirdLife International, Brussels.

Biryetega, S (2006) 2006 Country Reports – Uganda. Global Integrity website, http://www.globalintegrity.org/reports/2006/UGANDA/notebook.cfm (viewed 11. August 2008).

Bishop, J (2007) Trading habitat: the potential of biodiversity offsets. AV 33. *Biodiversity Offsets* **AV33**(May): p7.

Bishop, J K B and Rossow, W B (1991) Spatial and temporal variability of global surface solar irradiance. *Journal of Geophysical Research* **96**(C9): pp16839-58.

Blüthner, A (2004) *Welthandel und Menschenrechte in der Arbeit*. Lang, Frankfurt/M.

BMELV (Bundesministerium für Ernährung, Landwirtschaft und Verbraucherschutz) (2006) *Die EU-Agrarreform – Umsetzung in Deutschland*. BMELV, Berlin.

BMELV (Bundesministerium für Ernährung, Landwirtschaft und Verbraucherschutz) (2008) *Land und Forstwirtschaft in Deutschland. Daten und Fakten. Ausgabe 2008*. BMELV, Berlin.

BMU (Bundesministerium für Umwelt, Naturschutz und Reaktorsicherheit) (2007a) Beitrag der erneuerbaren Energien zur Energiebereitstellung in Deutschland 2006. BMU website, http://www.erneuerbare-energien.de/inhalt/39613/ (viewed 12. December 2007).

BMU (Bundesministerium für Umwelt, Naturschutz und Reaktorsicherheit) (2007b) Entwurf und Begründung einer Verordnung über Anforderungen an eine nachhaltige Erzeugung von Biomasse zur Verwendung als Biokraftstoff (Biomasse-Nachhaltigkeitsverordnung – BioNachV). BMU, Berlin.

BMU (Bundesministerium für Umwelt, Naturschutz und Reaktorsicherheit) (2008a) *Fördergeld 2008 für Energieeffizienz und erneuerbare Energien. Programme – Ansprechpartner – Adressen.* BMU, Berlin.

BMU (Bundesministerium für Umwelt, Naturschutz und Reaktorsicherheit) (2008b) *Weiterentwicklung der Strategie zur Bioenergie.* BMU, Berlin.

BMU (Bundesministerium für Umwelt, Naturschutz und Reaktorsicherheit) (2008c) *Ökologische Industriepolitik. Nachhaltige Politik für Innovation, Wachstum und Beschäftigung.* BMU, Berlin.

BMU (Bundesministerium für Umwelt, Naturschutz und Reaktorsicherheit) (2008d) *Bundesumweltminister stoppt Biosprit-Verordnung. Biokraftstoffstrategie wird fortgesetzt. BMU-Pressedienst Nr. 052/08.* BMU, Berlin.

BMU (Bundesministerium für Umwelt, Naturschutz und Reaktorsicherheit) (2008e) *Life Web. Global Initiative on Protected Areas to be Launched at CBD COP 9 in Bonn.* BMU, Berlin.

BMU (Bundesministerium für Umwelt, Naturschutz und Reaktorsicherheit) (2008f) Gesetz zur Änderung der Förderung von Biokraftstoffen (Entwurf und Begründung). BMU website, http://www.erneuerbare-energien.de/inhalt/42435/4593 (viewed 29. October 2008).

BMWi (Bundesministerium für Wirtschaft und Technologie) (2008) *Energiezahlen.* BMWi, Berlin.

BMZ (Bundesministerium für wirtschaftliche Zusammenarbeit und Entwicklung) (2008a) Die Rettung des Brasilianischen Regenwalds. BMZ website, http://www.bmz.de/de/themen/umwelt/projektschaufenster/brasilien.html (viewed 04. March 2008).

BMZ (Bundesministerium für wirtschaftliche Zusammenarbeit und Entwicklung) (2008b) *Tropenwaldprojekte FZ und TZ i.w.S. aus dem Epl. 23 des BMZ. BMZ-Datenauskunftssystem.* BMZ, Bonn.

BMZ (Bundesministerium für wirtschaftliche Zusammenarbeit und Entwicklung) (2008c) *Entwicklungspolitische Positionierung zu Agrartreibstoffen. Diskussionspapier.* BMZ, Berlin.

Boerrigter, H and van der Drift, A (2004) *Biosyngas – Description of R&D Trajectory Necessary to Reach Large-Scale Implementation of Renewable Syngas From Biomass.* Energy Research Centre of the Netherlands (ECN), Petten.

Bogner, J, Abdelrafie Ahmed, M, Diaz, C, Faaij, A, Gao, Q, Hashimoto, S, Mareckova, K, Pipatti, R and Zhang, T (2007) Waste management. In Metz, B, Davidson, O R, Bosch, P R, Dave, R and Meyer, L A (eds) *Climate Change 2007: Mitigation. Contribution of Working Group III to the Fourth Assessment Report of the Intergovernmental Panel on Climate Change* Cambridge University Press, Cambridge, New York, pp585-618.

Bolhár-Nordenkampf, M (2004) *Techno-Economic Assessment on the Gasification of Biomass on the Large Scale for Heat and Power Production. Dissertation.* Technical University, Vienna.

Bondeau, A, Smith, P C, Zaehle, S, Schaphoff, S, Lucht, W, Cramer, W, Gerten, D, Lotze-Campen, H, Müller, C, Reichstein, M and Smith, B (2007) Modelling the role of agriculture for the 20th century global terrestrial carbon balance. *Global Change Biology* **13**(3): pp679–706.

Borges Silva, M, Kanashiro, M, Yamaguishi Ciampi, A, Thompson, I and Sebbenn, A M (2008) Genetic effects of selective logging and pollen gene flow in a low-density population of the dioecious tropical tree Bagassa guianensis in the Brazilian Amazon. *Forest Ecology and Management* **255**: pp1548–58.

Bormann, F H and Likens, G E (1979) Catastrophic disturbance and the steady-state in northern hardwood forests. *American Scientist* **67**: pp660–9.

Bounoua, L, deFries, R, Collatz, G J, Sellers, P and Khan, H (2002) Effects of land cover conversion on surface climate. *Climatic Change* **52**: pp29–64.

BP (British Petroleum) (2007) Statistical Review of World Energy 2007. BP website, http://www.deutschebp.de/genericarticle.docategoryId=9003692&contentId=7034169 (viewed 19. August 2008).

BP (British Petroleum) (2008) *Statistical Review of World Energy 2008.* BP, London.

BR (Bundesregierung) (2007) *Das Integrierte Energie- und Klimaprogramm der Bundesregierung – Herausforderung und energie- und klimapolitische Zielsetzungen.* Bundesregierung, Berlin.

Brandon, K, Redford, K H and Sanderson, S E (ed) (1998) *Parks in peril. People, politics, and protected areas.* Island Press, Washington, DC.

Bringezu, S, Ramesohl, S, Arnold, K, Fischedick, M, von Geibler, J, Liedtke, C and Schütz, H (2007) *Towards a Sustainable Biomass Strategy. What we Know and What we Should Know. Wuppertal Papers 163.* Wuppertal Institute, Wuppertal.

Bringezu, S and Schütz, H (2008) Auswirkungen eines verstärkten Anbaus nachwachsender Rohstoffe im globalen Maßstab. *Technikfolgenabschätzung – Theorie und Praxis* **17**(2): pp12-23.

Brock, N (2008) Der Müll, die Stadt – und wer davon lebt. In Mexiko-Stadt ist Abfallbeseitigung ein Geschäft mit strengen Regeln. *welt-sichten* **4**.

Bruner, A G, Gullison, R E, Rice, R E and da Fonseca, G A B (2001) Effectiveness of parks in protecting tropical biodiversity. *Science* **291**: pp125–8.

Brüntrup, M (2008) *Steigende Lebensmittelpreise – Ursachen, Folgen und Herausforderungen für die Entwicklungspolitik. DIE Analysen und Stellungnahmen 4.* German Development Institute (DIE), Bonn.

Bryant, D, Nielsen, D and Tangley, L (1997) *The Last Frontier Forests. Ecosystems & Economies on the Edge.* World Resources Institute, Washington, DC.

BTG (Biomass Technology Group) (2008) *Sustainability Criteria & Certification Systems for Biomass Production. Final Report. Project No. 1386.* BTG, Enschede.

Buchmann, N and Schulze, E-D (1999) Net CO_2 and H_2O fluxes of terrestrial ecosystems. *Global Biogeochemical Cycles* **13**: pp751–60.

Butler, D (2008) Food Crisis spurs research spending. *Nature* **453**: p8.

BUWAL (Bundesamt für Umwelt Wald und Landschaft) (1999) Nationale Standards für die Waldzertifizierung in der Schweiz. BUWAL website, http://www.fsc-schweiz.ch/index.php?option=com_docman&task=doc_download&gid=40&lang=de (viewed 15. April 2008).

Calder, I R (1999) *The Blue Revolution: Land Use and Integrated Water Resources Management.* Earthscan, London.

Calfapietra, C, Gielen, B, Galema, A N J, Lukac, M, de Angelis, P, Moscatelli, M C, Ceulemans, R and Scarascia-Mugnozza, G (2003) Free-air CO_2 enrichment (FACE) enhances biomass production in a short-rotation poplar plantation. *Tree Physiology* **23**: pp805-14.

Campbell, J E, Lobell, D B, Genova, R C and Field, C B (2008) The global potential of bioenergy on abandoned agriculture lands. *Environmental Science & Technology* **42**(15): pp5791-4.

Canadell, J and Raupach, M R (2008) Managing forests for climate change mitigation. *Science* **320**: pp1456-7.

Canadell, J G, Le Quére, C, Raupach, M R, Fielde, C B, Buitenhuisc, E T, Ciaisf, P, Conwayg, T J, Gillettc, N P, Houghton, R A and Marlandi, G (2007) Contributions to accelerating atmospheric CO_2 growth from economic activity, carbon intensity, and efficiency of natural sinks. *PNAS* **early edition**: p5.

Carey, C, Dudley, N and Stolton, S (2000) *Squandering Paradise? The Importance and Vulnerability fo the Worlds Protected Areas.* WWF International, Gland.

Carroll, N, Fox , J and Bayon, R (ed) (2007) *Conservation and Biodiversity Banking: A Guide to Setting up and Running Biodiversity Credit Trading Systems. Environmental Market Insights.* Earthscan, London.

Castro-Arellano, I, Presley, S J, Saldanha, L N, Willig, M R and Wunderle jr., J M (2007) Effects of reduced impact logging on bat biodiversity in terra firme forest of lowland Amazonia. *Biological Conservation* **138**: pp269–85.

CBD (Convention on Biological Diversity) (2000) *Ecosystem Approach. Decision V/6.* CBD, Montreal.

CBD (Convention on Biological Diversity) (2002a) *Global Strategy for Plant Conservation. Decision VI/9.* CBD, Montreal

CBD (Convention on Biological Diversity) (2002b) *Strategic Plan for the Convention on Biological Diversity. Decision VI/26.* CBD, Montreal.

CBD (Convention on Biological Diversity) (2002c) *Alien Species That Threaten Ecosystems, Habitats or Species. Decision VI/23.* CBD, Montreal.

CBD (Convention on Biological Diversity) (2004a) *Ecosystem Approach. Decision VII/11.* CBD, Montreal.

CBD (Convention on Biological Diversity) (2004b) *Protected areas (Articles 8 (a) to (e)). Decision VII/28.* CBD, Montreal

CBD (Convention on Biological Diversity) (2004c) *Strategic Plan: Future Evaluation of Progress. Decision VII/30.* CBD, Montreal.

CBD (Convention on Biological Diversity) (2004d) *Addis Ababa Principles and Guidelines for the Sustainabe Use of Biodiversity. Decision VII/14.* CBD, Montreal.

CBD (Convention on Biological Diversity) (2006a) *Global Biodiversity Outlook 2.* CBD-Secretariat, Montreal.

CBD (Convention on Biological Diversity) (2006b) *Framework for Monitoring Implementation of the Achievement of the 2010 Target and Integration of Targets Into the Thematic Programmes of Work. Decision VIII/15.* CBD, Montreal.

CBD (Convention on Biological Diversity) (2007) *Review of Implementation of the Programme of Work on Protected Areas for the Period 2004–2007. UNEP/CBD/WG-PA/2/2.* CBD, Montreal.

CBD (Convention on Biological Diversity) (2008a) *Forest Biodiversity. Decision IX/5.* CBD, Montreal.

CBD (Convention on Biological Diversity) (2008b) *Ecosystem Approach. Decision IX/7.* CBD, Montreal.

Chape, S, Harrison, J, Spalding, M and Lysenko, I (2005) Measuring the extent and effectiveness of protected areas as an indicator for meeting global biodiversity targets. *Philosophical Transactions of the Royal Society* **360**: pp443–55.

Chaytor, B (2002) Negotiating further liberalization of environmental goods and services. An exploration of the terms of art. *Review of European Community and International Environmental Law* **11**: pp287–97.

Chemrec (2007) The BLGMF System. Chemrech website, http://www.chemrec.se. (viewed 28. July 2007).

Chivian, E and Bernstein, A (ed) (2008) *Sustaining Life: How Human Health Depends on Biodiversity.* Oxford University Press, Oxford, New York.

Chomitz, K M, Thomas, T S and Brandao, A S P (2004) *Creating Markets for Habitat Conservation when Habitats are Heterogeneous. World Bank Policy Research Working Paper No. 3429.* World Bank, Washington, DC.

Choren (2007) CHOREN-Verfahren zur Erzeugung von BtL-Kraftstoff aus Biomasse. Choren Industries website, http://www.choren.com (viewed 28. July 2007).

Clarkson, R and Deyes, K (2002) *Estimating the Costs of Carbon Emissions.* HM Treasury, London.

Clifton-Brown, J C, Breuer, J and Jones, M B (2007) Carbon mitigation by the energy crop. Miscanthus. *Global Change Biology* **13**: pp2296–307.

Cochrane, M A (2003) Fire science for rainforests. *Nature* **421**: pp913–9.

Collatz, G J, Ribas-Carbo, M and Berry, J A (1992) Coupled photosynthesis-stomatal conductance model for leaves of V4 plants. *Australian Journal of Plant Physiology* **19**(5): pp519–38.

Collier, P (ed) (2007) *The Bottom Billion: Why the Poorest Countries Are Failing and What Can Be Done About It.* Oxford University Press, Oxford, New York.

Collins, W D, Bitz, C M, Blackmon, M L, Bonan, G B, Bretherton, C S, Carton, J A, Chang, P, Doney, S C, Hack, J J, Henderson, T B, Kiehl, J T, Large, W G, McKenna, D S, Santer, B D and Smith, R D (2006) The Community Climate System Model Version 3 (CCSM3). *Journal of Climate* **19**(11): pp2122–43.

Colwell, R K, Brehm, G, Cardelús, C L, Gilman, A C and Longino, J T (2008) Global warming, elevational range shifts, and lowland biotic attrition in the wet tropics. *Science* **322**: pp258-60.

COMPETE (2008) Project Website. COMPETE Platform for Bioenergy in Arid and Semi-Arid Ecosystems in Africa website, http://www.compete-bioafrica.net/ (viewed 24. July 2008).

Conniff, R (2007) Bioenergy – The Cure for Our Oil Addiction? Yale School of Forestry & Environmental Studies website, http://forestry.yale.edu/pubs/Bioenergy-The-Cure-for-Our-Oil-Addiction/ (viewed 10. October 2008).

Constantin, A L (2008) *A Time of High Prices: An Opportunity For the Rural Poor?* Institute for Agriculture and Trade Policy, Minneapolis, Minnesota.

Cotula, I, Dyer, N and Vermeulen, S (2008) *Fuelling Exclusion? The Biofuels Boom and Poor Peoples Access to Land.* Food and Agriculture Organization of the United Nations (FAO), Rome.

Council of the European Union (2008) Presidency Suggestions for a Common Scheme of Sustainability Criteria for Biofuels. 9 September 2008, Brussels. Counicl of the European Union website, http://register.consilium.europa.eu/pdf/en/08/st12/st12157-re01ad01.en08.pdf (viewed 15. October 2008).

Cramer, W, Bondeau, A, Schapphoff, S, Lucht, W, Smith, B and Sitch, S (2004) Tropical forest and the global carbon cycle: impacts of atmospheric carbon dioxide, climate change and rate of deforestation. *Philosophical Transactions of the Royal Society of London* **B 359**: pp331-43.

Crews, T E and Peoples, M B (2005) Can the synchrony of nitrogen supply and crop demand be improved in legume and fertilizer-base agroecosystems? A review. *Nutrient Cycling in Agroecosystems* **72**: pp101–20.

Dasappa, S, Sridhar, G and Sridhar, H V (2003) *"Bio-residue gasification – Science and Technology" at Roundtable on Biomass Gasification Technologies, December 2003, Combustion, Gasification & Propulsion Laboratory, Dept. of Aerospace Engineering.* Indian Institute of Science, Bangalore.

Datta, S K (2002) *Recent Developments in Transgenics for Abiotic Stress Tolerance in Rice. JIRCAS Working Report.* Japan International Research Center for Agricultural Sciences (JIRCAS), Ibaraki, Japan.

DB (Deutsche Bahn Energie) (2008) DB Energie – Zusammensetzung des Strom-Mix der DB Energie GmbH. DB Energie GmbH website, http://www.dbenergie.de/site/dbenergie/de/start.html (viewed 18. September 2008).

DBV (Deutscher Bauernverband) (2004) *Der Biomasse gehört die energetische Zukunft. Pressemitteilung vom 10.02.2004.* DBV, Berlin, Brussels.

de Fraiture, C, Giordano, M A and Yonsong, L (2007) *Biofuels and Implications for Agricultural Water Use: Blue Impacts of Green Energy.* International Water Management Institute, Colombo, Sri Lanka.

de Fraiture, C, Giordano, M A and Liao, Y (2008) Biofuels and implications for agricultural water use: blue impacts of green energy. *Water Policy* **10**(Supplement 1): pp67–81.

de Koning, G H J, Veldkamp, E and López-Ulloa, M (2003) Quantification of carbon sequestration in soils following pasture to forest conversion in northwestern Ecuador. *Global Biogeochemical Cycles* **17**: p1098.

de La Torre Ugarte, D G (2006) *Developing Bioenergy: Economic and Social Issues. IFPRI Bioenergy and Agriculture: Promises and Challenges, Focus 14, Brief 2.* International Food Policy Research Institute (IFPRI), Washington, DC.

de Lacerda, A E, Kanashiro, M and Sebbenn, A M (2008) Effects of reduced impact logging on genetic diversity and spatial genetic structure of a Hymenaea courbaril population in the Brazilian Amazon forest. *Forest Ecology and Management* **2555**: pp1034–43.

de Santi, G (2008) *Biofuels in the European Context: Facts and Uncertainties.* European Commission Joint Research Centre (JRC), Brussels.

Debiel, T and Werthes, S (2006) Fragile Staaten und globale Friedenssicherung. In Debiel, T, Messner, D and Nuscheler, F (eds) *Globale Trends 2007.* Fischer, Frankfurt/M., pp81–103.

DEFRA (Department for Environment Food and Rural Affairs) (2007) *The Waste Strategy for England.* DEFRA, London.

DeFries, R S and Townsend, J R G (1999) Global land cover characterization from satellite data: from research to operational implementation? *Global Ecology and Biogeography* **8**: pp367–79.

DeFries, R S, Bounoua, L and Collatz, G J (2002) Human modification of the landscape and surface climate in the next fifty years. *Global Change Biology 8, 438-458.* **8**: pp438–58.

Delgado, C, Rosegrant, M, Steinfeld, H, Ehui, S and Courbois, C (1999) *Livestock to 2020 – The Next Food Revolution. Food, Agriculture and the Environment Discussion. Paper 28.* International Food Policy Research Institute (IFPRI), Washington, DC.

Delworth, T L, Broccoli, A J, Rosati, A, Stouffer, R J, Balaji, V, Beesley, J A, Cooke, W F, Dixon, K W, Dunne, J, Dunne, K A, Durachta, J W, Findell, K L, Ginoux, P, Gnanadesikan, A, Gordon, C T, Griffies, S M, Gudgel, R, Harrison, M J, Held, I M, Hemler, R S, Horowitz, L W, Klein, S A, Knutson, T R, Kushner, P J, Langenhorst, A R, Lee, H-C, Lin, S-J, Lu, J, Malyshev, S L, Milly, P C D, Ramaswamy, V, Russell, J, Schwarzkopf, M D, Shevliakova, E, Sirutis, J J, Spelman, M J, Stern, W F, Winton, M, Wittenberg, A T, Wyman, B, Zeng, F and Zhang, R (2006) GFDLs CM2 Global Coupled Climate Models. Part I: Formulation and Simulation Characteristics. *Journal of Climate* **19**(5): pp643–74.

Denevan, W M and Woods, W I (2004) Discovery and Awareness of Anthropogenic Amazonian Dark Earths. Edwardsville: University of Wisconsin-Madison, Southern Illinois University.

Denman, K L, Brasseur, G, Chidthaisong, A, Ciais, P, Cox, P M, Dickinson, R E, Hauglustaine, D, Heinze, C, Holland, E, Jacob, D, Lohmann, U, Ramachandran, S, da Silva Dias, P L, Wofsy, S C and Zhang, X H C U P, Cambridge, UK and New York, USA (2007) Couplings between changes in the climate system and biogeochemistry. In Intergovernmental Panel on Climate Change (IPCC) (ed) *Climate Change 2007: The Physical Science Basis. Contribution of Working Group I to the Fourth Assessment Report of the Intergovernmental Panel on Climate Change.* Cambridge University Press, Cambridge, New York, pp499–589.

Déry, P and Anderson, B (2007) Peak Phosphorus. Energy Bulletin online website, http://www.energybulletin.net/33164.html (viewed 16. May 2008).

Destatis (Statistisches Bundesamt) (2006) Land- und Forstwirtschaft, Fischerei. Statistisches Bundesamt website, http://www.destatis.de/jetspeed/portal/cms/Sites/destatis/Internet/DE/Navigation/Statistiken/LandForstwirtschaft/LandForstwirtschaft.psml (viewed 27. February 2008).

Deutsch, C A, Tewksbury, J J, Huey, R B, Sheldon, K S, Ghalambor, C K, Haak, D C and Martin, P R (2008) Impacts of climate warming on terrestrial ectotherms across latitude. *PNAS* **105**(18): pp6668–72.

Deutscher Bundestag (2008) *Unterrichtung durch die Bundesregierung. Bericht der Bundesregierung über das Ministertreffen der Welthandelsorganisation in Genf vom 21. bis 30. Juli 2008 (Doha-Runde) Zugeleitet mit Schreiben des Bundesministeriums für Wirtschaft und Technologie vom 23. August 2008. Drucksache 16/10171. 16. Wahlperiode 26.08.2008.* Deutscher Bundestag, Berlin.

DGE (Deutsche Gesellschaft für Ernährung) (2007) *Neue Richtwerte für die Energiezufuhr.* DGE, Bonn.

Diaz, R J and Rosenberg, R (2008) Spreading dead zones and consequences for marine ecosystems. *Science* **321**: pp926–9.

DIN (Deutsche Industrie Norm) (1990) *DIN 4702-8 – Heizkessel; Ermittlung des Norm Nutzungsgrades und des Norm Emissionsfaktors. DIN 4702-8:1990-03.* Beuth Verlag, Berlin.

DIN (Deutsche Industrie Norm) (2008a) *DIN EN 14511-3 – Luftkonditionierer, Flüssigkeitskühlsätze und Wärmepumpen mit elektrisch angetriebenen Verdichtern für die Raumbeheizung und Kühlung. Teil 3: Prüfverfahren. Deutsche Fassung EN 14511-3:2007.* Beuth Verlag, Berlin.

DIN (Deutsche Industrie Norm) (2008b) *DIN EN 255-3 – Luftkonditionerer, Flüssigkeitskühlsätze und Wärmepumpen mit elektrisch angetriebenen Verdichtern – Heizen – Prüfungen und Anforderungen an die Kennzeichnung von Geräten zum Erwärmen von Brauchwarmwasser. Deutsche Fassung prEN 255-3:2008.* Beuth Verlag, Berlin.

Djaja, K (2006) *Development of Biofuel / Green Energy in Indonesia: Plan and Strategy. Präsentation auf dem Workshop on Mainstreaming Policies and Investment in Low Carbon: Opportunities for New Approaches to Investment and Flexible Mechanism. Bangkok, 30-31.08.2006.* Coordinating Ministry of Economic Affairs, Bangkok.

DMFA (Dutch Ministry of Foreign Affairs) (2008) SNV Biogas Programme Shortlisted for Energy Prize. DMFA website, http://www.minbuza.nl/en/news/pressreleases,2007/03/SNV-biogas-programme-shortlisted-for-energy-prize.html (viewed 27. February 2008).

Donner, S D and Kucharik, C J (2008) Corn-based ethanol production compromises goal of reducing nitrogen export by the Mississippi River. *Proceedings of the National Academy of Sciences of America* **105**(11): pp4513–8.

Doornbosch, R and Steenblik, R (2007) *Biofuels: is the cure worse than the disease? Background paper for the OECD Round Table on Sustainable Development, Paris, 11–12 September 2007. OECD-Dokument SG/SD/RT(2007)3.* OECD, Paris.

Doyle, U, Vohland, K, Rock, J, Schümann, K and Ristow, M (2007) Nachwachsende Rohstoffe – eine Einschätzung aus Sicht des Naturschutzes. *Natur und Landschaft* **82**(12): pp529–353.

Dreier, T and Tzscheutschler, P (2000) *Ganzheitliche Systemanalyse für die Erzeugung und Anwendung von Biodiesel und Naturdiesel im Verkehrssektor. Technische Universität München, Energiewirtschaft und Anwendungstechnik.* Technical University Munich (TUM), Munich.

Droege, S (2001) Ecological labelling and the World Trade Organization. *Außenwirtschaft* **56**: pp99–122.

Dudley, N (2008) *Guidelines for Applying Protected Area Management Categories.* The World Conservation Union (IUCN), Gland.

Dudley, N and Stolton, S (1999a) *Conversion of Paper Parks to Effective Management: Developing a Target. Report to the WWF-World Bank Alliance from the IUCN/WWF Forest Innovation Project.* The World Conservation Union (IUCN), Gland.

Dudley, N and Stolton, S (1999b) *Threats to Forest Protected Areas. Summary of a Survey of 10 Countries Carried out in Association with the World Commission on Protected Areas.* IPCC, Geneva.

Dudley, N, Belokurov, A, Borodin, O, Higgins-Zogib, L and Hockings, M (2004) *How Effective Are Protected Areas? A Preliminary Analysis of Forest Protected Areas by WWF – The Largest Ever Global Assessment of Protected Area Management Effectiveness.* WWF International, Gland.

Dufey, A (2006) *Biofuels Production, Trade and Sustainable Development, Emerging Issues. Sustainable Markets Discussion Paper No. 2.* International Institute for Environment and Development (IIED), London.

Dupraz, C, Burgess, P, Gavaland, A, Graves, A, Herzog, F, Incoll, L D, Jackson, N, Keesman, K, Lawson, G, Lecomte, I, Liagre, F, Mantzanas, K, Mayus, M, Moreno, G, Palma, J, Papanastasis, V, Paris, P, Pilbeam, D J, Reisner, Y, van Noordwijk, M, Vincent, G and van der Werf, W (2005) *SAFE Final Report. Synthesis of the SAFE Project (August 2001–January 2005).* Silvoarable Agroforestry for Europe (SAFE), Brussels.

Durstewitz, M, Hahn, B, Lange, B, Rohrig, K and Wessel, A (2008) *Windenergie Report Deutschland 2007.* Institute for Solar Energy Technology, Kassel.

Dutschke, M, Kapp, G, Lehmann, A and Schäfer, V (2006) *Risks and Chances of Combined Forestry and Biomass Projects under the Clean Development Mechanism. CD4CDM Working Paper Series. Working Paper No. 1.* UNEP Risoe Centre on Energy, Climate and Sustainable Development, Roskilde.

EAA (Electric Auto Association) (2007) Frequently Asked Questions About Electrical Vehicles. EEA website, http://www.pluginamerica.com/faq.shtml (viewed 15. November 2007).

Easterling, W E, Aggarwal, P K, Batima, P, Brander, K M, Erda, L, Howden, S M, Kirilenko, A, Morton, J, Soussana,

J-F, Schmidhuber, J and Tubiello, F N (2007) Food, fibre and forest products. In Intergovernmental Panel on Climate Change (IPCC) (ed) *Climate Change 2007: Impacts, Adaptation and Vulnerability. Contribution of Working Group II to the Fourth Assessment Report of the Intergovernmental Panel on Climate Change.* Cambridge University Press, Cambridge, New York, pp273–313.

Economist (2008a) *A Special Report on the Future of Energy.* Economist, London.

Economist (2008b) Biofuels in India. Power plants. *Economist* (20.09.): p68.

EcoTrade (2008) *Comment on the Green Paper on Market-Based Instruments for Environmental and Related Policy Purposes by the European Research Team of the EcoTRADE Project.* Ecotrade Project at the Helmholtz Centre for Environmental Research Leipzig (UFZ), Leipzig.

Edelbrock, K (2005) *Initiative 2000plus. Präsentation Recyclingpapier.* Mimeo, Berlin.

EEA (European Environment Agency) (2006) *Urban Sprawl in Europe: The Ignored Challenge. EEA Report No 10.* EEA, Kopenhagen.

EEA (European Environment Agency) (2006a) *How Much Bioenergy can Europe Produce Without Harming the Environment? EEA Report no. 7.* EEA, Paris.

EEA (European Environment Agency) (2007a) *Environmentally Compatible Bio-Energy Potential From European Forests.* EEA, Copenhagen.

EEA (European Environment Agency) (2007b) *Estimating the Environmentally Compatible Bioenergy Potential From Agriculture. EEA Technical Report 12/2007.* EEA, Copenhagen.

EERE (US Department of Energy-Energy Efficiency and Renewable Energy) (2008) Federal Biomass Policy. EERE website, http://www1.eere.energy.gov/biomass/federal_biomass.html (viewed 11. June 2008).

Egenhofer, C (2007) Looking for the cure-all? Targets and the EUs New Energy Strategy. *CEPS Policy Brief* **118**: pp1–5.

EIA (Energy Information Administration) (2008) Official Energy Statistics From the US Government – Indonesia – Country Analysis Briefs. EIA website, http://www.eia.doe.gov/emeu/cabs/Indonesia/Background.html (viewed 7. March 2008).

Eide, A (2008) *The Right to Food and the Impact of Liquid Biofuels (Agrofuels).* Food and Agriculture Organization of the United Nations (FAO) Rome.

El Bassam, N (1998) *Energy Plant Species – Their Use and Impact on Environment and Development.* James & James, London.

Ellis, E C and Ramankutty, N (2008) Putting people in the map: anthropogenic biomes of the world. *Front Ecol Environ* **6**: doi:10.1890/070062.

Endres, A (1995) Zur Ökonomie internationaler Umweltvereinbarungen. *Zeitschrift für Umweltpolitik* **18**(2): pp143–78.

Engel, T (2007) *Plug-in Hybrids. Studie zur Abschätzung des Potentials zur Reduktion der CO_2-Emissionen im PKW-Verkehr bei verstärkter Nutzung von elektrischen Antrieben im Zusammenhang mit Plug-in Hybrid Fahrzeugen.* Deutsche Gesellschaft für Sonnenenergie, Munich.

Enkvist, P-A, Nucler, T and Rosander, J (2007) A cost curve for greenhouse gas reduction. *The McKinsey Quarterly* **1**: pp35–45.

EPIA (European Photovoltaic Industry Association) (2008) Cummulative Installed PV Capacities. EPIA website, http://www.epia.org/index.php?id=86 (viewed 27. February 2008).

Epiney, A (2000) Welthandel und Umwelt. *Deutsches Verwaltungsblatt* **115**: pp77–86.

Epiney, A and Scheyli, M (2000) *Umweltvölkerrecht. Völkerrechtliche Bezugspunkte des schweizerischen Umweltrechts.* Stämpfli, Bern.

Erbrecht, T and Lucht, W (2006) Impacts of large-scale climatic disturbances on the terrestrial carbon cycle. *Carbon Balance and Management* **1**: p7.

EU (European Union) (2003) *EU-Verordnung 1782/2003/EG: Verordnung (EG) Nr. 1782/2003 des Rates vom 29. September 2003 mit gemeinsamen Regeln für Direktzahlungen im Rahmen der Gemeinsamen Agrarpolitik. Amtsblatt der Europäischen Union v. 21.10.2003, L 270/1-L 270/69.* EU, Brussels.

EU Coherence (2008) Diverse Artikel zum Thema Biokraftstoffe und Entwicklungszusammenarbeit. EU Coherence website, http://www.eucoherence.org (viewed 07. October 2008).

EU Commission (2004) *Communication from the Commission to the Council and the European Parliament on the Future of the EU Energy Initiative and the Modalities for the Establishment of an Energy Facility for the ACP Countries. COM(2004) 711 final.* EU Commission, Brussels.

EU Commission (2005a) *Biomass Action Plan. Communication from the Commission. COM (2005) 628 final.* EU Commission, Brussels.

EU Commission (2005b) *The Support of Electricity From Renewable Energy Sources. Communication from the Commission. COM (2005) 627 final.* EU Commission, Brussels.

EU Commission (2005c) EU-Russia Energy Dialogue. Sixth Progress Report. Moscow/Brussels. EU Commission website, http://europa.eu.int/comm/energy/russia/joint_progress/doc/progress6_en.pdf (viewed 08. October 2008).

EU Commission (2005d) *Communication from the Commission to the Council and the European Parliament - Reporting on the Implementation of the EU Forestry Strategy [SEC(2005) 333].* EU Commission, Brussels.

EU Commission (2005e) *The European Consensus on Development. Joint Statement by the Council and the Representatives of the Governments of the Member States Meeting Within the Council, the European Parliament and the Commission.* EU Commission, Brussels.

EU Commission (2005f) *EU Strategy for Africa: Towards a Euro-African Pact to Accelerate Africa's Development.* EU Commission, Brussels.

EU Commission (2006a) *Biofuels Progress Report. Communication from the Commission to the Council and the European Parliament. COM (2006) 845 final.* EU Commission, Brussels.

EU Commission (2006b) *An EU Strategy for Biofuels. COM (2006) 34 final.* EU Commission, Brussels.

EU Commission (2007a) *Green Paper on Market-Based Instruments for Environment and Related Policy Purposes. COM (2007) 140 final.* EU Commission, Brussels.

EU Commission (2007b) *Commission Staff Working Document Accompanying the Green Paper on Market-Based Instruments for Environment and Energy: Related Policy Purposes. SEC(2007) 388.* EU Commission, Brussels.

EU Commission (2008a) *Directive of the European Parliament and of the Council on the Promotion of the Use of Energy From Renewable Sources. COM (2008) 19 final.* EU Commission, Brussels.

EU Commission (2008b) Trade in Agricultural Goods and Fishery Products. Agricultural Policy and Trade: Trade Partners. LDCs. EU Commission website, http://trade.ec.europa.eu/doclib/docs/2006/june/tradoc_120307.pdf (viewed 29. April 2008).

EU Commission (2008c) African Union Commission and European Commission Launch an Ambitious Africa-EU Energy Partnership. Press Release. EU Commission website, http://www.europafrica.org (viewed 20. October 2008).

EU Parliament (2006) *Directive 2006/32/EC of the European Parliament and of the Council on Energy End-Use Efficiency and Energy Services and Repealing Council Directive 93/76/EEC.* EU Parliament, Brussels.

EU Parliament (2008) More Sustainable Energy in Road Transport Targets. Pressemitteilung vom 10.9.2008. EU Parliament website, http://www.europarl.europa.eu/news/expert/infopress_page/064-36659-254-09-37-911-20080909IPR36658-10-09-2008-2008-false/default_en.htm (viewed 15. October 2008).

EUEI (European Union Energy Initiative) and GTZ (Deutsche Gesellschaft für Technische Zusammenarbeit) (2006) *The Significance of Biomass Energy Strategies (BEST) for Sub-Saharan Africa. Background Paper for 1st Regional Workshop on Biomass Energy Strategy (BEST) Development Held in Dar-es-Salaam, Tanzania, 12–14 September 2006.* EUEI, GTZ, Eschborn.

Europe Aid (2007) *Newsletter Europe Aid. ACP-EC Energy Facility. Seventeenth Issue.* Europe Aid Energy Facility, Brussels.

Europe Economics (2008) *A Comparison of the Costs of Alternative Policies for Reducing UK Carbon Emissions. Europe Economics Report for Open Europe.* Europe Economics, London.

Faaij, A (2008) Bioenergy and Global Food Security. Expertise for the WBGU Report "World in Transition: Future Bioenergy and Sustainable Land Use". WBGU website, http://www.wbgu.de/wbgu_jg2008_ex03.pdf

Fairless, D (2007) The little shrub that could – maybe. *Nature* **449**: pp652–3.

Fang, Q, Yu, Q, Wang, E, Chen, Y, Zhang, G, Wang, J and Li, L (2006) Soil nitrate accumulation, leaching and crop nitrogen use as influenced by fertilization and irrigation in an intensive wheat–maize double cropping system in the North China Plain. *Plant Soil* **284**: pp335–50.

FAO (Food and Agriculture Organization of the United Nations) (1990) *Soil Map of the World. Revised Legend.* FAO, Rome.

FAO (Food and Agriculture Organization of the United Nations) (1992) *Biogas Processes for Sustainable Development.* FAO, Rome.

FAO (Food and Agriculture Organization of the United Nations) (1996) *The State of the World's Plant Genetic Resources for Food and Agriculture. Background Documentation Prepared for ITCPGR.* FAO, Rome.

FAO (Food and Agriculture Organization of the United Nations) (1997) *Africover Land Cover Classification. Report.* FAO, Rome.

FAO (Food and Agriculture Organization of the United Nations) (2001a) *The State of Food Insecurity in the World 2001. Food Insecurity: When People Live With Hunger and Fear Starvation.* FAO, Rome.

FAO (Food and Agriculture Organization of the United Nations) (2001b) *Global Estimates of Gaseous Emissions of NH_3, NO and N_2O from Agricultural Land.* FAO, Rome.

FAO (Food and Agriculture Organization of the United Nations) (2003a) *World Agriculture: Towards 2015/2030.* Earthscan, London.

FAO (Food and Agriculture Organization of the United Nations) (2003b) *Compendium of Agricultural–Environmental Indicators 1989–91 to 2000.* FAO Statistics Analysis Service, Statistics Division, Rome.

FAO (Food and Agriculture Organization of the United Nations) (2004) Human Energy Requirements. Report of a Joint FAO/WHO/UNU Expert Consultation. Rome, 17–24 October 2001. FAO website, http://www.fao.org/docrep/007/y5686e/y5686e00.htm (viewed 23. July 2007).

FAO (Food and Agriculture Organization of the United Nations) (2005) *World Forest Assessment*. FAO, Rome.

FAO (Food and Agriculture Organization of the United Nations) (2006a) *The State of Food Insecurity in the World 2006. Eradicating World Hunger – Taking Stock Ten Years After the World Food Summit*. FAO, Rome.

FAO (Food and Agriculture Organization of the United Nations) (2006b) *World Agriculture: Towards 2030/2050. Interim Report*. FAO, Rome.

FAO (Food and Agriculture Organization of the United Nations) (2006c) *Global Forest Resources Assessment 2005. Progress Towards Sustainable Forest Management*. FAO, Rome.

FAO (Food and Agriculture Organization of the United Nations) (2007a) *State of the World's Forests 2007*. FAO, Rome.

FAO (Food and Agriculture Organization of the United Nations) (2007b) *The State of Food and Agriculture*. FAO, Rome.

FAO (Food and Agriculture Organization of the United Nations) (2007c) *A Review of the Current State of Bioenergy in G8+5 Countries*. FAO, Rome.

FAO (Food and Agriculture Organization of the United Nations) (2008a) *Soaring Food Prices: Facts, Perspectives, Impacts and Actions Required, Conference Document, High-Level Conference on World Food Security: The Challenges of Climate Change and Bioenergy, Rome 3–5 June 2008*. FAO, Rome.

FAO (Food and Agriculture Organization of the United Nations) (2008b) Crop Prospects and Food Situation. FAO website, http//www.fao.org/docrep/010/ai465e/ai465e01.htm (viewed 02. July 2008).

FAO (Food and Agriculture Organization of the United Nations) (2008c) *The State of Food and Agriculture 2008. Biofuels: Prospects, Risks and Opportunities*. FAO, Rome.

FAO (Food and Agriculture Organization of the United Nations) (2008d) *Climate Change, Water and Food Security*. FAO, Rome.

FAO (Food and Agriculture Organization of the United Nations) (2008e) FAO Unveils New Bioenergy Assessment Tool. FAO Newsroom. FAO website, http://www.fao.org/newsroom/en/news/2008/1000782/index.html (viewed 02. June 2008).

FAO (Food and Agriculture Organization of the United Nations) (2008f) International Commodity Prices Database. FAO website, http://www.fao.org/es/esc/prices/PreicesServlet.jsp?lang=en viewed 30. January 2008).

FAO-RWEDP (Food and Agriculture Organization of the United Nations - Regional Wood Energy Development Programme in Asia) (2008) Biomass Energy Technology – Wood Energy Data. FAO-RWEDP website, http://www.rwedp.org/d_technodc.html (viewed 13. October 2008).

FAOSTAT (Statistical Division of the Food and Agriculture Organization of the United Nations) (2006) FAOSTAT Agricultural Data. FAOSTAT website, http://faostat.fao.org/ (viewed 27. February 2008).

FAOSTAT (Statistical Division of the Food and Agriculture Organization of the United Nations) (2007) FAOSTAT Data on Production – ProdStat – Crops. FAOSTAT website, http://faostat.fao.org/site/567/default.aspx#ancor (viewed 09. October 2008).

FAOSTAT (Statistical Division of the Food and Agriculture Organization of the United Nations) (2008a) Data Archives. FAOSTAT website, http://faostat.fao.org (viewed 16. May 2008).

FAOSTAT (Statistical Division of the Food and Agriculture Organization of the United Nations) (2008b) Statistical Databases. FAOSTAT website, http://faostat.org/ (viewed 27. March 2008).

Farack, M (2007) *Ethanolgetreide – eine neue Qualitätsschiene für die Landwirtschaft. 9. Jahrestagung Thüringer Landwirtschaft*. Thüringer Landesanstalt für Landwirtschaft (TLL), Jena.

Fargione, J, Hill, J K, Tilman, D, Polasky, S and Hawthorne, P (2008) Land clearing and the biofuel carbon debt. *Science* **319**: pp1235-8.

Farnum, P, Lucier, A and Meilan, R (2007) Ecological and population genetics research imperatives for transgenic trees. *Tree Genetics & Genomes* **3**: pp119–33.

Farquhar, G D, Caemmerer, S V and Berry, J A (1980) A biochemical-model of photsynthetic CO_2 assimilation in leaves of C-3 species. *Planta* **149**(1): pp78–90.

Farwig, N, Brown, C and Böhning-Gaese, K (2007) Human disturbance reduces genetic diversity of an endangered tropical tree, Prunus africana (Rosaceae). *Conservation Genetics*: doi: 10.1007/s10592–007–9343–x.

Fatheuer, T (2007) Mit Agrotreibstoffen aus Brasilien gegen den Klimawandel? Kommentar von Thomas Fatheuer für die Heinrich-Böll-Stiftung. Heinrich-Böll-Stiftung website, http://www.boell.de/oekologie/klima/klima-energie-1557.html (viewed 20. October 2008).

Faurès, J-M, Hoogeveen, J and Bruinsma, J (2000) *The FAO Irrigated Area Forecast for 2030*. Food and Agriculture Organization of the United Nations (FAO), Rome.

Faust, J and Croissant, A (2007) *Staatlichkeit und Governance: Herausforderungen in Lateinamerika, Analysen und Stellungnahmen 1/2007*. German Development Institute (DIE), Bonn.

Fearnside, P M (2008) Amazon forest maintenance as a source of environmental services. *Annals of the Brazilian Academy of Sciences* **80**: pp101–14.

Feddema, J J, Oleson, K W, Bonan, G B, Mearns, L O, Buja, L E, Meehl, G A and Washington, W M (2005) The importance of land cover change in simulating future climates. *Science* **310**: pp1674–8.

Fedoroff, N V and Cohen, J E (1999) Plants and population: is there time? *Proceedings of the National Academy of Sciences USA* **96**: pp5903–7.

Fehrenbach, H (2007) *Kriterien zur nachhaltigen Bioenergienutzung im globalen Massstab. Zwischenstand aus einem Forschungsvorhaben im Auftrag des Umweltbundesamts. Vortragsfolien zu einem Vortrag im Rahmen der Tagung –Biomasseproduktion – ein Segen für die Land(wirt)schaft? BfN – INA, Insel Vilm, 12.–15. March 2007*. Bundesamt für Naturschutz Insel Vilm, Vilm.

Fehrenbach, H, Giegrich, J, Reinhardt, G, Schmitz, J, Sayer, U, Gretz, M, Seizinger, E and Lanje, K (2008) *Criteria for a Sustainable Use of Bioenergy on a Global Scale. UBA Texte 30/08*. Umweltbundesamt (UBA), Dessau.

Felton, A, Wood, J, Felton, A M, Hennessey, B and Lindenmayer, D B (2008) Bird community responses to reduced-impact logging in a certified forestry concession in lowland Bolivia. *Biological Conservation* **141**: pp545–55.

FES (Friedrich-Ebert-Stiftung) (2007) *The Creation of an International Renewable Energy Agency (IRENA). Veranstaltungsbericht, 21. June 2007*. FES Washington, Washington, DC.

FiBL (Forschungsinstitut für Biologischen Landbau) (2001) *Bio fördert Bodenfruchtbarkeit und Artenvielfalt. Erkenntnisse aus 21 Jahren DOK-Versuch (Dossier 1)*. FiBL, Frick, Switzerland.

Fichtner (2003) *Gutachten zur Berücksichtigung großer Laufwasserkraftwerke im EEG. Endbericht*. BMU, Fichtner, Berlin, Stuttgart.

Field, C B, Campbell, J E and Lobell, D B (2008) Biomass energy: the scale of the potential resource. *Trends in Ecology & Evolution* **23**(2): pp65–72.

Finon, D (2007) Pros and cons of alternative polices aimed at promoting renewables. *EIB papers* **12**(2): pp110-33.

Fischer, G, van Velthuizen, H, Shah, M and Nachtergaele, F O (2002) *Global Agro-Ecological Assessment for Agriculture in the 21st Century: Methodology and Results. Research Report RR-02-02*. International Institute for Applied Systems Analysis (IIASA), Laxenburg.

Fischlin, A, Midgley, G F, Price, J T, Leemans, R, Gopal, B, Turley, C, Rounsevell, M D A, Dube, O P, Tarazona, J and Velichko, A A (2007) Ecosystems, their properties, goods, and services. In Intergovernmental Panel on Climate Change (IPCC) (ed) *Climate Change 2007: Impacts, Adaptation and Vulnerability. Contribution of Working Group II to the Fourth Assessment Report of the Intergovernmental Panel on Climate Change*. Cambridge University Press, Cambridge, New York, pp211–72.

FNR (Fachagentur Nachwachsende Rohstoffe) (2005) *Leitfaden Bioenergie: Planung, Betrieb und Wirtschaftlichkeit von Bioenergieanlagen*. FNR, Gülzow.

FNR (Fachagentur Nachwachsende Rohstoffe) (2006a) *Handreichung Biogasgewinnung und -nutzung*. FNR, Gülzow.

FNR (Fachagentur Nachwachsende Rohstoffe) (2006b) *Biokraftstoffe – eine vergleichende Analyse*. FNR, Gülzow.

FNR (Fachagentur Nachwachsende Rohstoffe) (2006c) *Nachwachsende Rohstoffe in der Industrie*. FNR, Gülzow.

FNR (Fachagentur Nachwachsende Rohstoffe) (2006d) *Leitfaden – Bioenergie im Gartenbau*. FNR, Gülzow.

FNR (Fachagentur Nachwachsende Rohstoffe) (2006e) *Einspeisung von Biogas in das Erdgasnetz*. FNR, Gülzow.

FNR (Fachagentur Nachwachsende Rohstoffe) (2007a) *Bioenergie: Pflanzen, Rohstoffe, Produkte*. FNR, Gülzow.

FNR (Fachagentur Nachwachsende Rohstoffe) (2007b) *Handbuch – Bioenergie-Kleinanlagen*. FNR, Gülzow.

FOES (Förderverein Ökologische Steuerreform) (2008) *Schädliche Subventionen gegen die biologische Vielfalt. Eine Studie im Auftrag des DNR*. FOES, Munich.

Foley, J A (1995) An Equilibrium-Model of the Terrestrial Carbon Budget. *Tellus Series B-Chemical and Physical Meteorology* **47**(3): pp310–9.

Foley, J A, DeFries, R, Asner, G P, Barford, C, Bonan, G, Carpenter, S R, Chapin, F S, Coe, M T, Daily, G C, Gibbs, H K, Helkowski, J H, Holloway, T, Howard, E A, Kucharik, C J, Monfreda, C, Patz, J A, Prentice, I C, Ramankutty, N and Snyder, P K (2005) Global consequences of land use. *Science* **309**: pp570–4.

Foreign Policy (2008) The Failed State Index 2008. Foreign Policy July/August 2008. Foreign Policy website, http://www.foreignpolicy.com/story/cms.php?story_id=4350 (viewed 4. August 2008).

Fowles, M (2007) Black carbon sequestration as an alternative to bioenergy. *Biomass & Bioenergy* **31**: pp426-32.

Freibauer, A, Rounsevell, M D A, Smith, P and Verhagen, J (2004) Carbon sequestration in the acricultural soils of Europe. *Geoderma* **122**: pp1-23.

Fritsche, U R and Hennenberg, K (2008) *Stand der internationalen Prozesse zum Thema "nachhaltige Biomasse". Working Paper. Update 2*. Öko-Institut, Darmstadt.

Fritsche, U R and Wiegmann, K (2008) Ökobilanzierung der Umweltauswirkungen von Bioenergie-Konversionspfaden. Expertise for the WBGU Report "World in Transition: Future Bioenergy and Sustainable Land Use". WBGU website, http://www.wbgu.de/wbgu_jg2008_ex04.pdf

Fritsche, U R, Hennenberg, K J and Wiegmann, K (2008) *Bioenergy and Biodiversity: Potential for Sustainable Use of Degraded Lands. Briefing Paper for the Information Event at CBD-COP9 on May 27, 2008*. Öko-Institut, Darmstadt Office, Darmstadt.

Fritsche, U R, Dehoust, G, Jenseit, W, Hünecke, K, Rausch, L, Schüler, D, Wiegmann, K, Heinz, A, Hiebel, M, Ising, M, Kabasci, S, Unger, C, Thrän, D, Fröhlich, N, Scholwin, F, Reinhardt, G, Gärtner, S, Patyk, A, Bauer, F, Bemmann, U, Groß, B, Heib, M, Ziegler, C, Flake, M, Schmehl, M and Simon, S (2004) *Stromstoffanalyse zur nachhaltigen energetischen Nutzung von Biomasse. Verbundprojekt gefördert vom BMU im Rahmen des ZIP*. Bundesministerium für Umwelt, Naturschutz und Reaktorsicherheit (BMU), Darmstadt, Berlin.

Fritsche, U, Hünecke, K, Hermann, A, Schulze, F and Wiegmann, K (2006) *Sustainability Standards for Bioenergy*. WWF Germany, Frankfurt/M.

Fritz, T (2007) Zertifiziertes Raubrittertum. Wie Nichtregierungsorganisationen dem Welthandel auf die Sprünge helfen. *Lateinamerika Nachrichten* (396): pp31–5.

FSC (Forest Stewardship Council) (1996) *FSC International Standard. FSC Principles and Criteria for Forest Stewardship*. FSC Germany, Bonn.

Fürnsinn, S (2007) *Polygeneration – Strategie für eine dezentrale Energieerzeugung aus Biomasse. Internationale Tagung Thermo-chemische Biomasse-Vergasung für eine effiziente Strom-/Kraftstoffbereitstellung. Erkenntnisstand 2007*. IE Leipzig, TU Vienna, Leipzig, Vienna.

FWA (Fischer Weltalmanach) (2007) *Der Fischer Weltalmanach 2008: Zahlen, Daten, Fakten*. Fischer, Frankfurt/M.

FZK (Forschungszentrum Karlsruhe) (2007) Institut für Technische Chemie. FZK website, http://www.fzk.de/fzk/idcplg?IdcService=FZK&node=1429 (viewed 28. July 2007).

GBEP (Global Bioenergy Partnership) (2008) *A Review of the Current State of Bioenergy Development in G8+5 Countries*. GBEP, Food and Agriculture Organization of the United Nations (FAO), Rome.

GCP (Global Carbon Project) (2008) Carbon Budget and Trends 2007. GCP website, http://www.globalcarbonproject.org/ (viewed 30. September 2008).

GEF (Global Environment Facility) (2006) *GEF Resource Allocation Framework: Indicative Resource Allocations for GEF-4, for Biodiversity and Climate Change Focal Areas*. GEF, Washington, DC.

GEF (Global Environment Facility) (2007a) Focal Area Strategies and Strategic Programming for GEF-4. GEF/C.31/10. May 11, 2007. GEF, Washington, DC.

GEF (Global Environment Facility) (2007b) *Liquid Biofules in Transport: Conclusions and Recommendations of the Scientific and Technical Advisory Panel (STAP) to the Global Environment Facility (GEF)*. GEF/C.31/Inf.7. GEF Scientific and Technical Advisory Panel, New York.

GEF (Global Environment Facility) (2007c) Investing in Our Planet. GEF website, http://www.gefweb.org (viewed 30. September 2008).

GEF (Global Environment Facility) and UNDP – United Nations Development Programme (2006) *Environmentally Sustainable Transport and Climate Change. Experiences and Lessons From Community Initiatives. The GEF Small Grants Programme (SGF).* GEF, UNDP, Washington, DC, New York.

Geist, H J and Lambin, E F (2002) Proximate causes and underlying driving forces of tropical deforestation. *BioScience* **52**: pp143–50.

Gerbens-Leenes, P W, Nonhebel, S and Ivens, W P M F (2002) A method to determine land requirements relating to food consumption patterns. *Agriculture, Ecosystems and Environment* **90**(1): pp47–58.

Gerten, D, Schaphoff, S, Haberlandt, U, Lucht, W and Sitch, S (2004) Terrestrial vegetation and water balance - hydrological evaluation of a dynamic global vegetation model. *Journal of Hydrology* **286**(1-4): pp249–70.

Girardet, H (1996) *The Gaia Atlas of Cities: New Directions for Sustainable Urban Living.* Gaia Books Ltd., London.

Glastra, R, Wakker, E and Richert, W (2002) *Kahlschlag zum Frühstück. Palmöl–Produkte und die Zerstörung indonesischer Wälder: Zusammenhänge, Ursachen und Konsequenzen.* WWF Germany, Frankfurt/M.

Glowka, L, Burhenne–Guilmin, F and Synge, H (1994) *A Guide to the Convention on Biological Diversity.* The World Conservation Union (IUCN), Gland, Cambridge.

Gnansounou, E, Panichelli, L, Dauriat, A and Villegas, J D (2008) *Accounting for Indirect Land-Use Changes in GHG Balances of Biofules. Working Paper.* Ecole Polytechnique Federale de Lausanne, Lausanne, Switzerland.

Goeden, R D and Andres, L A (1999) Biological control of weeds in terrestrial and aquatic environments. Chapter 34. In Bellows, T S and Fisher, T W (eds) *Handbook of Biological Control: Principles and Applications of Biological Control.* Academic Press, London, pp871–90.

Govindasamy, B, Duffy, P B and Caldeira, K (2001) Land use changes and and northern hemisphere cooling. *Geophysical Research Letters* **28**: pp291–4.

Grain (2008) *Agrofuels in India.* Grain, Barcelona.

Grävingholt, J (2007) *Staatlichkeit und Governance: Herausforderungen in Zentralasien und im Südkaukasus. Analysen und Stellungnahmen 2/2007.* German Development Institute (DIE), Bonn.

Green, R E, Cornell, S J, Scharlemann, J P W and Balmford, A (2005) Farming and the fate of wild nature. *Science* **307**: pp550–5.

Greenpeace (2007) *Globale Energie-[r]evolution; Ein Weg zu einer nachhaltigen Enerige-Zukunft für die Welt.* Greenpeace International, Amsterdam, Brussels.

Greenpeace (2008) *Soja-Diesel im Tank.* Greenpeace Factsheet 4/2008. Greenpeace website, http://www.greenpeace.de/fileadmin/gpd/user_upload/themen/waelder/FSSojaDieselFINAL.pdf (viewed 23. June 2008).

Greenpeace and EREC (European Renewable Energy Council) (2007) *Globale Energie-(R)Evolution – Ein Weg zu einer nachhaltigen Energie-Zukunft für die Welt.* Greenpeace, EREC, Berlin.

Grieg-Gran, M (2006) *The Cost of Avoiding Deforestation. Report prepared for the Stern Review of the Economics of Climate Change.* International Institute for Environment and Development (IIED), London.

Grimm, S and Klingebiel, S (2007) *Staatlichkeit und Governance: Herausforderungen in Subsahara-Afrika. Analysen und Stellungnahmen 3/2007.* German Development Institute (DIE), Bonn.

Grogan, P and Matthews, R (2002) A modelling analysis of the potential for soil carbon sequestration under short rotation coppice willow energy plantations. *Soil Use and Management* **18**: pp175-83.

Grönkvist, S, Möllersten, K and Pingoud, K (2006) Equal opportunity for biomass in greenhouse gas accouniting of CO_2 capture and storage: A step towards more cost-effective climate change mitigation regimes. *Mitigation and Adaptation Srategies for Global Change* **11**: pp1083–96.

Grote, U (2002) *Eco-Labelling in the Agricultural Sector: An International Perspective. Paper anlässlich der High-level Pan-European Conference on Agriculture and Biodiversity: Towards Integrating Biological and Landscape Diversity for Sustainable Agriculture in Europe. Paris, Maison de lUnesco, 5.–7. June 2002.* UNESCO, Paris.

Grunert, M (2007) *Pflanzenöl als Kraftstoff. Möglichkeiten und Grenzen aus acker- und pflanzenbaulicher Sicht. Tagungsband der 6. Fachtagung Kraftstoff Pflanzenöl am 9.11. 2007 in Nossen.* Conference Secretariat, Nossen.

Grünzweig, J M, Gelfand, I, Fried, Y and Yakir, D (2007) Biogeochemical factors contibuting to enhanced carbon storage following afforestation of a semi-arid shrubland. *Biogeosciences* **4**: pp891–904.

GTZ (Deutsche Gesellschaft für Technische Zusammenarbeit) (2006) Liquid Biofuels for Transportation – Chinese Protential and Implications for Sustainable Agriculture and Energy in the 21st Century. GTZ website, http://www.gtz.de/de/dokumente/en-biofuels-for-transportation-in-china-2005.pdf (viewed 27. February 2008).

GTZ (Deutsche Gesellschaft für Technische Zusammenarbeit) (2007a) *Energiepolitische Rahmenbedingungen für Strommärkte und erneuerbare Energien. 23 Länderanalysen.* GTZ Sektorvorhaben TERNA Windenergie, Eschborn.

GTZ (Deutsche Gesellschaft für Technische Zusammenarbeit) (2007b) *The Voluntary Emission Reduction Market: Characteristics, Risks and Opportunities. Factsheet Commissioned by the Federal Ministry for Economic Cooperation and Development.* GTZ, Eschborn.

GTZ (Deutsche Gesellschaft für Technische Zusammenarbeit) (2007c) *Cooking Energy – Why it Really Matters if we Are to Halve Poverty by 2015.* GTZ, Eschborn.

GTZ-EAP (Deutsche Gesellschaft für Technische Zusammenarbeit Energy Advisory Project) (2007) *Energy Advisory Project Annual General Report 2007.* GTZ-EAP, Kampala.

Guo, L B and Gifford, R M (2002) Soil carbon stocks and land use change: a meta analysis. *Global Change Biology* **8**: pp345–60.

Gupta, S, Tirpak, D A, Burger, N, Gupta, J, Höhne, N, Boncheva, A I, Kanoan, G M, Kolstad, C, Kruger, J A, Michaelowa, A, Murase, S, Pershing, J, Saijo, T and Sari, A (2007) Policies, instruments and co-operative arrangements. In Metz, B, Davidson, O R, Bosch, P R, Dave, R and Meyer, L A (eds) *Climate Change 2007. Mitigation. Contribution of Working Group III to the Fourth Assessment Report of the Intergovernmental Panel on Climate Change.* Cambridge University Press, Cambridge, New York, pp745-808.

Gutman, P and Davidson, S (2007) *A Review of Innovative International Financial Mechanisms for Biodiversity Conservation. With Special Focus on the International Financing of Developing Countries Protected Areas.* WWF-MPO Macroeconomics Programme Office, New York.

GWEC (Global Wind Energy Council) (2008) *Global Wind 2008 Report.* GWEC, Brussels.

Haberl, H, Erb, K H, Krausmann, F, Gaube, V, Bondeau, A, Plutzar, C, Gingrich, S, Lucht, W and Fischer-Kowalski, M (2007) Quantifiying and mapping the human appropriation of net primary production in earth's terrestrial ecosystems. *PNAS* **104**(31): pp12942–7.

Hails, C (2006) *Living Planet Report 2006.* World Wildlife Fund, Global Footprint Network, Zoological Society London, Geneva, London.

Haines, D and Skinner, I (2005) *The Marketing of Mobility Services.* Institute for European Environmental Policy (IEEP), London.

Hakkila, P and Parikka, M (2002) Fuel resources from the forest. In Richardson, J, Björheden, R, Hakkila, P, Lowe, A T and Smith, C T (eds) *Bioenergy From Sustainable Forestry: Guiding Principles and Practice.* Kluwer, Dordrecht, pp19–48.

Halpin, P N (1997) Global climate change and natural-area protection: management responses and research directions. *Ecological Applications* **7**: pp828–43.

Hannah, L, Midgley, G F, Andelman, S J, Araújo, M B, Hughes, G, Martinez-Meyer, E, Pearson, R and Williams, P (2007) Protected area needs in a changing climate. *Frontiers in Ecology and the Environment* **5**(3): pp131–8.

Hansen, E A (1993) Soil carbon sequestration beneath hybrid poplar plantations in the north central United States. *Biomass and Bioenergy* **5**: pp431-6.

Hanstad, T, Haque, T and Nielsen, R (2008) Improving land access for India's rural poor. *Economic & Political Weekly* (3): pp49–55.

Harcombe, P A, Harmon, M E and Greene, S E (1990) Changes in biomass and production over 53 years in a coastal Picea-sitchensis – Tsuga-heterophylla forest. *Canadian Journal of Forest Research* **20**: pp1602–10.

Harmon, M-E, Ferrell, W K and Franklin, J F (1990) Effects on carbon storage of conversion of old-growth forests to young forests. *Science* **247**: pp699–702.

Hawn, A (2008) *Malua Wildlife Habitat Conservation Bank Launches in Sabah, Malysia. New Business Model Generates Innovative Product to Support Wildlife Conservation.* Malua BioBank, Sabah, Malaysia.

Heaton, E A, Clifton-Brown, J, Voigt, T B, Jones, M B and Long, S P (2004) Micanthus for renewable energy generation: European Union experience and projections for Illinois. *Mitigation and Adaptation Strategies for Global Change* **9**: pp433–51.

Hedde, M, Aubert, M, Decaens, T and Bureau, F (2008) Dynamics of soil carbon in a beechwood chronosequence forest. *Forest Ecology and Management* **255**: pp193–202.

Heistermann, M, Müller, C and Ronneberger, K (2006) Land in sight? Achievements, deficits and potentials of continental to global scale land-use modeling. *Agriculture Ecosystems & Environment* **114**(2-4): pp141–58.

Hepeng, J (2008) Chinese biofuel ,could endanger biodiversity'. SciDev Net online website, http://www.scidev.net/en/news/chinese-biofuel-could-endanger-biodiversity-.html (viewed 19. May 2008).

Hickler, T, Smith, B, Sykes, M T, Davis, M B, Sugita, S and Walker, K (2004) Using a generalized vegetation model to simulate vegetation dynamics in northeastern USA. *Ecology* **85**(2): pp519–30.

Hickler, T, Smith, B, Prentice, C I, Mjofors, C, Miller, P, Arneth, A and Sykes, M T (2008) CO_2 fertilization in temperate FACE experiments not representative of boreal and tropical forests. *Global Change Biology* **14**(7): pp1531–42.

Hilf, M and Oeter, S (2005) *WTO-Recht. Rechtsordnung des Welthandels.* Nomos, Baden-Baden.

Höhne, N, Wartmann, S, Herold, A and Freibauer, A (2007) The rules for land use, land use change and forestry under the Kyoto Protocol - lessons learned for the future climate negotiations. *Environmental Science & Policy* **10**(4): pp269–394.

Holmberg, N and Bülow, L (1998) Improving stress tolerance by gene transfer. *Trends in plant science* **3**(2): pp61–6.

Hoogwijk, M, Faaij, A, van den Broek, R, Berndes, G, Gielen, D and Turkenburg, W (2003) Exploration of the ranges of the global potential of biomass for energy. *Biomass & Bioenergy* **25**(2): pp119–33.

Hoogwijk, M, Faaij, A, Eickhout, B, de Vries, B and Turkenburg, W (2005) Potential of biomass energy out to 2100, for four IPCC SRES land-use scenarios. *Biomass & Bioenergy* **29**: pp225–57.

Hooijer, A, Silvius, M, Wösten, H and Page, S (ed) (2006) *PEAT-CO₂. Assessment of CO_2 Emissions From Drained Peatlands in SE Asia. Delft Hydraulics Report Q3943.* WL Delft Hydraulics, Delft.

Houghton, R A (2003) *Emissions (and Sinks) of Carbon from Land-Use Change (Estimates of National Sources and Sinks of Carbon Resulting From Changes in Land Use, 1950 to 2000.* Woods Hole Research Center, Falmouth, MA.

House, J I, Prentice, C I and Le Quére, C (2002) Maximum impacts of future reforestation or deforestation on atmospheric CO_2. *Global Change Biology* **8**: pp1047–52.

Huang, J, Pray, C and Rozelle, S (2002) Enhancing the crops to feed the poor. *Nature* **418**: pp678–84.

Huberman, D (2007) Scaling up financing for forest protected areas: developing international payments for ecosystem services. In Schmitt, C B, Pistorius, T and Winkel, G (eds) *A Global Network of Forest Protected Areas under the CBD: Opportunities and Challenges. Proceedings of an International Expert Workshop held in Freiburg, Germany, May 9–11, 2007.* Kessel, Remagen, pp75–81.

IAASTD (International Assessment of Agricultural Knowledge Science and Technology for Development) (2008) IAASTD Homepage. IAASTD website, http://www.agassessment.org/ (viewed 16. May 2008).

IADB (Inter-American Development Bank) (2008) Press Release: Inter-American Development Bank Announces Partnership to Develop Sustainable Biofuels. IDB website, http://www.iadb.org/news/articledetail.cfm?language=EN&artid=4507 (viewed 07. May 2008).

IBRD (International Bank for Reconstruction and Development) (2007) *Catalyzing Private Investment for a Low-Carbon-Economy – World Bank Group Progress on Renewable Energy and Energy Efficiency in Fiscal 2007.* World Bank, Washington, DC.

IE (Institute for Energy and Environment) (2007a) *Möglichkeiten einer europäischen Biogaseinspeisungsstrategie: Studie im Auftrag der Bundestagsfraktion Bündnis 90/Die Grünen.* IE, Berlin, Leipzig.

IE (Institute for Energy and Environment) (2007b) *Monitoring zur Wirkung des novellierten Erneuerbare-Energien-Gesetzes (EEG) auf die Entwicklung der Stromerzeugung aus Biomasse: Endbericht im Auftrag des Bundesministeriums für Umwelt, Naturschutz und Reaktorsicherheit (BMU).* IE, Berlin, Leipzig.

IEA (International Energy Agency) (2004) *Biofuels for Transport. An International Perspective.* IEA, Paris.

IEA (International Energy Agency) (2006a) *The Energy Situation in Brazil: An Overview.* IEA, Paris.

IEA (International Energy Agency) (2006b) *World Energy Outlook 2006.* IEA, Paris.

IEA (International Energy Agency) (2007a) *World Energy Outlook 2007.* IEA, Paris.

IEA (International Energy Agency) (2007b) IEA Energy Technology Essentials: Biomass for Power Generation and CHP. IEA website, http://www.iea.org/textbase/techno/essentials3.pdf (viewed 12. December 2007).

IEA (International Energy Agency) (2007c) *Renewables for Heating and Cooling*. IEA, Paris.

IEA (International Energy Agency) (2007d) *World Energy Outlook 2007: Zusammenfassung: China and India Insights*. IEA, Paris.

IEA (International Energy Agency) (2008a) *CHP/DHC Country Scorecard: Germany. The International CHP/DHC Collaborative*. IEA, Paris.

IEA (International Energy Agency) (2008b) *Combined Heat & Power and Emissions Trading: Options for Policy Makers IEA Information Paper*. IEA, Paris.

IEA (International Energy Agency) (2008c) Statistics and Balances. IEA website, http://www.iea.org/ (viewed 31. March 2008).

IEA (International Energy Agency) (ed) (2008d) *World Energy Outlook 2008. Global Energy Trends*. IEA, OECD, Paris.

IEA (International Energy Agency) and JREC (Johannesburg Renewable Energy Coalition) (2008) Global Renewable Energy – Policies and Measures. IEA, JREC website, http://www.iea.org/textbase/pm/grindex.aspx (viewed 01. April 2008).

IFEU (Institute for Energy and Environmental Research) (2007) *Nachwachsende Rohstoffe für die chemische Industrie: Optionen und Potenziale für die Zukunft*. IFEU, Heidelberg.

IFPRI (International Food Policy Research Institute (2008) *Hohe Nahrungsmittelpreise – Konzept für die Wege aus der Krise. IFPRI Policy Paper, May 2008*. IFPRI, Washington, DC.

Igelspacher, R, Antoni, D, Kroner, T, Prechtl, S, Schieder, D, Schwarz, W H and Faulstich, M (2006) Bioethanolproduktion aus Lignocellulose. Stand der Technik und Perspektiven. *BWK* **58**(3): pp50–4.

IGES (Institute for Global Environmental Strategies) (2008) Kyoto Protocol Related Information. CDM/JI Project Data. IGES website, http://www.iges.or.jp/en/cdm/report.html (viewed 13. October 2008).

IISc (Indian Institute of Science) (2006) Experience With Gasifiers in India. IISc website, http://cgpl.iisc.ernet.in/ (viewed 28. July 2007).

IISD (International Institute for Sustainable Development) (2007) *Summary of the Eighth Conference of the Parties to the Convention to Combat Desertification: 3–14 September 2007*. IISD, New York.

IISD (International Institute for Sustainable Development) (2008) *Summary of the Sixteenth Session of the Commission on Sustainable Development (CSD-16), New York 5–16 May 2008* IISD, New York.

IMF (International Monetary Fund) (2007) *World Economic Outlook. Globalization and Inequality*. IMF, Washington, DC.

Imhoff, M L, Bounoua, L, Ricketts, T, Lucks, C, Harriss, R and Lawrence, W T (2004) Global patterns in human consumption of net primary production. *Nature* **429**: pp870–3.

IMV (Institut for Miljøvurdering) (2002) *Assessing the Ecological Footprint. A Look at the WWF's Living Planet Report 2002*. IMV, Copenhagen.

Intelligent Energy (2007) *Energy Services for Poverty Alleviation in Developing Countries*. Intelligent Energy Europe, Brussels.

IPCC (Intergovernmental Panel on Climate Change) (2000) *Emissions Scenarios. A Special Report of Working Group III of IPCC*. Cambridge University Press, Cambridge, New York.

IPCC (Intergovernmental Panel on Climate Change) (2005) *Carbon Dioxide Capture and Storage. A Special Report of Working Group III of the IPCC*. Cambridge University Press., Cambridge, New York.

IPCC (Intergovernmental Panel on Climate Change) (2006) Guidelines for National Greenhouse Gas Inventories. IPCC website, http://www.ipcc-nggip.iges.or.jp/public/2006gl/index.htm (viewed 09. November 2007).

IPCC (Intergovernmental Panel on Climate Change) (2007a) *Climate Change 2007. The Physical Science Basis. Working Group I Contribution to the Fourth Assessment Report*. Cambridge University Press, Cambridge, New York.

IPCC (Intergovernmental Panel on Climate Change) (2007b) *Climate Change 2007: Impacts, Adaptation and Vulnerability. Contribution of Working Group II to the Fourth Assessment Report of the IPCC*. Cambridge University Press, Cambridge, New York.

IPCC (Intergovernmental Panel on Climate Change) (2007c) *Climate Change 2007. Mitigation of Climate Change. Working Group III Contribution to the Fourth Assessment Report*. Cambridge University Press, Cambridge, New York.

IPCC (Intergovernmental Panel on Climate Change) (2007d) *Climate Change 2007: The Fourth Assessment Report of the IPCC. Summary for Policymakers*. Cambridge University Press, Cambridge, New York.

IRENA (Initiative for an International Renewable Energy Agency) (2008) About IRENA. IRENA website, http://www.irena.org/irena.htm (viewed 01. September 2008).

ISAAA (International Service for the Acquisition of Agri-Biotech Applications) (2008) Globale Anbauflächen 2007: Weiter Zuwachs für gv-Pflanzen: Flächen steigen auf 114 Millionen Hektar. TransGen Wissenschaftskommunikation website, http://www.transgen.de/anbau/eu_international/531.doku.html (viewed 08. July 2008).

ISET (Institute for Solar Energy Technology) (2008) *Biogasaufbereitung zu Biomethan. 6. Hanauer Dialog. Tagungsband*. ISET, Kassel, Hanau.

ITADA (Grenzüberschreitendes Institut zur rentablen umweltgerechten Landbewirtschaftung) (2005) *Nachhaltige Maisproduktion am Oberrhein: Konzeption und vertiefte Auswertung von Anbausystemen. Abschlussbericht Projekt 3. Grenzüberschreitendes Institut zur rentablen umweltgerechten Landbewirtschaftung*. ITADA, Colmar, France.

ITTO (International Tropical Timber Council) (ed) (2006) *Status of Tropical Forest Management 2005*. ITTO, Yokohama, Japan.

Iturregui, P and Dutschke, M (2005) *Liberalisation of Environmental Goods & Services and Climate Change. Hamburg Institute of International Economics (HWWA) Discussion Paper 335*. HWWA, Hamburg.

IUCN (The World Conservation Union) (1994) *Guidelines for Protected Areas Management Categories*. IUCN, Gland.

IUCN (The World Conservation Union) (2003) *WPC Outputs. The Durban Accord. World Parks Congress 2003*. IUCN, Gland.

IWMI (International Water Management Institute) (ed) (2007) *Water for Food. Water for Life. A Comprehensive Assessment of Water Management in Agriculture*. Earthscan, London, New York.

IZT (Institute for Futures Studies and Technology Assessment) (2007) *Stoffliche oder energetische Nutzung? Nutzungskonkurrenz um die Ressource Holz. Holzwende Paper*. IZT, Berlin.

Jackson, R B, Jobbágy, E G, Avissar, R, Roy, S B, Barrett, D J, Cook, C W, Farley, K A, le Maitre, D C, McCarl, B A and

Murray, B C (2005) Trading water for carbon with biological carbon sequestration. *Science* **310**: pp1944-7.

Jagadish, K S (2004) *The Development and Dissemination of Efficient Domestic Cook Stoves and Other Devices in Karnataka. Department of Civil Engineering and Centre for Sustainable Technologies, Indian Institute of Science, Bangalore.* Indian Academy of Sciences, Bangalore.

James, A N, Gaston, K J and Balmford, A (1999) Balancing the Earths account. *Science* **401**: p323.

Jarnagin, S T (2004) Regional and global patterns of population, land use, and land cover change. *GIScience and Remote Sensing* **41**: pp207–27.

Jauhiainen, J, Takashi, H, Heikkinen, J E P, Martikainen, P J and Vasander, H (2005) Carbon fluxes from a tropical peat swamp forest floor. *Global Change Biology* **11**: pp1788-97.

JIKO (Joint Implementation Koordinierungsstelle im Bundesumweltministerium) (2007) The Bali CDM Agenda – Cleaning up the Leftovers? *JIKO* **4**: pp1-6.

JIKO (Joint Implementation Koordinierungsstelle im Bundesumweltministerium) (2008) Komplex und kostenintensiv in der Entwicklung: Eine Kommentierung der CDM-Methoden für den Verkehrssektor. *JIKO* **4**: pp1-10.

Johnson, F, Seebaluck, V, Watson, H and Woods, J (2006) Bio-ethanol from sugarcane and sweet sorghum in Southern Africa: agro-industrial development, import substitution and export diversification. In ICTSD Project on Trade and Sustainable Energy (ed) *Linking Trade, Climate Change and Energy. ICTSD Project on Trade and Sustainable Energy.* International Centre for Trade and Sustainable Development (ICTSD), Geneva, pp23–4.

Jonsell, M, Hansson, J and Wedmo, L (2007) Diversity of saproxilic beetle species in logging residues in Sweden – comparisons between tree species and diameters. *Biological Conservation* **138**: pp89-99.

Joos, F (2002) CO$_2$ Impulse Response Function of Bern SAR and Bern TAR Models. United Nations Framework Convention on Climate Change (UNFCCC) website, http://unfccc.int/resource/brazil/carbon.html (viewed 13. October 2008).

Jorgensen, J R, Deleuran, L C and Wollenberger, B (2007) Prospects of whole grain crops of wheat, rye and triticale under different fertilizer regimes for energy production. *Biomass and Bioenergy* **31**: pp308–17.

Jull, C, Redondo, P C, Mosoti, V and Vapnek, J (2007) *Recent Trends in the Law and Policy of Bioenergy Production, Promotion and Use.* Food and Agriculture Organization of the United Nations (FAO), Rome.

Jungbluth, N, Büsser, S, Frischknecht, R and Tuchschmid, M (2008) *Ökobilanz von Energieprodukten: Life Cycle Assessment of Biomass-to-Liquid Fuels. Final Report.* ESU-Services Ltd, Uster.

Jürgens, I, Schlamadinger, B and Gomez, P (2006) Bioenergy and the CDM in the emerging market for carbon credits. *Mitigation and Adaptation Strategies for Global Change* **11**: pp1051–81.

Kägi, T, Freiermuth Knuchel, R, Nemecek, T and Gaillard, G (2007) *Ökobilanz von Energieprodukten: Bewertung der landwirtschaftlichen Biomasse-Produktion (Draft).* Forschungsanstalt Agroscope, Bern.

Kaiser, M (2008) Entwaldung stoppen. In FUE (Forum Umwelt und Entwicklung) (ed) *Im Labyrinth der Labels. Nachhaltigkeit durch Zertifizierung? Rundbrief Forum Umwelt & Entwicklung 3/2008.* Projektstelle Umwelt & Entwicklung, Bonn, pp22-3.

Kaltner, F J, Azevedo, G F P, Campos, I A and Mundim, A O F (2005) *Liquid Biofuels for Transportation in Brazil. Poten-tial and Implications for Sustainable Agriculture and Energy in the 21st Century.* BMELV, GTZ, Berlin, Eschborn.

Kaltschmitt, M and Hartmann, H H (eds) (2003) *Energie aus Biomasse: Grundlagen, Techniken und Verfahren.* Springer, Berlin, Heidelberg, New York.

Kejun, J, Xiulian, H, Xianli, Z, Garg, A, Halsnaes, K and Qiang, L (2007) *Balancing Energy, Development and Climate Priorities in China: Current Status and the Way Ahead.* UNEP Risoe Centre on Energy, Climate and Sustainable Development, Roskilde.

Keyzer, M A, Merbis, M D, Pavel, I F P W and van Wesenbeeck, C F A (2005) Diet shifts towards meat and the effects on cereal use: Can we feed the animals in 2030? *Ecological Economics* **55**(2): pp187–202.

KfW Bankengruppe (2008) *Agrartreibstoffe und Entwicklung – Positionsbestimmung für die FZ.* KfW Bankengruppe, Frankfurt/M.

Kiers, T E, Leakey, R R B, Izac, A-M, Heinemann, J A, Rosenthal, E, Nathan, D and Jiggins, J (2008) Agriculture at crossroads. *Science* **320**: p320f.

Kirby, K R and Potvin, C (2007) Variation in carbon storage among tree species: Implications for the management of al small-scale carbon sink project. *Forest Ecology and Management* **246**: pp208–21.

Klein Goldewijk, K (2001) Estimating global land use change over the past 300 years: the HYDE database. *Global Biogeochemical Cycles* **15**: pp417–33.

Klein Goldewijk, C G M and Battjes, J J (1997) *A Hundred Year (1890–1990) Database for Integrated Environmental Assessments (HYDE Version 1.1). RIVM Report.* National Institute of Public Health and Environmental Protection (RIVM), The Hague.

Klein Goldewijk, K, Bouwman, A F and van Drecht, G (2007) Mapping contemporary global cropland and grassland distributions on a 5 by 5 minute resolution. *Journal of Land Use Science* **2**(3): pp167–90.

Klink, C A and Machado, R B (2005) Conservation of the Brazilian Cerrado. *Conservation Biology* **19**(3): pp707–13.

Knappe, F, Böß, A, Fehrenbach, H, Giegrich, J, Vogt, R, Dehoust, G, Fritsche, U, Schüler, D and Wiegmann, K (2007) *Stoffstrommanagement von Biomasseabfällen mit dem Ziel der Optimierung der Verwertung organischer Abfälle. Im Auftrag des Umweltbundesamtes. UBA Texte 04/07.* Institute for Energy and Environmental Research (IFEU), Heidelberg.

Knoeff, H (2005) *Handbook Biomass Gasification: BTG Biomass Technology Group.* Biomass Technology Group (BTG), Enschede.

Koh, L P and Wilcove, D S (2008) Is oil palm agriculture really destroying tropical biodiversity? *Conservation Letters* **1**(2): pp60–4.

Kojima, M and Johnson, T (2005) *Potential for Biofuels for Transport in Developing Countries.* The World Bank Energy Sector Management Assistance Programme (ESMAP), Washington, DC.

Koplow, D (2007) *Biofuels – At What Cost? Government Support for Ethanol and Biodiesel in the United States: 2007 Update.* International Institute for Sustainable Development (IISD), Geneva.

Krausmann, F, Erb, K-H, Gingrich, S, Lauk, C and Haberl, H (2007) Global patterns of socioeconomic biomass flows in the year 2000: a comprehensive assessment of supply, consumptions and constraints. *Ecological Economics* **65**: pp471-87.

KTBL (Association for Technology and Structures in Agriculture) (2006) *Energiepflanzen. Daten für die Planung des Energiepflanzenanbaus.* Association for Technology and

Structures in Agriculture, Darmstadt and Leibniz Institute for Agricultural Engineering Potsdam-Bornim.

Kulessa, M E (2007) Setting efficient climate policy targets: mission possible? *Intereconomics* **42**(2): pp64–71.

Kulessa, M E and Ringel, M (2003) Kompensationen als innovatives Instrument globaler Umweltschutzpolitik: Möglichkeiten und Grenzen einer Weiterentwicklung des Konzepts am Beispiel der biologischen Vielfalt. *Zeitschrift für Umweltpolitik und Umweltrecht* **83**: pp263–85.

Kumar, R, Lokras, S S and Jagadish, J S (1990) *Development, Analysis, Dissemination of a 3-Pan Cooking Stove*. Centre for the Application of Science & Technology to Rural Areas (ASTRA), Bangalore.

Kutas, G, Lindberg, C and Steenblik, R (2007) *Biofuels – At What Cost? Government Support for Ethanol and Biodiesel in the European Union: 2007 Update*. International Institute for Sustainable Development (IISD), Geneva.

La Rovere, E L, Pereira, A O, Simões, A F, Pereira, A S, Garg, A, Halsnaes, K, Dubeu, X and da Costa, R C (2007) Development First: Linking Energy and Emission Policies with Sustainable Development for Brazil. UNEP Risoe Centre on Energy, Climate and Sustainable Development website, http://www.developmentfirst.org/Publications/DevelopEnergyClimate_Brazil.pdf (viewed 20. October 2008).

Lal, R, Uphoff, N, Stewart, B A and Hansen, D O (ed) (2005) *Climate Change and Global Food Security*. Taylor & Francis, Boca Raton.

Lambin, E F, Turner, B L, Geist, H J, Agbola, S B, Angelsen, A, Bruce, J W, Coomes, O T, Dirzo, R, Fischer, G, Folke, C, George, P S, Homewood, K, Imbernon, J, Leemans, R, Li, X, Moran, E F, Mortimore, M, Ramakrishnan, P S, Richards, J F, Skaanes, H, Steffen, W, Stone, G D, Svedin, U, Veldkamp, T A, Vogel, C and Xu, J (2001) The causes of land-use and land-cover change: moving beyond the myths. *Global Environmental Change* **11**: pp261–9.

Lambin, E F, Geist, H J and Lepers, E (2003) Dynamics of land-use and land-cover change in tropical regions. *Annual Review of Environment and Resources* **28**: pp205–41.

Legutke, S and Voss, R (1999) *The Hamburg Atmosphere-Ocean Coupled Circulation Model ECHO-G. Technical Reports*. German High Performance Computing Centre for Climate and Earth System Research (DKRZ), Hamburg.

Lehmann, J (2007) A handful of carbon. *Nature* **447**: pp143–4.

Lehner, B and Doll, P (2004) Development and validation of a global database of lakes, reservoirs and wetlands. *Journal of Hydrology* **296**(1–4): pp1-22.

Lemus, R and Lal, R (2005) Bioenergy crops and carbon sequestration. *Plant Sciences Critical Reviews* **24**: pp1-21.

Lepers, E, Lambin, E F, Janetos, A C, de Fries, R, Achard, F, Ramankutty, N and Scholes, R J (2005) A synthesis of information on rapid land-cover change fort he period 1981–2000. *BioScience* **55**: pp115–24.

Leuschner, M (2008) Regelwerk für Biogaseinspeisung bis auf EEG-Bonus komplett. Netzbetreiber ist gerüstet. *BWK* **60**(6): pp21-3.

Levidow, L and Paul, H (2008) Land Use, Bioenergy and Agro-Biotechnology. Expertise for the WBGU Report "World in Transition: Future Bioenergy and Sustainable Land Use". WBGU website, http://www.wbgu.de/wbgu_jg2008_ex05.pdf

Levine, M, Ürge-Vorsatz, D, Blok, K, Geng, L, Harvey, D, Lang, S, Levermore, G, Mongameli Mehlwana, A, Mirasgedis, S, Novikova, A, Rilling, J and Yoshino, H (2007) Residential and commercial buildings. In Metz, B, Davidson, O R, Bosch, P R, Dave, R and Meyer, L A (eds) *Climate Change 2007: Mitigation. Contribution of Working Group III to the Fourth Assessment Report of the Intergovernmental Panel on Climate Change* Cambridge University Press, Cambridge, New York, pp387–446.

Lewandowski, I and Faaij, A P C (2006) Steps towards the development of a certification system for sustainable bioenergy trade. *Biomass & Bioenergy* **30**: pp83–104.

Lewandowski, I and Schmidt, U (2006) Nitrogen, energy and land use efficiencies of miscanthus, reed canary grass and triticale as determined by the boundary line approach. *Agriculture Ecosystems & Environment* **112**: pp335–46.

LfL Bayern (Bayerische Landesanstalt für Landwirtschaft) (2008) Basisdaten – für die Ermittlung des Düngebedarfs – für die Umsetzung der Düngeverordnung. LfL website, http://www.lfl.bayern.de/iab/duengung/mineralisch/10536/linkurl_0_9_0_0.pdf (viewed 16. June 2008).

Liberloo, M, Calfapietra, C, Lukac, M, Godbold, D, Luo, Z-B, Polle, A, Hoosbeek, M R, Kull, O, Marek, M, Raines, C, Rubino, M, Taylor, G, Scarascia-Mugnozza, G and Ceulemans, R (2006) Woody biomass production during the second rotation of a bio-energy Populus plantation increases in a future high CO_2 world. *Global Change Biology* **12**: pp1094–106.

Lieberei, R, Reisdorff, C and Franke, W (ed) (2007) *Nutzpflanzenkunde*. Thieme, Stuttgart, New York.

Lindlein, P (2007) *Bioenergy for Development in Africa*. international Consulting economists & engineers (iCee), Frankfurt/M.

Lloyd, J and Taylor, J A (1994) On the temperature-dependence of soil respiration. *Functional Ecology* **8**(3): pp315–23.

Loose, C and Korn, H (2008) Von Bonn nach Nagoya": Bewährungsprobe für die Biodiversitätskonvention. In Altner, G, Leitschuh-Fecht, H, Michelsen, G, Simonis, U E and von Weizsäcker, C (eds) *Jahrbuch Ökologie*. Hirzel, Stuttgart, pp57–67.

López-Claros, A (2006) *Global Competitiveness Report 2006–2007*. World Economic Forum, Geneva.

López-Hurtado, C (2002) Social labelling and WTO law. *Journal of International Economic Law* **5**: pp719–46.

López-Ulloa, M, Veldenkamp, E and de Koning, G H (2005) Soil carbon stabilization in coverted tropical pastures and forests depends on soil type. *Soil Science Society of America Journal* **69**: pp1110–7.

Low, T and Booth, C (2007) *The Weedy Truth About Biofuels*. The Invasive Species Council, Melbourne.

Loy, D (2007) *Energiepolitische Rahmenbedingungen für Strommärkte und erneuerbare Energien – 23 Länderanalysen*. GTZ, Eschborn.

Lucht, W, Prentice, I C, Myneni, R B, Sitch, S, Friedlingstein, P, Cramer, W, Bousquet, P, Buermann, W and Smith, B (2002) Climatic control of the high-latitude vegetation greening trend and Pinatubo effect. *Science* **296**(5573): pp1687–9.

Luhnow, D and Samor, G (2006) As Brazil fills up on Ethanol, it weans off energy imports. The Wall Street Journal, Washington, DC.

Lula da Silva, I (2007) Brazils President Lula on Trade, Agriculture, Poverty and Biofuels. Interview Europaen Parlimant, 05.07.2007. European Parliament website, http://www.europarl.europa.eu/sides/getDoc.do?language=EN&type=IM-PRESS&reference=20070703STO08738&secondRef=0 (viewed 02. July 2008).

Lundqvist, J, de Fraiture, C and Molden, D (2008) *Saving Water: From Field to Fork – Curbing Losses and Wastage in the Food Chain. SIWI Policy Brief*. Stockholm International Water Institute (SIWI), Stockholm.

Lunnan, A, Stupak, I, Asikainen, A and Raulund-Rasmussen, K (2008) Introduction to sustainable utilisation of forest energy. In Röser, D, Asikainen, A, Raulund-Rasmussen, K and Stupak, I (eds) *Sustainable Use of Forest Biomass for Energy. A Synthesis With Focus on the Baltic and Nordic Region*. Springer, Berlin, Dordrecht.

LUT (Lappeenranta University of Technology) (2008) International Bioenergy Technology in cooperation with University of Joensuu. LUT website, http://www.lut.fi/en/international_students/master_programmes/ente.html (viewed 07. June 2008).

Luyssaert, C, Schulze, E-D, Börner, A, Knohl, A, Hessen-möller, D, Law, B E, Ciais, P and Grace, J (2008) Old-growth forests as global carbon sinks. *Nature* **455**: doi:10.1038/nature07276.

LWF (Bayerische Landesanstalt für Wald und Forstwirtschaft) (2005) Anbau von Energiewäldern. LWF Merkblatt 19. LWF website, http://www.lwf.bayern.de/imperia/md/content/lwf-internet/veroeffentlichungen/lwf-merkblaeter/19/lwf_merkblatt_19.pdf (viewed 16. June 2008).

MA (Millennium Ecosystem Assessment) (2005a) *Ecosystems and Human Well-Being: Synthesis Report*. Island Press, Washington, DC.

MA (Millennium Ecosystem Assessment) (2005b) *Ecosystems and Human Well-Being. Biodiversity Synthesis*. World Resources Institute (WRI), Washington, DC.

MA (Millennium Ecosystem Assessment) (2005c) *Ecosystems and Human Well-Being: Current States & Trends*. Island Press, Washington, DC.

MA (Millennium Ecosystem Assessment) (2005d) *Ecosystems and Human Well-Being: Wetlands and Water. Synthesis*. Island Press, Washington, DC.

MA (Millennium Ecosystem Assessment) (2005e) *Ecosystems and Human Well-Being: Desertification Synthesis*. MA, Washington, DC.

Mack, R N, Simberloff, D, Lonsdale, W M, Evans, H, Clout, M and Bazzaz, F (2000) Biotic invasions: causes, epidemiology, global consequences and control. *Issues in Ecology* (5): p22.

Maeder, P, Fliessbach, A, Dubois, D, Gunst, L, Fried., P and Niggli, U (2002) Soil fertility and biodiversity in organic farming. *Science* **296**: pp1694-7.

Maier, J (2007) *CSD 2007 endet ergebnislos Analyse eines Scheiterns. Rundbrief Forum Umwelt und Entwicklung II*. Forum Umwelt und Entwicklung, Bonn.

Maier, J (2008) Nachhaltige Biokraftstoffe – gibts das überhaupt? In FUE – Forum Umwelt und Entwicklung (ed) *Im Labyrinth der Labels. Nachhaltigkeit durch Zertifizierung? Rundbrief Forum Umwelt & Entwicklung 3/2008*. Projektstelle Umwelt & Entwicklung, Bonn, pp10–1.

Mande, S and Kishore, V V N (2007) *Towards Cleaner Technologies: A Process Story on Biomass Gasifiers for Heat Applications in Small and Micro Enterprises*. The Energy and Resources Institute (TERI), New Delhi.

Manley, J, van Kooten, G C, Moeltner, K and Johnson, D W (2005) Creating carbon offsets in agriculture through no-till cultivation: a meat-analysis of costs and carbon benefits. *Climatic Change* **68**: pp41–65.

Mann, L and Tobert, V (2000) Soil sustainability in renewable biomass plantings. *Ambio* **29**: pp492-8.

Marris, E (2006) Black is the new green. *Nature* **442**: pp624–6.

Mathews, J A (2007) Biofuels: what a biopact between North and South could achieve. *Energy Policy* **35**: pp3550–70.

Matson, P A, Parton, W J, Power, A G and Swift, M J (1997) Agricultural intensification and ecosystem properties. *Science* **277**: pp504–9.

Matthews, E, Payne, R, Rohweder, M and Murray, S (2000) *Forest Ecosystems. Pilot Analysis of Global Ecosystems*. World Resources Institute (WRI), Washington, DC.

Maxwell, S (2008) *Downing Street Seminar on Food Prices. Power Point Presentation on 22 April*. Downing Street Seminar Group, London.

McCornick, P, Awulachew, S B and Abebe, M (2008) Water – food – energy – environment synergies and tradeoffs: major issues and case studies. *Water Policy* **10**(Supplement 1): pp23-36.

McNeely, J (2008) Protected areas in a world of eight billion. *GAIA* **17**(S1): pp104–6.

MDA (Ministério do Desenvolvimento Agrário) (2008) Selo Combustvel Social. Portal da Secretaria da Agricultura Familial. Governo Federal do Brasil website, http://www.biodiesel.gov.br/ (viewed 15. August 2008).

Meade, B and Rosen, S (1997) The influence of income on global food spending. *Economic Research Service/USDA: Agricultural Outlook* (July): pp14–7.

Melis, A and Happe, T (2001) Hydrogen production. Green Algae as a source of energy. *Plant Physiology* **127**: pp740-8.

MEMD (Ministry of Energy and Mineral Development) (2002) Energy Policy. MEMD website, http://www.energy-andminerals.go.ug/EnergyPolicy.pdf

MEMD (Ministry of Energy and Mineral Development) (2004) *Uganda Energy Balance*. MEMD, Kampala.

MEMD (Ministry of Energy and Mineral Development) (2007) *Renewable Energy Policy – 2007*. MEMD, Kampala.

Mendez, M A and Popkin, B (2004) Globalization, urbanization and nutritional change in developing world. In Food and Agriculture Organization of the United Nations (FAO) (ed) *Globalization of Food Systems in Developing Countries – Impact on Global Food Security and Nutrition. Food and Nutrition Paper*. FAO, Rome, pp55–80.

Meó Corporate Development (2008) International Sustainability and Carbon Certification Project. Meó Corporate Development website, http://www.iscc-project.org/projekt (viewed 30. April 2008).

Meyer, W B (1995) Past and present land use and land cover in the USA. *Consequences* **1**(1).

Mildner, S-A and Zilla, C (2007) Brasilien und Biokraftstoffe. Chancen und Stolpersteine für eine engere Zusammenarbeit mit der EU und Deutschland. *SWP-Aktuell* **A 60**(November): p4.

Ministry of Agriculture (2008) *Distribution of Agricultural Land by Different Usages in India From 1950–1951 to 1999–2000*. Ministry of Agriculture. Department of Statistics, New Delhi.

Ministry of Agriculture, Livestock and Food Supply (2006) Brazilian Agroenergy Plan 2006–2011. Brazilian Ministry of Agriculture Livestock and Food Supply, website, http://www.agricultura.gov.br/pls/portal/docs/PAGE/MAPA/PLANOS/PNA_2006_2011/PLANO%20NACIONAL%20DE%20AGROENERGIA%202006%20-%202011-%20INGLES_1_0.PDF (viewed 27. February 2008).

Ministry of Rural Development (2003) Wasteland Atlas. Ministry of Rural Development website, http://dolr.nic.in/fwaste-catg.htm (viewed 4. April 2008).

Misereor (2007) *"Bioenergie" im Spannungsfeld von Klimawandel und Armutsbekämpfung*. Bischöfliches Hilfswerk Misereor, Aachen.

Mitchell, D (2008) *A Note on Rising Food Prices*. World Bank, Washington, DC.

Mittermeier, R A, Myers, N, Gil, P R and Goettsch Mittermeier, C (ed) (1999) *Hotspots: Earths Biologically Richest and Most Endangered Terrestrial Ecoregions.* Cemex, Sierra Madre.

Mittermeier, R A, Mittermeier, C G, Brooks, T M, Pilgrim, J D, Konstant, W R, da Fonseca, G A B and Kormos, C (2003) Wilderness and biodiversity conservation. *Proceedings of the National Academy of Sciences of the United States of America* **100**(18): pp10309–13.

Mittermeier, R A, Robles Gil, P, Hoffmann, M, Pilgrim, J, Brooks, J, Goettsch Mittermeier, C, Lamoreux, J and Da Fonseca, G A B (2004) *Hotspots Revisited: Earths Biologically Riches and Most Endangered Terrestrial Ecoregions.* Conservation International, Arlington.

Mittler, D (2008) Schwach, schwächer, CSD? Die Kommission für nachhaltige Entwicklung der Vereinten Nationen 15 Jahre nach Rio. *Vereinte Nationen* **56**(1): pp16–9.

MME (Ministério de Minas e Energia) (2008) *Boletim Mensal Dos Combustveis Renováveis. Edição no. 7, Julho 2008.* MME, Brasilia.

MME (Ministério de Minas e Energia) and EPE (Empresa de Pesquisa Energética) (2007) *Balanço Energético Nacional 2007: Ano Base 2006. Relatório Final.* MME, EPE, Rio de Janeiro.

MoEF (Ministry of Environment and Forest) (2006) *Report of the National Forest Commission.* MoEF, New Delhi.

Molnar, A, Scherr, S and Khare, A (2004) *Who Conserves the World Forests? Community Drive Strategies to Protect Forests and Respect Rights. Forest Trends.* Forest Trends, Washington, DC.

Monfreda, C, Ramankutty, N and Foley, J A (2008) Farming the planet: 2. Geographic distribution of crop areas, yields, physiological types, and net primary production in the year 2000. *Global Biogeochemical Cycles* **22**: doi:10.1029/2007GB002947.

Montenegro, A, Brovkin, V, Eby, M, Archer, D and Weaver, A J (2007) Long term fate of anthropogenic carbon. *Geophysical Research Letters* **34**: doi:10.1029/2007GL030905.

Mooney, H A, Lubchenco, J, Dirzo, R and Sala, O E (1995) Biodiversity and ecosystem functioning: Ecosystem analysis. In Heywood, V H and Watson, R T (eds) *Global Biodiversity Assessment.* Cambridge University Press, Cambridge, New York, pp275–325.

Morgan, T (2007) *Energy Subsidies: Their Magnitude, How they Affect Energy Investment and Greenhouse Gas Emissions, and Prospects for Reform. Menecon Consulting. Final Report Commissioned by UNFCCC Secretariat Financial and Technical Support Programme.* UNFCCC Secretariat, New York.

Moritz, C, Patton, J L, Conroy, C J, Parra, J L, White, G C and Beissinger, S R (2008) Impact of a century of climate change on small-mammal communities in Yosemite National Park, USA. *Science* **322**: pp261-3.

Morton, D C, DeFries, R, Shimabukuro, Y E, Anderson, L O, Arai, E, del Bon Espirito-Santo, F, Freitas, R and Morisette, J (2006) Cropland expansion changes deforestation dynamics in the southern Brazilian Amazon. *PNAS* **103**: pp14637–41.

Müller, A (2008) Biofuels – driver of rural development? *Rural* **21**(3): pp12-5.

Müller-Langer, F, Perimenis, A, Brauer, S, Thrän, D and Kaltschmitt, M (2008) Technische und Ökonomische Bewertung von Bioenergie-Konversionspfaden. Expertise for the WBGU Report "World in Transition: Future Bioenergy and Sustainable Land Use". WBGU website, http://www.wbgu.de/wbgu_jg2008_ex06.pdf

Münch, J (2008) *Nachhaltig nutzbares Getreidestroh in Deutschland.* Institute for Energy and Environmental Research (IFEU), Heidelberg.

Murphy, S and Suppan, S (2003) *Introduction to the Development Box: Finding Space for Development Concerns in the WTOs Agriculture Negotiations.* Institute for Agriculture and Trade Policy, International Institute for Sustainable Development (IISD), Manitoba.

Murray, C C and López, A D (eds) (1996) *The Global Burden of Disease.* Harvard University Press, Harvard, MA.

Myers, N, Mittermeier, R A, Mittermeier, C G, de Fonseca, G A B and Kent, J (2000) Biodiversity hotspots for conservation priorities. *Nature* **403**: pp853-8.

Nabuurs, G J, Masera, O, Andrasko, K, Benitez-Ponce, P, Boer, R, Dutschke, M, Elsiddig, E, Ford-Robertson, J, Frumhoff, P, Karjalainen, T, Krankina, O, Kurz, W A, Matsumoto, M, Oyhantcabal, W, Ravindranath, N H, Sanz Sanchez, M J and Zhang, X (2007) Forestry. In Intergovernmental Panel on Climate Change (IPCC) (ed) *Climate Change 2007: Mitigation. Contribution of Working Group III to the Fourth Assessment Report of the Intergovernmental Panel on Climate Change.* Cambridge University Press, Cambridge, New York, pp541-84.

Nakicenovic, N and Swart, R (2000) *Emissions Scenarios. A Special Report of the Intergovernmental Panel on Climate Change.* Cambridge University Press, Cambridge, UK.

Namburete, S H E (2006) Mozambique Bio-Fuels: African Green Revolution Conference Oslo-Norway, 31 August–02 September 2006. Presentation by H.E. Salvador Namburete, Minister of Energy Republic of Moçambique (Power Point Presentation). Ministry fo Energy website, mediabase.edbasa.com/kunder/yaraimages/agripres/agripres/agripres/j2006/m09/t04/0000443_2.pdf (viewed 07. October 2008).

Neef, T, Eichler, L, Deecke, I and Fehse, J (2007) *Update on Markets for Forestry Offsets.* Centro Agronomico Tropical de Investigacion y Ensenanza (CATIE),

Nepstad, D C, Sticker, C M, Soares-Filho, B and Merry, F (2008) Interactions among Amazon land use, forests and climate: prospects for a near-term forest tipping point. *Philosophical Transactions of the Royal Society of London* **B 363**: pp1737–46.

New, M, Hulme, M and Jones, P (2000) Representing twentieth-century space-time climate variability. Part II: Development of 1901-96 monthly grids of terrestrial surface climate. *Journal of Climate* **13**(13): pp2217–38.

NFA (National Forestry Authority) (2006) NFA Should Not be Blamed For Forest Degradation. NFA website, http://www.nfa.org.ug/new.phpsubaction=showfull&id=116298811 4&archive=&start_from=&ucat=1& (viewed 4. April 2008).

Ng, F and Aksoy, M A (2008) *Who are the Net Food Importing Countries? Policy Research Working Paper 4457.* World Bank, Washington, DC.

NGA (National Governors Association) (2008) *Greener Fuels, Greener Vehicles: A State Resource Guide.* NGA, Washington, DC.

Nitsch, J (2007) *Leitstudie 2007 "Ausbaustrategie Erneuerbare Energien" Aktualisierung und Neubewertung bis zu den Jahren 2020 und 2030 mit Ausblick 2050.* DLR, BMU, Stuttgart, Berlin.

Nordic Ecolabel (2008) How to Apply for the Swan? Nordic Ecolabel website, http://www.ecolabel.nu/nordic_eco2/testing/how_to_app/ (viewed 05. August 2008).

ODI (Overseas Development Institute) (2008) *Rising Food Prices: A Global Crisis. Briefing Paper 37.* ODI, London.

ÖBB (Österreichische Bundesbahnen) (2008) Geschäftsbericht 2007. ÖBB-Holding AG website, http://www.railcargo.at/de/Ueber_uns/Geschaeftsbericht/Downloads/Geschaeftsbericht_2007.pdf (viewed 18. September 2008).

OECD (Organisation for Economic Co-operation and Development) (2002) *Aid Targeting the Objectives of the Rio Conventions 1998–2000. A Contribution by the DAC Secretariat for the Information of Participants at the World Summit for Sustainable Development in Johannesburg in August 2002.* OECD, Paris.

OECD (Organisation for Economic Co-operation and Development) (2003) Philanthropic foundations and development co-operation. Off-print of the *DAC Journal 2003, Volume 4, No. 3. Paris, OECD.*

OECD (Organisation for Economic Co-operation and Development) (2005) *Environmentally Harmful Subsidies: Challenges for Reform.* OECD, Paris.

OECD (Organisation for Economic Co-operation and Development) (2007a) *Agricultural Policies in OECD Countries. Monitoring and Evaluation 2007.* OECD, Paris.

OECD (Organisation for Economic Co-operation and Development) (2007b) *Agricultural Policies in Non-OECD Countries. Monitoring and Evaluation 2007.* OECD, Paris.

OECD (Organisation for Economic Co-operation and Development) (2008) *Economic Assessment of Biofuel Support Policies.* OECD Directorate for Trade and Agriculture, Paris.

OECD (Organisation for Economic Co-operation and Development) and FAO (Food and Agriculture Organization of the United Nations) (2005) *OECD–FAO Agricultural Outlook 2005–2014.* OECD, FAO, Paris, Rome.

OECD (Organisation for Economic Co-operation and Development) and FAO (Food and Agriculture Organization of the United Nations) (2006) *OECD–FAO Agricultural Outlook 2006–2015.* OECD, FAO, Paris, Rome.

OECD (Organisation for Economic Co-operation and Development) and FAO (Food and Agriculture Organization of the United Nations) (2008) *OECD–FAO Agricultural Outlook 2007–2016.* OECD, Paris.

Oeding, D and Oswald, B (2004) *Elektrische Kraftwerke und Netze.* Springer, Berlin, Heidelberg, New York.

Oelmann, Y, Wilcke, W, Temperton, V M, Buchmann, N, Roscher, C, Schumacher, J, Schulze, E-D and Weisser, W W (2007) Soil and plant nitrogen pools as related to plant diversity in an experimental grassland. *Soil Science Society of America* **71**: pp720-9.

Olaki, E (2008) Uganda: Cabinet Okays Oil, Gas Policy. New Vision website, http://allafrica.com/stories/200802041132.html (viewed 4. April 2008).

Oldeman, L R (1992) *Global Extent of Soil Degradation. ISRIC Bi-Annual Report 1991–1992.* International Soil Reference and Information Centre (ISCRIC), Wageningen.

Oldeman, L R, Hakkeling, R T A and Sombroek, W G (1991) *World Map of the Status of Human-Induced Soil Degradation. Global Assessment of Soil Degradation GLASOD.* International Soil Reference and Information Centre (ISRIC), Wageningen.

Olson, D M, Dinerstein, E, Wikramanayake, E D, Burgess, N D, Powell, G V N, Underwood, E C, D'Amico, J A, Itoua, I, Strand, H E, Morrison, J C, Loucks, C J, Allnutt, F, Ricketts, T, Kura, Y, Lamoreux, J F, Wettengel, W W, Hedao, P and Kassem, K R (2001) Terrestrial ecoregions of the world: a new map of life on earth. *BioScience* **51**(11): pp933–8.

Ong, C K, Black, C R and Muthuri, C M (2006) Modifying forestry and agroforestry to increase water productivity in the semi-arid tropics. *CAB Reviews: Perspectives in Agriculture, Veterinary Science, Nutrition and Natural Resources* **1**(65): p19ff.

Openshaw, K (2000) A review of Jatropha curcas: an oil plant of unfulfilled promise. *Biomass and Bioenergy* **19**: pp1–15.

Pacala, S and Socolow, R (2004) Stabilization wedges: solving the climate problem for the next 50 years with current technologies. *Science* **305**: pp968–72.

Pagiola, S (2002) Paying for water services in Central America: learning from Costa Rica. In Pagiola, S, Bishop., J and Landell-Mills, N (eds) *Selling Forest Environmental Services. Market-based Mechanisms for Conservation and Development.* Earthscan, London, pp37–62.

Palosuo, T, Peltoniemi, M, Mikhailov, A, Komarov, A, Faubert, P, Thürig, E and Lindner, M (2008) Projecting effects of intensified biomass extraction with alternative modelling approaches. *Forest Ecology and Management* **255**: pp1423–33.

Paquin, M (2007) *Advocacy for Sustainable Biofuels: Roles and Functions for UNCCD Actors in the Light of the Ten-Year Strategic Plan. Speech at the 8th VSK of UNCCD in Madrid, 6.9.07, UNISFERA.* UNISFERA, Montreal.

Parmesan, C and Yohe, G (2003) A globally coherent fingrprint of climate change impacts across natural systems. *Nature* **421**: pp37–42.

Parrish, D J and Fike, J H (2005) The biology and agronomy of switchgrass for biofuels. *Critical Reviews in Plant Sciences* **24**: pp423–59.

Pastowski, A, Fischedick, M, Arnold, K, Bienge, K, von Geibler, J, Merten, F and Schüwer, D (2007) *Sozial-ökologische Bewertung der stationären energetischen Nutzung von importierten Biokraftstoffen am Beispiel von Palmöl.* Wuppertal Institute, Wuppertal.

Paul, N (2008) Pflanzensprit mit Ökolabel? *Umwelt aktuell* **3**. Oekom Verlag, Munich.

Paustian, K, Six, J, Elliott, E T and Hunt, H W (2000) Management options for reducing CO_2 emissions from agricultural soils. *Biogeochemistry* **48**: pp147–63.

Pearce, D (2003) The social cost of carbon and its policy implication. *Oxford Review of Economic Policy* **19**(3): pp362–84.

Pearce, D (2004) Environmental market creation: saviour or oversell? *Portuguese Economic Journal* **3**(2): pp115–44.

Pearce, D (2007) Do we really care about biodiversity? *Environment and Resource Economics* **37**: pp313–33.

Peksa-Blanchard, M, Dolzan, P, Grassi, A, Heinimö, J, Junginger, M, Ranta, T and Walter, A (2007) *Global Wood Pellets Markets and Industry: Policy Drivers, Market Status and Raw Material Potential. IEA Bioenergy Task 40.* International Energy Agency (IEA), Paris.

Peoples Coalition (2008) Peoples Coalition in Biofuels. Open Letter to Minister For New and Renewable Energy for Pro-People Energy Policy, Febuary 2008. Deccan Development Society (DDS) website, http://www.ddsindia.com/www/people_coalition.html (viewed 03. July 2008).

Peregon, A, Maksyutov, S, Kosykh, N P and Mironycheva-Tokareva, N P (2008) Map-based inventory of wetland biomass and net primary production in western Siberia. *Journal of Geophysical Research* **113**: doi:10.1029/2007JG000441.

Pereira, R, Zweede, J, Asner, G P and Keller, M (2002) Forest canopy damage and recovery in reduced-impact logging and conventional selective logging in Eastern Para, Brazil. *Forest Ecology and Management* **168**: pp77–89.

Perrings, C and Gadgil, M (2005) Conserving biodiversity: reconciling local and global benefits. In Kaul, I, Coneicao, K, Le Goulven, K and Mendoza, R U (eds) *Providing Global Public Goods Managing Globalization.* United Nations Development Programme (UNDP), New York, pp532–55.

Perrot-Ma"tre, D and Davis, P (2001) *Case Studies: Developing Markets for Water Services from Forests.* Forest Trends, Washington, DC.

Pesambili, C, Magessa, F and Mwakabuta, N (2003) Sazawa Charcoal Stove Designed for Efficient Use of Charcoal. TU Delft website, http://www.io.tudelft.nl/research/dfs/ide-conference/papers/11_Mwakabuta.pdf (viewed 27. February 2008).

Peskett, L, Slater, R, Stevens, C and Dufey, A (2007) Biofuels, agriculture and poverty reduction. *ODI Natural Resource Perspectives* **107**: p6.

Peters, C J, Wilkins, J L and Fick, G W (2007) Testing a complete-diet model for estimating the land resource requirements of food consumption and agricultural carrying capacity – the New York State example. *Renewable Agriculture and Food Systems* **22**(2): pp145–53.

Pfahl, S, Oberthür, S, Tänzler, D, Kahlenborn, W and Biermann, F (2005) *Die internationalen institutionellen Rahmenbedingungen zur Förderung erneuerbarer Energien.* adelphi Research, Berlin.

Philibert, C (2004) *International Energy Technology Collaboration and Climate Change Mitigation. JT00165457. COM/ENV/EPOC/IEA/SLT(2004)1.* OECD Environment Directorate, Paris.

Phillips, A and Stolton, S (2008) Protected landscapes and biodiversity values: an overview. In Amend, T, Brown, J H, Kothari, A, Phillips, A and Stolton, S (eds) *Values of Protected Landscapes and Seascapes. Protected Landscapes and Agrobiodiversity Values.* Kasparek Verlag, Heidelberg, pp8–21.

Pickardt, T and de Kathen, A (2002) Verbundprojekt "Grundlagen für die Risikobewertung transgener Gehölze": Literaturstudie zur Stabilität transgen-vermittelter Merkmale in gentechnisch veränderten Pflanzen mit Schwerpunkt transgene Gehölzarten und Stabilitätsgene. Umweltbundesamt (UBA) website, http://www.umweltbundesamt.org/fpdf-l/2181.pdf (viewed 16. June 2008).

Pielke, R A, Marland, G, Betts, R A, Chase, T N, Eastman, J L, Niles, J O, Niyogi, D D S and Running, S W (2002) The influence of landuse change and landscape dynamics on the climate system: Relevance to climate-change policy beyond the radiative effect of greenhouse gases. *Philosophical Transactions of the Royal Society London Series A* **360**: pp1705–19.

Pilardeaux, B (2004) Entwicklungslinien internationaler Süßwasserpolitik. In Lozán, J L, Graßl, H, Hupfer, P, Menzel, L and Schönwiese, C-D (eds) *Warnsignal Klima: Genug Wasser für alle?* Wissenschaftliche Kooperationen, Hamburg, pp316-9.

Pilardeaux, B (2008) Konvention gegen Wüstenbildung: 8. Vertragsstaatenkonferenz 2007. *Vereinte Nationen* **56**(1): pp29–30.

Pingoud, K (2003) *Harvested Wood Products: Considerations on Issues Related to Estimation, Reporting and Accounting of Greenhouse Gases. Final Report Delivered to the UNFCCC Secretariat. January 2003.* UNFCCC, New York.

Pistorius, R, Schmitt, C B and Winkel, G (2008) *A Global Network of Forest Protected Areas Under the CBD – Analysis and Recommendation.* Institute of Forest and Environmental Policy (IFP), Freiburg.

Pitman, A, Pielke sr, R, Avissar, R, Claussen, M, Gash, J and Dolman, H (1999) The role of the land surface in weather and climate: does the land surface matter? *IGBP Newsletter* **39**: pp4–9.

Planning Comission (2003) *Report of the Committee on the Development of Biouel.* Government of India, New Delhi.

Poore, M E D and Fries, C (1985) *The Ecological Effects of Eucalyptus.* Food and Agriculture Organization of the United Nations (FAO), Rome.

Pope, V D, Gallani, M L, Rowntree, P R and Stratton, R A (2000) The impact of new physical parametrizations in the Hadley Centre climate model: HadAM3. *Climate Dynamics* **16**(2): pp123–46.

Popkin, B (2006) Global nutrition dynamics: the world is shifting rapidly towards a diet linked with noncommunicable diseases. *The American Journal of Clinical Nutrition* **84**(2): pp289–98.

Porter, M E, Schwab, K and Sala-i-Martin, X (2007) *The Global Competitiveness Report 2007–2008 World Economic Forum.* Palgrave Macmillan, Houndmills.

Portmann, F, Siebert, S, Bauer, C and Döll, P (2008) *Global data set of monthly growing areas of 26 irrigated crops.* Institute of Physical Geography, University of Frankfurt, Frankfurt/M..

Pressey, R L (1997) Priority conservation areas: Towards an operational definition for regional assessments. In Pigram, J J and Sundell, R C (eds) *National Parks and Protected Areas: Selection, Delimitation and Management.* University of New England, Centre for Water Policy Research, Armidale, Australia, pp337–57.

Puth, S (2003) *WTO und Umwelt. Die Produkt-Prozess-Doktrin.* Duncker & Humblot, Berlin.

Puth, S (2005) *Der Umweltschutz im Recht der WTO.* Nomos, Baden-Baden.

Qin, J (2007) WTO Panel decision in Brazil – Tyres supports safeguarding environmental values. ASIL Insight 11//23. American Society of International Law website, http://asil.org//insights//2007//09//insights070905.html (viewed 12. September 2007).

Raghu, S, Anderson, R C, Daehler, C C, Davis, A, Wiedenmann, R N, Simberloff, D and Mack, R N (2008) Adding biofuels to the invasive species fire? *Science* **313**: p1742.

Raison, R J (2005) Demonstrating the Sustainability of Forest Bioenergy Projects. Bioenergy Australia Conference (Melbourne). Sustainability Guide and Industry Code of Practice for Energy from Waste Projects. Waste Management Association of Australia website, http://www.wmaa.asn.au/efw/home.html (viewed 20. October 2008).

Ramankutty, N and Foley, J A (1999) Estimating historical changes in global land cover: croplands from 1700 to 1992. *Global Biogeochemical Cycles* **13**: pp997–1027.

Ramankutty, N, Evan, A T, Monfreda, C and Foley, J A (2008) Farming the planet: 1. Geographic distribution of global agricultural lands in the year 2000. *Global Biogeochemical Cycles* **22**: pGB1003.

Randall, R (2004) *Jatropha curcas (physic nut) Its Weed Potential in Western Australia and The Implications of Large Scale Plantations for Fuel oil Production.* Department of Agriculture,Western Australia, Perth.

Raskin, P, Hansen, E and Margolis, R (1995) *Water and Sustainability: Global Outlook.* Stockholm Environment Institute (SEI), Boston.

Reijnders, L (2008) Ethanol production from crop residues and soil organic carbon. *Resources, Conservation and Recycling* **52**: pp653–8.

Reinhardt, G, Rettenmaier, N and Gärtner, S (2007) *Regenwald für Biodiesel? Ökologische Auswirkungen der energetischen Nutzung von Palmöl.* WWF Germany, Frankfurt/M.

Reinicke, W and Reinicke, H (ed) (1998) *Global Public Policy. Governing Without Government.* Brookings Institution Press, Washington, DC.

Reisner, Y, de Filippi, R, Herzog, F and Palma, J (2007) Target regions for silvoarable agroforestry in Europe. *Ecological Engineering* **29**: pp401–18.

REN21 (Renewable Energy Policy Network for the 21st Century) (2006) *Global Status Report 2006 Update. Renewables.* REN21 Secretariat Paris.

REN21 (Renewable Energy Policy Network for the 21st Century) (2008) Renewables Global Status Report. REN21 Secretariat website, http://gsr.ren21.net/index.php?title=Main_Page (viewed 2. September 2008).

Ressortarbeitsgruppe Welternährungslage (2008) *Globale Ernährungssicherungdurch nachhaltige Entwicklung und Agrarwirtschaft. Bericht der Ressortarbeitsgruppe "Welternährungslage" an das Bundeskabinett.* Ressortarbeitsgruppe Bundesregierung "Welternährungslage", Berlin.

Reuters (2007) South Africa Biofuel Plan Upsets Monsanto, Maize Farmers. Reuters website, http://www.reuters.com/article/rbssConsumerGoodsAndRetailNews/idUSL1148171420071211 (viewed 17. January 2008).

RFA (Renewable Fuels Association) (2008) U.S. Fuel Ethanol Imports by Country. Industry Statistics. RFA website, http://www.ethanolrfa.org/industry/statistics/#F (viewed 12. August 2008).

Rhodes, J S and Keith, D W (2005) Engineering economic analysis of biomass IGCC with carbon capture and storage. *Biomass and Bionergy* **29**: pp440–50.

Ribeiro, H (2008) Sugar cane burning in Brazil: respiratory health effects. *Revista de Saúde Pública* **42**(2): pp370-6.

Richards, G P and Stokes, C (2004) A review of forest carbon sequestration cost studies: a dozen years of research. *Climatic Change* **63**: pp1–48.

Righelato, R and Spracklen, D V (2007) Carbon mitigation by biofuels or by saving and restoring forests? *Science* **317**: p902.

Ringel, M (2004) *Energie und Klimaschutz. Umweltökonomische Analyse der Klimaschutzmaßnahmen auf dem deutschen Elektrizitätsmarkt unter Berücksichtigung internationaler Erfahrungen.* Peter Lang, Frankfurt/M.

Rockström, J, Lannerstad, M and Falkenmark, M (2007) Assessing the water challenge of a new green revolution in developing countries. *PNAS* **104**: pp6253-60.

Rodrigues, A S L, Andelman, S J, Bakarr, M I, Boltani, L, Brooks, T M, Cowling, R M, Fishpool, L D C, da Fonseca, G A B, Gaston, K J, Hoffmann, M, Long, J S, Marquet, P A, Polgrim, J D, Pressey, R L, Schipper, J, Sechrest, W, Stuart, S N, Underhill, L G, Waller, R W, Watts, M E J and Yan, X (2004) Effectiveness of the global protected area network in representing species diversity. *Nature* **428**: pp641–3.

Roe, D (2006) *Biodiversity, Climate Change and Complexity: An Opportunity for Securing Co-Benefits?* International Institute for Environment and Development (IIED), London.

Roeckner, E, Bäuml, G, Bonaventura, L, Brokopf, R, Esch, M, Giorgetta, M, Hagemann, S, Kirchner, I, Kornblueh, L, Manzini, E, Rhodin, A, Schlese, U, Schulzweida, U and Tompkins, A (2003) *The atmospheric general circulation model ECHAM5 – Part I: Model description.* Max Planck Institute for Meteorology, Hamburg.

Rogner, H-H, Zhou, D, Bradley, R, Crabbé, P, Edenhofer, O, Hare, B, Kulipers, L and Yamaguchi, M (2007) Introduction. In Intergovernmental Panel on Climate Change (IPCC) (ed) *Climate Change 2007: Mitigation. Contribution of Working Group III to the Fourth Assessment report of the Intergovernmental Panel on Climate Change.* Cambridge University Press, Cambridge, New York, pp95–117.

Röhricht, C and Ruscher, K (2004) Anbauempfehlungen für schnellwachsende Baumarten. Fachmaterial Sächsische Landesanstalt für Landwirtschaft. Fachbereich Pflanzliche Erzeugung. Sächsische Landesanstalt für Landwirtschaft website, http://www.smul.sachsen.de/lfl/publikationen/download/858_1.pdf (viewed 16. June 2008).

Ronneburger, J-U (2008) Südamerika im Soja-Rausch – Rekorde bei Gewinn und Umweltzerstörung. dpa-Pressemeldung 24.04.2008. dpa website, http://www.klima-aktiv.com/article144_5953.html (viewed 23. June 2008).

Rösch, C, Raab, K, Skarka, J and Stelzer, V (2007) *Energie aus dem Grünland – eine nachhaltige Entwicklung? Wissenschaftliche Berichte FZKA 7333.* Forschungszentrum Karlsruhe in der Helmholtz-Gemeinschaft, Karlsruhe.

Rosebala, A, Morillo, E, Undabeytia, T and Maqueda, C (2007) Long-term impacts of wastewater irrigation on Cuban soils. *Soil Science Society of America Journal* **71**: pp1292–8.

Rosegrant, M W, Paisner, M S, Meijer, S and Witcover, J (ed) (2001) *Global Food Projections to 2020. Emerging Trends and Alternative Futures.* International Food Policy Research Institute (IFPRI), Washington, DC.

Rosegrant, M W and Cavalieri, A J (2008) Bioenergy and Agro-Biotechnology. Expertise for the WBGU Report "World in Transition: Future Bioenergy and Sustainable Land Use". WBGU website, http://www.wbgu.de/wbgu_jg2008_ex07.pdf.

Rosegrant, M W, Ewing, M, Msangi, S and Zhu, T (2008) Bioenergy and the Global Food Situation Until 2020/2050. Expertise for the WBGU Report "World in Transition: Future Bioenergy and Sustainable Land Use". WBGU website, http://www.wbgu.de/wbgu_jg2008_ex08.pdf.

Röser, D, Asikainen, A, Stupak, I and Pasanen, K I H S, Dordrecht, Niederlanden. (2008) Forest energy resources and potentials. In Röser, D, Asikainen, A, Raulund-Rasmussen, K and Stupak, I (eds) *Sustainable Use of Forest Biomass for Energy. A Synthesis With Focus on the Baltic and Nordic Region.* Springer, Berlin, Dordrecht.

Roy, M (2008) *Nachwärme aus Biogasverstromung ergänzt Holzhackschnitzel-Kessel. BWK Volume 60.* VDI-Springer-Verlag, Düsseldorf.

RSB (Roundtable on Sustainable Biofuels) (2008a) Principles and Criteria – Introduction. RSB, École Polytechnique Fédérale de Lausanne (EPFL) website, http://cgse.epfl.ch/page70341.html

RSB (Roundtable on Sustainable Biofuels) (2008b) *Global Principles and Criteria for Sustainable Biofuels Production. Version Zero.* RSB, École Polytechnique Fédérale de Lausanne (EPFL), Lausanne.

Rudloff, B (2008) *Nahrungsmittelkrisen und die falsche Angst vor der Globalisierung. SWP-Aktuell 45.* German Institute for International and Security Affairs (SWP), Berlin.

Russelle, M P, Morey, R V, Baker, J M, Porter, P M and Jung, H-J G (2007) Comment on –Carbon-negative biofuels from low-input high-diversity grassland biomass". *Science* **316**: p1567/b.

Sachs, J D (2008) Surging food prices mean global instability. Misguided policies favor biofuels over grain for hungry people. *Scientific American Magazine* **May**: p2.

Saffih-Hdadi, K and Mary, B (2008) Modeling consequences of straw residues export on soil organic carbon. *Soil Biology & Biochemistry* **40**: pp594–607.

Salvatore, M, Pozzi, F, Ataman, E and Bloise, M (2005) *Mapping Global Urban and Rural Population Distributions.* Food and Agriculture Organization of the United Nations (FAO), Rome.

Sanchez Blanco, L (2008) Key Challenges in Agro-Industrial Development in Africa, Presentation of the African Development Bank at the GAIF in New Delhi, April 2008. African Development Bank (AfDB) website, http://www.afdb.org/pls/portal/url/ITEM/4B1451DC39215FAEE040C00A0C3D3892 (viewed 07. May 2008).

Sanderson, E W, Jaiteh, M, Levy, M A, Redford, K H, Wannebo, A V and Woolmer, G (2002) The Human Footprint and the Last of the Wild. *Bioscience* **52**(10): pp891–904.

Sawyer, D (2008) Climate change, biofuels and eco-social impacts in the Brazilian Amazon and Cerrado. *Philosophical Transactions of the Royal Society B* **363**: pp1747-52.

SBGF (Swedish Biogas Association), SGC (Svenskt Gastekniskt Center AB) and Gasföreningen (2008) *Biogas From Manure and Waste Products – Swedish Case Studies*. SBGF, SGC, Gasföreningen, Stockholm.

Scanlon, B R, Jolly, I, Sophocleous, M and Zhang, L (2007) Global impacts of conversions fRom natural to agricultural ecosystems on water resources: Quantity versus quality. *Water Resources Research* **43**: ppW03437, doi: 10.1029/2006WR005486.

SCBD (Secretariat of the Convention on Biological Diversity) (2008) *The Potential Impacts of Biofuels on Biodiversity. UNEP/CBD/COP/9/26*. Secretariat of the CBD, Montreal.

Schaphoff, S, Lucht, W, Gerten, D, Sitch, S, Cramer, W and Prentice, I C (2006) Terrestrial biosphere carbon storage under alternative climate projections. *Climatic Change* **74**(1-3): pp97–122.

Schinninger, I (2008) Entwicklung der globalen Landnutzung. Expertise for the WBGU Report "World in Transition: Future Bioenergy and Sustainable Land Use". WBGU website, http://www.wbgu.de/wbgu_jg2008_ex09.pdf.

Schlamadinger, B, Grubb, M, Azar, C, Bauen, A and Berndes, G (2001) *Carbon Sinks and Biomass Energy Production. A Study of Linkages, Options and Implications. Project Initiation, Coordination and Dissemination by Climate Strategies*. UK Department for Environment, Food and Rural Affairs, London.

Schlamadinger, B, Faaji, A, Junginger, M, Woess-Gallasch, S and Daugherty, E (2005) *Options for Trading Bioenergy Products and Services. IEA Bioenergy Task 38, 40*. International Energy Agency (IEA), Paris.

Schlamadinger, B, Faaij, A and Daugherty, E (2006) *Should we Trade Biomass, Renewable Certificates, or CO₂ Credits. Study on Behalf of IEA Bioenergy Task 38 With Contributions of IEA Bioenergy Task 35 and Task 40*. International Energy Agency (IEA), Paris.

Schlamadinger, B, Bird, N, Johns, T, Brown, S, Canadell, J, Ciccarese, L, Dutschke, M, Fiedler, J, Fischlin, A, Fearnside, P, Forner, C, Freibauer, A, Frumhoff, P, Hoehne, N, Kirschbaum, M U F, Labat, A, Marland, G, Michaelowa, A, Montanarella, L, Moutinho, P, Murdiyarso, D, Pena, N, Pingoud, K, Rakonczay, Z, Rametsteiner, E, Rock, J, Sanz, M J, Schneider, U A, Shvidenko, A, Skutsch, M, Smith, P, Somogyi, Z, Trines, E, Ward, M and Yamagata, Y (2007) A synopsis of land use, land-use change and forestry (LULUCF) under the Kyoto Protocol and Marrakech Accord. *Environmental Science & Policy* **10**: pp271–82.

Schleich, J and Gruber, E (2008) Beyond case studies: Barriers to energy efficiency in commerce and the services sectors. *Energy Economics* **30**(2): pp449-64.

Schmer, M R, Vogel, K P, Mitchell, R B and Perrin, R K (2008) Net energy of cellulosic ethanol from Switchgrass. *Proceedings of the National Academy of Sciences of the United States of America* **105**: pp464-9.

Schmidhuber, J and Shetty, P (2005) *The Nutrition Transition to 2030. Why Developing Countries are Likely to Bear the Major Burden*. Plenary Paper Presented at the *97th Seminar of the European Association of Agricultural Economists. Reading: European Association of Agricultural Economists*. Food and Agriculture Organization of the United Nations of the United Nations (FAO), Rome.

Schmidt, J E U (2008) *Ökologische Risiken transgener Bäume unter Biodiversitätsgesichtspunkten. Gutachten im Auftrag des BfN*. Bundesamt für Naturschutz (BfN), Bonn.

Schmitt, C B, Pistorius, T and Winkel, G (2007) *A Global Network of Forest Protected Areas under the CBD: Opportunities and Challenges. Proceedings of an international Expert Workshop held in Freiburg, Germany, May 9–11, 2007*. Kessel, Remagen.

Schmitz, G and Schütte, G (2000) *Plants Resistant Against Abiotic Stress*. University Hamburg, Hamburg.

Schneider, L (2007) *Is the CDM Fulfilling its Environmental and Sustainable Development Objectives? An Evaluation of the CDM and Options for Improvement*. Worldwide Fund for Nature (WWF), Frankfurt/M.

Schuchhardt, F (2007) *Palmöl: Saubere Produktion eines natürlichen Energieträgers – Ein neuer Baustein zur umweltverträglichen Palmölproduktion. ForschungsReport 2*. Bundesforschungsanstalt für Landwirtschaft (FAL), Braunschweig.

Schulze, E-D, Freibauer, A and Matthes, F C (2007) *Kyoto-Protokoll: Untersuchung von Optionen für die Weiterentwicklung der Verpflichtungen für die 2. Verpflichtungsperiode, Teilvorhaben "Senken in der 2. Verpflichtungsperiode"*. UBA, Berlin.

Schütz, H and Bringezu, S (2006) *Flächenkonkurrenz bei der weltweiten Bioenergieproduktion. Kurzstudie im Auftrag des Forums für Entwicklung*. Forum für Entwicklung, Bonn.

Schwertmann, U, Vogl, W and Kainz, M (1987) *Bodenerosion durch Wasser: Vorhersage des Abtrags und Bewertung von Gegenmaßnahmen*. Ulmer, Stuttgart.

Searchinger, T, Heimlich, R, Houghton, R A, Dong, F, Elobeid, A, Fabiosa, J, Tokgoz, S, Hayes, D and Yu, T-H (2008) Use of U.S. croplands for biofuels increases greenhouse gases through emissions from land use change. *Science* **319**: pp1238–40.

Seemüller, M (2001) Ökologische bzw. konventionell-integrierte Landbewirtschaftung. *Zeitschrift für Ernährungsökologie* **2**(2): pp94–6.

SEKAB (Svensk Etanolkemi AB) (2008) The Worlds First Verified Sustainable Ethanol Comes to Sweden. SEKAB Press Release, 26. May 2008. SEKAB website, http://www.sustainableethanolinitiative.com/Eng/Standardsidor/Filer/080526%20-%20The%20worlds%20first%2026%205.pdf (viewed 24. June 2008).

Sell, M (2006) Trade, climate change and the transition to a sustainable energy future. Framing the debate. In International Centre for Trade and Sustainable Development (ICTSD) (ed) *Linking Trade, Climate Change and Energy. ICTSD Trade and Sustainable Energy Series*. ICTSD, Geneva, pp1–2.

Setyogroho, B (2007) Mondaq Topics – Energy and Environment – Indonesia Follows Suit in the Production and Use of Biofuels. Meldung vom 12.04.2007. Mondaq Environmental & Energy website, http://www.mondaq.com/article.asp?article_id=47580 (viewed 26. March 2008).

Sharman, J C (2008) International organizations and the implementation of new financial regulations by blacklisting. In Joachim, J, Reinalda, B and Verbeek, B (eds) *International Organizations and Implementation: Enforcers, Managers, Authorities?* Routledge, Oxford, pp48–62.

Shinozaki, K, Yoda, K, Hozumi, K and Kira, T (1964) A quantitative analysis of th eplant form – the pipe model theory. *Japanes Journal of Ecology* **14**(3): pp98–104.

Shiva, V (2008) Stolen harvest. *World Conservation* (May): p14.

Siegel, J (2008) Multilateral Financing for Biofuels. Präsentation im Rahmen der International Fund for Agricultural Development (IFAD), International Consultation on Pro-Poor Jatropha Development, 10-11 April 2008 Rome/Italy. International Fund for Agricultural Development (IFAD) website, http://www.ifad.org/events/jatropha/roundtable/J_Siegel.ppt (viewed 13. June 2008).

Silver, W L, Kueppers, L M, Lugo, A, Ostertag, R and Matzek, R (2004) Carbon sequestration and plant community dynamics following reforestation of tropical pasture. *Ecological Applications* **14**(4): pp1115-27.

Sims, R E H, Schock, R N, Adegbululgbe, A, Fenhann, J, Konstantinaviciute, I, Moomaw, W, Nimir, H B, Schlamadinger, B, Torres-Martnez, J, Turner, C, Uchiyama, Y, Vuori, S J V, Wamukonya, N and Zhang, X (2007) Energy supply. In Metz, B, Davidson, O, Bosch, P R, Dave, R and Meyer, L A (eds) *Climate Change 2007: Mitigation. Contribution of Working Group III to the Fourth Assessment Report of the Intergovernmental Panel on Climate Change.* Cambridge University Press, Cambridge, New York, pp251–322.

Singh, S (2005) *Environmental Goods Negotiations. Issues and options for ensuring win-win outcomes.* International Institute for Sustainable Development (IISD), Winnipeg.

Sirisomboon, P, Kitchaiya, P, Pholpho, T and Mahuttanyavanitch, W (2007) Physical and mechanical properties of Jatropha curcas L. fruits, nuts and kernels. *Biosystems Engineering* **97**: pp201–7.

Sitch, S, Smith, B, Prentice, I C, Arneth, A, Bondeau, A, Cramer, W, Kaplan, J O, Levis, S, Lucht, W, Sykes, M T, Thonicke, K and Venevsky, S (2003) Evaluation of ecosystem dynamics, plant geography and terrestrial carbon cycling in the LPJ dynamic global vegetation model. *Global Change Biology* **9**(2): pp161–85.

Six, J, Feller, C, Denef, C, Ogle, S M, de Moraes Sa, J C and Albrecht, A (2002) Soil organic matter, biota and aggregation in temperate and tropical soils – effects of no-tillage. *Agronomie* **22**: pp755–75.

Skysails (2008) Frachtschifffahrt & Umwelt. Berechnung von Treibstoffersparnis und Amortisationsdauer. Skysails website, http://www.skysails.info/index.php?id=89&L=0 (viewed 18. September 2008).

Smeets, E, Faaij, A P C and Lewandowski, I (2004) *A Quickscan of Global Bio-Energy Potentials to 2050. An Analysis of the Regional Variability of Biomass Resources for Export in Relation to the Underlying Factors.* Secretariaat programma Duurzame Energie Nederland (DEN), Utrecht.

Smeets, E M W, Faaij, A P C, Lewandowski, I M and Turkenburg, W C (2007) A bottom-up assessment and review of global bio-energy potentials to 2050. *Progress in Energy and Combustion Science* **33**(1): pp56–106.

Smith, P, Powlson, D S, Smith, J U, Falloon, P and Colemn, K (2000) Meeting Europes climate change committments: quantitative estimates of the potential for carbon mitigation by agriculture. *Global Change Biology* **6**: pp525–39.

Smith, P, Martino, D, Cai, Z, Gwary, D, Janzen, H, Kumar, P, McCarl, B, Ogle, S, O'Mara, F, Rice, C, Scholes, B and Sirotenko, O (2007a) Agriculture. In Intergovernmental Panel on Climate Change (IPCC) (ed) *Climate Change 2007: Mitigation. Contribution of Working Group III to the Fourth Assessment Report of the Intergovernmental Panel on Climate Change.* Cambridge University Press, Cambridge, UK, pp497–541.

Smith, P, Martino, D, Cai, Z, Gwary, D, Janzen, H, Kumar, P, McCarl, B, Ogle, S, OMara, F, Rice, C, Scholes, B, Sirotenko, O, Howden, M, McAllister, T, Pan, G, Romanenkov, V, Schneider, U and Towprayoon, S (2007b) Policy and technological constraints to implementation of greenhouse gas mitigation options in agriculture. *Agriculture, Ecosystems and Environment* **118**(1–4): pp6–28.

Smyth, A J and Dumanski, J (1993) *FESLM: An International Framework for Evaluating Sustainable Land Management. World Soil Resources Report 73.* Food and Agriculture Organization (FAO), Rome.

SNV (Netherlands Development Organisation) (2008) Biogas and Renewable Energy Programme. SNV website, http://www.snv.org.vn/ (viewed 27. February 2008).

Soussana, J-F, Loiseau, P, Vuichard, N, Ceschia, E, Balesdent, J, Chevallier, T and Arraouays, D (2004) Carbon cycling and sequestration opportunities in temperate grasslands. *Soil Use and Management* **20**: pp219–30.

SRU (German Advisory Council on the Environment) (2007) *Klimaschutz durch Biomasse. Sondergutachten.* SRU, Berlin.

SRU (German Advisory Council on the Environment) (2008) "Naturschutz im Umweltgesetzbuch". Offenes Schreiben des Vorsitzenden des SRU an Kanzleramtsminister de Maizière. SRU website, http://www.umweltrat.de/04presse/downlo04/hintgru/Brief_UGB_Naturschutz_2008_04.pdf (viewed 14. May 2008).

Staffhorst, M (2006) *The Way to Competitiveness of PV – An Experience Curve and Break-even Analysis. PhD-Thesis.* Kassel University Press GmbH, Kassel.

Statis (2008) Ein- u. Ausfuhr (Außenhandel): Deutschland, Monate, Warengruppen (EGW 2002: 3-Steller), Steinkohle, Braunkohle. Statistisches Bundesamt website, https://www-genesis.destatis.de/genesis/online/ (viewed 20. October 2008).

Stattersfield, A J, Crosby, M J, Long, A J and Wege, D C (ed) (1998) *Endemic bird areas of the world. Priorities for biodiversity conservation.* BirdLife Conservation Series. BirdLife International, Cambridge.

Stecher, K-H (2007) Neue Größenordnung: Perspektiven der Biokraft in Brasilien. website, http://www.entwicklungspolitik.org/home/12-007-02/ (viewed 13. May 2008).

Steenblik, R (2007) *Biofuels – At What Cost? Government Support for Ethanol and Biodiesel in Selected OECD Countries. A Synthesis of Reports Addressing Subsidies for Biofuels in Australia, Canada, the European Union, Switzerland and the United States.* The Global Subsidies Initiative (GSI), Geneva.

Steger, S (2005) *Der Flächenrucksack des europäischen Außenhandels mit Agrarprodukten – Welche Globalisierung ist zukunftsfähig?* Wuppertal Institute, Wuppertal.

Steinfeld, H, Gerber, P, Wassenaar, T, Rosales, M and de Haan, C (2006) *Livestocks Long Shadow. Environmental Issues and Options.* Food and Agriculture Organization of the United Nations (FAO), Rome.

Sterk, W and Arens, C (2006) *Die projektbasierten Mechanismen CDM & JI. Einführung und praktische Beispiele.* Bundesministerium für Umwelt, Naturschutz und Reaktorsicherheit (BMU), Berlin.

Stern, N (2006) *The Economics of Climate Change. The Stern Review.* HM Treasury, London.

Stern, N (2008) *Key Elements of a Global Deal on Climate Change.* The London School of Economics and Political Science, London.

Sterner, M (2007) *Technical Analysis and Assessment of Biomass-to-Liquid Technologies. Unveröffentlichte Masterarbeit.* Institute for Energy and Environment, Leipzig and Carl-von-Ossietzky University, Oldenburg.

Sterner, M, Schmid, J and Wickert, M (2008) *Effizienzgewinn durch erneuerbare Energien – der Primärenergiebeitrag von erneuerbaren Energien. BWK No. 60, 06/2008.* Springer-VDI-Verlag, Düsseldorf.

Stiens, H (2000) *Ermittlung des gesamtheitlichen Wirkungsgrades als Kennzahl zur rationellen Energienutzung in der Produktionstechnik. Dissertation.* Technische Universität, Shaker Verlag, Aachen.

Stolton, S, Maxted, N, Ford-Lloyd, B, Kell, S and Dudley, N (2006) *Food Stores: Using Protected Areas to Secure Crop Genetic Diversity. Arguments for Protection.* World Wide Fund for Nature (WWF), University of Birmingham, Gland, Birmingham.

Stolzenburg, K (2007) Chinaschilf (Miscanthus x giganteus) – Anbau und rechtliche Rahmenbedingungen. Merkblatt. Landwirtschaftliches Technologiezentrum Augustenberg. Aussenstelle Forchheim, Rheinstetten website, http://www.landwirtschaft-bw.info/servlet/PB/show/1199500/index.pdf (viewed 10. December 2007).

Stone, R (2007) Can palm oil plantations come clean? *Science* **317**: p1491.

Strassburg, B, Turner, K, Fisher, B, Schaeffer, R and Lovett, A (2008) *An Empirically-Derived Mechanism of Combined Incentives to Reduce Emissions from Deforestation.* The Centre for Social and Economic Research on the Global Environment (CSERGE), Norwich, UK.

Strauß, K (2006) *Kraftwerkstechnik: zur Nutzung fossiler, nuklearer und regenerativer Energiequellen.* Springer, Berlin, Heidelberg New York.

Sugathan, M (2006) Climate change benefits from liberalisation of environmental goods and services. In International Centre for Trade and Sustainable Development (ICTSD) (ed) *Linking Trade, Climate Change and Energy. ICTSD Trade and Sustainable Energy Series.* ICTSD, Geneva, pp8–9.

Sukhdev, P (2008) *The Economics of Ecosystems & Biodiversity. An Interim Report.* European Community, Brussels.

Sulzman, J and Ruhl, J B (2002) Paying to protect watershed services: wetland banking in the United States. In Pagiola, S, Bishop, J and Landell-Mills, N (eds) *Selling Forest Environmental Services. Market-Based Mechanisms for Conservation and Development.* Earthscan, London, pp77–90.

Sutherland, J W, Bailey, M J, Bainbridge, I P, Brereton, T, Dick, J T A, Drewitt, J, Dulvy, N K, Dusic, N R, Freckleton, R P, Gaston, K J, Gilder, P M, Green, R E, Heathwaite, A L, Johnson, S M, Macdonald, D W, Mitchell, R B, Osborn, D, Owen, R P, Pretty, J, Prior, S V, Prosser, H, Pullin, A S, Rose, P, Stott, A, Tew, T, Thomas, C D, Thompson, D B A, Vickery, J A, Walker, M, Walmsley, C, Warrington, S, Watkinson, A R, Williams, R J, Woodroffe, R and Woodroof, H J (2008) Future novel threats and opportunities facing UK biodiversity identified by horizon scanning. *Journal of Applied Ecology*: doi: 10.1111/j.365-2664.008.01474.x.

Swamy, S L and Puri, S (2005) Biomass production and C-sequestration of Gmelina arborea in plantation and agroforestry system in India. *Agroforestry Systems* **64**: pp181-95.

Swanson, T (1999) Why is there a biodiversity convention? The international interest in centralized development planning. *International Affairs* **75**(2): pp307–11.

Swearingen, J, Reshetiloff, K, B. Slattery, B and Zwicker, S (2002) Plant Invaders of Mid-Atlantic Natural Areas. National Park Service and U.S. Fish & Wildlife Service. National Park Service and U.S. Fish & Wildlife Service website, http://www.invasive.org/eastern/midatlantic/ (viewed 10. December 2007).

TAB (Büro für Technikfolgen-Abschätzung beim Deutschen Bundestag) (2000) *Risikoabschätzung und Nachzulassungs-Monitoring transgener Pflanzen. Sachstandsbericht, TA-Arbeitsbericht Nr. 68.* TAB, Berlin.

TAB (Büro für Technikfolgen-Abschätzung beim Deutschen Bundestag) (2005) *TA-Projekt Grüne Gentechnik – Transgene Pflanzen der 2. und 3. Generation. Endbericht.* TAB, Berlin.

TAB (Büro für Technikfolgen-Abschätzung beim Deutschen Bundestag) (2007) *Industrielle stoffliche Nutzung nachwachsender Rohstoffe. TAB-Arbeitsbericht 114.* TAB, Berlin.

ten Kate, K, Bishop, J and Bayon, R (2004) *Biodiversity Offsets. Views, Experience, and the Business Case.* IUCN, Gland.

Teplitz-Sembitzky, W (2006) *The Significance of Biomass Energy Strategies (BEST) for Sub-Saharan Africa. Background Paper for the First Regional Workshop of the Biomass Energy Strategy (BEST) Initative, a Joint Initiative of the EUEI Partnership Dialogue Facility (PDF) and GTZ (Energising Africa, Household Energy Programme [HERA]) Held in Dar-es-Salaam, Tanzania, 12–14 September 2006.* Deutsche Gesellschaft für Technische Zusammenarbeit (GTZ), Eschborn.

ter Heegde, W (2005) Domestic Biogas and CDM Financing – Support Project to the Biogas Programme for the Animal Husbandry Sector in Vietnam. Netherlands Development Organization (SNV) website, http://unapcaem.org/Activities%20Files/A01/The%20opportunities%20and%20challenges%20of%20the%20CDM%20for%20the%20financing%20of%20phase%20II%20of%20the%20Biogas%20Project%20in%20Vietnam.pdf (viewed 19. May 2008).

TERI (The Energy and Resources Institute) (2008) TERI Technologies – Biomass Gasifier – Based Power Generation System. TERI website, http://www.teriin.org/tech_power-generation.php (viewed 16. July 2008).

TERI (The Energy and Resources Institute) and GTZ (Deutsche Gesellschaft für Technische Zusammenarbeit) (2005) *Liquid Biofuels for Transportation: India Country Study on Potential and Implications for Sustainable Agriculture and Energy.* TERI, New Delhi.

Teufel, J (2005) Entwicklung stresstoleranter Nutzpflanzen im Zuge des Klimawandels: Überblick über den Forschungsstand und Perspektiven. In Korn, H, Schliep, R and Stadler, J (eds) *Biodiversität und Klima – Vernetzung der Akteure in Deutschland – Ergebnisse und Dokumentation des Auftaktworkshops an der Internationalen Naturschutzakadmie des Bundesamtes für Naturschutz, Insel Vilm, 29.09.-01.10.2004. BfN-Skripten 131.* Bundesamt für Naturschutz (BfN), Berlin, pp55–6.

TFZ (Technologie- und Förderzentrum des Bayrischen Staatsministeriums für Landwirtschaft und Forsten) (2008) Rutenhirse. Standortansprüche, Anbau und Ernte. TFZ website, http://www.tfz.bayern.de/rohstoffpflanzen/16592/ (viewed 13. October 2008).

The Royal Society (2008) *Sustainable Biofuels: Prospects and Challenges.* The Royal Society, London.

Thomas, C D, Cameron, A, Green, R E, Bakkenes, M, Beaumont, L J, Collingham, Y C, Erasmus, B F N, de Siqueira, M F, Grainger, A, Hannah, L, Hughes, L, Huntley, B, van Jaarsveld, A S, Midgley, G F, Miles, L, Ortega–Huerta, M A, Townsend Peterson, A, Phillips, O L and Williams, S E (2004) Extinction risk from climate change. *Nature* **427**: pp145–8.

Thonicke, K, Venevsky, S, Sitch, S and Cramer, W (2001) The role of fire disturbance for global vegetation dynamics: coupling fire into a Dynamic Global Vegetation Model. *Global Ecology and Biogeography* **10**(6): pp661–77.

Thrän, D M, Weber, A, Scheuermann, N, Fröhlich, J, Zeddies, A, Henze, C, Thoroe, J, Schweinle, U R, Fritsche, W, Jenseit, L, Rausch, L and Schmidt, K (2005) *Nachhaltige Biomassenutzungsstrategien im europäischen Kontext. Analyse im Spannungsfeld nationaler Vorgaben und der Konkurrenz zwischen festen, flüssigen und gasförmigen Bioenergieträgern. Kapitel 6: Biomassemärkte.* Institute for Energy and Environment, Leipzig.

Thrän, D, Seiffert, M, Müller-Langer, F, Plättner, A and Vogel, A (2007) *Möglichkeiten einer europäischen Biogaseinspeisungsstrategie.* Institute for Energy and Environment, Leipzig.

Thürer, D (2000) Soft law. In Bernhardt, R (ed) *Encyclopedia of Public International Law. Volume IV.* Elsevier, Amsterdam, pp452-60.

Tilman, D, Cassman, K G, Matson, P A, Naylor, R and Polasky, S (2002) Agricultural sustainability and intensive production practices. *Nature* **418**: pp671–7.

Tilman, D, Hill, J and Lehman, C (2006) Carbon-negative biofuels from low-input high-diversity grassland biomass. *Science* **314**(5805): pp1598–600.

Tilman, D, Hill, J and Lehman, C (2007) Response on comment on –Carbon-negative biofuels from low-input high-diversity grassland biomass". *Science* **316**: p1567c.

Toepfer International (2007) Statistische Informationen zum Getreide- und Futtermittelmarkt. Edition October 2007. Toepfer International website, http://www.acti.de/frameset. html (viewed 23. June 2008).

Tollefson, J (2008) Brazil goes to war against logging. *Nature* **452**: pp134–5.

Trewavas, A (2002) Malthus foiled again and again. *Nature* **418**: pp668–70.

TU Vienna (2005) *Energiezentrale zur Umwandlung von biogenen Roh- und Reststoffen einer Region in Wärme, Strom, BioSNG und flüssige Kraftstoffe: Endbericht.* Institut für Verfahrenstechnik, Vienna University of Technology (TU Vienna), Vienna.

Tuck, G, Glendining, M J, Smith, P, House, J I and Wattenbach, M (2006) The potential distribution of bioenergy crops in Europe under present and future climate. *Biomass and Bioenergy* **30**: pp183–97.

TUM – TU Munich (2000) *Raps – RME.* Technical University Munich.

Turner II, B L, Clark, W C, Kates, R W, Richards, J F, Mathews, J T and Meyer, W B (ed) (1990) *The Earth as Transformed by Human Action: Global and Regional Changes in the Biosphere Over the Past 300 Years.* Cambridge University Press, Cambridge, New York.

Turyareeba, P and Drichi, P (2001) Plan for the Development of Ugandas Energy Biomass Strategy. United Nations Environment Programme (UNEP) website, http://www.uneprisoe.org/SEAF/PlanDevelopBioEnergyStrategy.pdf (viewed 27. February 2008).

Tylianakis, J M, Rand, T A, Kahmen, A, Klein, A-M, Buchmann, N, Perner, J and Tscharntke, P (2008) Resource heterogeneity moderates the biodiversity-function relationship in real world ecosystems. *Public Library of Science Biology* **6**(e122): doi:10.1371/journal.pbio.0060122.

UBA (Umweltbundesamt) (2003a) *Reduzierung der Flächeninanspruchnahme durch Siedlung und Verkehr. Materialienband.* UBA, Berlin.

UBA (Umweltbundesamt) (2003b) *Klimaverhandlungen – Ergebnisse aus dem Kyoto-Protokoll, den Bonn-Agreements und den Marrakesh Accords.* UBA, Berlin.

UBA (Umweltbundesamt) (2004) Recycling von Phosphor verbessern. Presse-Information Nr. 103/2004. UBA, Berlin.

UBA (Umweltbundesamt) (2006a) *Monitoring und Bewertung der Förderinstrumente für Erneuerbare Energien in EU Mitgliedsstaaten. Climate Change Nr. 8.* UBA, Dessau.

UBA (Umweltbundesamt) (2006b) *Zertifikathandel für erneuerbare Energien statt Erneuerbare Energien-Gesetz? Hintergrundpapier zum Vorschlag des Verbands der Elektrizitätswirtschaft Dessau.* UBA, Dessau.

UBA (Umweltbundesamt) (2007) *Umweltbundesamt – Daten- und Rechenmodell TREMOD: Energieverbrauch und Schadstoffemissionen des motorisierten Verkehrs in Deutschland 1960–2030, Version 4, Heidelberg 2005, im Auftrag des Umweltbundesamtes 2006 Bundesministerium für Verkehr, Bau- und Wohnungswesen.* UBA, Dessau.

UBA (Umweltbundesamt) (2008a) *Bodenschutz beim Anbau nachwachsender Rohstoffe. Empfehlungen der "Kommission Bodenschutz beim Umweltbundesamt".* UBA, Dessau.

UBA (Umweltbundesamt) (2008b) *Wirtschaftliche Bewertung von Maßnahmen des integrierten Energie- und Klimaprogramms (IEKP): Wirtschaftlicher Nutzen des Klimaschutzes. Kostenbetrachtung ausgewählter Einzelmaßnahmen der Meseberger Beschlüsse zum Klimaschutz. Forschungsbericht 205 46 434. UBA-FB 001097.* UBA, Dessau.

UC Berkeley (University of California Berkeley) (2008) RAEL – Renewable and Appropriate Energy Laboratory. Understanding and Exploring the Future of our Worlds Energy. University of California website, http://rael.berkeley.edu (viewed 18. August 2008).

Ullrich, P (2008) Kraftstoffe aus Algen: Der Tank bleibt leer. *Umwelt aktuell* **June**: pp6-7.

Umsicht (Fraunhofer-Institut für Umwelt- Sicherheits- und Energietechnik) (2007) *Rechtsprobleme der Erzeugung von Biogas und der Einspeisung in das Erdgasnetz. Ergebnisse aus dem Workshop – Rechtsfragen der Einspeisung von Biogas in die Gasnetze. Fraunhofer-Institut für Umwelt-, Sicherheits- und Energietechnik Umsicht.* Umsicht, Oberhausen.

Umwelt aktuell (2008) Biokraftstoffe II: Wachsender Druck auf die EU-Beimischungsquote. *Umwelt aktuell* (6), p15.

UN (United Nations) (2006) *Cooperation Between the United Nations Environment Programme and the United Nations Development Programme: Note by the Executive Director: UNEP/GC/24/INF/19.* UN, New York.

UN (United Nations) (2008) *High Level Task Force on the Global Food Crisis: Elements of a Comprehensive Framework for Action (Draft as of June 3rd 2008).* UN, New York.

UNCCD (United Nations Convention to Combat Desertification) (2008) *An Introduction to the United Nations Convention to Combat Desertification. Fact Sheet.* UNCCD, New York.

UNCTAD (United Nations Conference on Trade and Development) (2006a) *Challenges and Opportunities for Developing Countries in Producing Biofuels.* UNCTAD, Geneva.

UNCTAD (United Nations Conference on Trade and Development) (2006b) *The Emerging Biofuels Market Regulatory. Trade and Development Implications.* UNCTAD, Geneva.

UNCTAD (United Nations Conference on Trade and Development) (2008a) The BioFuels Initiative of UNCTAD. UNCTAD website, http://r0.unctad.org/ghg/biofuels.htm (viewed 04. March 2008).

UNCTAD (United Nations Conference on Trade and Development) (2008b) Draft UNCTAD XII Negotiated Text. UNCTAD Twelfth Session, Accra, Ghana. 20–25 April 2008. UNCTAD website, http://www.unctad.org/Templates/Meeting.asp?intItemID=4287&lang=1 (viewed 13. June 2008).

UNECE (United Nations Economic Commission for Europe) (2005) *European Forest Sector Outlook Study, 1960–2000–2020. Main Report.* UNECE, Geneva

UN-Energy (2007a) UN-Energy Homepage. ESA website, http://esa.un.org/un-energy/index.htm (gelesen am 26. Oktober 2007).

UN-Energy (2007b) Sustainable Bioenergy: A Framework for Decision Makers. UN-Energy, New York.

UNEP (United Nations Environment Programme) (2002) *Global Environmental Outlook GEO-3. Past, Present and Future Perspectives.* UNEP, Nairobi.

UNEP (United Nations Environment Programme) (2007a) *Global Environment Outlook GEO-4. Environment for Development.* UNEP, Nairobi.

UNEP (United Nations Environment Programme) (2007b) Empowering Rural Communities by Planting Energy. Roundtable on Bioenergy Enterprise in Developing

Regions. Background Paper. UNEP website, http://www.unep.fr/energy/act/bio/index.htm (viewed 16. June 2008).

UNEP (United Nations Environment Programme) (2008) *UNEP Yearbook 2008. An Overview of Our Changing Environment*. UNEP, Nairobi.

UNEP Risoe Centre on Energy Climate and Sustainable Development (2008) CDM/JI Pipeline Analysis and Database. UNEP Risoe website, http://www.cdmpipeline.org (viewed 04. March 2008).

UNEP-WCMC (United Nations Environment Programme World Conservation Monitoring Centre) (2008) *World Database on Protected Areas*. UNEP-WCMC, Cambridge, UK.

UNESCO-MAB (United Nations Educational Scientific and Cultural Organization - Man and the Biosphere Programme) (1995) *Statutory Framework of the World Network of Biosphere Reserves*. UNESCO-MAB, Paris.

UNFCCC (United Nations Framework Convention on Climate Change) (2002) *Report of the Conference of the Parties on Its Seventh Session, Held at Marrakesh From 29 October to 10 November 2001. FCCC/CP/2001/13/Add.1*. UNFCCC, New York.

UNFCCC (United Nations Framework Convention on Climate Change) (2003) *Estimation, Reporting and Accounting of Harvested Wood Products. Techical Paper FCCC/TP/2003/7*. UNFCCC, New York.

UNFCCC (United Nations Framework Convention on Climate Change) (2006) CDM Executive Board 28th Meeting 12–15 December 2006. Annex 35 – Revisions to General Guidance on Leakage in Biomass Project Activities. UNFCCC website, https://cdm.unfccc.int/EB/028/eb28_repan35.pdf (viewed 20. May 2008).

UNFCCC (United Nations Framework Convention on Climate Change) (2007a) *Annual Report of the Executive Board of the Clean Development Mechanism to the Conference of the Parties Serving as the Meeting of the Parties. FCCC/KP/CMP/2007/3 (Part I+II) 6 November 2007*. UNFCCC, New York.

UNFCCC (United Nations Framework Convention on Climate Change) (2007b) *Investment and Financial Flows to Address Climate Change*. UNFCCC, New York.

UNFCCC (United Nations Framework Convention on Climate Change) (2007c) *Views on Issues Related to Further Steps Under the Convention Related to Reducing Emissions from Deforestation in Developing Countries: Approaches to Stimulate Action. FCCC/SBSTA/2007/MISC.14* UNFCCC, New York.

UNFCCC (United Nations Framework Convention on Climate Change) (2007d) *Views on the Range of Topics and Other Relevant Information Relating to Reducing Emissions from Deforestation in Developing Countries. FCCC/SBSTA/2007/MISC.2* UNFCCC, New York.

UNFCCC (United Nations Framework Convention on Climate Change) (2008a) CDM Home ó Project 0547: Facilitating Reforestation for Guangxi Watershed Management in Pearl River Basin. UNFCCC website, http://cdm.unfccc.int/Projects/DB/TUEV-SUED1154534875.41/view (viewed 04. March 2008).

UNFCCC (United Nations Framework Convention on Climate Change) (2008b) DM Project 529: Bunge Guará Biomass Project. UNFCCC website, http://cdm.unfccc.int/Projects/DB/SGS-UKL1152887084.53 (viewed 10. October 2008).

UNFCCC (United Nations Framework Convention on Climate Change) (2008c) Salto Grande Farmers Cooperative Self Consumption Biodiesel Plant. UNFCCC website, http://cdm.unfccc.int/Projects/Validation/DB W1ONENRIPBKD-J804YNXPDO86BSON75/view.html (viewed 10. October 2008).

UNFCCC (United Nations Framework Convention on Climate Change) (2008d) Fuel Switch from Petro-diesel to Biofuel for the Transport Sector in Bangalore Metropolitan Transport Corporation (BMTC), Karnataka, India. UNFCCC website, http://cdm.unfccc.int/Projects/Validation/DB/U5H6LUXR77B4DVODY5BHUUCB6E0IB7/view.html (viewed 10. October 2008).

Universität Hohenheim (2008) Bachelor Nachwachsende Rohstoffe und Bioenergie (NawaRo). Universität Hohenheim website, https://agrar.uni-hohenheim.de/60195.html (viewed 07. June 2008).

Universität Michigan (2008) ConsEnSus – Concentration in Environmental Sustainability. Programmbeschreibung. University Michigan website, http://www.engin.umich.edu/prog/consensus/ (viewed 18. August 2008).

UNPD (United Nations Population Division of the Department of Economic and Social Affairs) (2006) World Population Prospects: The 2006 Revision. Department of Economic and Social Affairs (UN DESA) website, http://esa.un.org/unpp (viewed 16. May 2008).

Uryu, Y, Mott, C, Foead, N, Yulianto, K, Budiman, A, Setiabudi, Takakai, F, Nursamsu, Sunarto, Purastuti, E, Fadhli, N, Hutajulu, C M B, Jaenicke, J, Hatano, R, Siegert, F and Stüwe, M (2008) *Deforestation, Forest Degradation, Biodiversity Loss and CO_2 Emissions in Riau, Sumatra, Indonesia*. WWF Indonesia, Jakarta.

U.S. Geological Survey – Earth Resources Observation and Science Center (2008) EDG Datensatz "MODIS/Terra Land Cover Types, Yearly L3 Global 0.05 Deg CMG". U.S. Geological Survey website, http://igskmnenwb001.cr.usgs.gov/modis/mod12c1v4.asp. (viewed 30. Dezember 2008).

USDA (U.S. Department of Agriculture) (2001) *Switchgrass, Panicum virgatum L. Plant Fact Sheet*. USDA, Washington, DC.

USDA (U.S. Department of Agriculture) (2008) *Secretary Ed Schafer Pre-Trip Media Availability for United Nations Food and Agriculture Organization June 3 Conference on World Food Security. Release No. 0140.08*. USDA, Washington, DC.

Valencia, A and Caspary, G (2008) *Hürden der ländlichen Stromversorgung mit erneuerbaren Energien. DIE Analysen und Stellungsnahmen*. German Development Institute (DIE), Bonn.

van Burgel, M (2006) *Biogas and Others in Natural Gas Operations (BINGO). A Project Under Development. 23rd World Gas Conference Amsterdam*. Nederlandse Gasunie, Groningen.

van Dam, D, Veldkamp, E and van Breemen, N (1997) Soil organic carbon dynamics: variability with depth in forested and deforested soils under pasture in Cosa Rica. *Biogeochemistry* **39**: pp343–75.

van Dam, J, Junginger, M, Faaij, A, Jürgens, I, Best, G and Fritsche, U (2008) Overview of recent developments in sustainable biomass certification. *Biomass and Bioenergy* **32**: pp749-80.

van Eijck, J and Romijn, H (2008) Prospects for Jatropha biofuels in Tanzania. An analysis with strategic niche management. *Energy Policy* **36**: pp311–25.

van Kooten, G C, Eagle, A J, Manley, J and Smolak, T (2004) How costly are carbon offsets? A meta-analysis of carbon forest sink. *Environmental Science & Policy* **7**: pp239–51.

Vavilov, N I (1926) Geographical regularities in the distribution of the genes of cultivated plants. *Bulletin of Applied Botany* **17**(3): pp411–28.

VDI (Verein Deutscher Ingenieure) (2000) *VDI Richtlinie 3986: Ermittlung des Wirkungsgrades von konventionellen*

Kraftwerken Düsseldorf. Verein Deutscher Ingenieure, Beuth Verlag GmbH, Berlin.

VDI (Verein Deutscher Ingenieure) (2002a) *Richtlinie VDI 2067: Wirtschaftlichkeit gebäudetechnischer Anlagen.* Verein Deutscher Ingenieure, Beuth Verlag GmbH, Berlin.

VDI (Verein Deutscher Ingenieure) (2002b) *Richtlinie VDI 6025: Betriebswirtschaftliche Berechnungen für Investitionsgüter und Anlagen.* Verein Deutscher Ingenieure, Beuth Verlag GmbH, Berlin.

VDI (Verein Deutscher Ingenieure) (2003) *VDI Richtlinie 4661: Energiekerngrößen – Definitionen, Begriffe, Methodik.* Beuth Verlag, Berlin.

VDI (Verein Deutscher Ingenieure) (2008) *VDI 4650 Blatt 1 – Berechnung von Wärmepumpen – Kurzverfahren zur Berechnung der Jahresarbeitszahl von Wärmepumpenanlagen – Elektro-Wärmepumpen zur Raumheizung und Warmwasserbereitung.* Beuth Verlag, Berlin.

Venetoulis, J and Talberth, J (2008) Refining the ecological footprint. *Environment Development and Sustainability* **10**: pp441–69.

Verburg, P H, Schot, P P, Dijst, M J and Veldkamp, A (2004) Land use change modelling: current practice and research priorities. *GeoJournal* **61**(4): pp309–24.

Vermeulen, S, Dufey, A and Vorley, B (2008) *Biofuels: Making Tough Choices. IIED Sustainable Development Opinion.* Inernational Institute for Environment and Development (IIED), London.

Vieira, S, Trumbore, S, Camargo, P B, Selhorst, D, Chambers, J Q, Higuchi, N and Martinelli, L A (2005) Slow growth rates of Amazonian trees: consequences for carbon cycling. *Proceedings of the National Academy of Sciences of America* **102**: pp18502–7.

Vogel, A (2006) *Synthetische Biokraftstoffe: Eine Analyse vorhandener Konzepte: Innovationsforum "Stoffumwandlung in Gase im Bereich der Energieverfahrenstechnik".* Institute for Energy and Environment (IE), Leipzig.

Vogel, A (2007) *Dezentrale Strom- und Wärmeerzeugung aus biogenen Festbrennstoffen: Eine technische und ökonomische Bewertung der Vergasung im Vergleich zur Verbrennung. Dissertation.* Technical University Hamburg-Harburg, Hamburg-Harburg, Leipzig.

Vogt, R, Gärtner, S, Münch, J, Reinhardt, G and Köppen, S (2008) *Optimierungen für einen nachhaltigen Ausbau der Biogaserzeugung und -nutzung in Deutschland.* Institute for Energy and Environmental Research (IFEU), Institute for Energy and Environment (IE), Institute for Applied Ecology – Darmstadt Bureau, Technical University Berlin, Berlin School of Economics and Law, Heidelberg, Leipzig, Berlin.

Voldoire, A (2006) Quantifying the impact of future land-use changes against increases in GHG concentrations. *Geophysical Research Letters* **33**: doi:1029/2005GL024354.

Voldoire, A, Eickhout, B, Schaeffer, M, Royer, J-F and Chauvin, F (2007) Climate simulation of the twenty-first century with interactive land-use changes. *Climate Dynamics*: doi 10.1007/s00382-007-0228-y.

von Braun, J (2007) *World Food Situation. New Driving Forces and Required Actions. Food Policy Report.* International Food Policy Research Institute (IFPRI), Washington, DC.

von Braun, J (2008a) *Rising Food Prices. What Should be Done? IFPRI Policy Brief April 2008.* International Food Policy Research Institute (IFPRI), Washington, DC.

von Braun., J (2008b) *Food and Financial Crises. Implications for Agriculture and the Poor, IFPRI Food Policy Report.* International Food Policy Research Institute (IFPRI), Washington, DC.

von Braun, J, Ahmed, A, Asenso-Okyere, K, Fan, S, Gulati, A, Hoddinott, J, Pandya-Lorch, R, Rosegrant, M W, Ruel, M, Torero, M, van Rheenen, T and von Grebmer, K (2008) *High Food Prices. The What, Who, and How of Proposed Policy Actions. Policy Brief.* International Food Policy Research Institute (IFPRI), Washington, DC.

von Drachenfels, C (2007) *Kurzanalyse und Vergleich ausgewählter Indikatoren zur Einstufung des Investitionsklimas und der Wettbewerbsfähigkeit von Volkswirtschaften, Stellungnahme für das BMZ.* German Development Institute (DIE), Bonn.

von Koerber, K, Kretschmer, J and Prinz, S (2008) Globale Ernährungsgewohnheiten und -trends. Expertise for the WBGU Report "World in Transition: Future Bioenergy and Sustainable Land Use". WBGU website, http://www.wbgu. de/wbgu_jg2008_ex10.pdf.

Wagner, H-J, Koch, M-K and Burkhardt, J (2008) *CO_2-Emissionen der Stromerzeugung. BWK-Energiefachmagazin 10/2007.* VDI-Springer Verlag, Düsseldorf.

Wagner, W, Scipal, K, Pathe, C, Gerten, D, Lucht, W and Rudolf, B (2003) Evaluation of the agreement between the first global remotely sensed soil moisture data with model and precipitation data. *Journal of Geophysical Research-Atmospheres* **108**(D19): p4611.

WBGU (German Advisory Council on Global Change) (1995a) *World in Transition: The Threat to Soils. Economica, Bonn.*

WBGU (German Advisory Council on Global Change) (1995b) *Scenario for the Derivation of Global CO_2 Reduction Targets and Implementation Strategies. Statement on the Occasion of the First Conference of the Parties to the Framework Convention on Climate Change in Berlin. Special Report 1995.* WBGU, Bremerhaven.

WBGU (German Advisory Council on Global Change) (1998) *The Accounting of Biological Sinks and Sources Under the Kyoto Protocol - A Step Forwards or Backwards for Global Environmental Protection? Special Report 1998.* WBGU, Bremerhaven.

WBGU (German Advisory Council on Global Change) (2000) *World in Transition: Strategies for Managing Global Environmental Risks.* Springer, Heidelberg, Berlin, New York.

WBGU (German Advisory Council on Global Change) (2001a) *World in Transition: Conservation and Sustainable Use of the Biosphere.* Earthscan, London.

WBGU (German Advisory Council on Global Change) (2001b) *World in Transition: New Structures for Global Environmental Policy.* London, Earthscan.

WBGU (German Advisory Council on Global Change) (2002) *Charging the Use of Global Commons. Special Report 2002.* WBGU, Berlin.

WBGU (German Advisory Council on Global Change) (2003) *Climate Protection Strategies for the 21st Century: Kyoto and beyond. Special Report 2003.* WBGU, Berlin.

WBGU (German Advisory Council on Global Change) (2004a) *World in Transition: Towards Sustainable Energy Systems.* Earthscan, London.

WBGU (German Advisory Council on Global Change) (2004b) *Renewable Energies for Sustainable Development: Impulses for renewables 2004. Policy Paper 3.* WBGU, Berlin.

WBGU (German Advisory Council on Global Change) (2005) *World in Transition: Fighting Poverty through Environmental Policy.* Springer, Berlin, Heidelberg, New York.

WBGU (German Advisory Council on Global Change) (2006) *The Future Oceans – Warming Up, Rising High, Turning Sour. Special Report 2006.* WBGU, Berlin.

WBGU (German Advisory Council on Global Change) (2007) *New impetus for climate policy: making the most of Germanys dual presidency. Policy Paper 5.* WBGU, Berlin.

WBGU (German Advisory Council on Global Change) (2008) *World in Transition: Climate Change as a Security Risk.* Earthscan, London.

WDPA (World Database on Protected Areas) (2008) World Database on Protected Areas. UNEP-WCMC website, http://www.wdpa.org (viewed 12. August 2008).

WEHAB Working Group (2002) *A Framework for Action on Biodiversity and Ecosystem Management.* WEHAB, Montreal.

Weigelt, A, Schumacher, J, Roscher, C and Schmid, B (2008) Does biodiversity increase spatial stability in plant community biomass? *Ecology Letters* **11**: pp338-47.

Weltsichten (2008) *"Im Müll liegen die Rohstoffe der Zukunft". Abfallmanagement in Entwicklungsländern muss den informellen Sektor einbinden. Gespräch mit Günter Wehenpohl.* Magazin für Globle Entwicklung und Ökumenische Zusammenarbeit, Bonn.

Weyerhaeuser, H, Tennigkeit, T, Yufang, S and Kahrl, F (2007) *Biofuels in China: An Analysis of the Opportunities and Challenges of Yatropha Curcas in Southwest China. ICRAF Working Paper No. 53.* International Centre for Research in Agroforestry (ICRAF) China, Nairobi, Peking.

White House (2006) *The National Security Strategy of the United States of America.* The White House, Washington, DC.

White, R P, Murray, S and Rohweider, M (2000) *Grassland Ecosystems. Pilot Analysis of Global Ecosystems.* World Ressources Institute (WRI), Washington, DC.

WHO (World Health Organisation) (2006) *Fuel for Life. Household Energy and Health.* WHO, Geneva.

WI (Worldwatch Institute) (2007) *Biofuels for Transport. Global Potential and Implications for Sustainable Energy and Agriculture.* Earthscan, London.

WI-IFEU (Wuppertal Institute for Climate, Environment and Energy-Institute for Energy and Environmental Research) (2007) *Elektromobilität und erneuerbare Energie. Arbeitspapier Nr. 5 im Rahmen des Projektes "Energiebalance – Optimale Systemlösungen für Erneuerbare Energien und Energieeffizienz".* WI and IFEU, Wuppertal, Heidelberg.

Wiegmann, K, Heintzmann, A, Peters, W, Scheuermann, A, Seidenberger, T and Thoss, C (2007) *Bioenergie und Naturschutz: Sind Synergien durch die Energienutzung von Landschaftspflegeresten möglich? Final Report.* Institute for Applied Ecology, Institute for Energy and Environment, Darmstadt, Leipzig.

Wiesenhütter, J (2003) *Nutzung von Purgiernuss (Jatropha curcas L.) zur Desertifikationsbekämpfung und Armutsminderung. Möglichkeiten und Grenzen technischer Lösungen in einem bestimmten sozio-ökonomischen Umfeld am Beispiel Kap Verde. Konventionsprojekt Desertifikationsbekämpfung (CCD Projekt).* Deutsche Gesellschaft für Technische Zusammenarbeit (GTZ), Eschborn, Bonn.

Wiggins, S and Levy, S (2008) *Rising Food Prices: A Global Crisis. ODI Briefing Paper 37.* Overseas Development Institute (ODI), London.

Willstedt, H and Bürger, V (2006) *Overview of Existing Green Power Labelling Schemes. A Report Prepared as Part of the EIE Project "Clean Energy Network for Europe (CLEAN-E)".* European Commission, Brussels.

Windfuhr, M (2008) Viele Initiativen, wenig Koordination. Die Welternährungskrise legt Defizite der internationalen Steuerung im Ernährungsbereich offen. *Welt-Sichten* **8**: pp36-8.

Wirsenius, S (2000) *Human Use of Land and Organic Materials. Doktorarbeit.* Chalmers University of Technology and Göteborg University, Göteborg.

Wissenschaftlicher Beirat Agrarpolitik beim BMELV (2007) *Nutzung von Biomasse zur Energiegewinnung – Empfehlungen an die Politik.* BMVEL, Bonn.

Wolf, J, Bindraban, P S, Luijten, J C and Vleeshouwers, L M (2003) Exploratory study on the land area required for global food supply and the potential global production of bioenergy. *Agricultural Systems* **76**(3): pp841–61.

World Bank (2002) *Project Appraisal Document on a proposed Grant from the Global Environment Facility Trust Fund in the Amount of SDR (USD 30 million equivalent) to the Fundo Brasileiro Para a Biodiversidad (FUNBIO) for an Amazon Region Protected Areas Project. Report No: 23756.* World Bank, Washington, DC.

World Bank (2003) *World Development Indicators.* The World Bank, Washington, DC.

World Bank (2004) *World Development Report 2004. Making Services Work for People.* World Bank, Washington, DC.

World Bank (2006a) BioCarbon Fund. World Bank website, http://carbonfinance.org/Router.cfm?Page=BioCF&FID=9708&ItemID=9708&ft=DocLib&CatalogID=6072 (viewed 06. June 2007).

World Bank (2006b) Carbon Finance for Sustainable Development – CFU Annual Report 2006. World Bank website, http://carbonfinance.org/Router.cfm?Page=DocLib&CatalogID=30716 (viewed 06. June 2007).

World Bank (2006c) *Strengthening Forest Law Enforcement and Governance. Addressing a Systemic Constraint to Sustainable Development.* World Bank, Washington, DC.

World Bank (2007) Brazil at a Glance. World Bank, Washington, DC.

World Bank (2008a) Mali: World Bank Approves Additional Funding for Energy Services Delivery in Rural Areas. News Release No: 2009/071/AFR. World Bank website, http://web.worldbank.org/WBSITE/EXTERNAL/NEWS/0,,contentMDK:21890422~pagePK:64257043~piPK:437376~theSitePK:4607,00.html (viewed 15. September 2008).

World Bank (2008b) Household Energy and Universal Access Project. World Bank website, http://web.worldbank.org/WBSITE/EXTERNAL/NEWS/0,,contentMDK:21890422~pagePK:64257043~piPK:437376~theSitePK:4607,00.html (viewed 16. September 2008).

World Bank (2008c) *World Development Report 2008. Agriculture for Development.* World Bank, Washington, DC.

World Bank (2008d) *Rising Food Prices Threaten Poverty Reduction. Press Release No:2008/264/PREM.* World Bank, Washington, DC.

World Bank (2008e) Biofuels. World Bank website, http://web.worldbank.org/WBSITE/EXTERNAL/COUNTRIES/LACEXT/EXTLACREGTOPENERGY/0,,contentMDK:20981118~menuPK:2717951~pagePK:34004173~piPK:34003707~theSitePK:841431,00.html (viewed 25. October 2007).

World Bank (2008f) Uganda Data Profile. World Bank website, http://devdata.worldbank.org/external/CPProfile.asp?CCODE=UGA&PTYPE=CP. (viewed 11. August 2008).

Worldwatch Institute (1999) *PaperCuts: Recovering the Paper Landscape. World Watch Paper 149.* Worldwatch Institute, Washington, DC.

Worldwatch Institute (2006) *Biofuels for Transport. Global Potential and Implications for Sustainable Agriculture and Energy in the 21st Century. Summary.* GTZ, BMELV, Berlin.

Worldwatch Institute (2007) *Biofuels for Transport. Global Potential and Implications for Sustainable Energy and Agriculture*. Earthscan, London.

Worldwatch Institute (2008) Carbon Markets Gain Momentum. Despite Challenges. Worldwatch Institute website, http://www.worldwatch.org/node/5597 (viewed 01. April 2008).

WRI (World Resources Institute) (2008) CAIT – Climate Analysis Indicators Tool. WRI website, http://cait.wri.org/ (viewed 13. August 2008).

WSSD (World Summit on Sustainable Development) (2002) *Plan of Implementation*. WSSD, Johannesburg.

Wuebbles, D J and Hayhoe, K (2002) Atmospheric methane and global change. *Earth-Science Revues* **57**: pp177–210.

Wunder, S (2005) *Payments for Environmental Services: Some Nuts and Bolts. CIFOR Occassional Paper No. 42*. Center for International Forestry Research (CIFOR), Bogor, Indonesia.

Wuppertal Institute and RWI (Rheinisch-Westfälisches Institut für Wirtschaftsforschung) (2008) *Nutzungskonkurrenzen bei Biomasse*. Wuppertal Institute and RWI, Wuppertal, Essen.

WWF (World Wide Fund for Nature) (2005a) Sugar and the Environment. Encouraging Better Management Practices in Sugar Production. Manuscript. WWF website, http://www. wwf.de/fileadmin/fm-wwf/pdf-alt/landwirtscgaft/WWF_ Action_for_Sustainable_Sugar_05.pdf (viewed 27. February 2008).

WWF (World Wide Fund for Nature) (2005b) WWF Action for Sustainable Sugar – Making it Sweeter for Nature. WWF Global Freshwater Programme. WWF website, http://assets. panda.org/downloads/sustainablesugar.pdf (viewed 27. February 2008).

WWF (World Wide Fund) and IUCN (The World Conservation Union) (1994) *Centres of Plant Diversity: A Guide and Strategy for their Conservation*. IUCN, Gland.

Yimer, F, Ledin, S and Abdelkadir, A (2007) Changes in soil organic carbon and total nitrogen contents in three adjacent land use types in the Bale Mountains, south-eastern highlands of Ethiopia. *Forest Ecology and Management* **242**: pp337–42.

Yu, V P (2007) *WTO Negotiating Strategy on Environmental Goods and Services for Asian Developing Countries. International Centre for Trade and Sustainable Development (ICTSD) Trade and Environment Series*. ICTSD, Geneva.

Zabarenko, D (2008) World Banks "Green" Energy Funding up 87 Percent. Pressemeldung Reuters. Reuters website, http://www.reuters.com/article/GCA-GreenBusiness/idUS-TRE49186B20081002 (viewed 03. October 2008).

Zah, R, Böni, H, Gausch, M, Hischier, R, Lehmann, M and Wäger, P (2007) *Ökobilanz von Energieprodukten: Ökologische Bewertung von Biotreibstoffen*. EMPA, St Gallen.

Zarrilli, S (2006) *The Emerging Biofuels Market: Regulatory, Trade and Development Implications*. UNCTAD, Geneva.

Annex I states

The group of countries listed in Annex I to the →United Nations Framework Convention on Climate Change (UNFCCC). It includes all OECD countries apart from Mexico and South Korea, as well as the eastern European states and Russia. Annex I countries have committed under the UNFCCC to adopt a leading role in the reduction of greenhouse gases. Moreover, most of the Annex I states have entered into binding reduction commitments under the Kyoto Protocol. These countries are listed in Annex B to the Protocol.

Biodiesel

Biodiesel (fatty acid methyl ester, FAME) is produced by esterification from →vegetable oils, at present mainly from rapeseed oil, soya oil and palm oil. 2nd-generation biofuels include →Fischer-Tropsch diesel.

Bioenergy

Bioenergy is the final or useful energy that can be released and made available from →biomass.

Bioethanol

Bioethanol is produced from →biomass with the aid of yeasts or bacteria, and subsequently purified and concentrated by means of distillation or rectification. Grain or sugar cane are the most common feedstock materials. Ethanol can also be produced from vegetable wastes, wood or straw (lignocellulose ethanol), but this process is still at the development stage. Bioethanol can be blended with petrol or used in its pure form as a fuel in vehicle engines, but also in →small-scale CHP units.

Biofuels

The term biofuels refers to fuels in liquid or gaseous form that are produced from →biomass and are used primarily as transport fuels but also to produce electricity and heat, e.g. in →small-scale CHP units. A distinction is made between 1st-generation and 2nd-generation biofuels. The 1st generation includes →vegetable oil, →biodiesel and →bioetha-

nol, obtained through established physical and chemical (pressing, extraction, esterification) or biochemical (alcoholic fermentation) processes. The 2nd generation includes synthetic biofuels such as →Fischer-Tropsch diesel, →biomethane and biohydrogen, produced using thermochemical processes (gasification, pyrolysis).

Biogas

Biogas is a generic term referring to gases from which energy can be recovered and that are created when →biomass decomposes under anaerobic conditions. Fermentation produces a mixture of the gases →methane (CH_4) and →carbon dioxide (CO_2). The methane is the fraction of the biogas that can be utilized for energy recovery, either directly through its combustion or through processing to →biomethane.

Biomass

Biomass comprises the organic substances of the biotic environment, either as living or dead biomass (e.g. fuelwood, charcoal and dung). Important conversion products of biomass include →biogas and →biofuels. In developing countries, →traditional biomass use is the dominant form.

Biomethane

Biomethane is →methane (CH_4) produced from →biomass. It can be produced from →biogas by separating →carbon dioxide and other impurities such as hydrogen sulphide. Biomethane can also be produced by gasifying solid or liquid biomass. This involves the initial production of a raw gas which is then converted to methane following cleaning and synthesis. Biomethane can be mixed with natural gas.

Blending quota

A blending quota is a prescribed minimum proportion of →biofuels within the overall quantity of fuel marketed, defined in terms of either the volume or the energy content of the fuels. The proportion can be attained by blending with petrol or diesel fuel, or by using pure biofuel.

BtL diesel
→Fischer-Tropsch diesel

Carbon dioxide (CO₂)

CO_2 is a naturally occurring →greenhouse gas. It is a product of burning fossil energy carriers and →biomass. CO_2 is also emitted as a result of deforestation and other land-use changes, as well as from industrial processes such as cement production.

Carbon dioxide equivalents (CO₂eq)

Carbon dioxide equivalents are a measure of the degree to which a mixture of gases contributes to global warming. A conversion factor, the Global Warming Potential, expresses the radiative forcing of other greenhouse gases as an equivalent quantity of CO_2. This makes it possible to register and compare the impacts of all greenhouse gases using a common metric.

Carbon sink

A carbon sink is a reservoir that absorbs and stores carbon in a temporary or permanent manner. The concept is not to be confused with that of the carbon store. While the store (or stock) is static, i.e. contains a certain quantity of carbon, sinks are dynamic, i.e. can experience incremental gains, as in the case of newly planted forests.

Carbon store

cf. →carbon sink

Cascading

Cascading refers to a strategy seeking to use resources or products made from them as long as possible within the economic cycle. The material passes through as many phases of use as possible, i.e. the same resource is utilized several times over in different functions, and is thus recycled comprehensively. This approach boosts overall value creation and improves environmental performance. In the case of →biomass, cascading can mean that the biomass is first used as an industrial feedstock and its energy content is then recovered at the end of the product cycle. For example, furniture or wood used in construction can be co-fired in a power plant at the end of its service life.

Certification

Certification is the process by which an independent body affirms that a product or a production or management process meets certain →standards. The certificate provides proof that a company or undertaking has subjected itself to such a certification process and meets certain standards. The certifying organization need not necessarily be the organization that has set the standards, but can also be accredited by

the latter. Certified products can be labelled as such so that they can be identified by the final consumer (labelling).

Clean Development Mechanism (CDM)

The CDM is one of the flexible mechanisms introduced by the Kyoto Protocol to the →UNFCCC, which allows an investor to carry out emissions-reducing projects in a developing or newly industrializing country and to receive tradable certificates for this, which an industrialized country can offset against its reduction commitments.

Cogeneration (combined heat and power, CHP)

Facilities with combined heat and power production not only generate electricity from the fuel consumed, but also make use of the waste heat. For instance, this heat can be used for space heating purposes (as in district heating systems). In industry, it can be used for heat-dependent production processes. →Small-scale CHP units are an example of cogeneration.

Combined-cycle power plant

A combined-cycle power plant is a plant fuelled by natural gas or biomethane, in which the principles of a gas turbine power plant and of a steam power plant are combined. With efficiencies of up to approx. 60 per cent, such plants are among the most efficient conventional power plants. Thanks to the rapid load changes which they permit, they can be deployed very flexibly in generation mix dispatching.

Convention on Biological Diversity (CBD)

The CBD is the key international regime pertaining to the biosphere. It was signed in 1992 at the UN Conference on Environment and Development, entered into force in 1993 and has since been ratified by 191 states. The parties to the CBD undertake (1) to conserve biological diversity, (2) to use its components sustainably and (3) to share the benefits arising from the utilization of genetic resources fairly and equitably. The CBD stresses the links between the conservation and use of biological diversity and seeks to reconcile the interests of industrialized and developing countries.

Cross compliance

Cross compliance rules make the provision of direct payments to agricultural holdings contingent upon compliance with rules in other spheres. In EU agricultural policy, for instance, cross compliance rules are being applied increasingly. They make direct payments dependent upon compliance with standards governing environmental performance, food and feed security, animal health or animal welfare.

Observance of the rules is monitored. If an infringement is found, payments are reduced.

Degraded areas
→marginal land

Ecosystem services
Ecosystem services are benefits gained by people from ecosystems. These include supply services such as food or water, regulation services such as protection against flooding or against the spread of disease, cultural services of a spiritual or recreational nature, and support services such as nutrient cycles that maintain Earth's life-support systems. The concept is here used synonymously with 'ecosystem products and services'.

Electromobility
Electromobility refers to the use of electricity in the transport sector, especially in road transport. Examples include the use of hybrid, battery-driven and fuel-cell vehicles.

Emissions trading
Emissions trading is an economic instrument for the limitation or reduction of environmentally harmful emissions. Generators of emissions are subject to reduction goals that they must either meet themselves or can have met in whole or part by other generators. To that end, reduction commitments can be traded among the participants in a trading system, which produces a cost-optimal allocation of the set overall reduction. The Kyoto Protocol to the →UNFCCC introduces this instrument at state level for those countries which have undertaken commitments. Moreover, some states or groups of states (such as the EU) have introduced emissions trading schemes in which companies can trade their emission rights among themselves.

Energy crops
Energy crops are cultivated for the purpose of extracting energy from their →biomass. This may involve using either a specific part of the crop (e.g. maize grain or →vegetable oil extracted from seeds), or the entire aboveground biomass (e.g. certain grass species or woody species such as poplar or willow).

Energy poverty
Energy poverty refers to the lack of sufficient access to energy services, in order to meet basic needs, that are affordable, reliable, high-quality and safe, and cause no undue health or environmental impacts. Countries where energy poverty is widespread are generally characterized by major development problems. Energy poverty affects around 38 per cent of the world's population. These people are dependent upon →traditional biomass use. Pollutants from open hearths cause more than 1.5 million deaths each year.

Energy system transformation
WBGU uses this term to describe the transformation of energy systems that will be required in the next decades in order to attain the sustainability goals of mitigating climate change and overcoming energy poverty. The transition involves a shift from fossil to renewable energy sources, and the provision of modern energy services for the entire population.

Fischer-Tropsch diesel
Diesel fuel produced from solid →biomass by means of the Fischer-Tropsch process. A synthesis gas is first extracted from straw, timber and similar biomass and then converted into liquid →biofuel. The process permits use of the entire aboveground biomass of →energy crops. Fischer-Tropsch diesel is also termed FT diesel or BtL (biomass-to-liquid) diesel.

Genetically modified organisms (GMOs)
GMOs are organisms whose genetic material has been modified by methods of molecular biology in a way that would not be possible naturally by means of crossing or natural recombination. Green genetic engineering is a term referring to the use of such techniques in plant breeding. The plants modified in this way are also termed transgenic plants.

Global Bioenergy Partnership (GBEP)
The GBEP is a high-level, intergovernmental discussion platform within the context of the G8+5 forum, with the aim of advancing renewable energies and developing a market for →bioenergy. It originated at the suggestion of the UK at the 2005 G8 economic summit in Gleneagles. In addition to the G8 and the five outreach countries (China, India, Mexico, Brazil, South Africa), several UN organizations are involved such as the FAO, UNEP and UNDP. Institutionally, the GBEP is located within the FAO in Rome.

Greenhouse gases
Greenhouse gases are those gaseous constituents of the atmosphere that, due to their selective absorption of thermal radiation, cause warming of the lower atmosphere. The primary anthropogenic greenhouse gases are →carbon dioxide, →nitrous oxide and →methane. Other greenhouse gases are industrial gases such as hydrofluorocarbons (HFCs), perfluorocarbons (PFCs) and sulphur hexafluoride (SF_6), and the ozone-depleting chlorofluorocarbons (CFCs).

Guard rail

Guard rails are quantitatively defined damage limits, exceedance of which is intolerable or would have catastrophic consequences. This is a concept introduced by WBGU. One example is the climate protection guard rail, which states that an increase in the global mean temperature by more than 2°C from the pre-industrial level should be prevented. Sustainable development pathways follow trajectories that lie within the range delimited by the ecological and socio-economic guard rails. This approach is based on the realization that it is virtually impossible to define a desirable, sustainable future in positive terms – that is, in terms of a goal or state to be achieved. It is, however, possible to agree on the boundaries of a range that is acknowledged to be unacceptable and that society seeks to avoid. Adherence to the guard rails is a necessary, but not sufficient criterion for sustainability.

Intergovernmental Panel on Climate Change

(IPCC) The IPCC was founded in 1988 by UNEP and WMO, and is the most influential international scientific institution for climate policy. The IPCC lays the scientific foundation for negotiations on the →UNFCCC and publishes regular assessment reports on global climate change. The 4th assessment report was published in 2007.

International Renewable Energy Agency (IRENA)

Established in 2009, the task of IRENA is to advance the worldwide use of renewable energies. Among other things the agency provides advice on policy settings for renewables, fosters technology and knowledge transfer, assists capacity building and advises member states on financing options. Its principal purpose is to increase the share of renewable energies worldwide.

Land-use changes

Land use is a term referring to the human use of an area of land for a certain purpose, while land-use changes refer to changes in such human use. These include logging, afforestation, sealing, drainage, the conversion of cropland to grassland (and vice versa) or the conversion of cropland to fallow. Land-use changes can take place directly, for instance when tropical forests are cleared and the land is used to cultivate →energy crops. Land-use changes induced by indirect mechanisms are more difficult to identify. When cropland is given over to the cultivation of energy plants, the agricultural production that previously took place on this land must now take place elsewhere. Through the world market for agricultural commodities these indirect displacement effects often acquire an international dimension.

Life-cycle assessment (LCA)

An LCA involves a systematic analysis of the environmental impacts of products across their entire life cycle. This includes all environmental impacts arising during production, in the use phase and in the disposal of the product, and all associated upstream and downstream processes (such as production of the raw, auxiliary and ancillary materials). The method involves compiling an inventory of all inputs along the product life cycle (such as metals, as well as fossil and renewable energy carriers) as well as the outputs such as emissions of substances hazardous to the environment or human health.

Marginal land

Marginal lands are (1) areas with little capacity for fulfilling a production or regulation function, and also (2) areas that have lost their production and regulation function, sometimes to a significant extent. Category 1 comprises areas whose productivity for agriculture or forestry is considered low. This category includes arid and semi-arid grasslands, desert fringes and areas of steep ground and structurally weak or erosion-prone soils, particularly in mountainous regions. Category 2 includes formerly productive areas; they may have lost their yield potential as a result of human-induced soil degradation (e.g. overused, degraded and therefore unproductive land, including both forests and pasture and arable land), or the land may have been deliberately taken out of production (e.g. set-aside land in central Europe that has been taken out of production for economic or political reasons). Marginal sites are generally highly susceptible to soil degradation.

Methane (CH_4)

CH_4 is a →greenhouse gas emitted mainly in rice cultivation and livestock management. It is the principal component of natural gas and biogas.

Nitrous oxide (N_2O)

N_2O is a persistent →greenhouse gas released above all through the use of nitrogen fertilizers in agriculture, and through the combustion of →biomass and fossil fuels.

Non-Annex I states

Non-Annex I states are countries not listed in Annex I to the →UNFCCC. The group is largely composed of developing and newly industrializing countries. The non-Annex I states do not have any quantified emissions reduction commitments under the Kyoto Protocol; cf. → Annex I states.

Roundtable on Sustainable Biofuels (RSB)

The RSB is a multilateral forum that has set itself the goal of developing global standards and a certification system (→certification) for →biofuels. The forum has its origins in an initiative of the Swiss Federal Institute of Technology, Lausanne, and brings together more than 300 different stakeholders from the realms of policymaking, business and society as well as experts and international organizations. The references it uses include the label scheme of the Forest Stewardship Council for wood products, as well as recent standard-setting processes (→standards) for →bioenergy.

Sequestration

Storage, by human action, of atmospheric carbon in terrestrial ecosystems, geological formations or the oceans. For instance, through new technologies the →carbon dioxide resulting from combustion processes can be captured, possibly liquefied and then pumped into underground repositories such as depleted gas and oil fields (carbon capture and storage, CCS).

Short-rotation plantations (SRPs)

Short-rotation plantation refers to the cultivation of fast-growing tree species (e.g. poplar, willow) on agricultural land to produce →biomass. SRPs originate in coppicing, a method which served in the past to produce firewood. The rotation period is the growth period until the trees are cut and depends on the use of the wood. For pulpwood or for woodchip production, the trees are harvested after 3–5 years. The below-ground root mass remains in the soil, enabling growth of coppice shoots the following year.

Small-scale CHP units

A small-scale CHP unit produces electricity and heat simultaneously. It is operated at the site of heat consumption or feeds useful heat into a local heat network. It operates on the principle of →cogeneration.

Standards

Standards are fixed criteria that products, manufacturing processes or management processes must meet in order to fulfil certain (quality) requirements. Compliance with a standard can be voluntary or mandatory. Standards can be initiated in voluntary form by private-sector actors, or set by the state. A mandatory minimum standard – such as a standard requiring compliance with certain sustainability criteria – can, moreover, be set as a precondition for the market approval of a product. By issuing a certificate (→certification), an independent body certifies that a standard has been complied with.

Technology transfer

Technology transfer refers to the set of processes involved in the exchange of knowledge, money and goods among different stakeholders that leads to the spread of technologies, e.g. for mitigating climate change or securing sustainable energy development. Transfer often has two meanings: dissemination of technologies and technological cooperation among and within countries.

Traditional biomass use

Form of energy production from →biomass, such as wood, dung, harvest residues, etc., above all for cooking and heating on open hearths. Worldwide about 2.4 billion people, predominantly in developing countries, depend upon traditional biomass and are thus often exposed to emissions-related health hazards due to inadequate combustion technologies.

UNCCD, Desertification Convention

The UNCCD (United Nations Convention to Combat Desertification in those Countries Experiencing Serious Drought and/or Desertification, Particularly in Africa) is, among the three Rio conventions (cf. →UNFCCC and →CBD), the one with the strongest development policy orientation. Besides conserving resources in drylands, poverty reduction is also a declared goal. The UNCCD entered into force in 1996 and has been ratified by 193 countries.

UNFCCC, United Nations Framework Convention on Climate Change

Adopted in 1992, the UNFCCC entered into force in 1994 and has been ratified by 192 states. Its ultimate objective is to stabilize →greenhouse gas concentrations in the atmosphere at a level that would prevent dangerous anthropogenic interference with the climate system ('dangerous climate change'). Such a level should be achieved within a timeframe sufficient to allow ecosystems to adapt naturally to climate change, to ensure that food production is not threatened and to enable economic development to proceed in a sustainable manner. The Kyoto Protocol adopted in 1997 sets out binding commitments to reduce greenhouse gas emissions.

Vegetable oil

Vegetable oils are produced by pressing the oil fruit or seeds of oleiferous plants. →Biodiesel can be produced from vegetable oil by means of esterification. Unrefined vegetable oil can also be used directly as a →biofuel in engines modified for that purpose. In Germany, vegetable oil is produced mainly from rape. Important tropical oleiferous plants include oil and coconut palms, as well as *Jatropha* and soya.